q	Charge on a molecule
r	Radius
rDNA	Recombinant DNA
Res	Resolution
RI	Refractive index
RPLC	Reversed-phase liquid chromatography
RSD	Relative standard deviation (= coefficient of variation [CV])
S/N	Signal-to-noise ratio
SDS	Sodium dodecyl sulfate
SDS-PAGE	Sodium dodecyl sulfate-polyacrylamide gel electrophoresis
SIM	Single ion monitoring
t	Time
T	Temperature
TBE	Tris–borate–EDTA
TCA	Trichloroacetic acid
TE	Terminating electrolyte
TEMED	*N,N,N′,N′*-Tetramethylethylenediamine
TLC	Thin-layer chromatography
TMAC	Trimethylammonium chloride
TMAPS	Trimethylammonium propylsulfonate
TR	Transfer ratio
Tricine	*N*-[Tris(hydroxymethyl)methyl] glycine
Tris	Tris(hydroxymethyl)aminomethane
UV	Ultraviolet
v_i	Velocity
w	Width
$w_{1/2}$	Width at half-height
x_i	Distance
zmol	Zeptomole (10^{-21} mol)

Δ	Difference
ΔP	Difference in pressure (pressure differential)
μ	Micro
μA	Microampere
μ_{app}	Apparent mobility
μ_{avg}	Average mobility of two solutes
σ	Standard deviation
σ^2	Variance
κ	Specific conductance
η	Solvent viscosity
Λ	Equivalent conductance

Handbook of

Capillary Electrophoresis

Handbook of

Capillary Electrophoresis

Edited by

James P. Landers

CRC Press
Boca Raton Ann Arbor London Tokyo

Cover figure by Huang, X. C., Quesada, M. A., and Mathies, R. A. reproduced with permission from *Anal. Chem.,* 64, 2149, 1992. Copyright © 1992 American Chemical Society.

Library of Congress Cataloging-in-Publication Data

Handbook of capillary electrophoresis / edited by James P. Landers.
p. cm.
Includes bibliographical references and index.
ISBN 0-8493-8690-X (alk. paper)
1. Capillary electrophoresis. I. Landers, James P.
[DNLM: 1. Electrophoresis—methods. QH 324.9.E4 H236 1993]
QP519.9.C36H35 1993
574.19′285—dc20
DNLM/DLC
for Library of Congress 93-11351
CIP

No claim to original U.S. Government works
International Standard Book Number 0-8493-8690-X
Library of Congress Card Number 93-11351
Printed in the United States of America 1 2 3 4 5 6 7 8 9 0
Printed on acid-free paper

DEDICATION

To Thomas C. Spelsberg for his vision identifying the potential biomedical applications of capillary electrophoresis.

FOREWORD

As evidenced by this book, capillary electrophoresis (CE) is a field coming of age. Practical solutions to real analytical and separation problems are demonstrated in this text. While the chapters focus on the well-known success of CE with biopolymer analysis (e.g., DNA, proteins, and carbohydrates) small molecules and ions are not neglected. For example, it is shown that capillary electrophoresis provides an effective means of analysis of drug and drug metabolites in biological fluids, where sample clean-up is often minimal prior to injection into the capillary. As another example, inorganic and small organic ion analysis using indirect detection is shown to be competitive with current technologies.

Electrophoresis in columns is not new; indeed, electrophoresis in tubes was established prior to slab gel electrophoresis. From the beginning, workers recognized that performance would increase with reduction of column diameter or slab gel thickness. Such narrow bore columns provide reduced current and a more effective means of Joule heat removal than wider diameter tubing, permitting high electric fields to be operated for rapid analysis and efficiency.

While capillary columns were utilized in the electrophoretic format prior to 1981, it is Jorgenson who is generally recognized as the first to assemble a workable format in CE. His experience in micro-liquid chromatography proved to be significant in this area. The 1980s saw the slow development of CE as more and more workers demonstrated high performance conditions. While many researchers contributed to the development of CE, several milestones in the late 1980s may be cited. Terabe introduced micellar electrokinetic chromatography in 1985; Lauer and McManigill demonstrated ultra-high performance for protein separations in open tubes in 1986; Hjertén established capillary isoelectric focusing in 1987; and our laboratory demonstrated the separating power of gel or sieving media in a capillary format in 1988.

Noteworthy as well in the establishment of the technique were the efforts of the instrument companies to introduce automated equipment that would permit a wide utilization of the technique. Bob Brownlee should be especially recognized as an early champion of capillary electrophoresis; his company, Microphoretics, was the first to introduce a full CE instrument. Unfortunately, Bob passed away several years ago and thus did not live to see his dream completely fulfilled. Today, approximately a dozen companies have instrumentation available on the market. As further evidence of the rapid development of capillary electrophoresis, the 5th International Symposium on CE was recently held in Orlando, Florida. Over 250 papers were presented with approximately 600 attendees.

Why is there such interest in capillary electrophoresis? First, it is a fully automatable technique that can provide separation and analysis of samples in a high performance format. It is suitable for samples that may be difficult to handle by liquid chromatography, or at the least complements this technique, since the principles of separation are different. It is important to note in this regard that capillary electrophoresis is fundamentally a homogeneous method of separation, not involving surface adsorption. Indeed, great efforts are expended in trying to minimize adsorption on the capillary wall, utilizing coatings or additives.

A second feature of CE is that ultra-trace analysis is feasible. The incorporation of laser spectroscopy, first demonstrated by Zare and co-workers, has enhanced the potential of CE in the trace area. In terms of mass detection, analysis at the attomole level or lower is possible. With appropriate volume utilization, sample concentration detection levels of 10^{-8} to 10^{-9} M are feasible, even with UV detection. The potential of on-line mass spectrometric detection of species migrating out of the column is also important. Today, this method is under development; however, we can anticipate its full utilization in the years ahead to complement LC/MS.

A third attribute of capillary electrophoresis is its miniature size. This leads to the possibility of separating and characterizing very small quantities of material. For example, solution binding constants of protein–protein or drug–protein interactions can be achieved. Moreover, the kinetics of this binding process can be directly studied. Furthermore, enzymatic reactions for analytical purposes can be conducted within the capillary column. Another advantage of the miniaturized system is that reagents that would normally be too costly or difficult to work with are cost effective. For example, labeled biospecific reagents can easily be employed in CE since only a minuscule amount is utilized for each assay.

The next few years should see a rapid expansion of capillary electrophoresis, particularly with respect to the number of users and the breadth of applications. In many respects the field is following the history of liquid chromatography. As more workers use capillary electrophoresis, those areas that need improvement are identified and the appropriate improvements developed.

It is thus clear that capillary electrophoresis is on its way to becoming a major analytical/separation tool. Further, it will likely emerge as a significant tool for biological measurements. By examining the chapters in this book, the reader will obtain a good understanding of the current status of capillary electrophoresis. The editor is to be congratulated on assembling such a broad list of CE experts to outline the scope of this field.

Barry L. Karger, Ph.D.
Barnett Institute
Northeastern University
Boston, MA

THE EDITOR

James P. Landers, Ph.D., is the Operations Director of the Capillary Electrophoresis Core Facility at the Mayo Clinic in Rochester, Minnesota.

Dr. Landers received his Bachelor of Science degree in Biochemistry and Biomedicine from the University of Guelph in Ontario (Canada) in 1984. He obtained his Ph.D. in Biochemistry in 1988 in the Department of Chemistry and Biochemistry at the same University. After a short postdoctoral fellowship at the Banting Institute in the School of Medicine at the University of Toronto (Ontario, Canada) he was awarded a Medical Research Council (Canada) postdoctoral fellowship to study with Dr. Thomas C. Spelsberg at the Mayo Clinic. Interest in exploring new state-of-the-art bioanalytical technologies for evaluation of certain disease states of both basic science and clinical importance led to his involvement with capillary electrophoresis. Major efforts at present involve applying capillary electrophoresis to solve separation/analytical problems in the biomedical sciences.

CONTRIBUTORS

Linda Benson, B.S.
Biomedical Mass Spectrometry Facility
Mayo Clinic/Foundation
Rochester, MN

Dean S. Burgi, Ph.D.
Genomyx Corporation
South San Francisco, CA

Norman J. Dovichi, Ph.D.
Department of Chemistry
University of Alberta
Edmonton, Alberta, Canada

F. M. Everaerts, Ph.D.
Laboratory of Instrumental Analysis
Eindhoven University of Technology
Eindhoven, The Netherlands

Andrew G. Ewing, Ph.D.
Department of Chemistry
Pennsylvania State University
University Park, PA

L. Liliana Garcia, Ph.D.
Department of Pathology
Bowman Gray School of Medicine
Wake Forest University
Winston-Salem, NC

David R. Goodlett, Ph.D.
Chemical Sciences Department
Battelle
Pacific Northwest Laboratories
Richland, WA

Andras Guttman, Ph.D.
Hafslund Nycomed Pharma
Linz, Austria

Carl R. Jolliff, S.M. (AAM)
Physician's Laboratory Services
Lincoln, NE

William R. Jones, Ph.D.
Millipore-Waters, Inc.
Milford, MA

Morteza G. Khaledi, Ph.D.
Department of Chemistry
North Carolina State University
Raleigh, NC

Ferenc Kilár, Ph.D.
Central Research Laboratory
University of Pécs, Medical School
Pécs, Hungary

Gerald L. Klein
Beckman Instruments
Fullerton, CA

Ira S. Krull, Ph.D.
Department of Chemistry
Northeastern University
Boston, MA

James P. Landers, Ph.D.
Department of Biochemistry
and Molecular Biology
Mayo Clinic/Foundation
Rochester, MN

Jeff R. Mazzeo, Ph.D.
Waters Chromatography Division
Millipore Corporation
Milford, MA

Randy M. McCormick, Ph.D.
Dionex Corporation
Sunnyvale, CA

Stephen Naylor, Ph.D.
Biomedical Mass Spectrometry Facility
Department of Biochemistry and
Molecular Biology/Pharmacology
Mayo Clinic/Foundation
Rochester, MN

Robert J. Nelson, Ph.D.
Thermo Separations Products
Fremont, CA

Judith A. Nolan, Ph.D.
Beckman Instruments
Fullerton, CA

Robert P. Oda, B.A.
Department of Biochemistry
and Molecular Biology
Mayo Clinic/Foundation
Rochester, MN

Joseph D. Olechno, Ph.D.
Applied Biosystems
Foster City, CA

Richard Palmieri, Ph.D.
Beckman Instruments
Fullerton, CA

Stephen Pentoney, Ph.D.
Beckman Instruments
Fullerton, CA

Charles Shaw, M.Sc.
R. W. Johnson Pharmaceutical
Research Institute
Raritan, NJ

Zak K. Shihabi, Ph.D.
Department of Pathology
Bowman Gray School of Medicine
Wake Forest University
Winston-Salem, NC

Charlotte Silverman, Ph.D.
R. W. Johnson Pharmaceutical
Research Institute
Raritan, NJ

Sandra Sloss, Ph.D.
Department of Chemistry
Pennsylvania State University
University Park, PA

Richard D. Smith, Ph.D.
Chemical Sciences Department
Battelle
Pacific Northwest Laboratories
Richland, WA

Thomas C. Spelsberg, Ph.D.
Department of Biochemistry
and Molecular Biology
Mayo Clinic/Foundation
Rochester, MN

Sally A. Swedberg, Ph.D.
Hewlett-Packard Company
Waldbronn Analytical Division
Waldbronn, Germany

Jonathan Sweedler, Ph.D.
Department of Chemistry
University of Illinois
Urbana, IL

Andrew Tomlinson, Ph.D.
Biomedical Mass Spectrometry Facility
Mayo Clinic/Foundation
Rochester, MN

Takao Tsuda, Ph.D.
Nagoya Institute of Technology
Nagoya, Japan

Kathi J. Ulfelder, A.B.
Beckman Instruments
Fullerton, CA

Jon H. Wahl, Ph.D.
Chemical Sciences Department
Battelle
Pacific Northwest Laboratories
Richland, WA

Bart J. Wanders, Ph.D.
Beckman Instruments
Fullerton, CA

Tim Wehr, Ph.D.
BioRad Laboratories
Hercules, CA

Mingde Zhu, Ph.D.
BioRad Laboratories
Hercules, CA

TABLE OF CONTENTS

CONCLUDING REMARKS

APPENDICES

Chapter 1

CAPILLARY ELECTROPHORESIS: HISTORICAL PERSPECTIVES

Tim Wehr and Mingde Zhu

It is not unusual in science that formulation of new concepts and analytical techniques predate their widespread acceptance in the scientific community by periods as long as decades. In the case of capillary electrophoresis (CE), the value of using the anticonvective properties of small internal diameter tubes to obtain improved resolution was already understood during the early development of electrophoresis. Although the first working prototype of a capillary electrophoresis system was constructed between 1958 and 1965, the CE literature did not begin to expand until 15 years later and the first commercial instruments were introduced only 4 years ago.

Acceptance of capillary electrophoresis required the convergence of three technologies: development of successful separation chemistries from conventional electrophoresis, development of inexpensive high-quality fused silica tubing from gas chromatography, and development of high-sensitivity optical detectors from high-performance liquid chromatography (HPLC). During the period from 1980 to 1990, as capillary electrophoresis instrumentation became technically feasible, it was becoming increasingly evident that requirements for reproducible and quantitative microscale analyses and reduced consumption of toxic reagents were not being met by conventional electrophoretic and chromatographic techniques. This chapter reviews the milestones in separation science that paved the way for the emergence of capillary electrophoresis and the important events in the development of CE as it is practiced today.

The roots of modern electrophoresis begin in the experiments of Arne Tiselius on moving boundary electrophoresis during the 1930s.[1] In these studies, which contributed to his receipt of the Nobel prize in 1948, protein mixtures were partially resolved in free solution as contiguous bands in U-shaped quartz tubes, and detected by ultraviolet (UV) absorbance. Although refinements were later made,[2,3] the approach was limited by the inherent weaknesses of the moving boundary method: incomplete separation of the sample proteins and the relatively large volume of sample (the length of the starting must be 8 to 15 cm).

Over the next two decades, research continued on development of moving boundary electrophoresis, zone electrophoresis, and isotachophoresis. In particular, major efforts were directed toward improving anticonvective media for zone electrophoresis. In the late 1960s, optimized polyacrylamide gels were developed with stacking and resolving buffer systems for high-resolution separations of native proteins and sodium dodecyl sulfate (SDS)–protein complexes.[4-8] Polyacrylamide gel electrophoresis (PAGE) and SDS-PAGE proved to be relatively rapid and inexpensive methods for characterizing protein mixtures and determining protein monomer molecular weights. Today they are ubiquitous tools found in virtually every biochemistry laboratory, and, coupled with blotting techniques, are used routinely for micropreparative isolation and immunoassay of polypeptides.

Concurrent with the development of PAGE systems for separating proteins on the basis of charge or size, the laboratory of Svensson/Rilbe had been investigating the use of gels containing stable pH gradients for resolving proteins on the basis of their isoelectric points.[9,10] By 1969, Vesterberg devised a method for synthesizing ampholyte mixtures suitable for high-resolution isoelectric focusing (IEF) of protein mixtures, and IEF was added to native and SDS-PAGE as routine tools in biomedical research and diagnosis.[11,12]

0-8493-8690-X/94/$0.00+$.50

In the 1970s IEF separations were combined with SDS-PAGE in the elegant two-dimensional system developed by O'Farrell.[13] With a resolving power of more than 50 components in each dimension, two-dimensional gels were shown to resolve complex physiological samples into several thousand spots;[14] in terms of resolution, two-dimensional PAGE is clearly the leader in "high-performance" separation techniques!

During this decade, while electrophoresis tanks and power supplies became the furniture of biochemistry laboratories, analytical chemists in the pharmaceutical and chemical industries were learning the virtues of high-performance liquid chromatography. The speed and resolution of HPLC made it the technique of choice for analysis of many low molecular weight compounds and for characterization of industrial polymers. Like gas chromatography, HPLC combined the advantages of high resolution, excellent quantitative precision, and automation. By the mid-1980s, the installed base of HPLC instruments was approaching that of gas chromatography.

Even before HPLC became fully established as a tool for analysis of drugs, the landscape in the pharmaceutical industry was changing. The techniques for gene cloning and expression developed by molecular biologists in the 1970s had enabled pharmaceutical companies to consider seriously the design and manufacture of recombinant protein-based biopharmaceuticals by the early 1980s. As bioanalytical chemists began to populate the ranks of pharmaceutical research and quality control laboratories, the limitations of gel electrophoresis in an industrial setting became obvious. Electrophoretic techniques could not meet the demands for quantitative analysis, facile preparative isolation, and automation. Throughout the first half of the 1980s, biochemists in industrial and academic research laboratories turned increasingly to HPLC for isolation and characterization of biomolecules, and commercial instrumentation and chromatographic media for HPLC of peptides, proteins, and nucleic acids became widely available. Although HPLC proved to be quite successful for chromatography of oligonucleotides, peptides, and small proteins, poor recovery and resolution were often observed with large molecules such as structural proteins.

Concurrent with the growth of the biotechnology industry, concern about the introduction of toxic substances into the environment was increasing. Closer scrutiny of laboratory waste management by regulatory agencies resulted in huge increases in disposal costs, and the organic solvent waste generated by HPLC was a major contributor. Several research groups had been investigating the feasibility of open tubular and packed capillary LC,[15-17] and escalating solvent costs sparked renewed interest in these "micro" LC systems. However, technical difficulties in the design of solvent delivery systems helped to deter the major instrument manufacturers from investing in development of micro-LC.

It was against this backdrop of increasing demands for quantitative precision in the analysis of biopharmaceuticals and increasing costs of waste management that capillary electrophoresis developed during the latter part of the 1980s.

The first capillary electrophoresis apparatus was described in 1967 by Hjertén.[3] This instrument (Figure 1), constructed in his laboratory at the University of Uppsala, consisted of a 1- to 3-mm I.D. quartz capillary immersed in a cooling bath. To minimize solute adsorption and electroosmotic flow, the internal surface of the capillary was coated with methylcellulose. In order to reduce convective mixing, the tube was continuously rotated about its axis. The rotating capillary and electrode reservoirs were mounted on a movable carriage; detection of separated components was accomplished by driving the capillary past a fixed UV monitor. Use of this "scanning" detector enabled the peak profiles to be monitored continuously during the separation. In this way, analyte mobilities could be accurately determined. Using this apparatus, Hjertén was able to demonstrate separations of inorganic ions, proteins, nucleic acids, and microorganisms by free zone electrophoresis and isoelectric focusing. Performance of the system was limited by variations in optical quality along the capillary axis and by the

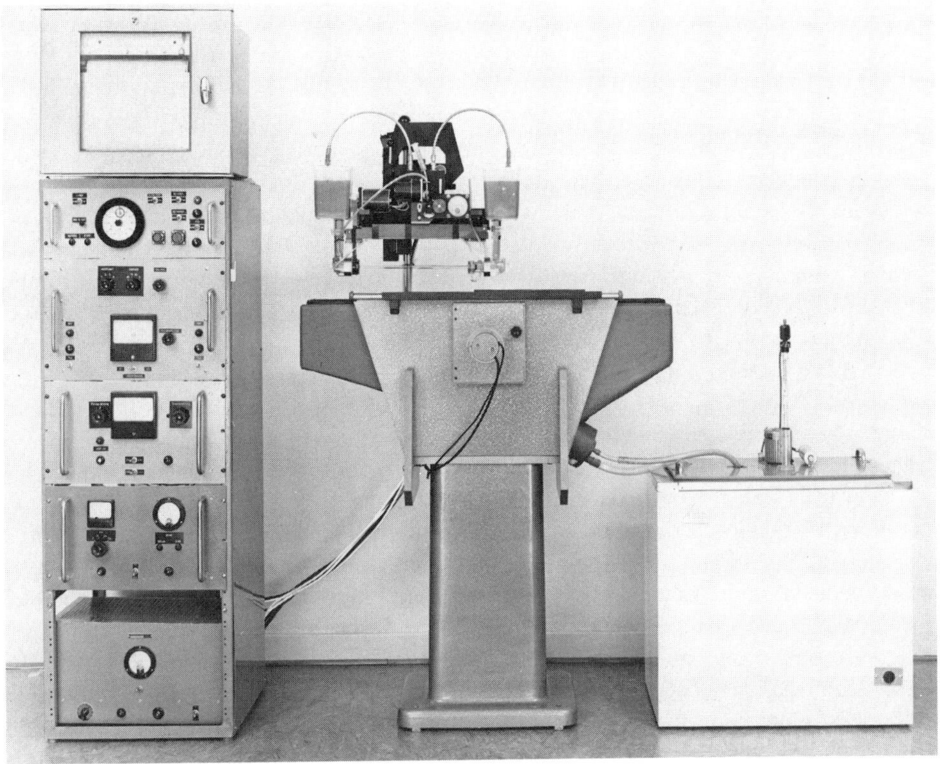

FIGURE 1. Photograph of the capillary electrophoresis system designed by Hjertén.[3] On the left are the high-voltage power supply and detector electronic components, on the right is the cooling reservoir, and in the center is the carriage with capillary and electrode vessels above the immersion bath.

large internal diameters of the tubes. Virtanen[18] later demonstrated that use of a tube of smaller inside diameter (0.2 mm) obviated the need for capillary rotation to eliminate convection, as predicted by Hjertén in 1967.[3]

In spite of the fact that Hjertén's instrument proved many of the concepts in capillary electrophoresis, the technique essentially lay dormant for over a decade. On the other hand, progress in isotachophoresis (ITP) was made during this period. Several groups investigated theoretical aspects of ITP and applied the technique to separations of inorganic and organic ions, proteins, and other substances;[19] at least two commercial instruments were introduced. However, ITP did not gain popular acceptance as an analytical technique, in part because the unusual data presentation was foreign to practitioners of both electrophoresis and chromatography. In addition, ITP was not a true microscale technique since large-diameter (0.2 to 0.4 mm) polytrifluoroethylene (PTFE) tubes were generally used. Currently ITP is enjoying a renaissance as a sample prefractionation and focusing step prior to CE separations.[20-22]

The modern era of capillary electrophoresis is considered by many to have commenced with publication of a series of papers by Jorgenson and Lukacs.[23-25] These papers described simple research instruments consisting of a fused silica capillary, electrode reservoirs, a high-voltage power supply, and (usually) a modified HPLC optical detector. Samples were easily introduced by dipping the capillary inlet into the sample solution and applying voltage or raising the level of the sample vial. Polyimide-clad fused silica capillaries similar to those used for gas chromatography were employed with inside diameters of 75 to 100 μm, and a section of the polyimide burned away to provide a detection "window." The high electroosmotic flow

(EOF) caused by the charge on the underivatized silica surface was used as the motive force to drive both anions and cations past the detection point. A rather simplistic theoretical treatment implied that resolution was limited only by diffusion and that efficiencies of hundreds of thousands of theoretical plates could be achieved due to the plug-flow characteristics of EOF. The simplicity of these early "home-made" systems and the excellent separations of both small and large molecules inspired many researchers in academic and industrial laboratories to build experimental systems to evaluate the technique. Chromatographers in the life sciences were particularly interested in CE, anticipating that free-solution CE or capillary IEF might eliminate the problems encountered in HPLC of large proteins. Major instrument companies began internal evaluation of CE technology as a market opportunity in a period of slackening gas chromatography (GC) and LC sales growth. By the end of the 1980s, the CE literature was expanding exponentially (at this writing there are over 1000 CE references [see Figure 1 of Chapter 27]) and 3 manufacturers had introduced CE systems (currently 15 commercial CE systems are available).

Since the landmark publications in the early 1980s, research in capillary electrophoresis has been vigorously pursued on several fronts. It was known from the work of Hjertén[25] that adsorption of biomolecules to fused silica surfaces necessitated the use of coated capillaries for successful CE separations of proteins, and Hjertén and others[26] have actively sought improvements in capillary coating techniques. Hjertén,[28] Hjertén et al.,[22] Cohen and Karger,[29] Cohen et al.,[30] and Drossman et al.[31] have developed techniques for casting polyacrylamide and agarose gels in capillaries and applied capillary gel electrophoresis to analysis of polymerase chain reaction (PCR) products and to DNA sequencing. Terabe et al.[32,33] and Otsuka and Terabe[34] pioneered the technique of micellar electrokinetic capillary chromatography (MECC), in which both neutral and charged species can be separated by partitioning in osmotically pumped micellar solutions. Hjertén and Zhu[35] and Hjertén et al.[22,36] have adapted isoelectric focusing to the capillary format, using novel chemical mobilization methods to move focused protein zones past the monitor point. Zhu et al. developed non-gel-sieving systems for size-based separation of biopolymers.[37] This technique avoids many of the problems encountered with capillary gel electrophoresis, and has been applied to separation of SDS–protein complexes by Ganzler et al.[38] Jandik et al.[39] have undertaken a comprehensive investigation of high-speed analysis of organic and inorganic ions, using indirect detection. Theoretical studies on the contributions to band broadening in capillary electrophoresis[40,41] and MECC[42] have been published by several investigators. In the area of instrumentation, progress has been made in the development of new detectors, including amperometric,[43] laser-induced fluorescence,[44,45] and CE-mass spectrometer interfaces.[46-48] The application of CE to analysis of the components of a single cell[43] is a dramatic illustration of the power of capillary electrophoresis as a microanalytical technique.

This brief overview of some of the major reserach directions underscores the great flexibility of capillary electrophoresis. Although CE has been an important topic in electrophoresis and chromatography meetings for the last 8 years, it only emerged as a separate entity in 1989 when the first International Symposium on High-Performance Capillary Electrophoresis was convened by B. Karger in Boston. This meeting drew over 400 scientists to attend 100 presentations on the theory and practice of CE. The fact that this symposium is now in its fifth year attests to the arrival of capillary electrophoresis as a viable member of the separation sciences.

REFERENCES

1. **Tiselius, A.,** The moving boundary method of studying the electrophoresis of proteins, Ph.D. thesis, *Nova acta regiae societatis scientiarum upsaliensis*, Ser. IV, Vol. 17, No. 4, Almqvist & Wiksell, Uppsala, Sweden, 1930, pp. 1–107.
2. **Tiselius, A.,** A new apparatus for electrophoretic analysis of colloidal mixtures, *Trans. Faraday Soc.*, 33, 524, 1937.
3. **Hjertén, S.,** Free zone electrophoresis, *Chromatogr. Rev.* 9, 122, 1967.
4. **Raymond, S. and Weintraub, L.,** Acrylamide gel as a supporting medium for zone electrophoresis, *Science,* 130, 711, 1959.
5. **Davis, B. J. and Ornstein, L.,** paper presented at the Society of the Study of Blood, N.Y. Acad. Med., 1959.
6. **Laemmli, U. K.,** Cleavage of structural proteins during the assembly of the head of bacteriophage T4, *Nature (London),* 227, 680, 1970.
7. **Neville, D. M.,** Molecular weight determination of protein-dodecyl sulfate complexes by gel electrophoresis in a discontinuous buffer system, *J. Biol. Chem.*, 246, 6328, 1971.
8. **Hjertén, S.,** as reported by Tiselius, A., in Quarterly Report No.1 to the European Research Office, U.S. Department of the Army, Frankfurt/Main, Germany, APO 757 U.S. Forces, 1960.
9. **Svensson, H.,** Isoelectric fractionation, analysis, and characterization of ampholytes in natural pH gradients. I. The differential equation of solute concentrations at a steady state and its solution for simple cases, *Acta Chem. Scand.*, 15, 325, 1961.
10. **Svensson, H.,** Isoelectric fractionation, analysis, and characterization of ampholytes in natural pH gradients. II. Buffering capacity and conductance of isoionic ampholytes, *Acta Chem. Scand.*, 16, 456, 1962.
11. **Vesterberg, O.,** Synthesis and isoelectric fractionation of carrier ampholytes, *Acta Chem. Scand.*, 23, 2653, 1969.
12. **Vesterberg, O.,** Twenty years of scientific work and development of isoelectric focusing, in *Electrophoresis '83*, Hirai, H., Ed., Walter de Gruyter, Berlin, 1984, pp. 467–479.
13. **O'Farrell, P. H.,** High resolution two-dimensional electrophoresis of proteins, *J. Biol. Chem.*, 250, 4007, 1975.
14. **Anderson, N. G. and Anderson, L.,** Analytical techniques for cell fractions. XXI. Two-dimensional analysis of serum and tissue proteins: multiple isoelectric focusing, *Anal. Biochem.*, 85, 331, 1978.
15. **Ishii, I. J.,** A study of micro-high-performance liquid chromatography. I. Development of technique for miniaturization of high performance liquid chromatography, *J. Chromatogr.*, 144, 157, 1978.
16. **Hirata, Y. and Novotny, M.,** Techniques of capillary liquid chromatography, *J. Chromatogr.*, 186, 521, 1979.
17. **Kucera, P. and Guiochon, G.,** Use of open-tubular columns in liquid chromatography, *J. Chromatogr.*, 283, 1, 1984.
18. **Virtanen, R.,** Zone electrophoresis in a narrow-bore tube employing potentiometric detection, *Acta Polytech. Scand. Chem.*, 123, 1, 1974.
19. **Everaerts, F. M., Beckers, J. L., and Verheggen, T. P.,** Isotachophoresis: theory, instrumentation and applications, in *Journal of Chromatography Library*, Vol. 6, Elsevier, Amsterdam, 1976.
20. **Dolnick, V., Cobb, K., and Novotny, M.,** Capillary zone electrophoresis of dilute samples with isotachophoretic preconcentration, *J. Microcolumn Sep.*, 2,127, 1990.
21. **Foret, F., Sustacek, V., and Boçek, P.,** On-line isotachophoretic sample preconcentration for enhancement of zone detectability in capillary zone electrophoresis, *J. Microcolumn Sep.*, 2, 229, 1990.
22. **Hjertén, S., Elenbring, K., Kilár, F., Liao, J. L., Chen, A. J. C., Siebert, C. J., and Zhu, M.,** Carrier-free zone electrophoresis, displacement electrophoresis and isoelectric focusing in a high-performance electrophoresis apparatus, *J. Chromatogr.*, 403, 47, 1987.
23. **Jorgenson, J. W. and Lukacs, K. D.,** Zone electrophoresis in open-tubular glass capillaries, *Anal. Chem.*, 53, 1298, 1981.
24. **Jorgenson, J. W. and Lukacs, K. D.,** High-resolution separations based on electrophoresis and electroosmosis, *J. Chromatogr.*, 218, 209, 1981.
25. **Jorgenson, J. W. and Lukacs, K. D.,** Capillary zone electrophoresis, *Science*, 222, 266, 1983.
26. **Hjertén, S.,** High-performance electrophoresis, elimination of electroendosmosis and solute adsorption, *J. Chromatogr.*, 347, 191, 1985.
27. **Wehr, T.,** Recent advances in capillary electrophoresis columns, *LC-GC Mag.*, 11, 14, 1993.
28. **Hjertén, S.,** High performance electrophoresis, in *Electrophoresis '83*, Hirai, H., Ed., Walter de Gruyter, New York, 1984, pp. 71–79.
29. **Cohen, A. S. and Karger, B. L.,** High performance sodium dodecyl sulfate polyacrylamide gel capillary electrophoresis of peptides and proteins, *J. Chromatogr.*, 397, 409, 1987.

30. **Cohen, A. S., Paulus, A., and Karger, B. L.,** High-performance capillary electrophoresis using open tubes and gels, *Chromatographia*, 24, 15, 1987.
31. **Drossman, H., Luckey, J. A., Kostichka, A., D'Cunha, J., and Smith, L. M.,** High-speed separations of DNA sequencing reactions by capillary electrophoresis, *Anal. Chem.*, 62, 900, 1990.
32. **Terabe, S., Otsuka, K., Ichikawa, K., Tsuchiya, A., and Ando, T.,** Electrokinetic separations with micellar solutions and open-tubular capillaries, *Anal. Chem.*, 56, 111, 1984.
33. **Terabe, S., Otsuka, K., and Ando, T.,** Band broadening in electrokinetic chromatography with micellar solutions in open-tubular capillaries, *Anal. Chem.*, 61, 251, 1989.
34. **Otsuka, K. and S. Terabe,** Effects of pH on electrokinetic velocities in micellar electrokinetic chromatography, *J. Microcolumn Sep.*, 1, 150, 1989.
35. **Hjertén, S. and Zhu, M.,** Adaptation of the equipment for high-performance electrophoresis to isoelectric focusing, *J. Chromatogr.*, 346, 265, 1985.
36. **Hjertén, S., Liao, J. L., and Yao, K.,** Theoretical and experimental study of high-performance electrophoretic mobilization of isoelectrically focused protein zones, *J. Chromatogr.*, 387, 127, 1987.
37. **Zhu, M., Hansen, D. L., Burd, S., and Gannon, F.,** Factors affecting free zone electrophoresis and isoelectric focusing in capillary electrophoresis, *J. Chromatogr.*, 480, 311, 1989.
38. **Ganzler, K., Greve, K. S., Cohen, A. S., Karger, B. L., Guttman, A., and Cooke, N. C.,** High performance capillary electrophoresis of SDS-protein complexes using UV-transparent polymer networks, *Anal. Chem.*, 64, 2665, 1992.
39. **Jandik, P., Jones, W. R., Weston, A., and Brown, P. R.,** Electrophoretic capillary ion analysis: origins, principles, and applications, *LC-GC Mag.*, 9, 634, 1991.
40. **Hjertén, S.,** Zone broadening in electrophoresis with special reference to high-performance electrophoresis in capillaries: an interplay between theory and practice, *Electrophoresis*, 11, 665, 1990.
41. **Grushka, E., McCormick, R. M., and Kirkland, J. J.,** Effect of temperature gradients on the efficiency of capillary zone electrophoresis separations, *Anal. Chem.*, 61, 241, 1989.
42. **Sepaniak, M. J. and Cole, R. O.,** Column efficiency in micellar electrokinetic capillary chromatography, *Anal. Chem.*, 59, 472, 1987.
43. **Wallingford, R. A. and Ewing, A. G.,** Capillary zone electrophoresis with electrochemical detection in 12.7 μm diameter columns, *Anal. Chem.*, 60, 1992, 1988.
44. **Gassmann, E., Kuo, J. E., and Zare, R. N.,** Electrokinetic separation of chiral compounds, *Science*, 230, 813, 1985.
45. **Kuhr, W. G. and Yeung, E. S.,** Indirect fluorescence detection of native amino acids in capillary zone electrophoresis, *Anal. Chem.*, 60, 1832, 1988.
46. **Smith, R. D. and Udseth, H. R.,** Capillary zone electrophoresis-MS, *Nature (London)*, 331, 639, 1988.
47. **Lee, E. D., Muck, W., Henion, J. D., and Covey, T. R.,** On-line capillary zone electrophoresis-ion spray tandem mass spectrometry for the determination of dynorphins, *J. Chromatogr.*, 458, 313, 1988.
48. **Caprioli, R. M., Moore, W. T., Martin, M., DaGue, B. B., Wilson, K., and Moring, S.,** Coupling capillary zone electrophoresis and continuous-flow fast atom bombardment mass spectrometry for the analysis of peptide mixtures, *J. Chromatogr.*, 480, 247, 1988.

PART I
MODES OF CAPILLARY ELECTROPHORESIS

Chapter 2

INTRODUCTION TO CAPILLARY ELECTROPHORESIS

Robert P. Oda and James P. Landers

TABLE OF CONTENTS

0-8493-8690-X-94/$0.00+$.50

I. INTRODUCTION

A critical prerequisite for any discipline in the chemical or biochemical sciences is the ability to separate the individual components of a mixture. Accordingly, it is often the case that only when a substance is available in pure form can an understanding of its structural and functional character be obtained. While this may appear to be a problem of recent times, this has, indeed, been a practical challenge in the chemical sciences for centuries. As far back as medieval times and until the turn of this century, the main separation methods available to the chemist for the isolation of the components in a mixture were filtration, distillation, and crystallization. However, the growth of a number of fields, most notably biochemistry, has invoked a demand for the development of new analytical separation technologies that can keep pace with the ever-evolving complexity of the systems studied. This has led to the development of a wide range of separation techniques, most of which can be categorized as involving either gas or liquid chromatography (GC or LC), centrifugation, or electromigration (electrophoresis).

In almost all areas of chemistry, including those involving biological systems, analytical- and preparative-scale separation techniques are required for the analysis of the multicomponent samples. In comparison with the separation technologies amenable to the study of purely chemical systems, the separation and analysis of biological mixtures is further complicated by the idiosyncratic characteristics inherent to biomolecules. The separation technologies must be amenable to (1) the strict pH and temperature requirements necessary to offset the often unstable nature of biomolecules, (2) the extremely complex nature of crude biological samples, where the desired component is usually present as a minor percentage of the total, and (3) distinguishing the desired component among other structurally similar components.

It is in this respect that electrophoresis has had such a tremendous impact on biomedical research. The pioneering experiments of Tiselius in 1937[1] showing the separation of serum proteins (albumin and α-, β-, and γ-globulin) by "moving boundary electrophoresis" provided the first intimation of the potential use of electrophoretic analysis for biologically active molecules. Several other forms of electrophoresis followed from the work of Tiselius, including paper electrophoresis, which was shown to be applicable to the analysis of a variety of molecules, and later, polymer and agarose gels for the analysis of peptides, proteins, DNA, and RNA.

From a historical perspective, capillary electrophoresis (CE) appears to have arrived on the analytical scene rather late. It was not until the late 1970s/early 1980s that CE was shown to be a viable analytical technique by Mikkers et al.[2] and Jorgenson and Lukacs.[3] However, it was the pioneering work of Hjertén[4] a decade earlier that laid the groundwork for the capillary electrophoretic analysis of a diverse array of analytes ranging from small molecules (inorganic ions, nucleotides) to macromolecular structures such as proteins and viruses. As discussed in

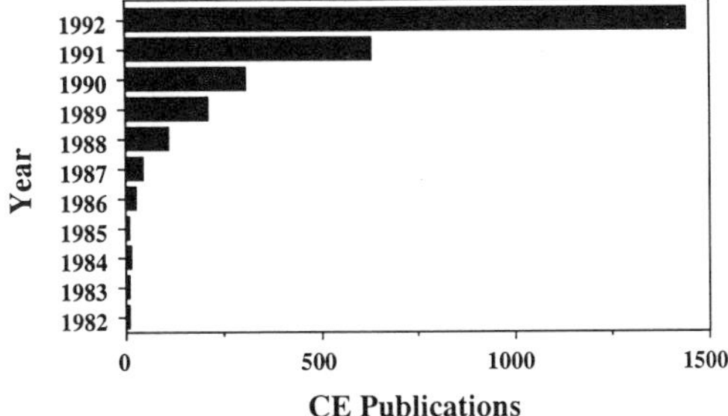

FIGURE 1. Exponential growth of CE literature.

Chapter 1, Hjertén had constructed a functional CE unit, albeit with 2-mm tubes, as early as 1959. The rejuvenation of CE following the 12-year dormancy appears to have resulted, at least in part, from advances in technology, particularly from the availability of narrow bore capillaries.

Interest in capillary electrophoresis has literally exploded over the past 5 years, and its matriculation to the modern analytical laboratory has begun. One very good indicator of this is the attendance at the annual International Symposium on High Performance CE. Initiated and organized by B. Karger in 1989, both attendance and the number of papers represented at this meeting since its inception 5 years ago is staggering. Papers presented are published in a special volume of the *Journal of Chromatography* dedicated specifically to the meeting proceedings.[5-7] With regard to CE literature, Figure 1 shows that there has been an exponential growth over the last decade, with predictions for 1993 exceeding 2000 publications. Several excellent reviews have been written that describe various aspects of the CE field.[8-10] Much of the recent focus in CE has been, and will continue to be, on methods development and the application of CE to a variety of modern-day analytical problems. The chapters to follow demonstrate that, unlike other forms of electrophoretic analysis, CE is not restricted to use by the biochemist. Its diversity is highlighted by its application to a wide spectrum of analyses in a variety of disciplines ranging from the detection and quantitation of priority pollutants in environmental samples, to the analysis of the human immunodeficiency virus (HIV) in blood.

Consistent with the theme of this handbook, this chapter focuses on the practical aspects of CE. It is not meant to be a comprehensive review of the general literature, but instead attempts to give the reader a flavor of the potential power of CE for solving real analytical/ separation problems. It aims to (1) provide the reader with an understanding of the basic principles governing CE, (2) introduce the parameters important to the CE practitioner, and (3) provide a logical approach to development of methods for CE analysis. As a result of the practice-oriented focus, a limited presentation of the theoretical aspects of CE is given in the text of this chapter, and a more complete discussion is provided in Appendix I at the end of the handbook. Appendix II describes, with examples, calculations of use to the CE practitioner. Finally, Appendix III provides the reader with a guide to troubleshooting typical problems that may be encountered with CE.

II. CAPILLARY ELECTROPHORESIS

A. WHY ELECTROPHORESIS IN A CAPILLARY?

Electrophoresis has been one of the most widely used techniques for the separation and analysis of ionic substances. Almost all modes of electrophoresis utilize some form of solid support for the electromigration of analytes. This has been in the form of paper (high-voltage separation of amino acids and other small organic molecules) or, more commonly, polyacrylamide or agarose gels, which have been used extensively for both protein and DNA/RNA analysis. There is little doubt that gel electrophoresis, in one form or another, has been the workhorse for modern research and an invaluable analytical tool. However, despite its ability for high-resolution analysis of complex systems, it suffers from several disadvantages, most of which scientists have endured for lack of an effective alternative. Perhaps the most obvious disadvantage is the speed of separation, which is ultimately limited by Joule heating (the heating of a conducting medium as current flows through it). The relatively poor dissipation of Joule heat in slab systems limits the use to low-potential electric fields. Moreover, from a methodological perspective, it is a series of cumbersome, time-consuming tasks from the casting of the gel, preparation and loading of samples, resolution of the ionic/molecular species, to the final stage where the gel is stained, and the results obtained. Other problems include (1) poor reproducibility, particularly with two-dimensional analysis, (2) analyte-dependent differences in staining, which makes quantitative accuracy difficult to achieve, e.g., glycosylated proteins have different dye-binding properties, and (3) the cumbersome methodology involved in modern gel electrophoresis has made it virtually impossible to automate the entire electrophoretic procedure.

The use of capillaries as an electromigration channel for separation of a diverse array of analytes, including biopolymers, presents not only a unique approach to separation, but also several advantages over the standard solid support approach. The physical characteristics of narrow bore capillaries make them ideal for electrophoresis. Fused silica capillaries employed in CE typically have internal diameters of 20 to 100 μm (375-μm O.D.), lengths from 20 to 100 cm, and are externally coated with the polymeric substance polyimide, which imparts tremendous flexibility to a capillary that would otherwise be very fragile (Figure 2). The high surface-to-volume ratio of capillaries with these dimensions allows for very efficient dissipation of Joule heat generated from applied fields. This is illustrated in Table 1, which compares the surface-to-volume ratio of a standard analytical slab gel system with standard capillaries used for CE. From this table it is clear why the slab gel is limited to electrical fields in the range of ≈15 to 40 V/cm resolving gel whereas up to 800 V/cm can be applied to a capillary containing the same type of gel matrix.[11] As a result of this tremendous ability to dissipate heat, electrophoretic separations can easily be performed at up to 30,000 V, with the external capillary temperature thermostatted at approximately room temperature.

In addition to efficient Joule heat dissipation, there are many advantages associated with the use of capillaries for electrophoresis. With typical CE capillaries, the small dimensions yield total column volumes in the microliter range, hence leading to the use of only minute quantities of buffer. Adhering to the rule of thumb restricting sample volume to 1 to 5% of the total capillary volume, sample volumes introduced into the capillary are in the nanoliter range (as low as 0.2 nl). As a result, as little as 5 μl of sample will suffice for repetitive analysis on some commercial units. Considering the small reagent (buffer) and sample requirements, as well as the rapid analysis times associated with the application of high fields (30,000 V), it is clear that CE has tremendous potential for extrapolation to a diverse number of analytical and separation problems.

B. THE FAMILY OF CAPILLARY ELECTROPHORESIS TECHNIQUES

In much the same way that standard gel electrophoretic techniques have diversified, so has CE. This has resulted in a family of specialized techniques that collectively constitute "capillary electrophoresis". The main modes of CE that have been developed and are presently being exploited include capillary zone electrophoresis (CZE), often referred to as free-solution CE (FSCE), micellar electrokinetic capillary chromatography (MECC; Chapter 3), capillary isoelectric focusing (CIEF; Chapter 4), capillary isotachophoresis (CITP; discussed in Chapter 5), and capillary gel electrophoresis (CGE; Chapter 6) (Table 2). While the CITP mode has

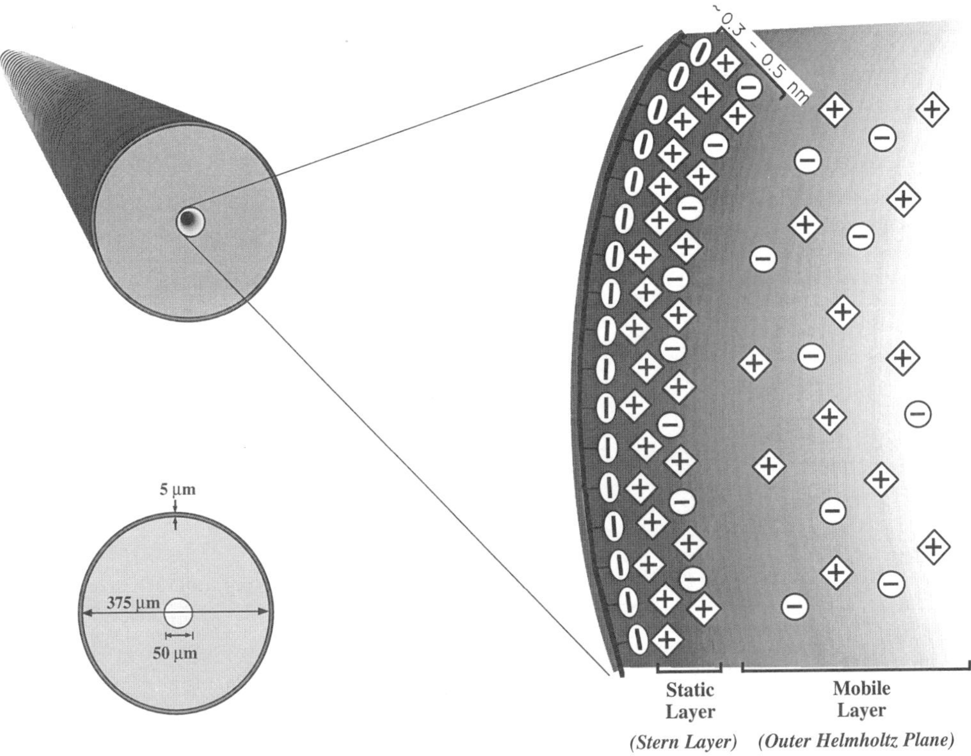

FIGURE 2. Diagram of the capillary and the ionic layer.

TABLE 1
A Comparison of Surface-to-Volume Ratios for a Slab Gel and 57-cm Capillaries with Various Internal Diameters

	Surface area (mm^2)	Volume (µl)	Surface-to-volume ratio
Slab gel (14 × 11.5 × 0.15 cm)	32,200	24,150	1.3
Capillary			
20 µm I.D.	35.81	0.179	200
50 µm I.D.	89.53	1.119	80
75 µm I.D.	134.3	2.518	53
100 µm I.D.	179.1	4.477	40
200 µm I.D.	358.1	17.907	20

the potential for being a separate analytical mode of CE, at the present time it is primarily used as an on-capillary preconcentration technique.

Capillary zone electrophoresis is the most universal of the techniques, having been shown useful for the separation of a diverse array of analytes varying in size and character. While a brief description of CZE is given in this chapter, various aspects of the technique will be discussed in a number of other chapters in this handbook (Chapters 8–13 and 15–23). The other CE modes are covered exclusively, and in detail, in separate chapters dedicated specifically to the theoretical and practical aspects of those modes, along with examples illustrating applications amenable to that particular mode. Table 3 provides a reference table categorizing the modes used for the analysis of various classes of samples/analytes.

TABLE 2
Different Modes of Capillary Electrophoretic Analysis

Mode	Principle of separation
Capillary zone electrophoresis (CZE)	
Continuous	Charge-to-mass ratio
Discontinuous	Charge-to-mass ratio
Capillary isoelectric focusing (CIEF)	Isoelectric point
Capillary isotachophoresis (CITP)	Mobility — separates analytes with same mobility as the buffer ions, which differ at inlet and outlet
Micellar electrokinetic capillary chromatography (MECC)	Charged species — Charge-to-mass ratio and partitioning into micelles based on hydrophobicity
	Neutral species — partitioning into detergent micelles based on hydrophobicity
Capillary gel electrophoresis	
Nondenaturing	Charge-to-mass ratio
Denaturing (SDS, urea)	Mass

TABLE 3
Mode of Capillary Electrophoresis Used for Analysis of Different Classes of Analytes

Small ions	Small molecules	Peptides	Proteins	Oligonucleotides	DNA
CZE	MECC	CZE	CZE	CGE	CGE
ITP	CZE	ITP	CGE	MECC	
	ITP	MECC	CIEF		
		CIEF	ITP		
		CGE			

C. CAPILLARY ZONE ELECTROPHORESIS

Capillary zone electrophoresis is not only the simplest form of CE, but also the most commonly utilized. Discussion of this mode allows for the presentation of the general design of instrumentation for CE. With the simple addition of specialized reagents to the separation buffer, the same instrumentation is used to perform other modes of CE: addition of (1) surfactants for MECC, (2) a sieving matrix (unpolymerized or polymerized) for CGE, and (3) ampholines for CIEF. A discussion of CZE also allows for analysis of some of the basic principles governing analyte separation by this technique.

1. Instrumention and Capillary Electrophoretic Analysis

The components of a CE system are given in Figure 3. The basic instrumentation involves a high-voltage power supply (0 to 60 kV), a capillary (externally coated with polyimide) with an internal diameter ≤200 μm, two buffer reservoirs that can accommodate the capillary and the electrodes connected to the power supply, and a detector. As is discussed later in this chapter, and in detail by Nelson and Burgi (Chapter 21), thermostatting of the capillary is critical for obtaining efficient separations and, hence, some type of capillary thermostatting system should be used. This is typically in the form of forced air convection or liquid. To perform a capillary zone electrophoretic separation, the capillary is filled with an appropriate separation buffer at the desired pH and sample is introduced at the inlet. Both ends of the capillary and the electrodes from the high-voltage power supply are placed into buffer reservoirs and up to 30,000 V applied to the system. The ionic species in the sample plug migrate with an electrophoretic mobility (direction and velocity) determined by their charge and mass, and eventually pass a detector where information is collected and stored by a data acquisition/analysis system.

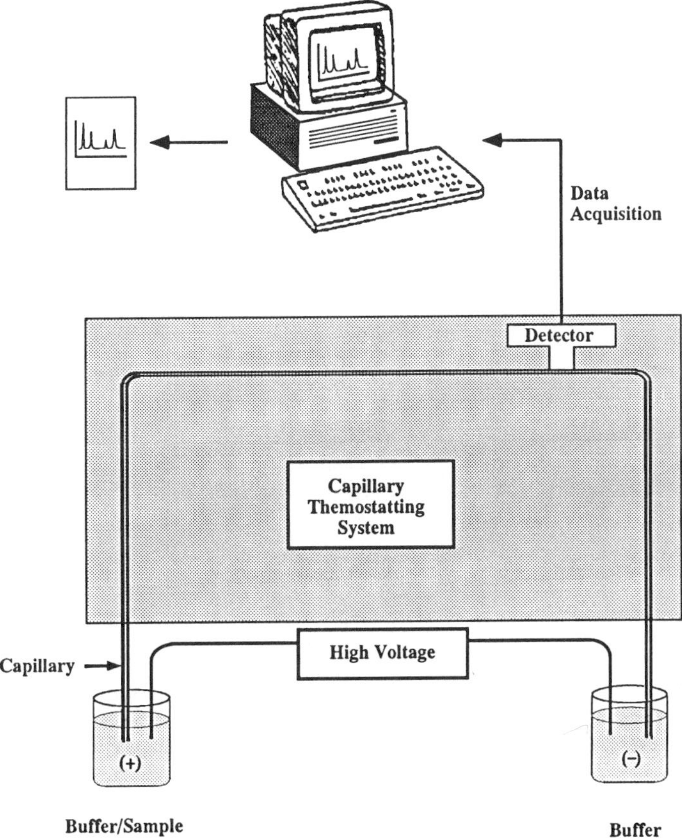

FIGURE 3. General schematic of a CE instrument.

Electrophoretic mobility (μ) of a charged molecular species can be approximated from the Debye–Huckel-Henry theory,

$$\mu = q/6\pi\eta r \tag{1}$$

where q is the charge on the particle, η is the viscosity of the buffer, and r is the Stokes' radius of the particle. The mass of the particle may be related to the Stokes' radius by $M = (4/3)\pi r^3 V$, where V is the partial specific volume of the solute. While one might infer the direct proportionality of mass and radius of the particle, empirical data suggest modifications of Equation 1 to allow for the nonspherical shape, counterion effects, and nonideal behavior of proteins and other biological molecules.[12]

With capillary zone electrophoresis, the "normal" polarity is considered to be (inlet [+] → detector → [–] outlet) as shown in Figures 3 and 4. As electrophoresis ensues under these conditions, the analytes separate according to their individual mobilities and pass the detector as *analyte zones* (hence, the term capillary *zone* electrophoresis). The fact that, under normal conditions, all charged species (net positive, negative, or neutral) pass the detector indicates that a force in addition to electrophoretic mobility is involved. If the applied field were the only force acting on the ions, net positively charged (cationic) substances would pass the detector while neutral components would remain static (i.e., at the inlet) and anionic components would be driven away from the detector. It is clear that, if this were the case, CE would be of limited use. Fortuitously, there is another force, *electroosmotic flow*, driving the

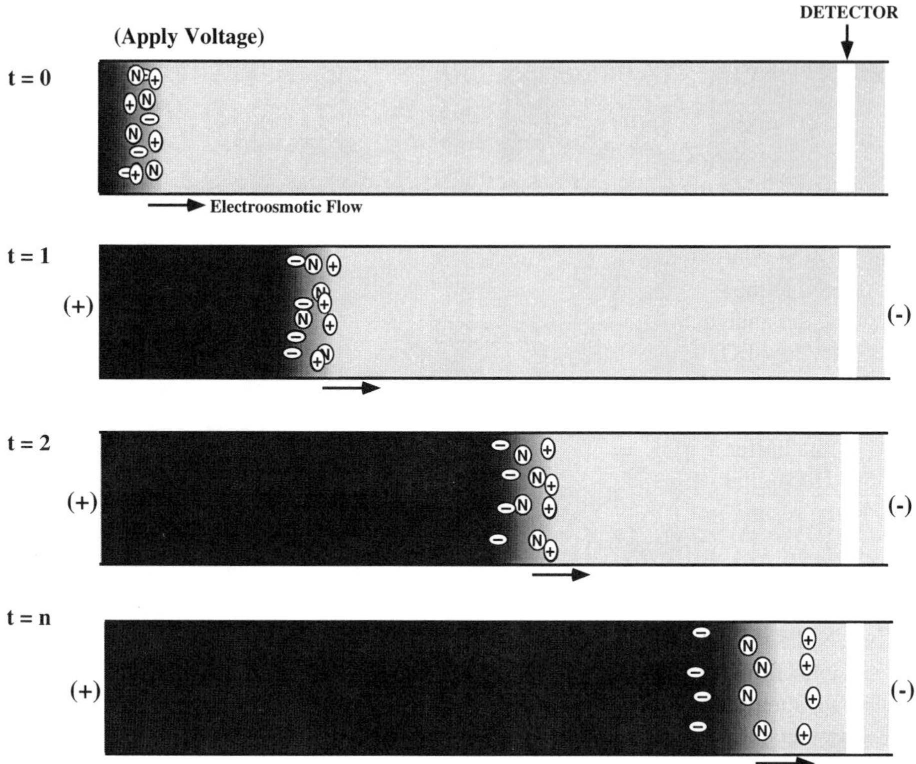

FIGURE 4. Mobility of charged and uncharged molecules in an applied field.

movement of all components in a capillary toward the detector when under an applied field (and a normal polarity, + → –). Electroosmotic flow, or EOF, plays a principal role in many of the modes of CE and most certainly in CZE. This is discussed briefly in the next section and in detail by Tsuda in Chapter 22.

2. Role of Electroosmotic Flow in Capillary Electrophoretic Analysis

Electroosmotic flow was first identified in the late 1800s when Helmholtz carried out experiments involving the application of an applied electrical field to a horizontal glass tube containing an aqueous salt solution.[13] Curious about the ionic character of the inner wall and the movement of ions, Helmholtz pointed out that the silica imparts a layer of negative charge to the inner surface of the tube, which, under an applied field, led to the net movement of fluid toward the cathode. A century later, this phenomenon plays the fundamental role in capillary electrophoretic analysis. Moreover, as is discussed in Chapter 22, the importance of the control of EOF has been realized and has become the focus of many research groups.

As a continuation of the pioneering work of Helmholtz, the basic principles governing EOF have been evaluated extensively (Chapter 22 and references therein). As shown by the expanded region of the inner wall of a capillary in Figure 2, the ionized silanol groups (SiO^-) of the capillary wall attract cationic species from the buffer. Obviously, the buffer pH will determine the fraction of the silanol groups that will ionized. The ionic layer that is formed has a positive charge density that decreases exponentially as the distance from the wall increases. The double layer formed closest to the surface is termed the *Stern layer* and is essentially static. A more diffuse layer formed distal to the Stern layer is termed the *outer Helmholtz plane* (OHP). Under an applied field, cations in the OHP migrate in the direction of the cathode carrying water of hydration with them. This EOF or *bulk flow* acts as a pumping mechanism to propel all molecules (cationic, neutral, and anionic) toward the detector with separation ultimately being determined by differences in the electrophoretic migration of the individual analytes. The magnitude of EOF is dependent on a number of parameters including pH (↑ pH → ↑ EOF) and ionic strength (↑ ionic strength → ↓ EOF). The importance of EOF in capillary electrophoretic analysis is highlighted in Figure 4. The buffer entering the capillary inlet behind the sample plug is represented by a graded shading. As electrophoretic migration occurs, all analytes are swept toward the detector by bulk flow. Provided that the EOF is adequate but not too strong, the respective electrophoretic mobilities of each of the analytes leads to the formation of zones by the time they pass the detector. If the EOF is slow, diffusion of the analyte zones could result in substantial band broadening and, under conditions of very low EOF, some of the analytes may not reach the detector within a reasonable analysis time.

As mentioned above, the EOF is pH dependent and can be quite strong. For example, in 20 m*M* borate buffer at pH 9.07 the EOF is ≈2 mm/s, which translates to a flow of ≈4 nl/s in a 50-μm I.D. capillary. The inclusion of electroosmotic flow in the calculation of velocity is essential and results in

$$v_i = \mu_{app}E = (\mu_{ep} + \mu_{eo})E \qquad (2)$$

where μ_{ep} is the mobility due to the applied electric potential and μ_{eo} is the mobility due to electroosmotic flow. A more complete theoretical treatment of EOF, with a discussion of experimental attempts to understand and control this phenomenon, is presented in Chapter 22.

From a practical perspective, it is important to have some idea of the magnitude of EOF under the conditions used for a given separation for two reasons. First, under conditions where EOF is very fast, components of the mixture may not have adequate "on-capillary time" for separation to occur. Second, it is useful to know where neutral compounds migrate in the obtained electropherogram. Knowing this, information about the charge character of the sample components is obtained based on whether they migrate faster (cationic) or slower (anionic) than EOF. The EOF marker is also useful as an internal standard for calculating "relative" migration times from the apparent (observed) migration times. Some compounds that adequately serve as neutral markers include dimethylformamide (DMF), dimethyl sulfoxide (DMSO), and mesityl oxide, although, due to rapid volatilization, we have found this last compound to be of limited use with samples maintained at ambient or higher temperatures. Typical working concentrations for use of these compounds as neutral markers is a 0.1% solution in water; a 1-sec hydrostatic (0.5 psi) loading of a 0.1% solution provides an adequate signal with ultraviolet (UV) detection at either 214 or 200 nm.

These markers are not of use in low-pH separations, where most analytes migrate markedly faster than the EOF (which is extremely slow). If a peak is still required as an internal standard for correction of migration time, any compound with a fast cathodic electrophoretic mobility will suffice as frontal marker. We have found that a synthetic peptide containing seven lysine residues and a single tryptophan (K_3WK_4) functions adequately for this purpose at pH 2.5. A frontal marker may also be useful at higher pH, e.g., when the neutral marker comigrates with a species of interest; under these conditions, cationic species such as normetanephrine can suffice.

3. Description of the Electrophoretic Process

In the early 1980s, Jorgenson and Lukacs discussed the theory and basis for electrophoretic separations in capillaries.[3] The following brief discussion develops, albeit at a basic level, the theoretical concepts describing the movement of a charged molecule in a buffer-filled capillary under an applied electrical field and the shape of the zone during the electrophoretic process. The separation process is discussed in terms of resolution (how adequately two components are separated) and efficiency (how long the separation takes). Finally, we discuss factors that affect the resolution. For more detailed evaluation of these processes, the reader is referred to several excellent discussions on this subject.[14,15]

a. Mobility of a Solute in an Applied Field

A charged particle in solution will become mobile when placed in an electric field. The velocity, v_i, acquired by the solute under the influence of the applied voltage H, is the product of μ_{app}, the apparent solute mobility, and the applied field E ($E = H/L$, where L is the length of the field), as described in Equation 2 above.

As the solute migrates a distance x from the origin after time t, the zone may be described by a Gaussian curve, with height proportional to the initial concentration, and peak width determined by dispersion. The width at the base is, by definition, 4σ. A fuller description of the analyte zone is found in Appendix 1.

b. Resolution and Efficiency

The simplest way to characterize the separation of two components is to divide the difference in migration distance by the average peak width:

$$\text{Res} = 2\,(x_2 - x_1)/(w_1 + w_2) \tag{3}$$

where x_i is the migration distance of the analyte i, subscript 2 denotes the slower moving component, and w is the width of the peak at the baseline. We can readily see that the position of a peak, x_i, is determined by the electrophoretic mobility. The peak width, w, is determined by diffusion and other dispersive phenomena.

From the equation describing a Gaussian curve, the two peaks touch at baseline when $\Delta x_i = 4\sigma$, and Res = 1, or

$$\text{Res} = \Delta x_i/4\sigma \tag{4}$$

or substituting from Equation 2, and defining the number of theoretical plates as $N = L^2/\sigma^2$,

$$\text{Res} = (1/4)(\Delta\mu_{app}/\mu_{avg})N^{1/2} \tag{5}$$

where $\Delta\mu_{app}$ is the apparent mobility of the two analytes, and μ_{avg} is the average mobility of the two analytes.

The utility of Equation 5 is that it permits one to independently assess the two factors that affect resolution, selectivity and efficiency. The selectivity is reflected in the mobility of the analyte(s), while the efficiency of the separation process is indicated by N. Another expression for N is derived from the definition, using the width at half-height of a Gaussian peak,

$$N = 5.54(L/w_{1/2})^2 \tag{6}$$

where 5.54 = 8 ln 2, and $w_{1/2}$ is the peak width at half-height.

At this point, it is important to point out that it is misleading to discuss theoretical plates in electrophoresis. The concept is a carryover from chromatographic theory, where a true partition equilibrium between two phases is the physical basis of separation. In electrophoresis, separation of the components of a mixture is determined by their relative mobilities in the applied electric field, which is a function of their charge, mass, and shape. The theoretical plate is merely a convenient concept to describe the analyte peak shape, and to assess the factors that affect separation.

While N is a useful concept to compare the efficiency of separation among columns, or between laboratories, it is difficult to use to assess the factors that affect that efficiency. This is because it refers to the behavior of a single component during the separation process, and is unsuited to describing the separation of two components or the resolving power of a capillary. A more useful parameter is the height equivalent of a theoretical plate (HETP),

$$\text{HETP} = L/N = \sigma^2_{tot}/L \tag{7}$$

The HETP might be thought of as the fraction of the capillary occupied by the analyte. It is more practical to measure HETP as an index of separation efficiency, rather than N, as the individual components that contribute to HETP may be individually evaluated and combined to determine an overall value. A consideration of all the factors influencing σ^2_{tot} should include not only diffusion, but also differences in mobility or diffusion generated by Joule heating, the reality that the sample is not introduced as a thin disk but as a plug of finite dimensions, and interaction of analytes with the capillary wall. Discussion of each of these factors is presented in Appendix 1. The variance of multiple dispersive phenomena on the analyte may be summed as

$$\sigma^2_{tot} = \sigma^2_{diff} + \sigma^2_T + \sigma^2_{\text{int}} + \sigma^2_{wall} \tag{8}$$

4. The Capillary

The capillary and its condition are key to successfully performing CE analysis efficiently and reproducibly. Hence, the treatment of the capillary prior to, during, and following electrophoretic separations is crucial. Microbore capillaries employed in CE can be purchased from a number of companies in either bare silica or coated format. As shown in Figure 2, the standard fused silica capillary typically has an internal diameter of 50 to 100 μm, although a range of internal and external diameters are commercially available (20- to 200-μm I.D.; 100- to 400-μm O.D.). The following sections describe some pertinent points regarding capillary preparation, conditioning prior to use, maintenance, and storage.

a. Preparation for Capillary Electrophoresis

For use in CE, the capillary is cut to the appropriate length with a ceramic knife, or an ampoule file, so that both ends are square and flat. This is of particular importance with the inlet end of the capillary where sample is to be introduced. A "window" is created in the polyimide coating so that on-line detection is possible. The window should ideally be approximately 0.2 cm in length (no longer than 1.0 cm) and can easily be made by burning off the polyimide with a flame and wiping the surface with an ethanol-soaked lens tissue. Another, more labor-intensive method is to scrape the polyimide surface off, being careful not to scratch the silica or break the capillary. Several mechanical devices have been devised to accomplish this.[16,17] Caution must be taken when handling the capillary once the window is created since the window area is extremely fragile. It is important to note that with internally coated capillaries, it is not advisable for detector windows to be created by burning off the polyimide, since the heat required may damage the internal coating. Alternatively, dropwise addition of 130°C sulfuric acid[18] or concentrated KOH[19] from a burette or glass pipette will remove the polyimide coating (create a window) without damaging the internal coating.

Prior to using new bare fused silica capillary for analysis, the capillary should be "preconditioned" by rinsing with 5 to 15 column volumes of 100 m*M* NaOH followed by 5 to 15 column volumes of water, before rinsing with 3 to 5 column volumes of separation buffer. If the capillary is coated, preconditioning should be performed as per the protocols recommended by the manufacturer or as deemed necessary by the surface coating.

When the capillary is installed, it may be useful to determine the "capillary fill time". This will be useful for determining the length of time for each rinse step (NaOH and separation buffer), as well as for diagnosing potential capillary problems such as complete or partial

TABLE 4
Capillary Fill times for Capillaries Typically Used for Capillary Electrophoresis[a]

Capillary dimensions	Capillary fill time (min)
27 cm × 50 μm	0.1
47 cm × 50 μm	0.2
57 cm × 50 μm	0.4
87 cm × 20 μm	6.0
87 cm × 50 μm	0.9
87 cm × 100 μm	0.1
107 cm × 50 μm	1.5

[a] Assuming a positive hydrostatic pressure of 20 psi.

obstruction. One approach for accomplishing this (with UV detection) is to pressure rinse the capillary with water, zero the detector, and follow with a rinse with 100 m*M* NaOH. The time required for a maximum change in absorbance to occur is the "fill time" to the detector. Factoring in the length between the detector and the capillary outlet yields the capillary fill time. Table 4 provides these values for capillaries of various dimensions, using a hydrostatic rinse pressure of 20 psi.

b. Conditioning

When using a newly preconditioned capillary or changing to a new separation buffer, the capillary should be adequately equilibrated with the separation buffer — this is termed "conditioning". This is particularly important when a phosphate-containing buffer is involved. When changing to a phosphate-containing buffer, equilibration with the new buffer should be a minimum of 4 h prior to use.[20] This process can be slightly accelerated by applying a separation-scale voltage to the system. When preparing to use a new separation buffer in a capillary that has been equilibrated with a phosphate-containing buffer, extensive equilibration of the capillary is typically required. The capillary should be equilibrated with the desired separation buffer for at least several hours prior to use and, if possible, equilibrated overnight. It is for this reason that "dedication" of capillaries to individual buffer systems is highly recommended. In this manner, no pre-equilibration time is required, since the capillary can be stored in the buffer to which it is dedicated.

c. Maintenance

Capillary maintenance plays a critical role in attaining reproducible results with CE. As with any untreated fused silica surface, ionized silanol groups are ideal for interaction with charged analytes, particularly peptides and proteins in neutral/basic pH buffers. Hence, following each separation, the capillary surface must be "regenerated" or "reconditioned", i.e., cleansed of any wall-adsorbed material. This is accomplished by following each run with a 3- to 5-column volume rinse with 100 m*M* NaOH, followed by flushing with 5 to 8 column volumes of fresh separation buffer.

When using acidic buffers, rinsing with NaOH may be disadvantageous since the drastic changes in pH may induce the requirement for extensive rinsing with the separation buffer for adequate pH re-equilibration. Under these conditions, it may be advisable to follow the NaOH rinse with a brief rinse with concentrated separation buffer (10× or more), followed by the normal rinse with the (1×) separation buffer. Alternatively, depending on the sample, the 10× separation buffer rinse alone may be adequate for regenerating the capillary and, hence, the NaOH rinse may be avoided completely. Peptide separation in phosphate buffer, pH 2.5, using 1.0 *M* phosphate buffer for regeneration is a good example.

d. Storage

Due to the small dimensions of the capillary, plugging is always a potential problem. This may occur as a result of solvent evaporation at the capillary ends, which leads to salt crystal formation. This is easily avoided in several ways. Dedicated capillaries to be stored for a short time (less than a week) should be rinsed with 100 m*M* NaOH, re-equilibrated with separation buffer, and stored with the ends immersed in buffer/distilled water, capped with silicone rubber stoppers, or inserted into crimped narrowbore teflon tubing. If the capillary is to be stored for longer periods of time (i.e., greater than 1 week), it should be washed with 100 m*M* NaOH, rinsed thoroughly with distilled water, flushed with 5 column volumes of methanol, dried, and stored. Alternatively, the capillary can be flushed with water, air dried (by pressure), and stored dry. If a capillary has been used with a detergent/surfactant-containing solution, all traces of the additive must be removed before storage. If this is not done, the capillary will require longer than normal re-equilibration on reuse due to the presence of the residual additive.

Before storing, coated capillaries should be thoroughly rinsed with the appropriate buffers or solvents. For example, we have found that it is best to rinse hydrophobically coated capillaries with 1 column volume of 100 m*M* NaOH, extensively with water (5 to 10 column volumes), and finally with methanol (5 to 10 column volumes), followed by drying and storage. With capillaries used with physical gels, the capillary should be rinsed and stored with fresh buffer, as described above.

e. Coating the Inner Surface

As will be discussed in detail by Mazzeo and Krull (Chapter 18) and elaborated on by Swedberg (Chapter 19), capillary coating for particular CE applications can be of paramount importance due to the affinity of some analytes, particularly proteins, for interaction with the wall. A number of methodologies have been described in the literature for coating the internal surface of the capillary (Chapters 18 and 19 and the references therein). The pH stability, as well as the effectiveness for preventing adsorption of analyte to the wall (i.e., separation efficiency), varies with the chemical nature of the coating. To circumvent the task of coating capillaries manually, precoated capillaries are commercially available from several companies. Table 5 provides a brief description of the manufacturers of capillaries that can be used with any type of CE instrumentation and the types of coatings available. Preconditioning and regeneration of commercially coated capillaries should be carried out as suggested by the manufacturer.

III. BUFFERS

A. SELECTION AND PREPARATION OF A SEPARATION BUFFER

A wide variety of electrolytes can be used to prepare buffers for CZE separations. When using absorbance detection, a major requirement of any component used in the buffer system is a low-UV absorbance at the wavelength used for detection. This restriction substantially limits the choices to a moderate number of non-UV-absorbing electrolytes. For low-pH buffers, phosphate and citrate have commonly been used, although the latter absorbs strongly at wavelengths <260 nm. For buffers in the basic pH range, Tris, Tricine, borate, and CAPS are acceptable electrolytes. For detection modes other than UV absorbance, a number of other electrolytes can be utilized. Table 6 presents a list of useful electrolytes for preparing separation buffers for CZE. In addition, an indication of their useful pH range and minimum functional wavelength (for absorbance detection) is given.

Perhaps one of the most important aspects of buffers used for CE is the requirement for pure reagents. It is recommended that the water used for buffer preparation be of Milli-Q (or equivalent)-purified quality and that the reagents be highly purified or ultrapure (Gold Label

TABLE 5
Commercially Available Coated Capillaries That Can Be Used with Any Instrumentation

Nature of coating	Isco	J&W	Supelco	Polymicro-technologies
Hydrophilic	CE-100 (C_{18}/Brij-35®) CE-200 (glycerol) CE-300 (sulfonic acid)	DB-1 (dimethyl-polysiloxane)	CE-elect™ H (C_1) CE-elect H1 (C_8)	
Hydrophobic		DB-17 (50% phenylpoly-siloxane//50% Methyl-polysiloxane) DB-wax (polyethylene glycol) μSil-Protein	P-1	
Gel-filled				
Chemical		μPAGE (3 or 5% poly-acrylamide)[a]		
Physical		DB-17 + HPMC[b]		
None (bare fused silica)	Yes	Yes	Yes	Yes (in various geometries and with various external coatings)

[a] Coated capillary containing "chemical gel".
[b] HPMC, hydroxypropylmethylcellulose — "physical gel". See Ref. 54.

TABLE 6
Commonly Used Capillary Electrophoresis Buffers and Their Associated Properties

Buffer	Useful pH range	Minimum usful λ (nm)
Phosphate	1.14–3.14	195
Citrate	3.06–5.40	260
Acetate	3.76–5.76	220
MES[a]	5.15–7.15	230
PIPES[a]	5.80–7.80	215
Phosphate	6.20–8.20	195
HEPES[a]	6.55–8.55	230
Tricine[a]	7.15–9.15	230
Tris	7.30–9.30	220
Borate	8.14–10.14	180
CHES	9.50	<190

[a] Zwitterionic buffers. MES, Morpholineethanesulfonic acid; PIPES, piperazine-*N,N'*-bis(2-ethanesulfonic acid); HEPES, *N*-2-hydroxyethlpiperazine-*N'*-2-ethanesulfonic acid; CHES, 2-[*N*-cyclohexylamino] ethane sulfonic acid.

Grade, Aldrich, Milwaukee, WI) reagents; the small amounts required for CE do not entail great expense. Passage of buffers through a 0.45 μm filter is recommended prior to use to remove any particulate matter. For buffers stored at 4°C, it is imperative that they be brought to room temperature and thoroughly degassed.

Buffer additives may also be used to obtain enhanced or differential selectivity in a separation. Several classes of additives have been identified as applicable to CE. Table 7

TABLE 7
Common Buffer Additives in Capillary Electrophoresis and Their Effects

Additive	Function
Inorganic salts	Protein conformational changes, protein hydration, modification of EOF
Organic solvents	Analyte solubilization, modification of EOF
Urea	Protein solubilization, denaturation of oligonucleotides, modification of EOF
Sulfonic acids	Analyte ion-pairing agents, hydrophobic interaction agents
Cationic surfactants	Charge reversal on capillary wall, hydrophobic interaction
Cellulose derivatives	Reduce EOF, provide sieving medium
Amines	Charge neutralization, analyte interaction

outlines some common buffer additives and their mode of action. As can be seen from Table 7, there are several classes of additives and many of them are multifunctional, not only suppressing analyte–wall interactions but also affecting analyte solubility. Some examples of the use of additives for enhancing/optimizing CE separations are given later in this chapter Section IV.B.5.d). An extensive list of examples of the use of additives in CZE and other modes of capillary electrophoresis is given in Appendix 4.

B. BUFFER CONCENTRATION: OHM'S LAW PLOT

Buffer concentration is restricted by several parameters, including the capillary length and internal diameter, the applied electrical field strength, and the efficiency of the capillary thermostatting/cooling system. The use of moderately high ionic strength buffers is desirable for suppression of ionic interactions between charged analyte ions and ionized silanol groups on the capillary wall. However, under high electrical fields, the current (Joule heat) associated with buffer concentrations greater than 100 m*M* may overcome the capillary thermostatting capability of the system. Excessive Joule heating can have undesirable effects on both resolution and analyte stability; this is discussed in detail in Chapter 21. Buffers that are particularly problematic in this respect include those containing high mobility electrolytes such as chloride, citrate, and sulfate. The excessive Joule heating associated with high-concentration buffers can be circumvented in three ways: (1) decrease the applied field, (2) increase capillary length, or (3) decrease capillary internal diameter. The easiest solution is to increase the capillary length so that a tolerable current is maintained. However, this will increase the on-capillary time and may result in band broadening. A reduction in cross-sectional area of the capillary will also reduce heating by decreasing the current density, and allowing for more effective dissipation of heat due to a greater surface-to-volume ratio. Alternatively, buffers that produce a relatively low current (and Joule heat) at high voltages can be used. One such buffer in the pH 7 to 9 range is borate buffer, which has been shown to be an excellent CE buffer at concentrations as high as 500 m*M*.[21] An added advantage of using higher ionic strength separation buffer is the increased sample-loading capacity that results due to the on-capillary stacking effect that will be described in a later section.

A simple method has been described by Nelson et al.,[22] termed the *Ohm's law plot*, that allows for easy determination of the "functional" buffer concentration and the maximum voltage that can be utilized with the particular buffer system (i.e., the functional limit for capillary thermostatting). After filling the capillary with the desired buffer and allowing for adequate equilibration, voltage is applied to the system for short intervals (e.g., 1 min) and the current recorded at each voltage. Linearity in a plot of observed current vs. applied voltage is an indication that the capillary temperature is being adequately maintained (i.e., the generated Joule heat is being effectively dissipated). At the point where linearity is lost, the thermostatting capacity of the system has been exceeded. One should strive for heat generation of <1 watt per meter of capillary (W/m) for optimum separation[23] and should not exceed 5 W/m. An example

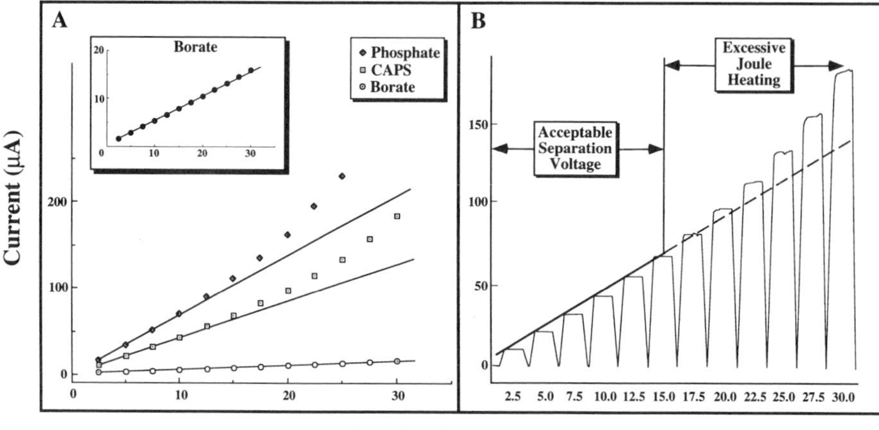

FIGURE 5. Ohm's law plot. (A) Plot of observed current vs. applied voltage for each of three buffers. The determinations were carried out on a Beckman (Fullerton, CA) P/ACE 2050. The voltage was incremented at 2.5 kV/min, and the current recorded through System Gold software, version 7.1. Buffers were 100 m*M* phosphate, pH 2.5, made by dilution of phosphoric acid and titration with NaOH; 100 m*M* borate, pH 8.3, made by titrating 25 m*M* sodium tetraborate with 100 m*M* boric acid; and 100 m*M* CAPS, pH 11.0, made by titration of the appropriate concentration dissolved in water with NaOH. The inset shows the borate data plotted on an expanded scale. (B) Direct plot of current vs. applied voltage for 100 m*M* CAPS, pH 11.0. A direct plot of observed current vs. applied voltage was obtained through System Gold version 7.1. Voltage is incremented by 2.5 kV/min. A straight line drawn through the front edge of the plateau illustrates the ability of the cooling system to dissipate the heat generated by the passage of current. The departure from linearity indicates the excessive increase in current at the applied voltage, and is a reflection of the increase in capillary temperature.

of the Ohm's law plot is given in Figure 5A for 100 m*M* concentrations of each of three buffers: phosphate, pH 2.5; borate, pH 8.3; and CAPS, pH 11.0.[24] At 20 kV, the current is lowest with borate (≈10 μA) while CAPS (≈100 μA) and phosphate (150 μA) are dramatically higher. The power associated with each buffer at 25 kV is 0.58, 10.07, and 5.88 W/m for borate, phosphate, and CAPS, respectively (see Appendix 2 for calculations). The recommended limits have clearly been exceeded with the latter two buffers. More importantly, linearity is lost with the phosphate and CAPS buffer at relatively low applied voltages in comparison with borate; the 100 m*M* phosphate and CAPS buffers should not be used at voltages greater than 10 and 15 kV, respectively. In contrast, borate buffer (inset in Figure 5A) exhibits a linear relationship between current and the applied voltage, even at 30 kV. This determination can be simplified if the software used for data acquisition and analysis has the ability to directly monitor/plot current as a function of time. Figure 5B shows a direct (on-screen) plot from System Gold version 7.1* for 100 m*M* CAPS buffer where voltage is incremented by 2.5 kV/min. This provides the same information without having to manually plot current vs. voltage.

C. BUFFER DEPLETION

Since the electrophoretic process occurring within a buffer-filled capillary under an applied electrical field involves the movement of ions, it is not surprising that the separation buffer can eventually be "depleted" if electrophoresis continues for extensive periods of time. Use of "depleted" buffer for CE analysis can result in deterioration of resolution and run-to-run reproducibility.[25] It is for this reason that buffer should be replenished after a limited number of runs (inlet and outlet every five to eight runs; 10- to 15-min analysis per run; 4- to 5-ml buffer reservoirs). One method that may be useful for extending the use of a buffer without adverse effects on reproducibility is a "limited" replenishment system in which two inlet

*System Gold 7.1 is a data acquisition/analysis software package available from Beckman Instruments, Fullerton, CA.

buffer reservoirs are used — one for electrophoresis and the other for rinsing. Rinsing the capillary with buffer that has not undergone changes in pH or composition due to electrophoresis, appears to substantially increase the number of runs (with the same buffer) before the onset of buffer-associated reproducibility problems.[24]

IV. INTRODUCTION OF SAMPLE INTO THE CAPILLARY

As a result of the small dimensions of capillaries used in CE, the total capillary volume is typically in the microliter range (see Table 1). Adhering to the "rule of thumb" restricting sample volume to 1 to 5% of the total capillary volume, sample volumes must be in the low-nanoliter range to avoid overloading of the capillary.[26] The technology for small sample volume introduction into a capillary has converged at three accepted methods. Introduction or, as it is sometimes referred to, "injection" or "loading" of sample into the capillary, can be accomplished hydrostatically, electrokinetically, and, in certain cases, by gravity. With hydrostatic injection, sample is introduced by immersing the capillary inlet into a vial containing the sample and either pressurizing the inlet vial containing the sample or applying a vacuum to the outlet vial (opposite end). Alternatively, sample introduction by gravity, which is more commonly used with noncommercial systems, relies on the siphoning of sample into the capillary by elevating the injection (inlet) end of the capillary relative to the outlet end.

With electrokinetic loading, the injection (inlet) end of the capillary is immersed in the sample, the outlet in the separation buffer, and a low voltage (1 to 10 kV) applied for durations of 1 to 20 s, depending on the capillary length and inner diameter. With this mode, the quantity of sample introduced into the capillary is dependent on a number of parameters, the most important of which are the electrophoretic mobilities of the sample components and the EOF (which is dependent on the characteristics of the capillary). The electrokinetic mode can "bias" sample introduction; sample components with the highest electrophoretic mobility will be preferentially introduced over those with lower mobilities. Therefore, one should bear in mind that the results obtained with electrokinetic loading may not be a quantitative representation of the sample components. In contrast, electrokinetic loading may be advantageous if the analyte of interest is a small percentage of the sample, but has a much higher electrophoretic mobility than other constituents. Under these conditions, electrokinetic loading provides a positive sample loading bias and may enhance the detectability of the component of interest. Other parameters that can affect this mode of loading include the buffer (type, ionic strength, pH) and the composition of the sample matrix. Sample matrices containing significantly high electrolyte concentrations are not conducive to efficient sample introduction.

This last point highlights a recurring theme in the chapters to follow; that is, the critical role that the sample matrix plays in obtaining efficient capillary electrophoretic separations. The sample matrix can have positive or negative effects on sample introduction and separation. The problems associated with high-salt sample matrices, and approaches to manipulating the matrix for adequate CE analysis, are discussed in detail in Chapter 20. In contrast, under the appropriate conditions, the sample matrix can have a positive effect on sample introduction, enhancing the detectability of the components of a dilute sample through on-capillary sample preconcentration. This is particularly important in light of the minute sample volume capacity inherent with CE and is discussed in the next section, as well as in Chapters 12 and 20.

V. ON-CAPILLARY SAMPLE CONCENTRATION TECHNIQUES

One of the main drawbacks of UV-detection CE is the low sensitivity resulting from the inherently small dimensions of the flow cell (i.e., the inner diameter of the capillary) and the sample volume capacity. As mentioned above, introduction of sample volumes larger than 1

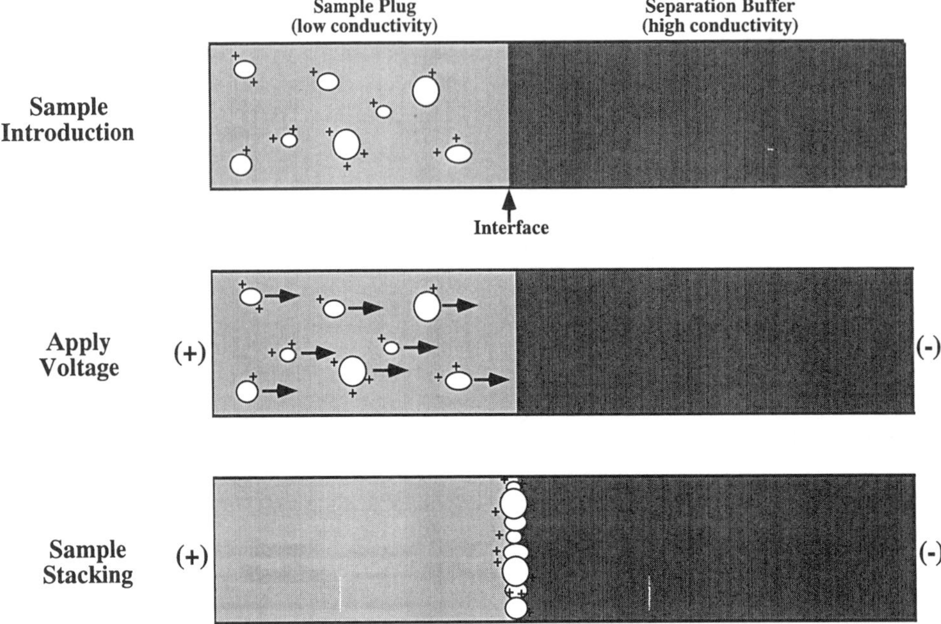

FIGURE 6A. The sample "stacking" mechanism. Sample ions have an enhanced electrophoretic mobility in a lower conductivity environment (i.e., elevated local field strength). When a voltage is applied, sample ions in the sample plug instantaneously accelerate toward the adjacent separation buffer zone where, on crossing the boundary, a higher conductivity environment (lower field strength) causes a decrease in electrophoretic velocity and "stacking" of the sample ions into a buffer zone smaller than the original sample plug.

to 5% of the total capillary volume can be detrimental to resolution. This limits the use of CE as an analytical technique to samples having nominally high concentrations (100 μg/ml or greater). It is for this reason that the development of several approaches for "on-capillary sample concentration" has been pivotal to the acceptance of CE as a microanalytical technique.

A. SAMPLE STACKING

One of the simplest methods for sample preconcentration is to induce "stacking" of the sample components and is easily accomplished by exploiting the ionic strength differences between the sample matrix and separation buffer.[27] As illustrated in Figure 6A, when the sample matrix is of lower ionic strength than the separation buffer (ideally tenfold), the sample components "stack" at the interface between the sample plug and the separation buffer. This results from the fact that sample ions have an enhanced electrophoretic mobility in a lower conductivity environment. When a voltage is applied to the system, sample ions in the sample plug instantaneously accelerate toward the adjacent separation buffer zone where, on crossing the boundary, the higher conductivity environment induces a decrease in electrophoretic velocity and subsequent "stacking" of the sample components into a smaller buffer zone than the original sample plug. Within a short time, the ionic strength gradient dissipates and the charged analyte molecules begin to move from the "stacked" sample zone toward the cathode. The illustration of stacking in Figure 6A is a simplified one, wherein all sample components are positively charged and migrate toward the sample plug–separation buffer interface. While this would not be an unrealistic representation for peptide separation at pH 2.5, at higher pH values (and with sample components that are both net positively and negatively charged) negatively charged components would accelerate toward the capillary inlet. Hence, it is advisable to follow sample plug introduction with an equal volume of separation buffer. This will result in the formation of more than one sample zone, each still substantially smaller than the original plug volume. Stacking can be utilized with either hydrostatic or electrokinetic

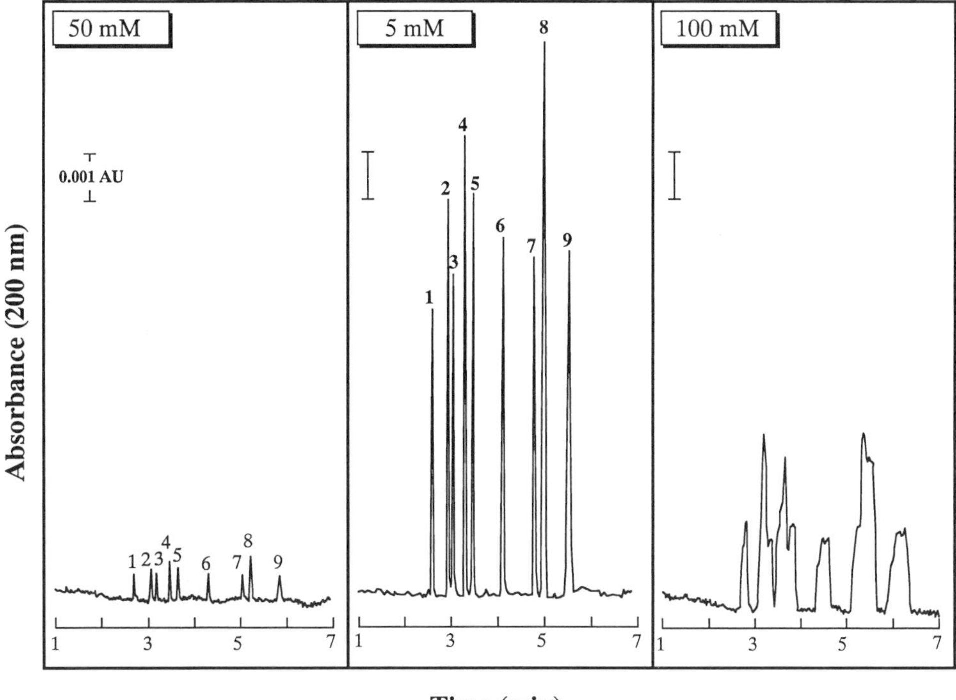

FIGURE 6B. "Stacking" of peptides in phosphate buffer. Separation of a standard peptide mixture (peptide calibration kit; Bio-Rad, Richmond, CA). Peptides were (1) bradykinin, (2) angiotensin, (3) α-melanin-stimulating hormone, (4) thyrotropin-releasing hormone, (5) leutinizing hormone releasing hormone, (6) leucine enkephalin, (7) bombesin, (8) methionine enkephalin, and (9) oxytocin, 50 μg/ml each. *Left*: Sample was dissolved in 50 m*M* phosphate buffer, pH 2.5. *Middle*: Sample was dissolved in 5 m*M* phosphate buffer, pH 2.5. *Right*: Sample was dissolved in 100 m*M* phosphate buffer, pH 2.5. Analyses were performed on a Beckman P/ACE System model 2100. Separation conditions were as follows: capillary, 50 μm × 20 cm (effective length), 27-cm total length, bare fused silica; *T*, 20°C; voltage, 10 kV; separation buffer, 50 m*M* phosphate, pH 2.5; 5-s hydrostatic injection; detection, 200 nm.

injection and can typically yield a tenfold enhancement in sample concentration and, hence, sensitivity. An example of this is given in Figure 6B, which shows separation of a standard peptide mixture in 50 m*M* phosphate separation buffer, pH 2.5. There is an obvious enhancement in detectability when the sample matrix is 5 m*M* in comparison with 50 m*M* (separation buffer). The undesirable effects of high ionic strength (100 m*M* phosphate buffer) sample matrix are clearly illustrated in the right panel of Figure 6. Detailed discussion of sample matrix effects on separation can be found in Chapter 20.

When possible, dissolving the sample in diluted separation buffer may be more advisable than in water since the dramatic differences in EOF between the sample plug and separation buffer may cause laminar flow within the capillary and, hence, lead to peak broadening.[28] Moreover, excessive heat produced in the sample plug may have additional effects, such as the possibility of denaturation of the sample components.[22,29,30] Therefore, it may be advisable to "stack" at lower applied voltages or "ramp" to the separation voltage over several minutes.

B. SAMPLE FOCUSING

An alternative approach to sample preconcentration by stacking is on-capillary "focusing", and is based on pH differences between the sample plug and separation buffer (Figure 7). This has been shown to be very useful for the analysis of peptides, mainly due to their relative stability over a wide pH range.[26] Focusing is easily accomplished by increasing the pH of the sample above that of the net p*I* (isoelectric point) of the sample. The high-pH sample plug is flanked between low-pH separation buffer zones (i.e., an equivalent volume of low-pH separation buffer

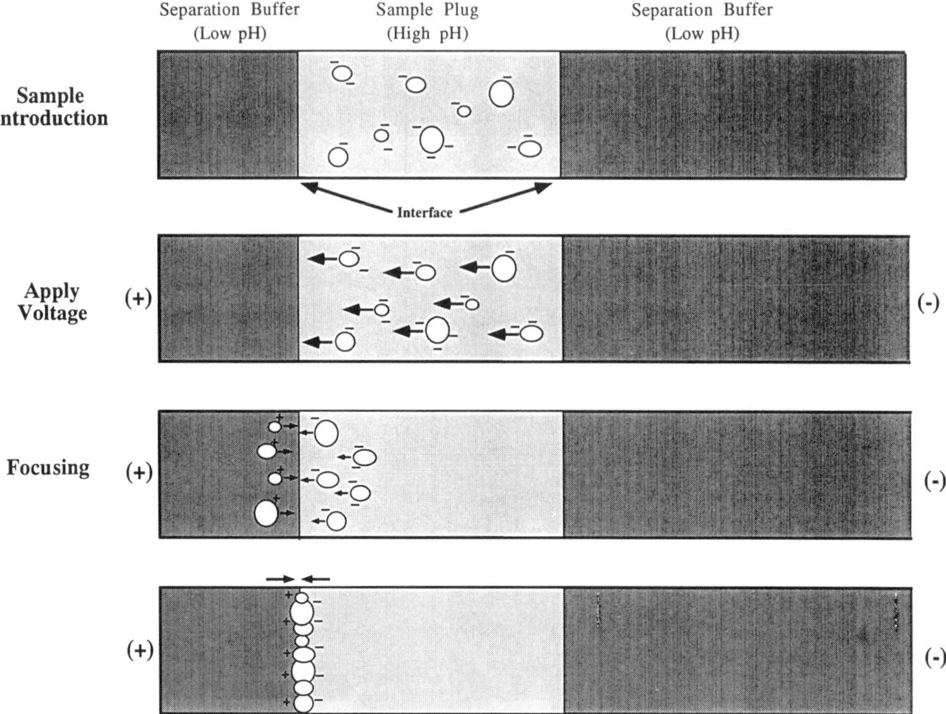

FIGURE 7. Sample-focusing mechanism. The high-pH sample plug is flanked between low-pH separation buffer zones. When a voltage is applied to the system, the negatively charged analytes in the sample zone migrate toward the anode. On entering the lower pH separation buffer, a pH-induced change in their charge state causes a reversal in the direction of mobility, causing the peptides to "focus" at the interface of the sample (high-pH) plug and low-pH buffer plug. After the pH gradient dissipates, the peptides, again positively charged, migrate toward the cathode as a sharp zone.

following introduction of the sample plug). Upon application of a voltage, the negatively charged peptides in the initial sample zone migrate toward the anode. Upon entering the lower pH separation buffer, a pH-induced change in their charge state causes a reversal in the direction of mobility, resulting in a "focusing" of the peptides at the interface of the sample (high pH) and low-pH buffer plugs. After the pH gradient dissipates, the peptides, again positively charged, migrate toward the cathode as a sharp zone. This approach, limited to samples that can withstand the inherent changes in pH without substantial denaturation, can yield a fivefold enhancement in sample concentration and, hence, sensitivity. The practical and some theoretical aspects of these techniques are addressed in some detail in Chapter 20.

C. ISOTACHOPHORETIC SAMPLE ENRICHMENT

Since Chapter 5 is dedicated solely to a discussion of the merits of ITP as an analytical technique, as well as for on-capillary sample concentration, a description of this technique will not be given here.

VI. CAPILLARY ELECTROPHORETIC METHODS DEVELOPMENT*

As with high-performance liquid chromatography (HPLC), obviously there will be no universal CE conditions that allow for the analysis of all types of samples. One of the first

* Modified from information given in the Beckman P/ACE 2000 customer Training Manual (with permission).

steps in designing a method is to determine, based on the type or class of sample involved, which of the CE modes is best suited to the sample (as outlined in Table 3). Once the appropriate CE technique has been identified, analysis is carried out and separation optimization initiated. Optimal separation of any sample requires a logical approach to (1) sample solubilization and/or dealing with diverse sample matrices and (2) the identification and utilization of the correct combination of CE operating parameters, each of which have distinct effects on the separation efficiency and resolution.

A. SAMPLE PARAMETERS TO CONSIDER

Clearly, for capillary electrophoretic analysis to occur, the sample must be adequately solubilized. Although it is not always possible to know the physicochemical properties of the sample involved, it is important to be familiar with at least some of the characteristics of the sample, in particular, those of the specific component of interest. If this is not known, some the physicochemical properties may ultimately have to be determined empirically. Different questions should be posed depending on whether the sample involved is lyophilized or already solubilized, as in a biological fluid. In either case, it is important to know something of the detectability, purity, and stability of the sample/component of interest. How complex is the sample? Is the substance of interest a major or minor component? What other substances in the sample might interfere with the detection or separation of the analyte of interest? What is its λ_{max}? Is it thermally stable? If structural information is available, what are the pK_a values of the ionizable groups? If the sample contains proteins or peptides, what pI values are involved? If the sample is not in solution, there are additional questions that must be entertained. Is it soluble in water at milligram per milliliter concentrations? Does it require extremes of pH for solubility and, if so, is it stable at this pH? Is solubility enhanced by buffer additives such as urea, methanol, acetonitrile, hexane sulfonic acid, sodium dodecyl sulfate (SDS), etc.? It is these types of questions that will allow for a sound methodological approach to methods development.

A final point to make in this section is the importance of the sample matrix. As discussed in Chapter 20, the constituents of the sample matrix will play an important role in attaining optimal separation of sample components. Figure 6B is a clear example of this. Unfortunately, the conditions required for solubilization may not provide an adequate sample matrix for introduction and analysis. Hence, compromises may have to be made between complete solubility and an ideal sample matrix. Moreover, it is crucial to view the sample matrix in relation to the separation buffer. Ideally the sample matrix should be approximately 100- to 200-fold less concentrated than the separation buffer, in order that the sample not contribute to the EOF. This will also enhance the possibility of adequate detection, since the lower ionic strength sample matrix will lead to sample stacking and, hence, on-capillary concentration.

Once solubility has been attained in a reasonable sample matrix, the sample should filtered through a 0.45-μm pore size filter to remove unsolubilized particulates. If this must be avoided to minimize loss of sample on the membrane, the sample should be centrifuged (13,000 rpm for 2 min) prior to analysis. A trial analysis is then carried out and the efficiency of separation optimized. This is accomplished by altering/adjusting a multitude of variables (discussed in the next section) in search of the combination that provides the best separation of the sample components within a reasonable analysis time.

B. INDEPENDENT AND DEPENDENT SEPARATION VARIABLES

The parameters that must be considered when developing a CE method are numerous and can essentially be separated into two groups: dependent or independent. The independent parameters are those under the direct control of the operator, e.g., instrumental settings (voltage, temperature, etc.) or choice of separation buffer (pH, ionic strength, etc.). The dependent parameters are those directly affected by the independent parameters.

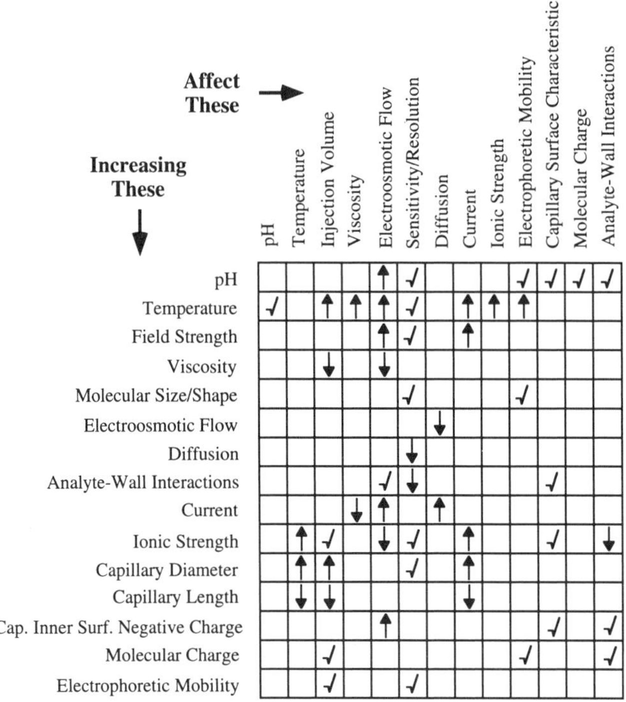

Increasing These ↓ / Affect These →	pH	Temperature	Injection Volume	Viscosity	Electroosmotic Flow	Sensitivity/Resolution	Diffusion	Current	Ionic Strength	Electrophoretic Mobility	Capillary Surface Characteristics	Molecular Charge	Analyte-Wall Interactions
pH					↑	✓				✓	✓	✓	✓
Temperature	✓		↑	↑	↑	✓		↑	↑	↑			
Field Strength					↑	✓		↑					
Viscosity			↓		↓								
Molecular Size/Shape						✓				✓			
Electroosmotic Flow							↓						
Diffusion						↓							
Analyte-Wall Interactions					✓	↓					✓		
Current				↓	↑		↑						
Ionic Strength		↑	✓		↓	✓		↑			✓		↓
Capillary Diameter		↑	↑			✓		↑					
Capillary Length		↓	↓					↓					
Cap. Inner Surf. Negative Charge					↑						✓		✓
Molecular Charge			✓							✓			✓
Electrophoretic Mobility			✓			✓							

FIGURE 8. A reference chart for evaluation of the relationships between variables influencing CE. ↑, ↓, Directional change; ✓, change in either direction.

Independent variables

- voltage or current
- temperature
- organic additives
- sample concentration
- pH
- capillary dimensions
- electrode polarity
- ionic strength

Dependent variables

- field strength (V/m)
- electroosmotic flow
- Joule heat production
- solution viscosity
- sample diffusion
- sample mobility
- sample charge
- sample shape
- sample interactions with capillary
- molar absorptivity

With the relatively large number of parameters to contend with in CE methods development, it is beneficial to understand how a change in one parameter will affect others. Figure 8 provides a quick reference chart for evaluation of the relationships between the variables outlined above. In this figure, the directional effects on the parameters along the horizontal axis are in response to increasing the magnitude of the parameters described on the vertical axis (✓ signifies the effect is not unidirectional). Parameters not covered in Figure 8 include the composition of the sample matrix and the presence of buffer additives that either interact with the sample components (e.g., surfactants) or affect the chemical nature of the wall (through dynamic/covalent modification). It is clear that with any given sample, each of these variables should be examined systematically for effective method development.

The following describes the qualitative effects on separation that result from alteration of the parameters described above. For examples illustrating the effects of many of these parameters on CE analysis of mixtures, the reader is referred to an excellent review by McLaughlin and co-workers.[31]

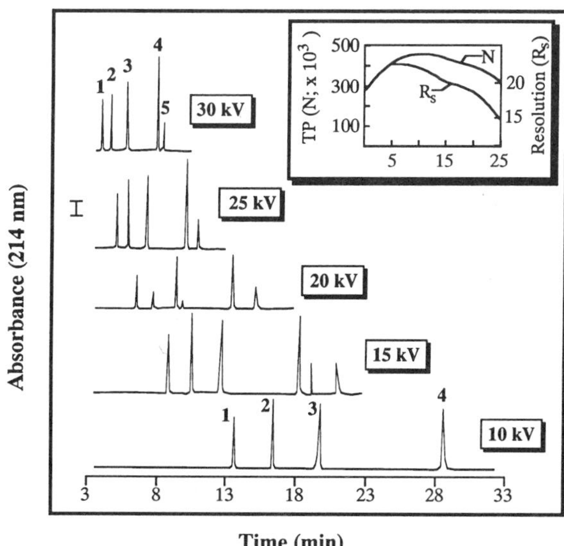

FIGURE 9. Effect of voltage on separation efficiency and resolution of a mixture of five vitamins: (1) niacinamide, (2) cyanocobalamine (vitamin B_{12}), (3) pyridoxine (vitamin B_6), (4) niacin, and (5) thiamine (vitamin B_1). Inset: Theoretical plates (TP) and resolution are plotted as a function of voltage. Separation conditions were as follows: capillary, 75 µm × 50 cm (effective length) fused silica; T, 25°C; buffer, 60 m*M* borate, 60 m*M* SDS, 15% methanol, pH 8.92; voltage varied as indicated to the right of each electropherogram. (Modified from McLaughlin, G. M., Nolan, J. A., Lindahl, et al., *J. Liq. Chromatogr.*, 15, 961, 1992. With permission.)

1. Electrode Polarity

Establishing the electrode polarity is of paramount importance in CZE and obviously one of the first considerations to be made before beginning analysis. As mentioned earlier, the normal polarity for CE is to have the anode (+) at the inlet and cathode (–) at the outlet. In this format, EOF is toward the cathode (detector/outlet). This is the standard polarity for most modes of CE utilizing a bare fused silica capillary. If set in reverse polarity (cathode at inlet; anode at outlet) by mistake, the direction of EOF is away from the detector and, hence, only negatively charged analytes with electrophoretic mobility greater than the EOF will pass the detector. This format is typically used with capillaries that are coated with substances that reverse the net charge of the inner wall (reverses EOF), or when analytes are all net negatively charged (e.g., DNA).

2. Applied Voltage

It is advisable to begin analysis with a midrange voltage (10 to 20 kV). Increasing the voltage will have a number of effects. While it will increase sample migration and electroosmotic flow, as well as shorten analysis time, it may increase the sharpness of the peaks and improve resolution. However, the advantages associated with increasing the voltage may be lost if the sample matrix ionic strength is much greater than the running buffer ionic strength and/or the increased production of Joule heat cannot be efficiently dissipated. Joule heating of the capillary results in a decreased solution viscosity. This leads to a further increase in EOF, ion mobility, and analyte diffusion, which may ultimately result in band broadening. An excellent example of the effects of increased voltage on the separation of five vitamins has been provided by McLaughlin et al.[31] and shown in Figure 9. Decreasing the applied voltage from 30 to 10 kV has the expected effect of increasing the migration time of all analytes in the mixture. At 10 kV the separation in inconveniently long and peak 5 has not reached the detector after 33 min. However, adequate separation at higher applied voltages is obtained at the expense of both resolution and efficiency (theoretical plates) (Figure 9, inset).

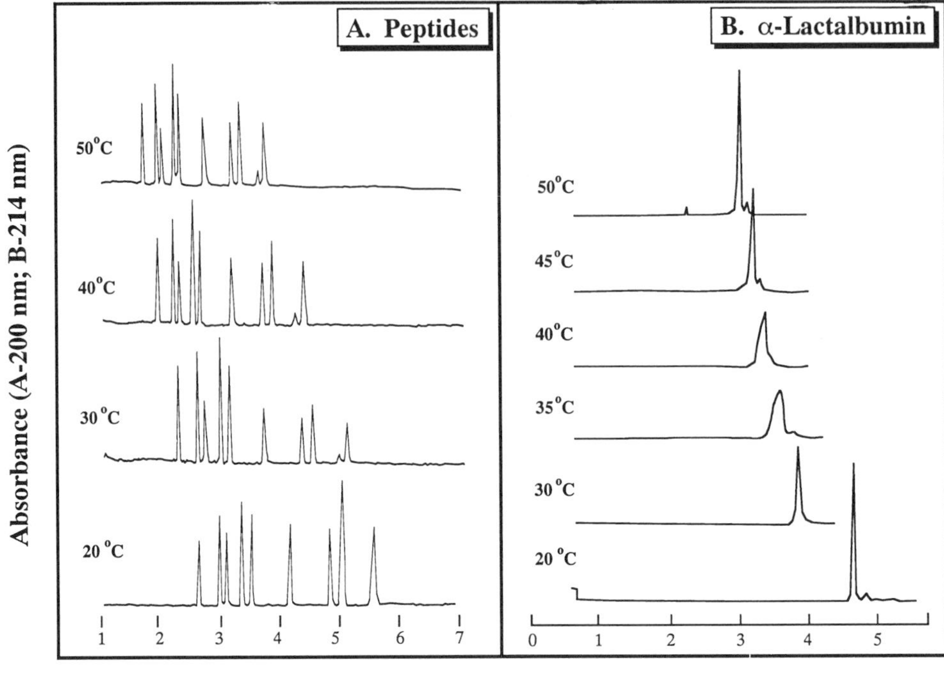

FIGURE 10. Effect of temperature on the electrophoretic behavior of peptides. Separation of a standard mix of peptides (50 μg/ml each, in 5 m*M* phosphate buffer, pH 2.5; 5-s pressure [0.5 psi] injection). Other separation parameters as in (B). (B) α-Lactalbumin (0.2 mg/ml, in water, 3-s pressure [0.5 psi] injection). Separation conditions: capillary, 75 μm × 50 cm (effective length) fused silica; buffer, 100 m*M* sodium tetraborate, pH 8.3; voltage, 350 V/cm; detection, 214 nm. (Adapted from Rush, R. S., Cohen, A. S., and Karger, B. L., *Anal. Chem.*, 63, 1346, 1991. With permission.)

3. Capillary Temperature

Separations should initially be attempted with the capillary thermostatted at close to ambient temperature. The capillary temperature can be increased on most commercial CE units to as high as 60°C without substantially increasing current. When this is done using the same applied voltage, decreased buffer viscosity leads to an increase in analyte electrophoretic mobility, thus decreasing separation times. Also, it is important to note that, when sample introduction is hydrostatic (same pressure/vacuum and time), increased capillary temperature will lead to an increase in the injected sample volume (not necessarily sensitivity) as a result of decreased buffer viscosity.[3,22,23,29,30,32-35] Undesirable effects include concurrent changes in buffer pH and band broadening due to increased diffusion and possible thermal denaturation of the sample.

Increasing the capillary temperature can have both positive effects and negative effects on separation. In some cases, the effects of elevated capillary temperatures are not only advantageous but necessary. A study by Guttman et al.[36] showed that separation of a mixture of five proteins in a physical gel was poorest at 20°C and optimal at 50°C. Apparently, higher temperatures were required to obtain the structural configuration of the physical gel required for adequate sieving of the proteins. Another positive effect of increasing capillary temperature is the substantial decrease in analysis time (which may or may not be associated with an increase in resolution). Figure 10A shows the effect of increasing temperature on a series of nine standard peptides. As the capillary temperature is increased, the time for analysis decreases with no apparent effect on the analytes.

Negative effects associated with elevated capillary temperature generally involve thermal effects on the sample. An example of this is given in Figure 10B with the temperature-dependent CE analysis of α-lactalbumin.[34] These separations appear to have identified 20 and 50°C as optimal temperatures for analysis of the particular protein; however, the authors demonstrate that the shift to a faster migration time is not solely due to increased EOF but also to a temperature-induced conformational change in the protein. Hence, while elevated capillary temperatures shorten analysis times, one should be cognizant of the potential adverse effects on analyte stability.

4. Capillary Dimensions

a. Internal Diameter

The main advantage resulting from increasing the capillary inner diameter is the enhancement in detection sensitivity (path length is longer). However, accompanying an increase in diameter is a decrease in the surface-to-volume ratio (see Table 1). This may lead to less efficient dissipation of Joule heat, which then results in a temperature gradient across the capillary and band broadening due to thermal effects. While a narrower capillary will accommodate a smaller sample volume (keeping $L_{int}/L_t < 3\%$), due to the concentrating phenomenon during electrophoresis, the narrower diameter enables one to detect a lower initial sample concentration.

b. Length

It is recommended that initial separations be performed on shorter capillaries (20 to 30 cm to detector), which provide short analysis times for determining sample introduction mode and time, buffer choice, etc. Once these are optimized, the capillary length can then be increased at a cost of increased migration (analysis) times. The increased length is often amenable to the separation of closely migrating species since an increased on-capillary time allows for subtle differences in analyte electrophoretic mobility to separate the components. Also, as the length increases, there with be a concomitant decrease in the electrical field strength at constant voltage and, hence, higher voltages may be applied.

5. Buffer

There are several parameters to consider when selecting an appropriate separation buffer and some of these (e.g., concentration limitations) were discussed in Section III. The sections below discuss pertinent details associated with other buffer parameters to be considered when designing a method.

a. pH

Assuming that information about the sample is available, it is advisable to choose a separation buffer pH that approximates the pK of the solute mixture. This choice is simple with pure or partially purified preparations. With crude biological mixtures, the pK_{avg} will typically be close to neutrality. Remember that increasing the separation buffer pH to 4–9 will result in an increase in the EOF. Also remember that the buffer pH may be altered in a secondary manner by other parameters such as temperature, ion depletion of the buffer (caused by repetitive use of the same separation buffer), and organic additives. Examples of the effect of buffer pH on CE separation is given in Figure 11 for both peptides and proteins. With the separation of the glycoprotein isoforms of ovalbumin, the expected decrease in migration time with increasing pH is observed, and optimal separation is observed at pH 9.0. Separation of a mixture of five peptide "isomers" is illustrated in the lower right of Figure 11. At pH 2.5, all five peptides, which are identical in amino acid composition but vary in sequence, comigrate as a broad peak (not shown). At pH 8.61, the peptides migrate as two groups while at pH 11.64, resolution of four of the five peptides is observed.

b. Ionic Species

As discussed in Section III, the choice of buffer for a particular CE separation is important from the perspective of Joule heat generation. Obviously, the pH required for the separation

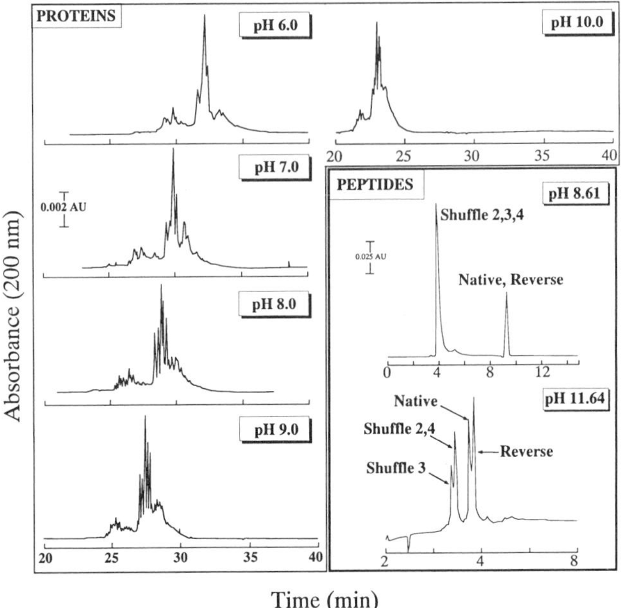

FIGURE 11. Dependence of electrophoretic separation and resolution on buffer pH. Proteins: Ovalbumin (1.0 mg/ml, in 100 m*M* borate, pH 8.3; 3-s pressure [0.5 psi] injection) was introduced into a preequilibrated capillary containing 100 m*M* borate buffer at the appropriate pH, indicated above each trace. Analysis performed on a Beckman P/ACE System 2050. Separation conditions: capillary, 50 µm × 80 cm (effective length), 87-cm total length bare fused silica; *T*, 28°C; voltage, 25 kV; detection, 200 nm. (Modified from Landers, J. P., Oda, R. P., Madden, B. J., et al., *Anal. Biochem.*, 205, 115, 1992. With permission.) Peptides: five peptides containing the same 12 amino acids but in varied sequence were analyzed under different pH conditions as labeled. Sequence of native peptide: KTNYCTKPQKSY. Separation conditions: capillary, 50 µm × 20 cm (effective length), 27-cm total length bare fused silica; *T*, 28°C; voltage, 25 kV; detection, 200 nm. Separation buffer, 100 m*M* borate, plus 10 m*M* diaminopentane, adjusted to the appropriate pH with NaOH.

will limit the candidate buffer systems amenable for use. However, it is important to note that, at a given pH, the type of buffer can have dramatic effects on resolution. The left panel of Figure 12 illustrates the importance of buffer choice on the CE analysis of recombinant human erythropoietin.[20] Although all buffers were of identical pH (4.0) and at a 100 m*M* concentration, resolution of the glycoforms was observed only with acetate-phosphate buffer and not with the acetate alone or acetate-sulfate solutions.

Other buffer systems that may have a dramatic effect on separations are those capable of specific interaction with the analyte(s). The ability of borate to form stable complexes with *cis*-diols has long been known and exploited in chromatography.[37] Several studies have shown this property to be particularly advantageous in CE since, at a slightly basic pH, borate complexation imparts an additional negative charge to the molecule (due to ionization of one of the free hydroxyl groups; $pK_a = 9.14$).[38-40] Figure 12 highlights the importance of borate for the separation of *cis*-diol-containing compounds from those similar in structure but lacking the functional group (i.e., cytidine vs. deoxycytidine).[41] The right panel of Figure 12 shows how complexation between buffer components and the analyte can affect separation. In Tricine at pH 8.3, dopamine is positively charged and migrates faster than EOF (mesityl oxide) while both tyrosine and dopa migrate slower than EOF (dopa more negatively charged than tyrosine). In borate buffer at the same pH, borate complexed with dopamine negates the positive charge on the amine, while dopa becomes more negatively charged. This example emphasizes the importance of buffer selection for enhancing selectivity in CE separations.

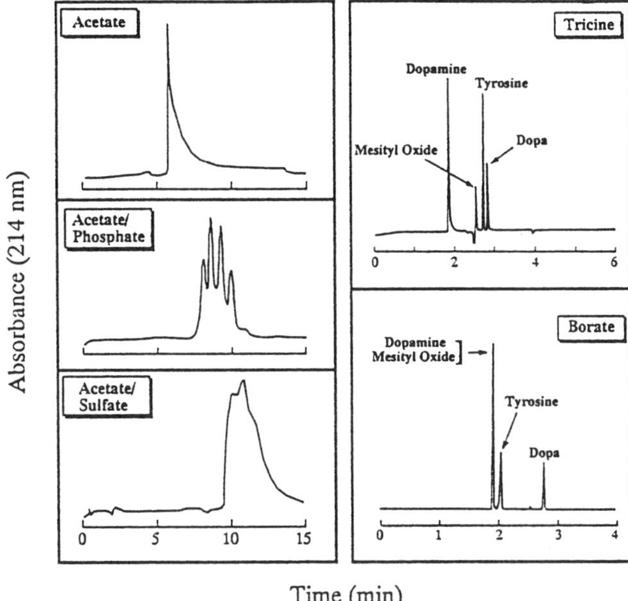

FIGURE 12. Effect of buffer ionic species on separation. (Left) CE analysis of recombinant human erythropoietin (r-hEPO). Separation conditions: capillary, 75 μm × 20 cm (effective length) fused silica; voltage, 10 kV; detection, 214 nm. Buffers were of identical pH (4.0) and 100 m*M* concentration. *Top*: Acetate buffer (30 μA). *Middle*: Acetate–phosphate (120 μA). *Bottom*: Acetate–sulfate (200 μA). (Modified from Tran, A. D., Park, S., Lisi, P. J., et al., *J. Chromatogr.*, 542, 459, 1991. With permission.) (Right) Separation of dopamine, tyrosine, dopa (0.5 mg/ml each, in water; 3-s pressure [0.5 psi] injection), and mesityl oxide (neutral marker, 3-s pressure [0.5 psi] injection). Analysis performed on a Beckman P/ACE System 2050. Separation conditions: capillary, 50 μm × 50 cm (effective length), 57 cm (total length) bare fused silica; *T*, 28°C; voltage, 25 kV; detection, 200 nm. *Top*: Separation in 100 m*M* tricine, pH 8.3. *Bottom*: Separation in 100 m*M* borate, pH 8.3. (Modified from Landers, J. P., Oda, R. P., Spelsberg, T. C., et al., *BioTechniques,* 14, 98, 1993. With permission.)

c. Ionic Strength

Increasing the ionic strength of the separation buffer has the effect of decreasing EOF and, hence, increasing the separation time. As mentioned in Section III, increasing the ionic strength will also increase the current at a constant voltage to the point where adequate thermostatting of the capillary becomes a concern. The advantages of increasing ionic strength, in addition to the obvious improvement in buffer capacity, will be to decrease analyte–wall interactions.[42] The net effect on the separation, therefore, will be to increase resolution, provided that capillary thermostatting capability is not overcome and that unwanted analyte dissociative processes (peptide/protein dimers) do not occur. On the other hand, an increase in ionic strength might improve resolution in mixtures by decreasing nonspecific analyte–analyte interactions. One example is with protein–DNA interactions. Figure 13 illustrates the effect of increased ionic strength on minimizing the nonspecific interaction of bovine serum albumin (BSA) with DNA restriction double-stranded DNA fragments.[43] The left panel shows that addition of BSA to the sample obliterates resolution of the double-stranded DNA fragments ranging in size from 72 to 310 base pairs. Simple addition of 50 m*M* NaCl to the 89 m*M* Tris–borate–EDTA separation buffer negates these effects (right panel).

d. Organic Additives

The addition of organic modifiers to the separation buffer will have differing effects, depending on the nature of the additive. One of the effects that often results is a change in EOF.

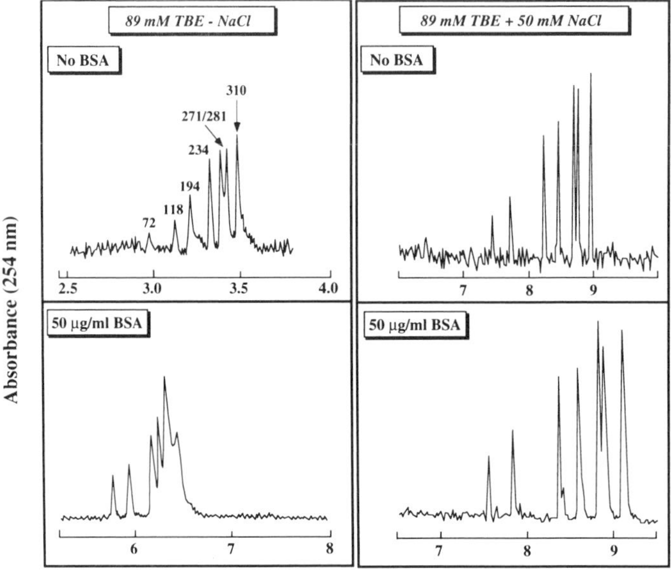

FIGURE 13. The effect of buffer ionic strength on the separation of low molecular weight DNA fragments. Analysis was carried out on a Beckman P/ACE System 2050. Restriction fragments from HaeIII digest of ϕX174 DNA (30 μg/ml, in water; 3-s pressure [0.5 psi] injection without BSA [top trace] and with 50 μg/ml BSA [bottom trace]). Separation conditions: capillary, 50 μm × 20 cm (effective length), 27-cm total length, DB-17-coated silica; *T*, 20°C; detection, 254 nm. *Left*: voltage, 7.0 kV; buffer, 89 m*M* Tris–borate-2 m*M* EDTA containing 0.1% HPMC and 10 μ*M* ethidium bromide, pH 8.6. *Right*: voltage, 4.0 kV; buffer, 89 m*M* Tris–borate–2 m*M* EDTA containing 0.1% HPMC and 10 μM ethidium bromide and 50 m*M* NaCl, pH 8.6. (Modified from Landers, J. P., Oda, R. P., Spelsberg, T. C., et al., *BioTechniques,* 14, 98, 1993. With permission.)

For example, the addition of 1,4-diaminobutane to the buffer has been proposed to enhance resolution by slowing EOF through a dynamic modification of the capillary wall.[44,45] Other additives have the effect of decreasing both the conductivity of the buffer and the EOF. In such cases, the subsequent enhancement in resolution may be a combination of the decreased EOF (i.e., increased on-capillary time), decreased thermal diffusion, and improved analyte solubility. An example is given in Figure 14 with the separation of pyrazoloacridine and two of its metabolites, 9-desmethyl-pyrazoloacridine, pyrazoloacridine-*N*-oxide.[46] These compounds have very low solubility in aqueous solutions and, hence, are separated in an acetate buffer (pH 4.0) containing 10% methanol. Increasing the methanol concentration to 30% in the separation buffer enhances the resolution of the parent compound and the methyl-metabolite, which are not resolved in 10% methanol.

e. *Modifiers of Electroosmotic Flow*

Two approaches can be taken to modify EOF in an attempt to enhance resolution: covalent or dynamic modification of the capillary wall. A study by Swedberg[47] provides an excellent example of the importance of capillary bonded phases for preventing protein–wall interactions during CE separations (Figure 15). A mixture containing seven proteins is clearly resolved in an arylpentofluoro (APF)-coated capillary, whereas separation in bare fused silica resulted in only a few broad peaks barely greater than baseline. As would be expected when protein–wall interactions are diminished, the migration times are markedly shorter.

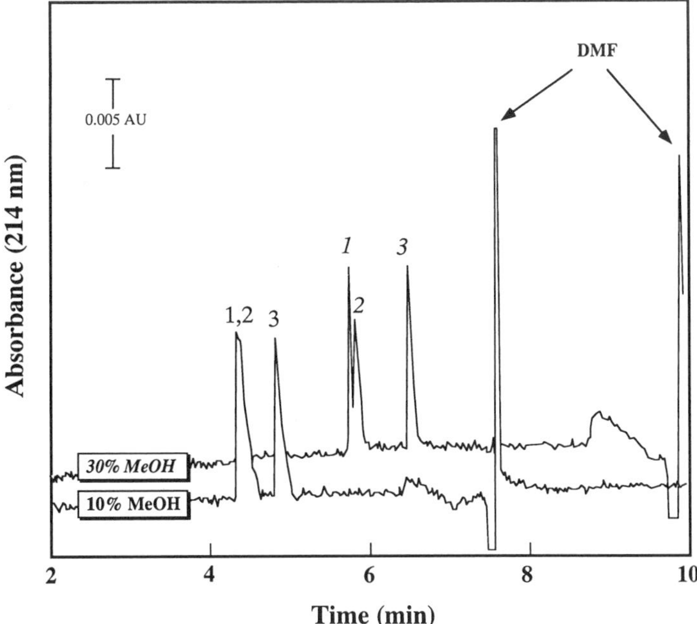

FIGURE 14. Effect of organic buffer additives on separation. The separation of pyrazoloacridine and two of its metabolites, 9-desmethyl-pyrazoloacridine and pyrazoloacridine-*N*-oxide (in methanol–water, 1:1 [v/v]; 2-s pressure [0.5 psi] injection). Separation conditions: capillary, 75 μm × 50 cm (effective length) bare fused silica; *T*, 40°C; voltage, 20 kV; detection, 214 nm. Separation buffer: lower trace, 20 m*M* ammonium acetate, 1% acetic acid, 10% methanol; upper trace, 20 m*M* ammonium acetate, 1% acetic acid, 30% methanol. (From Benson, L. M., Tomlinson, A. J., Reid, J. M., et al., *J. High Resolut. Chromatogr.*, 16, 324, 1993. With permission.)

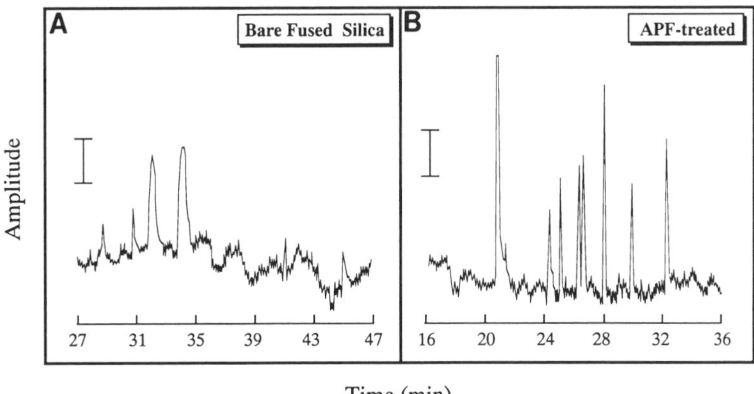

FIGURE 15. Effect of capillary coating on separation of proteins. Electropherograms of a mixture containing seven proteins and DMSO analyzed on (A) untreated fused silica, and (B) an APF-treated capillary. Peaks are (in order of elution) hen egg white lysozyme, DMSO, bovine ribonuclease A, bovine pancreatic trypsinogen, whale myoglobin, horse myoglobin, human carbonic anhydrase, and bovine carbonic anhydrase. Separation conditions: capillary, 20 μm × 100 cm (effective length); voltage, 250 V/cm; buffer, 200 m*M* phosphate, pH 7.0, containing 100 m*M* KCl; detection, 219 nm. (From Swedberg, S. A., *Anal. Biochem.*, 185, 51, 1990. With permission.)

Another approach to minimizing analyte–wall interactions through modification of the capillary wall is to use an additive that provides a dynamic (noncovalent) coating. Diaminoalkanes are one class of buffer additives that appear to enhance the resolution of protein mixtures, although the mechanism is poorly understood.[48-52] Nolan and Palmieri comment on this class of

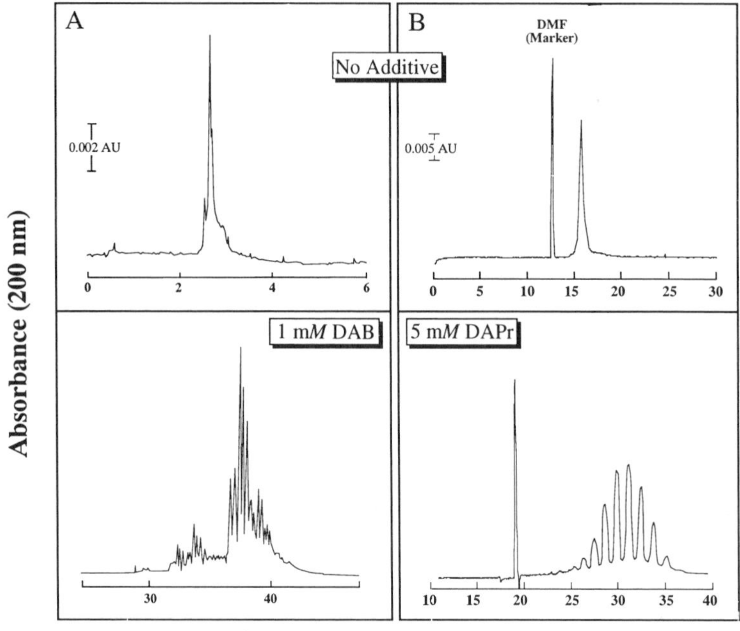

FIGURE 16. Effect of modifiers of EOF (α,ω-diaminoalkanes) on separation. (A) The effect of 1,4-diaminobutane (DAB) on the resolution of ovalbumin glycoforms. Analysis was carried out on a Beckman P/ACE System 2050. Ovalbumin (1 mg/ml, in water; 3-s pressure [0.5 psi] injection). Separation conditions: capillary, 50 μm × 80 cm (effective length), 87-cm total length bare fused silica; *T*, 28°C; voltage, 25 kV; buffer, 100 m*M* borate, pH 8.3; detection, 200 nm. *Top*: without additive. *Bottom*: buffer containing 1 m*M* DAB. (Modified from Landers, J. P., Oda, R. P., Madden, B. J., et al., *Anal. Biochem.*, 205, 115, 1992. With permission.) (B) The effect of 1,3-diaminopropane (DAPr) on human choriogonadotropin (hCG) isoforms. hCG (4 mg/ml, in water; 2-s vacuum injection). Analysis performed on an ABI 270A. Separation conditions: capillary, 50 μm × 80 cm (effective length), 100-cm total length bare fused silica; *T*, 28°C; voltage, 25 kV; buffer, 25 m*M* borate, pH 8.8; detection, 200 nm. *Top*: without additive. *Bottom*: buffer containing 5 m*M* DAPr. (Modified from Morbeck, D. E., Madden, B. J., Charlesworth, C. C., and McCormick, D. J., *Anal. Biochem.* [submitted]. With permission.)

additives in Chapter 13. Figure 16A shows the electropherograms resulting from the CE analysis of the glycoprotein, ovalbumin (known to be microheterogeneous with respect to its glycan content), with and without 1,4-diaminobutane.[44] The multitude of peaks observed in the presence of the diamino alkane are proposed to represent the glycoforms of the protein. A similar effect has been observed by Morbeck et al.[53] with the CE separation of urinary human chorionic gonadotropin (hCG), another glycoprotein known to be of microheterogeneous character (Figure 16B). The presence of diaminopropane in the buffer results in a dramatic reduction in EOF (as determined by reduced mobility of the neutral marker, DMF) and leads to resolution of eight hCG glycoforms that cannot be completely resolved by any other method.

C. STEPS IN DESIGNING A METHOD

Figure 17 shows a flow diagram that can be used as a general strategy for CE analysis of a sample. The first step is an obvious and important one: choosing the optimum wavelength for detection. This may not necessarily be the most sensitive detection wavelength (i.e., 200 nm), but more importantly, one that specifically enhances the chance of detecting the substance of interest and minimizing the absorbance of background components. Choosing a buffer system is of the utmost importance. The guidelines given earlier should be followed with strict attention to not only the pH requirements of the sample from a stability perspective,

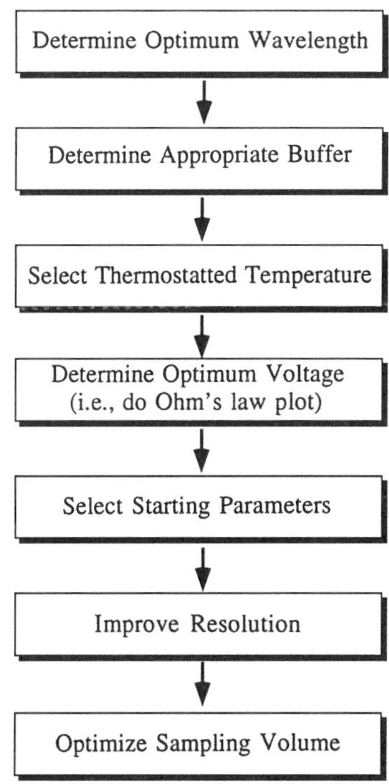

FIGURE 17. Flow diagram for CE method development.

but also to the wavelength restrictions of the buffer (i.e., λ_{min} for detection; see Table 6. If the use of a specific high ionic strength (high conductivity) buffer system is required (e.g., for preservation of bioactivity), the maximum working voltage should be determined by an Ohm's law plot to avoid capillary overheating problems. A review of the literature will provide a general idea of the basic parameters for an initial test run. The selected examples covered in the tables of Appendix 4 will act as a good starting point. A quick trial should establish sample introduction method, initial field strength, and capillary length. Optimization of the separation and resolution should include modification of the field strength, modification of the buffer by either altering pH or using additives to alter analyte mobility or EOF, and changing the length of the capillary if necessary. If peak resolution is poor, one might consider increasing the length of the capillary and/or the frequency of data collection. The remainder of the steps rely largely on the experience and intuition of the operator. Once adequate separation of the sample is attained, sample introduction should be optimized for a maximum signal-to-noise ratio. This step might include altering the sample concentration or the buffer composition (ionic strength, type, pH, etc.).

Typical starting conditions for analysis of unknowns in our laboratory are 50 μm × 47 cm capillary; the appropriate buffer at the appropriate pH (see the tables of Appendix 4) in the 10 to 100 m*M* range; 1- to 3-sec pressure injection of samples in the 100 μg to 1 mg/ml range (if possible); separation at 20 kV. Borate buffer (25 to 100 m*M*) in the neutral to slightly basic range has been found useful as a starting buffer for small molecules, peptides, and proteins. When possible, starting conditions should be tailored to the characteristics of the sample (e.g., pH needed for maximum solubility). We find 200 nm to be a good wavelength for ultraviolet detection of most organic compounds. For peptides/proteins, there is a three- to fivefold enhancement in signal over that at 280 nm due to the absorbance of the peptide bond. A reasonable starting concentration for protein/peptide samples is 100 μg/ml in buffer ($^1/_{10}$ × separation buffer) or water.

TABLE 8
Summary of the Analyte(s) Addressed in Each of the Chapters in this Book

Chapter	Small ions	Small molecules				Peptides	Proteins	Oligos	DNA
		General	Chiral	Drugs	Sugars				
2		✓				✓	✓		✓
3	✓	✓	✓	✓					
4							✓		
5	✓	✓				✓	✓		
6								✓	✓
7									
8		✓				✓	✓		
9	✓	✓							
10		✓	✓	✓					
11					✓		✓		
12						✓	✓		
13							✓		
14								✓	
15	✓	✓					✓		
16							✓		
17		✓		✓					
18							✓		
19							✓		
20		✓					✓		
21	✓								
22	✓								

VII. THE CHAPTERS TO FOLLOW

The chapters that follow are in-depth discussions of specific areas in the CE field. They have been chosen to be part of this handbook because they, in one form or another, are key to the practical use of CE for the analysis of a diverse array of analytes. The handbook is divided into five parts. Part I focuses on the various modes of CE that have been developed and applied to real analytical/separation problems. In some of the chapters, a theoretical development of the basic principles governing separation is presented in addition to the practical aspects. Part II deals with detection in CE. One chapter provides a comprehensive review of detector systems that have been successfully interfaced with CE, while the second focuses specifically on CE-mass spectrometry. In Part III, the application of CE to the analysis of a specific class of molecules/analytes or to solve specific separation problems is discussed, regardless of the mode involved. These include CE analysis of organic/inorganic ions, pharmaceuticals, glycoconjugates, peptides, proteins, DNA fragments, and the components of single cells, as well as specific discussions of the utility of capillary coatings in CE. Part IV is dedicated to discussions of two areas where the utility of CE is truly yet to be realized; hence the title, Specialized Applications of CE. These include a discussion of (1) the role of CE for solving separation problems in the clinical laboratory and (2) the potential for CE and CE-MS to advance analysis in the field of drug metabolism. Part V focuses on theoretical/practical aspects that are important to, and need to be addressed by, the practioner of CE. These include discussions of the sample matrix and its effects on separation, the need for controlling capillary temperature, and the present and future implications of EOF control. In the final chapter of the handbook, speculations are made regarding the future of CE and the specific areas where the exploitation of CE has only recently begun.

As a guide for the reader, Table 8 provides a quick reference chart describing the analyte(s) addressed in each of the specialty chapters.

REFERENCES

1. **Tiselius, A.,** A new apparatus for electrophoretic analysis of colloidal mixtures, *Trans. Faraday Soc.*, 33, 524, 1937.
2. **Mikkers, F. E. P., Everaerts, F. M., and Verheggen, T. P. E. M.,** High performance zone electrophoresis, *J. Chromatogr.*, 169, 11, 1979.
3. **Jorgenson, J. W. and Lukacs, K. D.,** Zone electrophoresis in open tubular glass capillaries, *Anal. Chem.*, 53, 1298, 1981.
4. **Hjertén, S.,** High performance electrophoresis, *Chromatogr. Rev.*, 9, 122, 1967
5. **Karger, B. L., Ed.,** 1st Int. Symp. on HPCE, *J. Chromatogr.*, 480, 1989.
6. **Karger, B. L., Ed.,** 2st Int. Symp. on HPCE, *J. Chromatogr.*, 516, 1990.
7. **Karger, B. L., Ed.,** 3st Int. Symp. on HPCE, *J. Chromatogr.*, 559, 1991.
8. **Gordon, M. J., Huang, X., Pentoney, S. L., Jr., and Zare, R. N.,** Capillary electrophoresis, *Science*, 242, 224, 1988.
9. **Karger, B. L., Cohen, A. S., and Guttman, A.,** High performance capillary electrophoresis in the biological sciences, *J. Chromatogr.*, 492, 585, 1989.
10. **Grossman, P. D., Colburn, J. C., Lauer, H. H., Nielsen, R. G., Riggin, R. M., Sittampalam, G. S., and Rickard, E. C.,** Application of free-solution capillary electrophoresis to the analytical separation of proteins and peptides, *Anal. Chem.*, 61, 1186, 1989.
11. **Rocheleau, M. J. and Dovichi, N. J.,** Separation of DNA sequencing fragments at 53 bases per minute by capillary gel electrophoresis, *J. Microcolumn Sep.*, 4, 449, 1992.
12. **Rickard, E. C., Strohl, M. M., and Nielsen, R. G.,** Correlation of electrophoretic mobilities from capillary electrophoresis with physicochemical properties of proteins and peptides, *Anal. Biochem.*, 197, 197, 1991.
13. **Helmholtz, H. Z.,** About electrical interfaces [translated title], *Ann. Phys. Chem.*, 7, 337, 1879.
14. **Wieme, R. J.,** Theory of electrophoresis, in *Chromatography: A Lab Handbook of Chromatography and Electrophoresis Methods*, 3rd ed., Heftmann, E., Ed., Van Nostrand, New York, 1984, pp. 228–278, chap. 10.
15. **Grossman, P. D. and Colburn, J. C., Eds.,** *Capillary Electrophoresis: Theory and Practice*, Academic Press, New York, 1992.
16. **Lux, J. A., Hausig, U., and Schomburg, G.,** Production of windows in fused silica capillaries for in-column detection of UV-absorption or fluorescence in capillary electrophoresis or HPLC, *J. High Resolut. Chromatogr.*, 13, 373, 1990.
17. **McCormick, R. M. and Zagursky, R. J.,** Polymide stripping device for producing detection windows on fused-silica tubing used in capillary electrophoresis, *Anal. Chem.*, 63, 750, 1991.
18. **Bocek, P. and Chambach, A.,** Capillary electrophoresis of DNA in agarose solutions at 40°C, *Electrophoresis*, 12, 1059, 1991.
19. **Bruno, A. E., Gassmann, E., Pericles, N., and Anton, K.,** On-column capillary flow cell utilizing optical waveguides for chromatographic application, *Anal. Chem.*, 61, 876, 1989.
20. **Tran, A. D., Park, S., Lisi, P. J., Huynh, O. T., Ryall, R. R., and Lane, P. A.,** Separation of carbohydrate-mediated microheterogeneity of recombinant human erythropoietin by free solution capillary electrophoresis, *J. Chromatogr.*, 542, 459, 1991.
21. **Chen, F. A., Kelly, L., Palmieri, R., Biehler, R., and Schwartz, H.,** Use of high ionic strength buffers for the separation of proteins and peptides with capillary electrophoresis, *J. Liq. Chromatogr.*, 15, 1143, 1992.
22. **Nelson, R. J., Paulus, A., Cohen, A. S., Guttman, A., and Karger, B. L.,** Use of Peltier thermoelectric devices to control column temperature in high-performance capillary electrophoresis, *J. Chromatogr.*, 480, 111, 1989.
23. **Sepaniak, M. J. and Cole, R. O.,** Column efficiencies in micellar electrokinetic capillary chromatography, *Anal. Chem.*, 59, 472, 1987.
24. **Landers, J. P., Oda, R. P., Madden, B. J., Sismelich, T. P., and Spelsberg, T. C.,** Reproducibility of sample separation using liquid or forced air convection themostated high performance capillary electrophoresis, *J. of High Resolut. Chromatogr.*, 15, 517, 1992.
25. **Strege, M. A. and Lagu, A. L.,** Studies of migration time reproducibility of CE protein separations, *J. Liq. Chromatogr.*, 16, 51, 1993.
26. **Aebersold, R. and Morrison, H. D.,** Analysis of dilute peptide samples by capillary zone electrophoresis, *J. Chromatogr.*, 516, 79, 1990.
27. **Chien, R.-L. and Burgi, D. S.,** On-column sample concentration using field amplification in CZE, *Anal. Chem.*, 64, 489A, 1992.
28. **Chein, R.-L. and Helmer, J. C.,** Electroosmotic properties and peak broadening in field-amplified capillary electrophoresis, *Anal. Chem.*, 63, 1354, 1990.
29. **Gobie, W. A. and Ivory, C. F.,** Thermal model of capillary electrophoresis and a method for counteracting thermal band broadening, *J. Chromatogr.*, 516, 191, 1990.
30. **Davis, J. M.,** Influence of thermal variations of diffusion coefficient on non-equilibrium plate height in capillary zone electrophoresis, *J. Chromatogr.*, 517, 521, 1990.

31. **McLaughlin, G. M., Nolan, J. A., Lindahl, J. L., Palmieri, R. H., Anderson, K. W., Morris, S. C., Morrison, J. A., and Bronzert, T. J.,** Pharmaceutical drug separations by HPCE: practical guidelines, *J. Liq. Chromatogr.*, 15, 961, 1992.
32. **Grushka, E., McCormick, R. M., and Kirkland, J. J.,** Effect of temperature gradients on the efficiency of capillary zone electrophoresis separations, *Anal. Chem.*, 61, 241, 1989.
33. **Terabe, S., Otsuka, K., and Ando, T.,** Electrokinetic chromatography with micellar solution and open-tubular capillary, *Anal. Chem.*, 57, 834, 1985.
34. **Rush, R. S., Cohen, A. S., and Karger, B. L.,** Influence of column temperature on the electrophoretic behavior of myoglobin and α-lactalbumin in high-performance capillary electrophoresis, *Anal. Chem.*, 63, 1346, 1991.
35. **Kurosu, Y., Hibi, K., Sasaki, T., and Saito, M.,** Influence of temperature control in capillary electrophoresis, *J. High Resolut. Chromatogr.*, 14, 200, 1991.
36. **Guttman, A., Horvath, J., and Cooke, N.,** Influence of temperature on the sieving effect of different polymer matrices in capillary SDS gel electrophoresis of proteins, *Anal. Chem.*, 65, 199, 1993.
37. **Moyer, T. P., Jiang, N.-S., Tyce, G. M., and Sheps, S. G.,** Analysis for urinary catecholamines by liquid chromatography with amperometric detection: methodological and clinical interpretation of results, *Clin. Chem.*, 25, 256, 1979.
38. **Wallingford, R. A. and Ewing, A. G.,** Separation of serotonin from catechols by capillary zone electrophoresis with electrochemical detection, *Anal. Chem.*, 61, 98, 1989.
39. **Lui, J., Shirota, O., and Novotny, M.,** Capillary electrophoresis of amino sugars with laser-induced fluorescence detection, *Anal. Chem.*, 63, 413, 1991.
40. **Honda, S., Makino, A., Suzuki, S., and Kakehi, K.,** Analysis of the oligosaccharides of ovalbumin by high performance capillary electrophoresis, *Anal. Biochem.*, 191, 228, 1990.
41. **Landers, J. P., Oda, R. P., and Schuchard, M. D.,** Separation of boron-complexed diol compounds using high-performance capillary electrophoresis, *Anal. Chem.*, 64, 2846, 1992.
42. **Bushey, M. and Jorgenson, J.,** Capillary electrophoresis of proteins in buffers containing high concentrations of zwitterionic salts, *J. Chromatogr.*, 480, 301, 1989.
43. **Landers, J. P., Oda, R. P., Spelsberg, T. C., Nolan, J. A., and Ulfelder, K. J.,** Capillary electrophoresis: a powerful microanalytical technique for biologically active molecules, *BioTechniques*, 14, 98, 1993.
44. **Landers, J. P., Oda, R. P., Madden, B. J., and Spelsberg, T. C.,** High-performance capillary electrophoresis of glycoproteins: the use of modifiers of electroosmotic flow for analysis of microheterogeneity, *Anal. Biochem.*, 205, 115, 1992.
45. **Landers, J. P., Oda, R. P., Madden, B. J., and Spelsberg, T. C.,** α,ω-Diamino alkanes as modifiers of electroosmotic flow: potential mechanisms for their role in high resolution capillary electrophoretic separations, *Anal. Chem.*, submitted.
46. **Benson, L. M., Tomlinson, A. J., Reid, J. M., Walker, D. L., Ames, M. M., and Naylor, S.,** Analysis of pyrazoloacridine drug metabolism by capillary electrophoresis in non-aqueous buffers, *J. High Resolut. Chromatogr.*, 16, 324, 1993.
47. **Swedberg, S. A.,** Characterization of protein behavior in high-performance capillary electrophoresis using a novel capillary system, *Anal. Biochem.*, 185, 51, 1990.
48. **Gordon, M. J., Lee, K.-J., Arias, A. A., and Zare, R. N.,** Protocol for resolving protein mixtures in capillary zone electrophoresis, *Anal. Chem.*, 63, 69, 1991.
49. **Wiktorowicz, J. E. and Colburn, J. C.,** Separation of cationic proteins via charge reversal in capillary electrophoresis, *Electrophoresis*, 11, 769, 1990.
50. **Towns, J. K. and Regnier, F. E.,** Capillary electrophoretic separation of proteins using nonionic surfactant coatings, *Anal. Chem.*, 63, 1126, 1991.
51. **Mazzeo, J. R. and Krull, I. S.,** Coated capillaries and additives for the separation of proteins by capillary zone electrophoresis and capillary isoelectric focussing, *BioTechniques*, 10, 638, 1991.
52. **Bolger, C. A., Zhu, M., Rodriguez, R., and Wehr, T.,** Performance of uncoated and coated capillaries in free zone electrophoresis and isoelectric focussing of proteins, *J. Liq. Chromatogr.*, 14(5), 895, 1991.
53. **Morbeck, D. E., Madden, B. J., Charlesworth, C. C., and McCormick, D. J.,** High resolution separation of human choriogonadotropin (hCG) isoforms using capillary electrophoresis, *Anal. Biochem.*, submitted.
54. **Schwartz, H. E., Ulfelder, K. J., Sunzeri, F. J., Busch, M. P., and Brownlee, R. G.,** Analysis of DNA restriction fragments and polymerase chain reaction products towards detection of the AIDS (HIV-1) virus in blood, *J. Chromatogr.*, 559, 267, 1991.

Chapter 3

MICELLAR ELECTROKINETIC CAPILLARY CHROMATOGRAPHY

Morteza G. Khaledi

TABLE OF CONTENTS

0-8493-8690-X/94/$0.00+$.50

I. INTRODUCTION

During the past decade, much attention has been given to broadening the scope of capillary electrophoresis (CE) to new classes of compounds and samples. The primary application of CE is the separation of charged solutes. In an effort to extend the enormous separation power of CE to uncharged solutes, Terabe and co-workers successfully introduced the use of ionic surfactants at concentrations above their critical micelle concentrations in the buffer solutions for CE.[1,2] This mode of CE separation is now known as micellar electrokinetic capillary chromatography (MECC). Several excellent review articles and book chapters on MECC have recently been published.[3-7]

Micellar electrokinetic capillary chromatography is regarded as a chromatographic technique mainly because the separation mechanism is due to the differential partitioning of solutes into a micellar pseudophase. The applications of the technique, initially intended for the separation of small uncharged molecules, has grown tremendously and continues to broaden into new areas that include a wide range of classes of charged and uncharged compounds. This has been accomplished through manipulation of the composition of micellar solutions and by incorporation of different types of chemical equilibria such as acid–base, complexation, ion exchange, or ion pairing.

Micellar electrokinetic capillary chromatography can be viewed as a hybrid of capillary zone electrophoresis (CZE) and reversed-phase liquid chromatography (RPLC) since electrokinetic migration, partitioning mechanisms, and hydrophobic interactions contribute to migration and separation of solutes. For separation of either uncharged solutes or mixtures of charged and uncharged compounds, the CE method of choice is MECC. In addition, for separation of compounds with identical electrophoretic mobilities, inclusion of some kind of organized media and/or chemical equilibrium can lead to the enhancement of selectivity.

In general, CE techniques (both CZE and MECC) offer higher efficiencies and faster analysis times than those achieved by RPLC. Reversed-phase liquid chromatography, on the other hand, offers the advantage of high selectivity, which is mainly due to the feasibility of controlling retention behavior of solutes through a careful selection of mobile-phase composition. The combination of high selectivity and efficiency has not only made RPLC the most widely used high-performance liquid chromatography (HPLC) technique but one that offers a wide scope of applications. In principle, it should be possible to manipulate selectivity in electrokinetic chromatography in a similar manner to that in RPLC.

Intuitively, one might argue that a higher degree of selectivity is achieved through differential partitioning of solutes from a bulk solvent into a solid (or alkyl-bonded) stationary phase than by distribution of solutes between the bulk solvent and micellar pseudophases in the same solution. However, this lower degree of selectivity, if any, can be largely compensated by a much higher efficiency in electrokinetic capillary chromatography. The smaller degree of band broadening in MECC, as compared to RPLC, is mainly due to the fact that faster rates of solute mass transfer into the micellar pseudophase are achieved in solution than across liquid–solid interfaces. Obviously, the nearly perfect flow profile in MECC as compared to the laminar profile in RPLC also contributes to higher efficiencies of the former technique.

A wider range of solutes, in terms of size and molecular weight, can be separated by CZE and RPLC in comparison with MECC. This is partly due to the existence of an elution window in MECC that limits the applicability of the technique. For example, RPLC is a versatile technique for the separation of highly hydrophobic compounds or large macromolecules such as peptides and proteins, both of which are difficult for MECC since they strongly interact with micelles and elute near or with micelles. The combination of hydro-organic mobile phases and alkyl-bonded phases in RPLC provides a wide range of elution strength and selectivity for the separation of broad classes of compounds. Both RPLC and MECC have shortcomings for separation of uncharged and highly polar molecules due to lack of adequate interaction with the hydrophobic alkyl-bonded phases and with micelles. An alternative LC technique is normal-phase HPLC, in which solutes partition from a nonpolar mobile phase into a more polar stationary phase. This option is not yet available for MECC.

In order to broaden the range of applications of electrokinetic chromatography and to control migration behavior of solutes for selectivity enhancement, the factors that influence migration behavior of solutes and resolution should be carefully examined. This chapter provides an overview of the fundamentals of the technique, with a major emphasis on migration mechanism of uncharged and charged solutes, as well as the factors that influence migration behavior and resolution in MECC. The importance of understanding the basic theoretical principles of MECC is illustrated through application of this technology for solving real separation and analytical problems. Accordingly, the figure legends contain methodological details of the separation.

A. TERMINOLOGY

Since the first report on incorporation of micelles in CE, the usefulness of several other chemical equilibria and organized media have been examined. Subsequently, it was suggested that a general terminology of electrokinetic chromatography with an abbreviation of EKC (instead of ECC) would be more appropriate[8] in order to emphasize the important role of electrokinetic phenomena that can be coupled with different types of interacting and/or organized media systems. This author agrees with these notions about the significance of the electrokinetic phenomena and the possibilities of using a wide range of organized media and equilibria to control selectivity. However, the abbreviation MECC is used in this chapter simply because it is the most widely known and utilized terminology for the technique.

B. MICELLES

The majority of reports on electrokinetic chromatography deal with using surfactants in the micelle-forming concentration ranges in the buffer solutions for CE. A surfactant consists of amphiphilic molecules, each with a hydrophobic tail and a polar or ionic head group. Surfactants can be classified as anionic, cationic, zwitterionic, and nonionic based on the head group charge, or as hydrocarbons, fluorocarbons, bile salts, etc., depending on the composition of the hydrophobic tail (Table 1).

At a certain concentration, known as critical micelle concentration (CMC), surfactants begin to form roughly spherical and dynamic aggregates that are at equilibrium with the monomer surfactant molecules in the solution. Spherical micelles have a diameter of 3 to 6 nm and are composed of 30 to 100 monomer surfactant molecules (known as the aggregation number, N).

The unique characteristics of micellar aggregates stem from their microscopically nonhomogeneous nature. They provide a microenvironment that is distinctly different from the bulk solvent. From the perspective of their applications in chemical separation, the availability of various sites of interaction with solutes in or on the micellar aggregates is their most important characteristic for providing selectivity in differential partitioning. Most notably, the hydrophobic core of micelles provides sites of interaction that greatly enhance the solubility of insoluble nonpolar compounds in aqueous media, a property that is known to "everyone who has washed dishes."[9] Other types of polar and electrostatic interactions may also exist, depending on the properties of the head group and/or hydrophobic tail.

TABLE 1
Typical Surfactants

Type	Name	Formula	c.m.c.[a]
Anionic	Sodium dodecylsulfate (SDS)	$C_{12}H_{25}OSO_3^-NA^+$	8.1
	Lithium perfluorooctane sulfonate (LiPFOS)	$C_8F_{17}SO_3^-$	6.72[b]
Cationic	Dodecyl trimethyl ammonium bromide (DTAB)	$C_{12}H_{25}N^+(CH_3)_3Br^-$	15
Nonionic	Polyoxyethylene(23) dodecanol (Brij35)	$C_{12}H_{25}(OCH_2CH_2)_{23}OH$	0.1
Zwitterionic	*N*-dodecylsultaine	$C_{12}H_{25}(CH_3)_2N^+CH_2CH_2CH_2SO_3^-$	1.2
Bile salts			

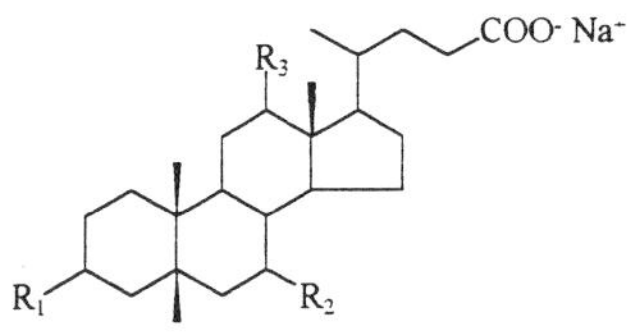

	R_1	R_2	R_3	c.m.c.[a]
Sodium cholate	OH	OH	OH	12.5
Sodium deoxycholate	OH	H	OH	6.4

[a] Data from Hinze, W. L., *Ordered Media in Chemical Separations,* Hinze, W. L. and Armstrong, D. W., Eds., American Chemical Society, Washington, D.C., 1987.
[b] Data from Yoda et al., *J. Colloid Interface Sci.,* 131(1), 282, 1989.

Micelles can organize or compartmentalize solutes in various sites with different microenvironmental properties such as polarity, fluidity, and acidity. It is important to recognize that the separation process in MECC is influenced by differences in the extent of interaction between solutes and micelles and not by micellar solubilization. All solutes are solubilized in the micelle-containing buffer solutions; however, solubility does not necessarily correlate with the quality of MECC separations. As a result, partition coefficients of solutes into micelles (or their binding constants to micelles), rather than solubility, should be regarded as the important physicochemical properties of solutes that provide information that is relevant to MECC separations. Solute–micelle interactions occur through different mechanisms such as surface adsorption, partitioning, and co-micellization. The exact mechanism depends on the chemical nature of both solute and surfactants. An excellent review on the use of surfactant assemblies in chemical separation has recently been published.[10]

II. THEORETICAL BASIS FOR SEPARATION IN MECC

Micelles provide both ionic and hydrophobic sites of interaction simultaneously, making MECC preferable to CZE for the separation of mixtures of charged and uncharged solutes. As mentioned above, solute interaction with micelles plays a major role in MECC separations for all solutes. There exist, however, significant differences in migration behavior of charged and uncharged compounds.

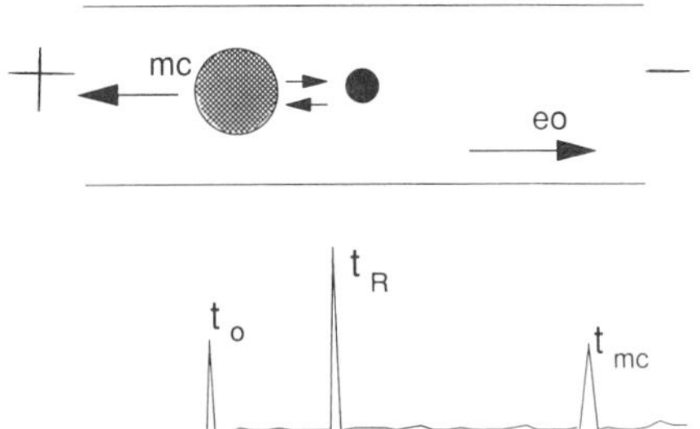

FIGURE 1. Migration of an uncharged compound in MECC. A typical chromatogram shows the elution of the uncharged solute within an elution window defined by t_o and t_{mc}.

A. UNCHARGED COMPOUNDS

Figure 1 illustrates the typical migration behavior of a neutral solute in MECC with an anionic surfactant in a fused silica capillary. The electrophoretic mobility of uncharged molecules is negligible. Therefore, their separation in MECC is solely due to differential partitioning into micelles. As in CZE, under the influence of an electric field, bulk buffer solution migrates from the positive electrode to the negative electrode. This is referred to as electroosmotic flow (EOF, with mobility equal to μ_{eo}) and is described in Chapter 2, Section II.C.2, and also discussed in detail in Chapter 22. Charged micellar aggregates possess their own electrophoretic mobility (μ_{mc}) and eventually "elute" from the capillary at a limited migration time, t_{mc}. The mobility of a neutrally charged compound would then depend on the extent of its interaction with micelles. The migration pattern of uncharged solutes in MECC falls within an elution window characterized by the elution of a highly polar and a highly hydrophobic compound. Highly polar solutes do not interact with micelles (i.e., they are "unretained" by micelles) and migrate at the velocity of the EOF. The migration time of polar, uncharged organic compounds such as methanol or acetonitrile, t_{eo}, can then be used as a marker for determining EOF velocity. Highly hydrophobic compounds, in contrast, have strong interactions with micelles and stay with micelles at all times (i.e., they are completely retained). The elution time of a very hydrophobic compound can then be regarded as a marker for determining the migration time of micelles, t_{mc}. Uncharged compounds of intermediate polarities would then elute within a window determined by the migration time of an unretained solute and that of micelles (Figure 1). This is considered a major drawback for MECC as separation will have to be achieved within this migration window.

The net migration velocity of micelles, v_{mc}, is then a vector sum of electroosmotic velocity, v_{eo}, and the electrophoretic velocity of micelles, v_{ep},

$$v_{mc} = v_{eo} + v_{ep}$$

The velocities are directly related to the electric field strength, E (defined as voltage/capillary length) with a proportionality constant equal to mobilities, μ (i.e., $v = \mu E$). Therefore, the following relation can be written for MECC systems:

$$\mu_{mc} = \mu_{eo} + \mu_{ep} \tag{1}$$

The migration behavior of uncharged solutes in MECC can be classified on the basis of the direction and relative magnitude of these mobilities.[2,11] The signs of the mobilities depend on their relative direction in the capillary. In unmodified fused silica capillaries, the direction of EOF is from the positive electrode (anode) to the negative electrode (cathode). This direction is usually used (by convention) as a reference and the mobility of species that migrate in this direction is considered as positive (i.e., $\mu > 0$). For example, for the case illustrated in Figure 1, the electrophoretic mobility, μ_{ep}, of the anionic micelle is negative (because the micelle moves toward the positive electrode); however, the net mobility of micelles is positive (i.e., $\mu_{mc} > 0$) because the mobility of the EOF is larger than the electrophoretic mobility of the micelle. Another possibility is when micelles migrate in the opposite direction of the EOF because their electrophoretic mobility is actually larger than the electroosmotic mobility, i.e., $\mu_{mc} < 0$ or $-\mu_{ep} > \mu_{eo}$, therefore the width of migration window (t_{mc}/t_o) is negative.[2,11] This situation is observed where the electroosmotic mobility is reduced to a point that it is unable to carry the anionic micelles toward the negative electrode. The EOF can be drastically reduced by a number of methods, such as using acidic buffers (pH < 5), chemical modification of the capillary, addition of an organic cosolvent to the aqueous micellar medium, or using cationic surfactants that dynamically modify the silica surface. Another obvious alternative is to choose a surfactant that forms micelles with a higher electrophoretic mobility. Finally, a special case can arise if the electrophoretic mobility of micelles is identical but in the opposite direction of the electroosmotic mobility ($-\mu_{ep} = \mu_{eo}$). Consequently, the net mobility of the micelles is zero ($\mu_{mc} = 0$), indicating that the micelles are stationary. This situation has been reported at pH 5 for sodium dodecyl sulfate (SDS) micelles.[12]

Due to the differential partitioning mechanism in MECC, Terabe et al. defined the capacity factor k' for MECC to represent retention as:[1,2]

$$k' = \frac{n_{mc}}{n_{aq}} = \Phi P \tag{2}$$

where n_{mc} and n_{aq} are the numbers of moles of solute in micelles (pseudostationary phase) and in the bulk aqueous phase (the mobile phase), Φ is the chromatographic phase ratio defined as the ratio of the volume of the micellar phase to that of the aqueous phase, and P is the distribution ratio (or partition coefficient) of a solute between aqueous phase and micelles. In this chapter the term *migration factor*, instead of capacity factor, is used to represent k' since the former better represents the migration process in MECC.

The migration factor can be calculated from the migration time values using Equation 3:[1,2]

$$k' = \frac{t_R - t_{eo}}{t_{eo}\left[1 - \left(t_R / t_{mc}\right)\right]} \tag{3}$$

This equation is very similar to that in conventional chromatography,

$$k' = \frac{t_R - t_o}{t_o} \tag{4}$$

The additional term in the denominator of Equation 3 accounts for the fact that the "stationary phase" in MECC actually moves. If micelles were made truly stationary, then Equation 3 equals Equation 4 as t_{mc} approaches infinity.

Migration behavior in MECC can also be described in terms of electrophoretic mobility.[11,13] An uncharged solute, however, does not possess its own electrophoretic mobility. One can define the mobility of a neutrally charged solute, μ_n, in terms of the fraction of solute that associates with micelles and migrates at the mobility of micelles as

$$\mu_n = (k'/k' + 1)\mu_{mc} \tag{5}$$

where μ_{mc} is the electrophoretic mobility of micelles and the ratio of $k'/k' + 1$ represents the fraction of the solute that is associated with micelles. One can then rearrange Equation 5 as

$$k' = \frac{\mu_n}{\mu_{mc} - \mu_n} \tag{6}$$

Electrophoretic mobility can be determined from the migration times as

$$\mu_n = \frac{L_t L_s}{V}\left(\frac{1}{t_R} - \frac{1}{t_{eo}}\right) \tag{7a}$$

$$\mu_{mc} = \frac{L_t L_s}{V}\left(\frac{1}{t_{mc}} - \frac{1}{t_{eo}}\right) \tag{7b}$$

where L_t and L_s are the total and separation lengths of the capillary, respectively, V is the applied voltage, and t_R is the migration time of solute (for calculation of μ_n) or micellar migration time, t_{mc} (for calculation of μ_{mc}). Substitution of Equations 7a and 7b into Equation 6 would result in Equation 3.

Using Equations 3 to 7, one can calculate a quantitative measure of migration in MECC from the migration time data, which can be obtained from MECC chromatograms. In contrast to migration time, the two migration parameters μ or k' are, at least in principle, independent of the system parameters such as field strength and EOF velocity. Compared to migration times, they better represent the extent of solute interactions with micelles and can be related to important variables (such as micelle concentration) that have a direct impact on resolution. These relationships can be used to achieve a better understanding of migration mechanisms and to facilitate optimization of MECC separations (see Sections IV.A and IV.E.l.a).

B. CHARGED COMPOUNDS

Migration behavior of ionic species in MECC is more complicated than those of neutral compounds for two reasons. First, charged solutes have an extra migration mechanism that influences the separation process. In addition to partitioning into (or associating with) micelles and moving at the velocity of micelles, ionic compounds possess their own electrophoretic mobility. Second, the involvement of additional chemical equilibria due to ionizable compounds (acid–base, ion pairing) would further add to the complexity of the system. As a result, the number of experimental parameters that influence the separation would be rather extensive, some examples being pH, surfactant type and concentration, and buffer concentration and type. These additional factors, if optimized effectively and efficiently, would provide additional tools to manipulate selectivity and thus enhance separation.[13]

1. Anions

The migration behavior of a weak acid (and that of an anion) is shown schematically in Figure 2a. An anionic compound is regarded as a special case of an acid that is fully dissociated. In contrast to neutral solutes, not all retention of charged solutes (both anions and cations) can be explained by their association with the micellar pseudo phase. This is because these compounds possess an electrophoretic mobility in the bulk aqueous phase. Uncharged solutes migrate with the bulk solvent (at electroosmotic velocity) during the time spent in the aqueous phase. Therefore, in MECC the mobility of an anion would be the weighted sum of the mobility of the micellar phase and its own mobility in the aqueous phase. This can be accounted for by the observed overall mobility in the absence of micelles, μ_o:

$$\mu_n = \left(\frac{k'}{1+k}\right)\mu_{mc} + \left(\frac{1}{1+k'}\right)\mu_o \tag{8}$$

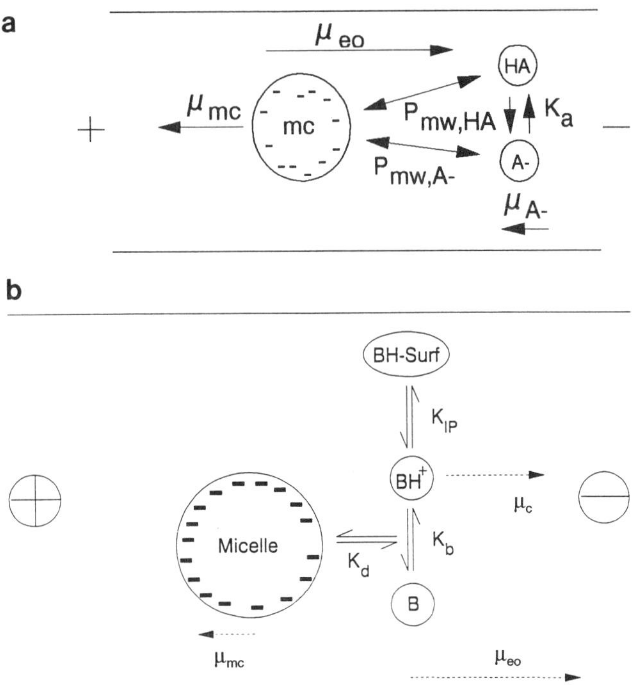

FIGURE 2. Migration of ionizable/charged compounds in MECC. (a) An acidic/anionic compound and (b) a basic/cationic compound in MECC. (From Khaledi, M. G., Smith, S. C., and Strasters, J. K., *Anal. Chem.*, 63, 1820, 1991; and Strasters, J. K. and Khaledi, M. G., *Anal. Chem.*, 63, 2503, 1991. With permission.)

As shown in Figure 2a, two migrating fractions can be discerned: the acid/conjugated base pairs associate with the micelles and migrate at a mobility of μ_{mc}, and the anionic (dissociated) form in the aqueous phase moving at a mobility of μ_o. Another important migration mechanism is through the EOF. All equations, however, are expressed in terms of electrophoretic mobilities because the electroosmotic mobility, μ_{eo}, is constant for all components in the system. It is important to note that in deriving these equations, several assumptions have been made. For example, it is assumed that the mobility of micelles is not altered when associated with analyte, secondary chemical equilibria with buffer components do not occur, and that the possible effects of the ionizable analytes and their equilibria on the ζ potential and, hence, EOF, are insignificant.

From Equation 8, one can derive the following equations for acidic/anionic compounds:

$$k' = \frac{\mu - \mu_o}{\mu_{mc} - \mu} \tag{9}$$

or

$$k' = \frac{t_R - t_o}{t_o\left(1 - t_R / t_{mc}\right)} \tag{10}$$

where t_o is the retention time of charged species in the bulk aqueous phase (i.e., in the absence of micelles). Again, using Equations 7 to 9, it is possible to calculate important migration parameters for anions that can be correlated with variables such as micelle concentration and pH (see Section IV.E.l).

2. Cations

It can be expected that the observed migration of these compounds will be more difficult to describe. This is due to the electrostatic interaction between the free surfactant ions in the mobile phase and the cations, which introduces an additional equilibrium (ion-pairing formation) into the equations. Cations may also interact with the silanol groups on the capillary walls. This will further complicate the equations. More importantly, however, it will lead to band broadening and loss of resolution. Therefore solute-wall interactions should be avoided. This can be done by adding a "competing" base to the buffer or by operating at higher micelle concentrations of an anionic surfactant.[14] Quang et al. observed a dramatic improvement in peak shapes of organic amines at higher concentrations of SDS. This was attributed to the effective competition of the anionic SDS micelles with the silanol groups on the capillary wall in interaction with the solutes.[14]

The various equilibria playing a role in the migration of a cationic solute in MECC are depicted in Figure 2b. First there will be a dissociation of the solute in the aqueous phase, represented by the equilibrium constant K_b, which also explains the major influence of the pH on the observed migration.[14,15] Both the base, B, and its conjugate acid, BH^+, will partition into (or associate with) the micellar phase, each according to its own distribution coefficient, $P^{m}_{BH^+}$ or P^{m}_{B}, respectively. Finally, there will be an interaction between the free ionized surfactant and the positively charged cations. This ion-pairing equilibrium is defined by the equilibrium constant K_{ip}. Since the concentration of free surfactant in micellar solutions is, in first approximation, constant and equal to the critical micelle concentration (CMC), we can define the constant K_I = (CMC)K_{ip}. The possible solute–wall interactions are not considered.[14,15]

The following equation has been derived for cationic solutes:[14,15]

$$k' = \frac{\mu - \left(\dfrac{1}{1 + K_a/[H^+] + K_I} \right) \mu_c}{\mu_{mc} - \mu} \tag{11}$$

Regarding Equation 12, a number of observations are immediately apparent:

1. As in all calculations with respect to k' in MECC, this equation shows that k' will approach infinity if the observed mobility of the solute becomes equal to the mobility of the micelles, i.e., complete partitioning of all species of a solute into the micelles. This is directly linked with the limited migration range in MECC.
2. In the case of complete complexation of the cations with the surfactants (i.e., K_I approaches infinity), Equation 11 will reduce to

$$k' = \frac{\mu}{\mu_{mc} - \mu} \tag{12}$$

   which is identical to the relation derived for uncharged solutes (Equation 7). This is in accordance with the expectations, since the cation–surfactant complex will not have an electrophoretic mobility of its own.
3. In the absence of complexation (i.e., K_I approaches zero) Equation 12 reduces to

$$k' = \frac{\mu - \mu_o}{\mu_{mc} - \mu} \tag{13}$$

which is identical to the relation derived for anionic solutes (Equation 9). The mobility is now completely defined by the mobility of the micellar phase and the mobility of the solute in the absence of micelles.

III. RESOLUTION IN MECC

The resolution equation for uncharged solutes in MECC was reported by Terabe and co-workers[2] as

$$\text{Res} = \left(\frac{N^{1/2}}{4}\right)\left(\frac{\alpha - 1}{\alpha}\right)\left(\frac{k_2'}{1 + k_2'}\right)\left[\frac{1 - \left(t_o/t_{mc}\right)}{1 + \left(t_o/t_{mc}\right)k_1'}\right] \tag{14}$$

where Res is the resolution between peaks 1 and 2, N is efficiency, and α is selectivity between the peak pair, and is equal to k_2'/k_1'.

The influence of efficiency on MECC separations is similar to those of chromatography and electrophoresis since resolution is directly proportional to the square root of the number of theoretical plates. Inherently, higher plate numbers can be generated in MECC than in HPLC. This is mainly due to the nearly perfect flow profile of electrokinetic migration and faster kinetics of solute mass transfer into micellar pseudostationary phases in solution as compared to that into solid stationary phases in HPLC. On the other hand, in comparing efficiency in CZE and MECC, inclusion of micelles in the buffer solutions of CZE causes additional band-broadening mechanisms that lead to the overall lower efficiency. In addition to various sources of intra- and extracolumn band broadening that exist in a CZE separation, the following mechanisms of band broadening have been discussed:[16-18] (1) axial diffusion, which is related to the diffusion coefficient of solutes in bulk solvent as well as to those associated with micelles, (2) micelle polydispersity, which accounts for the different migration rates of micelles of different sizes in a given solution, (3) mass transfer effect, and (4) thermal effects. Despite these sources of band dispersion, one can easily achieve a high number of theoretical plates, on the order of several hundred thousand. A more thorough discussion of the different band-broadening mechanisms can be found in Refs. 16 through 18.

The main difference between the resolution equation in MECC and that in chromatography is the last term in Equation 14, which represents the existence of the limited elution window in MECC. For a truly micellar stationary phase (i.e., t_{mc} = infinite), the last term would be equal to unity and Equation 14 would be identical to the fundamental resolution equation in chromatography. Better resolution can be achieved in MECC as the width of the elution window is increased (i.e., t_o/t_{mc} is reduced). This is shown in Figure 3, where the y axis, $f(\tilde{k}')$, is a product of the last two terms representing the influence of retention on resolution. For $t_o/t_{mc} = 0$, resolution increases with k' and reaches a plateau, which is the typical behavior in conventional chromatography. The limited elution window in MECC, however, creates a special situation where maximum resolution can be achieved in an optimum range of k' values at a given t_o/t_{mc}. Using a theory based on electrochemical parameters, Ghowsi and co-workers determined that the optimum range for the migration factor, for good resolution and resolution per unit time in MECC for uncharged solutes, is between 2 and 5 for the two cases where the net mobility of micelles is positive ($\mu_{mc} > 0$) and zero ($\mu_{mc} = 0$).[11] Foley has reported the following equations for optimum migration factors for best resolution and for best resolution per unit time:[19]

$$k'_{opt}(\text{maximum Res}) = \left(t_{mc}/t_o\right)^{1/2} \tag{15a}$$

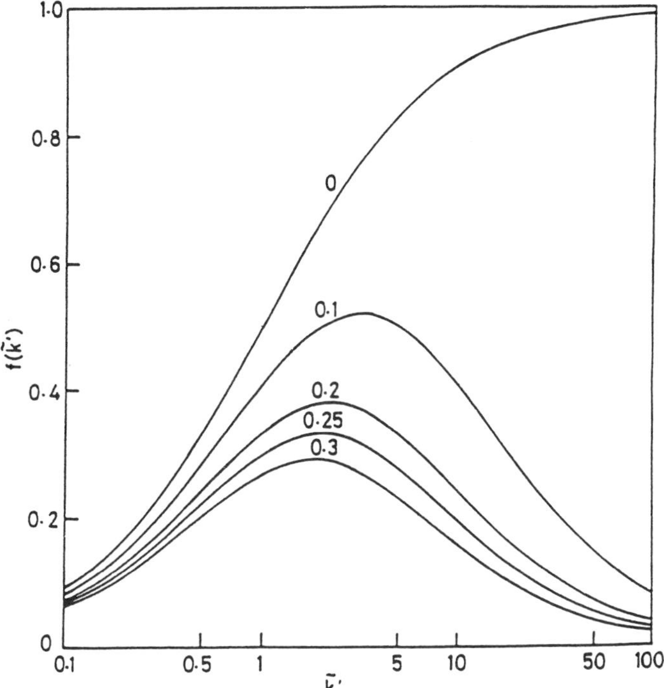

FIGURE 3. Effect of retention on MECC resolution. Dependence of the migration function, $f(k')$, on the migration factor k' at different t_o/t_{mc} ratios. Resolution in MECC is directly related to $f(k')$ that is, the product of the last two terms of the resolution equation (Equation 14). (From Terabe, S., Otsuka, K., and Ando, T., *Anal. Chem.*, 57, 834, 1985. With permission.)

and

$$k'_{opt}\left(\text{Res}/t_R\right) = \frac{-\left(1 - t_o/t_{mc}\right) + \left[\left(1 - t_o/t_{mc}\right)^2 + 16\left(t_o/t_{mc}\right)\right]^{1/2}}{4\left(t_o/t_{mc}\right)} \tag{15b}$$

Foley identified three distinct regions for migration factors of uncharged solutes in MECC. The most preferred region is where k' falls between the two optimum k' values for maximum resolution per time and maximum resolution, i.e., k'_{opt} (Res/t_R) < k' < k'_{opt} (Res). In the second most desirable region, k' < k'_{opt} (Res/t_R), resolution is not satisfactory due to low k' values (i.e., small interaction with micelles); however, analysis time is short. The least favorable region is k' > k'_{opt} (Res), where poor separations and long analysis times are observed due to large k'. The existence of the last k' region distinguishes MECC from conventional chromatography where no loss in resolution is observed due to prolonged retention.[19]

It is also important to note that the elution window in MECC limits the peak capacity (defined as maximum number of peaks that can be resolved within a specified range of retention time). Peak capacity, n, in MECC depends on the width of elution window, t_{mc}/t_o, as well as on the number of theoretical plates as,[2]

$$n = 1 + \frac{N^{1/2}}{4} \ln \frac{t_{mc}}{t_o} \tag{16}$$

IV. CONTROLLING MIGRATION BEHAVIOR IN MECC

Resolution in MECC can be enhanced through a careful manipulation of solute binding constants to micelles (which directly influences migration factor and selectivity) as well as extending the elution window, i.e., reducing t_o/t_{mc}. In general, the quality of MECC separations can be improved by increasing efficiency, enhancing selectivity, optimizing k', and widening the elution window. The composition of micellar solutions affects solute–micelle interactions, electrophoretic mobility, and EOF velocity. Consequently, the quality of MECC separations and analysis times depend on the selection of the composition of the micellar buffer solution and certain system parameters.

Method development in MECC is still based on trial and error as well as the researcher's intuition and experience. The number of independent variables is considerable, thus the number of possible variations of these variables is too large to screen all combinations for a given problem. For example, micelle concentration, type of surfactant, concentration and type of additives (such as organic co-solvents, cyclodextrins, other surfactants, urea, or other reagents that interact with solutes through complexation), ion exchange or other mechanisms, pH, type of buffer, ionic strength, as well as system parameters such as temperature, electric field strength, and modification of the capillary wall can be regarded as important parameters that influence MECC separations. In the following sections the role of some of the parameters having the largest influence on the three migration terms in the resolution equation — k', α, and t_o/t_{mc} — will be briefly discussed.

A. MICELLE CONCENTRATION

According to Equation 2, the migration factor is related to the partition coefficients of solutes between the aqueous phase and the micellar pseudophase as well as the phase ratio of the system defined as the volume of the micellar phase to that of the aqueous phase; the latter term depends on micelle concentration. The relationship between migration factor and micelle concentration can then be expressed as[2]

$$k' = \frac{V_{mc}}{V_{aq}} P = \frac{v(C_{sf} - \text{CMC})}{1 - v(C_{sf} - \text{CMC})} P \tag{17a}$$

where C_{sf} is the total surfactant concentration and v is the partial molar volume. Note that the micelle concentration is equal to C_{sf} – CMC. At low micelle concentrations, the second term in the denominator of Equation 17a becomes negligible and there exists a linear relationship between the migration factor and surfactant concentration as,[2]

$$k' = Pv(C_{sf} - \text{CMC}) \tag{17b}$$

This equation is valid for both charged and uncharged compounds; the only difference being the equation that is used for calculating k' for uncharged solutes, anions, and cations. Figure 4 shows the linear relationship between k' and C_{sf} for a group of neutral and charged substituted phenols.[13] Similar linear relationships have also been reported for cationic compounds.[14,15] The slope of Equation 17b is directly related to partition coefficients of solutes between the bulk aqueous solvent and micelles and the x intercept is the CMC. For acidic and basic compounds P is a function of the pH. One can then obtain information about solute interactions with micelles and CMC through MECC experiments.

Micelle concentration can be used to improve the quality of MECC separations through its influence on the migration factor, k'. As mentioned previously, the k' values should be adjusted

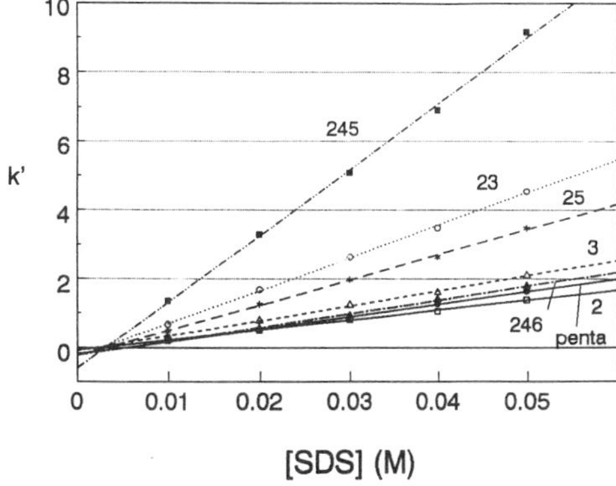

FIGURE 4. Effect of surfactant concentration on the migration factor. Charged and uncharged chlorinated phenols: 2, 2-chlorophenol; penta, pentachlorophenol; 246, 2,4,6-trichlorophenol; 3, 3-chlorophenol; 25, 2,5-dichlorophenol; 23, 2,3-dichlorophenol; 245, 2,4,5-trichlorophenol. (From Khaledi, M. G., Smith, S. C., and Strasters, J. K., *Anal. Chem.*, 63, 1820, 1991. With permission.)

within optimum ranges in order to achieve the maximum resolution and best separation per unit time (Equation 15).[19] Using Equations 15a and 15b, Foley[19] derived the following equations for calculating the optimum surfactant concentrations that provide the optimum k' value for maximum resolution between a pair of solutes and optimum k' for best resolution per unit time:

$$[\text{Surf}]_{\text{opt}}(\text{Res}) = \frac{\left(t_{\text{mc}}/t_{\text{o}}\right)^{1/2}}{Pv} + \text{CMC} \tag{18a}$$

and

$$[\text{Surf}]_{\text{opt}}\left(\text{Res}/t_{\text{R}}\right) = \frac{k'_{\text{opt}}(\text{Equation 15b})}{Pv} + \text{CMC} \tag{18b}$$

where P is the average partition coefficients of the critical pair of solutes.[19] Therefore, for a given MECC system and solutes of interest, the optimum micelle concentration is predetermined.[19] The optimum micelle concentration for a peak pair in a chromatogram can then be predicted using these equations, provided the partition coefficients of the solutes and other system values are known.

Another alternative for optimization of MECC parameters, in this case micelle concentration, is to predict the migration behavior of individual solutes in a mixture based on a limited number of initial experiments and with the help of a mathematical model that relates the migration factor to the experimental parameter. The linear relationship between k' and C_{sf} is particularly useful since predictions can be made through linear interpolation, using measured migration factors with as few as two micelle concentrations. Once the migration pattern of all solutes in a mixture over a wide range of micelle concentration is simulated, the optimum concentration that provides the best separation can be predicted.[20] The advantages of this method are that all peaks in a chromatogram are considered in the optimization process and no *a priori* knowledge about solute properties is required to identify the optimum. An example is illustrated in Figure 5. A linear model of k' vs. C_{sf} is derived on the basis of two initial values of surfactant concentrations. The top frame of Figure 5 shows the two initial chromatograms at 10 and 50 m*M* SDS concentrations. The parameter space for a given optimization is encompassed by the initial parameter values. Then the migration factor of the solutes in the mixture at surfactant concentrations other than those used in the actual measurements within the parameter space were predicted through interpolation of the assumed linear model of k' vs. [Surf] (shown in the second frame in Figure 5). From the

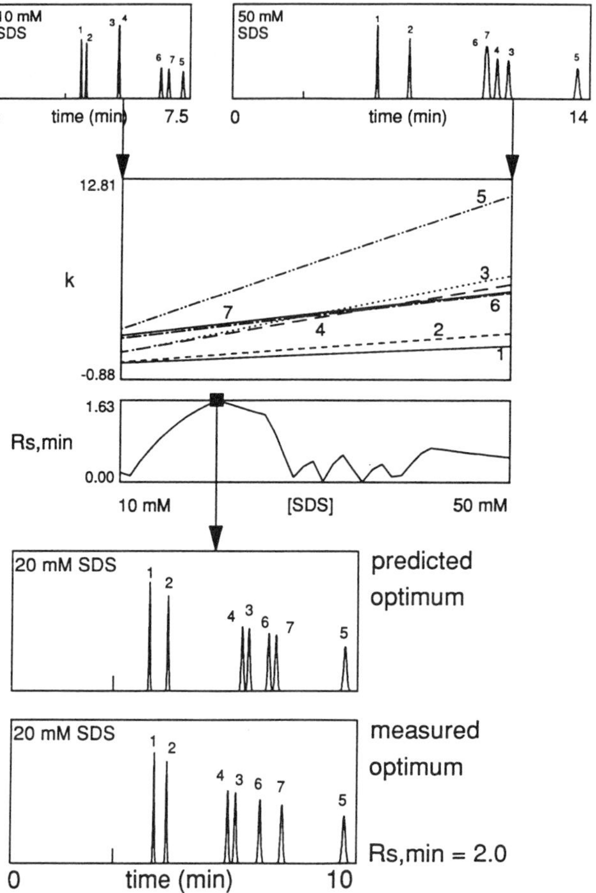

FIGURE 5. Optimization of MECC separation. Finding the optimum micelle concentration for the separation of chlorophenols. Experimental conditions: capillary, I.D. 50 μm, O.D. 375 μm, length 55 cm, separation length 38 cm; temperature, 35°C; voltage, 18 kV; 50 m*M* phosphate buffer, pH 7, sodium dodecyl sulfate (SDS) concentration varying between 10 and 50 m*M*. Solutes: (1) 2-chlorophenol, (2) 3-chlorophenol, (3) 2,3-dichlorophenol, (4) 2,5-dichlorophenol, (5) 2,4,5-trichlorophenol, (6) 2,4,6-trichlorophenol, and (7) pentachlorophenol. (From Strasters, J. K., et al., paper presented at the 15th Int. Symp. Column Liquid Chromatography, Basel, Switzerland, June 3–7, 1991. With permission.)

predicted migration values of all the compounds in the mixture, the quality of separation at all surfactant concentrations within the parameter space is calculated and an optimum is predicted (the third frame). Through computer modeling, electropherograms can be constructed based on these models in each point in the parameter space. The lower frame illustrates the excellent agreement between the predicted and measured separation. The success of predicting the optimum conditions based on a minimum number of experiments depends on the linearity of k' vs. parameter and, to some extent, on the reproducibility of migration behavior. If the linearity assumption is not accurate, additional experiments will be required to locate the optimum. This method, which is known as "iterative regression optimization strategy", was first developed for RPLC and has been applied in ion-pair chromatography and micellar liquid chromatography.[21-23]

It is important to note that in MECC, micelle concentration alters resolution primarily due to its influence on migration factor and to some extent on elution window width as micelle migration times change. Selectivity, however, is not affected by this factor as it is related to differential partitioning (or binding) of solutes into micelles. In order to improve the separation of solutes with identical partition coefficients into micelles, other parameters should be optimized. Micelle concentration can also have a dramatic effect on efficiency and consequently on resolution.[16]

B. TYPE OF SURFACTANT

Mobility of micelles and binding constants of solutes are influenced by the nature of the hydrophobic moiety, charged head group, and perhaps the counterion of surfactant-forming micelles.

The most widely used surfactant in MECC has been SDS, which is an anionic surfactant. Terabe and co-workers observed that migration factor values and micellar migration times can be significantly different for closely related surfactants such as sodium dodecyl sulfate, sodium tetradecyl sulfate, and sodium dodecyl sulfonate.[2] Polar compounds generally interact with micelles through surface adsorption and co-micellization. Therefore, their partition coefficients into micelles, and subsequently their migration factor in MECC, are more susceptible to changes in surfactant head group than to the hydrophobic moiety. Migration of hydrophobic solutes in MECC, on the other hand, should be more sensitive to changes in the hydrophobic interior of micelles such as the chain length.[2] In addition, the electrophoretic mobility, thus migration time of micelles, t_{mc}, changes with the hydrophobic tail and ionic head group, which in turn alters the elution window and ultimately MECC separations.

Burton and Sepaniak compared the influence of two anionic and two cationic surfactants with different chain lengths on migration behavior of a group of aromatic test solutes containing different acidic, basic, electron-withdrawing, and neutral functional groups.[24] The cationic surfactants were useful for solutes with limited aqueous solubility. Different elution orders were observed for several solutes which indicates the different selectivity for the cationic (alkyltrimethyl ammonium chlorides, C_{12}TAC and C_{16}TAC) and anionic surfactants (SDS, and sodium decyl sulfate, STS). It was concluded that migration behavior with the anionic SDS micelles is similar to that in RPLC with alkyl-bonded stationary phases. In contrast, Rasmussen et al. observed similar selectivity for SDS and STS micelles that differed from the selectivity obtained in RPLC for a group of aromatic compounds.[25]

Otsuka et al. observed completely different migration behavior for SDS and dodecyltrimethyl ammonium bromide (DTAB, a cationic surfactant) for the separation of PTH amino acids.[26] They reported an overall better resolution with SDS micelles; however, for certain PTH amino acids, CTAB provided better selectivity (Figure 6).

Morin et al.[27] reported the use of a cationic surfactant cetyltrimethyl ammonium bromide (CTAB) for the separation of anionic and neutral glucosinolates, which are a class of thioglucosides. The elution order was according to hydrophobicity of the solutes, while ion-pairing interactions also contributed to the migration of the anionic species.

Kaneta et al. also used a cationic surfactant (CTAC) for the separation of inorganic ions at concentrations below and above CMC.[28] At a submicellar concentration, ion association equilibria between the anionic solutes and cationic surfactant had an influence on migration behavior; however, the elution order of the ions was correlated with their molar conductivities. Above the CMC, partitioning into micelles also plays a role that leads to a different selectivity (Figure 7). The ion association constants and partition coefficients of the inorganic ions into the cationic micelles were determined from the MECC migration data.[28]

Bile salt surfactants have been used for the separation of hydrophobic compounds as well as for enantiomeric separations. Bile salts are biological surfactants with a hydrophobic steroid moiety and hydrophilic groups. Interaction between hydrophobic compounds and bile salt micelles is smaller than with hydrocarbon surfactants due to the more polar microenvironment of the bile salts. This leads to overall smaller MECC migration factors, which have

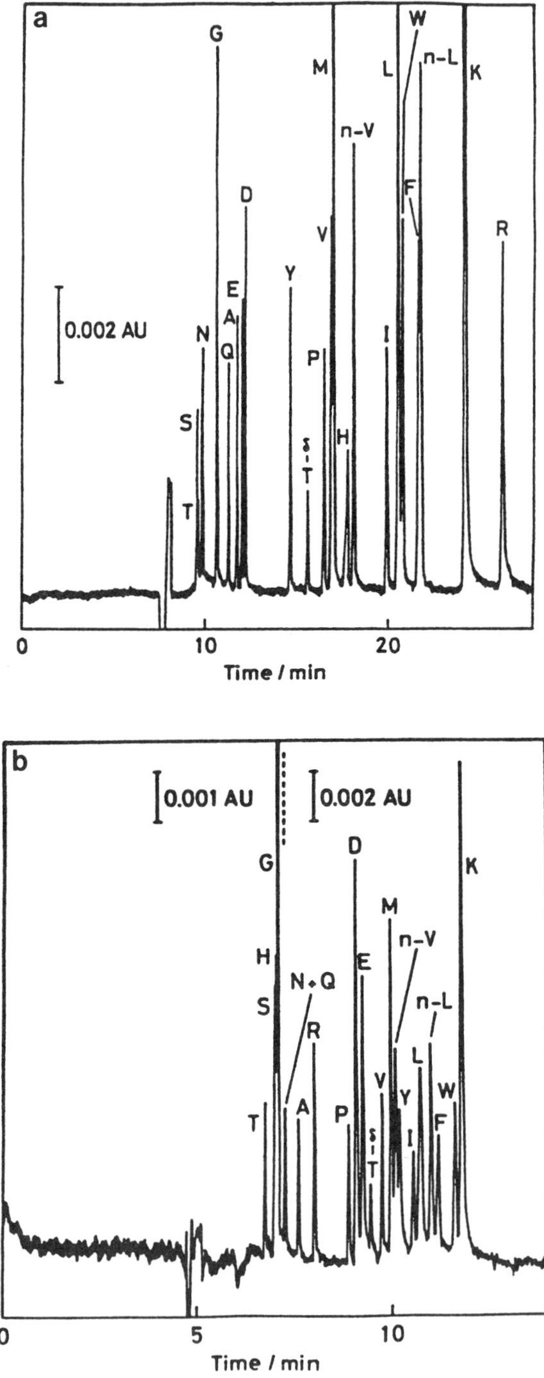

FIGURE 6. Anionic vs. cationic surfactant type: MECC separation of 22 PTH amino acids. (a) SDS (a C_{12} chain-length anionic surfactant), 0.05 *M* in phosphate-borate buffer (pH 7.0); capillary, I.D. 50 μm, total length 650 mm, separation length 500 mm; applied voltage, 10 kV; current, 14 μA; absorbance detection wavelength, 260 nm; capillary temperature, 35°C. (b) DTAB (a C_{12} chain-length cationic surfactant), 0.05 *M* in Tris-HCl buffer (pH 7.0); applied voltage, 15 kV; current, 37 μA, other conditions as in (a). (From Otsuka, K., Terabe, S., and Ando, T., *J. Chromatogr.*, 332, 219, 1985. With permission.)

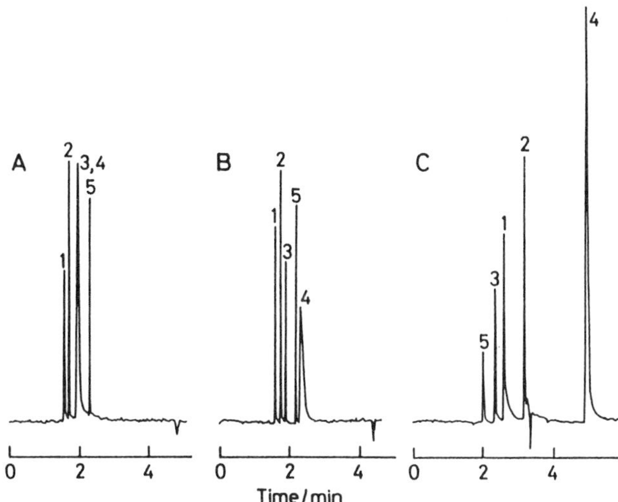

FIGURE 7. Effect of ion association interactions. Selectivity changes are observed due to the ion association interactions between the inorganic anions with DTAB, a cationic surfactant, at different concentrations: (A) 0.2 m*M*, (B) 1.0 m*M* (CMC), and (C) 25 m*M*. Key: (1) bromide, (2) nitrate, (3) bromate, (4) iodide, (5) iodate. Conditions: 20 m*M* phosphate-Tris buffer, pH 7.0; applied voltage, 15 kV; temperature, 22°C; fused silica capillary: I.D., 50 μm; total length, 500 mm; UV detection at 210 nm. (From Kaneta, T., Tanaka, S., Taga, M., and Yoshida, H., *Anal. Chem.*, 64, 798, 1992. With permission.)

been advantageously used for the separation of hydrophobic compounds.[29] Figure 8 illustrates the separation of ingredients of a cold medicine, using SDS micelles and two bile salt micelles, sodium cholate and sodium deoxycholate.[30] Note that in the SDS system, the hydrophobic compounds, peaks 10 to 14, coelute at the t_{mc} while they are less retained by the bile salts. In addition, different selectivity in migration is observed for these surfactant systems. Bile salt micelles are capable of chiral recognition, which makes them suitable for MECC separation of chiral compounds (see Section IV.E.2).

Another group of surfactants that has recently been used in MECC applications is fluorocarbon surfactants.[31] Figure 9 shows the plot of k' vs. micelle concentration for lithium dodecyl sulfate (LiDS) and for lithium perfluorooctane sulfonate (LiPFOS). An excellent linear relationship is observed for both systems. An important difference between the LiDS and LiPFOS surfactants is the degree of interaction with different solutes, represented by the solute partition coefficients into micelles or the slopes in Figure 9. For the LiDS system, the slope increases with the hydrophobicity of the solute, therefore k'(toluene) > k'(benzaldehyde) > k'(nitrobenzene). In LiPFOS, however, the order is reversed. This is interesting, in light of the fact that a $-CF_2$ group is about 1.5 times more hydrophobic than a $-CH_2$ group, yet hydrophobic hydrocarbon compounds have a lower affinity for the fluorocarbon micelles. For example, toluene, which is the most hydrophobic compound, has the smallest binding constant to fluorocarbon (FC) micelles in the group of the three compounds. This can be attributed to a phobicity effect between hydrocarbon and fluorocarbon compounds. The various affinity of compounds for these micellar systems should translate into different selectivity and migration behavior in MECC separations.

The dramatic variation in selectivity (and elution order) that is often observed for different surfactants indicates that the quality of MECC separations can be greatly influenced through the careful selection of surfactant type or even by "designing" surfactants with desirable structural properties.

An alternative approach to tailoring surfactant structure for achieving different interactions is to alter the micellar microenvironment through a judicious choice of mixed micelles.

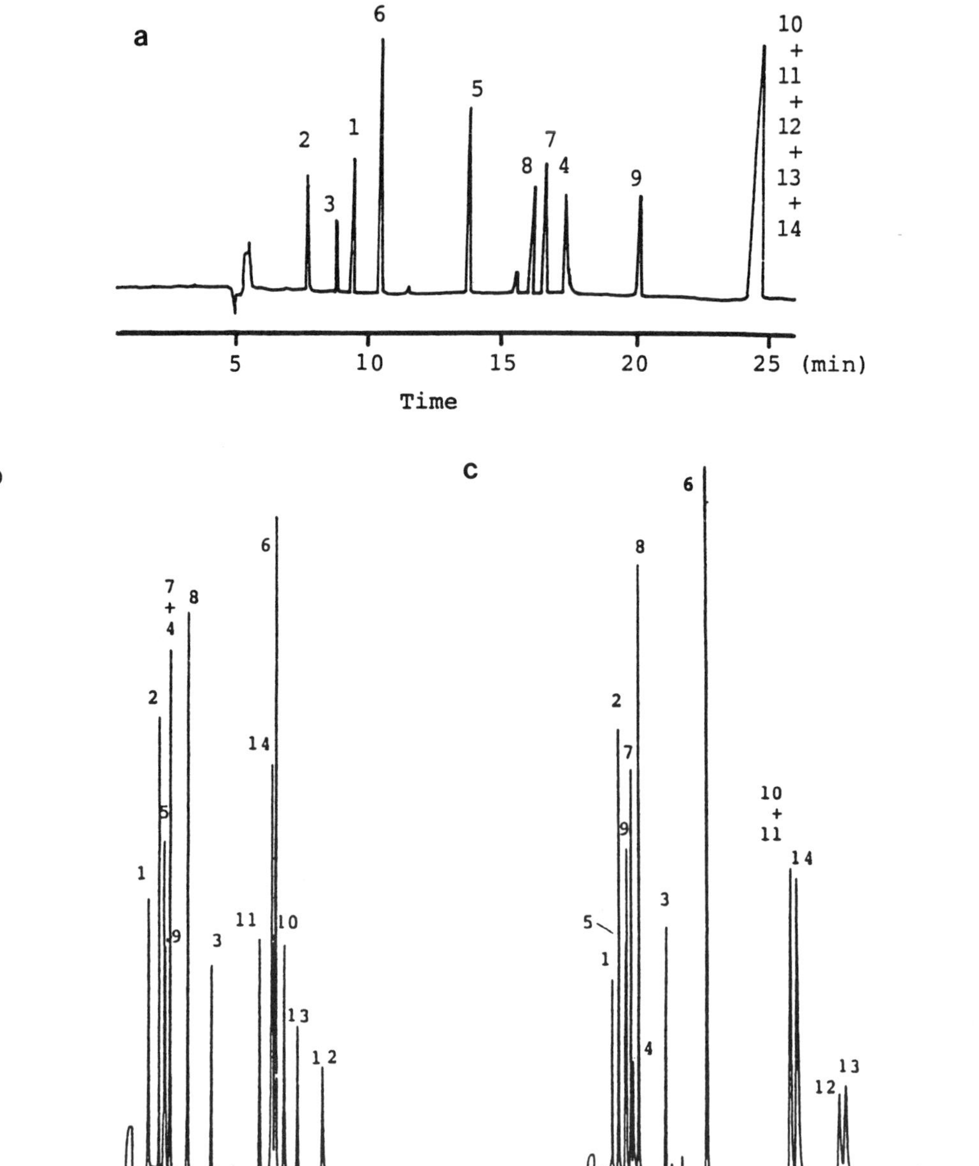

FIGURE 8. Hydrocarbon vs. bile salt surfactant type. Comparison of SDS and bile salt micelles for the MECC separation of ingredients of a cold medicine. (a) 100 m*M* SDS, 0.02 *M* phosphate-borate (pH 9.0), (b) 100 m*M* sodium cholate, and (c) 50 m*M* sodium deoxycholate. Applied voltage: 20 kV; fused silica capillary: I.D., 50 μm; total length, 650 mm; detection at 210 nm. Compounds: (1) Caffeine, (2) acetaminophen, (3) Sulpyrin (dipyrone), (4) trimethoquinol hydrochloride, (5) guaifenesin, (6) naproxen, (7) ethenzamide, (8) phenacetin, (9) isopropylantipyrine, (10) noscapine, (11) chlorpheniramine maleate, (12) Tipepidine hibenzate, (13) dibucaine hydrochloride, (14) triprolidine hydrochloride. (From Nishi, H., et al., *J. Chromatogr.*, 498, 313, 1990. With permission.)

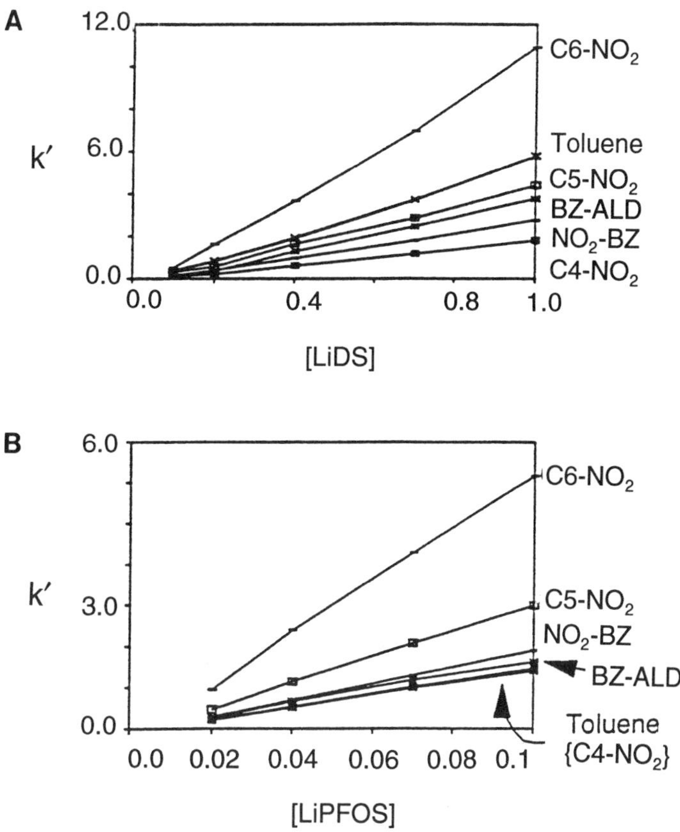

FIGURE 9. Hydrocarbon vs. fluorocarbon surfactant type. Variations in migration factor as a function of the surfactant concentration in MECC with (A) LiDS, a hydrocarbon surfactant, and (B) LiPFOS, a fluorocarbon micelle. (From Khaledi, M. G., et. al., *Anal. Chem.*, 1993, submitted. With permission.)

Mixtures of different surfactants in the micellar forms can be used to manipulate retention, selectivity, and t_o/t_{mc} in order to enhance resolution.

C. MIXED MICELLES

One can attain great control over MECC separations by mixing appropriate surfactants with different selectivity behavior.[31] An example is illustrated in Figure 10a and b for mixtures of the LiDS and LiPFOS system. The chromatograms shown in Figure 11 further illustrate this for uncharged compounds. As can be seen, by changing the mole fraction of hydrocarbon (HC) and FC surfactants (at a constant total concentration), the elution order of several peaks has changed.

The usefulness of mixed micelles in MECC has been demonstrated in a limited number of reports. Wallingford and co-workers observed greater selectivity and increased efficiency by using mixed SDS–sodium octyl sulfate for a group of borate-complexed catechols.[32] Rasmussen et al. reported selectivity changes between benzene and benzaldehyde on addition of a nonionic surfactant, Brij-35®, to SDS micelles.[25] The two compounds that coeluted at different concentrations of pure SDS micelles, were readily separated in mixed Brij-SDS micellar solutions. Foley reported a change in t_o/t_{mc} in Brij–SDS mixtures and also variations in selectivity for amino acids.[33] Chiral surfactants have also been mixed with nonchiral micelles to enhance selectivity of MECC separation for optical isomers.[34] Khaledi et al. have also shown the usefulness of mixed hydrocarbon–bile salts micelles.[35]

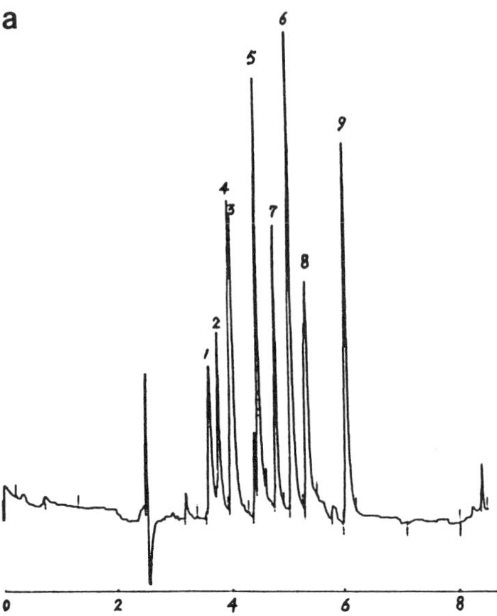

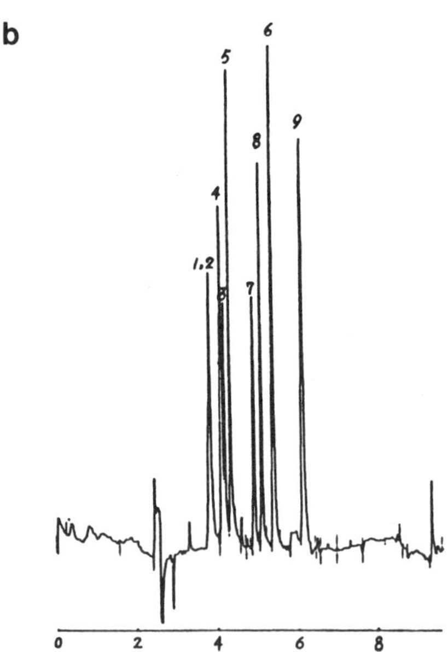

FIGURE 10. Selectivity changes in mixed micellar electrokinetic chromatography. Separation of a peptide mixture, using (a) 60 m*M* LiPFOS and 90 m*M* LiDS, (b) 90 m*M* LiPFOS and 60 m*M* LiDS. Conditions: applied voltage, 18 kV; fused silica capillary: I.D., 50 μm; total length, 520 mm; separation length: 355 mm. Peak I.D.: (1) Gly-Ala-Tyr, (2) Tyr-Tyr-Tyr, (3) Phe-Phe, (4) Tyr-Gly-Gly, (5) Leu-Trp, (6) Glu-Trp, (7) Phe-Gly-Gly, (8) Trp-Gly-Gly, (9) Leu-Trp. Capillary temperature, 40°C. (From Khaledi, M. G., et al., *Anal. Chem.*, 1993, submitted. With permission.)

As mentioned above, resolution in MECC can also be enhanced by increasing the size of the elution window (t_{mc}/t_o or reducing t_o/t_{mc}). The t_o/t_{mc} ratio changes with the composition of the mixed micellar solution. Therefore, in mixed systems one should consider the optimum compositions (both the total micelle concentration and mole fraction of individual surfactants) that would provide the optimum k', selectivity, and t_o/t_{mc} values. Although the t_o/t_{mc} ratio has

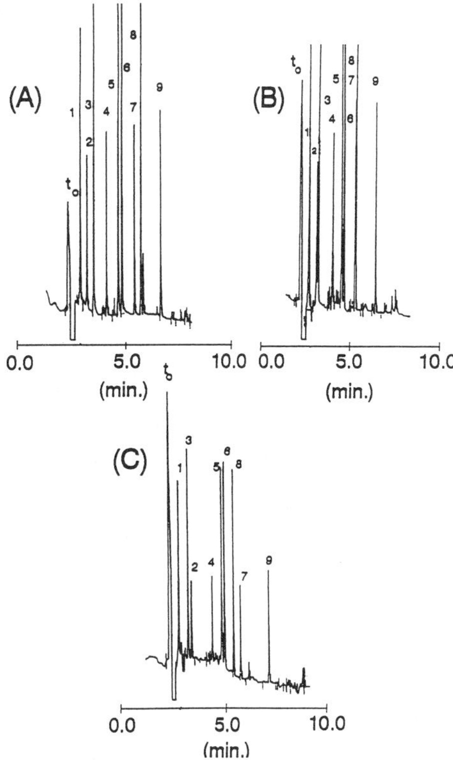

FIGURE 11. Effect of the surfactant ratio on selectivity in a mixed micellar system. Total surfactant concentration of LiDS and LiPFOS = 100 m*M*. Mole fraction of LiPFOS: (A) 0.2, (B) 0.4, and (C) 0.6. Other conditions same as in Figure 10. Solutes: (1) Resorcinol, (2) C_3NO_2, (3) phenol, (4) C_4NO_2, (5) benzaldehyde, (6) nitrobenzene, (7) C_5NO_2, (8) toluene, (9) C_6NO_2. (From Khaledi, M. G., *Anal. Chem.*, 1993, submitted. With permission.)

also changed in Figures 10 and 11, the overall separation times are relatively constant. This is a different situation than using modifiers such as organic solvents, where resolution is enhanced at the expense of analysis times (see Section IV.D.l). The examples shown for mixed micellar electrokinetic chromatography (MMEC) resemble those in HPLC, where selectivity can be easily manipulated at a constant solvent strength.

There are three types of mixed surfactant micelles:[36] (1) Mixed surfactants with similar head group charges: these are considered ideal mixed micelles whose critical micelle concentration can be predicted according to the ideal solution theory. (2) Mixtures of fluorocarbon and hydrocarbon surfactants that show a positive deviation from ideality: the critical micelle concentration of the latter mixture is larger than what is predicted according to an ideal solution. This has been attributed to a "phobicity" effect between $-CH_2$ and $-CF_2$ groups. It has been suggested in the literature that at a specific composition of FC and HC mixtures, two distinct types of micelles form, one that is rich in FC surfactants and one rich in HC surfactants. (3) Mixtures of ionic and nonionic surfactants: these mixtures exhibit a negative deviation from ideality, that is, the CMC of the mixture is smaller than that of the ideal solution. This is due to the reduction in electrostatic repulsion between the head groups of ionic surfactants due to the incorporation of the nonionic head groups.

Figure 12 shows the typical migration of an uncharged solute in MMEC. The upper frame corresponds to a situation where only one type of mixed micelle exists, while the lower frame illustrates the existence of two types of micelles with their corresponding mobilities and partition coefficients.

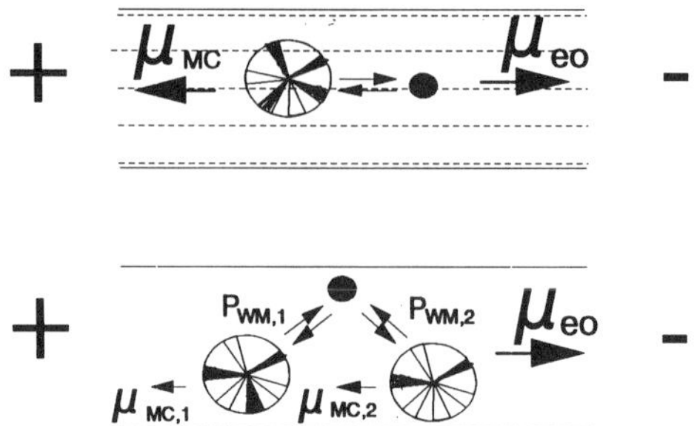

FIGURE 12. Mixed micellar electrokinetic chromatography. A schematic representation of migration of an uncharged compound in electrokinetic chromatography with mixed micelles. *Top*: only one type of mixed micelle exists in the solution. *Bottom*: two types of micelle in a mixed micellar solution. (From Khaledi, M. G., *Anal. Chem.*, 1993, submitted. With permission.)

D. INFLUENCE OF MODIFIERS

Modifiers are included in the micellar solutions of MECC for adjusting the migration factor, manipulation of the partitioning selectivity, and/or extension of the elution window. These additives influence partition coefficients of solutes into micelles, EOF velocity, and/or micellar migration times. The usefulness of organic solvents, cyclodextrins, and urea has been reported in the literature and these are discussed below.

1. Organic Solvents

The use of an organic cosolvent in the micellar solutions of MECC was first reported by Balchunas and Sepaniak[37] and later by Sepaniak et al.[38] and Gorse et al.[39] to improve the separation of highly hydrophobic compounds that elute near or with micelles (i.e., their migration time is about t_{mc}). They also reported the use of gradient elution (or solvent programming) in MECC (Figure 13).[40,41] Solute interaction with micelles and therefore the migration factor, k', will decrease in MECC upon the addition of an organic solvent to micellar solutions. Furthermore, the presence of an organic solvent would cause a reduction in EOF and subsequently an extension of the elution window. As a result, resolution of highly hydrophobic compounds in MECC can be greatly increased at the expense of longer separation times.

In many cases, the selectivity of partitioning into micelles (and MECC selectivity) may not actually change with the addition of an organic solvent. In other words, resolution enhancement can be due to the adjustment of migration factors and extension of the elution range and not selectivity enhancement, which is often the main effect of organic modifier in RPLC. It is important to differentiate between these two effects in order to better understand the role of organic modifier in MECC and its differences from that in RPLC. A major advantage of RPLC is the feasibility of manipulating selectivity through careful selection of the mobile

---

FIGURE 13. Effect of organic cosolvent and gradient elution in MECC: (A) 50 m*M* SDS in a purely aqueous buffer; (B) 50 m*M* SDS in 22.5% 2-propanol in water solution; (C) a stepwise gradient from 10 to 25% 2-propanol in 0.05 *M* SDS micellar solution. Other conditions: 0.01 *M* phosphate-borate buffer (pH 7.0). Laser fluorescence detection: excitation at 488 nm using an argon ion laser, emission at 525 nm. Solutes: alkyl amines derivatized with fluorescent reagent: 7-chloro-4-nitrobenz-2-oxa-1,3 diazole (NBD chloride). Peak I.D.: (a) NBD methyl amine, (b) ethyl amine, (c) dimethyl amine, (d) propyl amine, (e) diethyl amine, (f) *n*-butyl amine, (g) cyclohexylamine, (h) di-*n*-propyl, (i) *n*-hexyl, (j, k) Coumarin 153 and Coumarin 343. (From Balchunas, A. T. and Sepaniak, M. J., *Anal. Chem.*, 60, 617, 1988. With permission.)

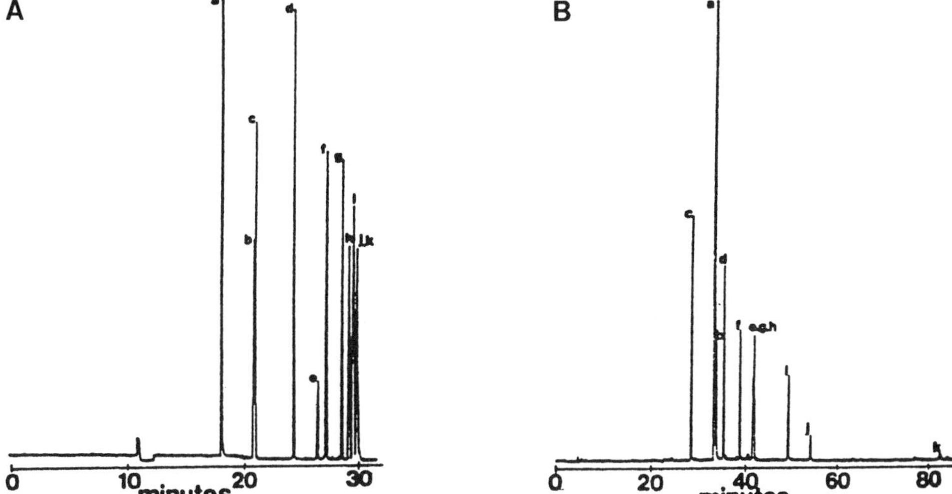
A
a
d
c
f
g
b
i
h
j,k
e
0
10
20
30
minutes
B
a
c
d
f
e,g,h
i
j
k
0
20
40
60
80
minutes

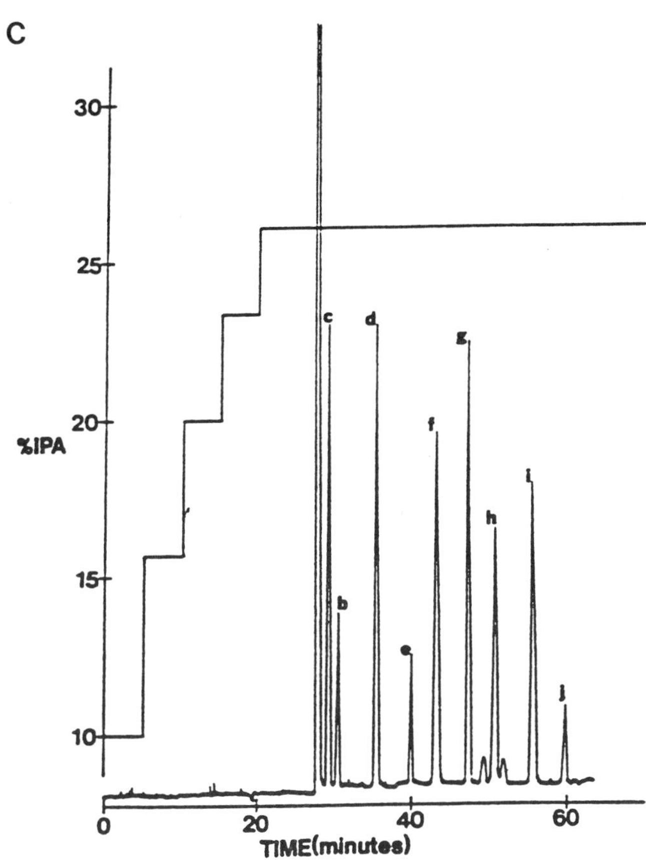
C
30
25
20
15
10
%IPA
c
d
g
f
b
e
h
i
j
0
20
40
60
TIME(minutes)

FIGURE 13.

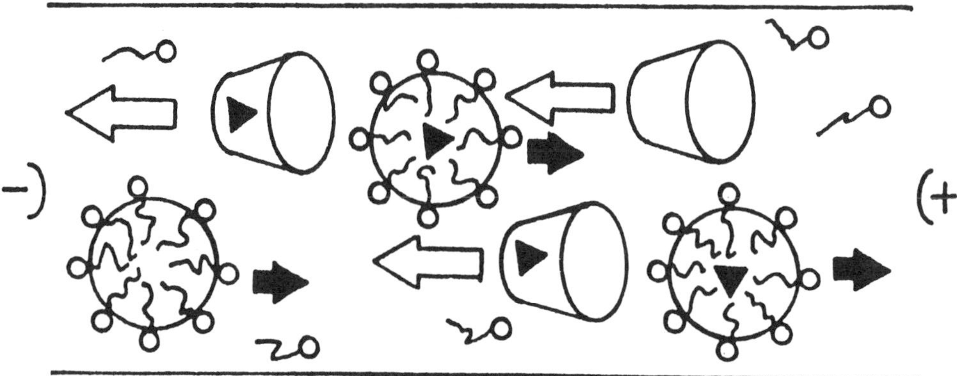

FIGURE 14. Cyclodextrin-modified MECC. Schematic illustration of the separation principle of including cyclodextrins in MECC. Filled arrows indicate the electrophoretic migration of the micelle and open arrows show the electroosmotic migration of CD. (From Terabe, S., et al., *J. Chromatogr.*, 516, 23, 1990. With permission.)

phase composition. The hydro-organic mobile phases composed of water, methanol, acetonitrile, and/or tetrahydrofuran offer a wide range of polarity and selectivity that facilitate RPLC separation of many classes of compounds.

The information on the effect of organic co-solvents on selectivity of partitioning into micelles is limited and yet to be fully explored. By definition, the solvent strength in MECC increases (i.e., partitioning into micellar pseudostationary phase and k' decrease) with the addition of an organic solvent. However, the overall separation time is increased as a result of a reduction in EOF velocity and micellar velocity. This is an opposite effect to what is often observed in RPLC, where an increase in solvent strength is accompanied by shorter retention times. Ideally, one should be able to enhance resolution in MECC without prolonging the overall separation time, as is possible in RPLC separations.

2. Cyclodextrins

Addition of cyclodextrins (α, β, or γ) to micellar solutions was first reported by Terabe and co-workers to improve the separation of very hydrophobic compounds by MECC.[8] Figure 14 shows the proposed migration model for cyclodextrin-modified MECC. An uncharged cyclodextrin migrates at the EOF velocity, which is in the opposite direction of the micellar migration. The hydrophobic interior of the cyclodextrins provides an alternative site to that of the micelles for interaction with solutes. Hydrophobic compounds are then distributed between micelles and cyclodextrins. As a result, their migration factor is reduced to within or near an optimal range that leads to enhancement in resolution.[8,42-44] Figure 15 shows a significant improvement in separation on addition of cyclodextrin to an SDS micellar solution. An important feature of this technique is that the inclusion complex between the cyclodextrins and compounds depends on the size and shape of molecules, as well as the cavity size of the cyclodextrins. As a result, cyclodextrins provide additional selectivity that makes it possible to separate geometrical, structural, and even optical isomers.[45-47]

3. Urea

Another method for extending the applications of MECC to hydrophobic compounds is the addition of urea to micellar media.[48] Urea increases the aqueous solubility of lipophilic compounds and concurrently reduces their interactions with micelles.[42,48] Taking advantage of these effects, Terabe et al. have shown that migration factors of hydrophobic compounds in MECC are dramatically decreased in the presence of urea.[48] In fact, linear relationships were

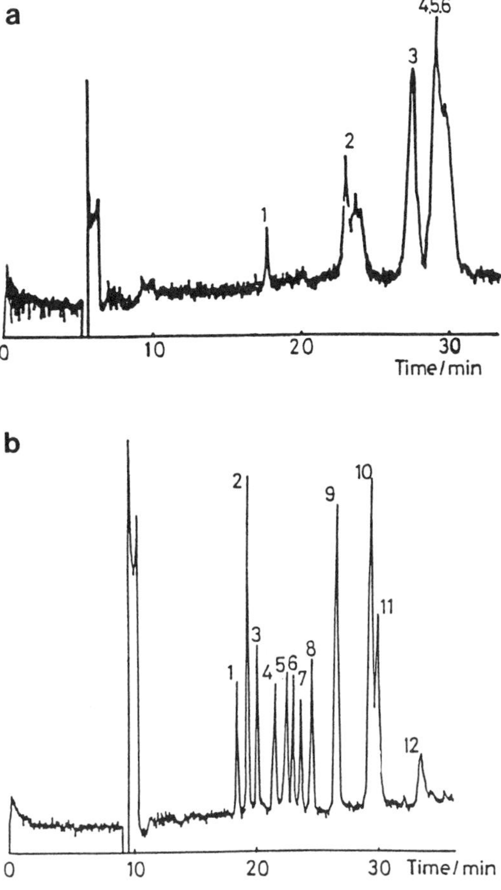

FIGURE 15. Separation of hydrophobic compounds by Cyclodextrin (CD)-MECC. Separation of chlorinated benzene congeners. (a) SDS (100 m*M*) in 100 m*M* borate buffer (pH 8.0) containing 2 *M* urea; applied voltage, 18 kV; current, 30 μA. Compounds: (1) mono-, (2) di-, (3) tri-, (4) tetra-, (5) penta-, (6) hexachlorobenzene. (b) j-CD (40 m*M*) added to the buffer in part (a); applied voltage, 15 kV; current, 23 μA. Capillary: I.D., 50 μm; total length, 700 mm; separation length, 500 mm. Compounds: (1) 1,2,3,5-tetra-, (2) 1,2,3-tri-, (3) 1,3,5-tri-, (4) 1,2-di-, (5) 1,2,4,5-tetra-, (6) mono-, (7) 1,3-di-, (8) 1,2,4-tri-, (9) 1,2,3,4-tetra-, (10) penta-, (11) 1,4-di-, and (12) hexachlorobenzene. (From Terabe, S., et al., *J. Chromatogr.*, 516, 23, 1990. With permission.)

observed between the logarithm of the migration factor and concentration of urea. As a result, one can adjust the MECC migration factor within the optimum range. Furthermore, the elution window was extended on the addition of urea. A combination of optimum k' range and wider elution window resulted in higher resolution of hydrophobic compounds (Figure 16).

E. SECONDARY CHEMICAL EQUILIBRIA

In HPLC, inclusion of secondary chemical equilibria (SCE) in the mobile phase (the primary equilibrium being partitioning into the stationary phase), has a great influence on selectivity. A variety of SCE have been reported, especially in RPLC, that have provided enhanced selectivity and have greatly extended the scope of HPLC applications.[49,50]

Similar approaches have been examined for MECC separations by incorporating a number of SCE in the micellar solutions in order to influence the primary equilibrium of solute partitioning into micelles. For example, prototropic (i.e., acid–base), ion pairing, and a variety of complexation equilibria between analytes with metal ions, cyclodextrins, borate

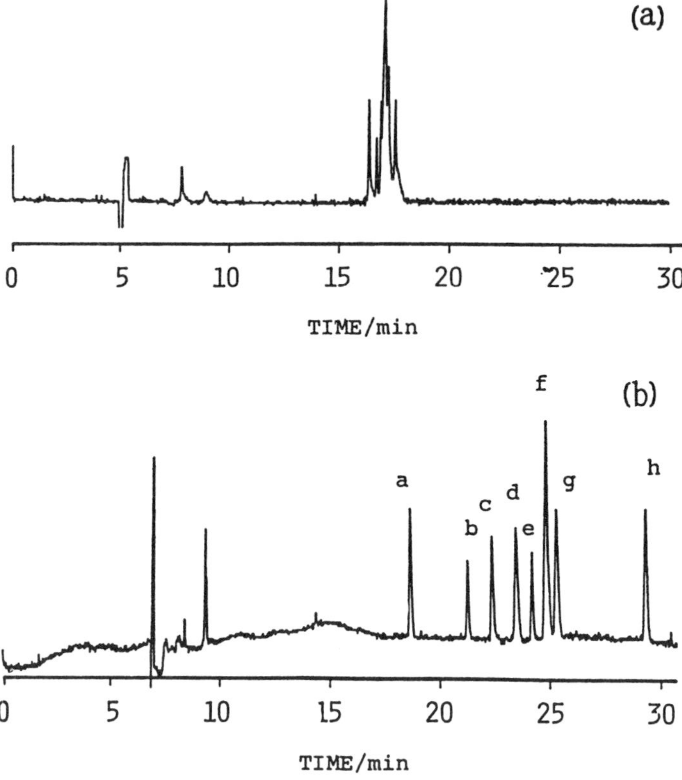

FIGURE 16. Effect of urea. Separation of eight corticosteroids. Conditions: (a) 50 m*M* SDS in 20 m*M* borate–20 m*M* phosphate buffer (pH 9.0); applied voltage, 20 kV. Capillary: I.D., 50 μm; total length, 650 mm; separation length, 500 mm. (b) Same solution as in (a) but containing 6 *M* urea. Compounds: (a) hydrocortisone, (b) hydrocortisone acetate, (c) betamethasone, (d) cortisone acetate, (e) triamcinolone acetonide, (f) fluocinolone acetonide, (g) dexamethasone acetate, (h) fluocinonide. (From Terabe, S., et al., *J. Chromatogr.*, 545, 359, 1991. With permission.)

ion, and chiral selectors have been used to enhance the capabilities and scope of MECC separations.

1. Acid–Base Dissociation Equilibria: pH Effects

As shown in Figure 2, acid–base equilibria have a great influence on migration of ionizable compounds in MECC. Therefore, the quality of MECC separations greatly depends on the pH since it determines the extent of solute ionization.[13-15,51-58] The migration behavior of a conjugate acid–base pair in MECC has been quantitatively described.[13-15] In general, the mobility μ of a conjugate acid–base pair is a weighted sum of mobility of the ionized fraction in bulk aqueous solvent and the fraction associated with micelles times the mobility of micelles:[15]

$$\mu_s = (F_{\text{acid, mc}} + F_{\text{base, mc}})\mu_{\text{mc}} + F_{\text{acid, aq}}\,\mu_{\text{aq, acid}} + F_{\text{base, aq}}\,\mu_{\text{aq, base}}$$

where the F values represent the molar fractions of a solute in conjugating acid–base form in aqueous and in micellar phase. A similar expression can also be written for the migration factor, k'. Substituting for the fractions, the following general expression can be derived for acidic and basic compounds in MECC with anionic surfactants. Using this method, similar equations can also be derived for other types of surfactants (e.g., cationic).

$$\mu_s = \frac{\left[K_{m,\,acid} + \frac{K_a}{[H^+]}K_{m,\,base}\right][M]\mu_{mc} + \mu_{aq,\,acid} + \frac{K_a}{[H^+]}\mu_{aq,\,base}}{1 + K_{m,\,acid} + K_{ip}CMC + \frac{K_a}{[H^+]}\left(1 + K_{m,\,base}[M]\right)} \tag{19}$$

where K_m values are binding constants of solutes to micelles in conjugating acidic and basic forms. Note that K is related to the partition coefficient P as: $K = Pv$, where v is the molar volume of surfactant.[58] K_{ip} is the ion-pair formation constant of ionic species with surfactant monomers, K_a is the acid dissociation constant, $[H^+]$ is the hydrogen ion concentration, [M] is micelle concentration, and CMC is the critical micelle concentration of the surfactant.

For an anionic solute (HA, A^-) in an anionic surfactant MECC system, both the ion-pair formation constant (K_{ip}) and $\mu_{aq,\,acid}$ equal zero, thus Equation 19 becomes

$$\mu_s = \frac{\left(K_{HA} + \frac{K_a}{[H^+]}K_{A^-}\right)[M]\mu_{mc} + \left(\frac{K_a}{[H^+]}\right)\mu_{aq,\,A^-}}{1 + K_{HA}[M] + \frac{K_a}{[H^+]}\left(1 + K_{HA}[M]\right)} \tag{20}$$

Similarly the electrophoretic mobility of a cationic solute (BH^+, B) in an anionic surfactant MECC system can be expressed as

$$\mu_s = \frac{\left(K_{m,\,BH^+} + \frac{K_a}{[H^+]}K_{m,\,B}\right)[M]\mu_{mc} + \mu_{aq,\,BH^+}}{1 + K_{m,\,BH^+}[M] + K_{ip}CMC + \frac{K_a}{[H^+]}\left(1 + K_{m,\,B}[M]\right)} \tag{21}$$

Likewise, a general equation can be derived for migration factor in MECC as

$$k' = \frac{\left[K_{m,\,acid} + K_{m,\,base}\frac{K_a}{[H^+]}\right]}{\left(1 + \frac{K_a}{[H^+]} + K_{ip}CMC\right)}[M] \tag{22}$$

Again, for an anionic solute (HA, A^-) in an anionic surfactant MECC system, the K_{ip} term is zero and the above equation is simplified to

$$k' = \frac{\left[K_{m,\,HA} + K_{m,\,A^-}\frac{K_a}{[H^+]}\right]}{\left(1 + \frac{K_a}{[H^+]}\right)}[M] \tag{23}$$

For a cationic solute (BH^+, B) the same equation applies:

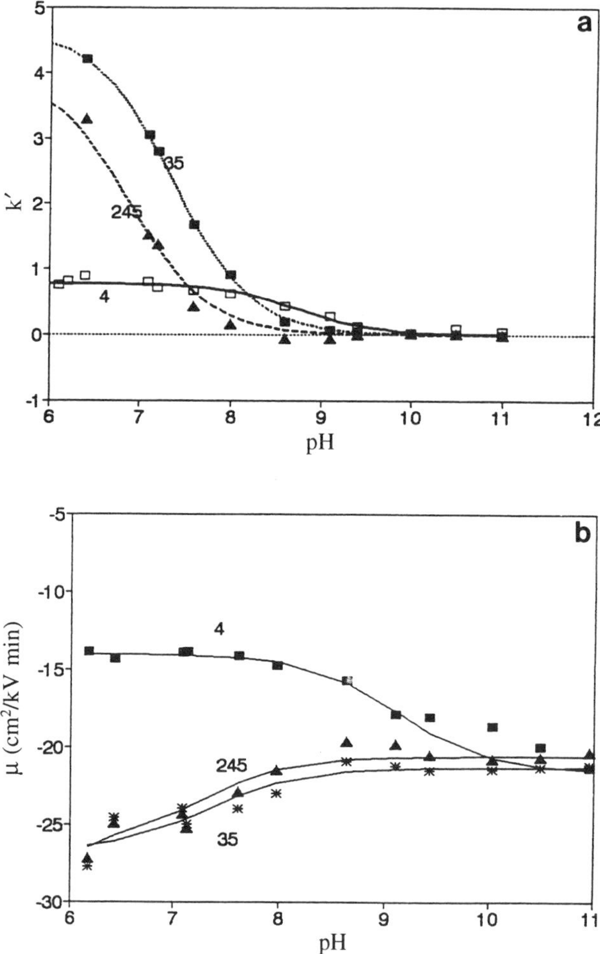

FIGURE 17. pH effect in MECC of acidic compounds. (a) The dependence of k' on pH in the presence of 20 m*M* SDS for three chlorophenols (4CP, 35CP, and 245CP). (b) The dependence of mobility on pH in the presence of 20 m*M* SDS. Solid line in (a) represents the behavior predicted by Equations 20 and 23. (From Khaledi, M. G., Smith, S. C., and Strasters, J. K., *Anal. Chem.*, 63, 1820, 1991. With permission.)

$$k' = \frac{K_{m,BH^+} + K_{m,B}\dfrac{K_a}{[H^+]}}{1 + \dfrac{K_a}{[H^+]} + K_{ip}CMC}[M] \tag{24}$$

In general, the migration of an ionic (or ionizable) solute described in terms of either mobility (μ) or the migration factor (k') is a function of

1. The controllable factors such as pH and micelle concentration.
2. The solute parameters, such as binding constants to micelles, dissociation constants, or ion mobility in bulk aqueous solvent.
3. The system constants (CMC and micellar mobility).

Using these equations, the variations in mobility and the migration factor as a function of pH and micelle concentration can be predicted on the basis of a few initial experiments.

For ionizable compounds, the pH of the buffer determines the net charge of molecules and greatly influences their migration behavior in MECC. As mentioned above, the charge of the surfactant head group is also important for MECC of acids and bases.

Migration behavior of ionizable solutes in MECC can be expressed in terms of the migration factor or mobility. The effect of pH on k' is, however, different from that on the mobility for acidic solutes. Using an anionic surfactant, for example, an increase in pH would enhance the dissociation of an acid, which reduces interaction of the solute with the anionic micelles. As a result, the migration factor, k', decreases with the pH of solution in a sigmoidal manner, as predicted by Equations 20 and 23 (i.e., $k'_{HA} > k'_{A^-}$) (Figure 17a). This is expected since the protonated acid, which is the dominant form at the lower pH values, interacts to a greater extent with micelles. This is as compared with the dissociated form A^-, which is electrostatically repelled from the micelles and, therefore, has a smaller partition coefficient into the micelles than the protonated form. The extent of the decrease in k' with pH increase is more pronounced for hydrophobic acids that have a stronger association with micelles.[13]

Unlike k', the trend of variations of mobility (μ) with the pH in micellar media depends on the type of acidic solutes (Figure 17b). For example, for monochlorophenols, the mobility increases with an increase in the pH; i.e., $\mu_{HA} < \mu_{A-}$. For the tri- and pentachlorophenols the trend is reversed; that is, the mobility of the more hydrophobic compounds decreases with an increase in pH, $\mu_{HA} < \mu_{A-}$. This is shown in Figure 17b for chlorophenols (4-, 3,5-, and 2,4,5-chlorophenol). For all three compounds, the trend for k' vs. pH is the same while that for μ vs. pH is different.

The different behavior in mobility can be explained by examining the relationship shown in Equation 25:

$$\mu = F_{mc,\,HA}\mu_{mc} + F_{aq,\,A^-}\mu_0 + F_{mc,\,A^-}\mu_{mc} \tag{25}$$

At the low pH values, the first term dominates and $\mu = \mu_{HA} = F_{mc,HA}\mu_{mc}$. The contribution of the second and the third terms to the overall mobility increase with the pH and at high pH values is

$$\mu = \mu_{A^-} = F_{aq,\,A^-}\mu_{A^-} + F_{mc,\,A^-}\mu_{mc}$$

The contribution of the third term is probably not significant due to the electrostatic repulsion of the dissociated solute from the anionic micelles, which results in small $F_{mc,A-}$ values. Therefore, the trend of the μ vs. pH curve is determined by the terms $F_{mc,HA}\mu_{mc}$ and $F_{aq,A-}\ \mu_{A-}$. The mobility of micelles is larger than any of the test solutes; however, the contribution of the protonated acid HA becomes larger than the dissociated form A^- only if it interacts strongly with micelles. This is the case for trichlorophenols. Therefore, for these compounds $F_{mc,HA}\ \mu_{HA} > F_{aq,A-}\ \mu_o + F_{mc,A-}\ \mu_{mc}$. The interaction of the monochlorophenols is considerably less, to the extent that the solute mobility in the dissociated form is more than that in the HA form.[13]

Note that the fractions of solute in micellar media are a function of solute–micelle interactions and micelle concentration. This is especially pronounced for protonated solutes. As a result, at higher micelle concentrations the difference between the mobilities of HA and A^- is reduced for more polar solutes while it is increased for the more hydrophobic ones. This can be seen in the predicted mobility vs. pH at two different micelle concentrations (Figure 18a and b) as well as the three-dimensional plots of mobility vs. pH/[SDS] (Figure 19a and b). In general, for less hydrophobic compounds and at low micelle concentrations, the difference between μ_{HA} and μ_{A-} is much larger than for the hydrophobic compounds. In other words, the changes in mobility with the pH variation is greater for the less hydrophobic monochlorophenols. The situation is different for the k' vs. pH function (Figure 19c and d). The difference in k'_{HA} and k'_{A-} increases with the solute hydrophobicity.

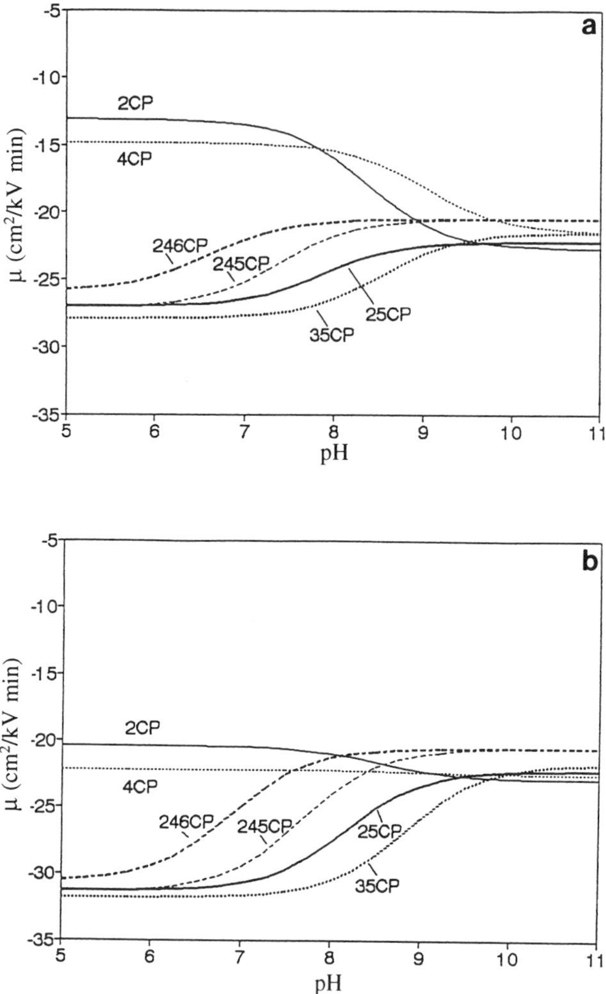

FIGURE 18. Effect of micelle concentration on μ vs. pH: (a) 20 m*M* SDS and (b) 40 m*M* SDS for chlorinated phenols. (From Khaledi, M. G., Smith, S. C., and Strasters, J. K., *Anal. Chem.*, 63, 1820, 1991. With permission.)

These two situations seem to contradict one another as far as the effect of pH on selectivity is concerned. Based on the mobility data, for more hydrophobic solutes, pH is not an effective parameter for selectivity manipulation at low micelle concentrations since there is not much difference in the mobilities of the protonated and the dissociated forms of an acid. At higher micelle concentrations, however, the changes in mobility of more polar compounds with pH is greatly reduced, while one observes a larger variation in mobility vs. pH for more hydrophobic compounds (Figure 18). On the other hand, one would observe dramatic changes in k' as the degree of the acid dissociation is changed through pH variations. Note, however, that k' represents the degree of solute interactions with micelles. The degree of this interaction does not necessarily indicate the true migration behavior of ionizable solutes in MECC; therefore, mobility is probably a more representative parameter for solute migration.

For basic compounds, lowering the pH would enhance the interaction of the protonated, positively charged base with the anionic micelles. Therefore k' increases as the pH is reduced (Figure 20). A similar behavior is observed for the variations of mobility with the pH, i.e., the larger mobilities are observed for bases at lower pH since the compounds migrate at the micellar mobility. This is also shown in Figure 21, which illustrates the influence of pH on

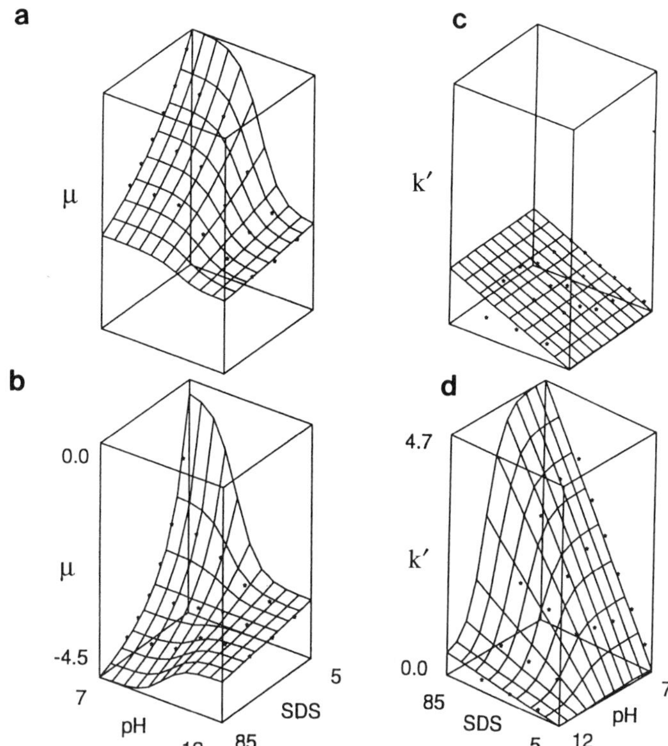

FIGURE 19. Migration surfaces for MECC of acids. Effect of pH and micelle concentration on mobility (a and b) and migration factor (c and d) for a moderately polar compound, *p*-methoxyphenol (a and c), and for a more hydrophobic compound (b and d) that has a stronger interaction with micelles. Note the different behavior of the mobility surfaces for the two compounds. (From Quang, C., Strasters, J. K., and Khaledi, M. G., *Anal. Chem*, 1993, submitted. With permission.)

the MECC separation of a mixture of organic bases at a constant micelle concentration. At pH 7.0, the bases are fully protonated and interact very strongly with micelles. As a result poor resolution is observed as most of the compounds elute near the micellar migration time, t_{mc}. The quality of the separation is greatly enhanced by increasing the pH, which reduces the solute–micelle interactions. Figure 21b shows the effect of increasing micelle concentration (at a constant pH) on the separation of the same mixture. Upon increasing the micelle concentration, the migration time of all solutes approach that of the micelles (t_{mc}) as k' of solutes are increased due to an increase in the phase ratio. Figure 22 shows a similar behavior for acidic solutes.

One important factor that should be considered in a study of the pH effect on the migration behavior and separation of ionizable compounds by MECC is the micellar-induced pK_a shift.[58-60] Consideration of micellar-induced shift of ionization constants is important in the comparison of the capabilities of MECC and CZE for the separation of ionizable compounds. The extent of ionization of acids in the MECC system depends on the apparent ionization constant in micellar solutions that is different from the aqueous ionization constants. More importantly, the magnitude of the pK_a shift is a function of the structural properties of the solutes.

This would create two interesting situations. The first is when two compounds with similar pK_a values in aqueous solutions have different pK_a values in micellar media, e.g., amino acids and small peptides.[58-60] As a result of this phenomenon, an enhanced separation selectivity is observed. Another possibility is that compounds with widely different pK_a values in aqueous media and different lipophilicities may have similar ionization constants in micellar solution.

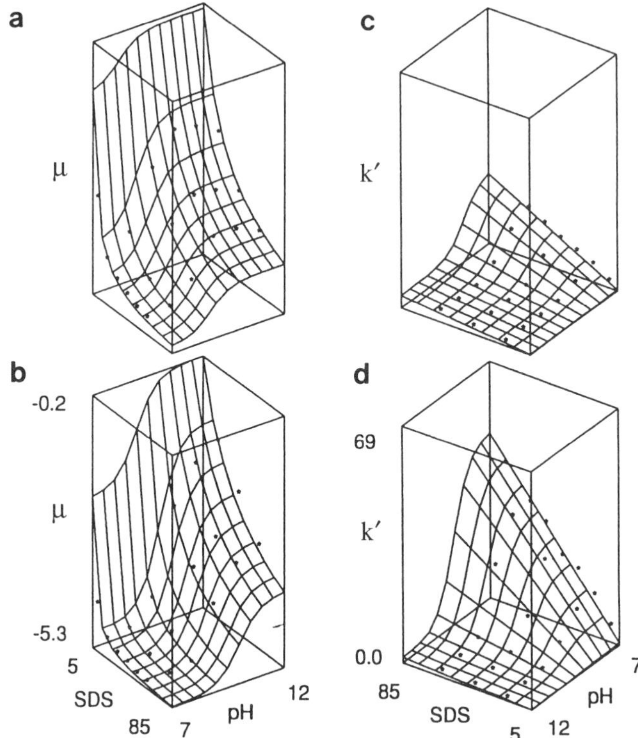

FIGURE 20. Migration surfaces for MECC of bases. Effect of pH and micelle concentration on mobility (a and b) and migration factor (c and d) for nicotine (a and c) and norephedrine (b and d). (From Quang, C., Strasters, J. K., and Khaledi, M. G., *Anal. Chem*, 1993, submitted. With permission.)

a. Simultaneous Optimization of pH and Micelle Concentration

Obviously, in MECC separations that include ionizable compounds, finding the optimum values for the pH and micelle concentration for maximum resolution is of primary importance. These two parameters have been used in a number of MECC applications.[13-15] The electropherograms shown in Figures 21 and 22 provide an example of the complexity of migration behavior of ionizable compounds in MECC, which makes the optimization by trial and error a difficult task. As shown in Figure 22, when micelles are not present, the elution order of the chlorophenols is according to their charge and size (or molecular weight). On the addition of micelles and with changing micelle concentration even at a constant pH, however, the elution order changes greatly. Large variations in migration as a function of pH and micelle concentration are also observed for organic bases, as shown in Figure 21. For the MECC separation of complex mixtures simultaneous optimization of pH and micelle concentration is required.

One important issue that should be addressed is whether these two parameters can be optimized independent of one another (i.e., sequentially) or simultaneously. Based on the observed results (see Figures 18 to 22) and the dependence of $pK_{a,\,app}$ on micelle concentration and solute type, one can conclude that the simultaneous multiparameter optimization would be a more effective strategy. Interpretive methods of optimization that are based on the use of a retention model would be quite appropriate.[61]

Using the models described above, one can predict the migration behavior as a function of the two parameters, based on a limited number of experiments. The initial experiments should be designed such that one would obtain an estimate of the important physicochemical parameters of the solutes in a mixture (e.g., for an acid: P_{HA}, P_{A^-}, and $K_{a,\,app}$). If these values are known, one can predict the mobility and migration factor of a compound at different pH values and micelle concentrations. In principle, or five or six initial measurements at different pH/[SDS] values are adequate to predict the migration behavior in MECC, using the nonlinear fitting of the models. It is important that the regression procedure requires no knowledge of different physicochemical properties of solutes such as ionization constants and partition coefficients. Once the migration surface (Figures 19 and 20) for each solute is known, finding an optimum composition that provides the best separation would not be difficult. This can be achieved through a plot of criterion (e.g., minimum resolution) vs. parameters.

An example is given for the MECC separation of organic bases (Figure 23). These organic amines have a pK_a range of 8 to 11. The use of pH to manipulate the HPLC separation of amines is not feasible due to the chemical instability of *n*-alkylsilicas at high pH. Consequently, these solutes exist almost in their protonated, positively charged forms (alkylammonium ion) in the operating pH range of HPLC with silica-based stationary phases. The CZE separation of amines in the basic pH range is complicated by the silanophilic interactions of amines with the capillary wall. In addition, the EOF is maximized in the basic pH range, which causes a fast elution of an unretained peak (i.e., short t_o). Since cationic solutes elute before t_o, the practical separation window (between time zero and t_o) is very short to achieve a reasonable separation of a complex mixture. Reducing the pH significantly improves the peak shape and reduces the EOF; however, selectivity is sacrificed as all amines have very similar charges. Despite the high efficiency of CZE, it is still difficult to separate a complex mixture of organic amines because of the limited separation window and nearly identical mobilities of solutes at low pH.

In the presence of micelles, one can operate in the basic pH range as the solute-micelles interactions compete effectively with the silanophilic interactions and, consequently, the peak shape is improved and pH can be manipulated to achieve an acceptable separation.

With a limited number of initial experiments, an estimate of the constants in Equations 21 and 22 can be obtained through a nonlinear fit of these equations. Consequently, the migration behavior of individual solutes can be predicted. Furthermore, with the parameters available for each compound, electropherograms at different buffer conditions (pH and SDS concentration) can be simulated. Figure 23a shows that the predicted electropherogram of 17 organic amines, based on the mobility model, agrees well with the observed MECC separation (Figure 23b). The quality of the separation can be improved by including 10% acetonitrile, which reduces the t_o/t_{mc} ratio (Figure 23c).

2. Complexation Equilibria

Inclusion of complexing reagents in the micellar solutions can have a dramatic influence on migration factor and selectivity in MECC separations.[62] During the past few years, the usefulness of a number of complexation equilibria between solutes and metal ions,[63] borate,[64] cyclodextrins,[8,42-47] chiral selectors,[65-68] and ion-pairing reagents[69] have been reported. These secondary equilibria greatly influence the primary solute-micelle equilibrium, which results in better separations. Additionally, they play an important role in broadening the scope of MECC applications.

Nishi et al. incorporated an ion-pair mechanism into the MECC separation of a mixture of antibiotics that contained both cationic and anionic functional groups.[69] Upon the addition of a tetraalkyl ammonium salt (TAAS) to an anionic micellar solution (SDS), the migration times of cationic solutes were reduced as the cationic ion-pair reagent competed with analytes for the available sites on SDS micelles. The migration times of the anionic solutes increased as TAAS interacted with anionic micelles and altered the character of micelles.

Cohen and co-workers enhanced the separation of oligonucleotides and bases by adding Zn ions to the SDS micellar solutions.[63] This was attributed to enhanced selectivity due to Zn–solute complexation on the surface of SDS micelles, as well as extension of the elution

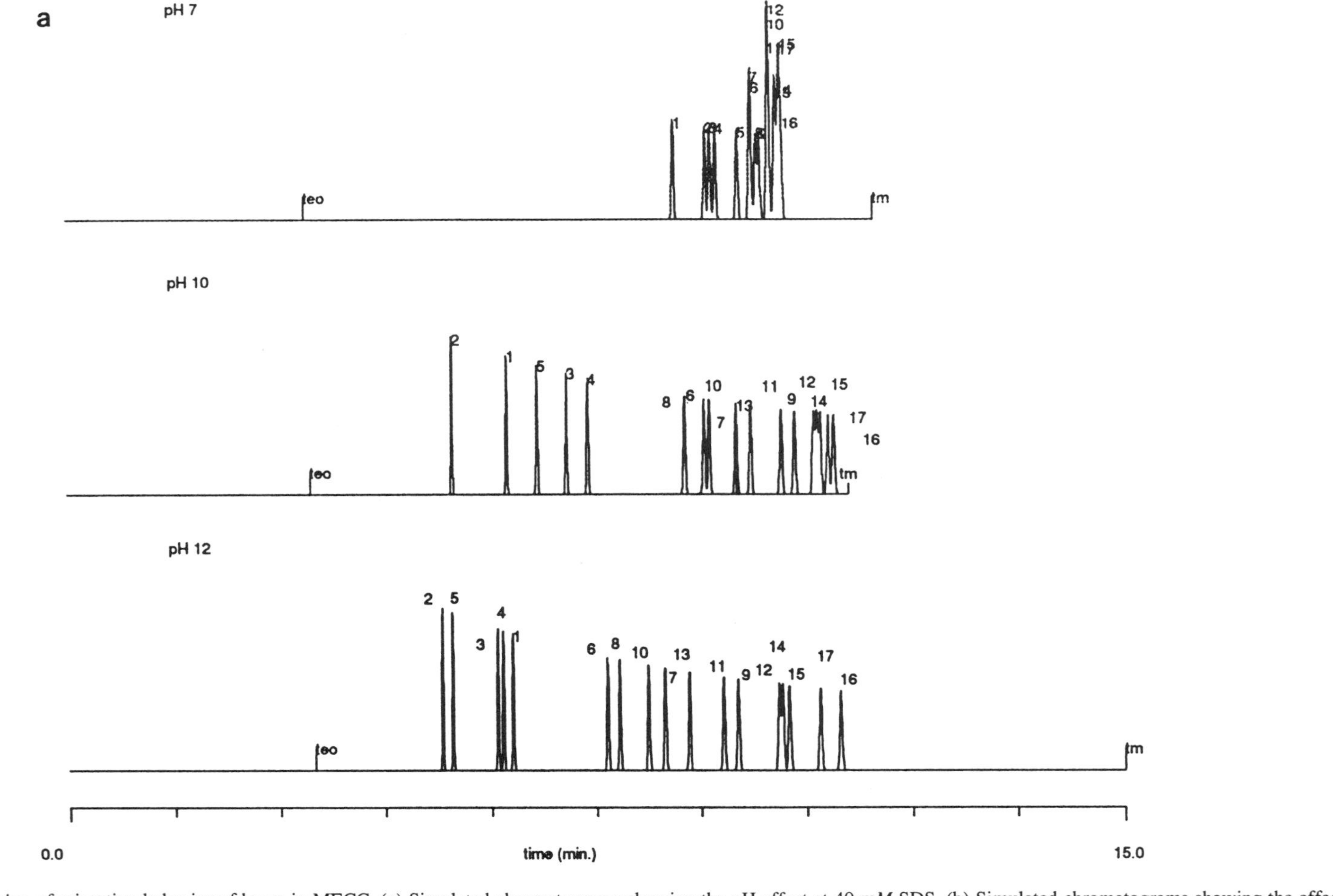

FIGURE 21. Prediction of migration behavior of bases in MECC. (a) Simulated chromatograms showing the pH effect at 40 m*M* SDS. (b) Simulated chromatograms showing the effect of increasing micelle concentration at pH 12 predicted by Equation 21. (From Quang, C., Strasters, J. K., and Khaledi, M. G., *Anal. Chem*, 1993, submitted. With permission.)

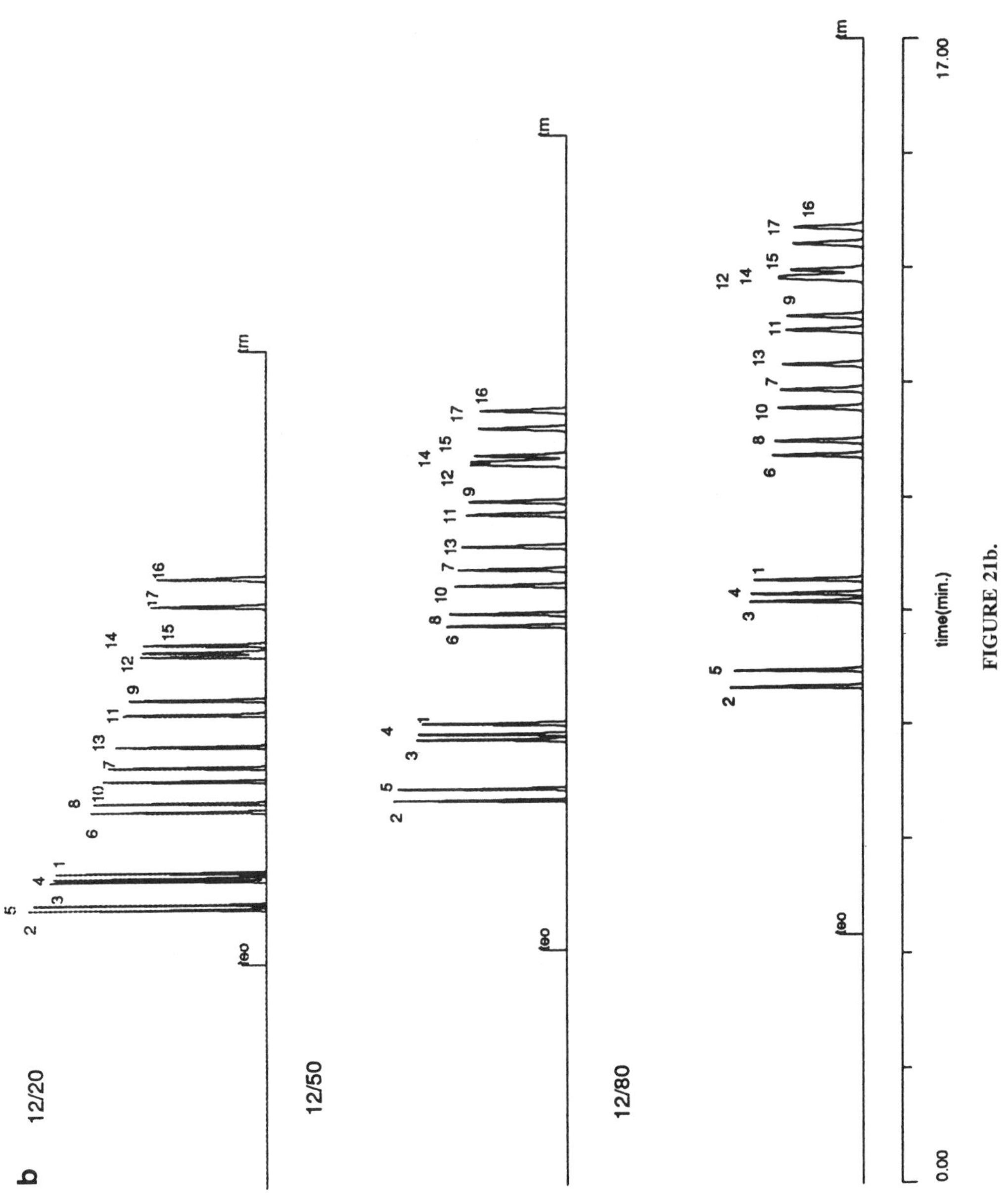

FIGURE 21b.

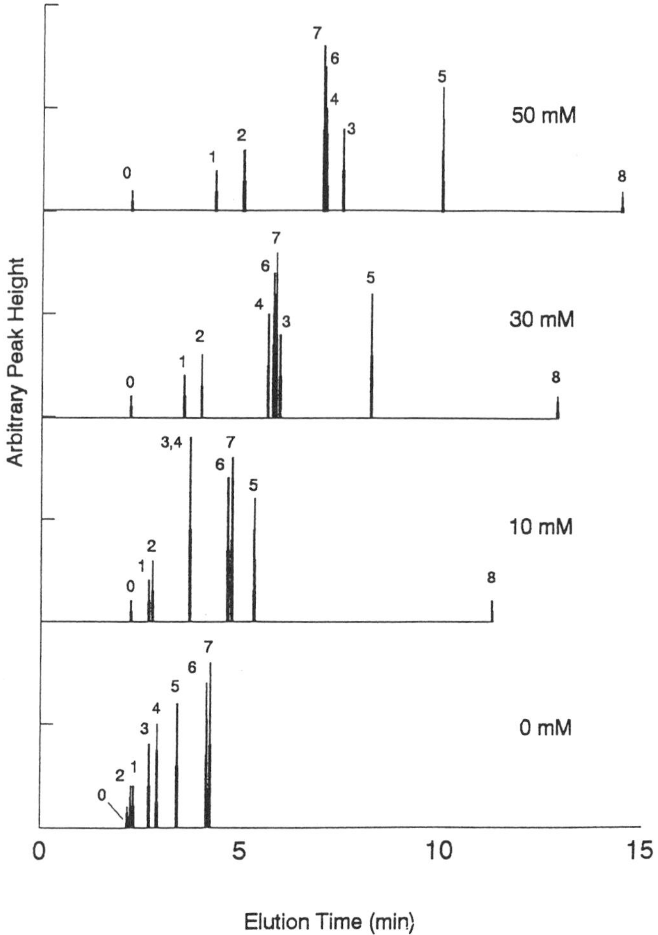

FIGURE 22. Changes in migration pattern of acidic compounds with micelle concentration. Peaks: (0) t_0, (1) 2-chlorophenol (2CP), (2) 3CP, (3) 23CP, (4) 25CP, (5) 245CP, (6) 246CP, (7) PCP, and (8) t_{mc} at 0, 10, 30, and 50 m*M* SDS in 50 m*M* phosphate buffer. Electropherograms were reconstructed from migration time data of individual experiments. Peak heights are arbitrary and vary to assist in peak identification. (From Khaledi, M. G., Smith, S. C., and Strasters, J. K., *Anal. Chem.*, 63, 1820, 1991. With permission.)

window. Wallingford and Ewing reported improved selectivity in the separation of catecholamines by using borate complexation.[64] The use of cyclodextrins is another example of enhancing selectivity of MECC separations for lipophilic solutes as well as optical isomers through inclusion complexes.[8,42-44]

Use of complexation equilibria is also an important part of chiral separations by CE (Figures 24 and 25). In order to extend the capability of CE to resolve optical isomers, a chiral

FIGURE 23. MECC separation of organic bases. (a) Predicted electropherogram of 17 organic amines, based on the mobility model at pH 11.5, 60 m*M* SDS. (b) Observed separation of 18 organic amines at pH 11.5, 60 m*M* SDS (0.05 *M* phosphate-carbonate mixed buffer). (c) Practical separation of 18 cationic amines in the same buffer as in (b), modified with 10% acetonitrite. Solute identification: (1) nicotine; (2) 4-nitrobenzylamine; (3) benzylamine; (4) 3-methyldopamine; (5) norephedrine; (6) ephedrine; (7) phenethylamine; (8) 4-nitrophenethylamine; (9) *N*-methylphenethylamine; (10) 4-chlorobenzylamine; (11) 2-methylpenethylamine; (12) phenylpropylamine; (13) 4-bromobenzylamine; (14) 4-chlorophenethylamine; (15) 2-(4-tolyl)-phenethylamine; (16) phenylbutylamine; (17) methylphenylbutylamine; (18) 4-fluorobenzylamine. All time in minutes. (From Quang, C., Strasters, J. K., and Khaledi, M. G., *Anal. Chem*, 1993, submitted. With permission.)

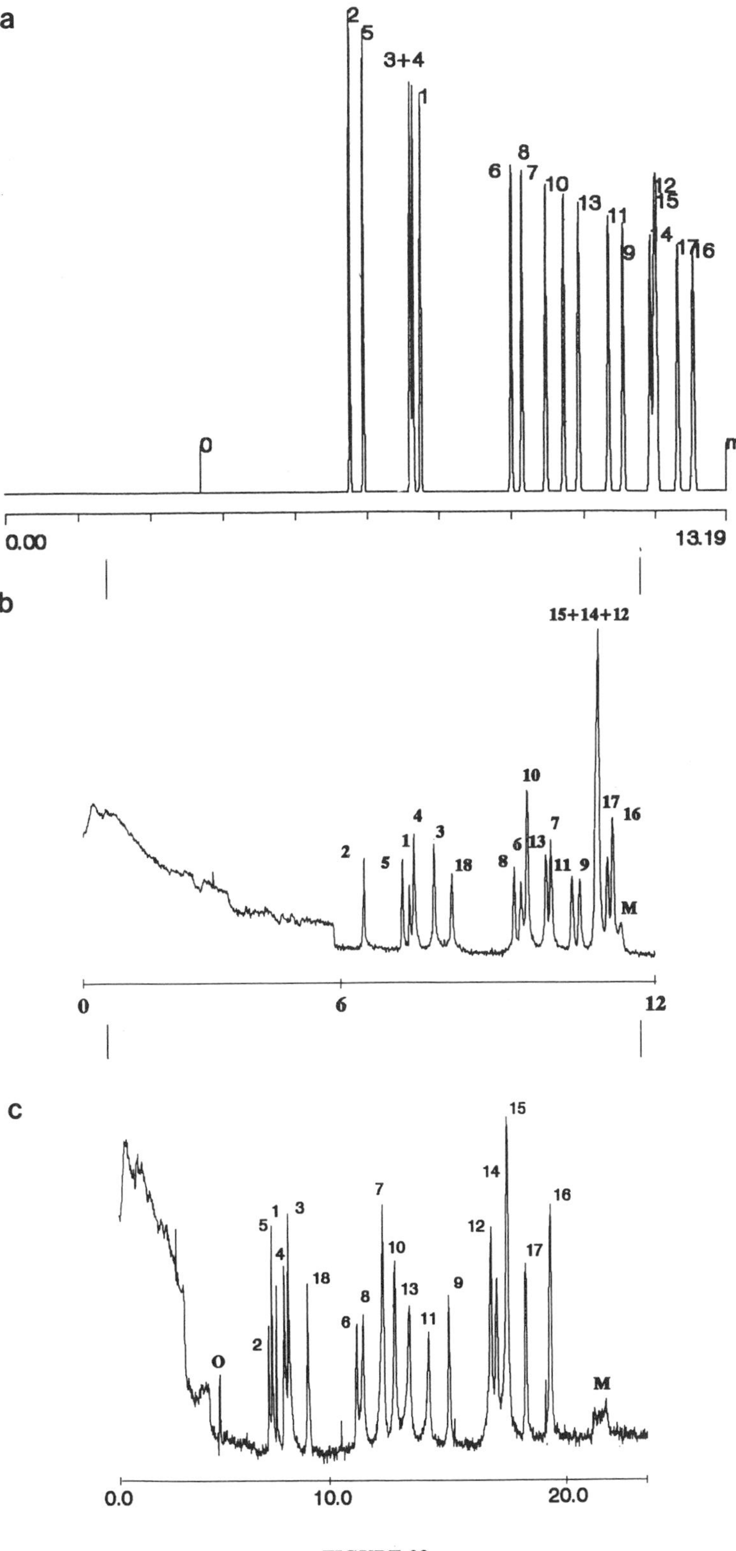

a
2
5
3+4
1
8
6
7
10
13
11
12
15
14
9
17
16
o
m
0.00
13.19
b
15+14+12
10
17
16
4
1
3
7
2
5
18
8
6
13
11
9
M
0
6
12
c
15
14
16
7
5
1
3
12
10
4
17
18
13
9
8
6
11
2
O
M
0.0
10.0
20.0

FIGURE 23.

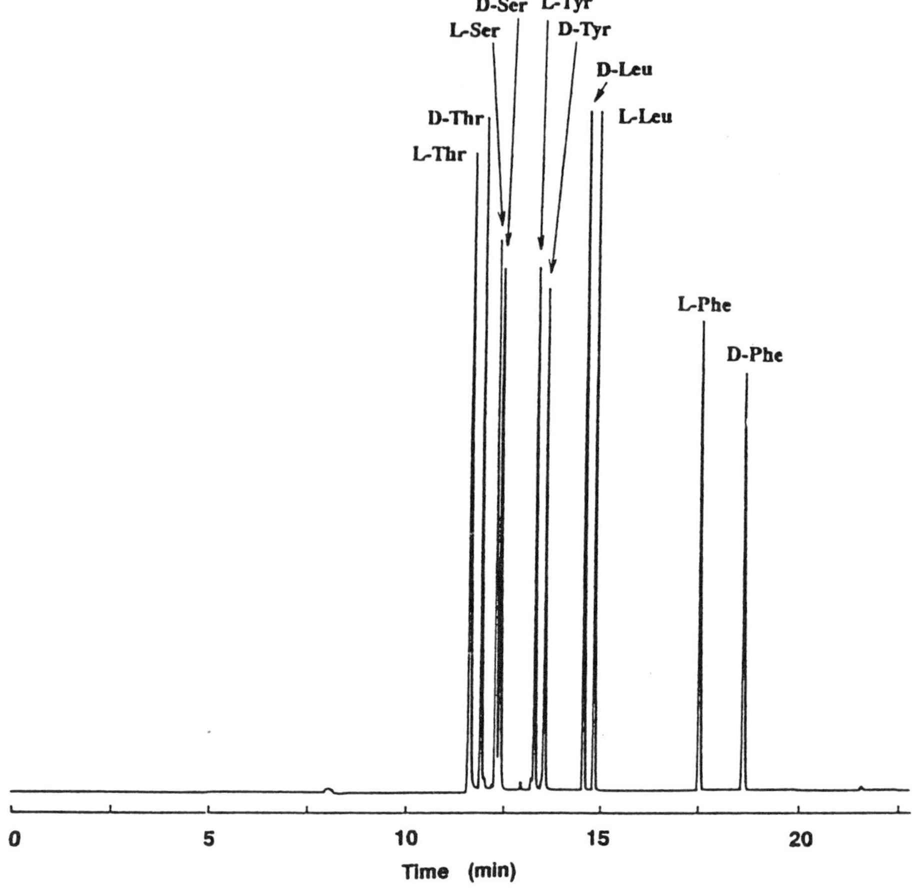

FIGURE 24. Chiral separation by CD-MECC. Electrolyte composition: 10 m*M* j-CD, 50 m*M* SDS, and 100 m*M* borate buffer, pH 9.0. A mixture of 1-cyano-2-substituted benz[*f*]isoindole-DL-amino acids. (From Ueda, T., et al., *Anal. Chem.*, 63, 2979, 1991. With permission.)

selector is incorporated in the buffer solution that forms enantioselective complexation with the isomers. To enhance enantioselectivity in MECC separations, the following approaches have been used:[45-47,65-68] (1) using chiral micelles such as bile salts and surfactants functionalized with L-amino acids residues, (2) addition of a chiral surfactant to nonchiral micelles (e.g., SDS) to form mixed micelles, (3) incorporation of cyclodextrins in SDS micellar solutions that enhances enantioselectivity through formation of inclusion complexes, and (4) derivatization of chiral molecules to form distereoisomers that are then separated by MECC.

V. CONSIDERATIONS IN METHODOLOGY

Despite certain similarities between experimental protocols in MECC and CZE, inclusion of micelles in the buffer solutions of CZE makes it necessary to consider a number of additional factors in MECC.

A. INSTRUMENTATION

As indicated in Chapter 2, the instrumentation and experimental procedure for MECC is identical to that in CZE. Absorbance detectors are the most widely used;[3-6] however, fluorescence,[70-73] electrochemical,[74,75] and indirect detection[73] have also been used in MECC. Sample can be introduced into the capillary via electromigration or hydrostatic (using vacuum or

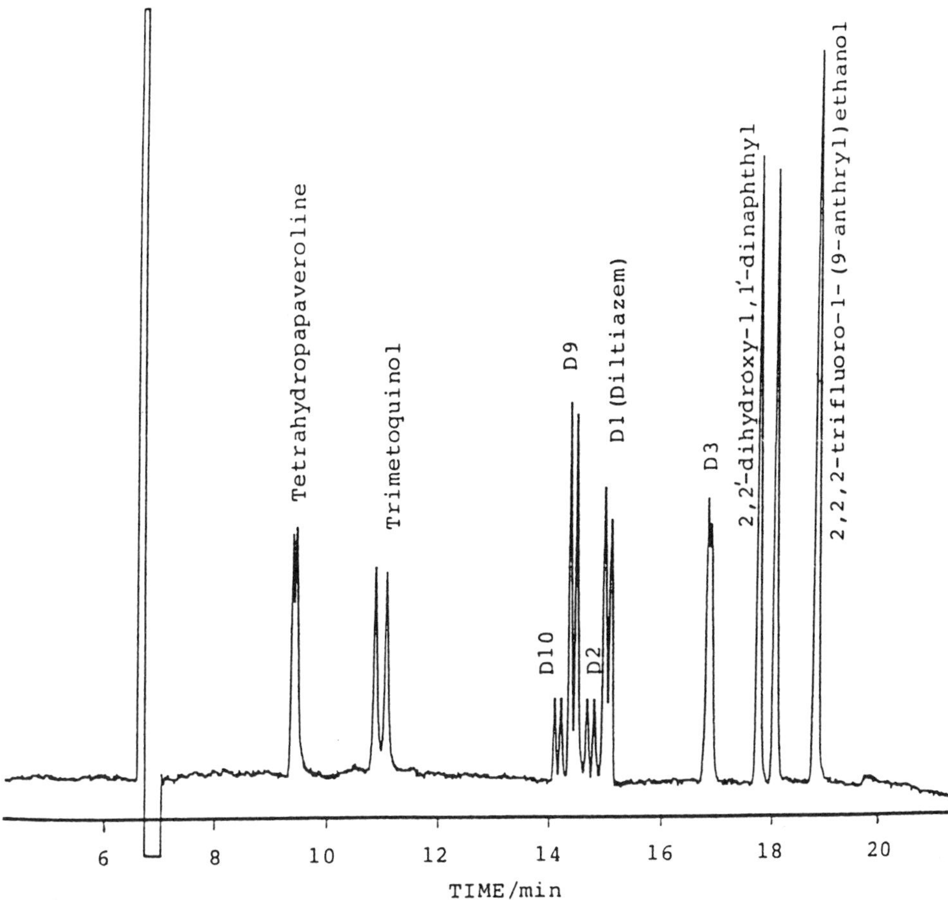

FIGURE 25. MECC chiral separation using bile salts. Electrolyte composition: 50 m*M* sodium taurodeoxycholate in 20 m*M* phosphate-borate buffer. Capillary: I.D., 50 μm; total length, 650 mm; separation length, 500 mm; applied voltage, 20 kV; UV detection at 210 nm. Compounds: Trimetoquinol hydrochloride, tetrahydropapaveroline, and five diltiazem-related compounds. (From Nishi, H., et al., *J. Chromatogr.*, 515, 233, 1990. With permission.)

gravity) injection.[17,76,77] Like CZE, the electromigration process discriminates between solutes based on their mobility, i.e., the injection amount is a function of solute mobility. This phenomenon also exists even for uncharged solutes in MECC since their mobility depends on the extent of their interaction with micelles (Equation 5). The method of internal standard is suitable for quantitation in MECC.[52,78]

B. SYSTEM PARAMETERS

In general, the experimental parameters for MECC such as applied voltage, capillary dimensions, and buffer concentrations should be selected such that the resultant electrical current is kept low, about 60 μA or less. Better results in terms of efficiency, reproducibility, and stability are achieved under these conditions as discussed in Chapters 1 and 13. High currents lead to excessive Joule heating in the CE system. Like CZE, Joule heating causes deterioration of efficiency in MECC separations. In addition, solute partitioning into micelles and solute migration in MECC are strongly dependent on temperature.[79] It is crucial to control the capillary temperature from the viewpoints of separation and reproducibility of migration behavior. This can easily and effectively be done by circulating a liquid (e.g., mineral oil) around the capillary at a constant temperature.

Increasing electric field strength (voltage/length) leads to shorter migration times and faster separations. Higher efficiencies may also be obtained by increasing the field strength, although excess Joule heating at higher voltages would be a limiting factor.

Ionic strength of the buffer also has a direct influence on the amount of the generated electrical currents and consequently the extent of Joule heating in a CE system. In addition to buffer salts, ionic surfactants contribute to the overall solution conductivity and to the electrical current in the system. However, their conductivity in the micellar form differs from that of monomer surfactants or a strong electrolyte. In micellar solutions, the rate of increase in conductivity with surfactant concentration is less than that in monomer surfactant solutions at concentrations below the CMC or a strong electrolyte. This is fortuitous because one would have more flexibility in adjusting micelle concentration without adversely generating current and heat in the system. The buffer concentration should be kept low in order to allow maximum flexibility in controlling micelle concentration. Buffer concentration does not usually have any significant influence on selectivity in MECC. Obviously, for the separation of ionizable compounds where the pH control is crucial, buffer capacity must be considered in selecting buffer concentration.

C. MICELLAR PARAMETERS

The lower end of the surfactant concentration range in MECC should be at least slightly higher than the CMC. The upper limit is basically dictated by factors such as surfactant solubility and electrical current. One should bear in mind that there is an optimum surfactant concentration range for best MECC separations. In addition, the linear relationship between k' and C_{sf} (Equation 17b), which facilitates predictability of migration behavior, is valid only at low concentrations. The sample solubility in a given micellar solution is also influenced by the micelle concentration. For SDS, which is the most widely used surfactant in MECC, a concentration range of 15 to 150 m*M* seems to be reasonable. The CMC of SDS in a purely aqueous solution at room temperature is about 8 m*M*.[9] Note, however, that CMC is not only a function of structural properties such as chain length and nature of the head group, but also depends on factors such as ionic strength, temperature, and presence of other reagents such as organic co-solvents.

As mentioned in Section IV.B, chemical characteristics of surfactants can have a dramatic influence on MECC separations. One should regard certain physical properties in selecting surfactants such as low CMC, surfactant solubility, purity, and detector compatibility. For example, sodium decyl sulfate (STS) exhibits greater mobility than SDS, which leads to a longer t_{mc} and a wider elution window, t_{mc}/t_o. The significance of this favorable electrophoretic behavior, however, is reduced by a higher CMC value of SDeS that is also due to the shorter chain length. Large CMC values are undesirable because high surfactant concentrations will be required to achieve the desired micelle concentration for the separation process. Under these conditions, the efficiency will be adversely influenced by the generation of large electrical currents and probably by mass transfer effects that occur at large surfactant concentrations.

It was mentioned in Section IV.D.1 that organic additives can be included in micellar solutions to influence MECC separations. The concentration of organic solvents should be kept low in order to maintain the integrity of micelles. Adding an organic solvent actually facilitates micelle formation at low concentrations (i.e., CMC decreases) since the electrostatic repulsions between the head groups in an ionic micelle are reduced. At higher concentrations, however, the hydrophobic interactions between the hydrocarbon chains, which are the main driving force in micelle formation, are decreased and formation of micellar aggregates becomes more difficult. The CMC variations change for different organic solvents and there are no determined concentration levels of organic solvents that are suitable for MECC. Based on the results reported for MECC and for micellar liquid chromatography, it is probably safe to stay below a 20% (v/v) concentration of organic solvent in MECC. Some workers have used up to 50% acetonitrile in MECC; however, no evidence for the existence of micelles was provided.[55] For bile salt micelles, higher levels of organic solvents can be used.[29]

D. ELECTROOSMOTIC FLOW AND MIGRATION DIRECTIONS

In a fused silica capillary anionic micelles migrate in the opposite direction of the EOF, toward the positive electrode. With normal buffer compositions and at pH > 5, the EOF velocity is stronger than the micellar velocity, thus the net direction of migration of micelles will be toward the negative electrode (downstream buffer). Cationic surfactants, on the other hand, migrate in a direction opposite to that of anionic surfactants. In addition, cationic head groups of the surfactants can interact with the silanol groups on the capillary wall, which results in a reduction of EOF. Adsorption of a large amount of cationic surfactants can actually lead to a change in the charge of the capillary surface and consequently to the reversal of the direction of EOF. Using cationic surfactants, it is often necessary to reverse the polarities of the two electrodes or alter the position of detection (and injection) with respect to electrode polarities. For example, with anionic surfactants, the high-voltage electrode (injection) should be positive and detection should be made at the negative electrode. Using the cationic surfactants, the high voltage should be negative and detection should be made at the positive electrode.[12,26]

Another important factor in MECC is the pH of the buffer solution, which should be controlled for the separation of ionizable compounds. In addition, pH can alter the overall MECC migration pattern through its influence on the EOF mobility, which drastically decreases with pH below 5.5.[12] Therefore, at lower pH values, EOF velocity is not strong enough to carry the anionic micelles, such as SDS, toward the negative electrode. Otsuka and Terabe[12] observed that below pH 5.0, the direction of migration for SDS micelles changed. The migration velocity of micelles is a sum of electroosmotic velocity, v_{eo}, and the electrophoretic velocity of micelles, v_{ep},

$$v_{mc} = v_{eo} + v_{ep} \tag{26}$$

The sign of each velocity depends on the migration directions. This influence of pH on these velocities is shown in Figure 26. The velocities were defined as positive when the migration was toward the negative electrode and as negative when the migration was toward the positive electrode.[12] Therefore, EOF velocity was positive at all pH values but decreases drastically below pH 5. Electrophoretic velocity of SDS micelles is negative over the entire pH range and is relatively constant, which shows that the pH variation has no effect on the micellar charge. The migration velocity of micelles, however, changes in sign and value according to Equation 26. It changes from positive values at pH > 5 (i.e., net micellar migration is toward the negative electrode because $v_{eo} > v_{ep}$) to zero at pH 5.0 (i.e., the electroosmotic velocity is canceled by the electrophoretic velocity of micelles) to negative values at pH < 5.0 (i.e., micelles migrate toward the positive electrode). This creates an interesting situation with regard to migration directions of solutes with different polarities (Figure 27a and b). At pH below 5, hydrophobic compounds (such as Sudan III) that interact strongly with micelles migrate toward the positive electrode while polar solutes (e.g., methanol) that have little interaction with micelles migrate with the EOF toward the negative electrode. Again, as in the case of using cationic surfactants, the polarity of the applied voltage has to be varied according to the migration direction of a solute due to the pH changes. For cases where solutes migrated toward the negative electrode, the positive voltage was applied to the injection end and vice versa.[12] Otsuka and Terabe[12] observed a poor reproducibility in migration times of uncharged solutes by operating in the low-pH range

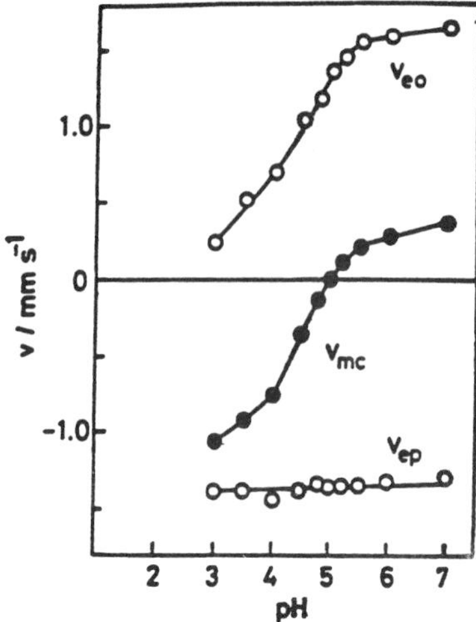

FIGURE 26. Dependence of electrokinetic velocities on pH. v_{eo}, Electroosmotic velocity; v_{mc}, migration velocity of micelles; v_{ep}, electrophoretic velocity of the micelle. Micellar solution: 100 m*M* SDS. (From Otsuka, K. and Terabe, S., *J. Microcolumn Sep.*, 1, 150, 1989. With permission.)

values; the reason for this was not clear. This problem was eliminated by rinsing with 0.10 *M* NaOH solution between each injection. This problem was also observed in this laboratory for ionizable compounds, which was improved but could not be avoided by a rigorous rinsing procedure.[13] This can partly be blamed on the strong dependence of v_{eo} on pH in this range. Any small shift in pH would lead to large changes in v_{eo} and consequently in migration times.

E. CAPILLARY RINSING

Reproducibility in migration behavior can be enhanced by properly rinsing the capillary, operating at low currents, controlling the capillary temperature, and filtering the micellar buffer solutions. As discussed in Chapter 2 (Section II.C.4.c), rinsing of the capillary is of the utmost importance. The results found in this laboratory indicated that the frequency of rinsing and the solutions used for rinsing had the greatest effect on migration reproducibility in both CZE and MECC.[80] A correlation was found between the stability of current and reproducibility of migration behavior, which shows that current might be used as an indicator for anomalous behavior.[80] Rinsing the capillary is essential to maintain "clean" reproducible capillary walls. Figure 28 shows the significance of the rinsing frequency in MECC. Large changes in selectivity and even elution reversal occurred when the capillary was not rinsed properly. Note that in CE techniques, resolution can vary with EOF mobility. Therefore, one can achieve better separations by controlling the EOF. This can also create reproducibility problems if EOF velocity changes from run to run or, even worse, during a run. Frequent rinsing of the capillary would help to maintain a "clean" and reproducible surface that would have an effect on EOF.

F. MODIFIED CAPILLARIES

Most MECC separations are performed in fused silica, untreated capillaries. Balchunas and Sepaniak[37] were the first to report the use of a chemically modified capillary in MECC. An alkyl functional group was bonded to the silica surface, which resulted in a decrease in EOF and broadening of the elution window. One problem with use of these modified capillaries is solute interactions with the hydrophobic moieties on the capillary walls, which can lead to additional band broadening. Lux and Schomberg[81] reported MECC separations on capillaries modified with polar and nonpolar functional groups (Figure 29).

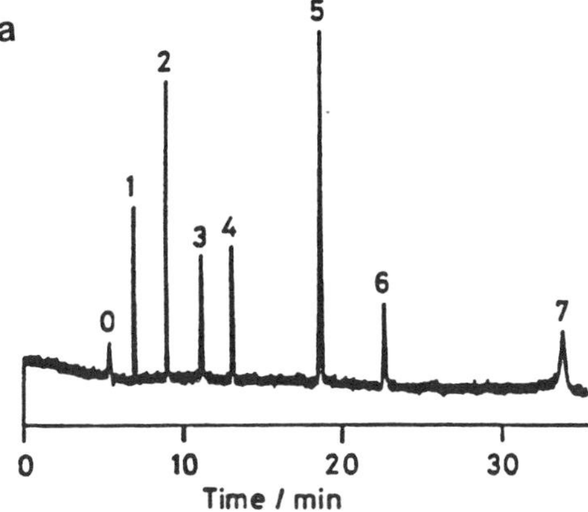

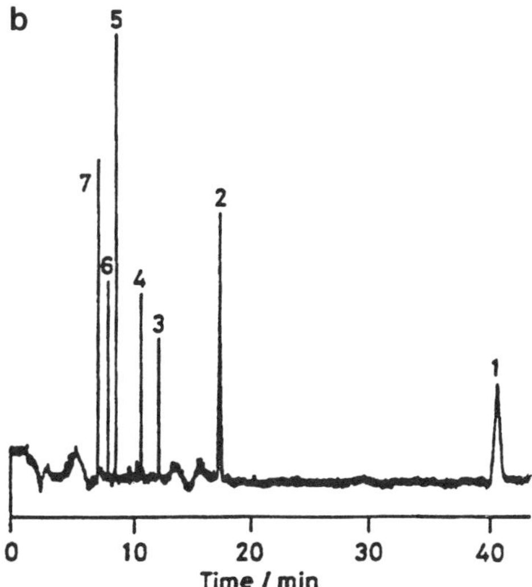

FIGURE 27. Influence of the electroosmotic flow direction on MECC separations. The direction of the electroosmotic flow is changed at pH below 5. Capillary: I.D., 50 μm; total length, 650 mm; separation length, 500 mm; voltage, 14.7 kV; current, 50 μA; UV detection at 220 nm. (a) pH 6.0; (b) pH 3.5. Micellar solution, 100 m*M*. Test solutes: (1) resorcinol, (2) phenol, (3) *p*-nitroaniline, (4) nitrobenzene, (5) toluene, (6) 2-naphthol, (7) Sudan III. (From Otsuka, K. and Terabe, S., *J. Microcolumn Sep.*, 1, 150, 1989. With permission.)

G. DETERMINATION OF t_o AND t_{mc}

Accurate determination of the migration factor, k', is crucial in systematic optimization of MECC separations. Error analysis of k' has revealed that the amount of relative error in k' depends on the t_o/t_{mc} ratio as well as the range of k'.[82] As shown in Figure 30a, an exponential increase in relative error is observed at "high" k' values. The range of acceptable k' values depends largely on the t_o/t_{mc} ratio. The best results are observed at larger elution windows, which can be achieved by the use of organic solvents or mixed micellar systems (Figure 30b).

This also emphasizes the necessity for accurate determination of t_o and t_{mc} in MECC. The direct measurement of these two quantities is not straightforward.

The retention of an unretained compound in MECC (which has no electrophoretic mobility and would not interact with micelles) is determined through the measurement of retention time of a polar organic solvent such as methanol or acetonitrile.[83] This is not feasible since these polar compounds do not absorb UV light, therefore their migration times should be measured from a baseline disturbance. It is often difficult to rationalize the correct position on the baseline

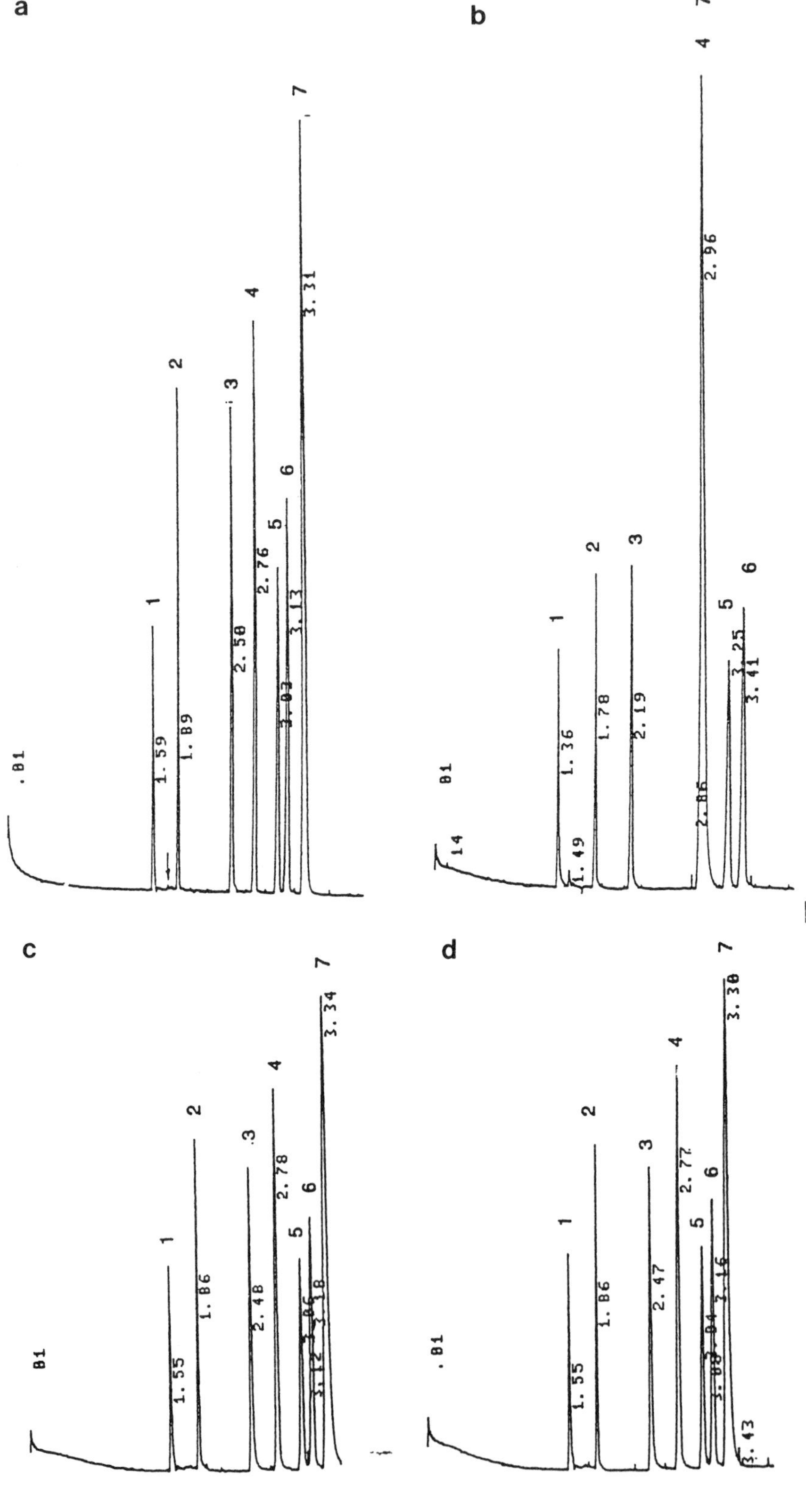

FIGURE 28.

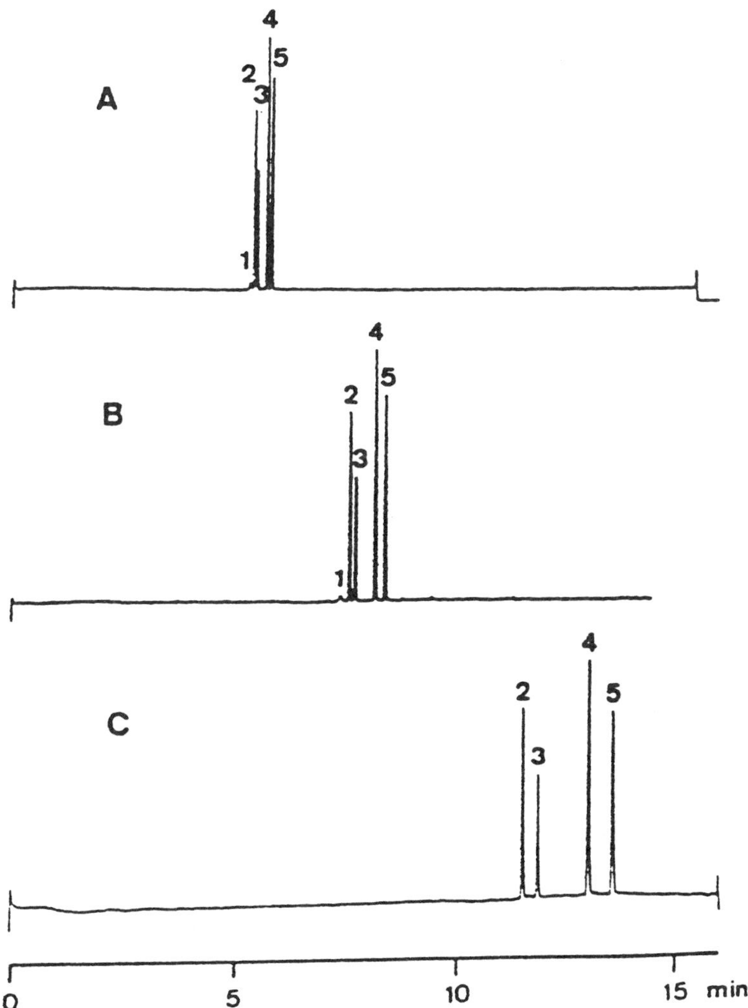

FIGURE 29. MECC separations in modified capillaries. Separations of nucleobases, using capillaries coated with polymers of different chemical structure. (A) Polymethylsiloxane OV-1 on fused silica, 50-mm I.D., film thickness 0.09 mm. (B) Uncoated fused silica capillary, 50-mm I.D. (C) Polyethylene glycol CW20M on fused silica, 50-mm I.D., film thickness 0.04 mm. Buffer: 50 m*M* SDS, 20 m*M* phosphate buffer. Applied voltage: 35 kV. Capillary: total length, 1200 mm; separation length, 800 mm. Sample: 0.2 mg/ml each: (1) water, (2) uridine, (3) cytidine, (4) guanosine, (5) adenosine. Injection: electrokinetic at 75 kV. Injection times: (A) 2 s, (B) 5 s, (C) 8 s. UV detection at 254 nm. Temp: 298 K. (From Lux, J. A. and Schomburg, G., *J. High Resolut. Chromatogr. Chromatogr. Commun.*, 13, 145, 1990. With permission.)

disturbance as an accurate measure of t_0. This is especially true considering the fact that the shape of the baseline disturbance is a function of buffer composition and is often irreproducible.

Hydrophobic compounds migrate at a velocity similar to that of micelles due to their strong association. As a result, the migration time of a highly hydrophobic compound (such as Sudan III) has been used as a measure of t_{mc}.[2] The direct measurement of migration time of Sudan

---

FIGURE 28. Rinsing effect on MECC reproducibility of migration. (a) Run 1, immediately after rinsing; (b) run 20; no rinsing was done before runs 2 to 20; (c) run 21, after rinsing before run; (d) run 30; capillary was rinsed before each run for runs 21 to 30. Conditions: 50 m*M* phosphate buffer (pH 7), 10 m*M* SDS; voltage, 15 kV; L_t = 40 cm, L_s = 26.5 cm. Peaks: (1) arterenol; (2) 2CP; (3) 23CP; (4) 2,4,6-trimethylbenzoic acid; (5) *p*-hydroxybenzoic acid; (6) *m*-methylbenzoic acid; (7) 245CP. Retention times are in minutes. (From Khaledi, M. G., Smith, S. C., and Strasters, J. K., *Anal. Chem.*, 63, 1820, 1991. With permission.)

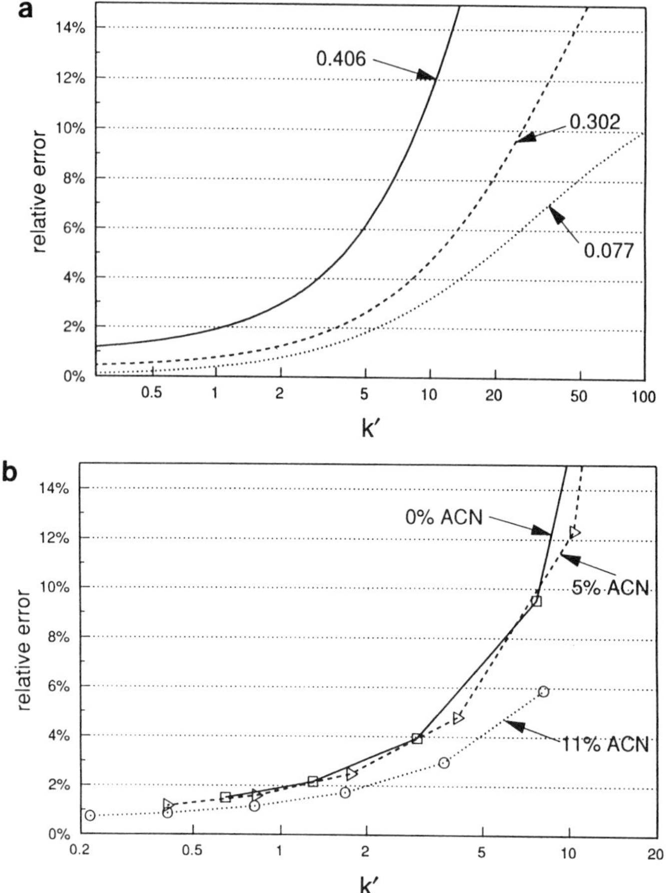

FIGURE 30. Error analysis of migration factor. (a) Relative error in k' in MECC as a function of k' at different t_o/t_{mc} ratios. (b) Relative error in k' in MECC as a function of k'. Buffer: 20 m*M* SDS and 0% (open squares), 5% (open triangles), or 11% (open circles) acetonitrile. (From Smith, D. M., Strasters, J. K., and Khaledi, M. G., manuscript in preparation. With permission.)

III (or other highly hydrophobic compounds) is not experimentally feasible. The low aqueous solubility of these compounds, their broad peaks due to long migration times, along with the low sensitivity of absorbance detectors in CE create serious problems in detecting these compounds. In addition, the assumption that Sudan III completely associates with micelles has not been verified experimentally. This assumption is especially doubtful for an MECC system that includes organic co-solvents in the micellar solution, since solute–micelle binding constants of compounds generally decrease with the addition of organic solvents to the aqueous media. In certain surfactant systems, such as fluorocarbon micelles, one would not observe a peak for Sudan III. Due to the difficulties in measuring t_{mc}, Ackermann et al. suggested the use of the solute mobility, instead of k', as a screening parameter in determination of drugs.[84]

One can determine t_{mc} and t_o values indirectly from the migration behavior of homologous series.[82] There is a linear relationship between the logarithm of k' and the number of $-CH_2$ groups, N_c

$$\ln k' = aN_c + b$$

The linearity of $\ln k'$ vs. N_c depends of the accuracy of t_{mc} and t_o. An iterative process has been used to search for the best t_{mc} value to provide the best linear relationship between $\ln k'$

and N_c. There was a good agreement between the predicted t_{mc} results and the observed values.[82]

VI. APPLICATIONS

Since the first introduction of this technique, numerous applications have been reported by many workers. Many of these have been illustrated in the preceding figures. For the sake of brevity, a detailed review of these applications is not possible; however, it is worth noting that a wide range of compounds and samples have been separated by MECC. These include amino acids and peptides,[26,85-87] chiral molecules,[65-68,86-88] organic acids,[13,51,58,89] nucleic acid constituents,[63,72,90,91] glucosides,[92,93] pharmaceuticals and clinical applications,[30,71,84,94-105] organic bases,[14,15,56,74,75] vitamins,[43,53,54,70] priority pollutants,[57,89] hydrophobic and structural isomers,[7,42] inorganic and organometallics,[28,106] and others.[107,108]

VII. CONCLUSIONS AND FUTURE TRENDS

The main focus of this chapter has been to examine the mechanisms for migration in MECC, the influence of different parameters on migration behavior and selectivity, the similarities and differences between RPLC and MECC separations, and appropriate strategies for optimization of migration parameters in method development. Despite its tremendous growth in the past few years, there is still much left to be explored in order to understand the full capabilities of the technique.

The existence of an elution window is a serious limitation of the technique; however, innovative approaches have greatly alleviated this shortcoming. An advantage of MECC is the feasibility of varying the composition of the micellar buffer solution to enhance separation. Exciting opportunities can be explored for broadening the range of MECC applications and controlling migration behavior by tailoring surfactant structure and by using novel chemistries. The microscale nature of MECC is an advantage that facilitates the use of exotic and expensive reagents.

Taking the high efficiency of the technique for granted, much effort has been expended to enhance resolution and the range of applications of MECC through modifications of the micellar solution compositions. This trend will undoubtedly continue. There is a need, however, for a better understanding of the band-broadening process in MECC in order to take full advantage of the separation power of the technique.

ACKNOWLEDGMENT

The author gratefully acknowledges funding from the National Institutes of Health (First Award, GM 38738), NIH-BMRG, and Glaxo, Inc.

REFERENCES

1. **Terabe, S., Otsuka, K., Ichikawa, K., Tsuchiya, A., and Ando, T.,** Electrokinetic separations with micellar solution and open-tubular capillaries, *Anal. Chem.*, 56, 111, 1984.
2. **Terabe, S., Otsuka, K., and Ando, T.,** Electrokinetic chromatography with micellar solution and open-tubular capillary, *Anal. Chem.*, 57, 834, 1985.
3. **Kuhr, W. G.,** Capillary electrophoresis, *Anal. Chem.*, 62, 403R, 1990.
4. **Kuhr, W. G. and Monning, C. A.,** Capillary electrophoresis, *Anal. Chem.*, 64, 389R, 1992.

5. **Jannini, G. M. and Issaq, H. J.,** Micellar electrokinetic capillary chromatography: basic considerations and current trends, *J. Liq. Chromatogr.*, 15, 927, 1992.
6. **Sepaniak, M. J., Powell, A. C., Swaile, D. F., and Cole, R. O.,** Fundamentals of micellar electrokinetic capillary chromatography, in *Capillary Electrophore sis: Theory and Practice*, Grossman, P. D. and Colburn, J. C., Eds., Academic Press, San Diego, 1992, p. 159.
7. **Terabe, S.,** Micellar electrokinetic chromatography, in *Capillary Electrophoresis Technology,* Guzman, N., Ed., Marcel-Dekker, New York, 1993, p. 65.
8. **Terabe, S., Miyashita, Y., Shibata, O., Barnhart, E. R., Alexander, L. R., Patterson, D. G., Karger, B. L., Hosoya, K., and Tanaka, N.,** Separation of highly hydrophobic compounds by cyclodextrin-modified micellar electrokinetic chromatography, *J. Chromatogr.*, 516, 23, 1990.
9. **Fendler, J. H.,** Membrane mimetic chemistry, *Chem. Eng. News*, 2nd January, 25, 1984.
10. **Hinze, W. L.,** Organized surfactant assemblies in separation science, in *Ordered Media in Chemical Separations*, Hinze, W. L. and Armstrong, D. W., Eds., American Chemical Society, Washington, D.C., 1987, p. 2.
11. **Ghowsi, K., Foley, J. P., and Gale, R. J.,** Micellar electrokinetic capillary chromatography theory based on electrochemical parameters: optimization for three modes of operation, *Anal. Chem.*, 62, 2721, 1990
12. **Otsuka, K. and Terabe, S.,** Effects of pH on electrokinetic velocities in micellar electrokinetic chromatography, *J. Microcolumn Sep.,* 1, 150, 1989.
13. **Khaledi, M. G., Smith, S. C., and Strasters, J. K.,** Micellar electrokinetic capillary chromatography of acidic solutes: migration behavior and optimization strategies, *Anal. Chem.*, 63, 1820, 1991.
14. **Strasters, J. K. and Khaledi, M. G.,** Migration behavior of cationic solutes in micellar electrokinetic capillary chromatography, *Anal. Chem.*, 63, 2503, 1991.
15. **Quang, C., Strasters, J. K., and Khaledi, M. G.,** Computer assisted modeling, prediction and multifactor optimization in micellar electrokinetic capillary chromatography of ionizable compounds, *Anal. Chem*, 1993, submitted.
16. **Sepaniak, M. G. and Cole, R. O.,** Column efficiency in micellar electrokinetic capillary chromatography, *Anal. Chem.*, 59, 472, 1987.
17. **Terabe, S., Otsuka, K., and Ando, T.,** Band broadening in electrokinetic chromatography with micellar solutions and open tubular capillaries, *Anal. Chem.*, 61, 251, 1989
18. **Davis, J. M.,** Random-walk theory of non equilibrium plate height in micellar electrokinetic capillary chromatography, *Anal. Chem.*, 61, 2455, 1989.
19. **Foley, J. P.,** Optimization of micellar electrokinetic chromatography, *Anal. Chem.*, 62, 1302, 1990.
20. **Strasters, J. K., Billiet, H. A. H., Smith, S. C., and Khaledi, M. G.,** Systematic method development in capillary zone electrophoresis, paper presented at the 15th Int. Symp. Column Liquid Chromatography, Basel, Switzerland, June 3–7, 1991.
21. **Billiet, H. A. H. and de Galan, L.,** Selection of mobile phase parameters and their optimization in reversed phase liquid chromatography, *J. Chromatogr.*, 485, 27, 1989.
22. **Strasters, J. K., Breyer, E. D., Rodgers, A. H., and Khaledi, M. G.,** Simultaneous optimization of variables influencing selectivity and elution strength in micellar liquid chromatography. Effect of organic solvent and micelle concentration, *J. Chromatogr.*, 511, 17, 1990.
23. **Strasters, J. K., Kim, S., and Khaledi, M. G.,** Multiparameter optimization in micellar liquid chromatography using iterative regression optimization strategy, *J. Chromatogr.*, 586, 221, 1991.
24. **Burton, D. E., Sepaniak, M. J., and Maskarinec, M. P.,** Evaluation of the use of various surfactants in micellar electrokinetic capillary chromatography, *J. Chromatogr. Sci.*, 25, 514, 1987.
25. **Rasmussen, H. T., Goebel, L. K., and McNair, H. M.,** Micellar electrokinetic chromatography employing sodium alkyl sulfates and Brij 35, *J. Chromatogr.*, 517, 549, 1990.
26. **Otsuka, K., Terabe, S., and Ando, T.,** Electrokinetic chromatography with micellar solutions. Separation of phenylthiohydantoin amino acids, *J. Chromatogr.*, 332, 219, 1985.
27. **Morin, P., Villard, F., Quinsac, A., and Dreux, M.,** Micellar electrokinetic capillary chromatography of glucosinolates and desulfoglucosinolates with a cationic surfactant, *J. High Resolut. Chromatogr. Chromatogr. Commun.*, 15, 271, 1992.
28. **Kaneta, T., Tanaka, S., Taga, M., and Yoshida, H.,** Migration behavior of inorganic anions in micellar electrokinetic capillary chromatography using a cationic surfactant, *Anal. Chem.*, 64, 798, 1992.
29. **Cole, R. O., Sepaniak, M. J., Hinze, W. L., Gorse, J., and Oldiges, K.,** Bile salt surfactants in micellar electrokinetic capillary chromatography. Application to hydrophobic molecule separations, *J. Chromatogr.*, 557, 113, 1991.
30. **Nishi, H., Tsukasa, F., Matsuo, M., and Terabe, S.,** Separation and determination of the ingredients of a cold medicine by micellar electrokinetic chromatography with bile salts, *J. Chromatogr.*, 498, 313, 1990.
31. **Khaledi, M. G., Hadjmohammadi, M. R., and Ye, B.,** Selectivity control in electrokinetic chromatography using mixed micelles, *Anal. Chem.*, 1993, submitted.

32. **Wallingford, R. A., Curry, P. D., and Ewing, A. G.,** Retention of catechols in capillary electrophoresis with micellar and mixed micellar solutions, *J. Microcolumn Sep.*, 1, 23, 1989.
33. **Little, E. L. and Foley, J. P.,** Optimization of the resolution of PTH-amino acids through control of surfactant concentration in micellar electrokinetic capillary chromatography, *J. Microcol. Sep.*, 4, 145, 1992.
34. **Otsuka, K., Kawahara, J., and Tatekawa, K.,** Chiral separations by micellar electrokinetic chromatography with sodium *N*-dodecanoyl-L-valinate, *J. Chromatogr.*, 559, 209, 1991.
35. **Khaledi, M. G., Hadjmohammadi, M. R., and Smith, D. M.,** Manuscript in preparation.
36. **Holland, P. M. and Rubingh, D. N.,** Mixed surfactant systems, in *ACS Symposium Series 501*, American Chemical Society, Washington, D.C., 1992.
37. **Balchunas, A. T. and Sepaniak, M. J.,** Extension of elution range in micellar electrokinetic capillary chromatography, *Anal. Chem.*, 59, 1466, 1987.
38. **Sepaniak, M. J., Swaile, D. F., Powell, A. C., and Cole, R. O.,** Capillary electrokinetic separations: influence of mobile phase composition on performance, *J. High Resolut. Chromatogr.*, 13, 679, 1990.
39. **Gorse, J., Balchunas, A. T., Swaile, D. F., and Sepaniak, M. J.,** Effects of organic mobile phase modifiers in micellar electrokinetic capillary chromatography, *J. High Resolut. Chromatogr. Chromatogr. Commun.*, 11, 554, 1988.
40. **Balchunas, A. T. and Sepaniak, M. J.,** Gradient elution for micellar electrokinetic capillary chromatography, *Anal. Chem.*, 60, 617, 1988.
41. **Sepaniak, M. J., Swaile, D. F., and Powell, A. C.,** Instrumental developments in micellar electrokinetic capillary chromatography, *J. Chromatogr.*, 480, 185, 1989.
42. **Nishi, H. and Matsuo, M.,** Separation of coticosteroids and aromatic hydrocarbons by cyclodextrin-modified micellar electrokinetic chromatography, *J. Liq. Chromatogr.*, 14, 973, 1991.
43. **Ong, C. P., Ng, C. L., Lee, H. K., and Li, S. F. Y.,** Separation of water- and fat-soluble vitamins by micellar electrokinetic chromatography, *J. Chromatogr.*, 547, 419, 1991.
44. **Imasaka, T., Nishitani, K., and Ishibashi, N.,** Cyclodextrin-modified micellar electrokinetic chromatography combined with semiconductor laser fluorimetry, *Analyst*, 116, 1407, 1991.
45. **Nishi, H., Fukuyama, T., and Terabe, S.,** Chiral separation by cyclodextrin-modified micellar electrokinetic chromatography, *J. Chromatogr.*, 553, 503, 1991.
46. **Ueda, T., Kitamura, F., Mitchell, R., Metcalf, T., Kuwana, T., and Nakamoto, A.,** Chiral separation of naphthalene-2,3-dicarboxaldehyde-labeled amino acid enantiomers by cyclodextrin-modified micellar electrokinetic chromatography with laser induced fluorescence detection, *Anal. Chem.*, 63, 2979, 1991.
47. **Fanali, S.,** Use of cyclodextrins in capillary zone electrophoresis. Resolution of terbutaline and propranolol enantiomers, *J. Chromatogr.*, 545, 437, 1991.
48. **Terabe, S., Ishihama, Y., Nishi, H. Fukuyama, T., and Otsuka, K.,** Effect of urea addition in micellar electrokinetic chromatography, *J. Chromatogr.*, 545, 359, 1991.
49. **Karger, B. L., Le Page, J. N., and Tanaka, N.,** "Secondary chemical equilibria in HPLC, in *High Performance Liquid Chromatography: Advances and Perspectives*, Vol. 1, Horvath, Cs., Ed., Academic Press, New York, 1980, p. 113.
50. **Foley, J. P. and May, W. E.,** Optimization of secondary chemical equilibria in liquid chromatography: theory and verification, *Anal. Chem.*, 59, 102, 1987.
51. **Otsuka, K., Terabe, S., and Ando, A.,** Electrokinetic chromatography with micellar solutions. Retention behavior and separation of chlorinated phenols, *J. Chromatogr.*, 348, 39, 1985.
52. **Fujiwara, S. and Honda, S.,** Determination of ingredients of antipyretic analgesic preparations by micellar electrokinetic capillary chromatography, *Anal. Chem.*, 59, 2773, 1987.
53. **Fujiwara, S., Iwase, S., and Honda, S.,** Analysis of water-soluble vitamins by micellar electrokinetic capillary chromatography, *J. Chromatogr.*, 447, 133, 1988.
54. **Nishi, H., Tsumagari, N., Kakimoto, T., and Terabe, S.,** Separation of water soluble vitamins by micellar electrokinetic chromatography, *J. Chromatogr.*, 465, 331, 1989.
55. **Vindevogel, J. and Sandra, P.,** Resolution optimization in micellar electrokinetic chromatography: use of Plackett-Burman statistical design for the analysis of testosterone esters, *Anal. Chem.*, 63, 1530, 1991
56. **Ong, C. P., Pang, S. F., Low, S. P., Lee, H. K., and Li, S. F. Y.,** Migration behavior of catechols and catecholamines in capillary electrophoresis, *J. Chromatogr.*, 559, 529, 1991.
57. **Ong, C. P., Lee, H. K., and Li, S. F. Y.,** Separation of phthalates by micellar electrokinetic chromatography, *J. Chromatogr.*, 542, 473, 1991.
58. **Smith, S. C. and Khaledi, M. G.,** Prediction of the migration behavior of organic acids in micellar electrokinetic capillary chromatography, *J. Chromatogr.*, 632, 177, 1993.
59. **El Seoud, O. A.,** Effects of organized surfactant assemblies on acid-base equilibria, *Adv. Colloid Interface Sci.*, 30, 1, 1989.
60. **Khaledi, M. G. and Rodgers, A. H.,** Micellar-mediated shifts of ionization constants of amino acids and peptides, *Anal. Chim. Acta*, 239, 121, 1990.

61. **Khaledi, M. G., Sahota, R. P., Strasters, J. K., Quang, C., and Smith, S. C.,** Controlling migration behavior in capillary electrophoresis: optimization strategy for method development, in *Capillary Electrophoresis Technology*, Guzman, N., Ed., Marcel Dekker, New York, 1993, p. 187.
62. **Snopek, J. and Smolkova-Keulemansova, E.,** Micellar, inclusion and metal-complex enantioselective pseudophases in high performance electromigration methods, *J. Chromatogr.*, 452, 571, 1988.
63. **Cohen, A. S., Terabe, S., Smith, J. A., and Karger, B. L.,** High performance capillary electrophoresis separation of bases, nucleosides, and oligonucleotides: retention manipulation via micellar solutions and metal additives, *Anal. Chem.*, 59, 1021, 1987.
64. **Wallingford, R. A. and Ewing, A. G.,** Retention of ionic and non-ionic catechols in capillary zone electrophoresis with micellar solutions, *J. Chromatogr.*, 441, 299, 1988.
65. **Terabe, S., Shibata, M., and Miyashita, Y.,** Chiral separation by electrokinetic chromatography with bile salts, *J. Chromatogr.*, 480, 403, 1989.
66. **Nishi, H., Fukuyama, T., Matsuo M., and Terabe, S.,** Chiral separation of diltiazem, trimetoquinol, and related compounds by micellar electrokinetic chromatography with bile salts, *J. Chromatogr.*, 515, 233, 1990.
67. **Dobashi, A., Ono, T., Hara, S., and Yamaguchi, J.,** Enantioselective hydrophobic entanglement of enantiomeric solutes with chiral functionalized micelles by electrokinetic chromatography, *J. Chromatogr.*, 480, 413, 1989.
68. **Otsuka, K. and Terabe, S.,** Enantiomeric resolution by micellar electrokinetic chromatography with chiral surfactants, *J. Chromatogr.*, 515, 221, 1990.
69. **Nishi, H., Tsumagari, N., and Terabe, S.,** Effect of tetraalkylammonium salts on micellar electrokinetic chromatography of ionic substances, *Anal. Chem.*, 61, 2434, 1989.
70. **Burton, D. E., Sepaniak, M. J., and Maskarinec, M. P.,** Analysis of B_6 vitamins by micellar electrokinetic capillary chromatography with laser excited fluorescence detection, *J. Chromatogr. Sci.*, 24, 347, 1986.
71. **Swaile, D., Burton, D., Balchunas, A., and Sepaniak, M. J.,** Pharmaceutical analysis using micellar electrokinetic capillary chromatography, *J. Chromatogr. Sci.*, 26, 406, 1988.
72. **Lee, T., Yeung, E. S., and Sharma, M.,** *J. Chromatogr.*, 565, 197, 1991.
73. **Amankwa, L. N. and Kuhr, W. G.,** Indirect fluorescence detection in micellar electrokinetic chromatography, *Anal. Chem.*, 63, 1733, 1991.
74. **Wallingford, R. A. and Ewing, A. G.,** Amperometric detection of catechols in capillary zone electrophoresis with normal and micellar solutions, *Anal. Chem.*, 60, 258, 1988.
75. **Wallingford, R. A. and Ewing, A. G.,** Separation of serotonin from catechols by capillary zone electrophoresis with electrochemical detection, *Anal. Chem.*, 61, 98, 1989.
76. **Burton, D. E., Sepaniak, M. J., and Maskarinec, M. P.,** Effect of injection procedures on efficiency in micellar electrokinetic capillary chromatography, *Chromatographia*, 21, 583, 1986.
77. **Otsuka, K. and Terabe, S.,** Extra-column effects in high performance capillary electrophoresis, *J. Chromatogr.*, 480, 91, 1989.
78. **Otsuka, K., Terabe, S., and Ando, T.,** Quantitation and reproducibility in electrokinetic chromatography with micellar solutions, *J. Chromatogr.*, 396, 350, 1987.
79. **Terabe, S.,** Personal communication, 1992.
80. **Smith, S. C., Strasters, J. K., and Khaledi, M. G.,** Influence of operating parameters on reproducibility in capillary electrophoresis, *J. Chromatogr.*, 559, 57, 1991.
81. **Lux, J. A. and Schomberg, G.,** Influence of polymer coating of capillary surfaces on migration behavior in micellar electrokinetic capillary chromatography, *J. High Resolut. Chromatogr. Chromatogr. Commun.*, 13, 145, 1990.
82. **Smith, D. S., Strasters, J. K., and Khaledi, M. G.,** manuscript in preparation.
83. **Ahuja, E. S., Little, E. L., and Foley, J. P.,** Selected organic solvents as electroosmotic velocity markers in micellar electrokinetic capillary chromatography, *J. Liq. Chromatogr.*, 15, 1099, 1992.
84. **Ackermann, M. T., Everaets, F. M., and Beckers, J. L.,** Determination of some drugs by micellar electrokinetic capillary chromatography. The pseudo-effective mobility as parameter for screening, *J. Chromatogr.*, 585, 123, 1991.
85. **Liu, J., Cobb, K. A., and Novotny, M.,** Capillary electrophoretic separations of peptides using micelle-forming compounds and cyclodextrins as additives, *J. Chromatogr.*, 519, 189, 1990.
86. **Ong, C. P., Ng, C. L., Lee, H. K., and Li, S. F. Y.,** Separation of Dns-amino acids and vitamins by micellar electrokinetic chromatography, *J. Chromatogr.*, 559, 537, 1991.
87. **Dobashi, A., Ono, T., Hara, S., and Yamaguchi, J.,** Optical resolution of enantiomers with chiral mixed micelles by electrokinetic chromatography, *Anal. Chem.*, 61, 1984, 1989.
88. **Tran, A. D., Blanc, T., and Leopold, E. J.,** Free solution capillary electrophoresis and micellar electrokinetic resolution of amino acid enantiomers and peptide isomers with L- and D-Marfey's reagents, *J. Chromatogr.*, 516, 241, 1990.

89. **Ong, C. P., Ng, C. L., Chong, N. C., Lee, H. K., and Li, S. F. Y.,** Retention of eleven priority phenols using micellar electrokinetic chromatography, *J. Chromatogr.*, 516, 263, 1990.
90. **Row, K. H., Griest, W. H., and Maskarinec, M. P.,** Separation of modified nucleic acid constituents by micellar electrokinetic capillary chromatography, *J. Chromatogr.*, 409, 193, 1987.
91. **Doinik, V., Liu, J., Banks, J. F., Novotny, M., and Bocek, P.,** Capillary zone electrophoresis of oligonucleotides. Factors affecting separation, *J. Chromatogr.*, 480, 321, 1989.
92. **Swedberg, S. A.,** Use of non-ionic and zwitterionic surfactants to enhance selectivity in high-performance capillary electrophoresis. An apparent micellar electrokinetic capillary chromatography mechanism, *J. Chromatogr.*, 503, 449, 1990.
93. **Pietta, P. G., Mauri, P. L., Rava, A., and Sabbatini, G. L.,** Application of micellar electrokinetic capillary chromatography to the determination of flavonoid drugs, *J. Chromatogr.*, 549, 367, 1991.
94. **Nishi, H., Tsumagari, N., Kakimoto, T., and Terabe, S.,** Separation of β-lactam antibiotics by micellar electrokinetic chromatography, *J. Chromatogr.*, 477, 259, 1989.
95. **Altria, K. D. and Smith, N. W.,** Pharmaceutical analysis by capillary zone electrophoresis and micellar electrokinetic capillary chromatography, *J. Chromatogr.*, 538, 506, 1991.
96. **Thormann, W., Meier, P., Marcolli, C., and Binder, F.,** Analysis of barbiturates in human serum and urine by high performance capillary electrophoresis-micellar electrokinetic capillary chromatography with on-column multi-wavelength detection. *J. Chromatogr.*, 545, 445, 1991.
97. **Meier, P. and Thormann, W.,** Determination of thiopental in human serum and plasma by high performance capillary electrophoresis-micellar electrokinetic capillary chromatography, *J. Chromatogr.*, 559, 505, 1991.
98. **Weinberger, R. and Lurie, I. S.,** Micellar electrokinetic capillary chromatography of illicit drug substances, *Anal. Chem.*, 63, 838, 1991.
99. **Fanali, S.,** Separation of optical isomers by capillary zone electrophoresis based on host-guest complexation with cyclodextrins, *J. Chromatogr.*, 474, 441, 1989.
100. **Nishi, H., Fukuyama, T., and Matsuo, M.,** Separation and determination of aspoxicillin in human plasma by micellar electrokinetic chromatography with direct sample injection, *J. Chromatogr.*, 515, 245, 1990.
101. **Weinberger, R., Sapp, E., and Moring, S.,** Capillary electrophoresis of urinary porphyrins with absorbance and fluorescence detection, *J. Chromatogr.*, 516, 271, 1990.
102. **Atamna, I. Z., Janini, G. M., Muschik, G. M., and Issaq, H. J.,** Separation of xanthines and uric acids by capillary zone electrophoresis and micellar electrokinetic capillary chromatography, *J. Liq. Chromatogr.*, 14, 427, 1991.
103. **Soini, H., Tsuda, T., and Novotny, M.,** Electrochromatographic solid phase extraction for determination of cimetidine in serum by micellar electrokinetic capillary chromatography, *J. Chromatogr.*, 559, 547, 1991.
104. **Wernly, P. and Thormann, W.,** Analysis of illicit drugs in human urine by micellar electrokinetic capillary chromatography with on column fast scanning polychrome absorption detection, *Anal. Chem.*, 63, 2878, 1991.
105. **Sakodinskaya, I. K., Desiderio, C., Nardi, A., and Fanali, S.,** Micellar electrokinetic chromatography study of hydroquinone and some of its ethers. Determination of hydroquinone in skin-toning cream. *J. Chromatogr.*, 596, 95, 1992.
106. **Saitoh, T., Hoshino, H., and Yotsuyanagi, T.,** Separation of 4-(2-pyridylazo)resorcinolato metal chelates by micellar electrokinetic capillary chromatography, *J. Chromatogr.*, 469, 175, 1989.
107. **Bushey, M. M. and Jorgenson, J. W.,** Separation of dansylated metylamine and dansylated methyl-d3-amine by micellar electrokinetic capillary chromatography with methanol modified mobile phase, *Anal. Chem.*, 61, 491, 1989
108. **Northrop, D. M., Martire, D. E., and McCrehan,W. A.,** Separation and identification of organic gunshot and explosive constituents by micellar electrokinetic capillary chromatography, *Anal. Chem.*, 63, 1038, 1991.

Chapter 4

ISOELECTRIC FOCUSING IN CAPILLARIES

Ferenc Kilár

TABLE OF CONTENTS

0-8493-8690-X/94/$0.00+$.50

I. BASIC PRINCIPLES OF CAPILLARY ISOELECTRIC FOCUSING

The separation method in which substances are separated on the basis of their isoelectric points (p*I* values) is called isoelectric focusing. Isoelectric focusing (IEF) experiments can be performed in gel matrices or free solution. The first isoelectric focusing experiments in capillaries were performed by Hjertén and co-workers in the mid-1980s.[1,2] Since that time several papers have appeared in the literature but the utilization of capillary isoelectric focusing is still rare in comparison with other CE techniques.

Isoelectric focusing has been commonly performed using slab and tube gel electrophoresis.[3] However, gel methods generally require tedious and time-consuming gel preparation and staining procedures. In the capillary format, IEF separations can be run with or without supporting gel.

Zwitterionic compounds with different p*I* values, such as proteins, peptides, amino acids, various drugs, etc., can be resolved by isoelectric focusing. Separation of these substances is achieved by a pH gradient that is generated by ampholytes (or other amphoteric buffer components) in an electric field. The high- and low-pH sides of the gradient are at the cathode and anode, respectively. The sample components focus according to their isoelectric points in the pH gradient between two electrolytes (e.g., sodium hydroxide as catholyte and phosphoric acid as anolyte). Any band broadening caused by thermal diffusion is quickly reduced by the existing pH gradient. If an analyte, for example, drifts toward the low-pH side of the capillary, it becomes positively charged and migrates back toward its isoelectric point. In gel electrophoresis the focused zones are then fixed and stained for visualization. In capillary electrophoresis, however, where no supporting material is available, other methods, such as mobilization or whole tube scanning, have to be used for the detection.

The most distinguishing phenomenon in capillary electrophoresis is whether the experiments are performed in the absence or in the presence of electroosmotic flow (EOF) (see Chapters 2 and 22 in this book for details on EOF). Unlike other types of capillary electrophoresis, isoelectric focusing can be performed under both modes. Since the experimental and theoretical principles governing these modes of capillary isoelectric focusing (CIEF) are different, they will be discussed separately.

In the absence of EOF (e.g., in non-cross-linked acrylamide-coated capillaries; see Sections II and IV.A.1), the focused zones should be mobilized either chemically or with hydrodynamic flow. During this mobilization, maintenance of the electric field serves as a stabilizer to prevent zone broadening.

In the presence of electroendosmosis, the mobilization of the focused zones is achieved by the EOF. However, under these conditions EOF must be controlled and, therefore, additives are used to obtain a dynamic coating of the capillary surface, providing a reduced bulk flow (see Sections III and IV.A.2). This mode of CIEF can also be combined with hydrodynamic elution.

In the following sections, techniques for performing isoelectric focusing in coated or uncoated capillaries in the absence or presence of additives will be discussed. Different electrophoretic and/or hydrodynamic mobilization procedures will also be considered.

II. ISOELECTRIC FOCUSING IN CAPILLARIES WITH MINIMIZED ELECTROENDOSMOSIS

Electroosmotic flow can be reduced and even eliminated through coating of the internal surface of the capillaries. For example, capillaries coated with methylcellulose or non-cross-

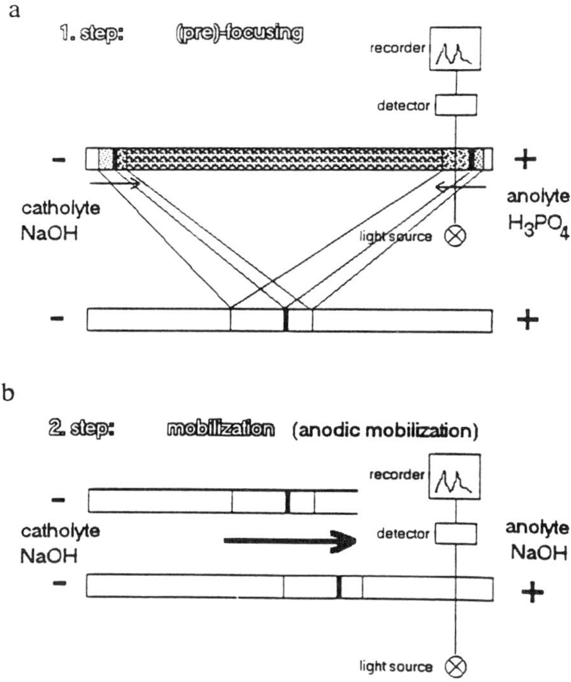

FIGURE 1. Schematic representation of isoelectric focusing in coated capillaries. Both patterns obtained in the (a) focusing and (b) mobilization steps are characteristic of the sample. Mobilization of the focused zones may occur introducing, e.g., Na^+ (shown here) or ethanolamine ions at the anodic end or introducing, e.g., Cl^- or phosphate ions at the cathodic end (not shown). The order of appearance of the components is opposite in the two steps.

linked acrylamide have negligible EOF.[4,5] As a result, the focusing of the substances will depend principally on the quality of the ampholytes and some other parameters (discussed below), but it is not influenced by electroendosmosis.

A. ISOELECTRIC FOCUSING WITH ELECTROPHORETIC MOBILIZATION

The schematic representation of isoelectric focusing in a system introduced by Hjertén and co-workers[1,2,6] is shown in Figure 1. An ultraviolet (UV) light beam is used for on-tube detection.

The substances to be separated are applied to the capillary together with the ampholytes. Typically, the entire capillary is filled with the ampholyte/sample mixture and CIEF is accomplished through two consecutive events: (pre)focusing and mobilization. The ampholytes will create the pH gradient within a short time on application of electric current. At the same time the substances are forced to move toward their isoelectric points in the pH gradient. During this movement the molecules are concentrated in migrating boundaries. The boundaries may be detected at the detection point and hence "peaks" can be recorded. These "peaks", however, do not feature the common Gaussian distribution pattern since they are not real electrophoretic zones. But the pattern obtained during this first step is characteristic of the sample. The focused zones will then be mobilized electrokinetically by replacing one electrolyte at one end of the tube. For instance, anodic mobilization occurs when the phosphoric acid at the anode is replaced by sodium hydroxide solution (see Figure 1).

Hjertén et al.[6] derived expressions that describe the theoretical basis of electrophoretic mobilization. The electroneutrality condition at steady state in the separation tube during focusing is

$$C_{H^+} + \Sigma C_{NH_3^+} = C_{OH^-} + \Sigma C_{COO^-} \quad (1)$$

where C_{H^+}, $\Sigma C_{NH_3^+}$, C_{OH^-}, and ΣC_{COO^-} are the concentrations in equivalents per liter (or coulombs per centimeter) of protons, hydroxyl ions, and positive and negative groups in the carrier ampholytes, respectively. Mobilization can be achieved by adding a positive term to the left side of Equation 1, which then takes the form

$$C_{X^{n+}} + C_{H^+} + \Sigma C_{NH_3^+} = C_{OH^-} + \Sigma C_{COO^-} \quad (2)$$

where X^{n+} (n is the valency) represents a cation. This equation illustrates one approach for accomplishing anodic mobilization, namely by replacing the anolyte used for focusing with a cation that can enter the tube electrophoretically.

The analogous expression for cathodic mobilization would be

$$C_{H^+} + \Sigma C_{NH_3^+} = C_{OH^-} + \Sigma C_{COO^-} + C_{Y^{m-}} \quad (3)$$

where Y^{m-} is an anion.

The above equations regarding the electroneutrality conditions indicate that the cations (anions) entering the separation tube will cause a pH change, but they do not reveal the course of events. The flux of the protons into the separation tube is affected by the composition of the anolyte (catholyte). At steady state in the focusing step the number of protons, N_{H^+}, from the anolyte passing electrophoretically the boundary between the anolyte and medium per time unit can be expressed by

$$N_{H^+} = v_{H^+} q n_{H^+} \quad (4)$$

where v_{H^+} is the migration velocity of the protons in the anolyte, q is the cross-sectional area of the tube, and n_{H^+} is the number per volume unit in the anolyte. Since $v_{H^+} = Eu_{H^+}$ and $E = I/q\kappa$ (where E is the field strength, u_{H^+} is the mobility of the proton in the anolyte, I is the current, and κ is the conductivity in the anolyte), Equation 4 takes the form

$$N_{H^+} = I u_{H^+} n_{H^+} / \kappa \quad (5)$$

For the mobilization we obtain a similar expression:

$$N'_{H^+} = I' u'_{H^+} n'_{H^+} / \kappa' \quad (6)$$

where κ' is the conductivity in the anolyte used for mobilization, n'_{H^+} is the number of protons in the same anolyte, and I' is the current in the tube (primed parameters refer to the mobilization step and nonprimed ones to the focusing step). Since in the initial phase of the mobilization I' has about the same value as I in the focusing step, and also because $u_{H^+} \approx n'_{H^+}$, a good approximation is

$$\frac{N_{H^+}}{N'_{H^+}} = \frac{\kappa' n_{H^+}}{\kappa n'_{H^+}} \quad (7)$$

If the number of protons is not changed in the anolyte during focusing and mobilization, then

$$N_{H^+} \kappa = N'_{H^+} \kappa' \quad (8)$$

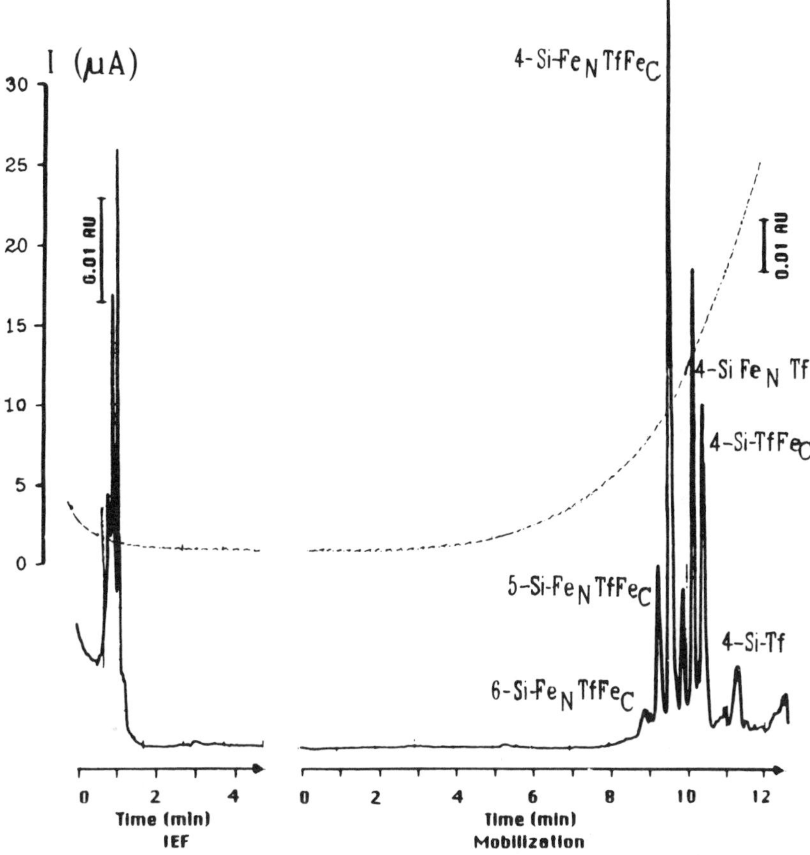

FIGURE 2. High-performance capillary isoelectric focusing experiment of a transferrin sample. Si, Sialo; Tf, transferrin; Fe_NTf and $TfFe_C$, monoferric transferrin forms containing iron at the N- or C-terminal lobe, respectively; Fe_NTfFe_C, diferric transferrin. The current (dotted line) decreases in the focusing step as a result of the immobilization of all the substances in the pH gradient. After a certain time the current reaches a "plateau value," which shows the formation of the steady state in the capillary. On replacing the electrolyte at one end of the capillary the current increases, indicating the electrophoretic migration of the mobilizing ion into the tube. Experimental conditions: tube length, 185 mm; detection point, 155 mm; detection, 280 nm; tube diameter, 0.1 mm; voltage, 5000 V; protein concentration, 1 mg/ml; ampholyte, 2% BioLyte 5/7; anolyte, 20 m*M* H_3PO_4 (focusing)–20 m*M* NaOH (mobilization); catholyte, 20 m*M* NaOH; the protein was dissolved in distilled, deionized water. (Reproduced from Kilár F., *J. Chromatogr.*, 545, 403, 1991. With permission of Elsevier Science Publishers.)

If we choose conditions where $\kappa < \kappa'$ (or $\kappa << \kappa'$), the ratio of N/N' will be > 1 (or >> 1), and one can therefore state that due to the increase in conductivity achieved by supplementing the anolyte with a cation, the number of protons entering the tube from the anolyte decreases, which gives rise to a pH increase in the tube. Analogous equations can be derived for the cathodic mobilization, where the introduction of an anion in the catholyte will cause a decrease in pH at the cathodic end of the capillary and, therefore, a mobilization of the pH gradient. Sodium or chloride ions are commonly used[6] for anodic or cathodic mobilization, respectively.

When the voltage is kept constant during isoelectric focusing, the current decreases in the focusing step due to the increasing resistance of the generated pH gradient (Figure 2). During the electrophoretic mobilization, the change in current is negligible at the beginning, gradually increases toward the end of the experiment, representing the entry of the mobilizing cation or anion in the whole tube.[7]

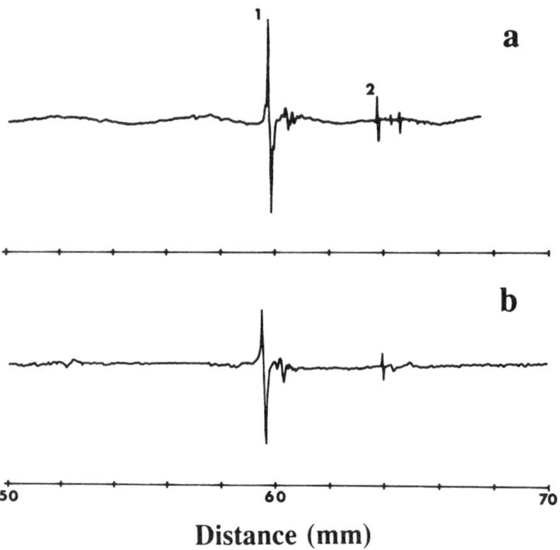

FIGURE 3. Mobilization electropherograms of phosphorylase *b* (peak 1) and ovalbumin (peak 2) focused (a) in a 12 cm × 20-μm I.D. capillary and (b) in a 12 cm × 10-μm I.D. capillary. The detection of the focused components was performed by a concentration gradient detector based on Schlieren optics. Concentrations of the introduced samples are 0.5 mg/ml for phosphorylase *b* and 0.1 mg/ml for ovalbumin, respectively, which correspond to 40 and 17 fmol, respectively, for absolute amounts of samples in the 10-mm I.D. capillary. (Reproduced from Wu, J. and Pawliszyn, J., *J. Chromatogr.*, 608, 121, 1992. With permission of Elsevier Science Publishers.)

Both the patterns obtained by monitoring the moving boundaries during the initial focusing step and those recorded in the mobilization step are characteristic of the sample analyses. The order of the appearance of the components is opposite in the two steps. The resolution and sensitivity are much lower in the former pattern although components in relatively large amounts could be identified (see Figure 2). However, the relative amounts of the components could not be estimated from this pattern.

The detecting UV light beam can be replaced by a concentration gradient detector based on Schlieren optics[8,9] (Figure 3). Electrophoretic mobilization has the advantage that it is also applicable to focusing performed in gel matrices.[1]

B. ISOELECTRIC FOCUSING WITH HYDRODYNAMIC MOBILIZATION

As an alternative to electrophoretic mobilization, the focused zones can be mobilized by pressure coupling an appropriate pump to one end of the capillary. In order to avoid distortion of the focused zones due to the hydrodynamic displacement and diffusion, the voltage must be maintained during the displacement. For hydrodynamic displacement of the focused substances, the anolyte (or the catholyte) is pumped into the capillary tube at a flow rate of about 60 to 70 nl/min (Figure 4a).

A quantitative isoelectric focusing method that exhibited linearity over a wide pH gradient was developed recently for the determination of RNase mutants.[10] The separation was done in a dimethyl polysiloxane (DB-1)-coated capillary in which EOF is sufficiently minimized. The mobilization was achieved by applying a precisely regulated vacuum, while maintaining the voltage. The stability of the coating was supported by the addition of methylcellulose to mask any exposed sites at the extremes of pH. The methylcellulose was always in the ampholyte mixture, and therefore the presumed loss of covalent coating did not reach the point at which resolution and pH gradient linearity became adversely affected.

C. ISOELECTRIC FOCUSING WITHOUT MOBILIZATION

The focused pattern can be recorded without mobilizing the pH gradient. The first report of isoelectric focusing in free solution using whole column absorbance detection[6] was performed in the "free zone electrophoresis apparatus" described by Hjertén.[4] In this case, a 2-mm I.D. rotating quartz tube was utilized, and the pattern of focused zones was scanned with high voltage applied by moving the whole tube past a 280-nm UV light beam. A similar system was adapted recently[11] where the fused silica capillary was moved through the detector at a speed of 0.30 mm/s. In this latter case, however, the voltage was not maintained during detection, which might result in band broadening. Since inhomogeneities of the capillary wall may cause relatively high noise levels, the method necessitates further improvements. (The free zone electrophoresis apparatus avoided this problem by using a special detection device.[4])

As mentioned earlier, a concentration gradient detector has been used successfully for CIEF. This system has been improved by using photodiode array detection,[12] in which a 3-mm segment is monitored simultaneously. The detection system is based on the use of lasers. The sensitivity is on the same order of magnitude as that of a UV absorbance detector, although the concentration gradient detectors possess the advantages that they have smaller detection volumes and are more universal since materials with no absorption can easily be detected without derivatization.

III. ISOELECTRIC FOCUSING IN CAPILLARIES IN THE PRESENCE OF ELECTROOSMOTIC FLOW

Provided that the electroosmotic flow rate is low enough to permit focusing of all substances to occur, isoelectric focusing can be performed in uncoated capillaries. This can be achieved by using a suitable additive in the buffer, e.g., hydroxypropylmethylcellulose (HPMC) or methylcellulose (MC), which forms a dynamic coating to reduce EOF and the interaction between analytes and the wall. Thormann et al., and Mazzeo and Krull, reported two different approaches for isoelectric focusing experiments in which the EOF served as mobilizer for the focused zones.[13-18] The main difference between the two techniques is that only a small plug of the sample–ampholyte mixture is introduced into a capillary filled with catholyte in one case[13,14] (see Figure 5), as opposed to filling the whole capillary with the sample–ampholyte mixture in the other case.[15-18] With this latter method, it is necessary to add a sufficient amount of *N,N,N′,N′*-tetramethylethylenediamine (TEMED), to block the region after the detection point so that substances would not focus in this region. The samples separate in the presence of EOF toward the cathode. Buffer additives reduce both the protein–wall interaction and the EOF, which allows the substances to be focused in the pH gradient, although no steady state conditions of the analytes can be achieved under these circumstances. However, the separation of the focused zones can be sufficient for analyses (Figure 6).

IV. METHODOLOGY FOR CAPILLARY ISOELECTRIC FOCUSING

Isoelectric focusing in capillaries offers several advantages over the conventional gel techniques. These include high-speed, two-step analysis in coated capillaries providing characteristic focusing and mobilization patterns and selective monitoring, using a suitable wavelength for detection. It must be emphasized that isoelectric focusing is a concentrating method, hence, the concentrations of the substances in the separated zones are higher than in the original sample solution. Therefore, isoelectric focusing has an advantage over other capillary

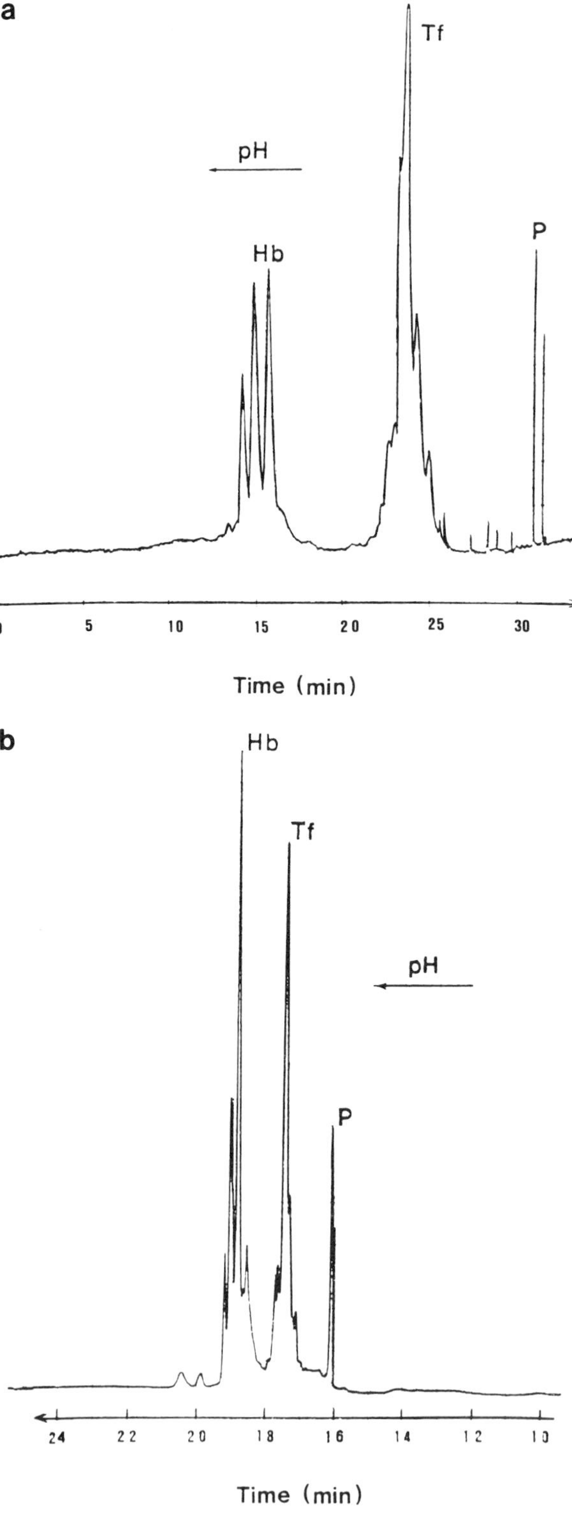
a
Tf
pH
P
Hb
0
5
10
15
20
25
30
Time (min)
b
Hb
Tf
pH
P
24
22
20
18
16
14
12
10
Time (min)

FIGURE 4.

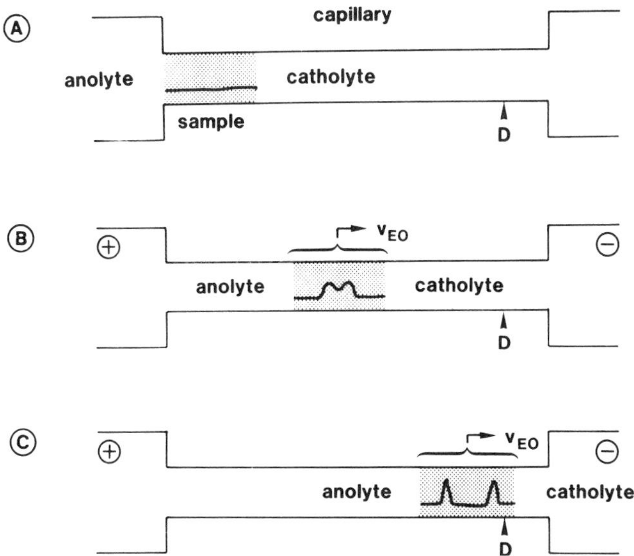

FIGURE 5. Schematic representation of (A) the initial configuration, (B) a transient state, and (C) detection for capillary IEF in the presence of electroendosmosis. D, The point of detection; v_{EO}, the electroosmotic displacement. Carrier ampholytes are represented by the stippled area and the protein distribution by the solid line within the stippled area. (Reproduced from Thormann, W., Caslavska, J., Molteni, S., et al., *J. Chromatogr.*, 589, 321, 1992. With permission of Elsevier Science Publishers.)

electrophoretic methods, with perhaps the exception of isotachophoresis, which may also be used to concentrate sample components. (See Chapter 5.)

Sample application can be done either by pressure or vacuum, but not electrophoretically, as in other capillary electrophoretic modes. The length of the sample plug in experiments performed in the presence of EOF must be carefully determined. Long sample zones may result a focusing step that will not be completed before the moving pH gradient reaches the detection point.[13] This may be circumvented by adding a strong base (e.g., TEMED) to the sample.[15,17,18]

The temporal behavior of the current during IEF provides information about the degree of focusing prior to sample detection at a specified location along the capillary (see Figure 2). In capillaries without EOF, the end of the focusing step can be estimated from the current. If the current is below 15 to 20% of its initial value then the mobilization can be started. The progress of the focusing in uncoated capillaries can be similarly followed by monitoring the current while the EOF transfers the pH gradient toward the detection point.

Isoelectric focusing in coated capillaries with minimized EOF can be performed within 15 to 30 min when electrophoretic mobilization is utilized.[10,19-24] The length of the mobilization

---

FIGURE 4. Isoelectric focusing of hemoglobin (Hb) and transferrin (Tf). (a) Isoelectric focusing performed in capillaries with "off-tube" detection system. Experimental conditions: tube length, 90 mm; tube diameter, 0.2 mm; voltage, 2000 V (10 min, focusing), 3500 V (hydrodynamic mobilization). Sample composition: protein concentration, 1 mg/ml and 10% (v/v) ethylene glycol; ampholyte, 2.5% Pharmalyte 3-10; anolyte, 10 m*M* H_3PO_4; catholyte, 0.5 *M* NaOH; the protein was dissolved in distilled, deionized water. H_3PO_4 (10 m*M*) was pumped into the capillary tube at a flow rate of about 67 nl/min for the hydrodynamic displacement. NaOH solution (0.5 m*M*) was delivered with an HPLC pump for transfer of the focused zones to the UV monitor as they left the capillary tube (0.13 ml/min). (Reproduced from Hjertén, S., Kilár, F., Liao, J.-L., et al., in *Electrophoresis '86*, Dunn, M. J., Ed., VCH, Weinheim, Germany, 1986. With permission of VCH Verlagsgesellschaft.) (b) Isoelectric focusing and hydrodynamic mobilization of the same sample with similar experimental conditions except that on-tube detection was performed at 280 nm and therefore no HPLC pump was necessary. P, precipitates of the protein samples.

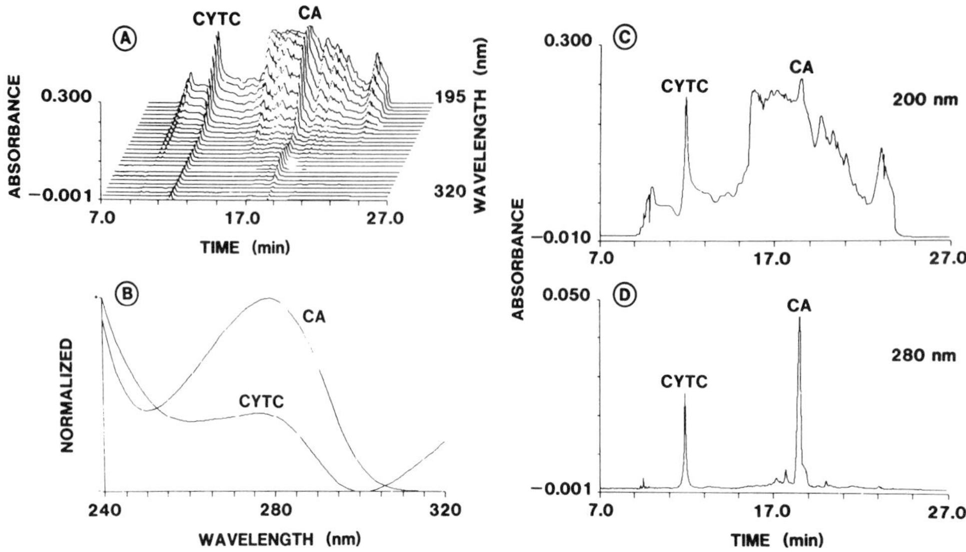

FIGURE 6. Capillary IEF of cytochrome *c* (CYTC; 0.16 mg/ml) and carbonic anhydrase (CA; 0.27 mg/ml) in 2.5% Ampholine. The HPMC concentration in the catholyte was 0.06%. Sampling occurred for 4 min at a height of 34 cm. The current values at the beginning, during protein detection, and after 30 min of power application were about 25, 2, and 4 μA, respectively. (A) Three-dimensional data plot; (B) background-corrected time slices (spectra) for the two proteins; (C) single-wavelength pherogram at 200 nm; (D) pherogram at 280 nm. (Reproduced from Thormann, W., Caslavska, J., Molteni, S., et al., *J. Chromatogr.*, 589, 321, 1992. With permission of Elsevier Science Publishers.)

depends on the applied voltage, length and diameter of the capillary, composition and ionic strength of the sample, ampholyte concentration, length of the focusing step, etc. Faster mobilization of the pH gradient can be obtained by increasing the concentration of, e.g., sodium chloride in the anolyte or catholyte according to Equations 2 or 3, provided that the voltage gradient is kept constant. However, to avoid thermal zone deformation caused by Joule heat, it is advisable to use a concentration of, e.g., NaCl not higher than 0.1 *M* (but not lower than 0.02 *M* to avoid long mobilization times). The speed of mobilization can also be increased with hydrodynamic mobilization, but distortion of the peaks by laminar flow band broadening may result and should be avoided by choosing an optimum flow rate.

In uncoated capillaries, CIEF separations can be carried out in 5 to 40 min,[13-18] depending on several parameters including the length of the capillary, the distance of the detection point from the sample application side, the composition of the electrolytes, voltage applied, etc. Since the dynamic coating of the capillary wall is not easily controlled, the migration times of the analytes may be strongly dependent on the quality of the capillary.

Selective monitoring using a suitable wavelength makes the detection of amphoteric compounds other than high molecular weight proteins possible, since no fixation and staining procedures are necessary. Since most ampholytes absorb strongly below 280 nm, low-UV detection is usually not possible in CIEF. However, the loss in sensitivity due to this detection problem is overcome by the concentrating power of isoelectric focusing itself. Meanwhile, a special benefit of capillary CIEF is that the quality of the ampholytes can be studied using low-wavelength detection (e.g., 225 nm; Figure 7). Possible "gaps" in the pH gradient, where the concentration of the respective ampholyte components is low, can easily be detected from the pattern. Fast scanning multiwavelength detection permits simultaneous monitoring of proteins and carrier ampholytes (Figure 6). In cases where proteins with extreme p*I* are to be separated, the pH gradient of the commercially available ampholytes (between pH 3 and 10) may be extended with certain additives, e.g., TEMED.[15,16,18,21]

In isoelectric focusing, the pH gradient is established using a mixture of synthetic carrier ampholytes. The experiment begins with a transient period (i.e., formation of the pH gradient

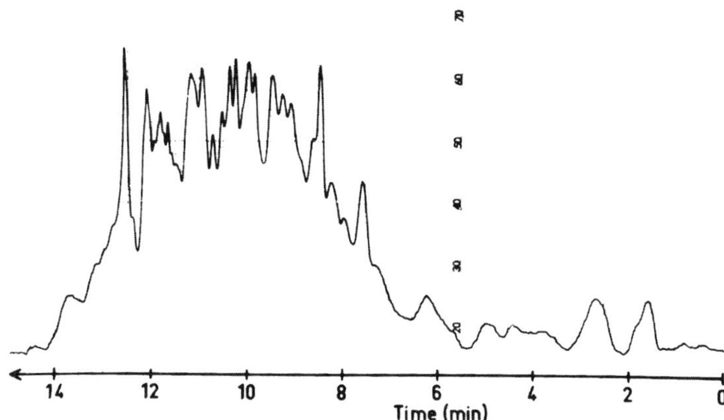

FIGURE 7. Monitoring of carrier ampholytes. The on-tube detection was made at 225 nm. BioLyte, pH 3 to 10, served as ampholyte. (Reproduced from Hjertén, S., Kilár, F., Liao, J.-L., et al., in *Electrophoresis '86*, Dunn, M. J., Ed., VCH, Weinheim, Germany, 1986. With permission of VCH Verlagsgesellschaft.)

and of the focused and concentrated zones) followed by a steady state, in which the amphoteric substances are in a position where the pH is equal to their isoelectric points. According to the theory, this steady state, once established, should remain invariant as long as the applied current density is constant. In practice, however, this has been shown not to be the case.[25] Both experimental and theoretical approaches showed that cathodic, anodic, as well as symmetrical drifts of the pH gradient occur after the formation of the pH gradient due to the loss of ampholytes at the acidic and/or cathodic end of the gradient.[25] The extent of drift in ampholyte systems strongly depends on the anolyte and catholyte concentrations. In experiments where capillaries with minimized EOF are used, an optimum ratio is $2.25[H_3PO_4] = [NaOH]$.[21] However, CIEF of acidic proteins in the presence of EOF showed poor separation due to anodic drift, which can be minimized by increasing the concentration of phosphoric acid.[18]

The two-step analysis of samples in IEF with electrophoretic mobilization is especially useful when proteins become so concentrated during the focusing step that they exceed their solubility in solution and precipitate. Hence, no true focusing pattern can be obtained. In such cases the (pre)focusing pattern is of particular importance (see Figure 2). The resolution in this pattern may be enhanced by increasing the distance between the detecting UV beam and the end of the capillary tube. Suppressing the tendency for precipitation may be achieved by supplementing the ampholytes with additives such as 10 to 50% (v/v) ethylene glycol or 0.5 to 2% (w/v) G 3707 (an efficient, non-UV-absorbing detergent), or a mixture of both.[6] The voltage during the mobilization step should be lowered compared to the focusing step since high current may also induce the blocking of the capillary by precipitate formation.

The metal complexation of ampholytes may cause unwanted results when iron-containing tools are used for handling the sample (e.g., syringes with metal needles, metal tubings, etc.). This was demonstrated with transferrin, a metal-binding protein, when iron-complexed forms of transferrin appeared in the mobilization pattern of an iron-free sample[19] on mixing the ampholytes and the protein solution with a Hamilton syringe equipped with a metal needle (Figure 8).

A. PROCEDURES OF CAPILLARY ISOELECTRIC FOCUSING

1. Typical Separation of Proteins in Coated Capillaries with Minimized Electroosmotic Flow

Ampholytes: Commercially available ampholyte solution (BioLyte, Pharmalyte, Servalyte, etc.), pH 3 to 10

Sample: 0.5- to 1.0-mg/ml solution mixed with 2% ampholytes (final concentration)

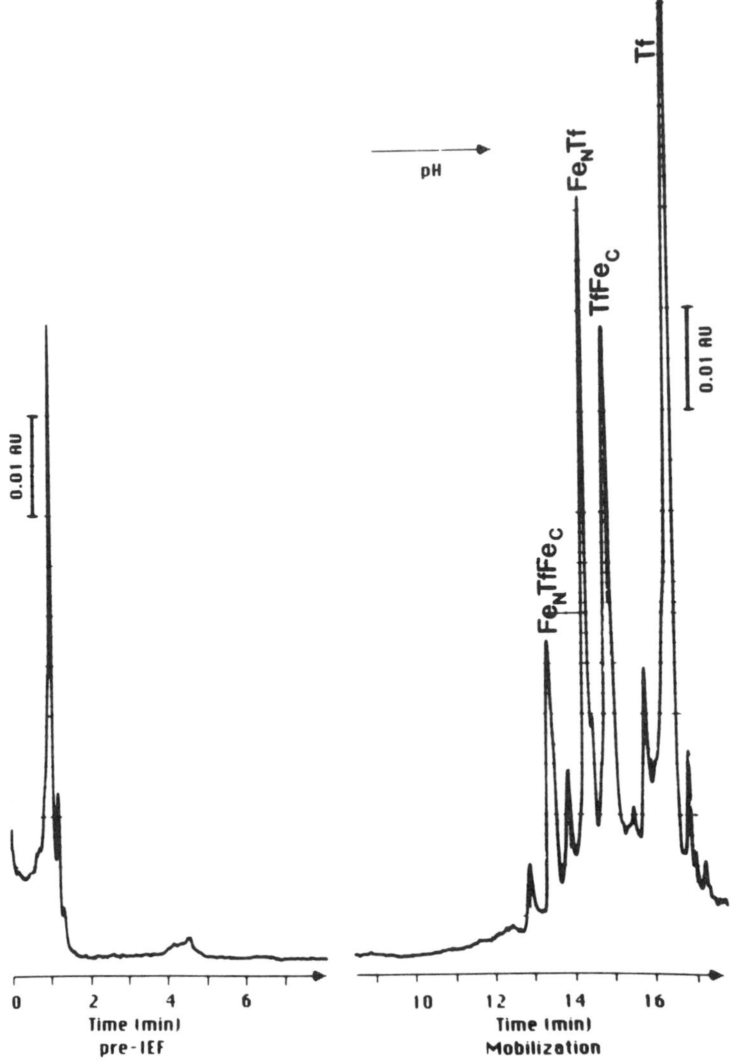

FIGURE 8. High-performance isoelectric focusing of iron-free transferrin sample after mixing the protein solution and carrier ampholytes, using a microsyringe with metal piston (20 mixing steps were performed before the experiment). Iron-complexed forms of transferrin appear in the pattern on the mixing steps. Experimental conditions: tube dimensions 0.1 (I.D.) × 0.3 (O.D.) × 150 mm; protein concentration, 0.76 mg/ml; carrier ampholytes, 3% Bio-Lyte 5/7; voltage, 5000 V; on-tube detection, 280 nm. Anodic mobilization. (Reproduced from Kilár, F. and Hjertén, S., *Electrophoresis*, 10, 23, 1989. With permission of VCH Verlagsgesellschaft.)

Capillary:	20 cm × 50 μm, coated, filled with sample
Anolyte:	10 m*M* H_3PO_4
Catholyte:	20 m*M* NaOH
Focusing:	8 kV, constant voltage, 10 min
Mobilizer:	80 m*M* NaCl, 20 m*M* NaOH (cathodic) or 20 m*M* NaOH (anodic)
Mobilization:	6 kV, constant voltage
Detection:	280 nm, near to the mobilizer
Washing:	After each run with, e.g., 10 m*M* H_3PO_4

Notes. All solutions should be degassed. To increase the resolution in short capillaries a smaller pH range of the ampholytes can be used. The ionic strength of the sample solution may influence dramatically the length of the focusing step; therefore, a low buffer ion concentration is preferable. The shorter the capillary the shorter the focusing step and the lower the resolution. The focusing step may be finished when the current decreases to 15 to 20% of its initial value, double peaks in the mobilization pattern might indicate incomplete focusing. The hydrolytic stability of coating is poor at alkaline pH; therefore, mobilization with NaOH may destroy the coating of the capillary wall after a few runs. Zwitterionic compounds may be used for mobilization as well. The voltage applied during mobilization should preferably be lower than in focusing to avoid high temperatures in the tube that might cause precipitation of the protein. (See other conditions in Refs. 1, 2, 6, 7, 19–24, and 26.)

2. Typical Separation of Proteins in Uncoated Capillaries in the Presence of Electroosmotic Flow

a. *Method 1*[13,14]

Ampholytes:	Commercially available ampholyte solution (BioLyte, Pharmalyte, Servalyte, etc.), pH 3 to 10
Sample:	0.5- to 1.0-mg/ml solution mixed with 2.5% ampholytes (final concentration)
Capillary:	70 cm × 75 μm, uncoated, filled with catholyte
Sample loading:	Filling 20 to 30% of effective column length
Anolyte:	10 m*M* H_3PO_4
Catholyte:	20 m*M* NaOH, 0.06% hydroxypropylmethylcellulose
Voltage:	10 kV
Detection:	280 nm
Washing:	Before each run with 0.1 *M* NaOH and catholyte

b. *Method 2*[15]

Ampholytes:	Commercially available ampholyte solution (BioLyte, Pharmalyte, Servalyte, etc.), pH 3 to 10
Sample:	0.5- to 1.0-mg/ml solution mixed with 5% ampholytes (final concentration), 0.1% methylcellulose, 1% TEMED
Capillary:	60 cm × 75 μm, uncoated, filled with sample
Anolyte:	10 m*M* H_3PO_4
Catholyte:	20 m*M* NaOH
Voltage:	20 kV
Detection:	280 nm
Washing:	Before each run with 0.1 *M* NaOH and water

Notes. The concentration of additives to the sample or catholyte should be carefully determined for the appropriate system. A high concentration of additives decreases the resolution. Conditioning of the capillary before the runs is essential for the performance. The length of the sample zone in method 1 influences the resolution: the shorter the length, the better the resolution. TEMED should be used in method 2 to extend the pH gradient. Thus TEMED can act as a blocking agent, focusing in the region past the detection window and allowing for the separation and detection of basic proteins. (See other conditions in Refs. 13–15, 17, and 18.)

V. APPLICATIONS OF CAPILLARY ISOELECTRIC FOCUSING

Capillary isoelectric focusing in free solution has been demonstrated in glass tubes,[1,2,6,7,19,20,26] in rectangular cross-section channels,[27-29] in Teflon tubings,[30] and in fused silica capillaries.[8-18,21-24,31,32] In those studies, capillary isoelectric focusing has been used for qualitative and quantitative investigation of proteins, and for the determination of the isoelectric points of macromolecules.

The following proteins have been investigated by capillary isoelectric focusing: hemoglobin,[1,22-24,30] transferrin,[1,6,7,9,14,19,20] RNase mutants,[10] plasminogen activator,[32] phosphorylase *b* and ovalbumin,[8,9,12] monoclonal antibodies,[9,21,31] myoglobin, carbonic anhydrase and cytochrome *c*,[13-18,30] albumin,[15,30] and chymotrypsinogen.[16] The studies show that most of these proteins have isoforms and the separation of those isoforms is easily performed in capillaries.

The determination of the p*I* of focused substances in isoelectric focusing experiments generally requires internal standards and their "migration time" values (i.e., the time parameter of the peaks in the electropherograms). Such experiments were done in coated capillaries and calibration curves obtained with electrophoretic[2] or hydrodynamic mobilization[10] were constructed. As an alternative the current measured during electrophoretic mobilization was used to determine the p*I* values without having internal standards in every run.[7] Keeping the experimental conditions unchanged, the current parameter characterizes the position of the respective substance in the pH gradient and, therefore, it can be used for the estimation of the isoelectric point. Experiments with transferrin isoforms show that the p*I* can be determined with an error of about 0.03 pH units or less.[7]

The application of off-tube detection methods has also been demonstrated.[1] Although it has somewhat lower resolution than the apparatus with on-tube monitoring (compare Figures 4a and b), it has the distinct advantage of permitting recovery of the separated proteins and can thus be used for micropreparative runs.

VI. CONCLUSIONS

Capillary isoelectric focusing is still in the developmental stage. Several applications using the two distinct methodologies described in this chapter have been reported, but further improvements are necessary to design fully automated systems, to improve column technology, to define the limits of resolution, and to describe the transport dynamics of substances in coated and uncoated capillaries.

ACKNOWLEDGMENT

This work was partially supported by Grants OTKA 1981 and T5218 (Hungary).

REFERENCES

1. **Hjertén, S. and Zhu, M.-D.,** Adaptation of the equipment for high-performance electrophoresis to isoelectric focusing, *J. Chromatogr.*, 346, 265, 1985.
2. **Hjertén, S., Kilár, F., Liao, J.-L., and Zhu, M.-D.,** Use of the high-performance electrophoresis apparatus for isoelectric focusing, in *Electrophoresis '86*, Dunn, M. J., Ed., VCH, Weinheim, Germany, 1986, p. 451.
3. **Righetti, P. G.,** *Isoelectric Focusing: Theory, Methodology and Applications*, Elsevier, Amsterdam, 1983.
4. **Hjertén, S.,** Free zone electrophoresis, *Chromatogr. Rev.*, 9, 122, 1967.
5. **Hjertén, S.,** High-performance electrophoresis. Elimination of electroendosmosis and solute adsorption, *J. Chromatogr.*, 347, 191, 1985.
6. **Hjertén, S., Liao, J.-L., and Yao, K.,** Theoretical and experimental study of high-performance electrophoretic mobilization of isoelectrically focused protein zones, *J. Chromatogr.*, 387, 127, 1987.
7. **Kilár, F.,** Determination of the pI with measuring the current in the mobilization step of high-performance capillary isoelectric focusing. Analysis of transferrin forms, *J. Chromatogr.*, 545, 403, 1991.
8. **Wu, J. Q. and Pawliszyn, J.,** High-performance capillary isoelectric focusing with a concentration gradient detector, *Anal. Chem.*, 64, 219, 1992.

9. **Wu, J. and Pawliszyn, J.**, Application of capillary isoelectric focusing with universal concentration gradient detector to the analysis of protein samples, *J. Chromatogr.*, 608, 121, 1992.
10. **Chen, S. M. and Wiktorowicz, J. E.**, Isoelectric focusing by free solution capillary electrophoresis, *Anal. Biochem.*, 206, 84, 1992.
11. **Wang, T. S. and Hartwick, R. A.**, Whole column absorbance detection in capillary isoelectric focusing, *Anal. Chem.*, 64, 1745, 1992.
12. **Wu, J. Q. and Pawliszyn, J.**, Universal detection for capillary isoelectric focusing without mobilization using a concentration gradient imaging system, *Anal. Chem.*, 64, 224, 1992.
13. **Thormann, W., Caslavska, J., Molteni, S., and Chmelik, J.**, Capillary isoelectric focusing with electroosmotic zone displacement and on-column multichannel detection, *J. Chromatogr.*, 589, 321, 1992.
14. **Molteni, S. and Thormann, W.**, Experimental aspects of capillary isoelectric focusing with electroosmotic zone displacement, *J. Chromatogr.*, 638, 187, 1993.
15. **Mazzeo, J. R. and Krull, I. S.**, Capillary isoelectric focusing of proteins in uncoated fused-silica capillaries using polymeric additives, *Anal. Chem.*, 63, 2852, 1991.
16. **Mazzeo, J. R. and Krull, I. S.**, Coated capillaries and additives for the separation of proteins by capillary zone electrophoresis and capillary isoelectric focusing, *BioTechniques*, 10, 638, 1991.
17. **Mazzeo, J. R. and Krull, I. S.**, Examination of variables affecting the performance of isoelectric focusing in uncoated capillaries, *J. Microcolumn. Sep.*, 4, 29, 1992.
18. **Mazzeo, J. R. and Krull, I. S.**, Improvements in the method developed for performing isoelectric focusing in uncoated capillaries, *J. Chromatogr.*, 606, 291, 1992.
19. **Kilár, F. and Hjertén, S.**, Fast and high-resolution analysis of human serum transferrin by high-performance isoelectric focusing in capillaries, *Electrophoresis*, 10, 23, 1989.
20. **Kilár, F. and Hjertén, S.**, Separation of the human transferrin isoforms by carrier-free high-performance zone electrophoresis and isoelectric focusing, *J. Chromatogr.*, 480, 351, 1989.
21. **Zhu, M. D., Rodriguez, R., and Wehr, T.**, Optimizing separation parameters in capillary isoelectric focusing, *J. Chromatogr.*, 559, 479, 1991.
22. **Bolger, C. A., Zhu, M., Rodriguez, R., and Wehr, T.**, Performance of uncoated and coated capillaries in free zone electrophoresis and isoelectric focusing of proteins, *J. Liq. Chromatogr.*, 14, 895, 1991.
23. **Zhu, M., Rodriguez, R., Wehr, T., and Siebert, C.**, Capillary electrophoresis of hemoglobins and globin chains, *J. Chromatogr.*, 608, 225, 1992.
24. **Zhu, M-D., Hansen, D. L., Burd, S., and Gannon, F.**, Factors affecting free zone electrophoresis and isoelectric focusing in capillary electrophoresis, *J. Chromatogr.*, 480, 311, 1989.
25. **Mosher, R. A. and Thormann, W.**, Experimental and theoretical dynamics of isoelectric focusing. 4. Cathodic, anodic and symmetrical drifts of the pH gradient, *Electrophoresis*, 11, 717, 1990.
26. **Hjertén, S., Elenbring, K., Kilár, F., Liao, J.-L., Chen, A. J. C., Siebert, C. J., and Zhu, M.-D.**, Carrier-free zone electrophoresis, displacement electrophoresis and isoelectric focusing in a high-performance electrophoresis apparatus, *J. Chromatogr.*, 403, 47, 1987.
27. **Mosher, R. A., Thormann, W., and Bier, M.**, Experimental and theoretical dynamics of isoelectric focusing. 2. Elucidation of the impact of the electrode assembly, *J. Chromatogr.*, 436, 191, 1988.
28. **Thormann, W., Firestone, A., Dietz, M. L., Cecconie, T., and Mosher, R. A.**, Focusing counterparts of electrical-field-flow fractionation electrical hyperlayer field flow fractionation and capillary isoelectric focusing, *J. Chromatogr.*, 461, 95, 1989.
29. **Mosher, R. A., Thormann, W., Kuhn, R., and Wagner, H.**, Experimental and theoretical dynamics of isoelectric focusing. 3. Transient multi-peak approach to equilibrium of proteins in simple buffers, *J. Chromatogr.*, 478, 39, 1989.
30. **Firestone, M. A. and Thormann, W.**, Capillary isoelectric focusing using an LKB 2127 Tachophor isotachophoretic analyzer, *J. Chromatogr.*, 436, 309, 1988.
31. **Silverman, C., Komar, M., Shields, K., Diegnan, G., and Adamovics, J.**, Separation of the isoforms of a monoclonal antibody by gel isoelectric focusing, high performance liquid chromatography and capillary isoelectric focusing, *J. Liq. Chromatogr.*, 15, 207, 1992.
32. **Yim, K. W.**, Fractionation of the human recombinant tissue plasminogen activator (rtPA) glycoforms by high-performance capillary zone electrophoresis and capillary isoelectric focusing, *J. Chromatogr.*, 559, 401, 1991.

Chapter 5

ISOTACHOPHORESIS IN CAPILLARY ELECTROPHORESIS

B. J. Wanders and F. M. Everaerts

TABLE OF CONTENTS

0-8493-8690-X/94/$0.00+$.50

I. INTRODUCTION

Isotachophoresis is an analytical technique founded on the same principle as zone electrophoresis, with the separation of ionic species based on the differences in their effective mobility in an electric field. It has been in existence since the early 1960s, but has never been widely accepted as a routine analytical tool. Different reasons have been given for this failure: parallel development with high-performance liquid chromatography (HPLC), strange staircase-shaped detector output, and the inability of some instrument companies to build a reliable, functional instrument. Isotachophoresis, however, is a valuable tool for small ion analysis, and is still used on a limited scale, mainly in Europe and Japan. It has detection limits comparable to capillary zone electrophoresis, but, to date, outperforms it in reproducibility.

In the last 2 to 5 years, the interest in isotachophoresis has increased again. This time, however, it is not for use as a separation technique by itself, but for use of the isotachophoretic principle as a sample-stacking method in capillary electrophoresis. This chapter focuses on this particular use of the isotachophoretic principle after explaining the basic principles of isotachophoresis itself.

II. HISTORIC OVERVIEW[1]

It was not until the early 1920s that a principle of electrophoresis other than zone electrophoresis was described by Kendall and Crittenden.[2] They successfully separated rare earth metals and some simple acids, by what they called the "ion migration method", which, in fact, was isotachophoresis. One of Kendall's observations was that the ions not only separated, but also adapted their concentration to the concentration of the first zone, as described by the Kohlrausch[3] regulating function. It took another decade before more relevant work was carried out on electrophoretic techniques other than zone electrophoresis. Martin,[4] in 1942, was able to separate chloride, acetate, aspartate, and glutamate by isotachophoresis, which he called "displacement electrophoresis" because of the similarity to displacement chromatography.

Another decade passed before Longsworth[5] realized the importance of Kendall's work. In a Tiselius moving boundary apparatus, he introduced a mixture of Ca^{2+}, Ba^{2+}, and Mg^{2+}, between two other zones, called the leading and the terminating electrolyte. Longsworth introduced a counter flow because the separation chamber in a Tiselius apparatus is very short, and adjusted it so that the zones remained in the detection region until they were separated. Once separated, the effective mobilities decrease from leading to terminating electrolyte. Longsworth also found that a steady state was reached once the components were separated.

In 1963, Everaerts,[6] together with Martin, performed an isotachophoretic separation (or as they called it then, "displacement electrophoresis") in a narrow-bore tube of Pyrex glass with an inner diameter of 0.5 mm and an outer diameter of 0.8 mm. To prevent hydrodynamic flow between the two electrode compartments they increased the viscosity of the electrolyte (up to 100 cP) by addition of a water-soluble linear polymer, hydroxyethylcellulose. A thermocouple was used as the detector. Based on the work of Everaerts[7] and Martin and Everaerts,[8] Verheggen and Everaerts built an instrument and introduced the technique in Bergstrom's department at the Karolinska Institute (Stockholm, Sweden) in 1968. This led to the first commercial production of an isotachophoretic instrument by LKB Produkter AB (Bromma, Sweden).

Prior to 1970 isotachophoresis was referred to as moving boundary method, displacement electrophoresis, steady state stacking, cons electrophoresis, and ionophoresis. In 1970 Haglund,[9] together with a group of researchers, introduced a new name for this method. Based on the important phenomenon of the electrophoretic technique, the identical velocities of the zones in the steady state, they termed it isotacho-electro-phoresis,* or isotachophoresis for short.

Since 1970, many scientists have made significant contributions to the development of isotachophoresis; two papers, in particular, represent milestones in the development of isotachophoresis. Arlinger and Routs[10] introduced an ultraviolet (UV) absorption detector in 1970, and Verheggen et al.[11] introduced an operational conductivity detector in 1972.

III. PRINCIPLE OF ISOTACHOPHORESIS

As already discussed in other chapters of this handbook, ions will move, under influence of an applied electric field (E), with a velocity (v) of

$$v = \mu E \tag{1}$$

where μ is the effective mobility of the ion ($m^2 V^{-1} s^{-1}$), which depends on several factors such as pH, temperature, and viscosity of the medium.

Differences in effective mobility will result in different velocities, which allows for ions to be separated. With only a single exception, electrophoretic separation techniques are based on this principle. The exception to this rule is isoelectric focusing, where the differences in p*I* separate the different components.

In the upcoming simplified model for isotachophoresis, we look at a separation of anions in a capillary. For the separation of anions, the capillary and the anode compartment are filled with an electrolyte called the leading electrolyte. The anions of the leading electrolyte must have a higher effective mobility than any of the sample anions. The cation of the leading electrolyte (called the counterion) must have a good buffering capacity at the pH at which the analysis will be performed. The cathode compartment is filled with an electrolyte called the terminating electrolyte, containing anions that must have an effective mobility lower than any of the sample anions. The sample is introduced between the leading and terminating electrolyte (Figure 1a). When a current is passed through the system, a uniform electric field occurs through the sample zone and all of the anions in the sample will migrate at different velocities according to their effective mobilities (Equation 1). The sample anions with the highest effective mobility will migrate forward, and the sample anions with lower effective mobilities will stay behind. This part of the separation, based on the moving boundary principle, will result in a series of mixed zones in front of and behind the original sample zone (Figure 1b). The fastest sample anions can never pass the leading anions, because the leading ions have a higher effective mobility. Similarly, the terminating anions can never pass the sample anions because of their lower effective mobility. In this way, the sample zones are sandwiched between the leading and terminating electrolyte. In the mixed zones of the sample, the separation continues until a steady state is reached. In the steady state, a series of zones is obtained, in which each zone only contains one anionic species of the sample, sandwiched between the leading and terminating electrolyte. The zones are ordered in mobility: the first zone has the anionic species with the highest effective mobility; the last zone the one with the lowest effective mobility.

It is in this respect that isotachophoresis differs greatly from zone electrophoresis. Once the steady state is reached, all zones must migrate connected together, in contrast to zone electrophoresis, in which the different zones separate from each other. In isotachophoresis, the zones cannot separate because there is no background electrolyte to conduct the electricity. Once the steady state is reached, all zones migrate with the same velocity, determined by the

* The name *isotachophoresis* is based on the Greek words ισο (equal), ταχοζ (velocity), and πορεεσθαι (to be dragged).

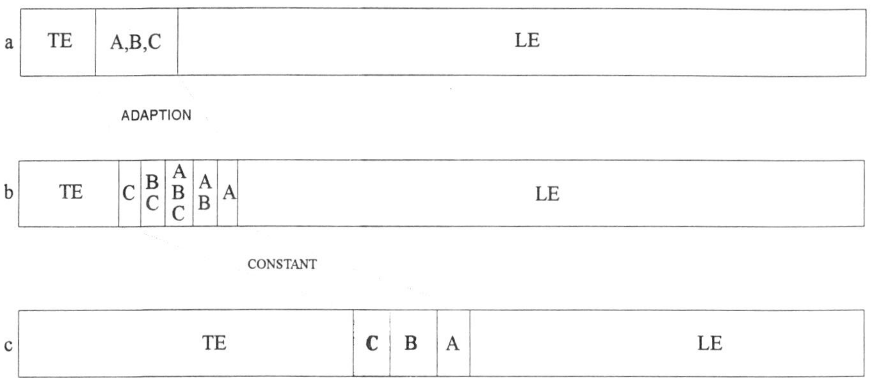

FIGURE 1. Isotachophoretic separation of a mixture of anions. A sample containing the anions A, B, and C is introduced between the leading electrolyte (with leading ion L) and terminating electrolyte (with terminating ion T) (a). After some time, some mixed zones are obtained according to the moving boundary principle (b). Finally, all anions are fully separated and the steady state is reached (c).

velocity of the anionic species in the leading electrolyte. If we consider a sample with three components, we can write

$$v_L = v_A = v_B = v_C = v_T$$

Using Equation 1 we can write

$$\mu_L E_L = \mu_A E_A = \mu_B E_B = \mu_C E_C = \mu_T E_T \tag{2}$$

Because the zones are ordered in decreasing effective mobility, the electric field strength increases from the leading to the terminating zone. Also, when running at a constant current, the heat production (= *EI*), and therefore the temperature, will increase from leading to terminating electrolyte. Hence, when using a conductivity or thermometric detection system, the resulting isotachopherogram will show the characteristic staircase (Figure 2).

Another important characteristic of isotachophoresis, induced by this staircase-shaped pattern of the electric field strength from the leading to the terminating zone, is the "self-correction" principle. In isotachophoresis, once the steady state is reached, the formed zones will not broaden further. This is in contrast to zone electrophoresis, in which zones will continue to broaden because of diffusion and, possibly, adsorption effects.

The reason for this self-correcting behavior is the fact that the zone is sandwiched between a zone with a lower (front) and higher (back) electric field strength. If an anion diffuses into the preceeding zone, the lower electric field of that zone will slow down the anion, and the anion will be overtaken by its own zone. The opposite effect occurs when an anion stays behind in a zone with a higher electric field strength. The anion will acquire a higher velocity (Equation 1) until its reaches its own zone.

A. SIMPLIFIED MODEL FOR ISOTACHOPHORESIS

For this simplified model for isotachophoresis, the following assumptions were made: (1) all ions are monovalent and fully ionized; (2) the influence of the presence of H^+ and OH^- can be neglected; (3) the influence of diffusion is negligible; and (4) the counterionic species, Q, are similar in all zones and have a constant mobility, μ_Q.

Working at a constant current, the following equations can be derived for two connected zones (1 and 2), containing the ionic species A and B with $\mu_A > \mu_B$. According to the principle of electroneutrality, the concentrations of positive and negative ions in both zones must be identical:

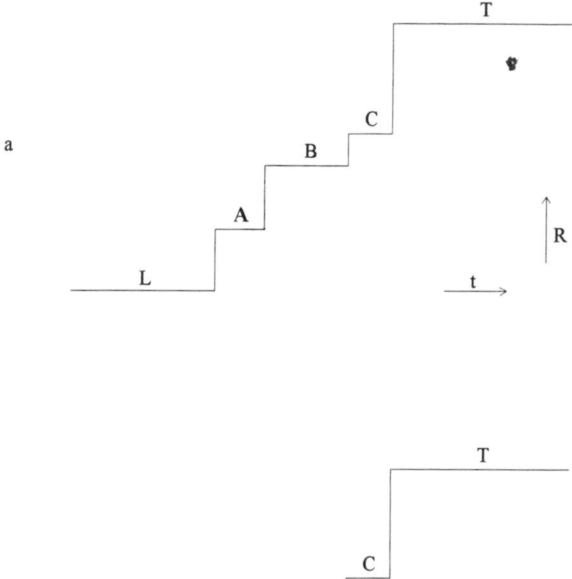

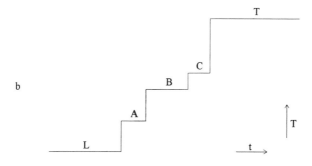

FIGURE 2. Isotachopherograms of the model separation of the anions A, B, and C shown in Figure 1, using a conductivity detector (a) or a thermometric detector (b).

$$c_{A,1} = c_{Q,1} \text{ and } c_{B,2} = c_{Q,2}$$

Once the steady state is reached, the zones are also migrating with the same velocity, therefore

$$v_1 = v_2$$

or

$$\mu_A E_1 = \mu_B E_2$$

or

$$E_1/E_2 = \mu_B/\mu_A \tag{2}$$

According to Ohm's law:

$$j = \text{constant} = E_1\sigma_1 = E_2\sigma_2 \tag{3}$$

in which j is the current density (A/m^2), σ_1 and σ_2 are the conductivities of zone 1 and zone 2 (S m^{-1}), which can be written as

$$\sigma_1 = c_{A,1}|\mu_A|F + c_{Q,1}|\mu_Q|F = c_{A,1}F\,(|\mu_A| + |\mu_Q|)$$

$$\sigma_2 = c_{B,2}|\mu_B|F + c_{Q,2}|\mu_Q|F = c_{B,2}F\,(|\mu_A| + |\mu_Q|)$$

in which $|\mu_A|$, $|\mu_B|$, and $|\mu_Q|$ are the absolute values of the mobilities μ_A, μ_B, and μ_C.

With Equation 3 this leads to

$$c_{A,1}E_1\,(|\mu_A| + |\mu_Q|) = c_{B,2}E_2\,(|\mu_A| + |\mu_Q|)$$

or

$$C_{B,2} = C_{A,1}\frac{E_1\left(|\mu_A| + |\mu_Q|\right)}{E_2\left(|\mu_B| + |\mu_Q|\right)}$$

Replacing E_1/E_2 with μ_B/μ_A (Equation 2):

$$C_{B,2} = C_{A,1}\frac{|\mu_B|\left(|\mu_A| + |\mu_Q|\right)}{|\mu_A|\left(|\mu_B| + |\mu_Q|\right)} \tag{4}$$

As can be seen from Equation 4, the concentration of all zones is determined by the concentration of the leading electrolyte, and depends on the mobilities of the ionic species of the sample. Therefore, when a sample is injected between the leading and terminating electrolyte, and the separation is initiated, the ionic species in the sample will adapt their concentration to that of the leading electrolyte. Because of this concentrating effect, isotachophoresis is very useful as a sample stacking method for use with other separation techniques like zone electrophoresis (see Section VI).

IV. CHOICE OF ELECTROLYTE SYSTEMS

In isotachophoresis, like most other electrophoretic techniques, ionic species are separated based on their difference in effective mobility, which is defined as

$$\mu = \sum_i \alpha_i \gamma_i m_i$$

in which α_i is the degree of dissociation (mainly dependent on pK value, pH, and temperature), γ_i is a correction factor (as described by Onsager, mainly dependent on ionic concentrations), and m_i is the absolute mobility (dependent on solvation, radius, charge of the ions, and the dielectric constant and viscosity of the solvents). The use of different solvent or buffer systems gives the analyst numerous possibilities for optimization of the separation.

A. CHOICE OF LEADING ELECTROLYTE

The leading electrolyte plays an important role in the isotachophoretic separation. It determines the pH at which the separation is performed (function of the counterion) and it also determines the final concentration of the different species of the sample when the steady state is reached (function of the leading ion; Equation 4).

1. Choice of Leading Ion

When choosing a leading ion for an isotachophoretic separation, the most important criterion is that the effective mobility of the leading ion be higher than that of all species of interest in the sample at the pH at which the separation is to be performed. Species in the sample with a higher effective mobility than the leading ion will migrate zone electrophoretically in the leading electrolyte and will not be part of the isotachophoretic train formed in the steady state. These species will also not adapt their concentration to that of the leading ion.

2. Choice of pH

As a general rule, the pH of the leading electrolyte is chosen so that there is a maximum difference in effective mobilities between the different species of interest in the sample. However, the pH should not be more then two or three units away from the p*K* values of the ionic species in the sample, for two different reasons:

1. If the pH of the leading elelctrolyte is much lower then the p*K* value of the ion, the effective mobility of that ion is so low that the resulting high electric field strength in the zone can result in thermal destruction of the species.
2. Because the buffering capacity of the counterion is optimal between $pK_C - 1$ and $pK_C + 1$, a p*K* of the sample ion differing more then 3 pH units from the p*K* of the buffering counterion will bring the pH of the zone outside of the buffering capacity of the counterion and, therefore, will no longer be fixed and well defined. A dramatic difference between isotachophoresis and zone electrophoresis is the concentration of the sample ions during the separation. In zone electrophoresis the concentration of the sample ions is generally two to three orders of magnitude lower then the buffering electrolyte, which means that the p*K* of the sample ions has little effect on the pH of the system. However, in isotachophoresis, where the sample ions adapt their concentration to that of the leading ion, the concentration of the ions in their zones is on the same order of magnitude as the buffering counterions. In this case, the p*K* of the sample ion will influence the pH of a zone, resulting in a zone pH between the p*K* of the sample ion and the p*K* of the counterion.

3. Choice of Counterion

The most important function of the counterion in an isotachophoretic separation is buffering. The counterion must have good buffering capacity at the pH where the analysis is to be performed. A constant pH fixes the effective mobilities of the different species in the different zones, and thus allows the maintainance of the "steady state." The counterion also acts, as its name indicates, as a counterion to fullfill the requirement of electroneutrality.

B. CHOICE OF TERMINATING ELECTROLYTE

The effective mobility of the terminating ion has to be lower than the effective mobility of all the species of interest in the sample at the pH at which the separation is to be performed. Species in the sample with a lower effective mobility than the terminating ion will migrate zone electrophoretically in the leading electrolyte and will not be part of the isotachophoretic train formed in the steady state.

C. PURITY OF CHEMICALS

Another requirement for both the preparation of leading and terminating electrolyte is that the chemicals used are of the highest quality. Impurities in the terminating electrolyte with a higher effective mobility than the terminating ion will migrate through the terminating zone and zero, one, or more sample zones (depending on their effective mobility) and create zones of impurities at the separation boundaries.These impurity zones become elongated in time, depending on the effective mobilities, concentrations of the impurities, and the analysis time. Similarly, impurities in the leading electrolyte with a lower effective mobility than the leading ion will form impurity zones at the separation boundaries.

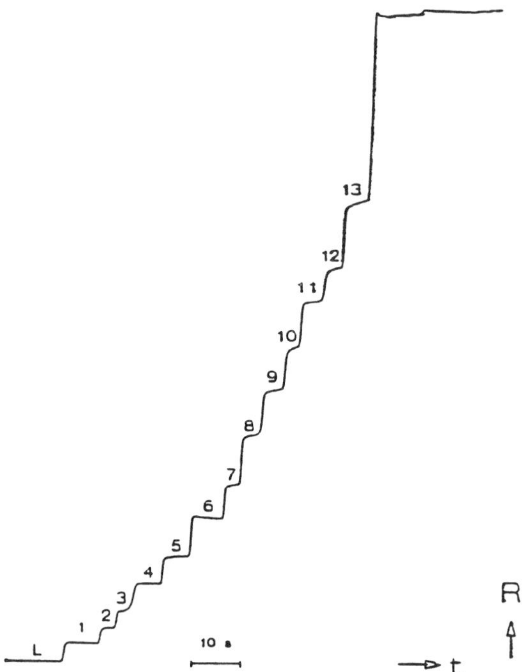

FIGURE 3. Isotachopherogram of a test mixture analyzed using the electrolyte system in Table 1. L, Leading ion; 1, sulfate; 2, chlorate; 3, chromate; 4, malonate; 5, pyrazole-3,5-dicarboxylate; 6, adipate; 7, acetate; 8, β-chloropropionate; 9, benzoate; 10, naphthalene-2-monosulfonate; 11, glutamate; 12, enanthate; 13, benzylaspartate; T, terminating ion; *t*, time; *R*, resistance. (From Wanders, B. J., Lemmens, A. A. G., Everaerts, F. M., et al., *J. Chromatogr.*, 470, 79, 1989. With permission.)

In contrast, impurities in the terminating electrolyte with lower effective mobilities than the terminating ion, and impurities in the leading electrolyte with higher effective mobilities than the leading ion will not affect the separation.

V. SOME APPLICATIONS

Figures 3 and 4 show examples of the separation power of isotachophoresis. These analyses were performed on a home-made instrument, using a conductivity detector. This detection technique is very useful in isotachophoresis, in contrast to zone electrophoresis, because of the much higher sample concentration levels in isotachophoresis.

The leading/terminating electrolyte system used in both of these separations is listed in Table 1.

VI. ISOTACHOPHORETIC SAMPLE STACKING IN CAPILLARY ELECTROPHORESIS

One of the biggest limitation for capillary electrophoresis is its relatively poor concentration detection limit (CDL), inherent when using small internal diameter capillaries. There are two obvious ways of improving the sensitivity in CE: (1) using larger inner diameter capillaries or (2) using a sample-stacking/concentration technique. The use of larger inner diameter capillaries, however, has some serious limitations, because of the increased heat production in the capillary. Other problems, like loss of resolution due to the increased radial temperature profile, limits the inner diameter of capillaries to around 100 to 150 μm for most applications. A better approach to increase the sensitivity of capillary electrophoresis is using a sample-

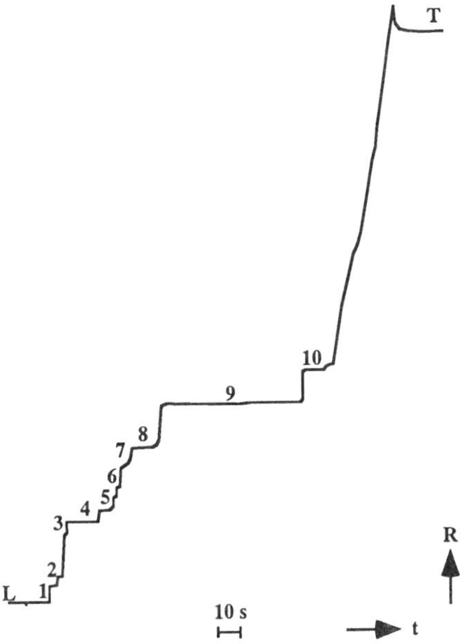

FIGURE 4. Isotachopherogram of a beer sample analyzed using the electrolyte system in Table 1. L, Leading ion; 1, sulfate; 3, formate; 4, malate; 5, citrate; 10, phosphate; 2, 6, 7, 8, and 9, not identified; T, terminating ion; *t*, time; *R*, resistance. (From Wanders, B. J., Lemmens, A. A. G., Everaerts, F. M., et al., *J. Chromatogr.*, 470, 79, 1989. With permission.)

TABLE 1
Electrolyte System for the Isotachophoretic Analysis of a Test Mixture[a] and Beer[b]

Leading ion	Chloride
Concentration	0.01 *M*
pH	6.0
Counterion	Histidine
Additive	Hydroxyethlcellulose (0.2%)
Terminating ion	Morpholinoethanesulfonic acid (MES)
Concentraiton	0.005 *M*

[a] See Figure 3.
[b] See Figure 4.

stacking or concentration technique. The traditional concentration method is the off-column preconcentration. The disadvantages of this technique are the nonspecificity (all species in the sample [ionic or not] are concentrated), quantitation problems, and its labor intensive nature. A more suitable technique is on-column sample stacking. Several techniques for on-column sample stacking have been described in the literature.

1. *Moving boundary stacking*: Moving boundary stacking is an easy and useful stacking technique based on the moving boundary principle, and can be used for samples with a lower conductivity matrix than the buffer system used for the separation.[12-16]
   When a voltage is applied over the capillary, the electric field strength in the injected sample plug will be higher than in the rest of the capillary because of the higher resistance. This discontinuous electric field strength results in a decrease in velocity when the sample ions cross the stationary sample/buffer boundary, leading to stacking of the ions. The resulting peaks will be assymetric because the focusing of the sample

species occurs only on one side of the zone and not on both sides, like the sandwiched zones in isotachophoresis. The peaks will show "heading" if they are stacked behind or "trailing" if they are stacked in front of the original sample plug. Figure 5 shows the effect of this moving boundary stacking technique on the separation of five proteins.

A problem with very low conductive sample matrices is the generation of excess heat in the sample zone, due to the high electric field. This excess heat generation can thermally degrade some or all of the sample components.

2. *Packed precolumns:* In this method, the first part of the capillary contains a packing material that will absorb/concentrate the sample when it is injected.[17,18] After injection, a solvent is introduced that will elute the compounds of interest from the packing, after which the voltage is applied to start the separation. This technique, however, shows considerable peak broadening in the elution step. Another way of eluting the absorbed sample is by electrodesorption.[19] Choosing the correct buffer system will create isotachophoretic conditions during the elution step that prevent the peak broadening described earlier.

   Both techniques require complex instrumentation and/or specialized capillaries.
3. *Column-coupled ITP-CZE:* This technique uses a coupling between two columns. The first column is used to concentrate the sample by isotachophoresis. After this step the concentrated sample is injected onto the second column and separated, using a CE buffer system.[20-24] Although this technique can attain concentration factors of $>10^4$, the technique requires complex instrumentation and, therefore, is currently not usable on commercial CE instruments.
4. *Single-column ITP-CZE:* By far the most interesting way to concentrate a sample in zone electrophoresis is by using a single-column ITP-CZE method. In this method a discontinuous buffer system is used to create isotachophoretic separation conditions at the beginning of the analysis. After a certain amount of time, the discontinuous buffer system disappears and the stacked zones are separated, using the now continuous CE buffer system.[25]

A. SAMPLE STACKING USING SINGLE-COLUMN ISOTACHOPHORESIS-CAPILLARY ZONE ELECTROPHORESIS

There are different approaches to create a discontinuous buffer system for isotachophoretic stacking at the beginning of the separation. The three methods that follow were described by Schwer et al.[25]

1. "Three Buffer" Stacking System

Figure 6 shows the system setup for a method similar to the classic stacking system used in sodium dodecyl sulfate-polyacrylamide gel electrophoresis (SDS-PAGE). In this system, an amphoteric buffer component first acts as a terminating electrolyte, but, because of a pH gradient, will eventually overtake the sample and create a continuous buffer system for the zone electrophoretic separation.

Figure 7 shows the CE separation of three basic peptides dissolved in the buffer electrolyte with and without the stacking method described above. The sample injected under stacking

FIGURE 5. Separation of five proteins: (1) lysozyme, 72 μg/ml (5×10^{-6} M); (2) cytochrome *c*, 40 μg/ml (3.3×10^{-6} *M*); (3) trypsin, 56 μg/ml (2.3×10^{-6} *M*); (4) ribonuclease A, 54 μg/ml (4×10^{-6} *M*); (5) α-chymotrypsinogen A, 56 μg/ml (2.4×10^{-6} *M*). Capillary, 27 cm × 75 μm (20 cm to detector); voltage, 8.2 kV; current, 22 μA; UV detection at 214 nm. BGE, 0.02 *M* triethylamine-acetic acid, pH 4.4. (a) Sample dissolved in water; sample volume, 38 nl. (b) Sample from (a) diluted fourfold in running buffer; sample volume, 38 nl. (c) Sample from (b) injected electrokinetically; voltage, 8 kV; injection time, 20 s. (From Foret, F., Szoko, E., and Karger, B. L., *J. Chromatogr.*, 608, 3, 1992. With permission.)

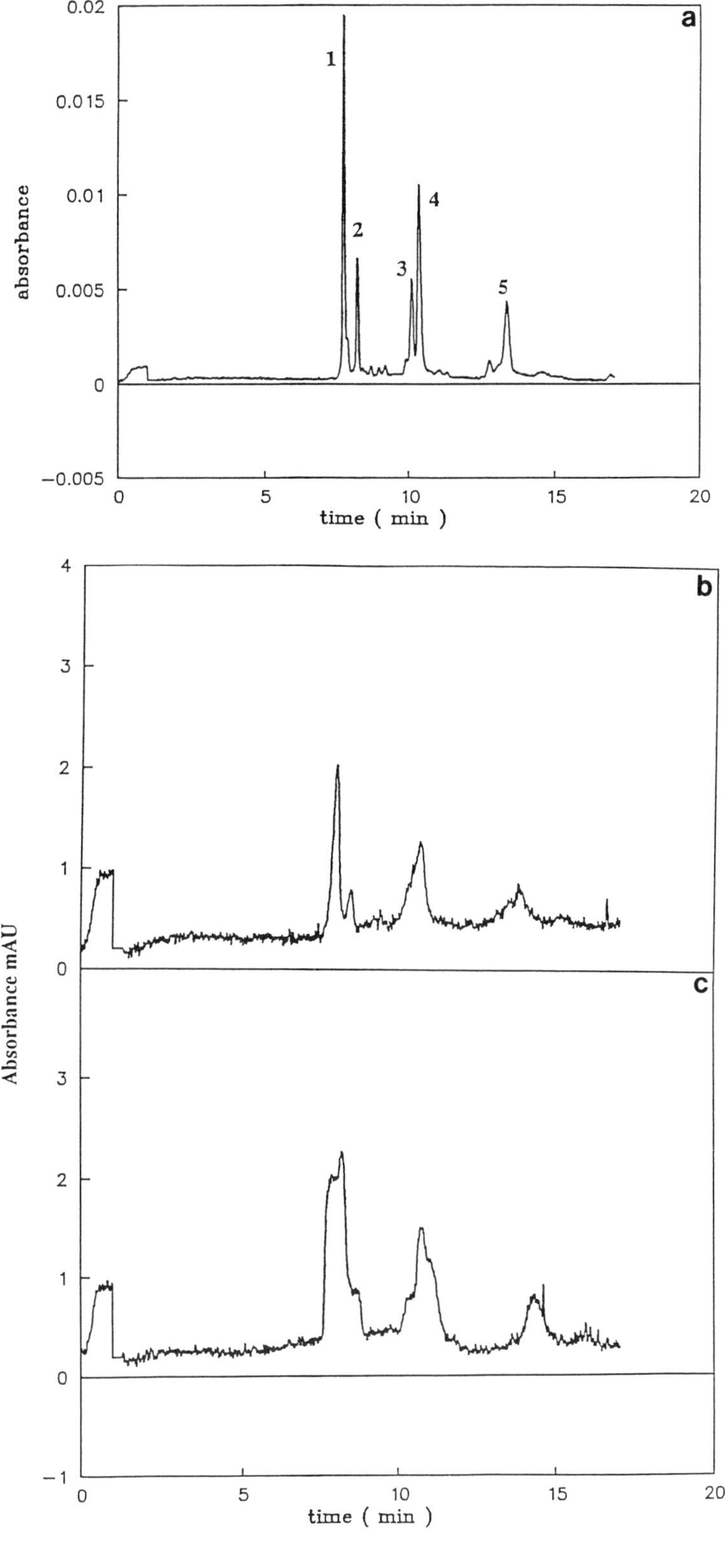

a
absorbance
0.02
0.015
0.01
0.005
0
−0.005
1
2
3
4
5
0
5
10
15
20
time (min)
b
c
Absorbance mAU
4
3
2
1
0
3
2
1
0
−1
0
5
10
15
20
time (min)

FIGURE 5.

TE (=BGE)	S	SE	LE

FIGURE 6. System setup for the "three buffer" stacking system. TE, Terminating electrolyte; BGE, background electrolyte; S, sample; SE, stacking electrolyte; LE, leading electrolyte.

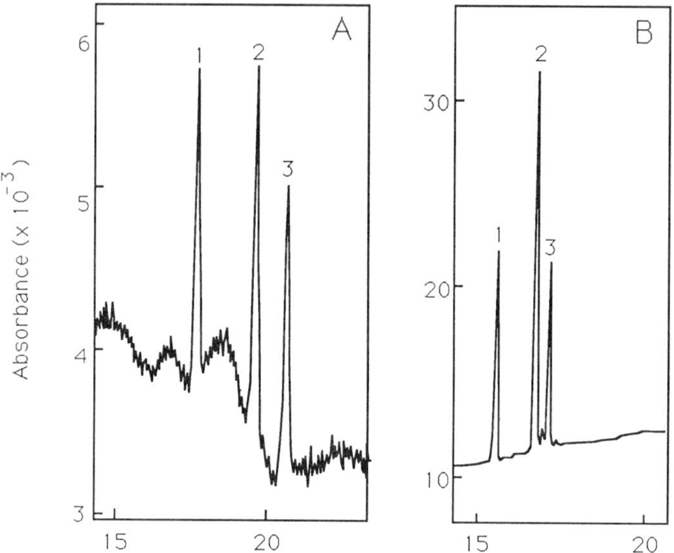

FIGURE 7. Separation of three basic peptides (1, Leu-Trp-Met-Arg; 2, Leu-Trp-Met-Arg-Phe; 3, Leu-Trp-Met-Arg-Phe-Ala): (A) without stacking; (B) using the "three buffer" stacking system. The peptides were dissolved in EACA buffer at a concentration of (A) 10 ng/ml and (B) 1 μg/ml. Injection time: (A) 3 s and (B) 30 s. Electrolyte system setup (Figure 6): LE, 0.05 *M* Tris-citrate, pH 4.8; SE, 0.05 *M* Tris-citrate, pH 6.5 (750 nl); TE, 0.05 *M* EACA-citrate, pH 4.8; S, sample solution (750 nl). (From Schwer, C. and Lottspeich, F., *J. Chromatogr.*, 623, 345, 1992. With permission.)

conditions was diluted 10-fold, but the injection time was ten times longer (injected volumes, 75 vs. 750 nl). In this separation, tris(hydroxyethyl)aminomethane (Tris) acts like the leading and stacking electrolyte (same concentration, but different pH), and the ε-aminocaproic acid (EACA) acts as a terminating electrolyte at high pH (>6) where its mobility is very low. The sample, injected between the stacking electrolyte and the terminating electrolyte, will initially separate under isotachophoretic conditions, and the different components of the sample will concentrate and adapt their concentration to that of the leading electrolyte. However, as soon as EACA enters a region of lower pH, it will gain a higher positive charge, and thus mobility, and overtake the sample zone. The different components in the sample will now be separated by zone electrophoresis in the continuous EACA buffer system. Figure 7 clearly shows the increase in sensitivity and resolution for the stacked sample, despite the large injection volume of 750 nl.

The disadvantage of this "three buffer" stacking system is the sometimes difficult search for a suitable terminating electrolyte. The mobility of the terminating electrolyte has to be higher at one pH and lower at another pH than the mobility of all the ions of interest (anions or cations) in the sample.

2. "Two Buffer Stacking" System

Another method for creating istoachophoretic stacking conditions at the beginning of a zone electrophoretic separation is by using the background electrolyte (BGE) as a leading or terminating electrolyte, as shown in Figure 8. By injecting a leading electrolyte plug in front

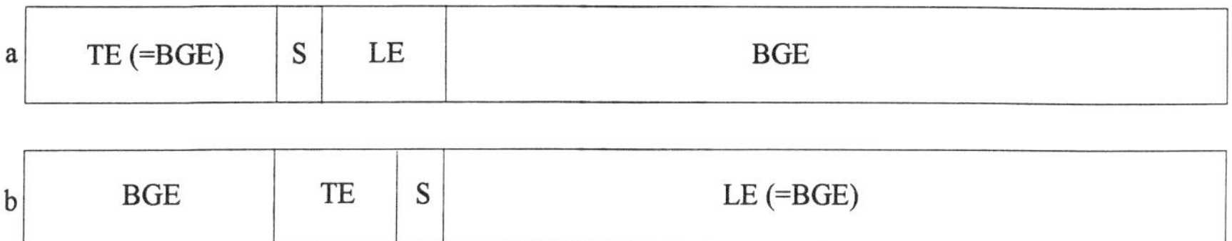

FIGURE 8. System setup for the "two buffer" stacking system. TE, Terminating electrolyte; BGE, background electrolyte; S, sample; LE, leading electrolyte.

of the sample plug (in case the BGE is used as the terminating electrolyte; Figure 8a) or a terminating electrolyte plug behind the sample plug (in case the BGE is used as the leading electrolyte; Figure 8b), isotachophoretic stacking conditions can be obtained. The sample is sandwiched between a leading and terminating electrolyte and will adapt its concentration to that of the leading electrolyte.

In the case where a leading electrolyte plug is injected before the sample plug, this leading electrolyte is preceded by the terminating electrolyte (= BGE), and will migrate zone electrophoretically in the BGE away from the sample. As soon as the leading zone separates from the istachophoretic sample train, the sample will also be separated by zone electrophoresis in the BGE. Instead of injecting a leading electrolyte plug before the sample zone, the sample can also be supplemented by a leading ion.[25] This option was confirmed mathematically.[26]

In the case where a terminating electrolyte plug is injected behind the sample plug, this terminating electrolyte is followed by the leading electrolyte (= BGE), and will migrate zone electropherotically in the BGE away from the sample, again ending the isotachophoretic stacking conditions. Instead of injecting the terminating electrolyte plug, the same effect can be obtained by replacing the inlet vial, after injection of the sample, by a vial containing the terminating electrolyte. After the isotachophoretic stacking is complete, the vial is replaced by a vial containing the BGE to start the actual zone electrophoretic separation.[24]

Figure 9 shows the separation of three basic peptides with and without a leading electrolyte injected in front of the sample zone (BGE acts like terminating electrolyte).

Figure 10 shows the separation of the same five proteins as shown in Figure 5, but now with the utilization of isotachophoretic stacking. by replacing the inlet vial by a vial containing the terminating electrolyte, before switching to the BGE vial.

A similar stacking effect can also be observed if a sample containing a high-concentration salt matrix is separated in a buffer of low mobility. The salt from the sample will now act as a leading electrolyte, resulting in isotachophoretic stacking conditions.

3. "One Buffer" Stacking System

A disadvantage of the stacking systems described above is that they do not work at extreme pH values. If a certain application requires an extreme pH to obtain an optimal separation, the following stacking method can be useful (Figure 11). In this stacking mechanism the injected sample is sandwiched between an OH^- and H^+ zone. When the voltage is applied the OH^- and H^+ zone will migrate toward each other, forming a region of low conductivity in the middle. This will lead to moving boundary stacking conditions, as described above. In special cases where the OH^- or H^+ ions can act like a terminator (H^+ can act as a terminating ion in the approximate pH range 4.0 to 5.2, and OH^- in the approximate pH range 9.0 to 10.0[1]), isotachophoretic separation conditions will further concentrate the sample.

When using this stacking mechanism, care has to be taken not to create OH^- and H^+ zones that are too long, as the generation of excess heat in the region of low conductivity may cause problems (see Moving boundary stacking, above).

Figure 12 shows the separation of four standard peptides (two basic, one neutral, and one acidic) in a phosphate buffer, pH 2.5, with and without the described stacking technique.

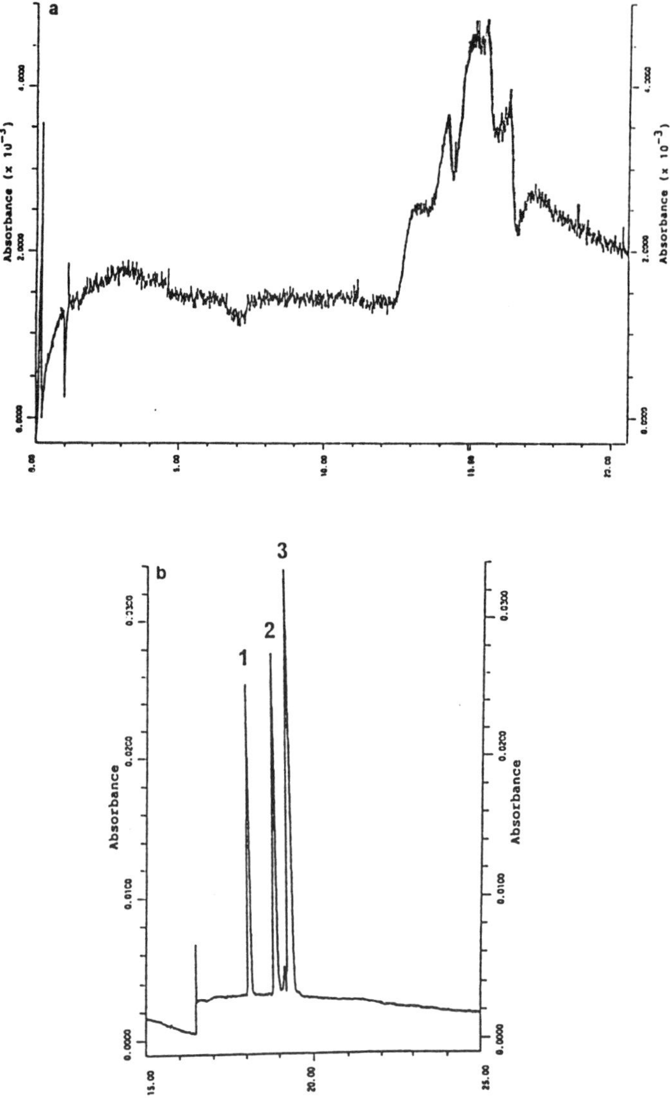

FIGURE 9. Separation of three basic peptides (for peak identification, see Figure 7): (a) without stacking; (b) using the "two buffer" stacking system. The peptides were dissolved in δ-alanine buffer at a concentration of 1 μg/ml. Injection time: 20 s. Electrolyte system setup (Figure 8): LE, 0.05 *M* Tris-citrate, pH 5.2 (500 nl); TE, 0.05 *M* δ-alanine-citrate, pH 4.8; S, sample solution (500 nl). (From Schwer, C. and Lottspeich, F., *J. Chromatogr.*, 623, 345, 1992. With permission.)

B. SUMMARY

In general, the moving boundary and single-column ITP-CZE are the two most useful sample-stacking techniques for capillary zone electrophoresis. Typical concentration factors are 2- to 10-fold for the moving boundary stacking and 10- to 1000-fold for the ITP-CZE stacking method. In addition to the already mentioned gain in sensitivity, the ability to stack samples and dramatically increase the sample load makes micropreparative work, like fraction collection, possible in small inner diameter capillaries. A potential problem, however, in using fraction collection in combination with a stacking technique is that the migration time is now also dependent on the length of the zone injected (which can be 20 to 30 cm long). The reason for this is that, because of the focusing during sample stacking, the effective length to the

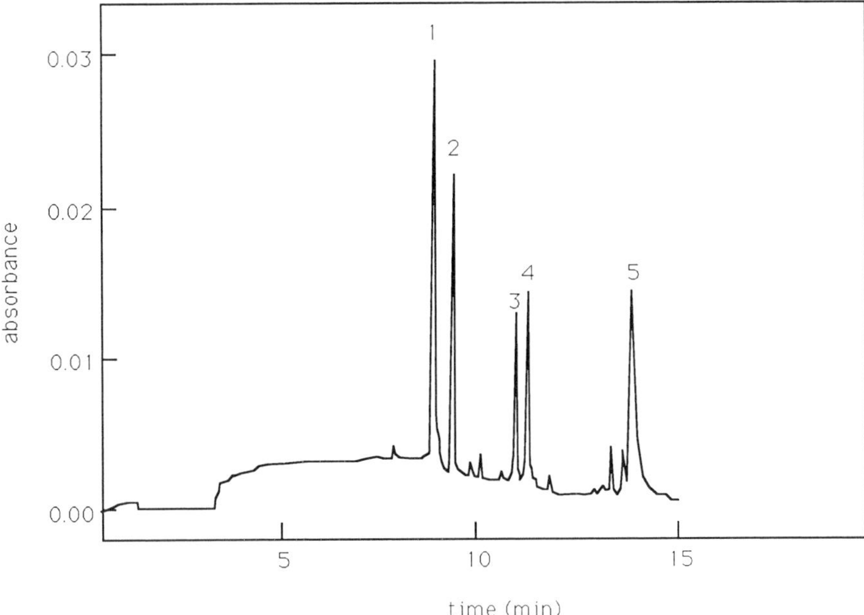

FIGURE 10. Separation of five proteins, using the "two buffer" stacking system (see text): (1) lysozyme, 13.5 μg/ml (1×10^{-6} *M*); (2) cytochrome *c*, 15 μg/ml (1.2×10^{-6} *M*); (3) trypsin, 7 μg/ml (3×10^{-7} *M*); (4) ribonuclease A, 7 μg/ml (5.2×10^{-7} *M*); (5) α-chymotrypsinogen A, 7 μg/ml (3×10^{-7} *M*). Capillary 47 cm × 75 μm (40 cm to detector); voltage 20 kV; current, 22 μA; UV detection at 214 nm. Sample dissolved in leading electrolyte; sample volume, 1 μl (57% of the effective length of the capillary); BGE (LE), 0.02 *M* triethylamine-acetic acid, pH 4.4; TE, 0.01 *M* acetic acid; focusing time, 2.5 min. (From Foret, F., Szoko, E., and Karger, B. L., *J. Chromatogr.*, 608, 3, 1992. With permission.)

BGE	H^+	S	OH^-	BGE

FIGURE 11. System setup for the "one buffer" stacking system. BGE, Background electrolyte; S, sample.

detector is dependent on the length of the injected zone. The average migration velocity of a zone, calculated from the migration time and length to detector, is not necessarily equal to the actual migration velocity when the zone passes through the detector. This can potentially lead to errors in calculating the times when the zones reach the end of the capillary. A dual-detector system can solve this problem by supplying the data needed to calculate the actual velocity of the sample.

VII. CONCLUSION

The relative simplicity of the method, together with the tremendous gain in sensitivity, make isotachophoretic stacking an important tool to increase the applicability of capillary zone electrophoresis. This is especially true for the analysis of peptides and proteins, where some spectacular improvements in sensitivity have been demonstrated.[24,25]

The usefulness of isotachophoresis as an analytical tool by itself among the currently available CE instruments is strongly dependent on the ability to suppress completely the electroosmotic flow (EOF).[27] With the coatings presently available to suppress the EOF and

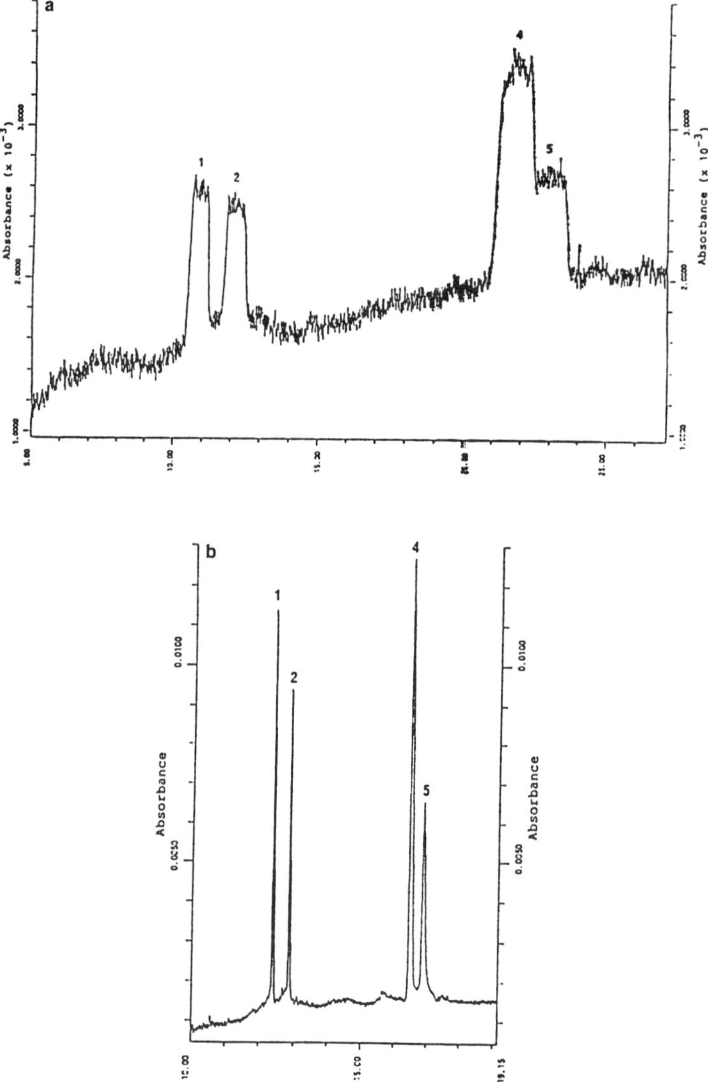

FIGURE 12. Separation of four standard peptides (1, Leu-Trp-Met-Arg; 2, Leu-Trp-Met-Arg-Phe; 4, Tyr-Gly-Gly-Phe-Leu; 5, Val-Leu-Ser-Glu-Gly) in a phosphate buffer of pH 2.5 (a) without stacking; (b) using the "one buffer" stacking system. The peptides were dissolved in the running buffer at a concentration of 2 μg/ml. Injection time: 20 s. Electrolyte system setup (Figure 11): BGE, 0.02 *M* phosphate, pH 2.5; H^+, 0.1 *M* phosphoric acid (650 nl); OH, 0.1 *M* sodium hydroxide (300 nl); S, sample solution (500 nl). (From Schwer, C. and Lottspeich, F., *J. Chromatogr.*, 623, 345, 1992. With permission.)

the availability of algorithms to analyze the staircase data,[28] ITP still has the potential to become a valuable tool in the analysis of small ions.

REFERENCES

1. **Everaerts, F. M., Beckers, J. L., and Verheggen, Th. E. P. M.,** *Isotachophoresis: Theory, Instrumentation and Applications*, Elsevier, Amsterdam, 1976, chap. 1.
2. **Kendall, J. and Crittenden, E. D.,** The separation of isotopes, *Proc. Natl. Acad. Sci. U.S.A.*, 9, 75, 1923.

3. **Kohlrausch, F.,** Über concentrations-verschiebungen durch electrolyse in inneren von lösungen und lösungsgemischen, *Ann. Physik Chemie*, 62, 209, 1897.
4. **Martin, A. J. P.,** Unpublished results.
5. **Longsworth, L. G.,** Moving boundary separation of salt mixtures, *Natl. Bur. Stand.*, 524, 59, 1953.
6. **Everaerts, F. M.,** Graduation report, University of Technology, Eindhoven, The Netherlands, 1964.
7. **Everaerts, F. M.,** Ph.D. thesis, University of Technology, Eindhoven, The Netherlands, 1968.
8. **Martin, A. J. P. and Everaerts, F. M.,** Displacement electrophoresis, *Anal. Chim. Acta*, 38, 233, 1967.
9. **Haglund, H.,** Isotachophoresis, a principle for analytical and preparative separation of substances such as proteins, peptides, nucleotides, weak acids, metals, *Science Tools*, 17, 21, 1970.
10. **Arlinger, L. and Routs, R. J.,** Boundary sharpness in capillary-tube isotachophoresis demonstrated by UV-detection, *Science Tools*, 17, 21, 1970.
11. **Verheggen, Th. E. P. M., Van ballegooijen, E. C., Massen, C. H., and Everaerts, F. M.,** Detection electrodes for electrophoresis, *J. Chromatogr.*, 64, 185, 1972.
12. **Mikkers, F. E. P., Everaerts, F. M., and Verheggen, Th. E. P. M.,** Concentration distributions in free zone electrophoresis, *J. Chromatogr.*, 169, 1, 1979.
13. **Moring, S. E., Colburn, J. C., Grossman, P. D., and Lauer, H. H.,** Analytical aspects of an automated capillary electrophoresis system, *LC-GC*, 8, 34, 1989.
14. **Burgi, D. S. and Chien, R. L.,** Optimization of sample stacking for high-performance capillary electrophoresis, *Anal. Chem.*, 63, 2042, 1991.
15. **Chien, R. L. and Burgi, D. S.,** Sample stacking of an extremely large injection volume in high performance capillary electrophoresis, *Anal. Chem.*, 64, 1046, 1992.
16. **Beckers, J. L. and Ackermans, M. T.,** Effect of sample stacking on resolution, calibration graphs and pH in capillary zone electrophoresis, *J. Chromatogr.*, 629, 371, 1993.
17. **Guzman, N. A., Trebilcock, M. A., and Advis, J. P.,** The use of a concentration step to collect urinary components separated by capillary electrophoresis and further characterization of collected analytes by mass spectrometry, *J. Liq. Chromatogr.*, 14, 997, 1991.
18. **Debets, A. J. J., Mazereeuw, M., Voogt, W. H., Van iperen, D. J., Lingeman, H., Hupe, K. P., and Brinkman, U. A. T.,** Switching valve with internal micro precolumn for online sample enrichment in capillary zone electrophoresis, *J. Chromatogr.*, 608, 151, 1992.
19. **Kasicka, V. and Prusik, Z.,** Isotachophoretic electrodesorption of proteins from an affinity adsorbent on a microscale, *J. Chromatogr.*, 273, 117, 1983.
20. **Kaniansky, D. and Marak, J.,** On-line coupling of capillary isotachophoresis with capillary zone electrophoresis, *J. Chromatogr.*, 498, 191, 1990.
21. **Dolnik, V., Cobb, K. A., and Novotny, M. V.,** Capillary zone electrophoresis of dilute samples with isotachophoretic preconcentration, *J. Microcolumn Sep.*, 2, 127, 1990.
22. **Stegehuis, D. S., Irth, H., Tjaden, U. R., and Van der Greef, J.,** Isotachophoresis as an on-line concentration pretreatment technique in capillary electrophoresis, *J. Chromatogr.*, 538, 393, 1991.
23. **Krivankova, L., Foret, F., and Bocek, P.,** Determination of halofuginone in feedstuffs by the combination of capillary isotachophoresis and capillary zone electrophoresis in a column-switching system, *J. Chromatogr.*, 545, 307, 1991.
24. **Foret, F., Szoko, E., and Karger, B. L.,** On-column transient and coupled column isotachophoretic preconcentration of protein samples in capillary zone electrophoresis, *J. Chromatogr.*, 608, 3, 1992.
25. **Schwer, C. and Lottspeich, F.,** Analytical and micropreparative separation of peptides by capillary zone electrophoresis using discontinuous buffer systems, *J. Chromatogr.*, 623, 345, 1992.
26. **Beckers, J. L. and Everaerts, F. M.,** Isotachophoresis with two leading ions and migration behaviour in capillary zone electrophoresis, *J. Chromatogr.*, 508, 19, 1990.
27. **Everaerts, F. M., Beckers, J. L., and Verheggen, Th. P. E. M.,** *Isotachophoresis: Theory, Instrumentation and Applications*, Elsevier, Amsterdam, 1976, chap. 9.
28. **Ackermans, M. T., Everaerts, F. M., and Beckers, J. L.,** Isotachophoresis in open systems: problems in quantitative analysis, *J. Chromatogr.*, 545, 283, 1991.
29. **Wanders, B. J., Lemmens, A. A. G., Everaerts, F. M., and Gladdines, M. M.,** Data acquisition in capillary isotachophoresis, *J. Chromatogr.*, 470, 79, 1989.

Chapter 6

SEPARATION OF DNA BY CAPILLARY ELECTROPHORESIS

Andras Guttman

TABLE OF CONTENTS

0-8493-8690-X/94/$0.00+$.50

I. INTRODUCTION

There is a great deal of interest in analytical biochemistry and molecular biology in the separation and identification of single- and double-stranded, and both short- and long-chain length DNA molecules.[1,2] For relatively short single-stranded DNA molecules (i.e., oligonucleotides) there is a necessity for separation at the level of a single base difference (for DNA sequencing)[3] or even for identical chain length with different sequences (identification of primers, probes, and antisense DNA molecules).[4] For double-stranded DNA molecules there is a significant interest in analyzing and identifying DNA molecules in the form of restriction fragments or polymerase chain reaction (PCR) products.

To solve these separation problems, the application of slab gel electrophoresis (polyacrylamide and agarose) rapidly became very popular in the 1960s and is still a widely used technique for DNA separation. Several samples can be run simultaneously on slab gels but the gel can be used only once. In addition to being time consuming, on-line detection is not possible and the analysis is not quantitative. These problems can easily be resolved using capillary electrophoresis.

After early attempts with open tubular capillary electrophoresis separation of single-stranded oligonucleotides and double-stranded DNA molecules,[5,6] it became evident that a sieving agent was required in order to improve resolving power. One of the major problems with the capillary electrophoretic separation of either nucleotides or DNA is their constant linear charge density. For example, a ten base pair and hundred base pair DNA molecule both have the same charge-to-mass ratio. As a result, a sieving medium is needed in order to separate molecules, not based on charge-to-mass ratio, but based on size. The sieving medium within the capillary can be either a high-viscosity gel or a low-viscosity polymer network solution. In capillary gel electrophoresis, cross-linked or non-cross-linked sieving matrices have been employed.[7-9] The cross-linked gels, i.e., chemical gels, have a well-defined pore structure and size. In contrast, non-cross-linked, or so-called physical gels, have a dynamic pore structure providing the non-cross-linked linear polymer networks with much higher flexibility compared to the cross-linked gels. With physical gels, one can operate at high temperatures (up to 50 to 70°C) and apply extremely high field strengths (up to 10^3 V/cm) without any damage to the linear polymer network formulations.[10] It is important to note that cross-linked (chemical) gels cannot withstand such extreme conditions.[11] The other main advantage of the linear polymer network system is that it can be easily replaced by simply rinsing the polymer network matrix through the capillary column by pressure or vacuum. Therefore, if the column becomes contaminated, the sieving matrix can be readily replaced, thereby extending the lifetime of the system. Moreover, employing the "replaceable sieving matrix" concept, the possibility exists for use of the pressure injection mode, in comparison to the cross-linked gels, where the electrokinetic injection mode is the only possibility.[12] It is important to note that, in addition to convenience, pressure injection permits more precise and quantitative analysis.

II. SPECIFIC APPLICATIONS

Capillary gel electrophoretic analysis of DNA molecules can be accomplished on any commercially available or home-made capillary electrophoresis (CE) system. When gel/polymer network-filled coated capillaries are used, the power supply of the CE system must be in reverse polarity mode, i.e., with the cathode on the injection side and the anode on the detection side. The separation is monitored on column at 254 to 260 nm for the DNA samples. If fluorescent detection is employed, the excitation/emission wavelength obviously depends on the dye used. The temperature of the gel/polymer network-filled capillary column is

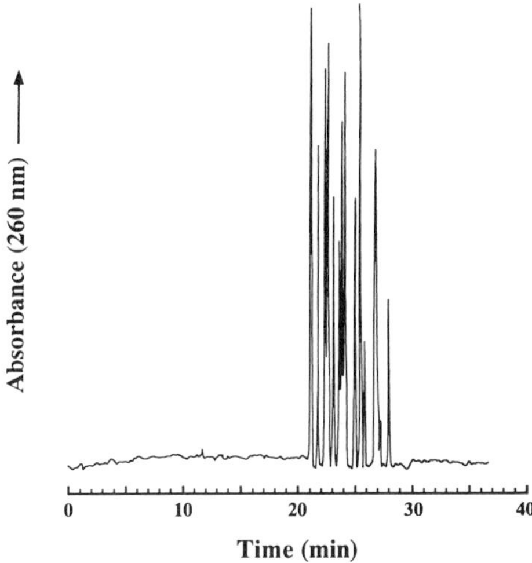

FIGURE 1. Separation of a mixture of 18 deoxyoligonucleotides, each with 8 bases. Buffer: 20 m*M* Tris, 5 m*M* Na_2HPO_4, 50 m*M* SDS, 7 *M* urea, and 3 m*M* Zn^{2+}; capillary length, 85 cm; E = 250 V/cm. (From Cohen, A. S., Terabe, S., Smith, J. A., and Karger, B. L., *Anal. Chem.*, 59, 1021, 1987. With permission.)

suggested to be maintained at constant temperature during the experiments (± 0.1°C). This is of particular concern at high field strengths, where it is recommended that a temperature controlling (cooling) system, such as a Peltier device,[13] be used in order to obtain results with high reproducibility. Use of a data-handling system is highly recommended for data acquisition and for storage of the electropherograms.

A. OPEN TUBULAR METHODS

Although in most laboratories gel or polymer network-filled capillary columns are used for DNA separations, application of open tubular methods (usually micellar electrokinetic capillary chromatography [MECC]) should also be described. The power of manipulating separation by the addition of complexing agents to the buffer system has been shown by Cohen et al.[14] Surfactants, such as sodium dodecylsulfate (SDS), above their critical micelle concentration (for SDS, > 8 m*M*) form micelles in which the interior is hydrophobic and the exterior is hydrophilic: in the case of SDS, the micelles are negatively charged. In this instance positively charged species can form ion pairs with negatively charged SDS micelles, causing changes in the migration/retention behavior. In addition, the chemistry of the surface of the micelles can also be manipulated.[15,16] This is discussed in detail in Chapter 3.

1. Metal Additives for Single-Stranded Oligonucleotide Separation

Metal ions can be electrostatically attracted to the micelle surface and are capable of complexing solutes. In Figure 1 a separation of 14 of 18 deoxyoligonucleotide octamers of varying sequence has been achieved, using a buffer system of 20 m*M* Tris, 5 m*M* Na_2HPO_4, 50 m*M* SDS, 7 *M* urea, and 3 m*M* Zn(II) in an untreated fused silica capillary column. Differential metal complexation of the deoxyoligonucleotides plays a significant role in the separation mechanism.[14]

2. Surfactant Additives for Double-Stranded DNA Separation

Several reports have recently appeared in the literature describing the employment of surfactant additives for separation of double-stranded DNA molecules, such as restriction fragments, by open tubular capillary electrophoresis.[5,6] Figure 2 shows a separation of a lambda DNA-*Hin*dIII/ϕX174 DNA *Hae*III mixture containing at least 18 fragments, from

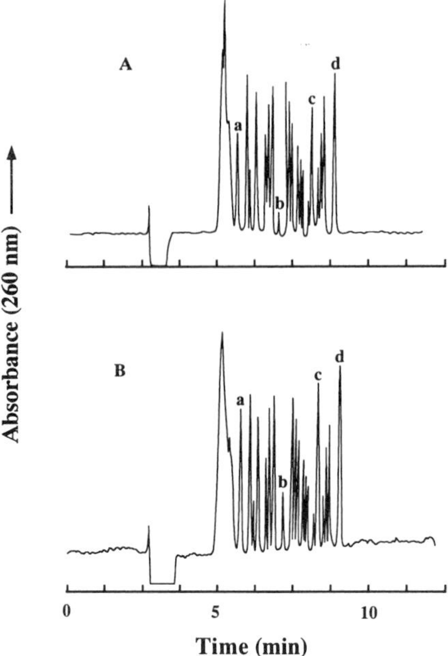

FIGURE 2. (A) Separation of DRIgest (lambda DNA-*Hin*dIII/ϕX174 DNA *Hae*III). Sample heated for 20 min at 60°C, injected hot by siphoning (3 to 4 nl). (B) Separation of DRIgest III sample spiked with four slab gel electrophoretically purified fragments: 72 bp (a), 564 bp (b), 4362 bp (c), 23,130 bp (d). Conditions in (A) and (B) buffer, 0.1 *M* Tris-borate, pH 8.1, 2.5 m*M* EDTA, 0.1% SDS, 7 *M* urea; capillary length, 30 cm; E = 500 V/cm. (From Cohen, A. S., Najarian, D., Smith, J. A., and Karger, B. L., *J. Chromatogr.*, 458, 323, 1988. With permission.)

72 base pairs (bp) up to 23,130 bp,[6] using a buffer system of 0.1 *M* Tris-borate (pH 8.1), 7 *M* urea, 2.5 m*M* ethylenediaminetetraacetic acid (EDTA), and 0.1% SDS. The samples are heated prior to injection to attain maximum resolution. Another approach to the separation of DNA restriction fragments is to add cationic surfactants to the buffer system.[5] In this case electrokinetic chromatography with micelles and ion pairing are both utilized for the separation.

B. POLYACRYLAMIDE GEL-FILLED CAPILLARY COLUMNS

As mentioned in the introduction, two types of gels commonly used in the separation of DNA molecules by capillary polyacrylamide gel electrophoresis are the cross-linked (chemical gel) and the non-cross-linked, linear polyacrylamide (physical gel). For the separation of single- and double-stranded DNA molecules, the polymerization of the cross-linked and linear (non-cross-linked) polyacrylamides is accomplished in fused silica capillary tubing (Polymicro Technologies, Inc., Phoenix, AZ).[9] For stabilization, the cross-linked and high-viscosity linear polyacrylamide gel is covalently bound to the wall of the column by means of a bifunctional agent, (3-methacryloxypropyl)-trimethoxysilane (Petrarch Systems, Bristol, PA).[17] When low-viscosity polyacrylamide solutions are used, the capillary wall should be coated, usually with linear polyacrylamide (up to 100–200 min).[17,18]

Cross-linked polyacrylamide gel is used mainly for single-stranded oligonucleotide separations of up to 200 bases, usually under denaturing conditions (7 to 9 *M* urea).[7,18,19] The recommended concentration of this gel is 3%T with 5%C and the column length depends on the resolving power required (effective length can be varied between 20 and 150 cm).[19] It should be mentioned that longer columns give higher resolution at the expense of longer separation time.

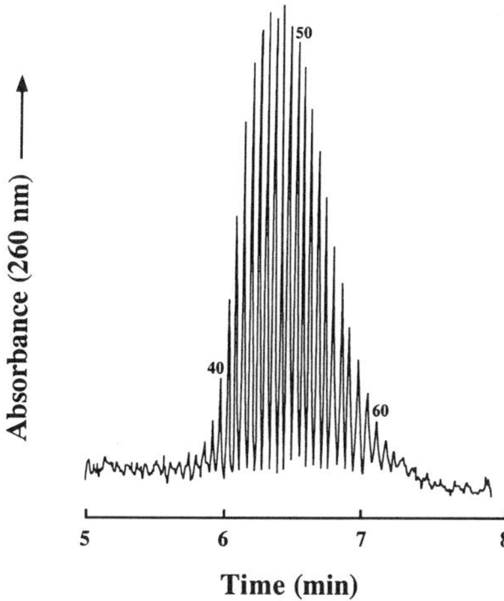

FIGURE 3. Separation of a deoxypolyadenylic acid mixture, $p(dA)_{40-60}$. The effective length of the polyacrylamide gel-filled capillary column (7.5%T/3.3%C) was 15 cm; E = 350 V/cm. Buffer: 100 m*M* Tris, 250 m*M* boric acid, pH 8.3, 7 *M* urea. (From Guttman, A., Paulus, A., Cohen, A. S., Karger, B. L., Rodriguez, H., and Hancock, W. S., in *Electrophoresis '88,* Schafer-Nielsen, C., Ed., VCH, Weinheim, Germany, 1988, p. 151. With permission.)

A second gel type is a low-viscosity linear polyacrylamide that is not covalently bound to the capillary wall. This is an attractive approach since it permits replacement of the gel–buffer system in the capillary (i.e., replaceable gel) by either the automated rinse mode of an electrophoretic apparatus or by means of a gas-tight syringe.[9,12,20-22] In this case, non-cross-linked polyacrylamide is used at low concentrations (<6%). For the best polymerization reproducibility it is advisable to prepare a higher concentration gel (9 to 12%) that can be diluted to the appropriate concentration prior to use.

Two types of buffer systems are commonly used in the separation of DNA molecules by gel electrophoresis: denaturing and nondenaturing systems. Denaturing polyacrylamide gel-filled capillary columns are utilized mainly for size separation of relatively short (up to several hundred bases) single-stranded DNA molecules, such as DNA primer and probe analysis, and in DNA sequencing (Figure 3). The most commonly used denaturing agent is urea, but several authors have recommended the use of other denaturants such as formamide.[23] As an alternative to self-made polyacrylamide gel columns, several companies offer commercial products for oligonucleotide separation (Beckman, Applied Biosystems, J&W). Figure 4 shows such an example, using a commercially available denaturing polyacrylamide gel-filled capillary column containing 7 *M* urea (eCAP ss100; Beckman Instruments, Inc., Fullerton, CA) for the purity check of a synthesized single-stranded oligodeoxyribonucleotide (102-mer). In addition to the peak of interest (the 102-mer), the electropherogram also shows full separation of all of the failure sequences. Figure 4 illustrates the very high resolving power of this technique (565,000 theoretical plates/m) as well as the rapid separation, both of which are important in molecular biology. These gels also have a micropreparative capability.[24] It is possible to collect any peak in Figure 3 in sufficient amount and purity to perform microsequencing for identification of possible failures of the oligonucleotide synthesis (Figure 5).

Rapid separations obtained with these gels suggest the possibility of developing a capillary gel electrophoresis-based automated DNA sequencer (see Chapter 14 for detailed discussion). Fluorescently labeled DNA fragments generated in enzymatic sequencing reactions are rapidly separated by capillary gel electrophoresis and detected at attomole levels within the gel-filled capillary. The application of this technology to automated DNA sequence analysis may permit the development of a second generation of automated sequencers capable of efficient and cost-effective sequence analysis on the genomic scale.[25-27]

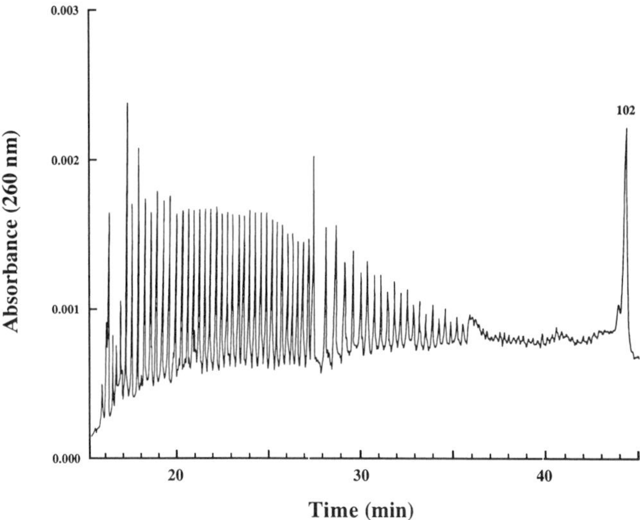

FIGURE 4. Denaturing capillary polyacrylamide gel electrophoresis separation of an in-house synthesized oligomer (102-mer). The Beckman eCAP U100P capillary gel electrophoresis kit was used for the separation, with a capillary length of 57 cm; E = 300 V/cm. (Courtesy of Beckman Instruments, Inc., Fullerton, CA.)

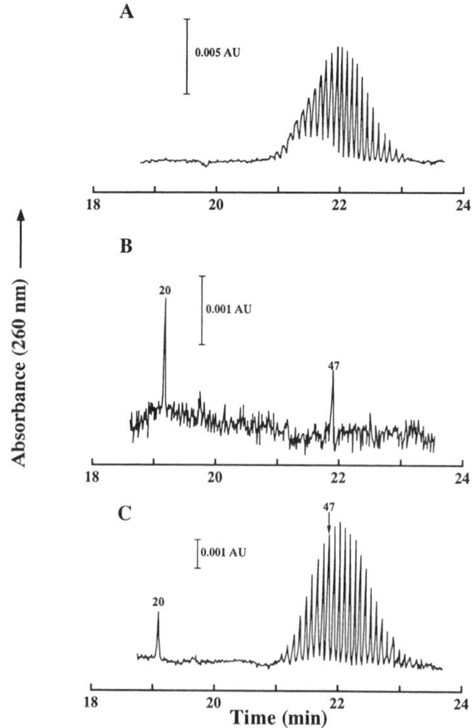

FIGURE 5. (A) Micropreparative high-performance capillary gel electrophoresis separation of the polydeoxyadenylic acid test mixture, $p(dA)_{40-60}$, (B) analytical run of the isolated $p(dA)_{47}$ spiked with $p(dA)_{20}$, and (C) analytical run of $p(dA)_{40-60}$ spiked with $p(dA)_{20}$. Capillary dimensions were 80 cm × 0.075 mm I.D. (effective length, 60 cm). Buffer: 0.1 *M* Tris, 0.25 *M* borate, 7 *M* urea, pH 8.3; E = 300 V/cm. Polyacrylamide gel: 3%T/5%T. (From Guttman, A., Cohen, A. S., Heiger, D. N., and Karger, B. L., *Anal. Chem.*, 62, 137, 1990. With permission.)

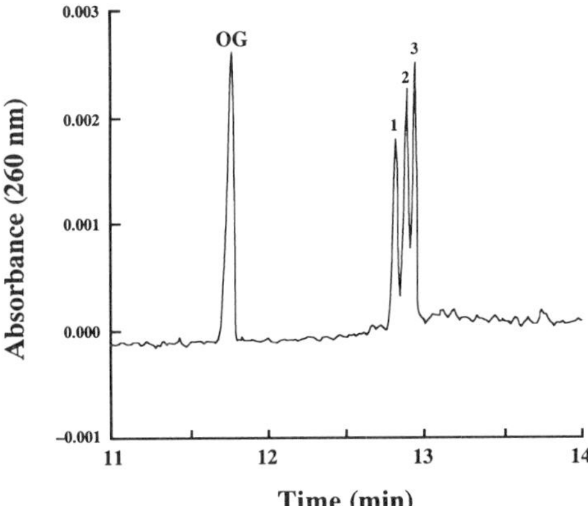

FIGURE 6. Nondenaturing capillary polyacrylamide gel electrophoresis (physical gel) separation of a human K-*ras* oncogene probe mixture (19-mers). Peaks: 1, dGTTGGAGCT-G-GTGGCGTAG; 2, dGTTGGAGCT-C-GTGGCGTAG; 3, dGTTGGAGCT-T-GTGGCGTAG. Capillary length was 47 cm; applied electric field, 400 V/cm. (From Guttman, A., Nelson, R. J., and Cooke, N., *J. Chromatogr.*, 593, 297, 1992. With permission.)

Nondenaturing polyacrylamide gels are utilized when separation is based on the shape, size, and charge of the molecules. Since DNA molecules have the same mass-to-charge ratio, regardless of chain length, nondenaturing conditions should be utilized for secondary structure difference recognition (restriction fragment length polymorphism, point mutation studies, etc.). Figure 6 shows a nondenaturing polyacrylamide capillary gel electrophoretic separation of three 19-mers (human K-*ras* oncogene probes). Orange-G was used as internal standard in the separation in order to increase precision of migration time measurement through standardization.[28,29] The three 19-mers in Figure 6 differ only by a single change in sequence in the middle (position 10) of the chains. This difference causes some secondary structure differences in the molecules. It is believed that these gel systems are able to separate these species based on those subtle differences.

The same type of nondenaturing media can be used at low concentration, without cross-linking, for the separation of longer chain double-stranded DNA molecules, such as restriction fragments. Figure 7A shows a separation of a ϕX174 DNA *Hae*III digest restriction fragment mixture, using a replaceable polyacrylamide gel matrix. It is important to point out that baseline separation of the two closest size fragments (271 and 281 bp) is achieved. This pair had proven difficult to separate by other types of sieving buffer systems, such as cellulose derivatives.[30] The separation of double-stranded DNA molecules can be improved by using special additives. The effect of an intercalator additive, ethidium bromide, on the separation of the previous test mixture can be seen in Figure 7B. Ethidium bromide is thought to intercalate between the two strands of the DNA double helix. Since it is oppositely charged it reduces the migration times of all the fragments. Because of this particular complexation phenomenon (one ethidium bromide molecule/5 bp) and the increasing rigidity of the complex, the larger DNA molecules migrate at a slower rate. Therefore, the separation time window opens up substantially, increasing peak capacity. Ethidium bromide is quite useful for manipulating migration time and separation, making it possible to separate even identical chain lengths of double-stranded DNA molecules[12] consisting of different sequences. As mentioned earlier this enhanced separation is based on secondary structure differences caused by different primary sequences.

Separation of DNA restriction fragments can also be improved by using different temperatures during the electrophoresis. Guttman and Cooke[10] evaluated the effect of temperature on the separation of DNA fragments on a physical gel. They concluded that migration

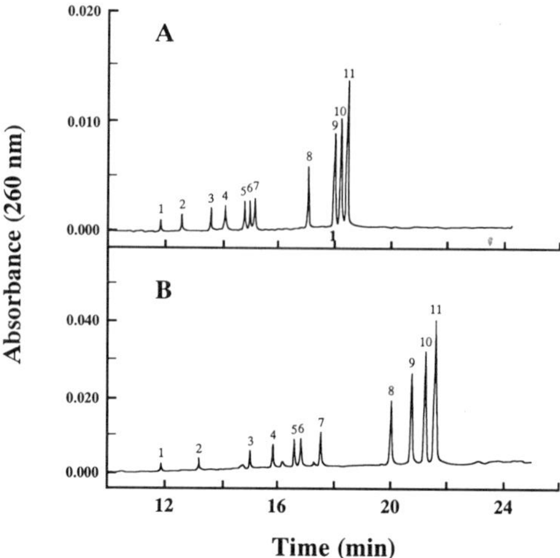

FIGURE 7. Effect of the intercalator additive, ethidium bromide, on the separation of the ϕX174 DNA-*Hae*III digest restriction fragment mixture. Separation without (A) and with (B) ethidium bromide (1 μg/ml) in the gel–buffer system (physical gel). Capillary length was 470 mm; applied electric field, 250 V/cm. (From Guttman, A. and Cooke, N., *Anal. Chem.*, 63, 2038, 1991. With permission.)

times decrease with increasing temperature in the isoelectrostatic (constant voltage) separation mode and show maximum migration time in the isorheic (constant current) separation mode. The resolution between the lower chain length DNA fragments (<300 bp) decreases in the isoelectrostatic separation mode and shows maxima in the isorheic mode at elevated temperatures. However, the efficiency in the higher molecular weight range (>100 bp) decreases in both modes with increasing temperature. Using different field manipulation methods, such as pulsed field electrophoresis,[9,31] analyte velocity modulation,[32] and field strength gradient separation techniques,[33] have also been shown to increase resolving power. These methods are useful in optimization of the separation time and improvement of selectivity and resolution.

C. AGAROSE GEL-FILLED CAPILLARY COLUMNS

Gels with agarose concentrations between 0.3 and 2.6% are suggested based on the method of Bocek et al.[20,34] for separation of DNA fragments. Problems usually encountered with polyacrylamide gel (i.e., air bubbles, volume expansion, etc.) are not observed with agarose solutions. In addition, agarose has the advantage over cellulose derivatives of being far more soluble and nonhydrophobic.[35,36]

DNA restriction fragments can be adequately separated by capillary electrophoresis using 0.3 to 2.0% agarose at 40°C. The advantage of employing liquefied agarose above its "gelling" temperature is that it is a replaceable gel (capillary can be easily filled, rinsed, and refilled). Another advantage of the agarose is that its background absorbance at 254 to 260 nm is sufficiently low that DNA detection at the nanogram level is possible. With an agarose sieving matrix, the inner surface of the capillary must be coated with linear polyacrylamide. Using this technique, the effective size range for separation of DNA can be extended up to 12 kb.[20] The log(bp) vs. mobility plot is biphasic for a 1.7% agarose solution, showing higher resolving power for DNA chains shorter than 1 kb in size in comparison with larger DNA. However, the resolving power of DNA larger than 1 kb is substantially enhanced when the agarose concentration is increased (up to 2.6%).[34]

D. USE OF CELLULOSE DERIVATIVES AS VISCOSITY ADDITIVES

It is well known that the electrophoretic mobility of double-stranded DNA in free solution is largely independent of molecular size.[37,38] Therefore, size separation is obtained by the use of various types of sieving media. Capillary electrophoretic separations of DNA molecules can be performed in different derivatized cellulose solutions such as hydroxyethylcellulose (HEC),[39,40] hydroxymethylcellulose (HMC),[41,42] hydroxypropylmethylcellulose (HPMC),[43] etc. Rheological studies confirmed that the polymer entanglement results in a mesh size smaller than that found in agarose gels.[40] It appears that low-viscosity polymer solutions not only provide a good sieving matrix, but also a good anticonvective medium, and are, therefore, suited for electrophoretic separations. DNA restriction fragments and polymerase chain reaction (PCR) products can be analyzed using this method. Ulfelder and co-workers[43] used a hydroxypropylmethylcellulose solution to achieve effective molecular sieving. According to the theory of MacCrehan et al.,[42] the addition of high concentrations of cellulose derivatives to the buffer system may also change the activity of water molecules available for association with the hydration sheath of the DNA, thereby modifying the effective Stokes' radius. In conventional gel sieving methods, a decrease in pore diameter improves resolution of small fragments at the expense of resolution of the larger ones, which are totally excluded. In contrast, capillary electrophoresis with derivatized cellulose sieving buffers gives improved resolution for both small and large fragments with increasing polymer concentration.

It is important to note that the use of derivatized cellulose additives forms an additional dynamic coating on the inner surface of the capillary wall. In this way, the use of this buffer additive significantly decreases the electroosmotic flow, even in an untreated fused silica capillary column.[35] It was shown recently by Bonn's group that the addition of nonbuffering alkaline salts such as Li, K, Na, Rb, and Cs improves the resolution of DNA fragments in uncoated fused silica capillaries by affecting the zeta potential of the capillary wall.[39]

Intercalating agents such as ethidium bromide,[35,39] thiazole orange,[44] etc., have also been used recently to increase selectivity in DNA fragment separation (usually by addition to the running buffer in a final concentration of 0.5 to 5 μg/ml). It is important to note that both of the above-mentioned intercalator additives are also fluorescent agents. As a result, very sensitive laser-induced fluorescent detection methods can also be employed. The most commonly used laser system at the present time is an argon-ion laser (excitation wavelength, 488 nm), which can be used for the excitation of thiazole orange. In this way the detection limit and selectivity are significantly improved for DNA restriction fragment and PCR analysis.[43,44]

E. SAMPLE INJECTION INTO GEL- OR POLYMER NETWORK-FILLED CAPILLARY COLUMNS

Samples must be injected electrokinetically (typically 0.015 to 0.15 Ws) with prepacked, cross-linked or high-concentration (>6%) linear polyacrylamide gel columns, and also into the columns containing high-viscosity derivatized cellulose additives or agarose gels. In contrast, samples can be injected by pressure (typically 2 to 20 sec, 0.5 psi) into the open tubular and the replaceable polyacrylamide gel-filled capillary columns. It is important to note that electrokinetic injection yields more efficient separation than pressure injection.[35] In electrokinetic injection from low ionic strength solutions [5 to 50 μg of DNA per milliliter, usually in water or very dilute buffers (>10 m*M*)], sample components migrate into the capillary where they are effectively stacked against the higher viscosity polymer network medium.[45] It is important to note that no sample bias occurs with electrokinetic injection.[46] Pressure injection is sometimes limited with highly viscous running buffers, such as higher concentration HPMC (>0.3%), since the volume of the sample injected is inversely proportional to the viscosity of the buffer.[47] In this instance, electrophoretic injection may be the preferred method.

III. METHODOLOGY

A. CAPILLARY SURFACE TREATMENT METHODS

The pretreatment of the inner surface of the fused silica capillary column is critical for electrophoresis of DNA and was first introduced by Hjertén et al.[48,49] In this method, a non-cross-linked monolayer of polyacrylamide is covalently attached to the inner surface of the capillary. Prior to starting the procedure, a short section of the outer polyimide coating (2 to 4 mm) is removed by a small torch or by a specially designed device.[50] It is important to note that this outside coating can also be removed after the inner surface treatment, but in such cases, special care should be taken not to destroy the existing inner coating.[20]

Procedure

To pretreat the surface of a 1-m long fused silica capillary column, 100 μl of 1 *M* HCl is flushed through the capillary for 30 min. This cleaning step is followed by a rinse with 100 μl of water. NaOH (1 *M*) is then flushed through the capillary for 30 min and the column is rinsed again with 100 μl of water. With this treatment the inner surface of the capillary column is activated for the reaction with a bifunctional silylation reagent. One hundred microliters of 0.4% 3-methacryloxypropyltrimethoxysylane in 0.4% acetic acid, pH 3.5, is rinsed through the capillary for 3 h at room temperature. After a final washing step with 100 μl of methanol and 100 μl of water, the capillary column is ready for the coating or gel polymerization procedure.

B. CAPILLARY COATING

It is important to understand that the buffer to be used for later separations should also be used in the coating procedures. Note that acrylamide is a neurotoxin and should be handled with caution.

Procedure

One milliliter of 6% electrophoretic grade acrylamide (Schwarz/Mann Biotech, Cambridge, MA)[51] solution in the appropriate buffer is vacuum degassed for at least 1 h at 100 mbar prior to polymerization. Polymerization is initiated by adding 2 μl of 10% ammonium persulfate and catalyzed by 4 μl of tetramethylethylenediamine (TEMED) at room temperature. After overnight polymerization the gel is removed form the capillary column by applying approximately 100 psi nitrogen pressure. Since this is a linear polymer (physical gel) only a covalently attached polymerized monolayer remains on the inner surface of the capillary. The coated capillary obtained in this way can be used for capillary electrophoresis studies, using various sieving matrices such as polyacrylamide, agarose, or cellulose derivatives.

Other researchers have used Grignard reactions to coat fused silica capillary columns. According to the procedure of Cobb et al.[52] the reaction of surface-chlorinated fused silica capillaries with vinyl magnesium bromide, followed by reaction of the vinyl group with acrylamide, results in an immobilized layer of polyacrylamide attached through hydrolytically stable Si–C bonds. Capillaries treated in this manner are claimed to be stable over a pH range of 2 to 10.5 without significant decomposition of the coating.

C. PREPARATION OF CROSS-LINKED POLYACRYLAMIDE GEL COLUMNS

Procedure

Thirty milligrams of electrophoretic grade acrylamide/*N,N′*-bisacrylamide mixture (1:19) (Schwarz/Mann Biotech) is dissolved in 1 ml of buffer solution (usually 100 m*M* Tris-borate, pH 8.3, with 7 *M* urea). Then the solution is carefully vacuum degassed for 1 h under 100 mbar with continuous stirring. The polymerization is initiated by adding 2 μl of 5% ammonium persulfate and catalyzed by 4 μl of 10% TEMED at room temperature. The pretreated capillary columns are immediately filled with this mixture by means of a gas-tight syringe, using a plastic sleeve to connect the needle to the capillary tubing. After overnight polymerization the gel-filled columns are ready to use. It is recommended that the capillary be checked for bubbles under a microscope before installing in the system.[53] If bubbles are found, the amount of ammonium persulfate, TEMED, and/or polymerization temperature should be decreased until the gel is polymerized bubble free. If no bubbles are found, an equilibration run is made following the installation of the column to condition the gel for the high applied electric field to be used. This is carried out by slowly increasing the applied electric field to the cross-linked gel filled column, e.g., 100 V/cm for 10 min, 200 V/cm for 10 min, and 300 V/cm for 10 min when 300 V/cm is the separation field strength. The recommended field strength for these types of columns is maximum 300 V/cm (11.1 kV for a 37-cm capillary) without temperature control and 500 V/cm (18.5 kV for a 37-cm capillary) with a capillary electrophoresis unit that has an adequate capillary thermostatting system.

D. PREPARATION OF LINEAR POLYACRYLAMIDE GEL COLUMNS

Procedure

Electrophoretic grade acrylamide (120 mg) is dissolved in 1 ml of buffer (usually 100 m*M* Tris-borate, pH 8.3, with 2 m*M* EDTA) and carefully vacuum degassed for 1 h under 100 mbar. The polymerization is initiated by adding 2 μl of 10% ammonium persulfate and catalyzed by 4 μl of concentrated TEMED at room temperature. The gel is polymerized overnight before use.

Depending on the application, different dilutions can be made from this "stock" gel. The most popular concentration is the 3% gel that is optimal for the most frequently separated base pair range of 100 to 1000. For other applications one should prepare lower or higher concentration solutions for longer or shorter chain length DNA molecules, respectively. For producing the 3% linear polyacrylamide gel, the above-manufactured 12% linear polyacrylamide should be diluted (3× volume) with the appropriate amount of buffer (same composition as above or as per the composition of the 12% gel if different than above) and stirred overnight. When no threadlike artifacts are observed in the solution the diluted polymer network solution is ready to use for separation, employing the coated capillary column described before. To remove any possible gas bubbles, sonication for 5 to 10 s is recommended. The temperature of the replaceable gel-filled capillary column has a significant effect on the pore structure (mesh size) of the polymer network. Resolving power therefore can be manipulated by changing the running temperature.[10]

E. MANUFACTURING OF AGAROSE GEL COLUMNS

Procedure

For DNA restriction fragment separations ranging from 100 to 1000 bp, 1.7% SeaPlaque GTG (FMC, Rockland, ME) agarose is dissolved in 1× TBE buffer (89 m*M* tris, 89 m*M* borate, 2.5 m*M* Na_2-EDTA) and heated to 50°C in a thermostatted oven. This will allow for easy refilling of the coated capillary (described earlier) with the melted gel. For DNA restriction fragment separations this agarose gel-filled capillary column should be thermostatted at 40°C. The upper limit of the size range of the DNA can be increased to 12 kbp by varying the concentrations of agarose solutions.[34]

F. SIEVING BUFFERS WITH CELLULOSE DERIVATIVES

It is known that linear hydrophilic polymers can be employed as viscosity additives to attain molecular sieving for the separation of biopolymers, such as DNA fragments, in coated fused silica capillary columns.[39-43] Similar results were published using uncoated fused silica capillaries having electroosmotic flow,[54] although with inferior resolving power. Use of derivatized cellulose solutions allows for the employment of linear polyacrylamide-coated capillary columns and commercially available coated capillaries such as nonpolar or intermediate polysiloxane coatings (OV-1, OV-17, and OV-225). These coated capillaries can be purchased from gas chromatography (GC) column suppliers (e.g., J&W, Sacramento, CA) with an inner diameter of 0.1 mm and a coating thickness of 0.05 to 0.2 μm. One disadvantage of using these precoated columns is that removal of the outer polyimide coating cannot be accomplished by means of a high-temperature flame without destroying (burning off) the inner coating. In this case, the procedure of Bocek and Crambach is recommended, using a 130°C sulfuric acid drop where the window is needed.[20] Although this procedure takes slightly longer, it is effective for making the window and does not destroy the inner coating.

Procedure

The derivatized cellulose (HEC, HMC, HPMC, etc.) is dissolved in the corresponding buffer system, usually the same or similar to that used with replaceable polyacrylamide gels, (50 to 100 m*M* Tris-borate, pH 8.3, 2.5 m*M* EDTA). The effective concentrations of the selected buffer additive polymers have been published by Ulfelder et al.[35] for sieving matrices such as HPMC, 4000 cP at 25°C, 0.1 to 0.7% (w/w), or HPMC, 100 cP at 25°C, in 0.5 to 1.0% (w/w). These concentrations can sufficiently separate the most frequently used range of double-stranded DNA molecules (100 to 1000 bp) and are not too viscous for filling and emptying the capillary. When cellulose additives are used, all buffers are filtered through a 1.2-μm pore size membrane filter and degassed by vacuum (100 mbar) with continuous stirring or by sonication (60-s maximum).

G. CHEMICALS

The single-stranded homooligomers and the DNA restriction fragment mixtures can be purchased from Pharmacia (Piscataway, NJ), New England BioLabs (Beverly, MA), or Bethesda Research Laboratories (Bethesda, MD). All DNA samples are diluted to 5 to 50 μg/ml with water or diluted buffers (>10 m*M*) before injection. The use of ultrapure electrophoresis grade acrylamide, Tris, boric acid, EDTA, urea, ammonium persulfate, and TEMED is strongly recommended in capillary gel electrophoresis experiments (Schwartz/Mann Biotech, Cambridge, MA). Orange G (Sigma, St. Louis, MO) is a good internal standard in capillary gel electrophoretic separations of DNA molecules in 0.01% concentration[28] since it migrates faster than DNA molecules of chain lengths longer than 10 bases.

All samples should be stored at –20°C or freshly used. Single-stranded DNA samples should be boiled for at least 3 min and then cooled by ice water prior to injection; this will prevent inter- and intrachain associations. The buffer and gel solutions should be filtered through a <1.2-μm pore size filter and carefully vacuum degassed at 100 mbar.

IV. THE FUTURE OF CAPILLARY GEL ELECTROPHORESIS

The methods currently used in capillary gel electrophoresis offer a diversity of analyses.[55] New techniques employing low-viscosity gel or polymer network systems with novel coatings for capillaries will have a great impact on the development of new applications. The introduction of capillary gel electrophoresis methods, such as size separation systems for DNA and protein analysis, will complement and perhaps replace traditional slab gel methods. Developments of novel coatings will permit the use of simpler gel–buffer systems and are postulated to improve the reproducibility of the technique. Capillary gel electrophoresis could easily adapt methods like detection of single-stranded DNA conformational polymorphism to determine point mutations in genetic sequences. Another possibility is to design various hybridization regimes to show DNA–DNA and DNA–protein interactions by a shift in electrophoretic mobility of the hybrid. Changes in double-stranded DNA melting temperature due to sequence variations can be determined by temperature gradient capillary gel electrophoresis. Based on the above, one can predict that capillary gel electrophoresis will be an essential part of novel automated analyzers designed for a variety of diagnostic, biomedical, and forensic research applications, as well as in the process control in the rapidly developing biotechnology industry.

REFERENCES

1. **Andrews, A. T.,** *Electrophoresis*, 2nd ed., Clarendon Press, Oxford, 1986.
2. **Chrambach, A.,** *The Practice of Quantitative Gel Electrophoresis*, VCH, Deerfield Beach, FL, 1985.
3. **Rickwood, D. and Hames, B. D., Eds.,** *Gel Electrophoresis of Nucleic Acids*, IRL Press, Washington, D.C., 1983.
4. **Maniatis, T., Fritch, E. F., and Sambrook, J.,** *Molecular Cloning: A Laboratory Manual*, Cold Spring Harbor Laboratory, Cold Spring Harbor, NY, 1982.
5. **Kasper, T. J., Melera, M., Gozel, P., and Brownlee, R. G.,** Separation and detection of DNA by capillary electrophoresis, *J. Chromatogr.*, 458, 303, 1988.
6. **Cohen, A. S., Najarian, D., Smith, J. A., and Karger, B. L.,** Rapid separation of DNA restriction fragments using capillary electrohoresis, *J. Chromatogr.*, 458, 323, 1988.
7. **Guttman, A., Paulus, A., Cohen, A. S., Karger, B. L., Rodriguez, H., and Hancook, W. S.,** High performance capillary gel electrophoresis: high resolution and micropreparative application, in *Electrophoresis '88*, Schafer-Nielsen, C., Ed., VCH, Weinheim, Germany, 1988, p. 151.
8. **Lux, J. A., Yin, H. F., and Schomburg, G.,** A simple method for the production of gel-filled capillaries for capillary gel electrophoresis, *J. High. Resolut. Chromatogr.*, 13, 436, 1990.
9. **Heiger, D. N., Cohen, A. S., and Karger, B. L.,** Separation of DNA restriction fragments by high performance capillary electrophoresis with low and zero crosslinked polyacrylamide using continuous and pulsed electric fields, *J. Chromatogr.*, 62, 137, 1990.
10. **Guttman, A. and Cooke, N.,** Effect of temperature on the separation of DNA restriction fragments in capillary gel electrophoresis, *J. Chromatogr.*, 559, 285, 1991.
11. **Tanaka, T.,** Gels, *Sci. Am.*, 244, 124, 1981.
12. **Guttman, A. and Cooke, N.,** Capillary affinity gel electrophoresis of DNA fragments, *Anal. Chem.*, 63, 2038, 1991.
13. **Nelson, R. J., Paulus, A., Cohen, A. S., Guttman, A., and Karger, B. L.,** Use of Peltier thermoelectric devices to control column temperature in high performance capillary electrophoresis, *J. Chromatogr.*, 480, 111, 1989.
14. **Cohen, A. S., Terabe, S., Smith, J. A., and Karger, B. L.,** High performance capillary electrophoretic separation of bases, nucleosides and oligonucleotides: retention manipulation via micellar solutions and metal additives, *Anal. Chem., 59,* 1021, 1987.
15. **Wallingford, R. A. and Ewing, A. G.,** Retention of ionic and nonionic catechols in capillary zone electrophoresis with micellar solutions, *J. Chromatogr.*, 441, 299, 1988.
16. **Nishi, H., Tsumagari, N., Kakimoto, T., and Terabe, S.,** Separation of water soluble vitamins by micellar electrokinetic chromatography, *J. Chromatogr.*, 465, 331, 1989.
17. **Cohen, A. S. and Karger, B. L.,** High performance sodium dodecyl sulfate polyacrylamide gel capillary electrophoresis of peptides and proteins, *J. Chromatogr.*, 397, 409, 1987.
18. **Cohen, A. S., Najarian, D. R., Paulus, A., Guttman, A. Smith, J. A., and Karger, B. L.,** Rapid separation and purification of oligonucleotides by high-performance capillary gel electrophoresis, *Proc. Natl. Acad. Sci. U.S.A..*, 85, 9660, 1988.

19. **Guttman, A., Cohen, A. S., Heiger, D. N., and Karger, B. L.,** Analytical and micropreparative ultrahigh resolution of oligonucleotides by polyacrylamide gel high performance capillary electrophoresis, *Anal. Chem.* 62, 137, 1990.
20. **Bocek, P. and Crambach, A.,** Capillary electrophoresis of DNA in agarose solutions at 40°C, *Electrophoresis,* 12, 1059, 1991.
21. **Sudor, J., Foret, F., and Bocek, P.,** Pressure refilled polyacrylamide columns for the separation of oligonucleotides by capillary electrophoresis, *Electrophoresis,* 12, 1056, 1991.
22. **Guttman, A.,** Effect of operating variables on the separation of DNA moleculaes by capillary polyacrylamide gel electrophoresis, *Appl. Theor. Electrophoresis,* 3, 91, 1992.
23. **Cohen, A. S., Gemborys, M. W., and Vilenchik, M.,** presented in the 5th Int. Symp. High Performance Capillary Electrophoresis, Orlando, FL, Jan. 25–28, 1993.
24. **Guttman, A. and Mazsaroff, I.,** Economical performance analysis in preparative capillary gel electrophoresis, in *New Approaches in Chromatography,* H. Kalàsz and E. L. Ehre, Eds., Intercongress, Budapest, Hungary, 1992.
25. **Drossmann, H., Luckey, J. A., Kostichka, J., D'Cunha, J., and Smith, L. M.,** High speed separations of DNA sequencing reactions by capillary electrophoresis, *Anal. Chem.,* 62, 900, 1990.
26. **Swerdlow, H., Zhang, J. Z., Chen, D. Y., Harke, H. R., Grey, R., Wu, S., and Dovichi, N. J.,** Three DNA sequencing methods using capillary gel electrophoresis and laser induced fluorescence, *Anal. Chem.,* 63, 2835, 1991.
27. **Swerdlow, H. and Gesteland, R.,** Capillary gel electrophoresis of DNA sequencing: laser induced fluorescence detection with the sheath flow cuvette, *Nucleic Acids Res.,* 18, 1415, 1990.
28. **Gaastra, W.,** Nucleic acids, in *Methods in Molecular Biology,* Vol. 2, Walker, J. M., Ed., Humana Press, Clifton, NJ, 1984, chap. 52.
29. **Guttman, A., Nelson, R. J., and Cooke, N.,** Prediction of migration behavior of oligonucleotides in capillary gel elecrophoresis, *J. Chromatogr.,* 593, 297, 1992.
30. **Schwartz, H. E., Ulfelder, K. J., Sunzeri, F. J., Busch, F. J., and Brownlee, R. G.,** Analysis of DNA restriction fragments and polymerase chain reaction products towards detection of the AIDS (HIV-1) virus in blood, *J. Chromatogr.,* 559, 267, 1991.
31. **Heiger, D. N., Carson, S. M., Cohen, A. S., and Karger, B. L.,** Wave form fidelity in pulsed field capillary electrophoresis, *Anal. Chem.,* 64, 192, 1992.
32. **Demana, T., Lanan, M., and Morris, M. D.,** Improved separation of nucleic acids with analyte velocity modulation capillary electrophoresis, *Anal. Chem.,* 63, 2795, 1991.
33. **Guttman, A., Wanders, B., and Cooke, N.,** Enhanced separation of DNA restriction fragments by capillary gel electrophoresis using field strength gradients, *Anal. Chem.,* 64, 2348, 1992.
34. **Bocek, P. and Crambach, A.,** Capillary electrophoresis in agarose solutions: extensions of size separation to DNA of 12 kb in length, *Electrophoresis,* 13, 31, 1992.
35. **Ulfelder, K., Anderson, K., and Schwartz, H. E.,** Analysis of PCR products and DNA restriction fragments to detect AIDS (HIV-1) virus in blood. Paper presented at HPCE '91, San Diego, Feb. 3–6, 1991.
36. **Bocek, P. and Crambach, A.,** Electrophoretic size separation in liquified agarose of polystyrene particles and circular DNA, *Electrophoresis,* 12, 623, 1991.
37. **Olivera, B. M., Baine, P., and Davidson, N.,** Electrophoresis of the nucleic acids, *Biopolymers,* 2, 245, 1964.
38. **Hermans, J. J.,** Structure and the mechanism of deformation in celluslose gels, *J. Polymer. Sci.,* 18, 257, 1953.
39. **Nathakarnkitkool, S., Oefner, P., Bartsch, G., Chin, M. A., and Bonn, G, K.,** High resolution capillary electrophoretic analysis of DNA in free solution, *Electrophoresis,* 13, 18, 1992.
40. **Grossmann, P. D. and Soane, D. S.,** Capillary electrophoresis of DNA in entangled polymer solutions, *J. Chromatogr.,* 559, 257, 1991.
41. **Zhu, M. D., Hansen, D. L., Burd, S., and Gannon, F.,** Factors affecting free zone electrophoresis and isoelectric focusing in capillary electrophoresis, *J. Chromatogr.,* 480, 311, 1989.
42. **MacCrehan, W. A., Rasmussen, H. T., and Northrop, D. M.,** Size selective capillary electrophoresis (SSCE) separation of DNA fragments, *J. Liq. Chromatogr.,* 15, 1063, 1992.
43. **Ulfelder, K., Schwartz, H. E., Hall, J. M., and Sunzeri, F. J.,** Restriction fragment length polymorphism analysis of ERBB2 oncogene by capillary electrophoresis, *Anal. Biochem.,* 200, 260, 1992.
44. **Schwartz, H. E. and Ulfelder, K.,** Capillary electrophoresis with laser induced fluorescence detection of PCR fragments using thiazole orange, *Anal. Chem.,* 64, 1737, 1992.
45. **Ornstein, L.,** Disc electrophoresis *Ann. N.Y. Acad. Sci.,* 121, 321, 1964.
46. **Huang, X., Gordon, M. J., and Zare, R. N.,** Bias in quantitative capillary zone electrophoresis caused by electrokinetic sample injection, *Anal. Chem.,* 60, 375, 1988.
47. **Harbaugh, J., Collette, M., and Schwartz, H. E.,** Determination of injection volumes with P/ACE system 2000, Technical Bulletin, TIBC-103, Beckman Instruments, Palo Alto, CA, 1990.

48. **Hjertén, S., Elenbring, K., Kilar, F., Liao, J. L., Chen, J. C., Siebert, C. J., and Zhu, M. D.,** Carrier free zone electrophoresis, displacement electrophoresis and isoelectric focusing in a high performance capillary electrophoresis apparatus, *J. Chromatogr.,* 403, 47, 1987.
49. **Hjertén, S.,** High performance electrophoresis, elimination of electroendosmosis and solute adsorption, *J. Chromatogr.,* 347, 191, 1985.
50. **Lux, J. A., Hausig, U., and Schomburg, G.,** Production of windows in fused silica capillaries for in-column detection of UV-adsorption of fluorescence in capillary electrophoresis or HPLC, *HRC,* 13, 373, 1990.
51. **Guttman, A., Arai, A., and Magyar, K.,** Influence of pH on the migration properties of oligonucleotides in capillary gel electrophoresis, *J. Chromatogr.,* 608, 175, 1992.
52. **Cobb, K. A., Dolnik, V., and Novotny, M.,** *Anal. Chem.,* 62, 2478, 1990.
53. **Yin, H. F., Lux, J. A., and Schomburg, G.,** Production of polyacrylamide gel filled capillaries for capillary gel electrophoresis (CGE): influence of capillary surface pretreatment on performance and stability, *HRC,* 13, 624, 1990.
54. **Biosystems Report,** No. 6, Applied Biosystems Newsl., October, 1989.
55. **Landers, J. P., Oda, R. P., Spelsberg, T. C., Nolan, J. A., and Ulfelder, K. J.,** Capillary electrophoresis: a powerful microanalytical technique for biologically active molecules, *BioTechniques,* 14, 98, 1993.

PART II
DETECTION IN CAPILLARY ELECTROPHORESIS

Chapter 7

OPTICAL DETECTION TECHNIQUES FOR CAPILLARY ELECTROPHORESIS

Stephen L. Pentoney, Jr. and Jonathan V. Sweedler

TABLE OF CONTENTS

0-8493-8690-X/94/$0.00+$.50

I. INTRODUCTION

Advances in detector technology have played a major role in the successful resurgence of capillary electrophoresis (CE) during the past 10 years. Although the roots of CE can be traced as far back as the 1930s with the work of Tiselius[1] and the 1970s with the pioneering works of Virtanen[2] and Mikkers et al.,[3] the full potential of capillary electrophoresis was not recognized until advances in miniature detector technology made routine analysis of nanoliter sample volumes a reality. Jorgenson's group was among the first to recognize the need for highly sensitive and selective detectors for CE.[4]

A wide variety of detection schemes have now been coupled to capillary electrophoresis with varying degrees of success. In general, efforts in the area of CE detector development have involved extending instrumentation and techniques originally developed for microcolumn high-performance liquid chromatography (HPLC) to the nanoliter regime of capillary electrophoresis. In this chapter we discuss a selected set of the more promising optical detection schemes that have been combined with capillary electrophoresis. The works represented here are drawn entirely from published information and as a result do not adequately cover advances made by instrument companies. Our emphasis is placed largely on the two most widely utilized optical detection schemes, ultraviolet-visible (UV-Vis) absorbance and fluorescence. We also survey several of the less widely employed optical detection schemes. A compilation of the various nonoptical detection techniques that have been reported may be found elsewhere.[5]

II. GENERAL PERFORMANCE AND DESIGN REQUIREMENTS

One of the most challenging areas of instrumentation development in the field of capillary electrophoresis is the design and development of the detector. In this section we describe the various parameters that must be considered in the design of any optical detection scheme for capillary electrophoresis.

A. APPLICATION

First and foremost, the requirements of the application to be addressed must be considered. If the instrument is to be devoted to a single application, then there exists the opportunity to tightly define the performance specifications and develop a fully optimized detector. If, on the other hand, the instrument is to be used for a wide variety of applications, then the detector design must be versatile and compromises must be made.

B. ELECTROPHORETIC RESOLUTION

Ideally the detector provides a means of visualizing the separation process occurring within the capillary without impacting the quality of the separation. Electrophoretic resolution should be determined solely by the separation process occurring within the capillary channel; the presence of a detector can either degrade or, at best, have no impact on resolution. The viewing length of the detector must be restricted to a fraction of the width of the narrowest bands expected to migrate through the capillary in order to avoid band broadening by the detector.

An example of how designing a general purpose CE system involves compromise follows. Detector performance (noise level and/or amount of signal generated) often improves as the viewing length is increased. However, in a general purpose CE system, the anticipated electrophoretic efficiencies and hence bandwidths vary greatly from one application to the next. As a result, the detector viewing region is generally set to about 10% of the width of the narrowest bands expected under any circumstances and resolution is maintained for the most demanding separations at the expense of sensitivity.

The following example illustrates one method of estimating the minimum acceptable detection length. The absolute minimum peak variance for any separation is given by broadening caused solely by longitudinal diffusion:

$$\sigma^2 = 2Dt$$

where σ^2 is the peak variance, D is the diffusion coefficient of the analyte in the separation medium, and t is the time required for the sample to reach the detector. For a separation time of 10 min and a diffusion coefficient of 1×10^{-6} cm²/s, σ^2 would be 1.2×10^{-3} cm², resulting in a peak base width of approximately 1.4 mm. Therefore an acceptable detection length would be 140 μm. Although this treatment is certainly an oversimplification[6,7] (this calculation assumes an injection variance of zero and considers only one contributing mechanism to zone broadening), it serves to illustrate the point that in order to maintain high resolution the detection length must be quite short. The maximum acceptable viewing length for common CE optical detectors is approximately 100 to 200 μm.

C. LIMITS OF DETECTION

One of the greatest challenges in the area of CE system design is the development of detection schemes that possess the required limits of detection. This difficulty is caused by the small capillary dimensions that must be maintained in order to fully realize the separation power of capillary electrophoresis. This situation is exacerbated because the trend in the field is toward faster separations achieved using smaller capillary internal diameters in order to limit resistive heating at large field strengths. The useful analytical working range of sample concentrations is bounded on the high end by the column capacity (assuming that no detector limitations such as fluorescence quenching exist) and on the low end by the limit of detection (LOD) for the particular detection scheme employed. Column capacity relates here to the relative conductivities of the sample plug and an equivalent size plug of the separation medium. Maximum resolution is achieved when the sample plug conductivity is much less than that of the supporting separation medium. In order to limit Joule heating of the separation medium, buffer concentrations employed for CE separations are generally in the 5 to 200 m*M* range, depending on the mobility of the particular buffer components selected. As a result, the upper sample concentration limit that the system will acceptably tolerate tends to be in the one-tenth millimolar range. With this established as a typical upper concentration limit, the need for detection schemes that are capable of at least submicromolar concentration limits of detection in order to achieve even a modest dynamic range is evident. Again, the specific detection level requirements are defined by the application to be addressed.

It is worth preceding our discussion of detection limits with several definitions. The terms *sensitivity* and *detection limit* are often used interchangeably but this is not strictly correct. The sensitivity of a detector refers to the minimum detectable concentration change that can be observed at a specified concentration ($\Delta C/C$). The limit of detection refers to the minimum mass or concentration of analyte that can be detected at an acceptable signal-to-noise ratio. Nearly all of the CE detectors reported in the literature have been characterized in terms of their minimum limit of detection. The limit of quantitation and the limit of detection should also be differentiated. The limit of quantitation is the analyte mass or concentration required to give an acceptable level of confidence in the measured analyte quantity and is always greater (usually a factor of ten) than the minimum limit of detection.

Detection limits reported in the literature have been determined and expressed in a variety of ways. For example, detection limits have been expressed both in terms of mass and in terms of concentration. Even more confusing, and often difficult to discern without careful reading, is the fact that mass detection limits are reported in terms of either the sample quantity injected or the sample quantity present within the detection volume as the peak transits the detector (generally no more than 10% of the total peak quantity). Reported sample quantities injected onto the capillaries have most often been calculated for electrokinetic sample introduction and are, at best, only estimates. In order to avoid confusion in this area we will, where possible, discuss detection limits in terms of the minimum detectable concentration (MDC) of sample that is injected onto the column.

There exist several sample concentration techniques that can be utilized to extend the dynamic range of any CE detection scheme. A detailed coverage of these sample concentration techniques is beyond the intended scope of this chapter, but they include stacking or field amplification,[8-10] isotachophoretic preconcentration,[11-14] and precolumn concentration.[15,16] It is important to be aware of these phenomena because they may allow the detection of a trace constituent that otherwise would be missed. In addition, these concentration mechanisms are often present and are seldom quantified. For example, electrokinetic injection from a low-conductivity solution into a relatively high-conductivity separation medium will produce significant sample stacking. This concentration can occur even when the user is unaware of the conductivity differences. Of course, this complicates injection quantity calculations and detector performance comparisons.

It is also important to bear in mind the difference between injected sample volumes and the sample solution volumes required to reliably make an injection. Typical sample injection volumes in CE are in the 0.1- to 10-nl range, while the minimum sample volume required to make reliable injections is generally in the 0.5- to 2-μl range. Often researchers or instrument manufacturers will report that a particular detection scheme is capable of detecting a certain quantity of analyte (e.g., 20 amol) without discussing the minimum sample volume; this often leads to confusion. For example, the chemist wishing to detect 20 amol of analyte present in 5 μl of solvent faces a much greater challenge than the chemist wishing to analyze the same sample at a concentration of 20 amol/nl with a minimum sample size of 2 μl. In order to remain consistent throughout this discussion, we report the concentration of the sample solution injected onto the capillary, wherever possible.

D. EASE OF USE

As the acceptance of capillary electrophoresis continues to grow, more and more practitioners will be "non-CE experts"; as a result, an increasing level of automation is required. Many of the approaches to low-level detection that have been reported in the literature are manually intensive and are simply too complicated to be of practical value to the average CE practitioner. Here we look to both the research community and instrument companies for creative engineering to bring the more sophisticated detection schemes into the realm of practical utility.

E. COST AND COMPLEXITY

Cost and complexity are always key issues in the design of a successful instrument. They have prevented many potentially useful detection schemes from making the transition from the academic instrumentation laboratory to a commercially viable product. Perhaps there is no better example of this than laser-induced fluorescence (LIF) detection for HPLC. For quite some time the academic community has been demonstrating significant detection limit improvements for HPLC by utilizing LIF detection.[17-20] However there are currently no commer-

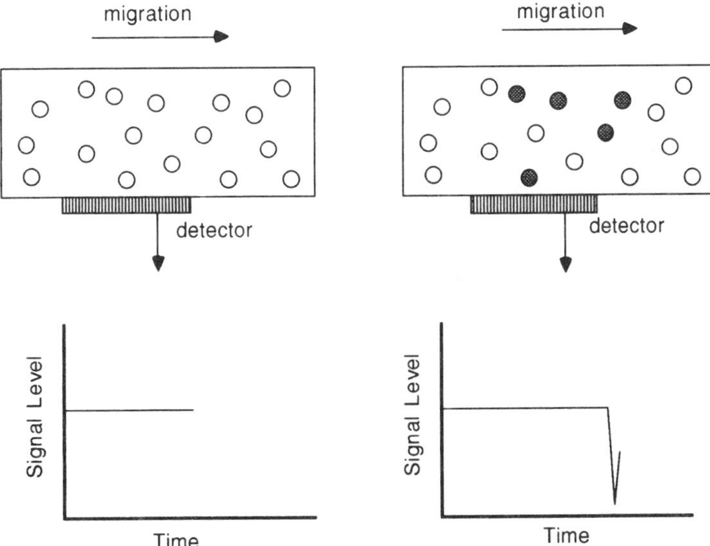

FIGURE 1. Diagram illustrating the principle of indirect detection. The open circles represent the signal-generating molecules that have been added to the separation medium. On the right, the analyte molecules are represented by shaded circles and have displaced some of the signal-generating molecules. This results in a loss of signal at the detector. (Adapted from Ref. 21.)

cially available HPLC/LIF systems, presumably because of the relatively high cost and complexity of lasers. In CE the increased demands placed on the detector have driven the commercial development of LIF detection.

III. INDIRECT VS. DIRECT DETECTION

In almost all instances, the analyst selects a detection technique based on physical and chemical properties of the analyte and the requirements of the application being addressed. Thus the analyst uses UV absorbance for molecules that are highly absorbing or fluorescence for easily derivatized analytes to be detected at low levels, etc. However, it is possible to indirectly detect analytes that lack desirable physical and chemical properties by the effect that the analyte has on another, more easily detected molecule. By incorporating a detectable molecule in the separation medium and monitoring the effect that the analyte has on the signal generated by this second molecule, one can indirectly observe the presence of the analyte. This indirect mode of detection has been successfully employed in the field of ion chromatography for years in the form of indirect absorbance.[21]

Indirect detection has also been applied to capillary electrophoresis for UV absorbance,[22-27] fluorescence,[28-33] and electrochemical[34] detection. Some applications of indirect detection are discussed in Chapter 11. As illustrated in Figure 1, the analyte of interest physically displaces some of the signal-generating molecules from the separation medium within the sample zone and a corresponding decrease in signal is observed. The mechanism of displacement may be as simple as conservation of volume (dilution) or as complex as conservation of charge or solubility alteration. For best performance (lowest limits of detection) indirect detection requires a well-characterized displacement mechanism such as charge displacement, a large transfer ratio (TR), and, most importantly, a very stable background. The transfer ratio refers to the number of signal-generating molecules displaced by an analyte molecule. Charge displacement refers, for example, to an anionic signal generator being displaced by an anionic analyte molecule.

Indirect detection is similar to absorbance detection in that as analyte concentration is decreased one is looking for a diminishingly small difference between two relatively large signals: the background signal and the background signal minus the signal arising from the displaced signal generators; hence the need for an extremely stable background. This background stability is achievable in the case of indirect absorbance detection in which very stable lamp sources are commonly employed. In the case of indirect fluorescence with laser sources, however, detection is limited by the "noisy" nature of lasers. As a result, indirect absorbance detection actually has lower concentration limits of detection than indirect fluorescence. The stability of the background signal in indirect detection is characterized by the dynamic reserve (DR). The dynamic reserve is the ratio of the background signal to the background noise. The overall MDC is given by

$$\text{MDC} = C_m/(\text{DR} \times \text{TR})$$

where C_m is the concentration of the signal-generating molecule and the other terms are as defined above.[21,28] This equation shows that the more stable the background (larger DR), the lower the detection limit. Additionally, the more efficient the transfer process, the lower the detection limit. Last, the lower the C_m, the larger the fractional change for a given number of analyte molecules. Thus if one is detecting a molecule present at low concentration, C_m can be set low in order to obtain the best possible detection limits. Of course, the effective dynamic range is also reduced by lowering C_m.

Dynamic range or limit of detection can be optimized by varying the concentration of the signal generator. While this allows one to optimize the detection conditions relative to the sample at hand (provided one knows enough about the sample), the dynamic range and limits of detection will not exceed (and in general will not approach) the detection limits of the corresponding direct technique. Although many have claimed that the instrumentation requirements are the same for both direct and indirect modes of detection, indirect detection does place a different set of requirements on the detection system. This is perhaps most apparent in the case of laser-induced direct fluorescence vs. indirect fluorescence. In the indirect mode of detection a high degree of precision is required in order to reliably detect small differences between two large signals; an extremely difficult task given the noise characteristics of common laser sources. External laser stabilizers have been used to reduce this problem in indirect laser-induced fluorescence.

Indirect detection is best suited for application-specific systems in which each of the critical parameters may be optimized and remain unchanged. In the case of indirect UV absorbance, one can truly optimize sensitivity and limit of detection by careful selection of the signal-generating chromophore; for example, one can choose a charge-transfer complex with a very high absorptivity for maximum sensitivity and appropriate charge to render a favorable TR for a group of analytes of like charge. In fact, Waters/Millipore (Bedford, MA) has commercialized this concept for ion analysis.

Over the last 10 years Yeung and co-workers have developed indirect fluorescence detection for CE and have discussed both theory and instrumentation.[21,28-33] One advantage of laser-based indirect fluorescence that this group has demonstrated is the ability to reduce the capillary inner diameter to as little as 5 µm; a very difficult task with lamp sources. Yeung's group has reported MDCs in the 10^{-7} *M* range for 5-µm I.D. capillaries, which corresponds to low attomole quantities of analyte.

IV. ABSORBANCE DETECTION

On-column UV absorbance detection is, by far, the most common means of monitoring CE separations. As is the case in HPLC, the primary reason for this is simply that the large

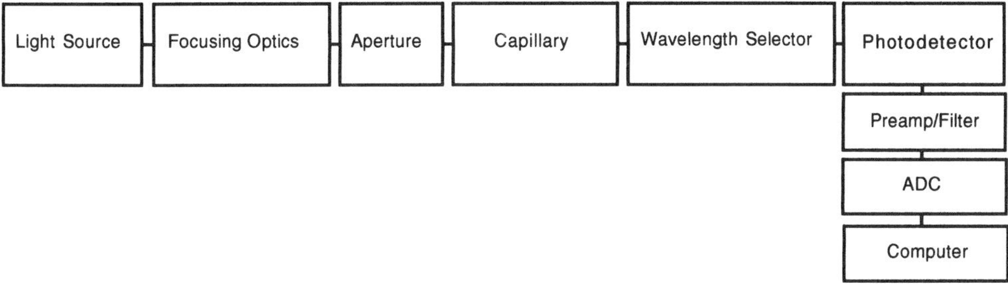

FIGURE 2. Functional block diagram of a UV-Vis absorbance detector for CE. A typical detector may also include a reference channel (not shown).

majority of compounds analyzed absorb somewhere in the UV region. Absorbance detection most often requires no chemical modification of the sample prior to analysis and is therefore an extremely convenient detection scheme.

A functional block diagram of a general CE/absorbance detector is illustrated in Figure 2. The output of a lamp source, most often UV radiation from a deuterium source, is directed either by focusing optics or optical fibers[35-37] through an aperturing means and into the detection region of a fused silica capillary where the polyimide coating has been removed. Wavelength selection is achieved using either optical filters or a diffraction grating and the transmitted radiation is directed onto a photodetector, usually a photodiode. The detector preamplifier converts the photocurrent to voltage and supplies the appropriate gain and electronic filtering. An analog-to-digital converter (ADC) digitizes the output signal for PC-based data collection. A second detector is often employed as a reference channel to eliminate source drift. It is important to bear in mind that the absorbance signal is derived from the ratio of the transmitted intensity when the capillary is filled with the separation medium alone (background transmittance) and the transmitted intensity when sample is migrating through the detection region. In a system like that depicted in Figure 2, the background transmittance value is measured at the onset of the run when no sample is present in the detection volume and this value is continuously ratioed against the run-time transmittance values.

When light passes through the separation medium containing absorbing molecules, the extent of absorption depends on the number of photon–absorber interactions. When an absorbing molecule absorbs a photon and is promoted to an excited electronic state, we observe a decrease in the intensity of the transmitted light reaching the photodetector. The number of photon–absorber interactions is a function of the sample concentration, the cell path length, and the illumination intensity. A fundamental limitation of the absorbance mode of detection is the fact that one is looking for a small difference between two relatively large signals; this problem becomes more severe as the concentration of absorbing analyte is reduced and generally places a lower boundary on the limit of detection.

The Lambert–Beer law indicates the role played by several design variable parameters in absorbance detection. The law states:

$$\mathrm{Log}(I_0/I) = A = \varepsilon bc$$

where I_0 is the intensity of radiation incident on the sample, I is the intensity of radiation emergent from the sample, A is the measured sample absorbance, ε is the molar extinction coefficient of the sample at the selected wavelength, b is the optical pathlength, and c is the molar concentration of the sample in the detection volume.

The primary disadvantage of absorbance detection for CE is the relatively high MDC, generally in the 10^{-6} M range, resulting in a linear dynamic range of two to three decades at best. The poor MDC is caused by geometric constraints imposed by the small capillary dimension, separation channel, and the inherently insensitive nature of absorbance detection. Much effort has been devoted toward increasing the utility of absorbance detection for CE by reducing the minimum detectable concentration. Most of the published works in this area have involved adapting approaches borrowed from the closely related field of microcolumn liquid chromatography[17-19] to the narrow fused silica capillaries employed for capillary electrophoresis separations. These approaches to improved detector performance can be divided into the following general categories.

1. Extension of absorbance pathlength
2. Reduction of detector noise
3. Maximization of separation efficiency
4. Optimum wavelength selection

A. EXTENSION OF ABSORBANCE PATH LENGTH

The limited sensitivity of UV absorbance detection in CE is expected because the optical pathlength is usually restricted by the capillary inner diameter, as illustrated in Figure 3A. The internal diameters of cylindrical capillaries employed for CE range from 2 to 200 μm, with 25 to 100 μm being most common. A dramatic decrease in separation efficiency is often observed above 100 μm due to resistive heating and the MDC is severely limited below 25 μm. The simplest and most commonly utilized approach to absorbance detection is transcolumn illumination, in which a spatially narrow band (detection length) of light of the selected wavelength is passed through the capillary walls in a direction orthogonal to the capillary axis (Figure 3A). Because the capillary is cylindrical, the average sample path length utilized for the transcolumn absorbance measurement is somewhat less than the capillary inner diameter. The mean effective path length for the transcolumn arrangement[38] may be approximated by

$$b = (\pi \times \text{column diameter})/4$$

Note that for a 50-μm I.D. capillary this would result in an effective mean path length of 39 μm and an absorbance value less than 1/250 of that measured on a spectrophotometer, using a standard 1-cm cuvette! Perhaps the most active area of absorbance detector improvement has been that of path length extension. Several different approaches to extending the effective path length have been reported in the literature and are described in the following sections.

Tsuda et al.[39] have extended the sample path length to as much as 1 mm while maintaining or even improving the heat dissipation capability of the capillary by employing rectangular cross-sectional channels. This work involved a transcolumn arrangement in which the source light was directed along the major axis of the rectangular capillary, as illustrated in Figure 3B. The authors reported the use of thin, rectangular borosilicate capillaries for absorbance measurements at 310 nm.

A trade-off between sample path length (absorbance) and optical throughput (noise $\propto$ [1/(throughput)$^{1/2}$]) is expected as the aspect ratio of these rectangular capillaries is increased. In a series of papers discussing general principles for the characterization of liquid phase flow through detectors, Poppe[40,41] has pointed out that simply increasing the absorbance path length while maintaining a fixed detection volume will not generally result in an increase in signal-to-noise ratio despite the increased absorbance. The primary reason for this is that the larger absorbance values are more than offset by an increase in detector noise caused by a decrease in light transmitted by the narrower cell (shot noise limit).

The approach taken by Tsuda et al. increased both the absorbance path length and the detection volume while maintaining a separation channel geometry that afforded efficient heat dissipation. The authors selected a 50 × 1000 μm rectangular geometry for application of this

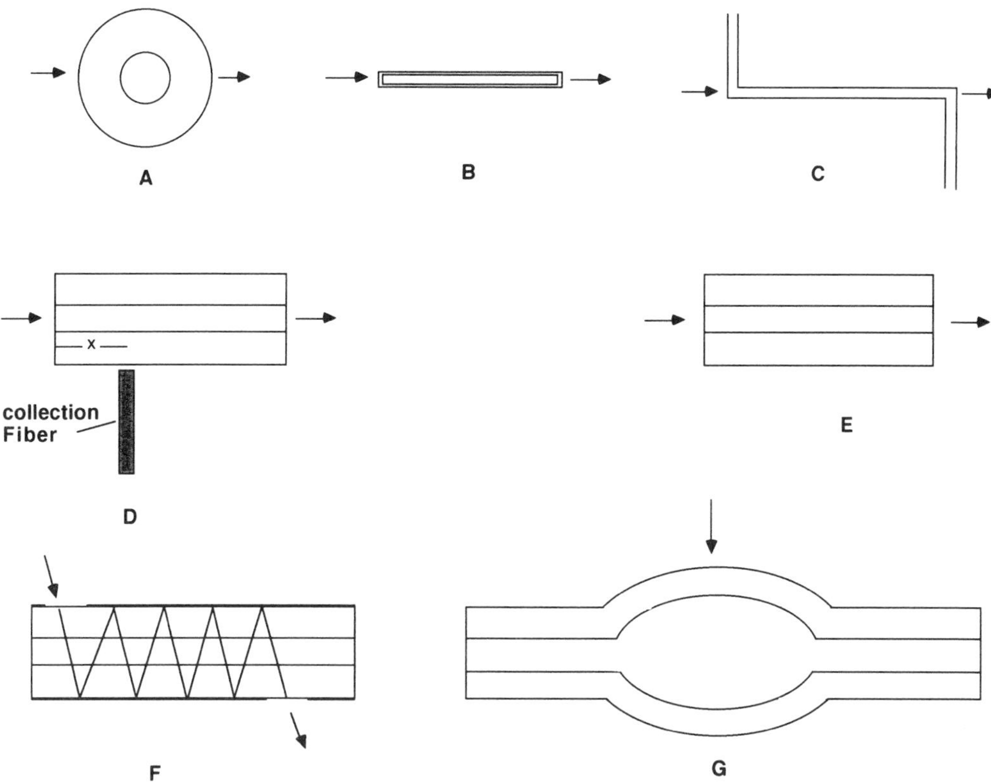

FIGURE 3. Several approaches to path-length extension that have been reported for CE. (A) The standard trans-column configuration; (B) the rectangular capillary; (C) the Z-type flow cell; (D) indirect absorbance determination by fluorescence monitoring; (E) axial illumination; (F) the multireflection cell; (G) the bubble cell. (Adapted from Refs. 38 to 49.)

approach to capillary electrophoresis. This represented a cell volume increase of greater than a factor of 25 per unit length and a 20-fold path length increase in comparison with a conventional 50-μm I.D. cylindrical capillary. For this 20-fold pathlength increase, the authors achieved approximately a 15-fold improvement in minimum detectable concentration, extending the MDC to the high 10^{-7} *M* range. One concern with the use of rectangular separation channels is the effect of the capillary corners on flow profile and electrophoretic resolution. The authors, however, found minimal peak asymmetry and band broadening caused by the nonconventional channel geometry.

Their work was limited to the use of borosilicate glass capillaries due to a lack of commercially available fused silica capillaries in a rectangular format. Thus the authors were precluded from operation at the most common UV wavelengths, which lie below the optical cutoff of borosilicate glass. The future commercial availability of high optical quality, fused silica, rectangular capillaries clad with polyimide should allow the use of rectangular capillaries to become a practical alternative.

Chervet et al.[42] and Bruin et al.[43] have reported the construction and evaluation of either Z- or U-shaped flow cells for CE. In the work of Chervet et al., formation of a Z-shaped flow cell resulted in extension of the sample path length to as much as 3 mm (Figure 3C) while still employing cylindrical capillaries. In their report the capillary was actually bent into a Z shape allowing axial, on-column detection along a 2- to 3-mm length of the capillary. No loss in electrophoretic resolution was observed as a result of the capillary bending. Again a tradeoff between sample path length and optical throughput is expected as the length of the flow cell region is increased and the column diameter is reduced. If the flow cell length is increased without a decrease in the cylindrical cross-section of the capillary there is the further concern with loss of electrophoretic resolution due to increased detector variance; this will ultimately limit the gains afforded by this approach. The authors reported a sixfold improvement in signal-to-noise ratio using the 3-mm Z-shaped flow cell at 254 nm and minimal efficiency loss caused by added detector volume. This approach has recently been commercialized and Z-shaped flow cells are currently available for a variety of different types of CE absorbance detectors.[44,45]

Another approach to path-length extension that has been reported is axial illumination,[38,46,47] in which the source light is focused into the capillary bore and directed along the capillary axis (Figure 3D). Grant and Steuer[38] first reported the axial illumination approach for indirectly measuring absorbance in CE. In their work a HeCd laser was selected as the axial UV light source and fluorescein was added to the electrophoresis buffer. In this arrangement, the fluorescence intensity at any point X along the length of the illuminated column is proportional to the light intensity (I_0 or I) at that point. If absorbing sample components are present within the monitored portion of the system (labeled x in Figure 3D) the absorbance magnitude may be derived from the resulting decrease in fluorescence intensity.

Using this approach the authors extended the absorbance path length to the width of the sample zones (2 to 3 mm in this instance), using a 50-µm I.D. capillary. The tradeoff between added detector volume and increased sample path length would be acceptable for some moderate resolution applications. Although absorbance values consistent with the extended path length were observed, the overall detection limits for this approach were poor. The primary limitation was the lack of a suitably stable, high-intensity UV light source and complications arising from photobleaching of the fluorophore in the illumination path.

Taylor and Yeung[46] reported "whole column" axial absorbance (Figure 3E) for capillary electrophoresis with direct absorbance. In this work the output beam of a red HeNe laser was focused into the inlet end of a 50-µm I.D. fused silica capillary and light exiting the outlet end of the capillary was imaged onto a photodetector. When the capillary was filled with standard aqueous buffer systems the refractive index of the separation medium was less than that of the surrounding fused silica capillary walls and only a fraction of the light passed along the capillary length by partial internal reflectance. The light exiting the capillary contained both wall-propagated and separation medium-propagated rays. The two were separated by spatial filtering and only the signal arising from the separation medium-propagated rays was monitored.

The incorporation of additives in the electrophoresis buffer that raised the refractive index of the separation medium above that of the capillary wall material eliminated the wall-propagated rays. In such cases the capillary acted like an optical fiber with high transmittance for the source light along the capillary "core." Taylor and Yeung were able to accomplish this by using PTFE capillary tubing (refractive index, 1.35 to 1.38) and adding ethylene glycol to the electrophoresis buffer.

With the axial illumination approach the authors reported a 15-fold improvement in detection limit over the transcolumn arrangement. The primary sources of noise limiting further improvement in detector performance were found to be capillary vibration caused by electrostatic motion in the applied electric field and the relatively unstable intensity of laser sources. Practical use of this system for CE separations would be limited by buffer restrictions and the relatively small number of analytes with absorbances suitable for the single-frequency red HeNe laser source.

Xi and Yeung also described an axial illumination absorbance system for CE, which incorporated a deuterium lamp source in place of laser excitation.[47] This system had the advantages of being wavelength variable and incorporating a much more stable illumination source, thereby making the system more suitable for absorbance measurements. This work was performed using a totally nonaqueous separation medium (dimethyl sulfoxide) in order to meet the criteria for total internal reflectance. Although the source intensity instability problem was apparently eliminated by replacement of the laser source with a deuterium lamp, the overall system performance was limited by capillary vibrations. The reported concentration limits of detection were 1.9×10^{-7} *M* for acridine and 1.1×10^{-6} *M* for 3-aminoquinoline. Again, restriction of the separation medium to solutions having refractive indices greater than that of the capillary wall material limit the practical applicability of this approach.

Hartwick et al. reported a 44-fold absorbance path-length extension and a 40-fold improvement in detection limits by developing a nanoliter scale multireflection cell.[48] The multireflection cell (Figure 3F) was fabricated on column by silver coating (mirroring) a short length of the detection region of the capillary. Two light windows were left in opposite sides of the capillary and served as the entrance and exit points for laser illumination. The distance between the entrance and exit extremes was 1.5 mm, resulting in a detection volume of 6.6 nl for a 75-μm I.D. capillary. A 5-mW red HeNe laser was used as the light source and a photomultiplier tube was selected for detection. The multireflection cell was compared with a conventional single pass cell using the same diameter capillary and brilliant green as the test analyte. The authors reported an extrapolated detection limit of 6.5×10^{-8} *M* for static measurements and a linear dynamic range of approximately 100. For CE separations, increased noise level and analyte dilution by zone broadening raised the extrapolated MDC to 3×10^{-7} *M*.

The multireflection approach is attractive as it imposes no major constraints on the composition of the separation medium and does not require major modifications to the capillary column — other than generation of the mirrored surfaces. The practical utility of this absorbance detection scheme for CE is, however, limited due to the use of a laser as the light source. Substitution of the laser source with a broadband lamp source greatly expands the utility of the multireflection approach, but is expected to be technically quite difficult in this arrangement due to the optical characteristics of lamp source radiation.

A particularly simple means of path length extension has been reported by Heiger.[49] This approach involves the expansion of the detection region of the capillary to form a "bubble cell" as depicted in Figure 3G. Expansion factors (or "bubble factors") as great as 4 have been utilized to improve detection limits in narrower bore fused silica capillaries. For example, a 25-μm I.D. capillary allows the application of relatively large fields to accomplish fast separations at lower current levels and a 100-μm bubble provides a fourfold path length extension.

B. REDUCTION OF DETECTOR NOISE

Considerable effort has also been directed toward reduction of the detector noise level in order to extend the MDC for absorbance detection in CE. In general, a compromise must be reached between the need to maximize the amount of light passing through the separation channel in order to reduce the effect of shot noise, and the need to limit the interrogated capillary volume in order to maintain maximum electrophoretic resolution. Ease of capillary replacement by the system operator is also an issue of major importance for design of commercial CE systems and this often impacts the optical throughput.

Capillary electrophoresis absorbance detectors are generally shot noise limited. Here the detector noise level is dominated by statistical fluctuations in photon flux reaching the photodetector. For absorbance detectors operating in this regime, the detector noise level will vary inversely with the square root of the number of photons reaching the photodetector at any given point in time. As a result, the detector noise level can be reduced by simply increasing the optical throughput of the system. Table 1 illustrates this point by considering two optical arrangements, the first of which provides a photon flux of 10^7 photons reaching the photodetector and the second providing 10^8 photons.

As illustrated in Table 1,[50,51] the tenfold increase in throughput for optical arrangement 2 results in roughly a threefold reduction in detector noise level. If this throughput increase is

TABLE 1
Effect of Optical Throughput on Detector Noise Level

	Optical arrangement 1	Optical arrangement 2
Photon flux:	10^7 photons	10^8 photons
Shot noise:	$(10^7)^{0.5} = 3162$ photons	$(10^8)^{0.5} = 10{,}000$ photons
σ_T	0.00044	0.00014
σ_{AU}	190 μAU	60 μAU

Note: The shot noise for each intensity measurement is simply the square root of the number of photons. This shot noise produces an uncertainty in the T reading given by $(2/I_0)^{1/2}$. At the detection limit, two measurements are made, ratioed, and the log taken. Using standard propagation of errors, the corresponding baseline noise expressed in absorbance units (AU) is given by $\sigma_{AU} = (0.434\sigma_T/T)$.[50,51]

achieved without a decrease in absorbance path length, the result would be a threefold improvement in signal-to-noise ratio. This treatment assumes that all of the light reaching the photodetector is transmitted through the sample in both optical arrangements and that the tenfold increase in throughput is not accompanied by an increase in spatially stray light. This is an important point as additional stray light can reduce the detector noise level but will not be absorbed, and can result in a nonlinear working curve.

Detector noise can be reduced in the following ways:

1. More tightly focusing the output of existing lamp sources to increase throughput
2. Employing stable lamp sources with greater light output
3. Maximizing the effective detection area by optimal aperturing for a given application
4. Employing effective electronic and digital filters

1. Focusing

Most absorbance detectors described for CE applications, whether they are modified HPLC detectors or detectors designed specifically for capillary electrophoresis, are very inefficient at coupling the source light with the capillary separation channel. For example, the area illuminated by the source light at the capillary plane is often several orders of magnitude larger than the capillary aperture, which defines the illumination target (Figure 4A). As a result only a small fraction of the available source energy passes through the detection cell and the rest of the light is blocked by the aperture. In some instances this is done intentionally, with the aperture attached to the capillary, in order to simplify capillary positioning during operator installation.

Weinberger[52] described the use of a spherical fused silica "ball lens," which served to increase throughput by focusing the source light down to a spot size comparable to the capillary internal diameter, as illustrated in Figure 4B. Ray trace analysis indicated that further focusing by the capillary wall reduced the spot size to just less than the capillary inner diameter, thus directing all rays to pass through the separation channel. Focusing of the source light in this manner resulted in a 100-fold reduction in detector noise level in comparison with an apertured detector arrangement. Good detector linearity was reported for over three orders of magnitude, indicating the presence of minimal stray light.

Fiber optic illumination/collection for CE/UV absorbance detection has also been reported by Foret et al.,[35] Bruno et al.,[36] and Ludi et al.[37] These groups discussed the potential of simultaneous multimode detection, i.e., fluorescence and absorbance or absorbance and conductivity. An on-column capillary flow cell design that was suitable for CE was reported

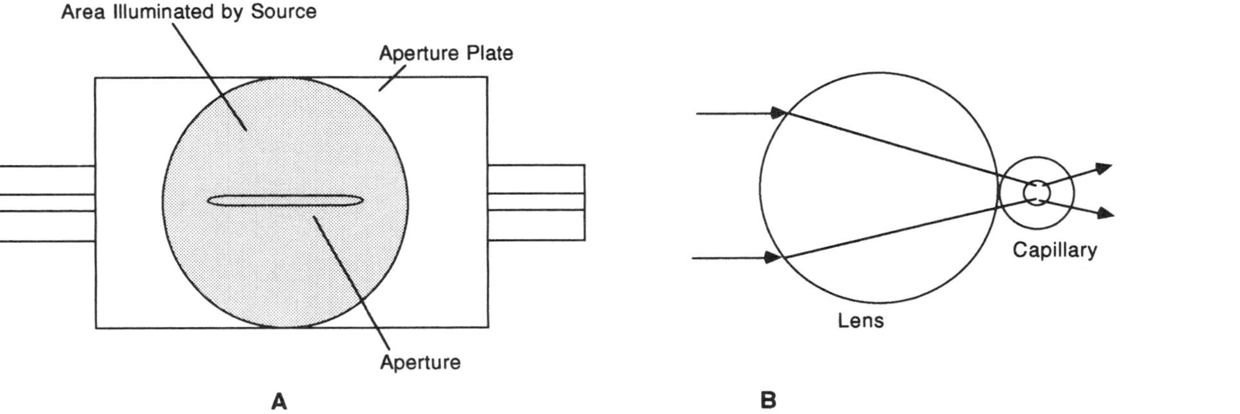

FIGURE 4. (A) An example of a large illuminated area superimposed on a relatively small-area aperture. This arrangement limits optical throughput but relieves capillary positioning requirements during column replacement. (B) Ball lens arrangement illustrating the ability of the lens to effectively couple light into the separation capillary. ([B] adapted from Ref. 52.)

by Bruno et al.[36] This cell used an optical fiber in combination with the focusing action of the capillary wall to achieve efficient illumination. A second, larger optical fiber was selected for collection of the transmitted radiation. The fiber optic interface is attractive because it has the potential for remote location of the source and detector, great ease of capillary replacement, and illuminating multiple capillaries from a single light source.

2. Lamp Sources

Lamp sources are generally selected on the basis of spectral output (wavelength and intensity) and stability. The stability of state-of-the-art lamp sources can be quite high. The spectral output requirements are determined by the application(s) to be addressed by the system. For an application-specific system, lamp sources delivering intense, discrete lines are an attractive option, especially in the UV spectral region. For a variable wavelength or multiwavelength system, broadband spectral distribution is preferable, making the deuterium lamp an attractive choice.

3. Aperturing

Aperturing or other means of physically defining the detector viewing length needs to be given a great deal of attention. With respect to reduction of detector noise, one should select the maximum tolerable detection length ($\sigma^2_{detector} = 0.1 \times \sigma^2_{peak\ min.}$) for the application to be addressed in order to maximize the optical throughput. For example, DNA sequencing separations performed using capillary gel electrophoresis tend to be very high resolution. Here theoretical plate numbers of several million are not uncommon and the detector length needs to be carefully limited in order to avoid resolution losses caused by detector variance. On the other hand, the separation of heterogeneous protein mixtures, such as the human serum proteins, tend to be much lower in resolution and a relatively large detection length can be used to increase optical throughput and reduce detector noise levels. For a general purpose CE system a variable detector viewing length is highly desirable.

4. Electronic and Digital Filtering

Nearly all CE detectors, whether home-made or commercial, employ some type of electronic and/or digital filtering to reduce noise. For repetitive analyses performed using an application-specific system in which peak widths are well defined and characterized, it becomes possible to optimize both electronic and digital filtering. For general purpose CE systems, variable (software selectable) filter settings are desirable in order to tailor the degree of filtering to the application at hand.

C. MAXIMIZING SEPARATION EFFICIENCY

The detectable sample concentration range can be extended by working to maximize separation efficiency.[53] Greater efficiency implies sharper, more concentrated sample zones and hence better detectability. The sample concentration techniques mentioned in Section II.C can be employed to improve the sample introduction process. Enrichment factors as large as 100 to 1000 have been reported in the literature. Careful selection of buffer systems and sample solvents and restriction of Joule heating also lead to improved efficiencies.

D. OPTIMUM WAVELENGTH SELECTION

Selecting the optimum wavelength for the analyte/separation medium combination at hand is also important. Simply taking the time to measure absorbance profiles on a spectrophotometer of both the separation medium and the sample dissolved in the separation medium (provided enough material is available) provides the CE practitioner with the necessary information to properly select the detector wavelength. The use of sub-200-nm illumination

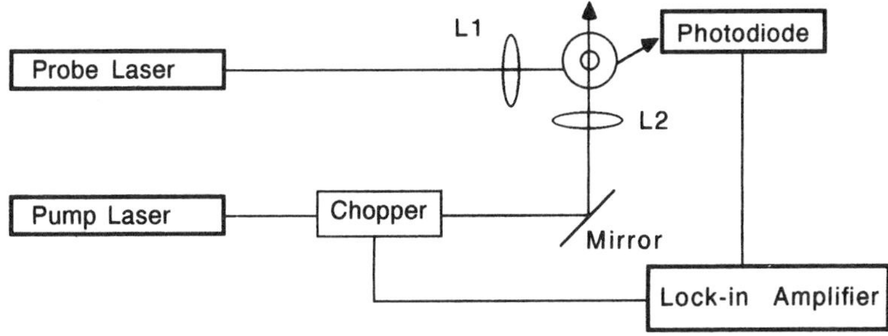

FIGURE 5. Thermooptical absorbance detection system for CE. The presence of an absorbing analyte is measured by heating the sample/separation medium using a pump beam, and then measuring the resulting refractive index change, using the probe beam. (Adapted from Ref. 62.)

has proven to be advantageous (though buffer selection is limited) when working with proteins and other biological molecules.[54] Several reports have been made of multiwavelength monitoring using either rapid scanning gratings or diode array detection (DAD),[55-58] and both of these options are now commercially available. These detectors provide additional spectral information on the fly that can be used to assist in analyte identification, peak purity assessment, or prerun screening to determine optimum wavelength monitoring settings.

V. THERMOOPTICAL DETECTION

Thermooptical detection is an absorbance-based detection scheme that has been successfully coupled with capillary electrophoresis by Yu and Dovichi,[59-61] Bornhop and Dovichi,[62] Waldron and Dovichi,[63] and Earle and Dovichi.[64] This detection scheme involves the use of two laser beams that intersect at the detection region of the separation capillary, as illustrated in Figure 5. The wavelength of the "pump" beam is selected to match the absorbance profile of the analyte of interest. The second beam, the "probe," is typically a red HeNe laser that monitors changes in the refractive index of the separation medium caused by thermal changes from the absorbance of the pump beam by the analyte. The analytical signal arises in this detection scheme when the probe beam reaching the photodiode is deflected on a refractive index change in the beam-crossing volume.

Dovichi et al. have described thermooptical absorbance systems for several different chromophore/laser combinations. A 4-mW HeCd laser (blue line, 442 nm) served as the pump source for the analysis of DABSYL-derivatized amino acids separated by CE[59] and thermooptical detection resulted in a mass LOD of 200 amol for DABSYL-glycine. Thermooptical absorbance sensitivity is proportional to the pump laser power (up to saturation) and higher power lasers can be useful for weakly absorbing or low-concentration analytes. In a subsequent report, Yu and Dovichi utilized a 130-mW argon-ion pump laser (458 nm) to achieve a mass detection limit of 37 amol (5×10^{-8} *M* concentration detection limit) for DABSYL-methionine.[60] In a more recent report, Waldron and Dovichi used an excimer laser (KrF, 248 nm, 5-mW average power) to detect phenylthiohydantoin (PTH) derivatized amino acids and were able to achieve a detection limit of 9×10^{-7} *M* for PTH-glycine.[63] Earle and Dovichi also reported a two-color thermooptical absorbance detector for capillary electrophoresis.[64] In this system, each of the two lasers served a dual role as both pump and probe, and the system was used to simultaneously monitor the absorbance of two different dye molecules.

Bruno et al. have also reported a thermooptical absorbance detector for capillary electrophoresis that used 25-µm I.D. capillaries and a frequency-doubled argon-ion laser to supply an ultraviolet (257-nm) pump beam.[65] This system has been applied to the analysis of dansylated amino acids.

VI. FLUORESCENCE DETECTION

Fluorescence detection is the most sensitive detection mode available for CE. The goals of most capillary-based detection methods are to maintain the concentration sensitivity of the much larger volume HPLC-based detection systems while the detector cell volume is reduced by orders of magnitude, thus yielding greatly superior mass sensitivities. In fluorescence, the detection limits of a well-designed laser-based system can be below 10^{-12} *M*, and mass limits of detection can be less than 1000 molecules. Until recently, the lack of commercial systems, problems with derivatizing chemistry, and the custom nature of the most sensitive systems all contributed to fluorescence being used by relatively few researchers. Recently, however, several commercial fluorescence systems have appeared (i.e., Beckman Instruments, Europhore, and Dionex). Unlike UV-Vis absorption, the ultimate sensitivity obtainable from fluorescence is not strictly path-length dependent, and laser-based systems can be miniaturized to the smallest diameter capillaries.

In fluorescence, analyte molecules absorb a photon, and a fraction of the electronically excited molecules emit a photon upon returning to the ground state. For a molecule to be a good fluorophore, it needs a high absorptivity at the excitation wavelength. After the molecule is excited, there are a number of possibilities, only one of which produces fluorescence: the molecule can return to the ground state by emitting a photon (fluorescence), can return to the ground state by a nonradiative process, can undergo a triplet state conversion, or can even react to form a nonfluorescent product. The most important parameters that determine the analytical utility of a fluorophore are the absorptivity, fluorescence quantum yield, and photostability. Just as in UV-Vis absorbance detection, having a higher absorptivity means that the molecule is more likely to be excited at a given illumination intensity. The fluorescence quantum yield is the fraction of the excited molecules that emit photons; therefore, having a high quantum yield is also important. For the most sensitive fluorophores, the quantum yields can approach unity, implying that for almost every photon absorbed, a photon is emitted.

One of the most important parameters is the photostability of a fluorophore. Photostability is commonly reported in terms of the average number of excitation/emission cycles a molecule can undergo before being destroyed. For fluorescein in water, Hirschfeld reported an average of 8000 cycles[66] and for the best rhodamine-based fluorophores, the number can be several hundred thousand cycles. Note that, in the absence of background, the absorptivity and quantum yield of a molecule are not the most important parameters determining the fluorescence signal from a molecule; the integrated fluorescence signal is dictated only by the photostability.[66] The photodestruction mechanisms are poorly understood for most fluorophores, and can be quite complex. As one example, the photodestruction rate for fluorescein depends on the amount of free oxygen in solution and the average number of cycles can be increased by the addition of oxygen scavengers.[67]

It is the task of the fluorescence system to efficiently excite the molecules and collect as large a fraction of the emitted photons as possible. While the intensity of the spectral background tends to be small, a number of background sources exist that restrict the detection limits obtainable for fluorescence measurements. The three most important background sources are reflections and Rayleigh scattering, Raman scattering, and background luminescence. Rayleigh scattered light is simply scattered light of the excitation wavelength; since the desired fluorescence signal will be at longer wavelengths, in principle, Rayleigh scattering can be removed by using the appropriate filters or monochromators. While it is well known that the Raman process is inherently weak, the solvent is present at many orders of magnitude

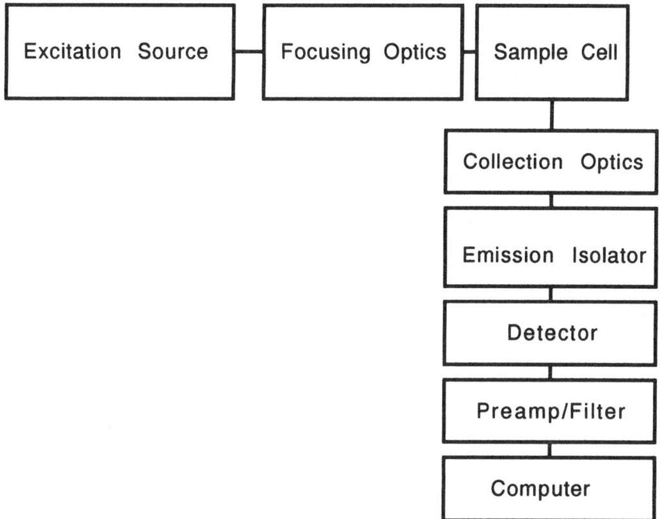

FIGURE 6. Functional block diagram of an LIF detector for CE.

higher concentration than the analyte and so Raman scattering becomes one of the most serious background sources at low fluorophore concentration. For aqueous solutions, the two major Raman emissions occur at 1650 cm^{-1} and in the range of 3100 to 3600 cm^{-1}; however, several much weaker transitions exist and often limit the ultimate sensitivity of the measurement. Other sources of background can include luminescence from the capillary walls and from background impurities in the separation medium. Even for the highest quality water, trace impurities are often the limiting background source for extremely low-concentration fluorescence detection. At the fluorescence detection limit, the background is often orders of magnitude more intense than the fluorescence signal, and, as in an absorbance measurement, the detection limit is determined by the ability to distinguish a small signal on a larger background. As the various geometries and instrumental configurations described below demonstrate, there are a number of methods to maximize fluorescence signal while minimizing the contributions of these background sources.

As mentioned above, fluorescence detection methods have some of the best performance characteristics of any CE detection mode in terms of linearity, sensitivity, and limits of detection; however, unless one is extremely fortunate, most samples of interest are not fluorescent. Thus, a large amount of the work published deals with methods to convert samples into fluorescent forms (i.e., derivatization, either pre-, on-, or postcolumn), or on combining the desirable properties of fluorescence with a less restrictive mode such as immunoassay or indirect detection. A brief overview of derivatization methods is presented in Section VI.D.

The basic fluorescence system is illustrated in Figure 6, and consists of an excitation source, optics to focus the excitation source onto the capillary or other sample cell, the sample cell itself, collection optics, and the detection system. Not surprisingly, there are a great variety of optical configurations that have been reported for fluorescence detection; this chapter does not attempt to review all configurations but concentrates on those that have demonstrated desirable performance characteristics and show promise of widespread applicability.

A. EXCITATION SOURCE

The first requirement of a fluorescence instrument is to excite the molecules of interest. Thus, an excitation source of the appropriate intensity, wavelength, and spatial characteristics

TABLE 2
Wavelength Range of Broad-Band Sources and Lasers

Source type	Wavelength (nm)
Lamp	
Deuterium	190–400 (continuous)
Xenon	190–2000 (continuous)
Hg-xenon	190–2000 (continuous)
Hg — low pressure	200–550 (line)
Laser	
Ar-ion — air cooled	457, 472, 476, 488, 496, 501, 514
Full frame and υ-doubled	Above visible and 275, 300, 305, 333, 351, 364, 385, 229, 238, 244, 248, 257
Ar/Kr	Above visible and 350–360, 521, 531, 568, 647, 752
HeCd	325, 354, and 442
HeNe	543, 594, 604, 612, 633
Excimer	
XeCl	308
KrF	248
Nitrogen	337
Nitrogen pumped dye	360–950 (tunable)
Solid state	
YAG (υ doubled)	532
YAG (υ quadrupled)	266
Diode lasers	620–infrared

becomes crucial. Excitation sources can be divided into two categories: coherent and incoherent. Coherent sources (lasers) offer the advantages of a spectrally narrow line source, and the ability to focus to very small beam diameters. The largest disadvantage is the limited number of emission lines available for reasonably priced lasers. Several of the most promising derivatizing agents for fluorescence detection such as fluorescamine have not been widely used for CE/LIF because of the lack of laser sources that match the excitation spectra of the label. Table 2 lists several commonly employed excitation sources with their excitation lines or wavelength range.

The first detector Jorgenson and co-workers described in their landmark CE work was a postcolumn fluorescence detector using a line source and various filters to isolate the excitation and emission wavelengths.[4,68-70] While not very sensitive by today's standards, such systems were easy to construct and demonstrated potential sensitivity gains offered by fluorescence detection in a capillary separation. Green and Jorgenson described an improved variable wavelength on-column fluorescence system that offered both wavelength tunability throughout the UV region and low stray light (by using a double monochromator) with an improved detection limit in the range of 0.1 to 1 μM.[70] Since that time, several reports have described further refinements, including a recent comparison of the performance of fluorescence systems based on a variety of broadband excitation sources ranging from deuterium, tungsten, and arc lamps to a system that used the 488-nm line of an Ar-ion laser.[71] For this study, a fiber optic-based collection system was used, which the authors claim reduces stray light from the lamp sources. The LOD for the Xe arc source for fluorescein-labeled glycine was 3×10^{-7} M; using the laser excitation system, the reported LOD was 5×10^{-10} M. While demonstrating a poorer LOD, the Xe arc lamp allowed detection of the native fluorescence of tryptophan-containing peptides, using ultraviolet excitation. Hernandez et al. described a fluorescence detector based on an epi-illuminated fluorescence microscope that achieved a limit of detection for riboflavin of 0.5 μM. Further refinements in this system, including a laser excitation source, greatly improved the detection limits. The LIF version is described in greater detail below.[72]

As excitation sources and focusing optics are optimized for CE, the performance offered by commercial lamp source fluorescence systems is expected to improve. Dovichi recently demonstrated the improvements possible with an incoherent source by designing a high collection efficiency filter fluorometer using a 75-W Xe arc lamp source and a sheath flow cuvette (to reduce scattering from the capillary walls). Using this system, concentration detection limits for fluorescein isothiocyanate (FITC)-labeled amino acids are approximately 10 p*M*, corresponding to 0.2 amol of tagged amino acid injected into the column.[73] While the performance of this system represents a large increase in sensitivity for a lamp-based system, the use of filters to select excitation wavelength and the non-UV-transmitting optics reduced one of the greatest advantages of using a broadband source — the increased flexibility in wavelength selection, especially in the ultraviolet. We expect that tunable, optimized broadband fluorescence systems will become available that will offer a performance in between the ever popular UV-Vis absorbance detector and the state-of-the-art laser-based fluorescence systems.

Lasers are particularly well suited for focusing into small capillaries and can be focused near the diffraction limit of light.[74] Another advantage that is often overlooked is that a monochromatic excitation source implies narrow Rayleigh, and narrower, simpler Raman features that are easier to filter or remove than the spectral background found with Xe arc lamps. A last advantage is that laser power can be varied over a wide range and can be optimized to the fluorescence system being studied; thus higher powers can be used for more photostable fluorophores and lower powers for less stable molecules. Gassman and co-workers described the first laser-based fluorescence detection system for CE.[75] In their original system, they focused a 325-nm HeCd laser into a fiber and used this fiber to illuminate a region of the capillary; a block diagram of this system is shown in Figure 7A. This first LIF system had detection limits of approximately 200 amol (0.1 μ*M*) for dansylated amino acids.

Since the original work, the Ar-ion and HeCd lasers have dominated laser-induced fluorescence (LIF) applications. In part, this is because the 488-nm emission of the Ar-ion laser almost exactly matches the excitation peak of 490 nm for one of the better fluorescing agents available today, fluorescein (available as a derivatizing agent in such forms as fluorescein isothiocyanate and fluorescein succinimidyl ester). The concentration detection limits obtainable with such systems will be discussed in detail below, but can be better than picomolar. The 325-nm emission of the HeCd laser works well with *o*-phthalaldehyde (OPA) and dansyl derivatizing chemistry, and the 442-nm line matches CBQCA and NDA. Also important, these are easy to use and relatively inexpensive lasers. Recently, the smaller and even less expensive green and yellow HeNe lasers have been used with various rhodamine derivatives such as tetramethyl rhodamine succinimidyl ester and Texas Red. For unmatched flexibility, the medium-power mixed Ar/Kr lasers offer "tunable" performance over a wide wavelength range. For example, the Innova Spectrum from Coherent has more than 20 emission lines over the 350- to 725-nm region.

Almost all LIF detection for CE has involved visible laser excitation. However, Swaile and Sepaniak[76] first demonstrated the use of UV excitation for native fluorescence detection, and Yeung's laboratory improved the detection limits several orders of magnitude by using the 275-nm line from a large Ar laser (this line is obtained only from expensive, 440-V full-frame lasers).[77] Using this emission line, they report nanomolar detection limits for tryptophan, and a 10^{-10} *M* detection limit for proteins that contain many tryptophan molecules. They have recently described the detection of native proteins in single red blood cells, using this system.[78] Additional work by the same group has demonstrated the detection of native fluorescence of several nucleotides, using UV excitation, with detection limits in the best cases of 10^{-8} *M*.[79] Several other groups have looked at pulsed UV excimer lasers for CE detection; for example, Chan and co-workers demonstrated nanomolar sensitivity for tryptophan and bovine serum albumin, using a 248-nm KrF excimer laser.[80]

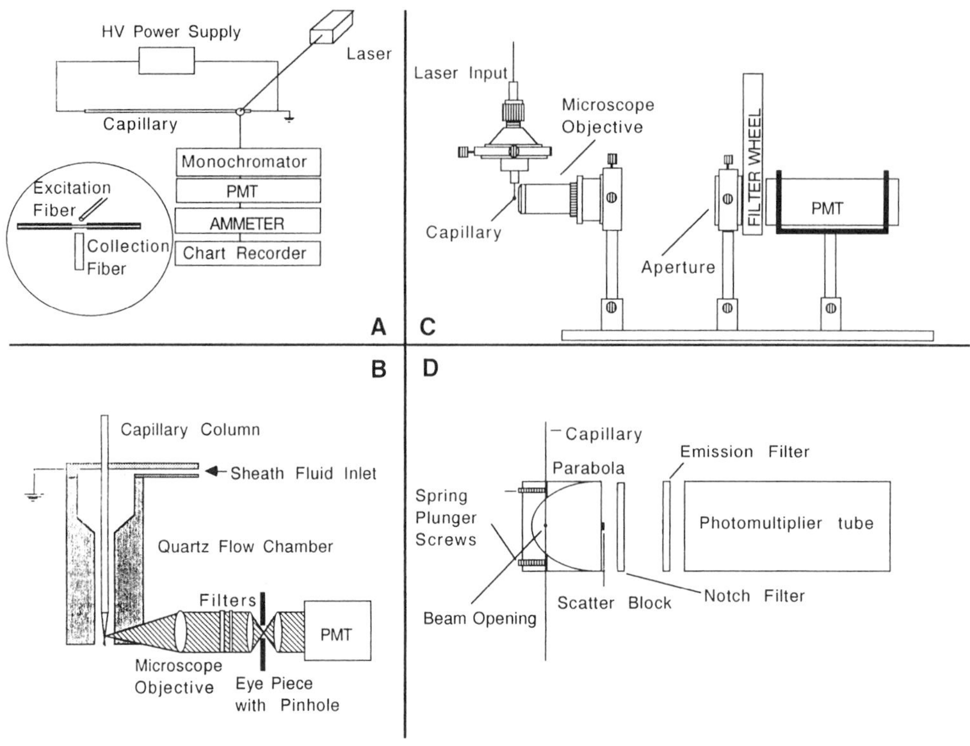

FIGURE 7. Various optical arrangements reported for CE/LIF. (A) The first reported CE/LIF interface, utilizing optical fibers; (B) the sheath flow cuvette interface; (C) a typical CE/LIF arrangement, using a microscope objective and scatter-eliminating aperture for collection; (D) parabolic reflector with scatter blocking mask. (Adapted from Refs. 75, 91, and 106.)

Another area of considerable interest is in the use of near-infrared (IR) excitation, especially as the availability of high-quality solid-state red and near-IR lasers improves. These lasers are attractive because of the reduced background, potentially lower cost, reduced size and superior noise performance. Red and near-IR diode laser detection have recently been reported by Chen et al.[81] and Williams et al.[82] In the near-IR regions, very few native materials are fluorescent. While this can be an advantage because of a simpler fluorescence background, it makes finding the appropriate derivatizing chemistry a challenge. While substantial progress has been made and more is sure to follow on developing near-IR tagging chemistry,[83,84] the currently available chemistries do not have the necessary combination of optical and chemical characteristics to make near-IR LIF a practical solution to general fluorescence problems. Higashijima et al.[85] have reported the use of the second harmonic (415 nm) of a near-IR semiconductor laser for CE/LIF analysis of DCCS (7-diethylaminocoumarin-3-carboxylic acid succinimidyl ester)-labeled amino acids. While a number of alternatives to visible lasers have been mentioned, it should be noted that the overwhelming majority of all LIF publications use visible laser excitation.

B. DETECTION CELL

One of the most critical of all components in a fluorescence detection system involves the detection cell. Replacing the standard fluorescence detection cell, the cuvette, with an on-column or postcolumn detection cell having a volume in the nanoliter to picoliter range poses a considerable challenge. Approaches that have been successfully implemented include using the capillary itself as the flow cell, using a postcolumn detection system such as a sheath flow cell, or using a constrained flow system. Each of these approaches are briefly described below.

Just as in UV-Vis absorbance, the most straightforward detection cell involves using the capillary as the flow cell. This obviously has the advantages of simplicity (the cell does not require any capillary connectors and does not degrade the separation efficiency). Guthrie and Jorgenson demonstrated one of the first on-column fluorescence detectors for modern CE,[69] and Gassman et al. demonstrated the first on-column LIF system.[75] As illustrated in Figure 7A, this first LIF system used optical fibers to both illuminate the capillary and collect the emission. Since this time, the majority of reported fluorescence detection systems have been based on on-column detection. While the majority of on-column methods use orthogonal illumination, two reports describe the use of illumination directed up the capillary — termed axial illumination.[86,87]

In addition to on-column detection, a variety of postcolumn detector cells have been reported. The most impressive in terms of performance is the sheath flow system developed by Dovichi and co-workers over a number of years.[88-91] Disadvantages of the on-column approach involve the broad-band luminescence background and scattering from the fused silica capillary, both of which are greatly minimized in a sheath flow system. The scattering and fluorescence background may be even higher in capillary gel electrophoresis, especially if UV excitation is used, and so alternative, off-column detection is important in this application. A sheath flow system is illustrated in Figure 7B, and shows the sheath fluid, the capillary, and the observation zone. Briefly, the end of the capillary is inserted into the sheath flow cell in which buffer is flowing at a rate that prevents mixing of the two flowing streams. In a properly operating system, the laser is focused at the output of the capillary and fluorescence is detected using a microscope objective. As there is no refractive index change at the junction between the two streams, little scattering occurs in the viewing field of the microscope objective. A pinhole placed at the image plane of the microscope objective ensures that only emission from the capillary flow stream is collected, futher minimizing scattered light. As the outlet of a CE capillary in a sheath flow system is slightly pressurized, the sheath flow cell does interact with the CE separation by producing a back pressure; this interaction has been explored in detail by Cheng et al.[92] While the pressure effect influences the amount injected unless the sheath flow is turned off during injection, the sheath flow cell allows high-efficiency separations with a million theoretical plates reported in best cases. The most impressive feature of this approach is detection limits in the range of 10^{-13} M (corresponding to ten molecules) for rhodamine-based fluorophores.[88,93] Recently, Dovichi et al. have duplicated Keller's single-molecule detection[94,95] in a sheath flow system[96], and we expect this to be coupled to a capillary separation in the near future. While single-molecule detection can be important in some applications, this is accomplished using the photon burst detection method, which severely decreases the dynamic range of the fluorescence measurement, and so this approach may have limited analytical utility.

Another postcolumn method recently reported is the deposition of the capillary output onto a membrane, followed by detection using a variety of techniques.[97-99] Huang and Zare demonstrated a proof of concept of this method; using a porous glass frit near the capillary outlet for grounding, they deposited the capillary effluent on filter paper and detected several fluorescently labeled amino acids.[99] While several other postcolumn methods have been explored, none have gained wide acceptance at this time.

C. COLLECTION OPTICS AND OPTICAL DETECTION

The goal of the collection optics, wavelength selector, and detector are obvious: to collect and detect as much light as possible with maximum discrimination against unwanted background sources. While the goals are straightforward, there is a wide variety of fluorescence implementations. The two most common collection methods include fiber optics and microscope objectives. Wavelength discrimination is almost always performed with a series of filters. Not surprisingly, the photomultiplier tube (PMT), whether operated in current mode or in a counting mode, has been used for almost all fluorescence detection to date, with four recent articles describing charge-coupled device (CCD) detection. Because the choices of collection optics, wavelength selector, and detector are linked, instead of covering each category separately, several representative approaches are described.

The microscope objective is the collection method that has yielded the lowest reported limits of detection; one embodiment is illustrated in Figure 7C. A microscope objective is designed to collect a fraction of light from a small spot and bring this spot back into focus; these goals exactly match the requirements of a CE fluorescence detection system. It is possible to collect nearly 50% of the light emission from a capillary with the highest numerical aperture objective; however, most objectives collect a much smaller fraction. Hernandez et al. have compared a number of objectives for LIF detection,[72,100-101] and, in a readable series of papers, Chen et al.[88] and Swerdlow et al.[89] have described the considerations one should take into account when designing an LIF system. As mentioned previously, the combination of a sheath flow cell and high collection efficiency microscope objective allows LODs of fewer than 100 molecules. An extension on the use of the microscope objective is to use an epi-illumination microscope as the illumination, collection, and wavelength-filtering system. Hernandez and co-workers have described such a system and have reported outstanding detection limits — below picomolar concentration (100 molecules) for fluorescently tagged analytes.[72,100-101]

Huang et al.[102,103] have described a laser-excited confocal fluorescence detection system for CE that allows rapid sequential monitoring of 25 capillaries run in parallel. This system has been reported for both single[102] and two[103] color DNA sequencing schemes and was directed toward high-throughput sequencing. The 25-capillary array was mounted on an *xy* translation stage, which was rapidly scanned beneath the confocal objective. This optical train coupled with argon-ion laser excitation allowed the detection of rhodamine at the low picomolar range and, more importantly, seems to have sufficient performance characteristics to meet the application being addressed.

It is important to note that while the fraction of fluorescence emission collected is often reported for optical systems, these numbers refer to the emission that leaves the capillary; a significant fraction is totally internally reflected in the capillary and the emission does not necessarily have a uniform spatial distribution.[104] Several reports have dealt with reducing the effect of internal reflection and scattering by placing the capillary into an index-matching fluid.[105] Of course, the sheath flow cuvette avoids this problem by having matched index liquids for both the capillary and sheath fluid.

These microscope-based systems generally include a spatial filter and wavelength filter located before the detector. The requirements of the wavelength filter are to pass a large fraction of the fluorescence emission while rejecting as much of the Rayleigh and Raman scattering as possible. While absorptive glass filters were used in many of the original LIF studies because of their high absorptivities, they tend to luminesce, which can interfere with low-level fluorescence measurements. If one is able to commit to a particular fluorophore/emission wavelength, one can purchase optical filters with extremely good Rayleigh and Raman blocking, and high fluorescence wavelength transmission. However, if a large number of fluorophores requiring different excitation lines will be investigated, having specially designed custom filters for each line becomes impractical. In addition to wavelength filtering, spatial filtering is important; the spatial filter consists of an aperture placed at the microscope objective imaging plane. The aperture size should be just slightly smaller than the size of the capillary inner diameter image formed by the objective. Such a spatial filter improves performance as only the light originating in the aqueous capillary channel passes through the filter. If a spectrometer is used instead of a wavelength filter, the spectrometer slit can be aligned with the capillary image. By choosing the spectrometer slit width to be just less than the size of the capillary channel image size, the system has an automatic spatial filter.[87]

Starting with the original LIF work of Zare and co-workers, many researchers have described fiber-based systems. As illustrated in Figure 7A, the use of collection and excitation fibers allows a compact and easily aligned system. However, a fiber does not have the same collection efficiency as a well-designed microscope objective and does not allow the same degree of spatial filtering. Therefore, the detection limits obtained with a fiber optic-based fluorescence detector tend to be higher than the microscope objective-based systems.

While microscope and fiber systems are among the most common approaches, other methods have been explored, including a single lens on up to a moderately complicated cylindrical lens and mirror system.[87] While such systems are generally more difficult to design, they can achieve very low detection limits. Perhaps the highest collection efficiency system described is the use of a highly efficient parabolic reflector. Pentoney et al. described a very simple system based on an inexpensive lamp reflector that has a collection efficiency approaching 50%.[106] A schematic of this system is shown in Figure 7D; by combining a high-quality laser blocking filter and a good bandpass filter with a scatter mask that blocks only a very small fraction of the fluorescence emission, most of the scattered light can be eliminated. In their system, the concentration limit of detection is in the low picomolar range. The first commercially available CE/LIF system, introduced by Beckman Instruments, incorporates a similar reflective collector design.

Until this point, only single-channel detection methods have been described. A natural extension is to use a multichannel detector to observe either multiple capillaries or to acquire the complete emission spectra simultaneously. Currently, one of the most robust multichannel detector arrays available for low light level spectroscopy or imaging is the CCD detector.[107-109] There have been several reports of using CCDs for LIF detection in CE. The first application of a CCD to CE used the CCD as a camera at the focal plane of the spectrometer.[110] The camera, complete with a shutter, was used to take a series of pictures of the fluorescence emission from a point on the capillary as a function of time. Limits of detection in the nanomolar range (attomole amounts) for FITC-labeled amino acids have been reported using this system. Sweedler and co-workers[87] demonstrated a CCD-based system that used a 2-cm observation zone (compared to the <100 μm of typical systems). Figure 8 shows a diagram of the system, where the optics image this section of capillary onto the entrance slit of an imaging spectrograph. During the separation (whether chromatographic or electrophoretic), the analyte band moves down the capillary and its image moves across the CCD. If the CCD readout rate is synchronized with the analyte band movement, then the effective integration time is the entire time the band is in the illuminated zone. Unlike nonimaging systems, this increase in observation zone does not correspond to a decreased spatial resolution (separation efficiency). Using this synchronization approach (which is called the time-delayed integration method), LODs of 10^{-12} *M* (zeptomole amounts) FITC-labeled amino acids have been reported.[87,111] Figure 9 shows a wavelength-resolved separation of approximately 3 amol (injected sample quantity) of fluorescein and sulforhodamine-101.[87] When using time-delayed integration, the spatial information is lost, and the output of the system is an infinite series of fluorescent spectra, each of which corresponds to a several hundred micron-wide band migrating through the observation region. Using rhodamine-based derivatizing agents, the LOD of this system was extended to 1000 tagged molecules. Karger and co-workers reported a CCD system specifically designed for DNA sequencing with the ability to distinguish the four bases, using four different fluorescent tags.[112]

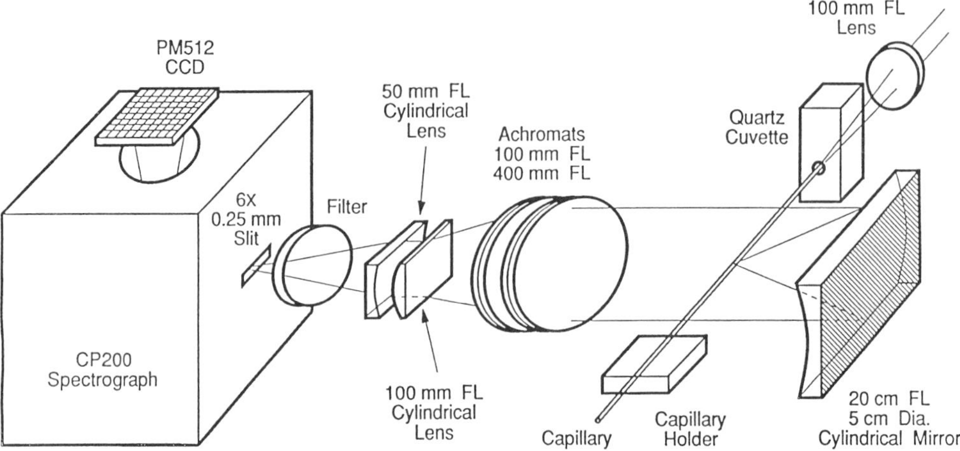

FIGURE 8. Detailed schematic diagram of the optical system for a multichannel CCD fluorescence detector. (From Sweedler, J. V., Shear, J. B., Fishman, H. A., et al., *Anal. Chem.,* 63, 496, 1991. With permission.)

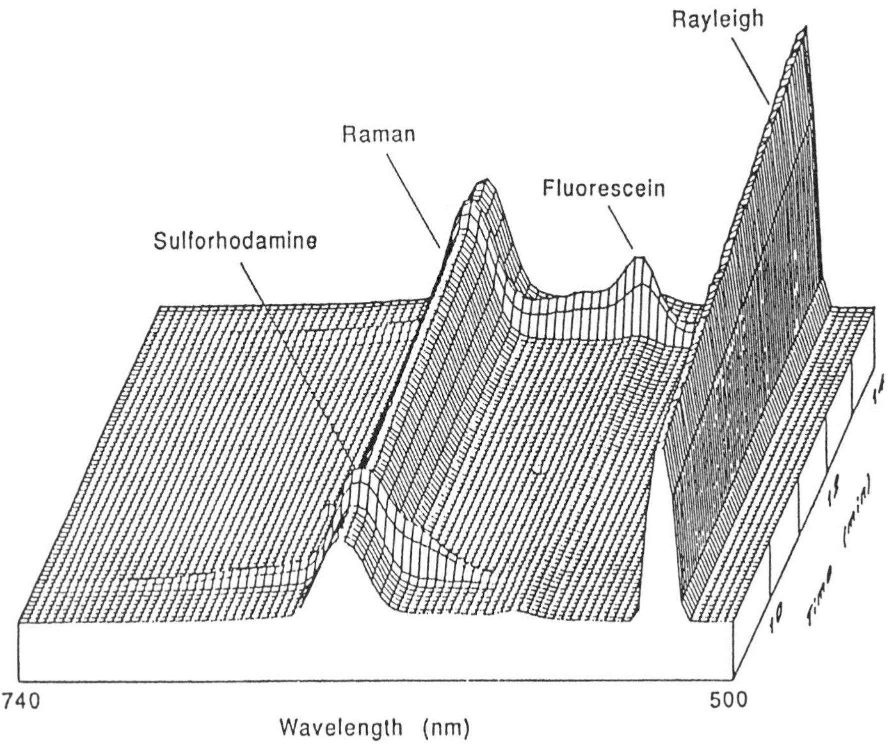

FIGURE 9. Wavelength-resolved electropherogram of sulforhodamine-101 and fluorescein. The Rayleigh and Raman backgrounds are apparent. (From Sweedler, J. V., Shear, J. B., Fishman, H. A., et al., *Anal. Chem.,* 63, 496, 1991. With permission.)

D. A BRIEF SURVEY OF DERIVATIZING METHODS

While fluorescence detection of a good fluorophore offers some of the most impressive performance characteristics of any CE detection mode, most analytes are not fluorescent. This is not a problem limited to CE, and there are thousands of publications and many books on the subject of fluorescence derivatization. For the latest chemistries and bibliographic infor-

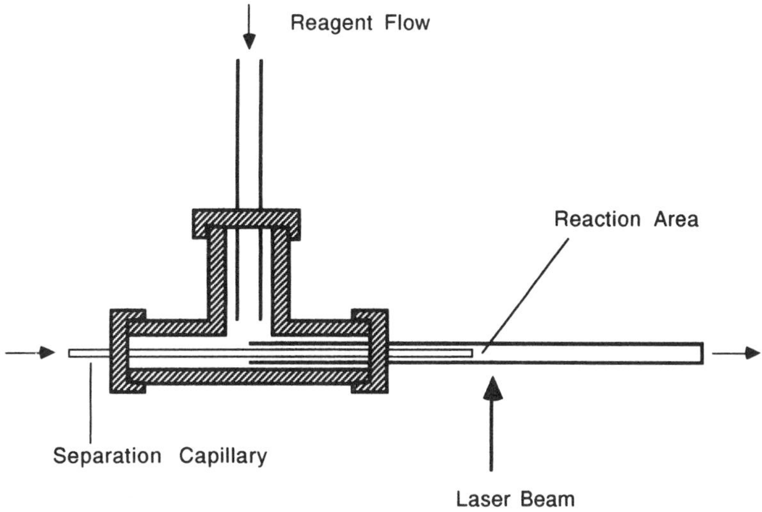

FIGURE 10. Postcolumn derivatization, using a coaxial sheathed capillary reactor. (Adapted from Ref. 129.)

mation on fluorescence derivatization, the Molecular Probes catalog (Eugene, OR) is an excellent resource containing over 5000 references to tagging and detection of a number of analytes.[113] While fluorescence tagging chemistry is not unique to CE, the small size of the separation capillary and sample make derivatization reactions a challenge. In this section, several instrumental approaches for pre-, post-, and on-column derivatization are reviewed.

Precolumn derivatization has been the most common method used with CE; in part, this is because most LIF reports have dealt with characterizing systems and proof-of-concept studies, and optimizing a reaction is easiest when working with the largest volumes possible with precolumn methods. A large number of reports have examined the use of various derivatizing chemistries for CE, including fluorescamine,[114] FITC,[7,90,91,115] CBQCA,[116,117] OPA,[114,118-121] NDA,[122-124] and NBD.[125,126] Nickerson and Jorgenson compared the relative sensitivities of OPA, NDA, and FITC,[124] and Albin et al. compared FITC, OPA, and FMOC.[71]

Almost all off-line precolumn derivatization schemes involve large volumes of analyte and reactants, even though only a small fraction is injected into the column. However, Oates and co-workers have demonstrated a 25-nl NDA derivatization of a single *Helix aspersa* cell using open tubular liquid chromatography; their method involves transferring the cell and nanoliter amounts of reagent to a reaction vial consisting of a small hole drilled in a quartz substrate.[127] They also described the use of a double internal standard to account for the incomplete tagging and adsorption of analyte on the vial walls.

When examining the techniques for postcolumn reactors for CE, an obvious extension is to adapt HPLC methods of introducing the required reagents using separate pumps at a capillary connector. Tsuda et al. demonstrated such a system, using three pumps and a four-way connector; not surprisingly, the separation efficiency was reduced and a nonlinear calibration was obtained for putrescine labeled with fluorescamine.[128] Rose and Jorgenson developed an on-line postcolumn tagging scheme that did not require any pumps.[129] As illustrated in Figure 10, a coaxial capillary reactor was used to introduce *o*-phthalaldehyde (OPA); the system achieved 10^{-9} *M* concentration (attomole mass) detection limits without causing excessive band broadening. Rose has also described a free-solution reactor for postcolumn fluorescence detection in CE.[130] An elegant post (separation)-column system has been described by Albin et al., in which two capillaries are separated by a small gap, as illustrated in Figure 11.[71] By the selection of a larger second capillary and careful alignment, reagent can be drawn into the capillary while maintaining a high separation efficiency.

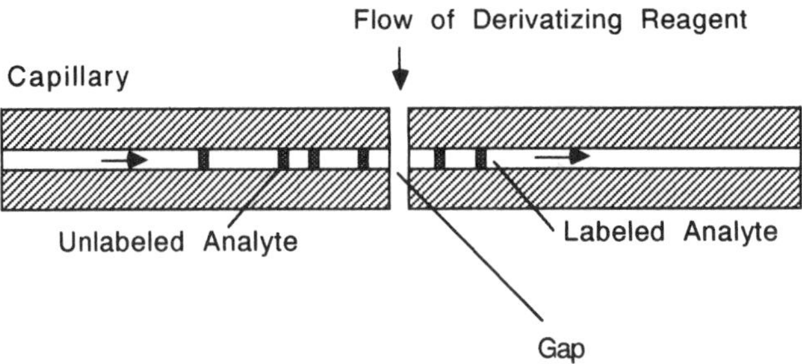

FIGURE 11. Off-column derivatization, using the gap junction reactor. Analyte is confined within the CE flow path by the electric field across the gap. (Adapted from Ref. 71.)

Several on-column methods have been developed, ranging from an on-column connector to capillary tip derivatization methods. Pentoney and co-workers demonstrated an extremely low-volume connector made by laser-drilling holes into the separation capillary and constructing the on-column T illustrated in Figure 12,[120] through which the derivatization reagent was introduced by gravity. Using OPA, the on-column derivatization of femtomole amounts of amino acids was accomplished with little zone spreading due to the connection. Both Shear and co-workers, and Honda and co-workers, have recently presented information on a unique method for on-column derivatization; in their approaches, they preload a small plug of derivatizing reagent into the capillary and then perform a normal injection.[131,132] After such an injection, they wait for the reaction to take place, and then proceed with a separation. While the results presented were preliminary and the linearity and detection limits obtainable with this method have not yet been determined, such an approach has high potential in sample-limited situations.

Yeung has developed several methods to detect nonfluorescent molecules by laser-induced fluorescence. While an earlier section described indirect capillary detection methods, Garner and Yeung have also reported the detection of absorbing but nonfluorescing analytes by LIF, using fluorescence energy transfer.[133] This method involves the transfer of the energy of the excited molecule to a fluorescent molecule. Using fluorescein as the fluorescent moiety and several highly absorbing test molecules (i.e., cresol red, orange G, and DABSYL-labeled amino acids), detection limits in the low attomole range were obtained. Although this approach produced a linear enhancement in fluorescence signal at low analyte concentrations, at high concentrations the emission reached a steady state value. While an interesting approach, the general utility of this method is yet to be demonstrated. On the other hand, the use of an absorbing molecule to shift the emission spectrum for a particular tagging chemistry to a more useful wavelength can be valuable.

Although a large variety of fluorescence implementations have been described, and a brief introduction to tagging chemistry presented, the goal of this section has not been to describe every system reported but rather to illustrate the design criteria of importance in a fluorescence measurement. Many of the approaches described are capable of 10^{-10} *M* (attomole) detection limits, and several extend below the 10^{-12} *M* (zeptomole) level. These detection limits indicate that the CE/LIF systems are capable of detecting extremely small amounts of tagged analytes; however, the detection limits have usually been determined under the artificial conditions of conjugating the fluorescent probe to the analyte at relatively high concentration and then diluting orders of magnitude before injection. While this is a valid approach to characterizing the detection system, the detection limits obtained in this fashion obscure the very large challenges faced when a complex sample must be analyzed. When the derivatization reactions are carried out at the low concentrations required for subcellular sampling of neurotransmitters, for example, nothing is detected! Unfortunately, the fluorescent tagging reactions are

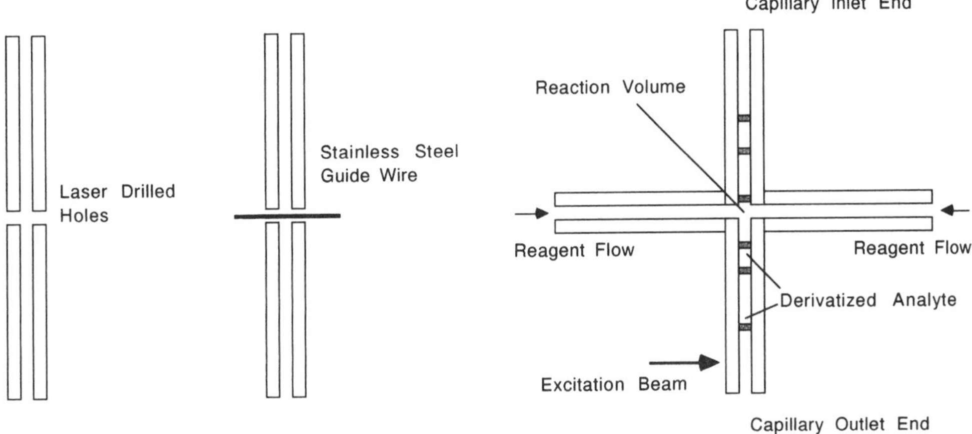

FIGURE 12. On-column derivatization, using a T connector. Derivatization reagent is introduced by gravity through the two side channels and fluor-labeled analyte is detected within the separation capillary, just downstream of the reaction volume. (Adapted from Ref. 120.)

slow and the large number of competing side reactions prevent the transmitters from being tagged and detected at concentrations many orders of magnitude above the limits predicted by instrument sensitivity. Thus, while the next decade is expected to bring more robust and easier to use fluorescence systems, we are hopeful that there will also be significant advances in fluorescence derivatization chemistry and methodology.

VII. CHEMILUMINESCENCE

Chemiluminescence is produced when a chemical reaction yields an electronically excited species that emits light as it returns to the ground state. Chemiluminescence is often encountered in biological systems (termed *bioluminescence*); however, since there are few reactions that produce chemiluminescence, the application of chemiluminescent techniques is limited. In certain instances the chemiluminescent moiety can be connected to derivatizing groups such as isothiocyanate or succinimidyl ester, and therefore can be used for the detection of a wide range of chemicals. For these, the simplicity and sensitivity offered by chemiluminescence are attractive characteristics. However, in most cases, the derivatizing chemistry described under fluorescence detection restricts the linearity, sensitivity, and detection limits.

When comparing chemiluminescence and fluorescence detection modes, the most notable difference is the simpler chemiluminescent instrumentation. The instrumentation for chemiluminescence may consist of only the capillary, a light collecting element (e.g., parabolic mirror), and a detector. Often, since the only source of radiation is the reaction, no wavelength-filtering device is required. Because of the lack of a wavelength selector (i.e., filters), higher efficiency collection optics are possible. At first glance, this would imply that chemiluminescence may have a lower limit of detection. However, it is important to remember that a good fluorophore can absorb and emit photons thousands of times before it is destroyed, while in most chemiluminescent reactions, the analyte may be consumed after only a single cycle. Thus, an important question in terms of relative sensitivity is, can the single photon emitted in the chemiluminescent experiment be detected more easily in the low background situation than the thousand emitted in the fluorescence experiment — even in the presence of background? Of course, the simpler instrumentation and different background sources also play important roles in the selection of a capillary electrophoresis detection mode.

To date, only two reports describe chemiluminescence detection in capillary electrophoresis.[134,135] The first, from Zare's research group, demonstrated the detection of luminol and isoluminol;[134] the second, by Ruberto and Grayeski, demonstrated the reaction of acridinium esters.[135] Both reports show LODs in the nanomolar range; however, the only electropherograms shown involve the separation of related chemiluminescence reagents (i.e., a mixture of luminol derivatives or of a mixture of acridinium esters). The ability to tag an analytically useful molecule with the chemiluminescent reagent and then perform a separation is needed to further demonstrate this detection mode.

Several other chemiluminescent reagents may be applicable to CE. These include luciferase, an enzyme that emits light in the presence of ATP, oxygen, luciferin, and magnesium. It may be possible to measure the ATP content of a cell by using luciferase-based luminescent detection after a CE separation (for such an application see Ng et al.[136]). Another important reagent is Tris(2,2′-bipyridyl)ruthenium(II) ($Ru(bpy)_3 2^+$).[137] This reagent has a relatively high chemiluminescent quantum yield and can be recycled many times. A limitation of many chemiluminescence approaches is the requirement to continuously deliver the reagent into the detection zone because the chemiluminescent reagent is consumed. The $Ru(bpy)_3 2^+$ can be regenerated electrochemically with a nafion electrode.[137] By using such a nafion-coated electrode at the outlet end of the capillary, it may be possible to make a simple, regeneratable chemiluminescence detection system for CE.

VIII. REFRACTIVE INDEX DETECTION

Not surprisingly, the search for a universal detection system for capillary electrophoresis has led to the development of refractive index (RI) detection. Potentially, the most attractive feature of this detector is its wide applicability; even samples without UV chromophores should be detectable with an RI detector. In fact, Tiselius described the use of a refractive index detector based on Schlieren optics for analyzing components separated using the moving boundary method in some of the original electrophoretic work.[1]

The Joule heating generated during electrophoresis poses major problems in the design of a CE-based RI detector; the change in refractive index of the electrolyte is on the order of 10^{-4} RI units/°C, and a minimum detectable RI change of at least 10^{-6} RI units is desired for CE analysis. It is for this reason that most of the published RI detection studies have concentrated on methods to reduce or eliminate thermal effects. Synovec has presented an excellent theoretical overview of the factors affecting RI detection.[138] Three approaches to a CE-based RI detector are described below.

Bruno and co-workers[139] have modified the approach demonstrated by Bornhop and Dovichi for HPLC detection.[62,140] Their method takes advantage of the light interferences produced on side illumination of the liquid-filled capillary tube by a laser. The resulting scattered fringe pattern is characteristic of the capillary dimensions, the RI of the capillary wall and inside medium, and the light polarization. For a 50-μm capillary and a 1-Hz time constant, the inherent sensitivity of their system is 3×10^{-8} RI units; however, Joule heating produced during a run significantly degrades the LOD. In addition to the characterization of the system and verification of their theoretical model, they demonstrated several separations. For sugars, the linear dynamic range was from approximately 10 μ*M* to almost 100 m*M*; this represents an attractive method for analyzing these difficult-to-detect carbohydrates.

In a series of papers, Chen and Morris[141] and Chen and co-workers[142] have reported a novel method for reducing the background thermal and capillary vibration drifts that limit the performance of an RI detector. In their system, they modulate the electrophoretic flow by superimposing an AC signal onto the DC separation voltage. They use a diode laser focused across the capillary in combination with a position-sensitive detector. By using a lock-in amplifier to detect only the component of the output that is modulated because of the modulated flow, the baseline drift is greatly reduced. The detection limits for their system are in the 10^{-8} RI unit range; this is similar in performance to the system described above.

Pawliszyn,[143] Wu and Pawliszyn,[144,145] and Mcdonnell and Pawliszyn[146] have demonstrated the use of an RI detector based on Schlieren optics for capillary isoelectric focusing (IEF), isotachophoresis, and moving boundary electrophoresis. Because this detection system is sensitive to concentration gradients, those CE methods that focus the sample and produce sharp bands and hence large concentration gradients are well suited to this detection scheme. Their systems consist of a few, relatively inexpensive components such as a diode laser or a HeNe laser, a position-sensitive detector, and a simple optical system. Pawliszyn et al. have reported LODs for sucrose below the micromolar range in a flow injection capillary system, and concentration limits of detection for proteins in the micromolar range in an IEF system. Because of the large concentrating effect of IEF, it is difficult to compare these results with those reported above for the other refractive index detection schemes.

IX. RADIONUCLIDE DETECTION

Radionuclide detection has also been successfully coupled with capillary electrophoresis. This mode of detection is both highly sensitive and selective. Selectivity arises from the fact that only radiolabeled sample components yield a response at the detector. The concentration limits of detection for on-line monitoring of radiolabeled analytes separated by CE are in the 10^{-9} *M* (low nano-Curie) range and are determined by the half-life of the isotope being monitored, the overall detection efficiency of the system, and the available measurement time. The integrated number of counts observed over a peak in the electropherogram is given by:

$$\begin{aligned} \text{NOC} &= (\text{DPM}_{\text{peak}})(\text{residence time})(\text{efficiency}) \\ &= (\text{DPM}_{\text{peak}})(\text{detector length/zone velocity})(\text{efficiency}) \end{aligned}$$

where NOC represents the number of observed counts integrated over a peak, DPM represents the number of radioactive transformations occurring each minute in the injected analyte zone, residence time is the amount of time (in minutes) a radioactive molecule within a given sample zone spends in the detection volume, and efficiency is the fractional number of events sensed by the detector. Literature reports to date have focused on maximizing both the detector efficiency and the residence time of the analyte in the detection volume. Several different approaches to radioisotope detection have been reported, including both optical- and nonoptical-based detection schemes.[147-153] We describe here the reported CE/radioisotope detection schemes that are largely optical in nature.

All of the reported optical-based radioisotope detection schemes involve the use of a scintillator that acts as a transducer, converting the decay particle energy into light that is then collected by an optical train and sensed by a photodetector. Radioisotope detection of ^{32}P, ^{14}C, and ^{99}Tc was reported by Kaniansky et al.[147,148] for isotachophoretic separations performed using 300-μm I.D. fluorinated ethylene-propylene copolymer capillary tubing. Either a Geiger–Mueller tube or a plastic scintillator/photomultiplier tube combination was used to detect emitted β particles and the reported detector efficiency was 13 to 15%. A minimum detection limit of 0.44 nCi was reported for a 212-nl cell volume. Altria et al.[149] reported the CE separation and detection of radiopharmaceuticals containing ^{99m}Tc, a γ emitter with a 6-h half-life. In their design a short section of capillary was passed through a block of solid scintillator and the light generated as technetium-labeled sample zones traversed the detection volume was detected. Detection limits and detector efficiency were not reported for this system.

Pentoney and co-workers[150-153] described the design of an on-line radioisotope detector for ^{32}P-labeled analytes that utilized parabolic plastic scintillator material completely surrounding

(360°) the detection region of the capillary. In this design the energetic ^{32}P β particles first passed through the capillary walls and then encountered the scintillator material. The light generated in the scintillator was directed off the reflective surface of the parabola and toward a cooled photomultiplier tube operated in the photon-counting mode. The efficiency of this particular design was determined to be 65% and the concentration LODs for ^{32}P-labeled nucleotides were in the 10^{-9} *M* range.

An improved optical-based ^{32}P detector for CE, which involved the use of two room-temperature photomultiplier tubes operated in the coincidence mode, was subsequently described by Pentoney et al.[153] In this work, the cell compartment of a commercially available HPLC radioisotope detector was modified to accommodate fused silica capillaries surrounded by a thin disk (0.5 to 1.0 mm) of plastic scintillator material. In the coincidence mode, only counts registered simultaneously (20-ns gate in this work) at both of the photomultiplier tubes are accepted and all noncoincident counts are rejected. This results in an extremely low background noise level, even though room temperature PMTs are used. Virtually any light generated in either the separation medium/fused silica (Cerenkov radiation) or the scintillator material is detected in this arrangement and the efficiency was found to be nearly 100%.

As predicted from the above expression, increasing the residence time decreases the minimum amount of analyte that may be accurately quantified by the detector. Residence time extension was accomplished by Zare's group[151-153] by employing "flow programming," in which the electrophoresis voltage was reduced as labeled sample passed through the detection volume in order to reduce the zone velocity. The electrophoresis voltage was then returned to the initial run voltage until the next labeled analyte was sensed, at which time the flow programming was repeated. It is possible to actually reduce the zone velocity to zero by removing the applied field; the measurement time may then be extended arbitrarily. However there is an obvious tradeoff between measurement time and resolution loss caused by diffusional broadening. The latter may be significantly reduced by freezing the capillary contents. This allows an autoradiographic snapshot of the separation channel by exposing directly the frozen capillary to X-ray film.[153]

In the radioisotope detection approaches described by Pentoney and co-workers the emission being monitored must be sufficiently energetic to pass through the capillary walls and then interact with an external transducer that converts the particle energy into light, which is collected and sensed by the optical system. Extension of CE/radioisotope detection to other lower energy emitting isotopes would greatly expand the utility of this detection scheme. Sweedler et al.[154] have described a postcolumn detection approach that allows the detection of lower energy β emitters such as ^{35}S. In this system, the effluent from the capillary is continuously deposited directly onto the surface of a moving substrate that incorporates a scintillator. A computer-controlled *xy* translation stage is used to record a series of lines on the substrate and detection is then made, postrun, by imaging the entire surface with a cooled CCD array detector.

In addition to its applicability to other isotopes, this system is more properly suited to take advantage of extended measurement periods than the on-line approaches described above. Once a record of the separation has been deposited on the substrate surface, the separation and detection processes are decoupled and the measurement interval may be arbitrarily selected without compromising electrophoretic resolution.

X. OTHER OPTICAL DETECTION MODES

Our emphasis in this chapter has been to describe the detection systems and chemistries that have had the largest impact on the field of CE. Several unique methods have been published that have not been mentioned, and in general, only a single reference to the technique has been published. Notable examples include the interesting work of Chen and Morris, demonstrating

Raman detection in CE[155] and the laser-induced fluorescence-based circular dichroism detector reported by Christensen and Yeung.[156] Both of these techniques have low micromolar detection limits; their impact will arise from the unique properties they are detecting. As an example in the case of the Raman system, the structural information available from Raman scattering complements the resolution offered by the CE separation, and potentially offers a powerful means of identifying the class of compound separated. Interestingly, we have yet to read of an attempt to couple FTIR with CE even though both flow cell and solvent elimination interfaces have been reported for microcolumn HPLC.[157-159] Time and further refinements will determine the ultimate impact that these additional optical detection schemes will have on the field of CE.

XI. CONCLUSIONS

There is an extremely wide range of CE detection modes that have been published. As is demonstrated in the following chapter, electrochemical detection and mass spectrometric detection offer unique characteristics and advantages compared to optical detection in many important applications. Electrochemical detection is also very useful in certain applications. It is important to realize that the vast majority of CE systems in the world today, whether home-built or commercial, use UV-Vis absorbance. As we have demonstrated in this chapter, the figures of merit for other optical detection modes can be impressive, and the impact that these additional detection schemes will have on the field of CE greatly depends on the success of transferring them from the academic research laboratory into the commercial marketplace.

ACKNOWLEDGMENTS

J.V.S. acknowledges the support of a National Science Foundation Young Investigator Award, a Dreyfus New Faculty Award, and a David and Lucille Packard Fellowship. S.L.P. wishes to thank Terry Pentoney and the Beckman Technical Library staff for assistance in preparing this manuscript.

REFERENCES

1. **Tiselius, A.,** A new apparatus for electrophoretic analysis of colloidal mixtures, *Trans. Faraday Soc.*, 33, 524, 1937.
2. **Virtanen, R.,** Zone electrophoresis in a narrow-bore tube employing potentiometric detection. Theoretical and experimental study, *Acta Polytech. Scand. Chem. Incl. Metall. Ser.*, 123, 1, 1974.
3. **Mikkers, F. E. P., Everaerts, F. M., and Verheggen, Th. P. E. M.,** High-performance zone electrophoresis, *J. Chromatogr.*, 169, 11, 1979.
4. **Jorgenson, J. W. and Lukacs, K. D.,** Capillary zone electrophoresis, *Science*, 222, 266, 1983.
5. **Kuhr, W. G. and Monnig, C. A.,** Capillary electrophoresis, *Anal. Chem.*, 64, 389R, 1992.
6. **Huang, X., Coleman, W. F., and Zare, R. N.,** Analysis of factors causing peak broadening in capillary zone electrophoresis, *J. Chromatogr.*, 480, 95, 1989.
7. **Otsuka, K. and Terabe, S.,** Extra-column effects in high-performance capillary electrophoresis, *J. Chromatogr.*, 480, 91, 1989.
8. **Chien, R. L. and Burgi, D. S.,** On-column sample concentration using field amplification in CZE, *Anal. Chem.*, 64, 489A, 1992.
9. **Chien, R. L. and Burgi, D. S.,** Sample stacking of an extremely large injection volume in high-performance capillary electrophoresis, *Anal. Chem.*, 64, 1046, 1992.
10. **Vinther, A., Everaerts, F. M., and Soeberg, H.,** Differences in absorbance levels as found in capillary zone electrophoresis under stacking conditions, *J. High Resolut. Chromatogr.*, 13, 639, 1990.
11. **Stegehuis, D. S., Irth, H., Tjaden, U. R., and Van der Greef, J.,** Isotachophoresis as an on-line concentration pretreatment technique in capillary electrophoresis, *J. Chromatogr.*, 538, 393, 1991.

12. **Stegehuis, D. S., Tjaden, U. R., and van der Greef, J.,** Analyte focusing in capillary electrophoresis using on-line isotachophoresis, *J. Chromatogr.,* 591, 341, 1992.
13. **Krivankova, L., Foret, F., and Bocek, P.,** Determination of halofuginone in feed stuffs by the combination of capillary isotachophoresis and capillary zone electrophoresis in a column-switching system, *J. Chromatogr.,* 545, 307, 1991.
14. **Foret, F., Szoko, E., and Karger, B. L.,** On-column transient and coupled column isotachophoretic preconcentration of protein samples in capillary zone electrophoresis, *J. Chromatogr.,* 608, 3, 1992.
15. **Swartz, M. E. and Merion M.,** On line sample preconcentration using a packed inlet capillary for improving the sensitivity of capillary electrophoretic analysis, *J. Chromatogr.,* 632, 209, 1993.
16. **Hoyt, A. M., Jr., Beale, S. C., Larmann, J. P., Jr., and Jorgenson, J. W.,** Preparation and evaluation of an on-line preconcentrator for capillary electrophoresis, *J. High Resolut. Chromatogr.,* 1993, submitted.
17. **Yeung, E. S. and Synovec, R. E.,** Detectors for liquid chromatography, *Anal. Chem.,* 58, 1237A, 1986.
18. **Yeung, E. S.,** Advances in optical detectors for micro-HPLC, *J. Chromatogr. Sci.,* 45, 117, 1989.
19. **Diebold, G. J. and Zare, R. N.,** Laser fluorimetry: subpicogram detection of aflatoxins using high-pressure liquid chromatography, *Science,* 196, 1439, 1977.
20. **Zare, R. N.,** Laser chemical analysis, *Science,* 226, 298, 1984.
21. **Yeung, E. S.,** Indirect detection methods: looking for what is not there, *Acc. Chem. Res.,* 22, 125, 1989.
22. **Hjertén, S., Elenbring, K., Kilar, F., Llao, J. L., Chen, A. J. C., Siebert, C. J., and Zhu, M. D.,** Carrier-free zone electrophoresis, displacement electrophoresis and isoelectric focusing in a high-performance electrophoresis apparatus, *J. Chromatogr.,* 403, 47, 1987.
23. **Foret, F., Fanali, S., Ossicini, L., and Bocek, P.,** Indirect photometric detection in capillary zone electrophoresis, *J. Chromatogr.,* 470, 299, 1989.
24. **Foret, F., Fanali, S., Nardi, A., L., and Bocek, P.,** Capillary zone electrophoresis of rare earth metals with indirect UV absorbance detection, *Electrophoresis,* 11, 780, 1990.
25. **Oefner, P. J., Vorndran, A. E., Grill, E., Huber, C., and Bonn, G. K.,** Capillary zone electrophoretic analysis of carbohydrates by direct and indirect UV detection, *Chromatographia,* 34, 308, 1992.
26. **Weiss, C. S., Hazlett, J. S., Datta, M. H., and Danzer, M. H.,** Determination of quaternary ammonium compounds by capillary electrophoresis using direct and indirect UV detection, *J. Chromatogr.,* 608, 325, 1992.
27. **Bruin, G. J. M., van Asten, A. C., Xu, X., and Poppe, H.,** Theoretical and experimental aspects of indirect detection in capillary electrophoresis, *J. Chromatogr.,* 608, 97, 1992.
28. **Kuhr, W. and Yeung, E. S.,** Indirect fluorescence detection of native amino acids in capillary zone electrophoresis, *Anal. Chem.,* 60, 1832, 1988.
29. **Kuhr, W. and Yeung, E. S.,** Optimization of sensitivity and separation in capillary zone electrophoresis with indirect fluorescence detection, *Anal. Chem.,* 60, 2642, 1988.
30. **Gross, L. and Yeung, E. S.,** Indirect fluorometric detection and quantification in capillary zone electrophoresis of inorganic anions and nucleotides, *J. Chromatogr.,* 480, 169, 1989.
31. **Gross, L. and Yeung, E. S.,** Indirect fluorometric detection of cations in capillary zone electrophoresis, *Anal. Chem.,* 62, 427, 1990.
32. **Garner, T. W. and Yeung, E. S.,** Indirect fluorescence detection of sugars separated by capillary zone electrophoresis with visible laser excitation, *J. Chromatogr.,* 515, 639, 1990.
33. **Hogan, B. L. and Yeung, E. S.,** Indirect fluorometric detection of tryptic digests separated by capillary zone electrophoresis, *J. Chromatogr. Sci.,* 28, 15, 1990.
34. **Olefirowicz, T. M. and Ewing, A. G.,** Capillary electrophoresis with indirect amperometric detection, *J. Chromatogr.,* 499, 713, 1990.
35. **Foret, F., Deml, M., Kahle, V., and Bocek, P.,** On-line fiber optic UV detection cell and conductivity cell for capillary zone electrophoresis, *Electrophoresis,* 7, 430, 1986.
36. **Bruno, A. E., Gassmann, E., Pericles, N., and Anton, K.,** On-column capillary flow cell utilizing optical waveguides for chromatographic applications, *Anal. Chem.,* 61, 876, 1989.
37. **Ludi, H., Gassmann, E., Grossenbacher, H., and Marki, W.,** Analysis of peptides synthesized by recombinant DNA-technology using capillary zone electrophoresis, *Anal. Chem. Acta,* 213, 215, 1988.
38. **Grant, I. H. and Steuer, W.,** Extended path length UV absorbance detector for capillary zone electrophoresis, W. *J. Microcolumn Sep.,* 2, 74, 1990.
39. **Tsuda, T., Sweedler, J. V., and Zare, R. N.,** Rectangular capillaries for capillary zone electrophoresis, *Anal. Chem.,* 62, 2149, 1990.
40. **Poppe, H.,** The performance of some liquid phase flow-through detectors, *Anal. Chem. Acta,* 145, 17, 1983.
41. **Poppe, H.,** Characterization and design of liquid phase flow-through detector systems, *Anal. Chem. Acta,* 114, 59, 1980.
42. **Chervet, J. P., Van Soest, R. E. J., and Ursem, M.,** Z-Shape flow cell for UV detection in capillary electrophoresis, *LC Packings, Tech. Commun.,* 1990.
43. **Bruin, G. J. M., Stegeman, G., Van Asten, A. C., Xu, X., Kraak, J. C., and Poppe, H.,** Optimization and evaluation of the performance of arrangements for UV detection in high-resolution separations using fused-silica capillaries, *J. Chromatogr.,* 559, 163, 1991.

44. LC Packings Technical Data Sheet, 1990.
45. **Moring, S. E., Pairaud, C., and Albin, M.,** Sensitivity enhancement for capillary electrophoresis, Poster M203, in HPCE '93, Orlando, FL, January, 1993.
46. **Taylor, J. A. and Yeung, E. S.,** Axial-beam absorbance detection for capillary electrophoresis, *J. Chromatogr.*, 550, 831, 1991.
47. **Xi, X. and Yeung, E. S.,** Axial-beam absorption detection for capillary electrophoresis with a conventional light source, *Appl. Spectrosc.*, 45(7), 1199, 1991.
48. **Wang, T., Aiken, J. H., Huie, C. W., and Hartwick, R. A.,** Nanoliter-scale multireflection cell for absorption detection in capillary electrophoresis, *Anal. Chem.*, 63, 1372, 1991.
49. **Heiger, D. N.,** Sensitivity improvements in HPCE, paper presented at HPCE '93, Orlando, FL, 1993.
50. **Skoog D. A. and Leary, J. J.,** *Principles of instrumental Analysis,* 4th ed., Harcourt Brace Jovanovich, Orlando, FL, 1992, chap. 8.
51. **Rothman, L. D., Crouch, S. R., and Ingle, J. D., Jr.,** Theoretical and experimental investigation of factors affecting precision in molecular absorption spectrophotometry, *Anal. Chem.*, 47, 1226, 1975.
52. **Weinberger, S. R.,** Optical considerations for "on-the-fly" UV-visible high performance capillary electrophoresis, Linear Instruments Data Sheet, 1989.
53. **Sustacek, V., Foret, F., and Bocek, P.,** Selection of the background electrolyte composition with respect to electromigration dispersion and detection of weakly absorbing substances in capillary zone electrophoresis, *J. Chromatogr.*, 545, 239, 1991.
54. **Fuchs, M., Timmoney, P., and Merion, M.,** Sub-200 nm detection for capillary electrophoresis — the case for fixed wavelength detection, Poster PM-24, in HPCE '91, San Diego, CA, 1991.
55. **Kobayashi, S., Ueda, T., and Kikumoto, M.,** Photodiode array detection for high-performance capillary electrophoresis, *J. Chromatogr.*, 480, 179, 1989.
56. **Sepaniak, M. J., Swaile, D. F., and Powell, A. C.,** Instrumental developments in micellar electrokinetic capillary chromatography, *J. Chromatogr.*, 480, 185, 1989.
57. **Yeo, S. K., Lee, H. K., and Li, S. F. Y.,** Separation of antibiotics by high-performance capillary electrophoresis with photodiode-array detection, *J. Chromatogr.*, 585, 133, 1991.
58. **Hjertén, S.,** High-performance electrophoresis: the electrophoretic counterpart of high-performance liquid chromatography, *J. Chromatogr.*, 270, 1, 1983.
59. **Yu, M. and Dovichi, N. J.,** Sub-femtomole determination of DABSYL-amino acids with capillary zone electrophoresis separation and laser-induced thermo-optical absorbance detection, *Mikrochim. Acta,* III, 27, 1988.
60. **Yu, M. and Dovichi, N. J.,** Attomole amino acid determination by capillary zone electrophoresis with thermooptical absorbance detection, *Anal. Chem.*, 61, 37, 1989.
61. **Yu, M. and Dovichi, N. J.,** Attomole amino acid analysis: capillary-zone electrophoresis with laser-based thermo-optical detection, *J. Appl. Spectrosc.*, 43, 196, 1989.
62. **Bornhop, D. and Dovichi, N. J.,** Simultaneous laser-based refractive index and absorbance determinations within micrometer diameter capillary tubes, *Anal. Chem.*, 59, 1632, 1987.
63. **Waldron, K. C. and Dovichi, N. J.,** Sub-femtomole determination of phenylthiohydantoin-amino acids: capillary electrophoresis and thermooptical detection, *Anal. Chem.*, 64, 1396, 1992.
64. **Earle, C. W. and Dovichi, N J.,** Simultaneous two-color thermooptical absorbance detector for capillary zone electrophoresis, *J. Liq. Chromatogr.*, 12(13), 2575, 1989.
65. **Bruno, A. E., Paulus, A., and Bornhop, D. J.,** Thermo-optical absorption detection in 25-μm-i.d. capillaries: capillary electrophoresis of dansyl-amino acids mixtures, *Appl. Spectrosc.*, 45, 462, 1991.
66. **Hirschfeld, T.,** Quantum efficiency independence of the time integrated emission from a fluorescent molecule, *Appl. Optics*, 15, 3135, 1976.
67. **Giloh, H. and Sedat, J. W.,** Fluorescence microscopy: reduced photobleaching of rhodamine and fluorescein protein conjugates by *n*-propyl gallate, *Science*, 217, 1252, 1982.
68. **Jorgenson, J. W. and Lukacs, K. D.,** Zone electrophoresis in open-tubular glass capillaries, *Anal. Chem.*, 53, 1298, 1981.
69. **Guthrie, E. J. and Jorgenson, J. W.,** On-column fluorescence detector for open-tubular capillary liquid chromatography, *Anal. Chem.*, 56, 483, 1984.
70. **Green, J. S. and Jorgenson, J. W.,** Variable-wavelength on-column fluorescence detector for open-tubular zone electrophoresis, *J. Chromatogr.*, 352, 337, 1986.
71. **Albin, M., Weinberger, R., Sapp, E., and Moring, S.,** Fluorescence detection in capillary electrophoresis: evaluation of derivatizing reagents and techniques, *Anal. Chem.*, 63, 417, 1991.
72. **Hernandez, L., Marquina, R., Escalona, J., and Guzman, N.,** Detection and quantification of capillary electrophoresis zones by fluorescence microscopy, *J. Chromatogr.*, 502, 247, 1990.
73. **Arriaga, E., Chen, D. Y., Cheng, X., and Dovichi, N. J.,** High efficiency filter fluorometer for capillary electrophoresis: zeptomole detection limits for fluoresceinthiocarbamyl amino acids, *J. Chromatogr.*, 1993, in press.
74. *Melles Griot Optics Guide 5*, Melles Griot, Irvine, CA, 1993.

75. **Gassman, E., Kuo, J. E., and Zare, R. N.,** Electrokinetic separation of chiral compounds, *Science*, 230, 813, 1985.
76. **Swaile, D. F. and Sepaniak, M. J.,** Laser-based fluorimetric detection schemes for the analysis of proteins by capillary zone electrophoresis, *J. Liq. Chromatogr.*, 14, 869, 1991.
77. **Lee, T. and Yeung, E. S.,** High-sensitivity laser-induced fluorescence detection of native proteins in capillary electrophoresis, *J. Chromatogr.*, 595, 319, 1992.
78. **Lee, T. and Yeung, E. S.,** Quantitative determination of native proteins in individual human erythrocytes by capillary zone electrophoresis with laser-induced fluorescence detection, *Anal. Chem.*, 64, 3045, 1992.
79. **Milofsky, R. E. and Yeung, E. S.,** Native fluorescence detection of nucleic acids and DNA restriction fragments in capillary electrophoresis, *Anal. Chem.*, 65, 153, 1993.
80. **Chan, K. C., Janini, G. M., Muschik, G. M., and Issaq, K. L.,** Pulsed UV laser-induced fluorescence detection of native peptides and proteins in capillary electrophoresis, *J. Chromatogr.*, 1993, in press.
81. **Chen, F. A., Pentoney, S. L., Jr., Tusak, A., Koh, E. V., and Sternberg, J. C.,** Laser induced fluorescence detection in capillary electrophoresis using a cyanine-based fluorophore and a red diode laser, Poster M208, in HPCE '93, Orlando, FL, January, 1993.
82. **Williams, S. J., Bergstrom, E. T., Goodall, D. M., and Evans, K. P.,** Development of a diode laser based detector for use in capillary electrophoresis, Poster M213, in HPCE '93, Orlando, FL, January, 1993.
83. **Green, M. D., Patonay, G., Thilivhali, N., and Warner, I. M.,** Spectroscopic effects of organized media on a cyanine dye/phenanthrene derivative, *Appl. Spectrosc.*, 46, 1724, 1992.
84. **Williams, R. J., Malgorzata, L., Patonay, G., and Strekowski, L.,** Comparison of covalent and non-covalent labeling with near infrared dyes for the high performance liquid chromatographic determination of human serum albumin, *Anal. Chem.*, 65, 601, 1993.
85. **Higashijima, T., Fuchigami, T., Imasaka, T., and Ishibashi, N.,** Determination of amino acids by capillary zone electrophoresis based on semiconductor laser fluorescence detection, *Anal. Chem.*, 64, 711, 1992.
86. **Taylor, J. A. and Yeung, E. S.,** Axial-beam laser-excited fluorescence detection in capillary electrophoresis, *Anal. Chem.*, 64, 1741, 1992.
87. **Sweedler, J. V., Shear, J. B., Fishman, H. A., Zare, R. N., and Scheller, R. H.,** Fluorescence detection in capillary zone electrophoresis using a charge-coupled device with time-delayed integration, *Anal. Chem.*, 63, 496, 1991.
88. **Chen, D. Y., Swerdlow, H. P., Harke, H. R., Zhang, J. Z., and Dovichi, N. J.,** Low-cost, high-sensitivity laser-induced fluorescence detection for DNA sequencing by capillary gel electrophoresis, *J. Chromatogr.*, 559, 237, 1991.
89. **Swerdlow, H., Wu, S., Harke, H., and Dovichi, N. J.,** Capillary gel electrophoresis for DNA sequencing; laser-induced fluorescence detection with the sheath flow cuvette, *J. Chromatogr.*, 516, 61, 1990.
90. **Wu, S. and Dovichi, N. J.,** High-sensitivity fluorescence detector for fluorescein isothiocyanate derivatives of amino acids separated by capillary zone electrophoresis, *J. Chromatogr.*, 480, 141, 1989.
91. **Cheng, Y. F. and Dovichi, N. J.,** Subattomole amino acid analysis by capillary zone electrophoresis and laser-induced fluorescence, *Science*, 242, 562, 1988.
92. **Cheng, Y. F., Wu, S., Chen, D. Y., and Dovichi, N. J.,** Interaction of capillary zone electrophoresis with a sheath flow cuvette detector, *Anal. Chem.*, 62, 496, 1990.
93. **Swerdlow, H., Zhang, J. Z., Chen, D. Y., Harke, H. R., Grey, R., Wu, S., Fuller, C., and Dovichi, N. J.,** Three DNA sequencing methods using capillary gel electrophoresis and laser-induced fluorescence, *Anal. Chem.*, 63, 2835, 1991.
94. **Shera, E. B., Seitzinger, N. K., Davis, L. M., Keller, R. A., and Soper, S. A.,** Detection of single fluorescent molecules, *Chem. Phys. Lett.*, 174, 553, 1990.
95. **Soper, S. A., Shera, E. B., Martin, J. C., Jett, J. H., Hahn, J. H., Nutter, H. L., and Keller, R. A.,** Single-molecule detection of rhodamine-6G in ethanolic solutions using continuous wave laser excitation, *Anal. Chem.*, 63, 432, 1991.
96. **Dovichi, N. J., Chen, D. Y., Harke, H., Zhao, J. Y., Cheng, X. L., and Bay, S.,** High sensitivity fluorescence detection with capillary electrophoresis for determination of DNA sequencing fragments, amino acids and monosaccharides, paper presented at HPCE '93, Orlando, FL, January, 1993.
97. **Eriksson, K. O., Palm, A., and Hjertén, S.,** Preparative capillary electrophoresis based on adsorption of the solutes (proteins) onto a moving blotting membrane as they migrate out of the capillary, *Anal. Biochem.*, 201, 211, 1992.
98. **Cheng, Y. F., Fuchs, M., Andrews, D., and Carson, W.,** Membrane fraction collection for capillary electrophoresis, *J. Chromatogr.*, 608, 109, 1992.
99. **Huang, X. and Zare, R. N.,** Continuous sample collection in capillary zone electrophoresis by coupling the outlet of a capillary to a moving surface, *J. Chromatogr.*, 516, 185, 1990.
100. **Hernandez, L., Escalona, J., Narahari, J., and Guzman, N.,** Laser-induced fluorescence and fluorescence microscopy for capillary electrophoresis zone detection, *J. Chromatogr.*, 559, 183, 1991.

101. **Hernandez, L., Joshi, N., Escalona, J., and Guzman, N.,** Attomolar concentration sensitivity in CZE coupled with LIF detection, Abstract 15, Pittcon, Chicago, IL, March, 1991.
102. **Huang, X. C., Quesada, M. A., and Mathies, R. A.,** Capillary array electrophoresis using laser-excited confocal fluorescence detection, *Anal. Chem.*, 64, 967, 1992.
103. **Huang, X. C., Quesada, M. A., and Mathies, R. A.,** DNA sequencing using capillary array electrophoresis, *Anal. Chem.*, 64, 2149, 1992.
104. **Abromson, D. and Bickel, W. S.,** Fluorescent angular scattering emissions from dye-filled fibers, *Appl. Optics*, 30, 2980, 1991.
105. **Kurosu, Y., Sasaki, T., and Saito, M.,** Fluorescence detection with an immersed flow cell in capillary electrophoresis, *J. High Resolut. Chromatogr.*, 14, 186, 1991.
106. **Pentoney, S. L., Jr., Konrad, K. D., and Kaye, W. I.,** A single-fluor approach to DNA sequence determination using high performance capillary electrophoresis, *Electrophoresis*, 13, 467, 1992.
107. **Sweedler, J. V., Bilhorn, R. B., Epperson, P. M., Sims, G. R., and Denton, M. B.,** High-performance charge transfer device detectors, *Anal. Chem.*, 60, 282A, 1988.
108. **Epperson, P. M., Sweedler, J. V., Bilhorn, R. B., Sims, G. R., and Denton, M. B.,** Applications of charge transfer devices in spectroscopy, *Anal. Chem.*, 60, 327A, 1988.
109. **Sweedler, J. V.,** Charge transfer device detectors and their applications to chemical analysis, *Crit. Rev. Anal. Chem.*, 24, 59, 1993.
110. **Cheng, Y. F., Piccard, R. D., and Vo-Dinh, T.,** Charge-coupled device fluorescence detection for capillary-zone electrophoresis (CCD-CZE), *Appl. Spectrosc.*, 44, 755, 1990.
111. **Sweedler, J. V., Shear, J. B., Fishman, H. A., Zare, R. N., and Scheller, R. H.,** Analysis of neuropeptides using capillary zone electrophoresis with multichannel fluorescence detection, *Proc. SPIE-Int. Soc. Opt. Eng.*, 1439, 37, 1992.
112. **Karger, A. E., Harris, J. M., and Gesteland, R. F.,** Multiwavelength fluorescence detection for DNA sequencing using capillary electrophoresis, *Nucleic Acids Res.*, 19, 4955, 1991.
113. **Haugland, R.,** *Handbook of Fluorescent Probes and Research Chemicals,* Molecular Probes, Inc., Eugene, OR, 1992.
114. **Wright, B., Ross, G. A., and Smith, R. D.,** Capillary zone electrophoresis with laser fluorescence detection of marine toxins, *J. Microcolumn Sep.*, 1, 85, 1989.
115. **Otsuka, K., Terabe, S., and Ando, T.,** Electrokinetic chromatography with micellar solution separation of phenylthiohydantoin-amino acids, *J. Chromatogr.*, 332, 219, 1985.
116. **Liu, J., Shirota, O., Wiesler, D., and Novotny, M.,** Ultrasensitive fluorometric detection of carbohydrates as derivatives in mixtures separated by capillary electrophoresis, *PNAS*, 88, 2302, 1991.
117. **Liu, J. P., Hsieh, Y. Z., Wiesler, D., and Novotny, M.,** Design of 3-(4-carboxybenzoyl)-2-quinolinecarboxaldehyde as a reagent for ultrasensitive determination of primary amines by capillary electrophoresis using laser fluorescence detection, *Anal. Chem.*, 63, 408, 1991.
118. **Nickerson, B. and Jorgenson, J. W.,** Characterization of a postcolumn reaction laser-induced fluorescence detector for capillary zone electrophoresis, *J. Chromatogr.*, 480, 157, 1989.
119. **Rose, D. J. and Jorgenson, J. W.,** Characterization and automation of sample introduction methods for capillary zone electrophoresis, *Anal. Chem.*, 60, 642, 1988.
120. **Pentoney, S. L., Jr., Huang, X., Burgi, D., and Zare, R. N.,** On-line connector for microcolumns: application to the on-column *o*-phthaldialdehyde derivatization of amino acids separated by capillary zone electrophoresis, *Anal. Chem.*, 60, 2625, 1988.
121. **Liu, J., Cobb, K. A., and Novotny, M.,** Separation of precolumn ortho-phthalaldehyde-derivatized amino acids by capillary zone electrophoresis with normal and micellar solutions in the presence of organic modifiers, *J. Chromatogr.*, 468, 55, 1989.
122. **Ueda, T., Mitchell, R., Kitamura, F., Metcalf, T., Kuwana, T., and Nakamoto, A.,** Separation of naphthalene-2,3-dicarboxaldehyde-labeled amino acids by high-performance capillary electrophoresis with laser-induced fluorescence detection, *J. Chromatogr.*, 593, 265, 1992.
123. **Guthrie, E. J., Jorgenson, J. W., and Dluzneski, P.,** On-column helium cadmium laser fluorescence detector for open-tubular capillary liquid chromatography, *J. Chromatogr. Sci.*, 24, 347, 1984.
124. **Nickerson, B. and Jorgenson, J.,** High sensitivity laser induced fluorescence detection in capillary zone electrophoresis, *J. High Resolut. Chromatogr. Chromatogr. Commun.*, 11, 878, 1988.
125. **Swaile, D. F., Burton, E. D., Balchunas, A. T., and Sepaniak, M. J.,** Pharmaceutical analysis using micellar electrokinetic capillary chromatography, *J. Chromatogr. Sci.*, 26, 406, 1988.
126. **Balchunas, A. T. and Sepaniak, M. J.,** Extension of elution range in micellar electrokinetic capillary chromatography, *Anal. Chem.*, 59, 1466, 1987.
127. **Oates, M. D., Cooper, B. R., and Jorgenson, J. W.,** Quantitative amino acid analysis of individual snail neurons by open tubular liquid chromatography, *Anal. Chem.*, 62, 1573, 1990.

128. **Tsuda, T., Kobayashi, Y., Hori, A., Matzumoto, T., and Suzuki, O.**, Post-column detection for capillary zone electrophoresis, *J. Chromatogr.*, 456, 375, 1988.
129. **Rose, D. J. and Jorgenson, J. W.**, Post-capillary fluorescence detection in capillary zone electrophoresis using *o*-phthaldialdehyde, *J. Chromatogr.*, 447, 117, 1988.
130. **Rose, D. J.**, Free-solution reactor for post-column fluorescence detection in capillary zone electrophoresis, *J. Chromatogr.*, 540, 343, 1991.
131. **Shear, J. B., Fishman, H. A., Gelber, S., Scheller, R. H., and Zare, R. N.**, Microscale labeling strategies for capillary electrophoresis in the analysis of *Aplysia californica* neurons, poster presentation, HPCE '93, Orlando, FL, January, 1993.
132. **Honda, S., Taga, A., and Ohta, Y.**, Analysis of biological substances based on diffusion in free solution, paper presented at HPCE '93, Orlando, FL, January, 1993.
133. **Garner, T. W. and Yeung, E. S.**, Absorption detection in capillary electrophoresis by fluorescence energy transfer, *Anal. Chem.*, 62, 2193, 1990.
134. **Dadoo, R., Colon, L. A., and Zare, R. N.**, Chemiluminescence detection in capillary electrophoresis, *J. High Resolut. Chromatogr.*, 15, 133, 1992.
135. **Ruberto, M. A. and Grayeski, M. L.**, Acridinium chemiluminescence detection with capillary electrophoresis, *Anal. Chem.*, 64, 2758, 1992.
136. **Ng, M., Blascke, T. F., Arias, A. A., and Zare, R. N.**, Analysis of free intracellular nucleotides using high-performance capillary electrophoresis, *Anal. Chem.*, 64, 1682, 1992.
137. **Downey, T. M. and Nieman, T. A.**, Chemiluminescence detection using regenerable Tris(2,2′-bipyridyl)ruthenium(II) immobilized in nafion, *Anal. Chem.*, 64, 261, 1992.
138. **Synovec, R. E.**, Refractive index effects in cylindrical detector cell designs for microbore high-performance liquid chromatography, *Anal. Chem.*, 59, 2877, 1987.
139. **Bruno A. E., Krattiger, B., Maystre, F., and Widmer, H. M.**, On-column laser-based refractive index detector for capillary electrophoresis, *Anal. Chem.*, 63, 2689, 1991.
140. **Bornhop, D. J. and Dovichi, N. J.**, Simple nanoliter refractive index detector, *Anal. Chem.*, 58, 504, 1986.
141. **Chen, C. Y. and Morris, M. D.**, Raman spectroscopic detection system for capillary zone electrophoresis, *Appl. Spectrosc.*, 42, 515, 1988.
142. **Chen, C. Y., Demana, T., Huang, S. D., and Morris, M. D.**, Capillary zone electrophoresis with analyte velocity modulation. Application to refractive index detection, *Anal Chem.*, 61, 1590, 1989.
143. **Pawliszyn, J.**, Nanoliter volume sequential differential concentration gradient detector, *Anal. Chem.*, 60, 2796, 1988.
144. **Wu, J. and Pawliszyn, J.**, High-performance capillary isoelectric focusing with a concentration gradient detector, *Anal. Chem.*, 64, 219, 1992.
145. **Wu, J. and Pawliszyn, J.**, Universal detection for capillary isoelectric focusing without mobilization using a concentration gradient imaging system, *Anal. Chem.*, 64, 224, 1992.
146. **Mcdonnell, T. and Pawliszyn J.**, Capillary isotachophoresis with concentration gradient detection, *Anal. Chem.*, 63, 1884, 1991.
147. **Kaniansky, D., Rajec, P., Svec, A., Havasi, P., and Macasek, F.**, On-line radiometric detection in capillary electrophoresis. I. Preliminary experiments, *J. Chromatogr.*, 258, 238, 1983.
148. **Kaniansky, D., Rajec, P., Svec, A., Marak, J., Koval, M., Lucka, M., Franko, S., and Sabanos, G.**, On-column radiometric detector for capillary isotachophoresis, *J. Radioanal. Nucl. Chem.*, 129(2), 305, 1989.
149. **Altria, K. D., Simpson, C. F., Bharij, A., and Theobald, A. E.**, Radiopharmaceutical analysis by high voltage capillary electrophoresis, paper presented at the 1988 Pittsburgh Conference and Exposition, Abstract 642, New Orleans, LA, February, 1988.
150. **Gordon, M., Huang, X., Pentoney, S. L., Jr., and Zare, R. N.**, Capillary electrophoresis, *Science*, 242, 224, 1988.
151. **Pentoney, S. L., Jr., Zare, R. N., and Quint, J. F.**, On-line radioisotope detection for capillary electrophoresis, *Anal. Chem.*, 61, 1642, 1989.
152. **Pentoney, S. L., Jr., Zare, R. N., and Quint, J. F.**, Semiconductor radioisotope detector for capillary electrophoresis, *J. Chromatogr.*, 480, 259, 1989.
153. **Pentoney, S. L., Jr., Zare, R. N., and Quint, J. F.**, On-column radioisotope detection for capillary electrophoresis, in *Analytical Biotechnology, ACS Symposium Series,* Horvath, C. and Nikelly, J. G., Eds., American Chemical Society, Washington, D.C., 1990, chap. 4, p. 60.
154. **Sweedler, J. V., Fuller, R., Tracht, S., Timperman, A., Toma, V., and Khatib, K.**, Novel detection schemes for trace analysis of neuropeptides using capillary electrophoresis, *J. Microcol. Sep.*, 5(5), 1993, in press.
155. **Chen, C. and Morris, M. D.**, On-line multichannel Raman spectroscopic detection system for capillary zone electrophoresis, *J. Chromatogr.*, 540, 355, 1991.

156. **Christensen, P. and Yeung, E. S.,** Fluorescence-detected circular dichroism for on-column detection in capillary electrophoresis, *Anal. Chem.,* 61, 1344, 1989.
157. **Norton, K. L., Lange, A. J., and Griffiths, P. R.,** A unified approach to the chromatography-FTIR interface: GC-FTIR, SFC-FTIR, and HPLC-FTIR with subnanogram detection limits, *J. High Resolut. Chromatogr.,* 14, 225, 1991.
158. **De Haseth, J. A. and Robertson, R. M.,** Magic (monodisperse aerosol generation interface combining)-LC/FT-IR: a viable interface for HPLC and FT-IR spectrometry, *J. Microchem.,* 40(1), 77, 1989.
159. **Fujimoto, C. and Jinno, K.,** Microcolumn high-performance liquid chromatography with Fourier transform infrared spectrometric detection, *Trends Anal. Chem.,* 8(3), 90, 1989.

Chapter 8

CAPILLARY ELECTROPHORESIS-MASS SPECTROMETRY

Richard D. Smith, David R. Goodlett, and Jon H. Wahl

TABLE OF CONTENTS

0-8493-8690-X/94/$0.00+$.50

I. INTRODUCTION

Of all capillary electrophoresis (CE) detection methods reported to date, mass spectrometry (MS) clearly has the greatest potential. The advantages of MS detection in conjunction with various forms of chromatography are well recognized, and include the capability for determining both molecular weight and structural information.[1] The mass spectrometer provides the equivalent of up to several thousand discrete selective "detectors" functioning in parallel. Capillary electrophoresis is based on the differential migration of ions in solution, while MS analyzes ions by their mass-to-charge ratio (m/z) in the gas phase. These two highly orthogonal analytical methods, however, exploit ion motion in two quite different environments: moderately conductive liquid buffers and high vacuum, respectively. Capillary electrophoresis places significantly different demands on an MS interface than does liquid chromatography[2] because CE flow rates are quite low or negligible and electrical contact must be maintained with both ends of the capillary so as to define the CE field gradient. In addition, if the high separation efficiencies possible with CE are to be realized, any extra capillary broadening due to laminar flow from pressure differences between the capillary termini or to dead volumes associated with detection needs to be minimized.

This chapter emphasizes experimental considerations, methods, and selected applications for CE-MS based on electrospray ionization (ESI) interfaces. This emphasis reflects the explosive growth in the availability of ESI/MS instrumentation, as well as the authors' belief that the ESI interface is currently the method of choice for CE-MS due to its sensitivity, versatility, and ease of implementation. We describe the status of CE-MS, as well as the present limitations due to the sensitivity and scan speed constraints of current MS technology. Although MS is still one of the most complex and expensive CE detectors, costs continue to decrease, and the amount of information that can be gained from this combination increasingly mitigates these disadvantages. Recent developments involving tandem-MS methods (i.e., MS/MS) and higher order-MS methods (i.e., MS^n, where $n \geq 3$) are discussed, which promise future instrumentation providing even greater selectivity and structural information.

II. INSTRUMENTATION FOR CAPILLARY ELECTROPHORESIS-MASS SPECTROMETRY

This chapter is focused on on-line CE-MS using ESI interfaces. However, the earlier reports of off-line CE-MS combination should be noted,[3,4] as well as the more promising recent reports based on plasma desorption mass spectrometry[5] and matrix-assisted laser desorption methods.[6] A considerable number of reports also describe on-line CE-MS based on either "liquid junction" fast atom bombardment (FAB) interfaces,[7-15] or "coaxial" continuous flow FAB interfaces.[16-22] While this is a viable approach to CE-MS, current results indicate that electrospray CE-MS interfaces have numerous advantages, and we limit this chapter to that approach.

The electrospray ionization process is based on the electrostatically induced nebulization of liquid flowing from a capillary in a high electric field at near atmospheric pressure. At the proper electric field strength and liquid flow rate, the liquid is dispersed as nearly uniform, micron-sized droplets. Each droplet is highly charged, with a polarity that depends on the voltage bias of the capillary relative to the counterelectrode toward which the charged droplets will drift. The droplets will have a charge close to the Rayleigh limit, i.e., the maximum extent of charging feasible. Next, the droplets are exposed to conditions that cause solvent evaporation, and the droplets shrink rapidly to the point at which either evaporation must seize, or charge must be lost. Although the subsequent details remain uncertain, these conditions result

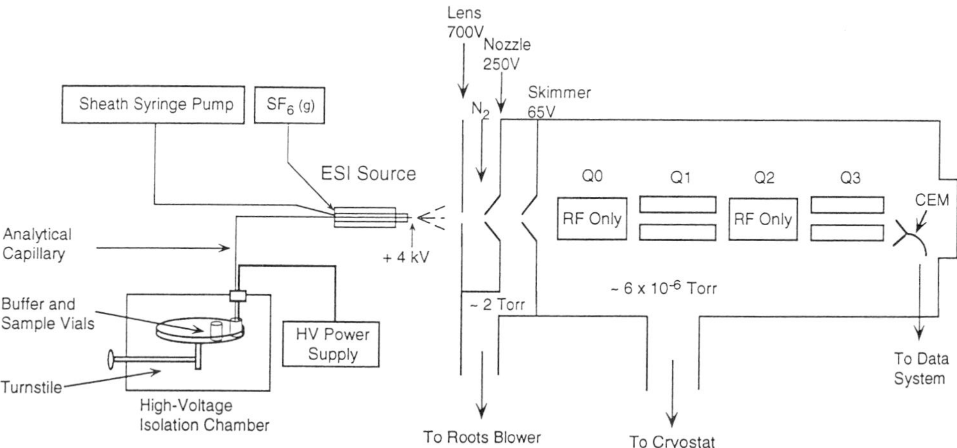

FIGURE 1. Schematic illustration of the experimental arrangement for CE-MS using an electrospray interface. The sheath syringe pump provides a low flow (<5 µl/min) of liquid around the CE capillary terminus. Sulfur hexafluoride is used as a sheath gas to suppress electric discharges. Q0 and Q2 are rf-only quadrupoles and Q1 and Q3 are quadrupole mass filters (i.e., rf/DC).

in transfer of ions from solution to the gas phase. It is these ions that can ultimately be detected after transport into the high vacuum of the MS. Ions of molecular species of molecular weights >1000 for a given class of compounds generally have more than one charge on average, and the extent of multiple charging increases nearly linearly with molecular weight.

Figure 1 shows a schematic illustration of a typical arrangement of instrumentation used for CE-MS based on ESI. Considerations for interfacing generally derive from some limitations on CE buffer composition and the desire to position the ESI source (i.e., the point of charged droplet formation) as close as possible to the analytical capillary terminus, avoiding lengthy transfer lines. The ESI "source" operates at a voltage difference of typically 3 to 5 kV relative to the sampling orifice, where ions are entrained in a flow of gas that enters the mass spectrometer. Some ESI sources allow the liquid effluent to be at ground potential, and others require the capillary terminus to be operated 3 to 5 kV relative to ground potential. Many of the practical constraints and considerations for CE-MS interfacing derive from the voltage bias between these components. The ESI liquid nebulization process can be pneumatically assisted, using a high-velocity annular flow of gas at the capillary terminus, and it sometimes is referred to as "ion spray."[23] For the liquid–junction interface variation,[24] electrical contact with the capillary terminus is established through a liquid reservoir that surrounds the junction of the analytical capillary and a transfer capillary. The gap between the two capillaries is typically adjusted to 10 to 20 µm, a compromise resulting from the need for sufficient make-up liquid being drawn into the transfer capillary while avoiding analyte loss by diffusion into the reservoir. The flow of make-up liquid arises from a combination of gravity-driven flow, due to the height of the make-up reservoir, and flow induced in the transfer capillary by a mild vacuum generated by the Venturi effect of the nebulizing gas used at the ESI source.

In the sheath flow (or coaxial) electrospray interface,[25,26] an organic liquid (typically pure methanol, methoxyethanol, or acetonitrile, but frequently modified with as much as 10 to 20% formic acid, acetic acid, water, or other reagents) flows through the annular space between the ~200-µm O.D. CE capillary and a fused silica or stainless steel capillary (generally >250 µm I.D.). Enhanced stability can be obtained by degassing the organic solvents used in the sheath or by including gas trapping volumes for the sheath liquid, and by minimizing heating (e.g., from the ion source or a "countercurrent" gas flow).

The performance in terms of spectral quality for the sheath flow and liquid junction interfaces is similar, as are most other MS-related considerations; however, some differences have been noted. Thibault et al.[27] have compared the liquid junction variation of Lee and coworkers[24] with the sheath flow interface using a pneumatically assisted ESI interface and have noted that the latter provided generally better and more convenient operation. In a comparison for a separation of marine toxins, where care was taken to obtain comparable conditions, they found significantly improved signal-to-noise ratios, and slightly improved separations were obtained with the sheath flow interface.

The dependence of ESI ion current on solution conductivity is relatively weak, generally 0.1 to 0.3 μA at atmospheric pressure, and about 10 to 100 pA of integrated ion current (i.e., the sum of all ions) is actually focused into the MS and transmitted for detection. Electrospray ionization currents for a typical water/methanol/5% acetic acid solution are in the range of 0.1 to 0.3 μA, only a small portion of which ($<10^{-3}$) is generally transmitted for MS analysis. An electron scavenger is often used to inhibit electrical discharge at the capillary terminus, particularly for ESI of aqueous solutions. Sulfur hexafluoride has proved particularly useful for suppressing corona discharges and improving the stability of negative ion ESI. Introduction of this gas is most effectively accomplished using a gas flow (~100–250 ml/min) through an annular volume surrounding the sheath liquid.

An important attribute of a CE-MS interface based on ESI is the efficiency of sampling and transport of ions into the MS. One commercially available instrument utilizes a 100- to 130-μm pinhole sampling orifice to a vacuum region maintained by a single stage of high-speed cryopumping (Sciex, Thornhill, Ontario, Canada). Charged droplets formed by ESI drift against a countercurrent flow of dry nitrogen, which serves to speed desolvation and exclude high-*m/z* residual particles and solvent vapor. As the ions pass through the orifice into the vacuum region, further desolvation is accomplished as the gas density decreases due to collisions as the ions are accelerated by the ion optics of the mass spectrometer. Also, differentially pumped interfaces are widely used,[28] which are available in versions compatible with many mass spectrometers. A countercurrent flow of nitrogen bath gas is often used to assist solvent evaporation, similar to the approach described above. Ions migrate toward the sampling orifice where some small fraction is entrained in the gas flow entering a glass capillary (metalized at both ends to establish well-defined electric fields). Ions emerge from the capillary, in the first differentially pumped stage of the MS (~1 torr), as a component of a free-jet expansion. A fraction of the ions are then transmitted through a "skimmer" and additional ion optics into the MS. The electrically insulating nature of the glass capillary provides considerable flexibility because ion transmission does not depend strongly on the voltage gradient between the conducting ends of the capillary. This approach has the advantage that the CE capillary terminus can be at ground potential, simplifying current measurements and voltage manipulation during injection.

An alternative approach to droplet desolvation for ESI relies solely on heating during droplet transport through a heated metal or glass capillary.[29,30] A countercurrent gas flow is not essential and, in fact, will decrease obtainable MS ion currents. The charged droplets from the ESI source are swept into the heated capillary, which heats the gas sufficiently to provide effective ion desolvation, particularly when augmented by a voltage gradient in the capillary–skimmer region. The electrospray source can be closely positioned to the sampling capillary orifice (typically 0.3 to 1.0 cm) because the spacing is not used for desolvation, resulting in more efficient transport into the MS.

The ESI/MS technique is an extremely "soft" ionization technique, and will yield, under appropriate conditions, intact molecular ions. Molecular weight measurements for large biopolymers that exceed the MS mass range can be obtained because the ESI process generally yields a distribution of molecular ion charge states, without contributions due to dissociation unless induced during transport into the MS vacuum.[28,31,32] The envelope of charge states for biopolymers, generally arising from protonation for positive ion ESI, yields a distinctive pattern of peaks due to the discrete nature of the electronic charge; i.e., adjacent peaks vary by addition or subtraction of one charge. In addition, while mass spectra of noncovalently bonded species, such as multimeric proteins, typically show spectra containing only the individual subunits, and the detection of such labile species is feasible under appropriate conditions.[32] Although the range of application has not been completely defined, ESI appears amenable to most charge-carrying species in solution, a situation that makes it an ideal detector for CE.

III. A MODEL FOR CAPILLARY ELECTROPHORESIS-ELECTROSPRAY IONIZATION/MASS SPECTROMETRY DETECTION SENSITIVITY

The efficiency of the ESI detection process will be considered in terms of two distinct regimes. In the first, and generally encountered regime, the mass flow rate of the various electrolyte species to the ESI source exceeds that capable of being transferred to the gas phase by the ESI process. In the second regime, the low mass flow rate of the various electrolytes constrains the ESI current.

For simple electrospray systems, using capillaries without additional sheath liquid flows and incorporating simple strong electrolytes, reasonable agreement between results and theory for ESI efficiency have been obtained. For example, Kebarle and co-workers showed that the analyte signal intensity detected in ESI/MS was highly dependent on the solution conditions.[33-35] They invoke a simple model that qualitatively rationalizes the functional dependence of mass spectrometrically detected analyte ion intensity on analyte concentration in the presence of a background electrolyte.[35] In this model, the initial ESI droplet formation process is viewed as an electrophoretic charge separation process, where droplets have either an excess of positive ions for positive mode ESI or an excess of negative ions for negative mode ESI. Further, this simple model assumes that the analyte and other charged species in solution compete for transfer to the gas phase.

The electrospray interfaces developed and used at our laboratory for CE-MS analytical applications use a coaxial sheath flow that provides an additional flow to the electrospray source that can incorporate charge carrying species. All CE-ESI/MS methods currently used involve this or a similar method, which serves to complicate any treatment of ESI efficiency. The sheath liquid (or make-up flow from a liquid junction) provides electrical contact with the CE terminus. However, uncertainties in the efficiency of mixing between the liquid sheath and the CE effluent prevent a simple accounting of its contributions to the ESI current. Therefore, it is probably unreasonable to expect an exact quantitative agreement between experimental results and the simple model discussed below.

Most work with CE-MS is conducted under conditions where this competition between charged species in the ESI process should apply, and we first consider this regime. In an adaptation of the approach of Tang and Kebarle,[35] the analyte signal intensity, $I\,(\mathrm{A}^+)$, can be expressed as a function of the mass flow rate of the analyte, $V_\mathrm{M}(\mathrm{A}^+)$, and background, $V_\mathrm{M}(\mathrm{B}^+)$, constituents as follows:

$$I\left(\mathrm{A}^+\right) = IPf\,\frac{k_\mathrm{A} V_\mathrm{M}\left(\mathrm{A}^+\right)}{k_\mathrm{A} V_\mathrm{M}\left(\mathrm{A}^+\right) + k_\mathrm{B} V_\mathrm{M}\left(\mathrm{B}^+\right)} \tag{1}$$

where I is the total electrospray current, P is a proportionality constant relating the sampling efficiency of the system and is assumed constant, f is a proportionality constant representing the fraction of droplet charge that is converted into gas-phase ions, and k_A and k_B are rate constants that describe the relative efficiencies for the transfer of ions from a droplet to the gas phase. Consequently, the analyte signal intensity is dependent on both the solute and background mass flow rates, all else being constant.

According to Equation 1, in this regime two limiting detection regions are predicted. First, when the analyte mass flow rates are relatively high and substantially greater than the background mass flow rates, Equation 1 reduces to the following expression:

$$I(\mathrm{A}^+) = IPf \tag{2}$$

where, according to Pfeifer and Hendricks,[36] the total electrospray current is a weak function of the solution conductivity, σ:

$$I \propto \sigma^n \quad n \approx 0.2\text{–}0.4 \tag{3}$$

By substituting Equation 3 into Equation 2, the analyte signal intensity is expected to be a weak function of the analyte concentration because the solution conductivity is a function of concentration. In this analyte-rich region, the analyte is not efficiently ionized by the electrospray process, and any further increase in the analyte concentration produces a smaller gradual increase, not a proportional increase in the signal intensity. Here the ESI charge separation step during droplet formation is less efficient (i.e., the number of droplet charges from the ESI process becomes substantially less than the number of electrolyte species in the droplet), and the overall ionization efficiency for conversion from solution to the gas phase is decreased. Further deviations in analyte signal intensity from that predicted by Equation 2 might also result from a decrease in f at high concentrations, where the ESI process may become hindered by slow droplet evaporation and, possibly, aggregation or precipitation of the various solute species.

Second, when the background electrolyte mass flow rate is assumed constant and much greater than the analyte mass flow rate, Equation 1 reduces to the following form:

$$I(\mathrm{A}^+) = IPf\frac{k_\mathrm{A}V_\mathrm{M}(\mathrm{A}^+)}{k_\mathrm{B}V_\mathrm{M}(\mathrm{B}^+)} \tag{4}$$

In this high-background electrolyte region the analyte signal intensity is predicted to be directly proportional to the analyte mass flow rate. Moreover, the analyte sensitivity is predicted to be inversely proportional to the background electrolyte mass flow rate in this detection region. Consequently, analyte sensitivity can be improved by decreasing the amount of background electrolyte mass flow rate.

These two detection regions are illustrated in Figure 2, where the relative analyte signal intensity is shown as a function of the analyte mass flow rate according to Equation 2. For curve A, the maximum and minimum analyte mass flow rate is presumed to be 10 and 0.1 times that of the background mass flow rate, k_A and k_B are assumed equal, and product Pf is assumed constant. Also, a constant ESI current, I, is assumed for the present discussion, because it is observed to be nearly constant, which in part may be attributed to the coaxial sheath flow arrangement. From Figure 2, a constant analyte sensitivity is then expected within the dynamic range of the detector when the mass flow rate of the analyte is small compared to the background electrolyte.

According to Equations 1 and 4 an increase in analyte sensitivity is predicted when the mass flow rate of the background electrolyte is decreased. This increase in sensitivity is illustrated in Figure 2 where curve B represents the signal intensity predicted when both the analyte and background mass flow rates are both reduced by a factor of 10 relative to curve

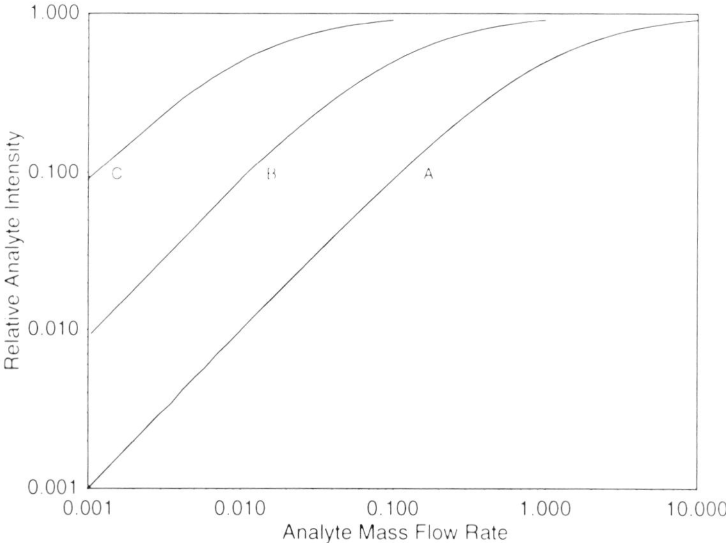

FIGURE 2. Electrospray analyte signal response from Equation 1, where curve B represents the signal intensity predicted when both the analyte and background electrolyte mass flow rates are reduced by a factor of 10 relative to curve A, and curve C corresponds to a concurrent reduction of 100, i.e., where the background electrolyte delivery rate to the ESI source is reduced by factors of 10 (B) and 100 (C).

A, and curve C corresponds to reduction of mass flow rates of 100 for both components. From Figure 2, when the relative analyte mass flow rate is 0.001, the analyte signal intensity is two orders of magnitude greater when both the analyte and background ions are reduced by the same magnitude (curve C) compared to when only the analyte mass flow rate is decreased (curve A). Additionally, as shown in Figure 2, this model predicts that the onset of the analyte-rich region occurs at lower mass flow rates when both the analyte and background are reduced. The increase in sensitivity arises directly from the decrease in background electrolyte, which allows a greater fraction of the analyte to be converted into gas-phase ions.

It is clear that the simple relationship described by Equation 1 must fail at sufficiently low background electrolyte amounts. Equations 1 and 4 predict the ion signal intensity will remain constant as the analyte and background electrolyte mass flow rates are decreased in an indefinite manner. Consequently, a second regime of detection occurs when charge-carrying species are no longer supplied to the ESI source at a rate sufficient to sustain the maximum electrospray current. The ionization efficiency for the system in this regime is maximized because there is no longer a competition between analyte and buffer for ionization. That is,

$$I(A^+) \propto V_M(A^+) \tag{5}$$

and

$$I(B^+) \propto V_M(B^+) \tag{6}$$

In CE, separations are generally performed under conditions where the analyte concentration is much less than the buffer concentration, and thus either Equation 4 or 5 should most generally apply. In both these situations, it is predicted that the observed MS signal will be proportional to analyte concentration; however, one important and somewhat subtle distinction exists. For Equation 4, a change in background electrolyte mass flow rate will affect analyte signal intensity, and in the course of a given CE experiment, the signal intensity directly reflects analyte concentration. Changes in flow rate are generally irrelevant, and hence the ESI/MS emulates a concentration-sensitive detector. In the case of Equation 5, encountered at the lower limits of electrolyte mass flow rate or for very small capillary diameters, the electrospray current is limited by electrolyte flow to the electrospray source and ionization efficiency will be optimum. Thus, in this regime, the ESI/MS detector appears to function as a mass-sensitive detector. The transition between the regimes described by Equations 4 and 5 should be evident when further decreases in the background electrolyte do not lead to additional gain in analyte sensitivity.

It is not clear, however, if the increased CE sensitivity predicted by this model will be realized in practice with an electrospray interface using a coaxial sheath flow. Questions arise because of the unknown mixing efficiency of the CE effluent with the sheath liquid. In addition, if the background electrolyte contribution to ESI arising from the CE buffer system is small compared to that present in the sheath flow, an increase in solute sensitivity would not be expected if good mixing was obtained.

These simple considerations, predictions, and related uncertainties indicate that analyte sensitivity in CE-ESI/MS may be increased by reducing the mass flow rate of the background components. This decrease in background flow rates can be experimentally accomplished by using a more dilute solvent system (i.e., decreasing the concentration of the supporting electrolyte); however, low-concentration buffer systems can lead to poor separation efficiencies.[37] In addition, for very low conductivity buffers, the maximum analyte concentration at which linear response will be obtained will decrease, leading to a reduced dynamic range. This second regime of detection should occur when the CE current is less than the normal ESI current. The lower limits to CE current for larger capillary diameters are generally defined by the small amounts of trace ionic contaminants in aqueous solutions that are nearly impossible to eliminate; consequently, no further reduction in the relative concentration of the background electrolyte occurs. Thus, in CE the use of highly dilute buffer systems is not generally an option. The mass flow rate from the analytical capillary, however, can be reduced by (1) performing the entire separation with a reduced electric field, which decreases ion migration rates and electroosmotic flow and causes a proportionality longer analysis time, (2) selectively reducing the electric field strength only when the solutes of interest are migrating into the electrospray source, which presumes that the solute migration time is previously known,[38] and (3) using small inner diameter capillaries for the separation, which reduces the mass flow rate of both the analyte and the background ions, but provides a constant analysis time and does not require prior knowledge of solute migration times. The prediction that the use of small inner diameter capillaries offers increased analyte sensitivity in CE-ESI/MS has been supported by experimental studies, as discussed in the next section.

IV. EXPERIMENTAL METHODS

Both the liquid junction and sheath flow ESI interfaces allow a wide range of CE buffers to be successfully electrosprayed due to the effective dilution of the low CE elution flow by a much larger volume of liquid, providing considerable flexibility despite some constraints (e.g., high salt and surfactant concentrations are often problematic). Both aqueous and mixed aqueous/organic buffer solutions can be effectively utilized. Buffer concentrations of at least 0.1 *M* can be used for CE-MS due to dilution by the sheath liquid or liquid junction buffer. However, because of practical sensitivity constraints, CE buffer concentrations are generally minimized, and it has also been found that sensitivity can vary significantly with buffer composition. In general, the best sensitivity is obtained with the use of volatile buffer components, such as acetic acid at the lowest practical concentration, and by minimizing nonvolatile components. In addition, buffer components that interact strongly with the sample (e.g., denaturants) degrade sensitivity. For example, surfactants generally give rise to intense signals in both positive and negative ion ESI, presenting a major barrier for sufficiently sensitive micellar electrokinetic capillary chromatography-MS applications.

Proper CE capillary surface coatings and conditioning are essential for many applications. As an example, our initial CE-MS of proteins in acidic buffers (pH 3 to 5) with uncoated capillaries produced poor separations for proteins such as myoglobin and cytochrome *c* due to interactions with capillary surfaces.[39] However, excellent separations of myoglobin mixtures have been obtained at higher pH (>8) due to the net negative charge of the protein, and reasonable detection limits (<100 fmol) obtained using multiple ion monitoring in 10 m*M* Tris.[40] In this case, methanol–water sheath liquid containing 5 to 10% acetic acid was used to obtain the acidic conditions for ESI-MS. The use of the Tris buffer component, however, led to reduced sensitivity compared to that obtained for the same separations conducted in volatile acidic buffers.[40] Thibault et al.[27] have demonstrated that protein separations conducted in acidic solution (e.g., 10 m*M* acetic acid, pH 3.4), using reversible amino-based coated capillaries, result in substantially improved sensitivity by CE-MS.

Probably the most widely used buffers to date for CE-ESI/MS are acetic acid, ammonium acetate, and formic acid systems, choices made primarily due to their volatility. Nonvolatile buffer systems may be used for CE-MS, but often present difficulties. For example, Tris and phosphate buffers at higher concentrations produce cluster ions extending to at least *m/z* 1000 with sufficient abundance to substantially limit CE-MS applicability. Conversely, acetic acid buffers often produce a broad range of background ions, presumably due to solution impurities, which illustrates the need to use high-purity buffer systems in CE-MS. For example, Figure 3 shows mass spectra obtained during elution of a substituted [Ala^2]Met-enkephalin from a CE-MS separation in a 20-μm I.D. capillary using a 10 m*M* acetic acid buffer. The molecular weight of this compound is sufficiently small such that the singly charged (protonated) molecular ion dominates the mass spectrum. The top panel in Figure 3 shows the "raw" mass spectrum (above *m/z* 300) for which a major peak arises at *m/z* 589 due to the protonated molecule from an 8-fmol injection in a binary mixture, with Leu-enkephalin. A large "background" exists below *m/z* ~700, which is likely due to impurities and degradation products in the acetic acid and "clustering" involving solvent association with trace ionic impurities. A subtraction of the "background" using spectra obtained just before peak elution gives the mass spectrum shown in Figure 3 (bottom), which is dominated by the eluting analyte.

The major considerations relevant to MS detection generally arise due to the nature and complexity of the particular sample and pragmatic constraints due to MS detection sensitivity, resolution, and the related scan-speed compromises. For quadrupole mass spectrometers, single- or multiple-ion monitoring, sometimes referred to as selected ion monitoring (SIM), leads to significantly enhanced detection limits compared to MS scanning operation due to the greater "dwell time" at specific *m/z* values. One of the advantages of CE-MS is that mixture components strongly discriminated against by direct infusion often show much more uniform response after CE separation. For samples where analyte molecular weights are known, and *m/z* values can be predicted, SIM detection is an obvious choice. If sufficient sample is available, direct infusion can be used to produce a mass spectrum of the unseparated mixture and the results used to guide selection of specific *m/z* values for SIM detection.

The sensitivity gain with SIM detection vs. scanning detection with conventional quadrupole MS instruments can be substantial. Figure 4 shows SIM profiles for a separation of the same simple pentapeptide mixture as used for obtaining the spectrum in Figure 3. Figure 4 shows three SIM plots (from a total of 7 *m/z* values selected) for injection of samples from serial dilution covering more than two orders of magnitude lower concentration than shown in Figure 3. For the 5- to 10-fmol injection (Figure 4, top), comparable to Figure 3, SIM detection provides excellent signal-to-noise ratios, and the CE performance and peak shape is

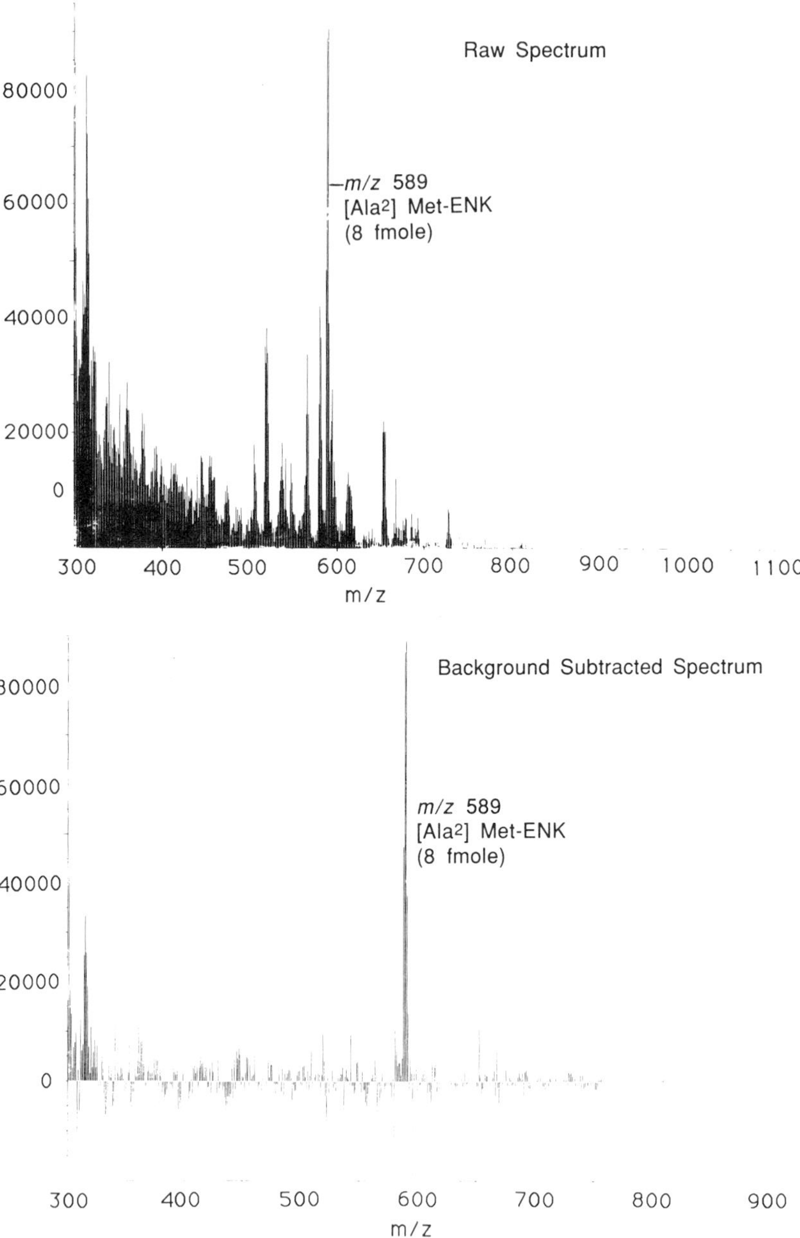

FIGURE 3. Capillary electrophoresis-mass spectrum obtained during elution of [Ala^2]Met-enkephalin from an 8-fmol injection, using a 1 m × 20-μm I.D. capillary with a 10 m*M* acetic acid buffer. *Top*: Raw spectrum (16.15 min) with large "background" contributions arising from the buffer. *Bottom*: "Background-subtracted" spectrum.

captured by the MS detector in near-ideal fashion. For injection sizes corresponding to 0.5 to 1 fmol (Figure 4, middle), the excellent peak shape is retained, but increased baseline "noise" due to buffer impurities is evident. (A shift in elution times is also noted that becomes greater with capillary use, due to a decrease in electroosmotic flow.) For injections of 40 to 75 amol (Figure 4, bottom) good peak shape was retained, but a noisy and elevated baseline suggests detection limits of ~10 amol, unless a reduction of background "chemical noise" can be obtained. This noise results from "real" solution components or other ions derived from these

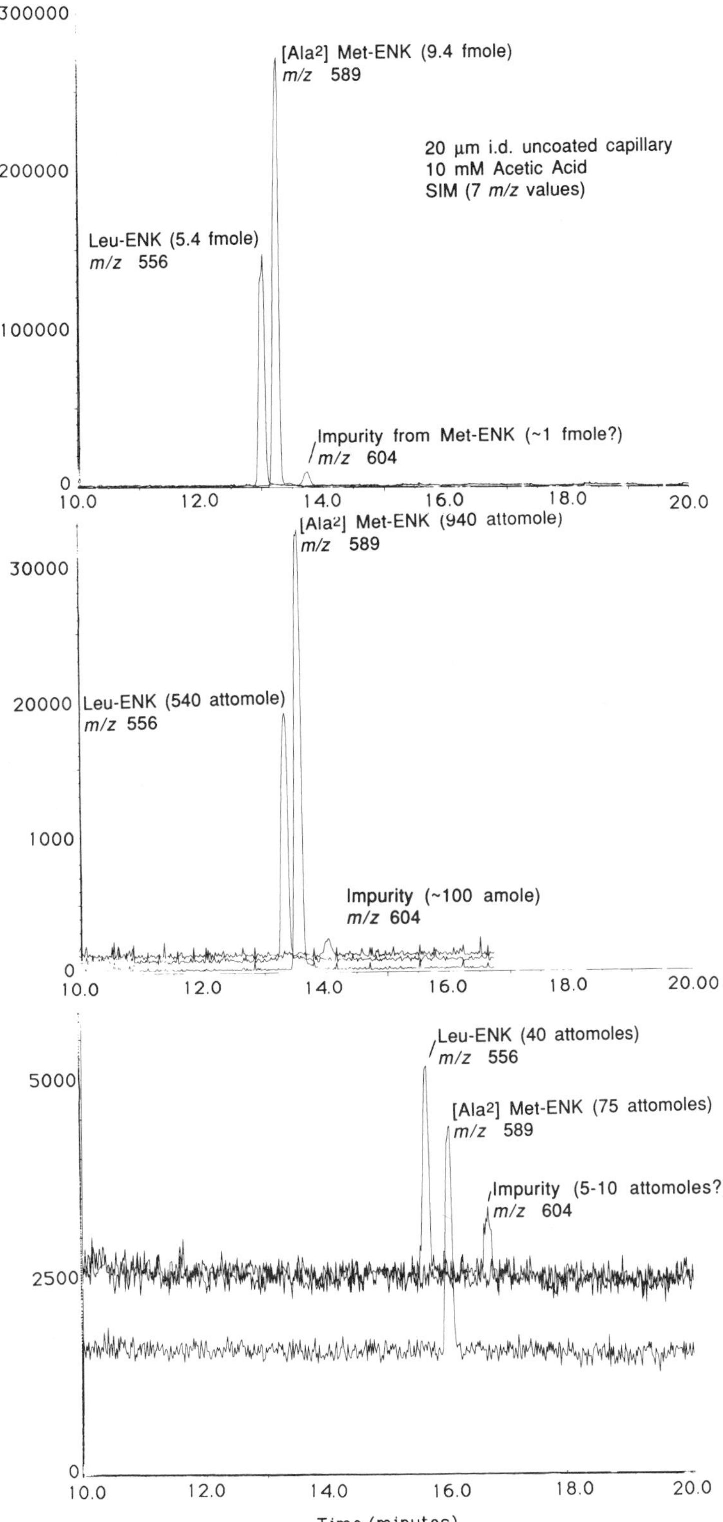

FIGURE 4. Selected ion electropherograms (of seven ions monitored) for injections of a pentapeptide mixture of Leu-enkephalin and [Ala^2]Met-enkaphalin for sample dilutions covering approximately two orders of magnitude in sample size, using the same conditions for the mass spectrum shown for Figure 3.

species due to the ESI process; thus, the use of cleaner buffers may provide a basis for an extension to subattomole detection limits.

An advantage of MS relative to other detectors is its high specificity. As shown in Figure 4, the elution times for the analytes change somewhat with time due to (often unavoidable) capillary surface modification. With less selective detectors, great care is generally necessary to establish reproducible migration times to identify eluents. In contrast, highly reproducible elution times are far less important for MS detection due to its much greater selectivity. Thus, relative migration times are usually sufficient, and one is generally more concerned with sensitivity, resolution, and signal/noise. Obtaining the maximum number of theoretical plates possible with combined CE-MS is rarely required unless closely related mixture components have nearly similar molecular weights.

Sensitivity for larger biomolecules (e.g., proteins) is often substantially less with ESI than for smaller peptides, and even greater needs arise for the use of tandem-MS (i.e., MS/MS methods) for obtaining structural information, as in polypeptide sequencing, due to the much lower signal intensities obtained for dissociation products with conventional tandem instruments. These limitations are a major driving force for the implementing of improved MS instrumentation.

There are several possible approaches to resolving these sensitivity limitations. One is to increase the efficiency of ion transport from the ESI source into the mass spectrometer, now generally only 0.1 to 0.01% in overall efficiency.[32] A second approach is to analyze all or a greater portion of the ions that enter the mass spectrometer, using either ion-trapping methods or array detection. Quadrupole ion trap mass spectrometers (ITMS) can trap and accumulate ions over a wide *m/z* range, followed by a rapid swept ejection/detection.[41] Thus, the fraction of total ions detected compared to that entering the ITMS is potentially much greater.

The best MS performance would, in principle, be obtained with full-range array detection, but practical instrumentation for this purpose is not yet available, and would likely be much more expensive than either the ITMS or existing instrumentation. Alternatively, the orthogonal (i.e., perpendicular ion extraction) time-of-flight mass spectrometer has the potential of obtaining high ion utilization efficiency, and initial results with an electrospray ion source have demonstrated efficiency of ~2.5% for ions transmitted into the orthogonal ion draw-out region.[42] Other developments are also promising. McLafferty and co-workers have pioneered the combination of ESI with Fourier transform-ion cyclotron resonance mass spectrometry (FT-ICR), and shown that very high resolution is obtainable.[43] Very recently our laboratory has demonstrated that the combination of CE with FT-ICR can provide both high MS resolution (>50,000) and high sensitivity (low femtomole for proteins) simultaneously.[44] The potential for higher order mass spectrometry, MS^n ($n \geq 3$), using FT-ICR or ITMS instruments, also appears promising.

The small solute quantities in CE require highly sensitive detection methods. The low signal intensities generally produced by ESI/MS, typically resulting in maximum analyte ion detection rates of no greater than 10^5 to 10^6 ion counts per second, effectively limit the maximum practical scan speeds with quadrupole mass spectrometers. Thus, depending on the desired *m/z* range, solute concentration, and other factors related to the nature of the solute and buffer species, maximum *m/z* scan speeds are often insufficient to exploit the high-quality separations feasible with CE when coupled to quadrupole or other scanning mass spectrometers.

As discussed in Section III, the maximum electrospray ion current is a weak function of solution conductivity. When the amount of charge-carrying solute entering the ESI source exceeds the capability of the electrospray process, then the efficiency of solute ionization decreases. Thus, at higher analyte concentrations or flow rates, ESI/MS signal strengths become relatively insensitive to analyte mass flow rate. At very low flow rates or concentrations the ESI signal strength becomes limited by the number of charge-carrying species in solution. In this regime, optimum sensitivity will be obtained. However, most ESI work, and nearly all CE-MS, has been conducted in the former regime where the efficiency of solute ionization is substantially limited. Capillary electrophoretic separations generally incur higher currents (5 to 50 μA) than typical ESI currents (0.1 to 0.5 μA), and therefore deliver charged species to the ESI source at a rate where ionization is necessarily inefficient. Any contaminant present may also compete with the analyte for available charge, thus decreasing the solute ionization efficiency. Ideal buffer components have characteristics that include being able to facilitate CE separation, volatility, being discriminated against during ESI, and forming minimal gas-phase contributions to the mass spectra arising from ionized clusters of solvent and buffer constituents.

The most obvious approach to improving sensitivity is to modify the injection step to load more sample onto the capillary. Such methods are of general interest to CE techniques and the reader is referred to Chapters 5, 6, 20, and 22. Such techniques have already been used for CE-MS; a particularly useful approach involving isotachophoretic sample preconcentration has been reported by Tinke et al.[45] and Thompson and co-workers.[46] Our discussion in the remainder of this section is from the viewpoint of subsequent steps that can be taken to enhance CE-MS sensitivity.

Another approach developed in our laboratory is based on reduced elution speed (RES) detection, which aids in alleviating both the sensitivity and scan speed limitations of CE-MS with scanning mass spectrometers.[38] Involving only step changes in the CE electric field strength, the technique is simple and readily implemented. Prior to elution of the first analyte of interest into the ESI source, the electrophoretic voltage is decreased and elution of solutes is slowed, allowing more scans to be recorded without a significant loss in ion intensity. Under conditions where the amount of solute entering the ESI source per unit of time exceeds its ionization capacity, no substantial decrease in maximum ion intensity is expected when the electric field strength is decreased. As a result, a greater fraction of the analyte ions can be transferred to the gas phase during an RES CE-MS experiment than during a normal constant electric field strength CE-MS experiment.[38]

Some of the greatest challenges for CE-MS involve analysis of complex mixtures of biopolymers. An important goal is to decrease the quantity of protein required for sequencing using methods based on an initial enzymatic digestion of a protein. Comparison of constant electric field strength (A) and RES (B) CE-MS for a 40-fmol injection of peptides produced by digestion of bovine serum albumin with trypsin is shown in Figure 5 and results in only a small decrease (20%) in ion intensity when the electrophoretic voltage was decreased to slow elution. Over 100 tryptic fragments can be readily extracted from the MS data obtained during this separation (Figure 5B).[38] The reduced complexity of individual scans aids data interpretation of complex mixtures due to the greater number of scans obtained during elution of a given component and the reduced likelihood of other components eluting during the same scan.

The utility of RES CE-MS is particularly useful for larger polypeptides and proteins, where broad *m/z* range spectra can be useful in molecular weight determination. For a mixture of standard proteins, Goodlett et al. observed that only a 14% reduction in the ion intensity is observed for the strongest response (myoglobin), without observable loss in separation quality.[38] Each protein elutes five times slower in the RES CE-MS analysis than in the constant electric field strength CE-MS separation. Quality of separation was not affected and actually appeared greater due to the increased number of scans recorded during solute elution. The quality of mass spectral data for a scan from 600 to 1200 *m/z* was sufficient to allow molecular weights to be calculated to 0.02 to 0.1% from 60 fmol/protein.

Reduced elution speed CE-MS provides an increase in the efficiency of mass spectrometric scanning compared to conventional CE-MS methods. The prolonged analyte elution into the electrospray source can be exploited by (1) increasing the *m/z* range scanned, (2) increasing the number of scans recorded during migration of a given solute, and (3) enhancing sensitivity

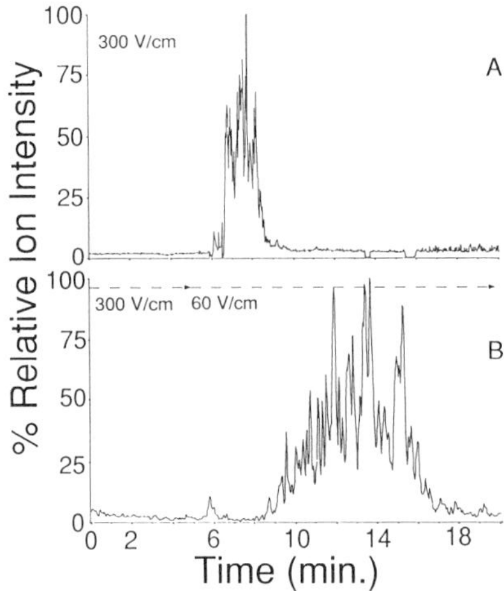

FIGURE 5. Comparison of constant field strength (A) and reduced elution speed (B) CE-ESI/MS analysis of bovine serum albumin after digestion by trypsin. The constant field strength CE-ESI/MS analysis was conducted at 300 V/cm. The reduced elution speed CE-ESI/MS analysis was conducted at 300 V/cm until 1 min prior to elution of the first peptide, when the electric field strength was reduced to 60 V/cm.

and signal intensity for a given solute. The method does not increase solute consumption, provides improved sensitivities for peptide and protein analyses extending into the low-femtomole regime, and incurs very little loss in ion intensity, which is particularly important for tandem MS methods and their potential application to peptide sequencing.

An alternative approach described by Wahl et al.[47-49] for obtaining greater sensitivity involves the use of smaller diameter capillaries than conventionally used for CE. As discussed in Section III, optimum ESI/MS sensitivity is generally obtained at low CE currents, where the rate of delivery of charged species to the ESI source is minimized. An optimum CE capillary diameter ideally meets several criteria: it should (1) be available commercially or readily prepared, (2) be amenable to alternative detection methods, and (3) provide the necessary detector sensitivity. For the last criterion, one would expect that optimum sensitivity would be obtained for CE currents approximately equal to or less than the ESI current. Even for a relatively low-conductivity acetic acid buffer system, CE capillary diameters greater than 40 μm will generally have currents that exceed that of the ESI source.[49] A series of CE-MS separations obtained using capillary inner diameters of 100, 50, 20, and 10 μm are shown in Figure 6 on the same absolute intensity scale. For these comparisons, all the experimental conditions were the same among the separations, and duplicate separations showing good agreement were obtained for each capillary diameter. The injected amounts of sample were proportional to the capillary cross-sectional area, where for example the relative amount injected for the 10-μm I.D. capillary is two orders of magnitude smaller compared to the 100-μm I.D. capillary. The relative ESI/MS signal intensity, however, decreased by only approximately half for melittin in comparing the 100- and 10-μm I.D. capillaries, and the corresponding sample injection sizes decreased from 800 to 8 fmol.[49]

To examine more closely the effect of solute concentration as a function of decreasing capillary inner diameter, a series of separations were performed with dilution of the peptide mixture at ratios of 1:2, 1:5, 1:10, 1:20, and 1:50. The results for one solute, tryptophan, from these CE-MS separations are shown in Figure 7, where the solute zone area measured is shown as a function of the injected amount for the four different capillary inner diameters. The results are convoluted by instabilities in the electrospray process while each solute was eluting to the electrospray source, and the longer day-to-day variation in instrumental performance over the course of the studies. For the lower analyte concentrations, the observed results may be summarized as suggesting that roughly comparable sensitivities are obtained with the 10- and 20-μm I.D. capillaries.

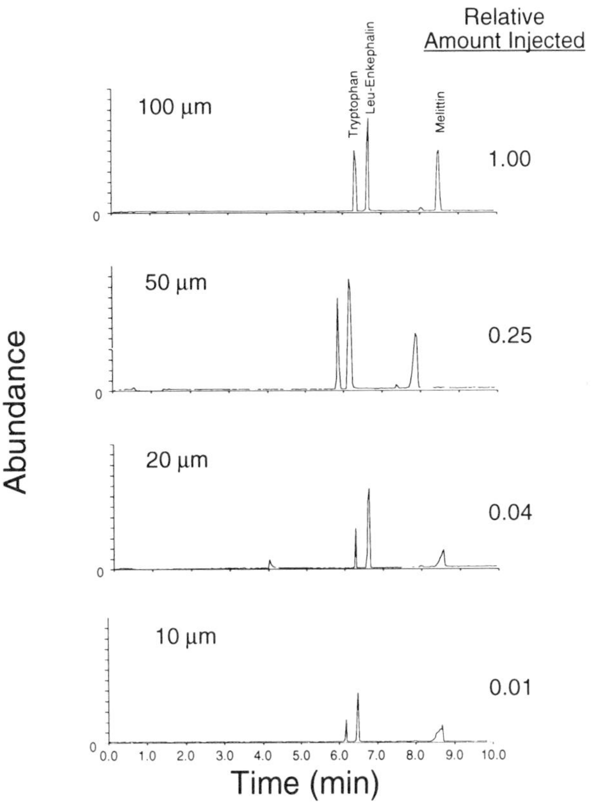

FIGURE 6. Comparison of the CE-MS total ion electropherograms for separation of tryptophan, Leu-enkephalin, and melittin for four different capillary diameters. Electromigration injection and separation conditions were identical for all experiments and signal intensities are shown on the same scale. The relative amount of analyte injected is a factor of 100 lower for the 10-μm capillary compared to the 100-μm capillary. The vertical axis is the same for all separations (1.2×10^6 count/s full scale).

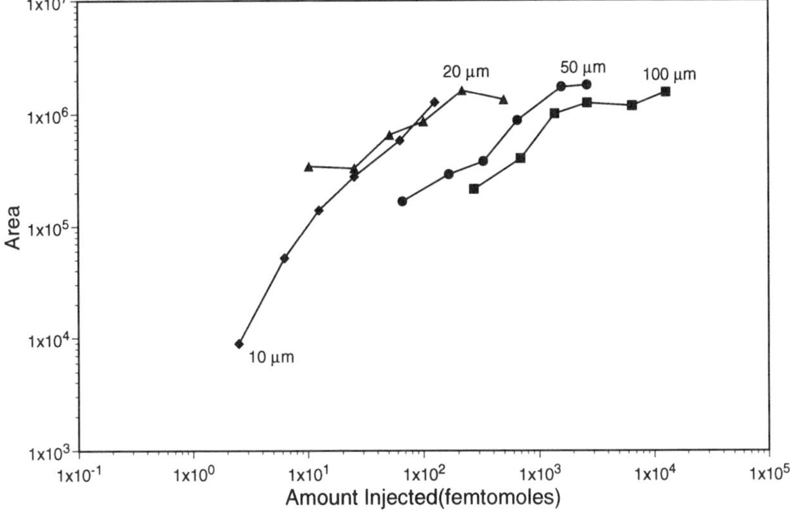

FIGURE 7. CE-MS sensitivity comparison for tryptophan from the mixture examined in Figure 6. The solute peak area is shown as a function of the injected amount for the four different capillary inner diameters. In general, for all solutes, the sensitivity increased with decreasing capillary diameter.

The prediction of the simple model discussed in Section III (Equations 1 and 4), that sensitivity will increase indefinitely as the mass flow rate through the CE capillary decreases, must fail at some point. This point corresponds to the regime where Equation 5 should apply. In the absence of a sheath flow, the CE current will ultimately reach a condition where the rate of analyte and background electrolyte delivered to the ESI source causes a decrease in the ESI current. The electrophoretic current is less than the electrospray current for both the 10- and 20-μm I.D. capillaries. As a result, the mass flow rate of ions to the ESI source from the analytical capillary may be inadequate to account for all the ions necessary for the electrospray process. If the contribution from the sheath liquid is small, the ionization efficiency for the analyte ions will approach its maximum, and further reduction in the CE background electrolyte delivery rate or concentration will have little effect.

The high analyte delivery rate limit of the ESI sensitivity model is given by Equation 2, where the ESI ion signal is expected to show no significant dependence on sample concentration. In this region, ESI sensitivity is poor and the high solute band concentration suggests that poor separation quality will often result. As the analyte concentrations are reduced, the regime of linear response is entered (Equation 4). The effect of capillary diameter reduction is apparent, with an increase in sensitivity observed as the capillary diameters are reduced. For the smallest capillary diameters, the results indicate that the amount of analyte delivered to the ESI region is now the primary determinant of the mass spectral response (Equation 5), and little change in sensitivity is expected with further reduction in capillary diameter. Uncertainties concerning the sheath liquid contribution to the observed results, however, do not allow a quantitative determination of the onset of such behavior. These considerations may at least partially account for variations in relative response between 10- and 20-μm I.D. capillaries for different analytes. Of particular significance, however, is the observation that the contribution to the total ESI current from the liquid sheath is substantial for the small inner diameter capillaries. Again, if the ESI current was dominated by the sheath liquid contributions, no sensitivity gain would be expected for smaller inner diameter capillaries. Similar gains in CE-MS sensitivity for protein mixtures have been reported by Wahl et al.[47] in a study of 50- to 5-μm I.D. capillaries. As with the peptide mixture discussed above, the relative signal intensity decreased by only a factor of two to four between the 50- and 5-μm I.D. capillaries, and the amount of each protein injected decreases from approximately 60 fmol to 600 amol for the 50- and 5-μm I.D. capillaries, respectively. A gain of 25 to 50 in sensitivity was observed. These results show that CE-MS of proteins is feasible at subfemtomole levels, and that sensitivity for the SIM mode of MS detection should extend to the low-attomole range for the proteins studied because generally a sensitivity improvement of at least 10 is noted relative to scanning detection with quadrupole instruments.

These results suggest the goal of ultrasensitive peptide and protein analysis at low-attomole and even subattomole levels is obtainable. However, a number of methodological problems remain to be addressed. Improved procedures for sample and buffer preparation and handling are also needed to prevent capillary plugging, a more common problem with small inner diameter capillaries. In addition, the ESI interface sensitivity must be improved further because the efficiency of ion transport from the ESI source, which is at atmospheric pressure, to the MS detector remains relatively low. The results also indicate that the sheath liquid makes only a modest contribution to the ESI current, and that ions delivered to the ESI source from the CE analytical capillary are transferred to the gas phase with very high efficiency for the smaller ($\leq$20-μm) diameter capillaries.

V. APPLICATION OF CAPILLARY ELECTROPHORESIS-MASS SPECTROMETRY

As yet, routine applications of CE-MS are few, and most reports have aimed at evaluation of its use for specific applications. A particular strength or weakness of CE and CE-MS, depending on one's viewpoint, of CE and CE-MS is the extremely small sample sizes generally used. Thus, the emphasis of CE-MS applications would appear to focus most effectively on those situations where sample sizes are inherently limited. Some of the most demanding applications involve the characterization of gene products and, specifically, proteins and polypeptide mixtures generated by enzymatic means.

A. POLYPEPTIDES AND ENZYMATIC DIGESTS

Application of CE-MS to complex polypeptide mixtures (e.g., tryptic digests) is potentially an extremely important area of application due to the significance of peptide mapping and sequencing using ever smaller sample sizes. Complex mixtures of peptides generated from tryptic digestion of large proteins present a difficult analytical challenge because the large number of fragments cover an extensive range of both isoelectric points and hydrophobicity. Because trypsin specifically cleaves peptide bonds on the C-terminal side of lysine and arginine residues, the resulting peptides generally form doubly charged as well as singly charged molecular ions by positive ion ESI. Such doubly charged tryptic peptides generally fall within the *m/z* range of modern quadrupole mass spectrometers. The resolving power of these methods has been demonstrated by the comparison of UV electropherograms obtained in conjunction with a CE-MS analysis[40] of tryptic digests of cytochrome *c* (in a 50 m*M* acetate buffer [at pH 6.1] mixed with an equal volume of acetonitrile). In this work,[40] a commercial CE instrument (Beckman [Fullerton, CA] P/ACE 2000) was modified to allow UV detection of the separation at one half of the MS detection time, effectively providing a "preview" of ESI/MS detection. The individual peaks indicate separation efficiencies of up to ~4×10^5 theoretical plates. It was noted that the apparent efficiency for MS detection is significantly greater than for ultraviolet (UV) detection (1.2×10^5 theoretical plates for the same peak), a fact that can be largely attributed to the longer separation time and the very small effective dead volume or residence time in the ESI interface.[40]

The sensitivity of CE-MS for such applications can be improved by the use of small inner diameter capillaries. The total ion electropherograms obtained from tryptic digests of bovine, *Candida krusei,* and horse cytochrome *c* by Wahl and Smith[48] are shown in Figure 8. For each separation, a 10 m*M* ammonium acetate/acetic acid buffer system, pH 4.4, was used. The separation capillary (10-μm I.D.) was chemically modified with 3-aminopropyltrimethoxysilane, and was 50 cm in length. The injection size corresponded to approximately 30 fmol of protein before digestion. The mass spectrometer was scanned from *m/z* 600 to 1200 in two *m/z* steps at 0.6 s/scan. In each case, the separation was complete within 6 min. The MS detection allows each of the tryptic fragments to be identified. For example, shown in Figure 9 are the extracted electropherograms for the individual tryptic fragments YIPGTK (*m/z* 678), which is a tryptic fragment common to all three proteins, EDLIAYLK (*m/z* 964), which is common to bovine and horse cytochrome *c*, and MAFGGLK (*m/z* 723), which is specific to *Candida krusei* cytochrome *c*. In addition, the three extracted electropherograms show additional solute zones due to other components producing ions at nearly the same *m/z* as these tryptic fragments. These additional species may arise due to incomplete digestion, adduction of ion species, noncovalent complexes, or fragmentation occurring within the interface of the mass spectrometer.

Experience to date suggests that nearly all tryptic fragments can be effectively detected by ESI-MS methods if first separated; infusion of the unseparated digest often results in dramatic discrimination against some components. Fragments observed by the UV detector are almost always detected by ESI/MS. However, detection has sometimes been problematic for very small fragments (i.e., amino acids and dipeptides) due to difficulties in obtaining optimum ESI/MS conditions over a sufficiently wide *m/z* range. Excessive internal excitation or large solvent-related background peaks at low *m/z* appear to be the origin of some difficulties. There are also indications that some ESI interface designs discriminate against low-*m/z* ions.

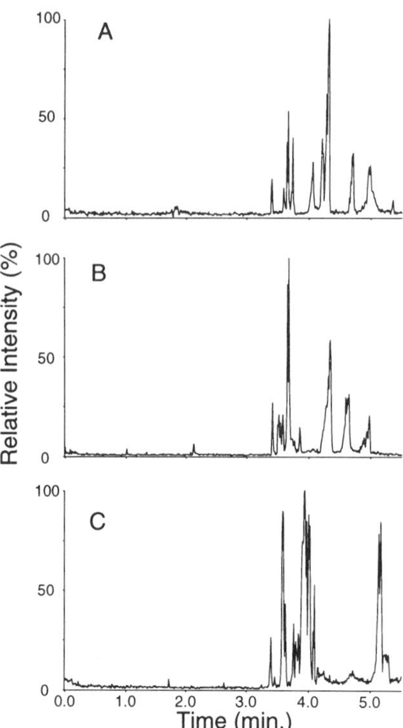

FIGURE 8. Total ion electropherograms obtained from tryptic digests of bovine (A), *Candida krusei* (B), and horse cytochrome *c* (C). Separation conditions: 10 m*M* ammonium acetate/acetic acid buffer system, pH 4.4; capillary, 10-μm I.D., 50 cm in length, and chemically modified with 3-aminopropyltrimethoxysilane.

Available evidence, however, indicates that tryptic fragments not detected by CE-MS are generally "lost" prior to the separation, a problem particularly evident with "nanoscale" sample handling. The full realization of more sensitive analytical methods depends on appropriate care in sample handling and preparation.

B. PROTEINS

The ESI/MS interface provides the basis for detection of molecules having molecular weights exceeding 100,000. The analysis of proteins by CE-MS is challenging due to the well-established difficulties associated with protein interactions with capillary surfaces. As discussed in Chapters 18 and 19, capillaries must often be coated to minimize, or prevent, wall interactions. It is unlikely that one capillary or buffer system will be ideal for all proteins, but rather, as for liquid chromatography (LC) separations, procedures optimized for specific classes of proteins will be developed. Most CE separations require the ionic strength of the buffer to be about 100-fold greater than that of the sample to prevent degradation of the separation due to perturbation of the local electric field in the capillary, although much lower relative buffer concentrations are often used in CE-MS to enhance sensitivity. Detection sensitivity for proteins is generally lower than for small peptides due to the greater number of charges per molecule and the greater number of charge states. Our laboratory was the first to explore CE-MS of proteins,[39] but impressive results have been recently reported by Thibault et al.[27] and Moseley et al.,[50] among others.

An important attribute of the ESI/MS interface is the capability to obtain structurally related information on large molecules by dissociation in the interface by a "collisional heating" process induced by a simple electric field gradient. This capability was first demon-

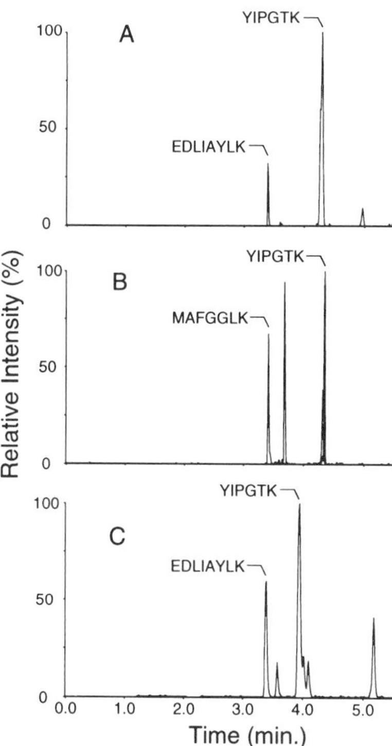

FIGURE 9. Extracted ion electropherograms obtained from Figure 8 for the individual tryptic fragments YIPGTK (*m/z* 678), which is a tryptic fragment common to all three proteins, EDLIAYLK (*m/z* 964), which is common to bovine (A) and horse (C) cytochrome *c*, and MAFGGLK (*m/z* 723), which is specific to *Candida krusei* (B) cytochrome *c*. The additional solute zones are probably due to incomplete protein digestion and/or fragmentation occurring within the mass spectrometer interface.

strated for large molecules by our laboratory in 1988,[51] and has the advantage, compared to collisionally induced dissociation (CID) MS/MS methods, of much greater selectivity. However, it is also associated with the disadvantage of greater ambiguity because only one MS step is utilized and a good separation is generally required. The complexity of large-molecule collisional dissociation spectra precludes obtaining sequence information with current MS quadrupole technology, but its potential for "fingerprinting" has been established.[31] Extensive sequence-related fragmentation can be obtained for molecules as large as serum albumins (~66 kDa), yielding spectra suitable for fingerprinting purposes.

The use of small-diameter capillaries and optimized buffer systems allows full-scan mass spectrometric data to be obtained for more concentrated protein samples. For example, Figure 10 shows the separation and mass spectrum obtained by injection of ~30 fmol of the protein carbonic anhydrase I, using a 20-μm I.D. capillary. The high-quality mass spectrum allows molecular weight determination with a precision of approximately 0.02%. Much improved performance has recently been obtained using FTICR detection.[44]

The potential of CE-MS for protein characterization has only begun to be explored and is expected to grow as more sensitive and powerful MS detection methods become available. Cases where the combination of CE and MS provide unique information, or provide answers to characterization problems much more rapidly, are expected to become increasingly common. For example, Tsuji et al. demonstrated that CE-MS allowed recombinant somatotropins (molecular mass ~22 kDa) to be characterized as well as the detection of both mono- and dioxidized homolog that could not be unambiguously detected by either CE or low-resolution

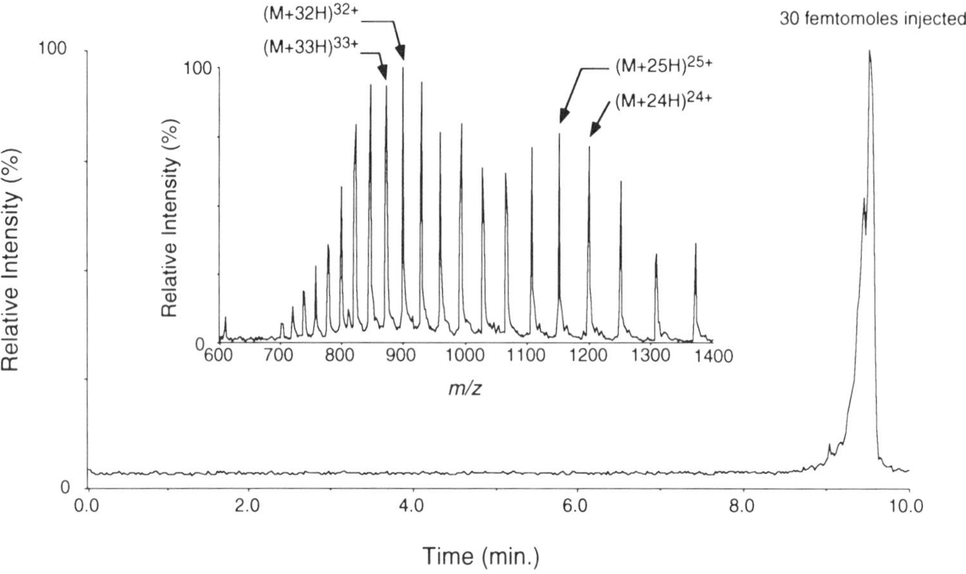

FIGURE 10. CE-MS of a 50-µm carbonic anhydrase I sample, showing the separation and mass spectra obtained for a 30-fmol injection with a 20-µm capillary that was chemically treated with 3-aminopropyltrimethoxysilane.

MS.[52] Recent results suggest that the powerful combination of CE with FTICR can provide extremely high performance MS and MS^n capabilities.[44]

VI. FUTURE PROSPECTS

A major development anticipated in the near future will be the use of high performance mass spectrometric detectors for CE-MS. The use of orthogonal time of flight instrumentation promises to facilitate both high speed CE-MS separations (since $>10^3$ spectra/s are obtainable), and much better (zeptomole to attomole range) detection limits should be achievable for well-designed instruments. Alternatively, as demonstrated by our initial work, CE using FTICR mass spectrometers provides the basis for very high MS resolution ($>10^6$!) and multidimensional mass spectrometry (i.e., MS^n for structural studies.[44] The combination of better MS performance, control of the ESI process, and the development of more compatible buffer systems will also aid the use of alternative CE formats (MECC, capillary gel electrophoresis, and isoelectric focusing). Finally, the introduction of relatively inexpensive, but highly sensitive quadrupole ion trap mass spectrometers should serve to make many of these advanced capabilities much more widely available. For the next few years, the use of CE-MS will clearly be driven by both the continued growth in the application of CE and the desire for more sensitive and information-rich detection methods in research.

ACKNOWLEDGMENTS

This research was supported by internal exploratory research and the Director, Office of Health and Environmental Research, U.S. Department of Energy. Pacific Northwest Laboratory is operated by Battelle Memorial Institute for the U.S. Department of Energy, through Contract DE-AC06-76RLO 1830. We thank C. J. Barinaga for helpful discussions and contributions to the work described in this chapter.

REFERENCES

1. **McCloskey, J. A., Ed.,** Mass spectrometry, in *Methods in Enzymology*, Vol. 193, Academic Press, San Diego, 1990.
2. **Yergey, A. E., Edmonds, C. G., Lewis, I. A. S., and Vestal, M. L.,** Liquid chomatography/mass spectrometry: techniques and applications, in *Modern Analytical Chemistry*, Hercules, D., Ed., Plenum Press, New York, 1990.
3. **Kenndler, E. and Kaniansky, D.,** Off-line combination of isotachophoresis and mass spectrometry, *J. Chromatogr.*, 209, 306, 1981.
4. **Kenndler, E., Haidl, E., and Fresenius J.,** Mixed zone analysis in isotachophoresis with selective detection by mass spectrometry applied to the quantitation of hydrogenation products of aromatic quaternary ammonium compounds, *Anal. Chem.*, 322, 391, 1985.
5. **Takigiku, R., Keough, T., Lacey, M. P., and Schneider, R. E.,** Capillary-zone electrophoresis with fraction collection for desorption mass spectrometry, *Rapid Commun. Mass Spectrom.*, 4, 24, 1990.
6. **Castoro, J. A., Chiu, R. W., Monnig, C. A., and Wilkins, C. L.,** Matrix assisted laser desorption/ionization of capillary electrophoresis effluents by Fourier transform mass spectrometry, *J. Am. Chem. Soc.*, 114, 7571, 1992.
7. **Caprioli, R. M., Moore, W. T., Martin, M., and DaGue, B. B.,** Coupling capillary zone electrophoresis and continuous-flow fast-atom mass spectrometry for the analysis of peptide mixtures, *J. Chromatogr.*, 480, 247, 1989.
8. **Reinhoud, N. J., Schroder, E., Tjaden, U. R., Niessen, W. M. A., Ten Noever de Brauw, M. C., and van der Greef, J.,** Static and scanning array detection in capillary electrophoresis-mass spectrometry, *J. Chromatogr.*, 516, 147, 1990.
9. **Suter, M. J.-F., DaGue, B. B., Moore, W. T., Lin, S.-N., and Caprioli, R. M.,** Recent advances in liquid chromatography-mass spectrometry and capillary zone electrophoresis-mass spectrometry for protein analysis, *J. Chromatogr.*, 553, 101, 1991.
10. **Moore, W. T. and Caprioli, R. M.,** Monitoring peptide synthesis stepwise by mass spectrometry, in *Techniques in Protein Chemistry*, Vol. II, Academic Press, San Diego, 1991, p. 511.
11. **Suter, M. J.-F. and Caprioli, R. M.,** An integral probe for capillary zone electrophoresis continuous-flow fast-atom bombardment mass spectrometry, *J. Am. Soc. Mass Spectrom.*, 3, 198, 1992.
12. **Minard, R. D., Luckenbill, D., Curry, R., Jr., and Ewing, A. G.,** Capillary electrophoresis-mass spectrometry, in Proc. 37th ASMS Conference on Mass Spectrometry and Allied Topics, 1989, p. 950.
13. **Wolf, S. M., Vouros, P., Norwood, C., and Jackim, E.,** Identification of deoxynucleoside-polyaromatic hydrocarbon adducts by capillary zone electrophoresis continuous-flow fast-atom bombardment mass spectrometry, *J. Am. Soc. Mass Spectrom.*, 3, 757, 1992.
14. **Reinhoud, N. J., Niessen, W. M. A., and Tjaden, U. R.,** Performance of a liquid-junction interface for capillary electrophoresis mass spectrometry using continuous-flow fast-atom bombardment, *Rapid Commun. Mass Spectrom.*, 3, 348, 1989.
15. **Verheij, E. R., Tjaden, U. R., Niessen, W. M. A., and Van Der Greef, J.,** Pseudo-electrochromatography-mass spectrometry: a new alternative, *J. Chromatogr.*, 554, 339, 1991.
16. **deWit, J. S. M., Deterding, L. J., Moseley, M. A., Tomer, K. B., and Jorgenson, J. W.,** Design of a coaxial continuous-flow fast-atom bombardment probe, *Rapid Commun. Mass Spectrom.*, 2, 100, 1988.
17. **Moseley, M. A., Deterding, L. J., Tomer, K. B., and Jorgenson, J. W.,** Coupling of capillary zone electrophoresis and capillary liquid chromatography with coaxial continuous-flow fast-atom bombardment tandem sector mass spectrometry, *J. Chromatogr.*, 480, 197, 1989.
18. **Moseley, M. A., Deterding, L. J., Tomer, K. B., and Jorgenson, J. W.,** Capillary-zone electrophoresis fast-atom bombardment mass spectrometry: design of an on-line coaxial continuous flow interface, *Rapid Commun. Mass Spectrom.*, 3, 87, 1989.
19. **Moseley, M. A., Deterding, L. J., Tomer, K. B., and Jorgenson, J. W.,** Determination of bioactive peptides using capillary zone electrophoresis/mass spectrometry, *Anal. Chem.*, 63, 109, 1991.
20. **Deterding, L. J., Moseley, M. A., Tomer, K. B., and Jorgenson, J. W.,** Nanoscale separations combined with tandem mass spectrometry, *J. Chromatogr.*, 554, 73, 1991.
21. **Deterding, L. J., Perkins, J. R., and Tomer, K. B.,** Nanoscale separations of biomolecules in combination with coaxial continuous-flow-FAB mass spectrometry, Proc. of the 40th ASMS Conference on Mass Spectrometry and Allied Topics, 1992, 392.
22. **Deterding, L. J., Parker, C. E., Perkins, J. R., Moseley, M. A., Jorgenson, J. W., and Tomer, K. B.,** Capillary liquid chromatography-mass spectrometry and capillary zone electrophoresis-mass spectrometry for the determination of peptides and proteins, *J. Chromatrogr.*, 554, 329, 1991.
23. **Bruins, A. P., Covey, T. R., and Henion, J. D.,** Ion spray interface for combined liquid chromatography/atmospheric pressure ionization mass spectrometry, *Anal. Chem.*, 59, 2642, 1987.

24. **Lee, E. D., Mück, W., Henion, J. D., and Covey, T. R.,** Liquid junction coupling for capillary zone electrophoresis/ion spray mass spectrometry, *Biomed. Environ. Mass Spectrom.*, 18, 844, 1989.
25. **Olivares, J. A., Nguyen, N. T., Yonker, C. R., and Smith, R. D.,** On-line mass spectrometric detection for capillary zone electrophoresis, *Anal. Chem.*, 59, 1230, 1987.
26. **Smith, R. D., Olivares, J. A., Nguyen, N. T., and Udseth, H. R.,** Capillary zone electrophoresis-mass spectrometry using an electrospray ionization interface, *Anal. Chem.*, 60, 436, 1988.
27. **Thibault, P., Paris, C., and Pleasance, S.,** Analysis of peptides and proteins by capillary electrophoresis/mass spectrometry using acidic buffers and coated capillaries, *Rapid Commun. Mass Spectrom.*, 5, 484, 1991.
28. **Fenn, J. B., Mann, M., Meng, C. K., Wong, S. F., and Whitehouse, C. M.,** Electrospray ionization–principles and practice, *Mass Spectrom. Rev.*, 9, 37, 1990.
29. **Chowdhury, S. K., Katta, V., and Chait, B. T.,** An electrospray-ionization mass spectrometer with new features, *Rapid Commun. Mass Spectrom.*, 4, 81, 1990.
30. **Rockwood, A. L., Busman, M., and Smith, R. D.,** Thermally induced dissociation of ions from electrospray mass spectrometry, *Rapid Commun. Mass Spectrom.*, 5, 582, 1991.
31. **Smith, R. D., Loo, J. A., Edmonds, C. G., Barinaga, C. J., and Udseth, H. R.,** New developments in biochemical mass spectrometry: electrospray ionization, *Anal. Chem.*, 62, 882, 1990.
32. **Smith, R. D., Loo, J. A., Ogorzalek Loo, R. R., Busman, M., and Udseth, H. R.,** Principles and practice of electrospray ionization-mass spectrometry for large polypeptides and proteins, *Mass Spectrom. Rev.*, 10, 359, 1991.
33. **Ikonomou, M. G., Blades, A. T., and Kebarle, P.,** Investigations of the electrospray interface for liquid chromatography/mass spectrometry, *Anal. Chem.*, 62, 957, 1990.
34. **Ikonomou, M. G., Blades, A. T., and Kebarle, P.,** Electrospray-ion spray: a comparison of mechanisms and performance, *Anal. Chem.*, 63, 1989, 1991.
35. **Tang, L. and Kebarle, P.,** Effect of conductivity of the electrosprayed solution on the electrospray current. Factors determining analyte sensitivity in electrospray mass spectrometry, *Anal. Chem.*, 63, 2709, 1991.
36. **Pfeifer, R. J. and Hendricks, C. D., Jr.,** Parametric studies of electrohydrodynamic spraying, *AIAA J.*, 6, 496, 1968.
37. **Hjertén S.,** Zone broadening in electrophoresis with special reference to high performance electrophoresis in capillaries: an interplay between theory and practice, *Electrophoresis*, 11, 665, 1990.
38. **Goodlett, D. R., Wahl, J. H., Udseth, H. R., and Smith, R. D.,** Reduced elution speed detection for capillary electrophoresis-mass spectrometry, *J. Microcolumn Sep.*, 5, 57, 1993.
39. **Loo, J. A., Jones, H. K., Udseth, H. R., and Smith, R. D.,** Capillary zone electrophoresis-mass spectrometry with electrospray ionization of peptides and proteins, *J. Microcolumn Sep.*, 1, 223, 1989.
40. **Smith, R. D., Udseth, H. R., Barinaga, C. J., and Edmonds, C. G.,** Instrumentatin for high-performance capillary electrophoresis-mass spectrometry, *J. Chromatogr.*, 559, 197, 1991.
41. **Van Berkel, G. J., Glish, G. L., and McLuckey, S. A.,** Electrospray ionization combined with ion trap mass spectrometry, *Anal. Chem.*, 62, 1287, 1990.
42. **Boyle, J. G. and Whitehouse, C. M.,** Time-of-flight mass spectrometry with an electrospray ion beam, *Anal. Chem.*, 64, 2084, 1992.
43. **Henry, K. D., Quinn, J. P., and McLafferty, F. W.,** High-resolution electrospray mass spectra of large molecules, *J. Am. Chem. Soc.*, 113, 5447, 1991.
44. **Hofstadler, S. A., Wahl, J. H., Bruce, J. E., and Smith, R. D.,** On-line capillary electrophoresis with fourier transform ion cyclotron resonance mass spectrometry, *J. Amer. Chem. Soc.*, 115, 6983, 1993.
45. **Tinke, A. P., Reinhoud, N. J., Niessen, W. M. A., Tjaden, U. R., and van der Greef, J.,** *Rapid Commun. Mass Spectrom.*, 6, 560, 1992.
46. **Thompson, T. J., Foret, F. P., Vouros, P., and Karger, B. L.,** Capillary electrophoresis/electrospray ionization mass spectrometry: improvement of protein detection limits using on-column transient isotachophoretic sample preconcentration, *Anal. Chem.*, 65, 900, 1993.
47. **Wahl, J. H., Goodlett, D. R., Udseth, H. R, and Smith, R. D.,** Attomole level capillary electrophoresis-mass spectrometric protein analysis using 5-µm-i.d. capillaries, *Anal. Chem.*, 64, 3194, 1992.
48. **Wahl, J. H. and Smith, R. D.,** Unpublished results.
49. **Wahl, J. H., Goodlett, D. R., Udseth, H. R., and Smith, R. D.,** Use of small-diameter capillaries for increasing peptide and protein detection sensitivity in capillary electrophoresis-mass spectrometry, *Electrophoresis*, 14, 448, 1993.
50. **Moseley, M. A., Shabanowitz, J., Hunt, D. F., Tomer, K. B., and Jorgenson, J. W.,** Optimization of capillary zone electrophoresis/electrospray ionization parameters for the mass spectrometry and tandem mass spectrometry analysis of peptides, *J. Am. Soc. Mass Spectrom.*, 3, 289, 1992.
51. **Loo, J. A., Udseth, H. R., and Smith, R. D.,** Collisional effects on the charge distribution of large molecules from electrospray ionization-mass spectrometry, *Rapid Commun. Mass Spectrom.*, 2, 207, 1988.
52. **Tsuji, K., Baczynskyj, L., and Bronson, G. E.,** Capillary electrophoresis-electrospray mass spectrometry for the analysis of recombinant bovine and porcine somatotropins, *Anal. Chem.*, 64, 1864, 1992.

PART III
GENERAL CAPILLARY ELECTROPHORESIS APPLICATIONS

Chapter 9

ELECTROPHORETIC CAPILLARY ION ANALYSIS

William R. Jones

TABLE OF CONTENTS

0-8493-8690-X/94/$0.00+$.50

I. INTRODUCTION

A. DEFINITION AND ORIGIN

Electrophoretic capillary ion analysis (CIA), often referred to as capillary ion electrophoresis (CIE) or inorganic capillary electrophoresis (ICE), is a capillary electrophoretic technique optimized for the rapid determination of low molecular weight inorganic and organic ions. Sensitive detection is based predominantly on indirect ultraviolet (UV) since the majority of the ions lack specific chromophores. The bulk flow of the solution within the capillary, electroosmotic flow (EOF), is controlled by electrolyte chemistry, power supply polarity, and applied potential. The EOF is directed toward the detector, resulting in short analysis times of 5 min or less through augmentation of the migration velocity of the analyte.

Introduced at the beginning of this decade, CIA methodologies are rapidly expanding to encompass the classes of ions that were once the domain of ion chromatography (IC).[1,2] Table 1 lists 147 ionic species that have been characterized to date. Although there are exceptions, Figure 1 maps the general domains of CIA, capillary electrophoresis (CZE), and micellar electrokinetic capillary chromatography (MECC) with respect to analyte mobility as reflected in the CZE literature. Capillary ion analysis effectively separates anions and cations with ionic equivalent conductance (IEC) values between the range of 30 to 110. Above an IEC of plus or minus 110 there are very few ions with higher conductivity (mobility). The exceptions are the ions responsible for pH, hydroxide (–198), deutronium (298), and hydronium (350). Although detection of the hydronium ion has yet to be reported, the hydroxide ion has been qualitatively determined.[16,23]

Separation of ions with IEC values generally less than 30 is best done by classic CZE and MECC. To represent the classes of analytes in the CZE/MECC according to class, a logarithmic IEC scale would be required to allow room for the large variety of ionic and neutral species.

The earliest CZE report of inorganic cation analysis can be traced to Hjertén in 1967[3] and to Mikkers et al. in 1979.[4] Several papers dealing with various approaches to separating low molecular weight ions are found dating from 1979 to 1989.[5-12] The first paper on indirect UV detection using a chromate electrolyte containing an EOF modifier for anion analysis was reported in 1990 by Jones and Jandik.[13] Modifications of the chromate electrolyte for controlled analyte selectivity and sensitivity are found in References 15, 16, 30, and 36. Weston et al. have described optimization of detection of cations and factors controlling selectivity.[27,33] A good review of the origins and principles of CIA/CZE can be found in the paper by Jandik et al.[17]

B. WHY CAPILLARY ION ANALYSIS OVER ION CHROMATOGRAPHY?

Capillary ion analysis to date has separation efficiencies approaching 1 million theoretical plates, with a very high peak capacity. Figure 2 shows a separation of 36 anions achieved in 2.9 min, where all the peaks are baseline resolved in an 83-s section.[32] In Figure 3A, separation of 19 cations consisting of alkali, alkaline earths, and lanthanides by CIA is compared with 1 cation by IC, using a weak cation-exchange column that separates alkali and alkaline earths isocratically in 20 min. The CIA cation separation has reached completion before the first IC peak elutes off the column. For anions, the 36-peak electropherogram shown in Figure 2 is compared with a chromatogram obtained from a conventional anion-exchange column (Figure 3B). Only three anions in the chromatogram are eluted within the same time frame. Increased sample throughput is achieved with the open tubular capillaries in CIA because the analysis is complete when the last analyte of interest has passed through the detector. Slower components remaining in the capillary that are not required for analysis are removed by automated purge routines included with most CZE instruments. This 2-min procedure flushes the undesired components out of the capillary with fresh electrolyte in preparation for the next sample to be analyzed. In IC, the separation is over when all of the components have eluted off the column.

TABLE 1
Ions Characterized by Capillary Ion Analysis

Inorganic anions
- Arsenate
- Arsenite
- Azide
- Borate
- Bromate
- Bromide
- Carbonate
- Chlorate
- Chloride
- Chlorite
- Chromate
- Cyanide
- Fluoroborate
- Fluoride
- Hypochlorite
- Iodide
- Metasilicate
- Metavanadate
- Molybdate
- Monofluorophosphate
- Nitrate
- Nitrite
- Orthovanadate
- Perchlorate
- Persulfate
- Phosphate
- Phosphite
- Selenate
- Selenite
- Sulfate
- Sulfide
- Sulfite
- Thiocyanate
- Thiosulfate
- Tungstate

Organic anions
- Acetate
- *trans*-Aconitate
- Ascorbate
- *dl*-Aspartate
- Benzoate
- Butanesulfonate
- Butyrate
- Caproate
- Caprylate
- 4-Carboxybenzaldehyde
- Chloroacetate
- Citrate
- Crotonate
- Decanesulfonate
- Dodecanesulfonate
- Dichloroacetate
- Ethanesulfonate

Organic anions (contd.)
- Formate
- Fumarate
- Galactarate
- *d*-Galacturonate
- *d*-Gluconate
- Glucuronate
- *l*-Glutamate
- Glutarate
- Glycerate
- Glycolate
- Glyphosate
- Heptanesulfonate
- Hexanesulfonate
- α-Hydroxybutyrate
- Hydroxyethyldiphosphonate
- Hydroxymethylbenzoate
- 2-Hydroxyvalerate
- *dl-myo*-Inositol-1-monophosphate
- *dl-myo*-Inositol-2-monophosphate
- *dl-myo*-Inositol-1,4-biphosphate
- *dl-myo*-Inositol-1,4,5-triphosphate
- Isocitrate
- α-Ketoglutarate
- Lactate
- Maleate
- Malonate
- Methanesulfonate
- Nonanesulfonate
- Octanesulfonate
- Orotate
- Oxalacetate
- Oxalate
- Pentanesulfonate
- *o*-Phthalate
- Phytate
- Propanesulfonate
- Propionate
- Pyridinedicarboxylate
- Pyruvate
- Quinate
- Salicylate
- Shikimate
- Sorbate
- Succinate
- Tartarate
- Terephthalate
- Trichloroacetate
- Trifluoroacetate
- Trimesate
- *p*-Toluate
- Valerate

Alkali metals
- Lithium
- Sodium
- Potassium
- Rubidium
- Cesium

Alkaline earths
- Beryllium
- Magnesium
- Calcium
- Strontium
- Barium

Transition metals
- Scandium
- Chromium
- Manganese
- Iron
- Cobalt
- Nickel
- Copper
- Zinc
- Yttrium
- Cadmium
- Tin
- Mercury
- Lead

Lanthanides
- Lanthanum
- Cerium
- Praseodymium
- Neodymium
- Samarium
- Europium
- Gadolinium
- Terbium
- Dysprosium
- Holmium
- Erbium
- Thulium
- Ytterbium
- Lutetium

Nonmetal cation
- Ammonium

Organic cations
- Dimethylamine
- Trimethylamine
- Diethylamine
- Triethylamine
- Diethanolamine
- Triethanolamine

Note: This table lists the anions and cations that have been characterized by CIA. The anions are divided into two categories, inorganic and organic, and are listed alphabetically. The cations are listed according to their class and are listed in order of increasing atomic number. Ultraviolet detection was used throughout.

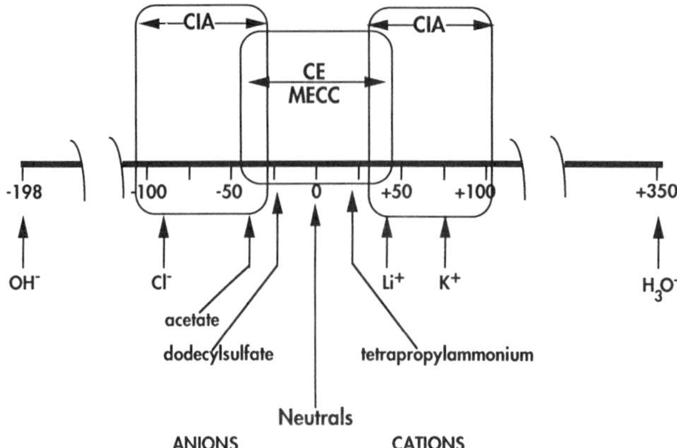

FIGURE 1. Depiction of the domains of ions as determined by CIA, CE, and MECC on the basis of ionic equivalent conductance. Capillary ion analysis is optimized for ions with conductivity values greater than 30.

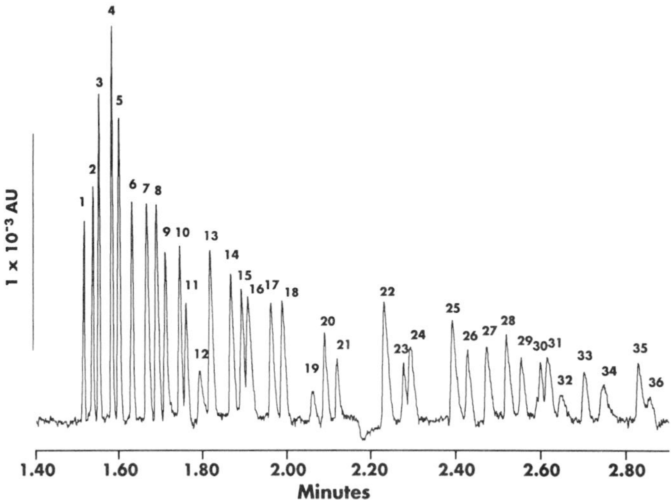

FIGURE 2. Peak identities and concentrations (ppm) for 36 anions displayed in an 85-s section of an electropherogram. Peaks: 1, thiosulfate (1.3); 2, bromide (1.3); 3, chloride (0.7); 4, sulfate (1.3); 5, nitrite (1.3), 6, nitrate (1.3); 7, molybdate (3.3); 8, azide (1.3); 9, tungstate (3.3); 10, monofluorophosphate (1.3); 11, chlorate (1.3); 12, citrate (0.7); 13, fluoride (0.3); 14, formate (0.7); 15, phosphate (1.3); 16, phosphite (1.3); 17, chlorite (1.3); 18, glutarate (1.7); 19, *o*-phthalate (0.7); 20, galactarate (1.3); 21, carbonate (1.3); 22, acetate (1.3); 23, chloroacetate (0.7); 24, ethanesulfonate (1.3); 25, propionate (1.3); 26, propanesulfonate (1.3); 27, *dl*-aspartate (1.3); 28, crotonate (1.3); 29, butyrate (1.3); 30, butanesulfonate (1.3); 31, valerate (1.3); 32, benzoate (1.3); 33, *l*-glutamate (1.3); 34, pentanesulfonate (1.7); 35, *d*-gluconate (1.7); 36, *d*-galacturonate (1.7). The electrolyte is a 5 m*M* chromate and 0.4 m*M* OFM Anion-BT adjusted to pH 8.0. Applied potential is 30 kV (negative polarity). Capillary dimensions are 60 cm (L_t), 52 cm (L_d), and 50-μm I.D. Injection by electromigration at 1 kV for 15 s. (From Jones, W. R. and Jandik, P., *J. Chromatogr.*, 608, 385, 1992. With permission.)

The versatility of CIA is evident from the list of ions in Table 1, which represents a nearly 180% increase over the 53 ions first reported in 1990.[13] The key to the versatility is the open tubular fused silica capillary, which provides a user-definable separation medium that is solely dependent on the electrolyte chemistry and polarity of the power supply. Ion chromatography, however, achieves selectivity through various types of analytical columns that are generally dedicated either for anions or cations. Selectivity of the column is based on the composition

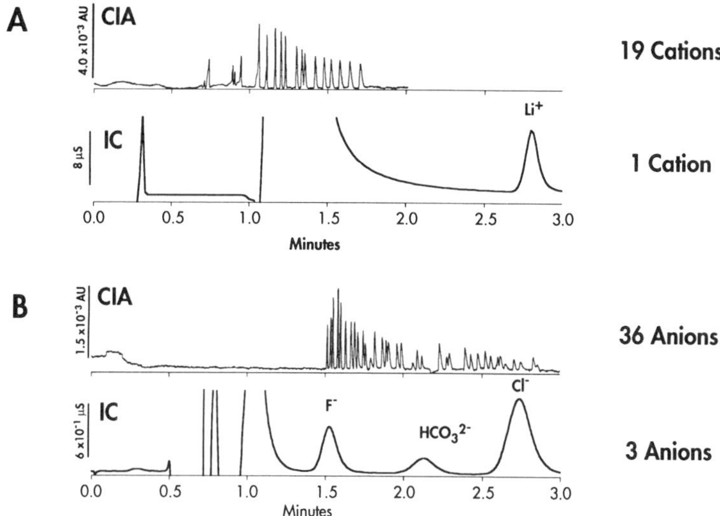

FIGURE 3. Comparison of peak capacity for IC separations and CIA separations. (A) Comparison of a 19-cation CIA separation with a 1-cation IC separation. The CIA separation is composed of alkali, alkaline earth, and lanthanide cations. Capillary, 36.5 cm × 75-µm fused silica; electrolyte, 10 m*M* UVCat-1, 4.0 m*M* α-hydroxyisobutyric acid (adjusted to pH 4.4 with acetic acid); voltage, +30 kV; injection, hydrostatic, 10 cm for 20 s; detection, indirect UV, 214 nm. The IC cation separation uses a Waters IC-Pak™ C M/D column, nitric acid/EDTA eluent with conductivity detection. (B) Comparison of a 36-anion separation (conditions given in Figure 2) with a 3-anion separation, using a Waters IC-Pak™ anion column, borate/gluconate eluent, and conductivity detection.

of the stationary phase and to the ion-exchange moieties covalently bound or dynamically coated on its surface. Further refinements in selectivity are achieved with eluent composition.[38] Detection in IC is primarily conductometric, but detection of all of the ions listed in Table 1 frequently requires more than one detection mode. This is due to the lower efficiency of the ion-exchange columns, which range between 1000 and 10,000 theoretical plates. Some of the more specific detection schemes include amperometric detection for electroactive species (e.g., sulfide and cyanide), and postcolumn reaction with photometric detection for transition metals and lanthanides. All of the ions in Table 1 have been visualized by indirect or direct UV detection.

Unlike IC, CIA migration order is predicted from IEC values readily found in the literature.[15] Since the analyte IEC is inversely related to the analyte migration time, a linear relationship exists between adjusted IEC values for the analyte ion and its reciprocal migration time (Figure 4). Conversely, the migration time of an unidentified peak in a CIA separation is used to facilitate identity.

It is worth noting that capillary ion analysis has lower operational costs mainly derived from the capillary, which can be purchased in spools at costs of less than $8 U.S./m. Daily electrolyte consumption is measured in milliliters reducing disposal costs. In contrast, IC columns cost up to 100 times more than the capillary with daily IC eluent consumption measured in liters.

C. CAPILLARY ION ANALYSIS APPLICATIONS IN THE LITERATURE

Rapid separations with high plate counts do not serve any purpose if the methodology cannot be applied to real samples. The availability of commercial CZE instrumentation and narrow bore fused silica capillaries, and advances in electrolyte chemistry, have led to solutions to a number of difficult and challenging sample matrix problems. Sulfur speciation of Kraft black liquors found in the pulp and paper industry was reported by Salomon and Romano.[23] Bondoux et al.

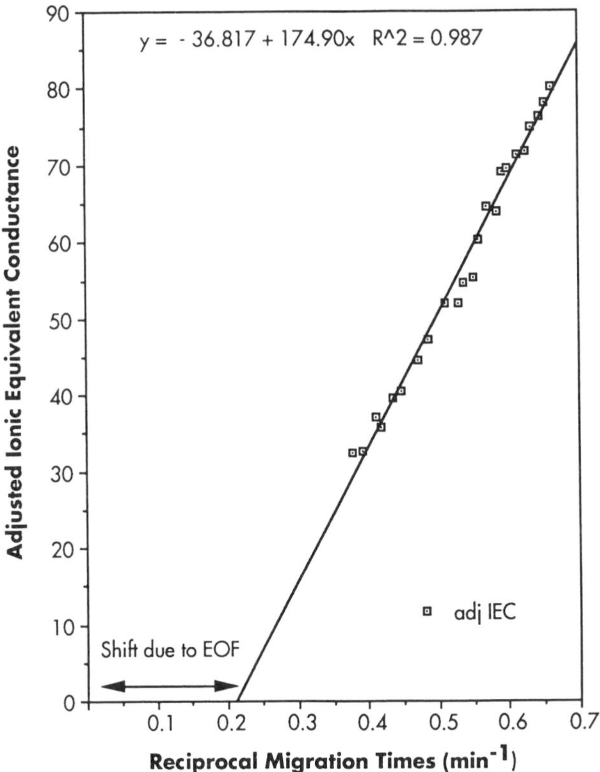

FIGURE 4. Reciprocal migration times of some anions from Figure 2 plotted against their respective values of ionic equivalent conductance for the prevalent ionic form at pH 8.0. The ionic equivalent conductances were also adjusted according to valence state as described in Ref. 15. (From Jones, W. R. and Jandik, P., *J. Chromatogr.*, 608, 385, 1992. With permission.)

were able to detect low parts per billion (ppb)* corrosive anions in the presence of 1000 parts per million (ppm) boron as boric acid.[22] The inherent simplicity of the hardware with low electrolyte consumption has lead to investigations evaluating CIA as an analyzer for in-flight ionic contamination monitoring of Space Station *Freedom* potable water.[29] Other applications for anions and cations are listed in Refs. 18-20, 22-23, 25-28, and 32-34.

II. CAPILLARY ION ANALYSIS PRINCIPLES

Discussed in this section are the principles governing CIA electrolyte composition and sample handling. These principles are a result of the higher mobility of the analytes.

A. THE ELECTROOSMOTIC FLOW FACTOR

In traditional CZE and MECC, it is common to use counter-EOF to drag anionic species or neutral species partitioned in anionic micelles to the cathodically positioned detector. This is possible because the ions are quite large (i.e., biomolecules) and do not migrate rapidly. The velocity of the cathodically directed EOF is greater than that of the anions migrating slowly in the opposite direction. Although the separations may take 30 min or longer, cations, neutral species, and anions are resolved in a single run.

The simultaneous separation of anions and cations found in the CIA domain (Figure 1) is very attractive, but this type of resolution is very difficult to achieve without significant

* The American billion (10^9).

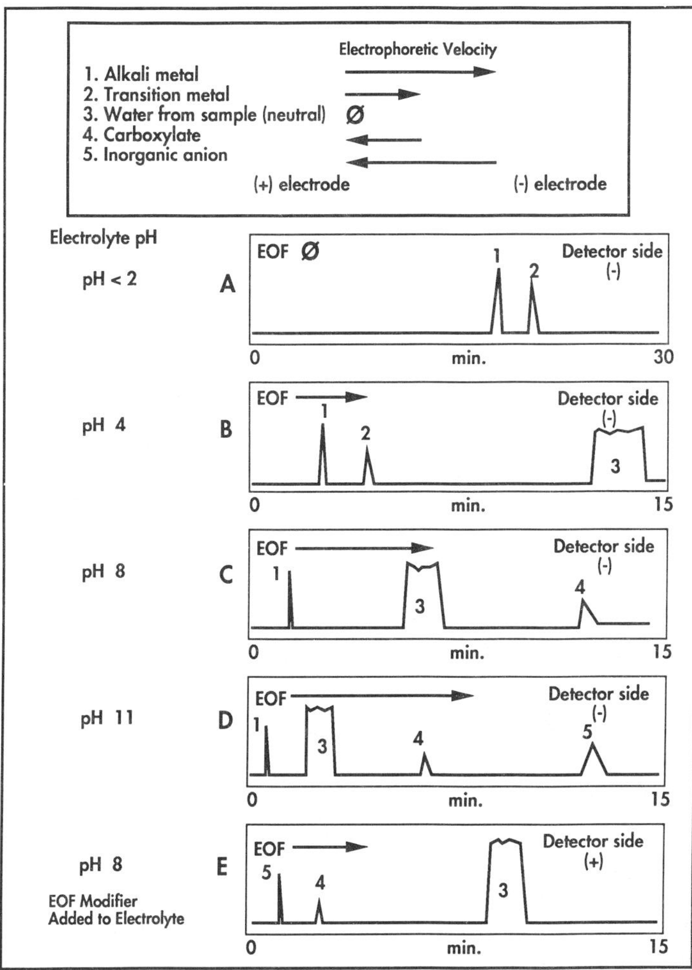

FIGURE 5. The migration order and relative migration times for four classes of ions (an alkali metal, transition metal, carboxylate, and an inorganic anion) under different EOF velocities and directions. (A–D) EOF controlled with electrolyte pH; (E) EOF controlled with an alkyl quaternary amine.

compromises. This is true not only for CIA but also for IC.[44-46] Capillary ion analysis separations require that the EOF moves in the same direction as the analytes, which is referred to as a Co-EOF condition. Under these conditions, the polarity of the system must be adjusted according to the ions to be analyzed. Anion analysis then requires the anode (positive electrode) to be at the detector whereas cation analysis requires the reverse situation. The benefit of Co-EOF is illustrated in Figure 5A–D. Four different classes of ions plus a neutral marker are injected into electrolytes with different EOF velocities with detection at the cathodic end. In Figure 5A there is no EOF since the silanol groups of the capillary are fully protonated (pH<2). This results in long migration times for the alkali metal (1) and the alkaline earth (2). In Figure 5B there is a cathodic EOF induced by the elevated electrolyte pH of 4. Peaks 1 and 2 have reduced migration times and the neutral peak 3, which represents water from the sample, is brought to the detector at the rate of the EOF. Under these conditions, the anions still have a greater mobility than the EOF and are not seen. In Figure 5C, the EOF is greater than the mobility of the carboxylate ion (peak 4) and is seen after the neutral marker. Peak 2 does not appear since most alkaline earths and transition metals are insoluble in

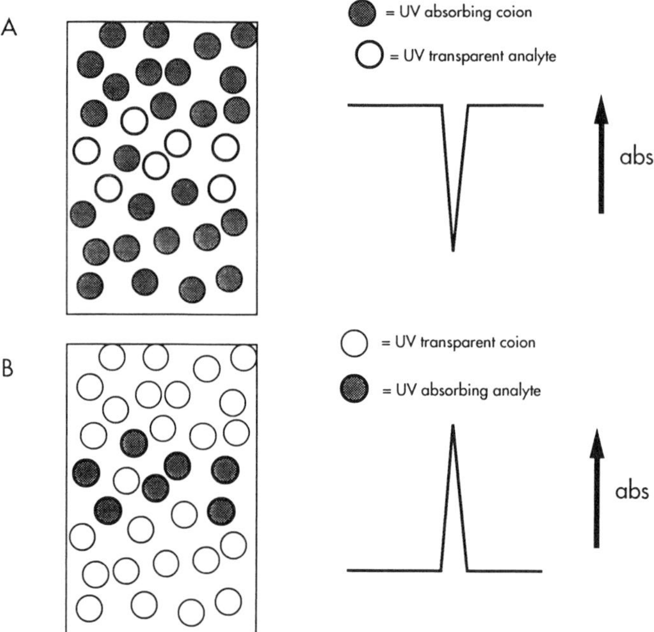

FIGURE 6. (A) Indirect UV detection, where analyte is UV transparent and electrolyte co-ion is UV absorbing. (B) Direct UV detection, where analyte is UV absorbing and electrolyte co-ion is UV transparent.

alkaline electrolytes. In Figure 5D, the EOF is greater than the mobility of all the analytes migrating in the opposite direction and, thus, the inorganic anion is brought to the detector. This electrolyte is not practical for cations but has shown limited use for the analysis of reactive ions such as hypochlorite and persulfate[32] and is discussed in more detail in Section V. Anions are resolved using the Co-EOF principle, illustrated in Figure 5E. Under these conditions, the polarity of the power supply is reversed so the detector is at the anode and an EOF modifier, consisting of an aliphatic quaternary amine, is added to the electrolyte. The amine dynamically coats the inner wall of the capillary through electrostatic attraction to the silanol groups and forms a bilayer that exposes the positively charged quaternary group to the electrolyte. A stable anodic EOF, independent of pH, is established permitting fast analysis times of anions. This allows the use of electrolyte pH as a way to change the selectivity of weakly acidic anions without affecting the magnitude of the EOF.

B. CO-ION REQUIREMENTS: ULTRAVIOLET-ABSORBING CHARACTER

For indirect photometric detection[41] the electrolyte, sometimes called the background electrolyte (BGE), contains a UV-absorbing salt. The portion of the salt that is UV absorbing is the co-ion, which has the same polarity as the analyte ion. The co-ion must possess at least one unique UV-absorbing maximum in a region that is not shared by the analytes. Depending on the UV lamp source energy of the detector, a wavelength is selected on or near the UV maxima of the co-ion for optimal absorption differences between the analyte and co-ion.[39,40] The detector measures the absence of the co-ion, which is physically displaced charge-for-charge by the analyte ions. The analyte zone in the detector path permits more UV energy from the lamp source to be transmitted to the photodiode. In the electropherogram, this is visualized as a peak opposite in polarity to that obtained for direct UV detection where the analyte has more absorbance than the electrolyte ion. To make the negative absorbance peaks appear positive the detector output polarity is reversed. Figure 6 illustrates the principle of indirect UV detection.

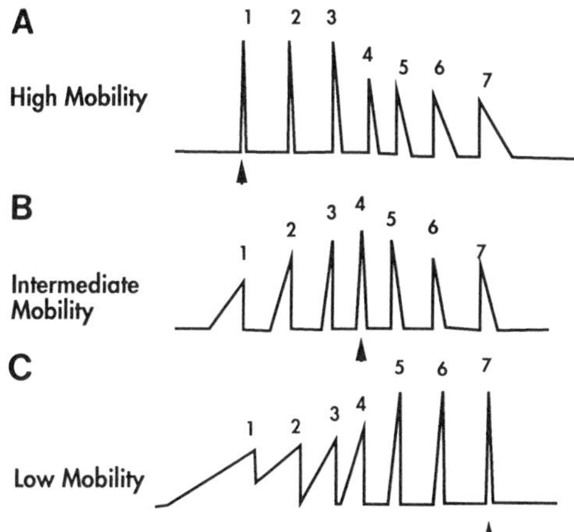

FIGURE 7. Illustration of the peak symmetry obtained for an (A) high-mobility, (B) intermediate-mobility, and (C) low-mobility electrolyte for anion analysis. Peaks 1 and 2 represent inorganic anions; peaks 3 to 5 represent short-chain carboxylates; peaks 6 and 7 represent short-chain linear alkyl sulfonates. The co-ions for the high-, intermediate-, and low-mobility electrolytes are chromate, phthalate, and *p*-hydroxybenzoate, respectively.

C. CO-ION REQUIREMENTS: MOBILITY

As the UV requirement of the electrolyte co-ion is important, so too is the mobility requirement. The rule states that the co-ion must have a mobility similar to that of the analytes. The closer the match the better the peak symmetry. The faster the mobility of the analyte in comparison with the co-ion, the more the analyte fronts. Conversely, the slower the analyte is compared to the co-ion, the more the analyte tails. Mikkers, Everaerts, and Verheggen were the first to discuss the rules governing this process,[4] while Hjertén described the peak asymmetry with respect to conductivity.[42] More recently, Sustacek, Foret, and Bocek describe the peak asymmetry in terms of electromigrative dispersion and suggest not only matching the mobilities of analyte and co-ion, but increasing the ionic strength of the electrolyte.[43] For 75-µm I.D. capillaries, Jones and Jandik show that electrolyte co-ion chromate (high mobility) shows optimal peak symmetry with inorganic anions; phthalate (intermediate mobility) is best for carboxylates; and *p*-hydroxybenzoate (low mobility) is best for linear alkyl sulfonates and alkyl sulfates.[13] Figure 7A–C represents separations of anions obtained using the high, intermediate and low mobility electrolytes just discussed. Peaks 1 and 2 are inorganic anions, peaks 3 through 5 are carboxylates while peaks 6 and 7 are alkylsulfonates. The mobility for each electrolyte co-ion is indicated by an arrow and is the region where analyte peak shapes will be the most symmetrical. As shown in Figure 7A, the high-mobility electrolyte effectively separates all seven anions while in Figure 7C, the conditions are optimized only for the lower mobility anions. The broader efficiency range is attributed to the higher conductivity of the co-ion, which aids in sample stacking. Using 50-mm I.D. capillaries, Jones and Jandik show that a chromate electrolyte, which is optimized for highly mobile inorganic anions, can be used to achieve efficient separations of ions that are outside the range of optimum matched mobility as shown in Figure 2.

II. INJECTION MODES

A. HYDROSTATIC INJECTION (PPM TO PERCENT CONCENTRATIONS)

The hydrostatic injection mode introduces small volumes of sample (measured in nanoliters) into the capillary. Applying the co-ion principles discussed earlier, detection limits range from 50 to 500 ppb, depending on the ion, using a 75-µm I.D. fused silica capillary. The quantitative concentration range is from 100 ppb to a few hundred parts per million. Increasing the concentration of the electrolyte co-ion extends the linear calibration range.[16] Concentrations exceeding a few hundred parts per million are diluted in water prior to injection to bring analytes concentrations into the linear calibration range.

The hydrostatic injection mode introduces sample into the capillary via a pressure differential due to the force of gravity. The injection end of the capillary is immersed into the sample and raised a predetermined height (e.g., 10 cm) for a period of time. The sample volume entered into the capillary is calculated by determining the rate of sample loading from the Poiseuille equation:

$$\text{Flow rate (nl/s)} = \Delta P D^4 \pi / 128\ \eta L \tag{1}$$

where ΔP is the pressure drop, D is the diameter of the capillary, η is the viscosity, and L is the length of the capillary.

For gravity-based injections, the pressure drop is determined as follows:

$$\Delta P = \rho g \Delta h \tag{2}$$

where ρ is the density of water, g is the gravitational constant (980 g/cm^2), and Δh is the height of the elevated capillary.

The sample flow rate for a 75 μm × 60 cm capillary at a elevated height of 9.8 cm is calculated to be 1.24 nl/s or 37.2 nl total volume injected in a 30-s injection from Figure 8A. This translates to a sample zone length of 0.842 cm or 1.4% of the capillary length. This exceeds the generally recommended rule that the sample zone length should not exceed 1% of the capillary length, yet the separation in Figure 8A achieves higher than expected efficiency. The higher efficiency is obtained through utilization of the "electrostacking" principle (discussed in Chapter 2), which is a function of a higher electric field existing in the sample zone. To better understand how an ion moves in an electric field, a review of a few equations is in order. The migration velocity of the analytes varies with the applied electric field E shown in Equation 3:

$$E = V/L \tag{3}$$

where V is the applied voltage and L is the length of capillary.

The migration velocity ν is dependent on the electric field as shown in Equation 4:

$$\nu = (\mu_e + \mu_{eo})E \tag{4}$$

where μ_e is the electrophoretic mobility and μ_{eo} is the contribution of the electroosmotic flow. Since ν moves proportionately to the electric field, higher E produces a higher ν. Higher electric fields are induced if the ionic strength in a particular zone (such as the sample) is lower than its surroundings (the electrolyte). It is then important to dilute samples whenever possible in high-purity water rather than in electrolyte. For CIA and CZE this approach is contrary to common laboratory practices with IC, where samples are sometimes diluted in eluent.

If dilution of the sample is not possible, the field strength can be increased if the electrolyte ionic strength is increased. Increasing the electrolyte concentration should be done in small increments, so as to prevent other dispersion effects such as excessive Joule heating or selectivity changes.[15]

B. ELECTROMIGRATION (ppt TO ppb CONCENTRATIONS)

The hydrostatic mode of sample injection relies on stacking of ions in nanoliter-sized size volumes introduced into the capillary, with detection limits in the low to mid-ppb range. To attain

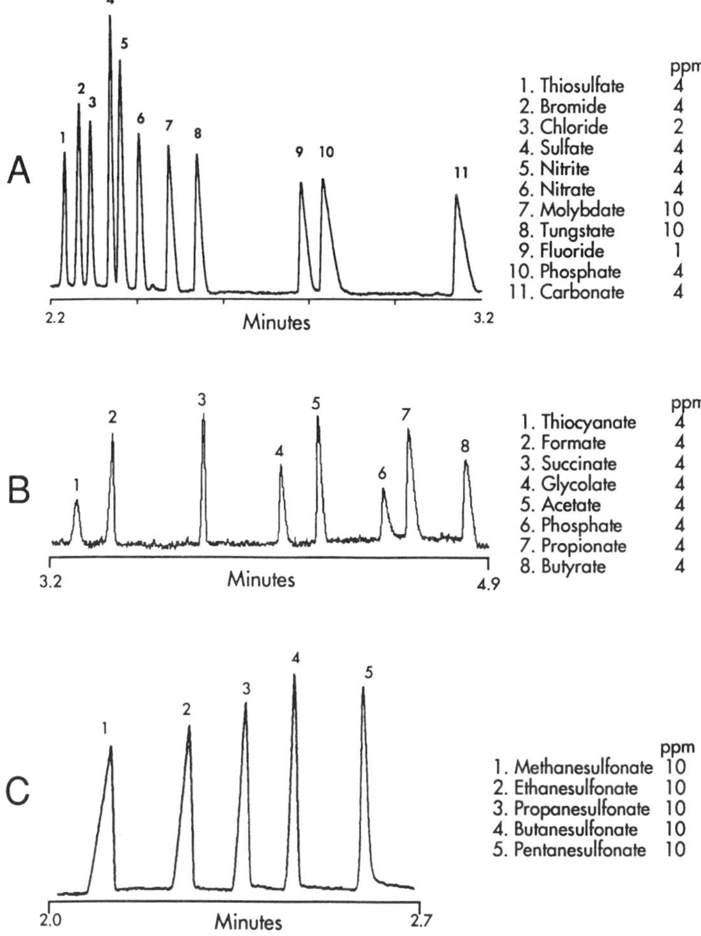

FIGURE 8. (A) Representative separation of 11 inorganic anions, using a high-mobility chromate electrolyte. (B) Separation of one inorganic and seven organic anions, using a phthalate electrolyte. (C) Separation of five linear alkyl sulfonates, using a *p*-hydroxybenzoate electrolyte. See Table 2 for operating conditions and refer to Methodology for electrolyte composition.

even lower detection limits, in the parts per trillion (ppt) range, the sample analytes must be preconcentrated prior to analysis. Traditional techniques of enriching ions with a small ion-exchange concentrator cartridge and eluting the accumulated analytes with an eluent is not required as is the case in IC.[48] In the electromigration mode of sample introduction, the analyte ions can be enriched from larger sample volumes (2 to 4 ml) by applying a low-kilovolt potential through the sample. Jandik and Jones first reported electromigrative enrichment of inorganic anions with ppt detection limits.[16] With this approach the Kohlrausch regulation function[47] was used to optimize the isotachophoretic stacking that occurs during sample loading.

The Kohlrausch regulation function was described in Chapter 5 on isotachophoresis. As given in Equation 5:

$$c_x/c_l = \mu_x(\mu_x + \mu_c)^{-1}(\mu_l + \mu_c)\mu_l^{-1} \tag{5}$$

c_x is the sample ion concentration in an isotachophoretic zone, c_l is the concentration of the leading electrolyte ion, and μ is the ionic mobility. The subscripts x, c, and l stand for sample, counter-, and leading electrolyte ions, respectively. The chromate electrolyte in this situation is

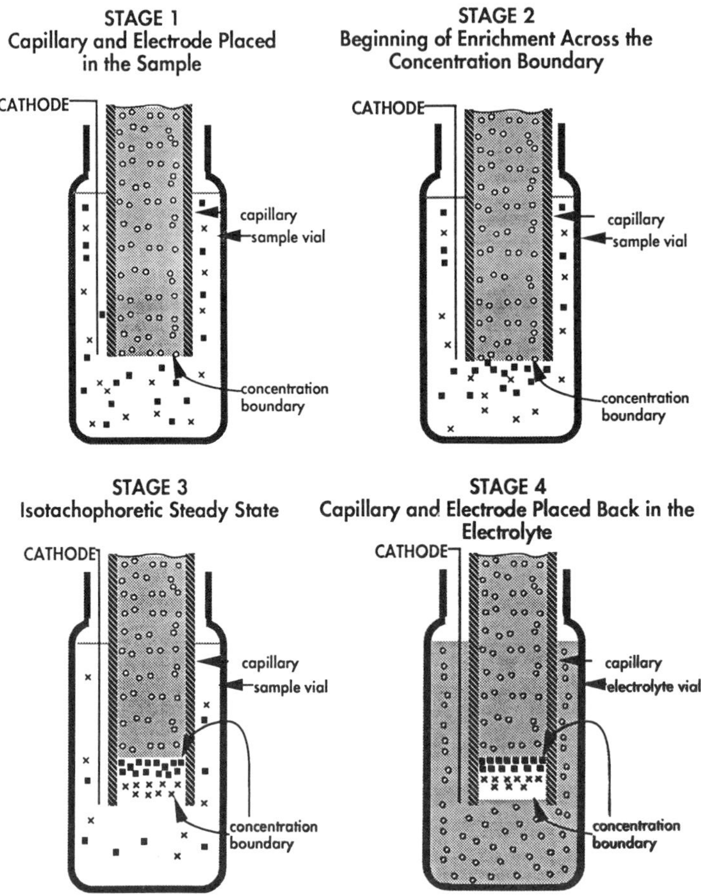

FIGURE 9. The four stages of electromigrative enrichment according to the Kohlrausch regulation function. An anionic additive (sodium octanesulfonate) is added to make the sample a terminating electrolyte. ■, Analyte anions; ❍, carrier electrolyte anions; X, anionic additive. See text for details.

the leading electrolyte while the sample is made into the terminating electrolyte through the addition of 40 to 80 μ*M* sodium octane sulfonate, NaOS.[22] The addition of this linear alkyl sulfonate is advantageous in that it will not be seen in the electropherogram due to the ion-pairing/hydrophobic tail-to-tail mechanism with the alkyl quaternary amine used for EOF modification. The NaOS also reduces the resistance of high-purity samples such as 18-MΩ water, since a certain level of background conductance is required to permit current to travel through the sample during loading. The electromigrative enrichment process is summarized in four stages in Figure 9. The capillary and the electrode are placed in the sample (Figure 9; stage 1) where a voltage of approximately 5 kV is applied to the sample (Figure 9, stage 2). The ions in the sample, including the anionic additive (NaOS), begin migrating into the capillary. Since the sample is significantly less conductive than the electrolyte, a higher electric field is found in the sample region. This increases the mobility of the analytes relative to the electrolyte. As the analyte ions reach the interface of the electrolyte zone, termed the concentration boundary, they accumulate and begin stacking due to the lower field at the boundary. Stacking continues until the isotachophoretic steady state is reached (Figure 9, stage 3). The steady state occurs when the conductivity of the analyte zones approaches that of the electrolyte where the electric fields are equalized. Beyond this point no more enrichment takes place. The capillary and sample are then placed back into the electrolyte to begin the analysis (stage 4).

TABLE 2
Standard Protocols for Running Capillary Ion Analysis Separations for Anions and Cations

Capillary:	Waters AccuSep fused silica, 60 cm × 75 μm I.D.[a]
Applied voltage:	20 kV
Power polarity:	Anion methods (negative) Cation method (positive)
Hydrostatic injection:	10 cm for 30 s
Detector wavelength:	Anion method (254 nm) Cation method (185 or 214 nm)
Detector output polarity:	Negative
Detector time constant:	Anion methods (0.1 s) Cation method (0.3 s)
Data acquisition rate:	Anion methods (20 pts/s) Cation method (10 pts/s)
Temperature:	Ambient
Electrolyte purge:	2.0 min at 15 psi vacuum[b]

Note: Methods were developed and optimized using the Waters Quanta™ 4000 CZE system. See electrolyte recipes for further details.

[a] If capillary is to be prepared manually from an alternative source, the outside diameter of the capillary is 375-mm O.D., with a capillary length to detector (L_d) of 52 cm.

[b] Fresh electrolyte must be flushed through the capillary between sample runs. This procedure brings the capillary back to original conditions and removes any ions, neutrals, and water that remain in the capillary when the analysis is complete.

IV. METHODOLOGY

Three recipes for anion analysis and one for cations are presented in this section. The anion analysis is divided into three mobility classes (high, intermediate, and low) as discussed earlier. The cation recipe is optimized for alkali metals and alkaline earth cations, although a number of transition metals can also be analyzed. To simplify operating conditions each method is run under a standardized format listed in Table 2. Figure 10 can be used in determining the appropriate mobility electrolyte for use in anion analysis.

A. CARE AND USE OF CAPILLARIES FOR CIA

As discussed in Chapter 2, capillaries are quite inexpensive and, as a result, it is recommended that the capillaries be dedicated for either anion or cation analysis. This saves time in converting the capillary from one form to another. If conversion of a capillary from anion to cation analysis or vice versa is necessary, a four-step rinse procedure is recommended. First, flush the existing electrolyte from the capillary with a 10% (v/v) methanol and Milli-Q mixture for 2 min followed by 0.5 *M* hydroxide for 5 min. Flush the base from the capillary with Milli-Q* water for 5 min, then for 5 min with the new electrolyte. Capillaries should be flushed with water prior to long-term storage. To change a capillary from one anion electrolyte to a different type, use 0.1 *M* hydroxide for 2 min followed by Milli-Q water for 1 min prior to flushing with the next electrolyte composition.[36] Preconditioning a new capillary with base and water prior to using an anion electrolyte is not necessary. However, when using the same capillary on consecutive days a start-up procedure of flushing with 0.1 *M* hydroxide for 5 min followed by 1 min with water before rinsing with electrolyte can help to ensure day-to-day reproducibility.

* The hydrostatic height of the Milli-Q® water (or 18-MΩ equivalent) should be higher than the hydroxide solution (5 mm). This ensures that the base is rinsed from the outside of the capillary and does not contaminate the running electrolyte.

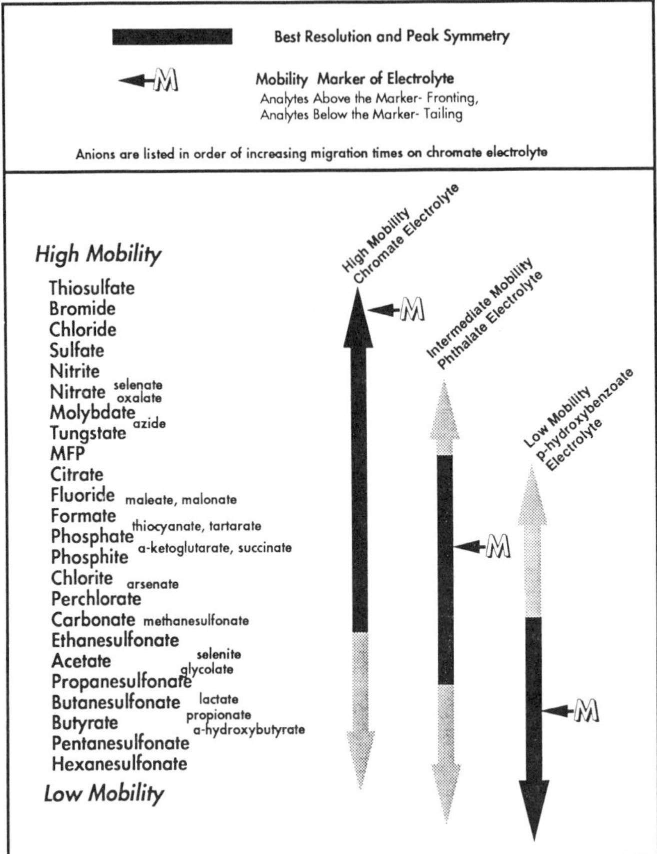

FIGURE 10. Chart assisting in the selection of the proper mobility background electrolyte for anion analysis. By selecting the proper electrolyte, optimal speed, sensitivity and resolution are achieved. In all cases, 0.5 m*M* OFM Anion-BT is included.

B. CHROMATE (HIGH-MOBILITY) ELECTROLYTE

A representative electropherogram is shown in Figure 8A.

Electrolyte composition

Sodium chromate (5 m*M*),* 0.5 m*M* CIA-Pak™ † OFM Anion-BT, pH 8.0

Chemicals required

Sodium chromate tetrahydrate (AR grade; Mallinckrodt, St. Louis, MO)
Sulfuric acid (Ultrex grade; J. T. Baker, Phillipsburg, NJ), CIA-Pak™ OFM Anion-BT (20 m*M* concentrate; Waters Div. of Millipore Corp.)

Chromate (100 m*M* concentrate) preparation: To a 1-l volumetric flask, add in the following order:

1. Milli-Q water (~500 ml)
2. Sodium chromate tetrahydrate (23.41 g)
3. Sixty-eight milliliters of 10 m*M* sulfuric acid (the 10 m*M* sulfuric acid is prepared by placing 560 μl of sulfuric acid into a clean 1-l flask and filling to the mark)

* Electrolyte is covered under U.S. patents.[48-50]

† OFM anion-BT is a solution containing an alkyl quaternary amine and is used as an EOF modifier.

Fill the flask to the mark with Milli-Q water or 18-MΩ equivalent and mix thoroughly. This concentrate may be stored in a volumetric or sealed glass container for up to 1 year. The 1-l concentrate makes 20 l of electrolyte.

Electrolyte preparation: To a clean 100-ml volumetric add in the following order:

1. Chromate concentrate (100 m*M*; see recipe above) (5 ml)
2. Waters CIA-Pak™ OFM Anion-BT solution (2.5 ml)

Fill the flask to the mark with Milli-Q water, mix thoroughly, and filter electrolyte through a 0.45-μm pore size Millicup-HV filter unit that is prerinsed with Milli-Q water prior to use. This results in a chromate electrolyte of pH 8.

Note 1. The electrolyte should be prepared the day of use.
Note 2. If fluoride and phosphate begin to comigrate, replace spent electrolyte with unused electrolyte.
Note 3. Running current at 20 kV is 18 to 20 μA.
Note 4. For further information for controlling selectivity see Refs. 15 and 30.

C. PHTHALATE (INTERMEDIATE-MOBILITY) ELECTROLYTE

A representative electropherogram is shown in Figure 8B.

Electrolyte composition
Potassium hydrogen phthalate (5 m*M*), 0.5 m*M* CIA-Pak™ OFM Anion-BT, pH 5.6

Chemicals required
Potassium hydrogen phthalate (Aldrich Chemical Co., Milwaukee, WI)
CIA-Pak™ OFM Anion-BT (20 m*M* concentrate, Waters Div. of Millipore Corp.)
Lithium hydroxide monohydrate (Aldrich Chemical Co.)

Electrolyte preparation: To a clean 200-ml volumetric add in the following order:

1. CIA-Pak™ OFM Anion-BT (2.5 ml)
2. Potassium hydrogen phthalate (0.204 g)

Fill the flask to the mark with Milli-Q water and mix thoroughly. Transfer the solution to a 250-ml beaker and adjust the pH of the electrolyte to 5.6, using a 50 m*M* lithium hydroxide solution. Filter the electrolyte through a 0.45-μm pore size Millicup-HV filter unit that is prerinsed with Milli-Q water prior to use.

Note 1. The electrolyte should be prepared the day of use.
Note 2. Because of the acidic pH of electrolyte, fluoride analysis is not recommended.
Note 3. Running current at 20 kV is 15 μA.

D. *p*-HYDROXYBENZOATE (LOW-MOBILITY) ELECTROLYTE

A representative electropherogram is shown in Figure 8C.

Electrolyte composition
p-Hydroxybenzoate (5 m*M*) 0.5 m*M* CIA-Pak™ OFM Anion-BT, pH 6.0

Chemicals required

p-Hydroxybenzoate acid (Aldrich Chemical Co.)
CIA-Pak™ OFM Anion-BT (20 m*M* concentrate; Waters Div. of Millipore Corp.)
Lithium hydroxide monohydrate (Aldrich Chemical Co.)

Electrolyte preparation: To a clean 200-ml volumetric add in the following order:

1. CIA-Pak™ OFM Anion-BT (2.5 ml)
2. *p*-Hydroxybenzoic acid (0.138 g)
3. Milli-Q water (~100 ml)

Fill the flask to the mark with Milli-Q water and mix thoroughly. Transfer the electrolyte to a 250-ml beaker and adjust the pH of the electrolyte to 6.0, using 50 m*M* lithium hydroxide. Filter the electrolyte through a 0.45-µm pore size Millicup-HV filter unit that is prerinsed with Milli-Q water prior to use.

Note 1. The electrolyte should be prepared the day of use.
Note 2. Running current at 20 kV is 7 µA.

E. ALKALI/ALKALINE EARTH CATION ELECTROLYTE

A representative electropherogram is shown in Figure 11A, using 214 nm for detection, and in Figure 11B, which is the same separation detected at 185 nm.

Electrolyte composition

UV-Cat1 (5 m*M*),* 6.5 m*M* HIBA, pH 4.4

Chemicals required

UV-Cat1 (Waters Div. of Millipore Corp.)
Hydroxyisobutyric acid (HIBA); (Aldrich Chemical Co.)
18-Crown-6 ether (optional) (Sigma Chemical Co., St. Louis, MD)

Electrolyte preparation: Place precisely 100 ml of Milli-Q water into a clean 250-ml plastic beaker (polypropylene or polymethylpentene) with a clean magnetic stirring bar. Perform the following steps in order.

1. Place the beaker with stir bar and water on a stirring plate.
2. Create a vortex with the water, such that the vortex just touches the stir bar.
3. Add 0.068 g of HIBA.
4. Pipette 64 µl of UV-Cat1 and deliver to the center of vortex
5. Filter the electrolyte through a 0.45-µm pore size Millicup-HV filter unit that is prerinsed with Milli-Q water prior to use.

The natural pH of the electrolyte is 4.4; no adjustment is required.

Note 1. Prepare the electrolyte fresh daily.
Note 2. Potassium and ammonium comigrate. To resolve the two cations add 0.0528 g of 18-crown-6 ether to the electrolyte.
Note 3. Running current at 20 kV is 4 µA.
Note 4. Improved sensitivity is obtained by using 185 nm for detection in place of 214 nm.

* Electrolyte is covered under U.S. patents.[49-50]

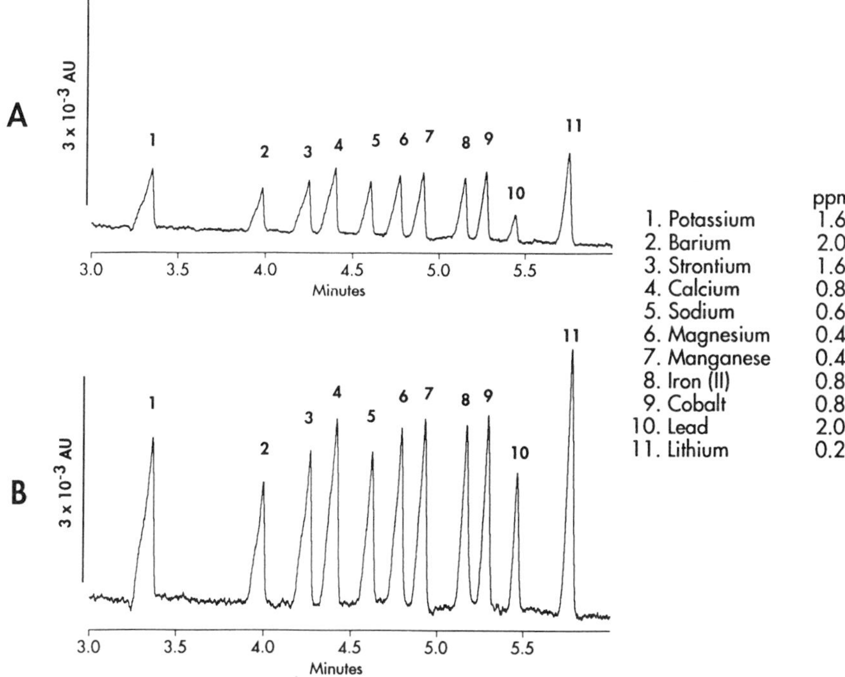

FIGURE 11. (A) Capillary ion electrophoretic analysis of a standard consisting of alkali, alkaline earth, and some transition metals. Indirect detection at 214 nm. (B) Same separation as in (A), except for indirect detection at 185 nm. See text for details. (From Weston, A., Brown, P. R., Jandik, P., et al., *J. Chromatogr.*, 608, 395, 1992. With permission.)

F. REFERENCE PEAKS INCORPORATED INTO THE ELECTROLYTE

In some applications it is desirable to include a reference in the sample to aid in peak identification. Situations where this may be advisable include difficulty in identifying peaks in an electropherogram where there is a high density of peaks or migration time shifts due to differences in sample composition. Rather than spiking a unique ion with an appropriate migration time into each sample, that unique ion can be added to the electrolyte instead. Adding a small concentration of a reference ion (25 to 50 ppm) into the electrolyte will induce a disturbance in the baseline precisely where that ion would be if it were placed into the sample except that the polarity of the peak is reversed. It is important to note that the reference component placed into the electrolyte is used for peak identification only in terms of a migration time ratio (migration time of analyte peak/migration time of reference peak). It cannot be used as an internal standard for quantitation since the magnitude of the peak is dependent on sample composition. The benefit of this technique is easier reference peak identification due to its reversed polarity and elimination of spiking multiple samples. The EOF modifier (OFM) used in the chromate electrolyte is a bromide salt, which can be used as a reference peak when the EOF modifier concentration is increased to 3.0 m*M*. Figure 12 shows the negative bromide peak induced by the increased OFM concentration. An added benefit in this separation is a change in selectivity for bromide, sulfate, nitrate, and iodide.

G. SAMPLE-HANDLING CONSIDERATIONS

The most important rule to observe is the electrostacking rule described earlier in Section

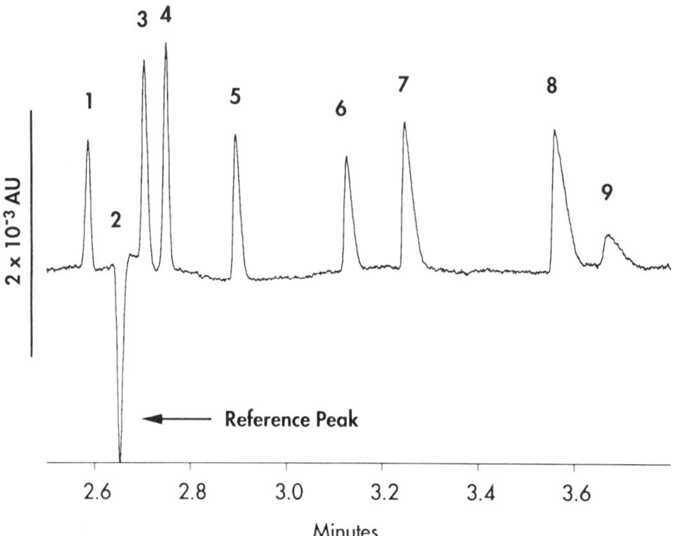

FIGURE 12. Peak identities and concentrations (ppm) for separation of nine anions. Peaks: 1, chloride (2); 2, bromide (–); 3, nitrite (4); 4, sulfate (4); 5, nitrate (4); 6, fluoride (1); 7, phosphate (4); 8, carbonate (4); 9, iodide (5).The electrolyte is a 5 m*M* chromate and 3 m*M* OFM Anion-BT adjusted to pH 8.0. Applied potential is 20 kV (negative polarity). Capillary dimensions are 60 cm (L_t), 52 cm (L_d), and 75-µm I.D. Hydrostatic injection at 10 cm for 30 s.

III.A. When possible, samples should be diluted in water instead of electrolyte to enhance stacking. It is sometimes advantageous to measure the conductivity both of the electrolyte and the sample(s) to determine if dilution is required. However, diluting a sample is often not an option. This situation occurs when the analyte of interest is present in trace amounts with respect to a major constituent (i.e., trace chloride and sulfate in food-grade citric acid), at which point dilution would bring the trace analyte below detection limits. In these situations it is important to select the optimum mobility electrolyte that best matches the analyte of interest. Also, the ionic strength of the running electrolyte should be increased if possible. Any minor constituent, such as the pH-adjusting acid (sulfate from sulfuric acid) used in the chromate concentrate, will incorporate into the sample zone only if the sample conductivity is higher than the electrolyte. This occurs because the electrolyte has a higher electric field than does the sample. This can be circumvented by eliminating or replacing the electrolyte constituent with a noninterfering ion.

V. APPLICATIONS

Of the various types of samples, Kraft Black liquor from pulp and paper production is one of the more difficult samples for ion chromatography. Kraft Black liquor is the spent white liquor used to separate lignin from wood fibers. The high-pH matrix containing reactive sulfur species and polyphenolic compounds is renowned for poisoning IC columns.[14,18] In contrast, fused silica capillaries for CIE do not contain a stationary phase to poison, hence, only a dilution in water is required as sample preparation. The chromate electrolyte permits the simultaneous analysis of inorganic anions and organic acids (Figure 13), where two separate IC methods would otherwise be required.[23]

Although fused silica capillaries do not have a packing material that can be poisoned, some sample excipients can adhere to the inner wall of the capillary. This will affect stability of the

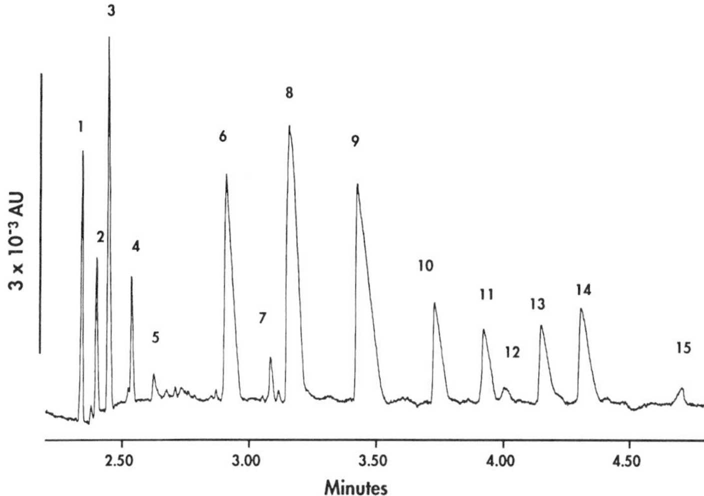

FIGURE 13. Analysis of Kraft Black liquor. Peak identity is as follows: 1, thiosulfate; 2, chloride; 3, sulfate; 4, oxalate; 5, sulfite; 6, formate; 7, unknown; 8, carbonate; 9, acetate; 10, propionate; 11, butyrate; 12–15, unidentified organic acids. The sample was diluted 1 to 1000 in 10 m*M* mannitol and filtered. Separation was performed at 20 kV with a 75 μm × 60 cm capillary, using 5 m*M* chromate, 0.5 m*M* OFM anion-BT at pH 10.0. Hydrostatic injection at 10 cm for 20 s. (From Jones, W. R., Jandik, P., and Pfeifer, R., *Am. Lab.*, 23, 40, 1991. With permission.)

EOF if it changes the charge on the wall (ζ potential). A good example of this is brewed coffee. Consecutive injections of brewed coffee into a chromate electrolyte (containing a compound that dynamically coats the inner wall; OFM) resulted in increased migration times with each injection (Figure 14A). An automated three-stage rinse cycle (consisting of a dilute base, water, and electrolyte) performed between injections efficiently removed the excipients and reestablished the EOF modifier on the inner wall, as evidenced by the consistent migration times (Figure 14B).

The analysis of trace levels of anions in the presence of high concentrations of other ions can be challenging, especially if the trace levels are below hydrostatic injection detection limits. A secondary water (SW) sample from a nuclear power plant pressurized water reactor (PWR) contains inorganic contamination in the low-ppb range. Anions, such as chloride and sulfate, promote intergranular stress corrosion which leads to cracking in pipes. A corrosion inhibitor, morpholine, is added in the 3-ppm concentration range to keep the SW alkaline. Addition of morpholine contributes degradation contamination products that are a result from the high temperatures and pressures applied to the SW system. These contaminants are found in the form of amines and carboxylates along with extractables from ion-exchange resins. Using electromigrative enrichment, both inorganic anions under 10 ppb and organic acids in SW are separated in a single run (Figure 15).

There are times when indirect photometric detection is not advantageous. The monovalent aromatic impurities found in 99.9% pure terephthalic acid (TA) are such an example. Since the analytes are all UV absorbers, better sensitivity is obtained with direct UV detection. Terephthalic acid is used to make plastics such as those in soft drinks containers. The quality of the plastic is dependent on how efficient the polymerization process is. Trace monovalent impurities in TA, such as benzoate, act as chain terminators affecting the degree of polymerization. Figure 16 shows separation of TA (present at 5000 ppm) migrating before the monovalent impurities found in concentrations of less than 1 ppm. Ultraviolet-transparent hexanesulfonate electrolyte was used at 185 nm.

Occasionally the EOF modifier used for the co-EOF condition of anions can interfere with the analysis and requires electrolyte conditions as depicted in Figure 5E. Reactive ions such

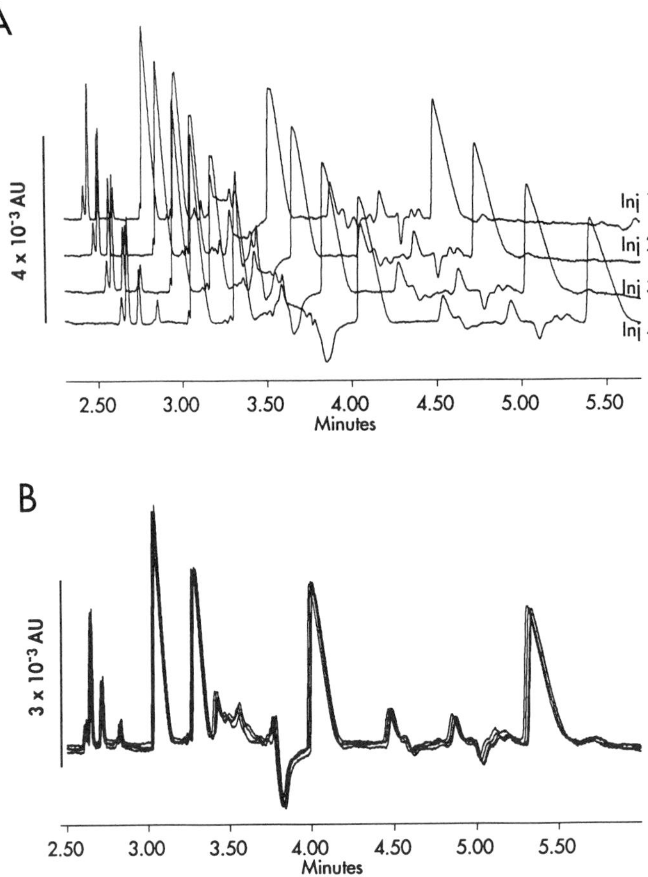

FIGURE 14. (A) Four consecutive injections (overlaid) of a brewed coffee sample, using only a 2-min electrolyte purge before each sample loading. (B) Four consecutive injections (overlaid) of a brewed coffee sample, using a three-stage rinse cycle consisting of 100 m*M* lithium hydroxide for 2 min, 18-MΩ water (Milli-Q® water) for 1 min, and running electrolyte for 2 min, performed between sample injections. The electrolyte is 5 m*M* chromate and 0.5 m*M* OFM Anion-BT, pH 8. Applied potential is 20 kV (negative polarity). Capillary dimensions are 60 cm (L_t), 52 cm (L_d) and 50-μm I.D. Injection is hydrostatic (10 cm for 30 s). (From Jones, W. R. and Jandik, P., *J. Chromatogr.*, 608, 385, 1992. With permission.)

as persulfate interact with the EOF modifier and are found late in the electropherogram as a broad, diffused baseline disturbance. To analyze for persulfate the EOF modifier is omitted from the electrolyte. Using a positive power supply, a strong counter-EOF condition must be attained to draw the highly mobile persulfate anion to the cathode. Figure 17 shows persulfate as the last peak of the electropherogram, using a chromate electrolyte at pH 11 with no EOF modifier. This electrolyte was also shown to be useful for hypochlorite analysis, which cannot be separated on ion-exchange columns as it will "bleach" the resin.[32]

VI. FUTURE PROSPECTS

As CIA/CZE technology evolves, the advances can be categorized into two basic areas: chemistry and hardware. From a chemistry perspective, electrolyte methodology should continue to expand into different classes of ions since the separations are rapid and equilibration of the capillary is essentially the time required to fill the tube. Further work is certainly needed to bring cation separations up to the peak capacity achieved for anions. This will most likely be achieved with the use of different types of complexing agents. Indications of

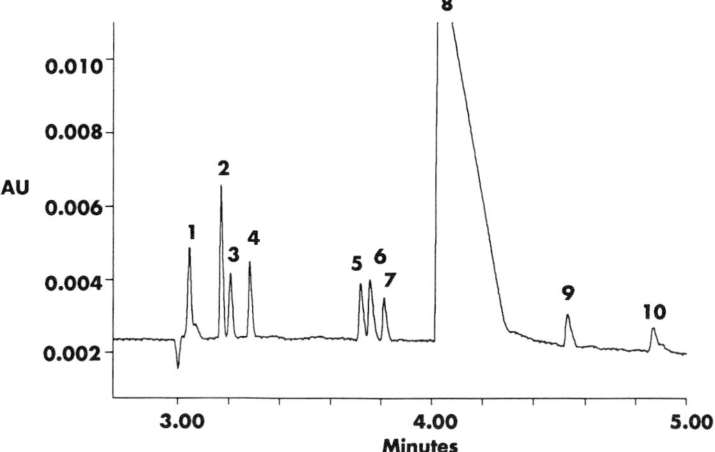

FIGURE 15. Trace anions in the presence of 3-ppm morpholine. The separation was generated after electromigrative trace enrichment, 45 s at 5 kV and with 75 μ*M* octanesulfonate additive in the sample. Peak identity and concentrations (ppb) injected are as follow: 1, chloride (7); 2, sulfate (9.6); 3, nitrate (12); 4, oxalate (10); 5, fluoride (3.8); 6, formate (10); 7, phosphate (6.2); 8, carbonate (not quantitated); 9, acetate (10); 10, propionate (10). The electrolyte consisted of 10 m*M* sodium chromate and 0.5 m*M* CIA-Pak OFM Anion-BT, adjusted to pH 8 with dilute sulfuric acid. The separation voltage was –15 kV. Capillary dimensions 75 μm × 60 cm. (From Bondoux, G., Jandik, P., and Jones, W. R., *J. Chromatogr.*, 602, 79, 1992. With permission.)

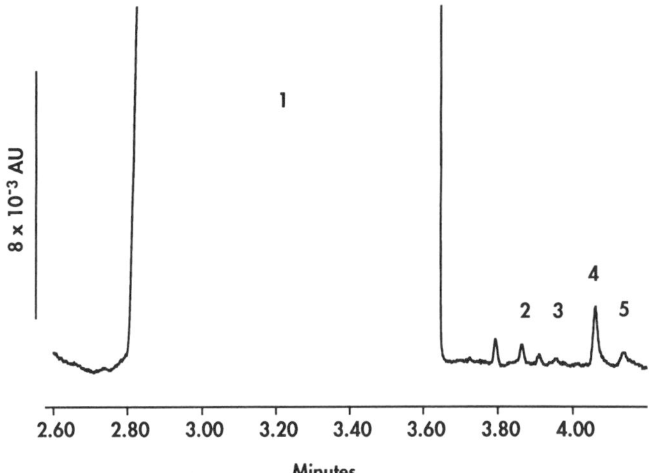

FIGURE 16. Electropherogram of a 99.9% pure terephthalic acid sample scaled to show the trace impurities. Peak identity and concentrations (ppm) injected are as follow: 1, terephthalate (5000); 2, unknown, 3, benzoate (not quantitated); 4, unknown, 5, toluate (0.56). The electrolyte is a 25 m*M* hexanesulfonate, 0.5 m*M* OFM anion-BT (converted to chloride form), adjusted to pH 10 with lithium hydroxide. Separation voltage, 25 kV (negative polarity), detection at 185 nm using 60 cm (L_t), 52 cm (L_d), and 75-μm I.D. fused silica capillary. Hydrostatic injection (10 cm for 30 s). (From Jones, W. R. and Jandik, P., *J. Chromatogr.*, 608, 385, 1992. With permission.)

improvement have been reported by Fritz and co-workers, with the resolution of 23 cations in a single run.[52] The commercially available, chemically modified capillaries used for capillary GC have been shown to be unstable for use under CZE operating conditions. We should, therefore, expect to see improved bonding techniques for covalently modified capillaries. Various functional groups covalently bound to the inner wall of the capillary may be used to control EOF and place less dependence on electrolyte chemistry for EOF control.

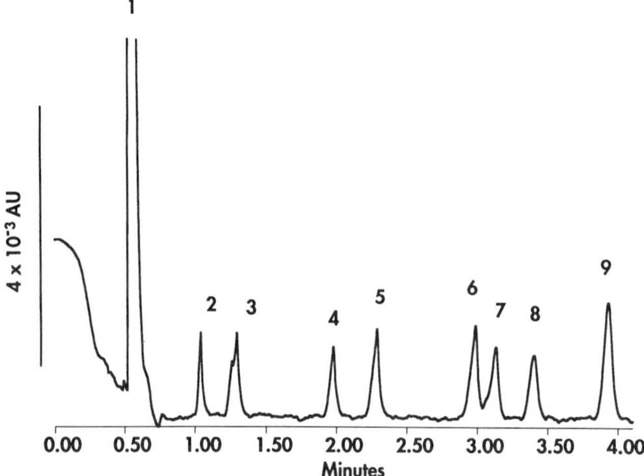

FIGURE 17. A separation of a sample including the reactive ion persulfate. Peak identity and concentrations injected (ppm) are as follow: 1, water peak (not quantitated); 2, fluoride (1); 3, carbonate (not quantitated); 4, nitrate (4); 5, nitrate (4); 6, sulfate (4); 7, chloride (2); 8, bromide (4); 9, persulfate (10). The electrolyte is 5 m*M* chromate, adjusted to pH 11.0 using lithium hydroxide. Separation voltage, 30 kV (positive polarity), detection at 254 nm using 40 cm (L_t), 32 cm (L_d), and 75-µm I.D. fused silica capillary. Hydrostatic injection (10 cm at 30 s). (From Jones, W. R. and Jandik, P., *J. Chromatogr.*, 608, 385, 1992. With permission.)

Unfortunately, these added value products will come at a much higher price than the generic fused silica capillary.

With respect to hardware, we may expect to see more focus on end-of-capillary detectors, such as conductometric or suppressed conductometric detectors, as reported by Dasgupta.[53,54] Other advances to be expected may include isomigration time methodology, reported by the author, which eliminates migration time changes due to sample conductivity.[55]

REFERENCES

1. **Small, H., Stevens, T., and Bauman, W. C.,** Novel ion exchange chromatographic method using conductimetric detection, *Anal. Chem.*, 47, 1801, 1975.
2. **Gjerde, D. T., Fritz, J. S., and Schmuckler, G.,** Anion chromatography with low conductivity eluents, *J. Chromatogr.*, 186, 509, 1979.
3. **Hjertén, S.,** Free zone electrophoresis, *Chromatogr. Rev.*, 9, 122, 1967.
4. **Mikkers, F. E. P., Everaerts, F. M., and Verheggen, T. P. M.,** Concentrations distributions in free zone electrophoresis, *J. Chromatogr.*, 169, 1, 1979.
5. **Gebauer, P., Deml, M., Kahle, V., Bocek, P., and Janak, J.,** Determination of nitrate, chloride and sulphate in drinking water by capillary free-zone electrophoresis, *J. Chromatogr.*, 267, 455, 1983.
6. **Foret, F., Deml, M., Kahle, V., and Bocek, P.,** On-line fiber optic UV detection cell and conductivity cell for capillary zone electrophoresis, *Electrophoresis*, 7, 430, 1986.
7. **Hjertén, S., Elenbring, K., Kilar, F., Liao, J., Chen, A. J. C., Siebert, C. J., and Zhu, M.,** Carrier-free zone electrophoresis, displacement electrophoresis and isoelectric focusing in a high-performance electrophoresis apparatus, *J. Chromatogr.*, 403, 47, 1987.
8. **Beckers, J. L., Verheggen, T. P. E. M., and Everaerts, F. M.,** Use of double-detector system for the measurement of mobilities in zone electrophoresis, *J. Chromatogr.*, 452, 591, 1988.
9. **Kuhr, W. G. and Yeung, E. S.,** Optimization of sensitivity and separation in capillary zone electrophoresis with indirect fluorescence detection, *Anal. Chem.*, 60, 2642, 1988.
10. **Foret, F., Fanali, S., Ossiscini, L., and Bocek, P.,** Indirect photometric detection in capillary zone electrophoresis, *J. Chromatogr.*, 470, 299, 1989.
11. **Gross, L. and Yeung, E. S.,** Indirect fluorimetric detection and quantification in capillary zone electrophoresis of inorganic anions and nucleotides, *J. Chromatogr.*, 480, 169, 1989.

12. **Huang, X., Gordon, M. J., and Zare, R. N.,** Effect of electrolyte and sample concentration on the relationship between sensitivity and resolution in capillary zone electrophoresis using conductivity detection, *J. Chromatogr.*, 480, 285, 1989.
13. **Jones, W. R. and Jandik, P.,** New method for chromatographic separation of anions, *Am. Lab. (Fairfield, Conn.)*, 22, 5, 1990.
14. **Jones, W. R., Jandik, P., and Pfeifer, R.,** Capillary ion analysis: an innovative technology, *Am. Lab. (Fairfield, Conn.)*, 23, 40, 1991.
15. **Jones, W. R. and Jandik, P.,** Controlled changes of selectivity in the separation of ions by capillary electrophoresis, *J. Chromatogr.*, 546, 445, 1991.
16. **Jandik, P. and Jones, W. R.,** Optimization of detection sensitivity in the capillary electrophoresis of inorganic anions, *J. Chromatogr.*, 546, 431, 1991.
17. **Jandik, P., Jones, W. R., Weston, A., and Brown, P. R.,** Electrophoretic capillary ion analysis: origins, principles, and applications, *LC-GC*, 9, 634, 1991.
18. **Romano, J., Jandik, P., Jones, W. R., and Jackson, P.,** Optimization of inorganic capillary electrophoresis for the analysis of anionic solutes in real samples, *J. Chromatogr.*, 546, 411, 1991.
19. **Kenney, B. F.,** Determination of organic acids in food samples by capillary electrophoresis, *J. Chromatogr.*, 546, 423, 1991.
20. **Wildman, B. J., Jackson, P., Jones, W. R., and Alden, P. G.,** Analysis of anion constituents of urine by inorganic capillary electrophoresis, *J. Chromatogr.*, 546, 459, 1991.
21. **Weston, A., Brown, P. R., Jandik, P., Jones, W. R., and Heckenberg, A. L.,** Factors affecting the separation of inorganic metal cations by capillary electrophoresis, *J. Chromatogr.*, 593, 289, 1992.
22. **Bondoux, G., Jandik, P., and Jones, W. R.,** New approaches to the analysis of low level anions in water, *J. Chromatogr.*, 602, 79, 1992.
23. **Salomon, D. R. and Romano, J.,** Applications of capillary ion electrophoresis in the pulp and paper industry, *J. Chromatogr.*, 602, 219, 1992.
24. **Chen, M. and Cassidy, R. M.,** Bonded-phase capillaries and the separation of inorganic ions by high-voltage capillary electrophoresis, *J. Chromatogr.*, 602, 227, 1992.
25. **Koberda, M., Konkowski, M., Youngberg, P., Jones, W. R., and Weston, A.,** Capillary electrophoretic determination of alkali and alkaline-earth cations in various multiple electrolyte solutions for parenteral use, *J. Chromatogr.*, 602, 235, 1992.
26. **Hargadon, K. A. and McCord, B. R.,** Explosive residue analysis by capillary electrophoresis and ion chromatography, *J. Chromatogr.*, 602, 241, 1992.
27. **Weston, A., Brown, P. R., Heckenberg, A. L., Jandik, P., and Jones, W. R.,** Effect of electrolyte composition on the separation of inorganic metal cations by capillary ion electrophoresis, *J. Chromatogr.*, 602, 249, 1992.
28. **Grocott, S. C., Jefferies, L. P., Bowser, T., Carnevale, J., and Jackson, P. E.,** Applications of ion chromatography and capillary ion electrophoresis in the alumina and aluminum industry, *J. Chromatogr.*, 602, 257, 1992.
29. **Mudgett, P. D., Shultz, J. R., and Sauer, R. L.,** Evaluation of capillary electrophoresis for in-flight ionic contamination monitoring of SSF potable water, in 22nd Int. Conf. Environ. Systems, 1992, p. 1.
30. **Buchberger, W. and Haddad, P.,** Effects of carrier electrolyte composition on separation selectivity in capillary zone electrophoresis of low-molecular-mass anions, *J. Chromatogr.*, 608, 59, 1992.
31. **Nielen, M. W. F.,** Indirect time-resolved luminescence detection in capillary zone electrophoresis, *J. Chromatogr.*, 608, 85, 1992.
32. **Jones, W. R. and Jandik, P.,** Various approaches to analysis of difficult sample matrices of anions using capillary ion electrophoresis, *J. Chromatogr.*, 608, 385, 1992.
33. **Weston, A., Brown, P. R., Jandik, P., Heckenberg, A. L., and Jones, W. R.,** Optimization of detection sensitivity in the analysis of inorganic cations by capillary ion electrophoresis using indirect photometric detection, *J. Chromatogr.*, 608, 395, 1992.
34. **Henshall, A., Harrold, M. P., and Tso, J. M. Y.,** Separation of inositol phosphates by capillary electrophoresis, *J. Chromatogr.*, 608, 413, 1992.
35. **Carpio, R. A., Mariscal, R., and Welch, J.,** Determination of boron and phosphorous in borophosphosilicate thin films on silicon substrates by capillary electrophoresis, *Anal. Chem.*, 64, 2123, 1992.
36. **Jones, W. R.,** Methods development approaches for capillary ion electrophoresis, *J. Chromatogr.*, 640, 387, 1993.
37. **Beck, W. and Engelhardt, H.,** Capillary electrophoresis of organic and inorganic cations with indirect UV detection, *Chromatographia*, 33, 313, 1992.
38. **Haddad, P. R. and Jackson, P. E.,** *Ion Chromatography—Principles and Applications*, Journal of Chromatography Library, Vol. 46, Elsevier, Amsterdam, 1990, pp. 22–25.
39. **Baumann, W.,** Optical detectors for liquid chromatography, *Fresenius Z. Anal. Chem.*, 284, 31, 1977.
40. **Li, J. B.,** Signal-to-noise optimization in HPLC UV detection, *LC-GC*, 10, 856, 1992.

41. **Yeung, E. S.,** Indirect detection methods: looking for what is not there, *Acc. Chem. Res.*, 22, 125, 1989.
42. **Hjertén, S.,** Zone broadening in electrophoresis with special reference to high-performance electrophoresis in capillaries: an interplay between theory and practice, *Electrophoresis*, 11, 665, 1990.
43. **Sustacek, V., Foret, F., and Bocek, B.,** Selection of the background electrolyte composition with respect to electromigration dispersion and detection of weakly absorbing substances in capillary zone electrophoresis, *J. Chromatogr.*, 545, 239, 1991.
44. **Hayakawa, K. and Miyazaki, M.,** Novel improvements in cation determination by indirect photometric ion chromatography, *LC-GC*, 6, 508, 1988.
45. **Tarter, J. G.,** Eluent selection criteria for the simultaneous determination of anions and cations, *J. Chromatogr. Sci.*, 27, 462, 1989.
46. **Saari-Nordhaus, R., Nair, L., and Anderson, J. M., Jr.,** Dual-column techniques for the simultaneous analysis of anions and cations, *J. Chromatogr.*, 602, 127, 1992.
47. **Kohlrausch, F.,** Ueber concentrations-Verschiebungen durch electrolyse im innerem von lösungen und lösungsgemischen, *Ann. Phys. Chem. N. F.*, 62, 209, 1897.
48. **Jones, W. R., Jandik, P., and Swartz, M. T.,** Automated dual column coupled system for simultaneous determination of carboxylic acids and inorganic anions, *J. Chromatogr.*, 473, 171, 1989.
49. **Jones, W. R., Jandik, P., and Merion, M.,** Method for Analyzing Ionic Species by Capillary Electrophoresis, U.S. Patent 5,104,506, April 14, 1992.
50. **Jones, W. R., Jandik, P., Merion, M., and Weston, A.,** Method for Analyzing Ionic Species by Capillary Electrophoresis, U.S. Patent 5,128,005, July 7, 1992.
51. **Jones, W. R., Jandik, P., Merion, M., and Weston, A.,** Method for Analyzing Ionic Species by Capillary Electrophoresis, U.S. Patent 5,156,724, October 20, 1992.
52. **Shi, Y. and Fritz, J.,** Separation of metal ions by capillary electrophoresis using a complexing electrolyte, Oral Lecture 33, *Int. Ion Chromatogr. Symp.*, Linz, Austria, September 21–24, 1992.
53. **Dasgupta, P.,** Ion chromatography: quo vadis domine?, Plenary Lecture 1, *Int. Ion Chromatogr. Symp.*, Denver, Colorado, October 6–9, 1991.
54. **Dasgupta, P.,** Suppressed conductometric detection in capillary electrophoretic separations, Poster 129, *Int. Ion Chromatogr. Symp.*, Linz, Austria, September 21–24, 1992.
55. **Jones, W. R.,** Methods development approaches for capillary ion analysis, Oral Lecture 34, *Int. Ion Chromatogr. Symp.*, Linz, Austria, September 21–24, 1992.

Chapter 10

CAPILLARY ELECTROPHORESIS IN THE EVALUATION OF PHARMACEUTICALS

Charlotte Silverman and Charles Shaw

TABLE OF CONTENTS

0-8493-8690-X/94/$0.00+$.50

I. INTRODUCTION

A. INTRODUCTION TO PHARMACEUTICAL ANALYSIS

Pharmaceutical researchers are designing increasingly complex drug molecules and targeted drug formulations. These new drugs present the analytical chemist with a multiplicity of unique challenges. The analytical objectives are not only to collect information, but to design methods that will detect and quantitate contaminants at very low levels, typically down to 0.1% of the concentration present in the drug substance, and to define guidelines for quality assurance. The supplied methods must provide (1) purity determination and characterization of drug substance, (2) stability data for both drug substance and formulated material, (3) quantitative analysis of excipients, (4) resolution of enantiomers, and (5) quantitation and detection of drugs in biological fluids.

The development of methods for indicating purity and stability is a necessary and often difficult part of obtaining drug approval from the Food and Drug Administration (FDA).

Separation methods such as high-performance liquid chromatography (HPLC), gas chromatography (GC), and traditional electrophoresis have long been established in the pharmaceutical industry as a means of conducting the required drug analyses. The analytical methods that utilize these techniques follow certain system suitability guidelines, such as specificity, sensitivity down to 0.1% vs. active ingredient, overall accuracy of at least 95%, reproducibility with a relative standard deviation (RSD) of less than 2%, and a sufficient detector response linearity.

With increasingly complex molecules and more requirements from the regulatory agencies, new techniques that can satisfy these requirements in a faster and cost-effective manner are of potential interest. One such novel technique is capillary electrophoresis (CE). This high-efficiency separation technique has been considered primarily a technique for the separation of proteins. In contrast, there has been comparatively little interest in the analysis of pharmaceuticals. This chapter addresses the utility of capillary electrophoresis in pharmaceutical analysis. Enantiomeric separations will be covered as well as comparison of CE with HPLC.

B. COMPARISION BETWEEN HIGH-PERFORMANCE LIQUID CHROMATOGRAPHY AND CAPILLARY ELECTROPHORESIS

It is clear that analytical and preparative scale separation of pharmaceuticals has been dominated by HPLC. Gradient HPLC offers high peak capacity and efficiencies around 50,000 theoretical plates. Detection in HPLC provides high concentration sensitivity. These features make HPLC very useful for the determination of drugs in biological matrices, in drug stability investigations, for the determination of drug purity, and for the analysis of biotechnological products. An apparent weakness is the lack of a simple, universally sensitive detection method equivalent to flame ionization detection in GC.[1]

Capillary electrophoresis is capable of enormous efficiency, which allows separation methods to be rapidly developed. The separation efficiency is related directly to the solute mobility and the voltage used for the separation. Under normal operating conditions, between 20 and 30 kV, 100,000 to 200,000 theoretical plates can be routinely achieved for small pharmaceutical compounds.[1] There have been some reports of up to 1 million theoretical plates.[2] The ability to separate a large number of closely related materials in a very short time is one of the major benefits of CE. In a single CE run there is essentially zero solvent consumption and the sample injection volumes are on the order of 5 to 10 nl.[2] All these factors makes CE very suitable for pharmaceutical analysis. It is particularly useful for the separation of protein digests due to its enormous peak capacity.[1]

The main drawbacks of CE are the poor concentration sensitivity and the fact that the separation buffer and the sample diluent have to be similar.[1] For most applications the concentration sensitivity must be improved if CE is to compete with HPLC.[1] Several highly sensitive spectrophotometric detection methods, including direct and indirect laser-induced fluorescence (LIF), have been used to improve sensitivity[3-9] (see Chapters 7, 12, and 14 for detailed discussion of this topic). However, CE provides better mass sensitivity than other chromatographic and electrophoretic methods, and it can quantitate minute amounts of sample.[10] Sampling volumes on some instrumentation can be reduced to less than 1 nl. Highly sensitive detection methods, such as the capillary vibration-induced laser (CVL) method, have been proposed.[10] The CVL method was applied as a detector for capillary zone electrophoresis (CZE) and femtomole amounts of underivatized amino acids were detected.

II. PHARMACEUTICAL APPLICATIONS

To use CE routinely for the analysis of pharmaceutical products, the reproducibility of sample injection, solute migration time, and detection limits of the solute need to be considered. In general, quantitation in capillary electrophoresis is less precise than is found in HPLC.[11] Depending on the separation conditions, one can routinely achieve about 1% RSD (equivalent to percent coefficient of variation, CV) by peak area when the concentration of the test mixture is 1 mg/ml and about 3 to 6% RSD when the concentration is 100 μg/ml.[2] There are many possible explanations for this observation. Specifically, capillary electrophoresis instrumentation has been available commercially for only the last few years and the sample injection/loading precision of the instruments needs to be improved. Since temperature control of the capillary has proven to be essential for successful separations in capillary electrophoresis, temperature control of the sample vial tray may represent a significant means of improving precision. Further discussion of causes for this variation are presented in Section II.B.1.

In addition to ionic strength of the buffer, reproducible migration times are dependent on additional experimental conditions: capillary age, previous capillary treatments, frequency of the capillary treatments, applied voltage, and external capillary temperature. For example, reproducibility of migration times can vary from 1.5 to 4.5% for a vitamin mixture over the course of ten injections.[2] A better understanding and control of the factors that influence migration is expected to improve the reproducibility.

Although capillary electrophoresis is best known for separations of proteins and other large biomolecules, success has been achieved in the analysis of low molecular weight pharmaceuticals.[2] The majority of these small molecules are charged, making them ideal candidates for CZE. The use of CE as a tool for quantitative or purity determination depends on the initial parent drug concentration. Concentrations of impurities below 1% may not be readily detectable in CE due to the short pathlength in the ultraviolet (UV) on-column detection cell.[2] For more discussion concerning optical detection in CE, the reader is referred to Chapter 7.

A. QUALITATIVE

1. Micellar Electrokinetic Capillary Chromatography with Sodium Dodecyl Sulfate as Surfactant

Although MECC is discussed in detail in Chapter 3, separations pertinent to the evaluation of pharmaceuticals will be discussed. Most of the CE separations reported to date have been developed for water-soluble compounds. In a study by Fujiwara and co-workers,[12] seven water-soluble vitamins were separated using micellar electrokinetic capillary chromatography (MECC) with sodium dodecyl sulfate (SDS) as the surfactant. This method was used for quantitation of hydrophilic vitamins in a commercial parenteral solution. Good recovery and peak area reproducibility (RSD, 2.1%) were obtained using ethyl *p*-aminobenzoate as an internal standard. MECC has been investigated as an alternative technique to HPLC for the analysis of lipophilic drugs and pharmaceutical formulations.[13-15]

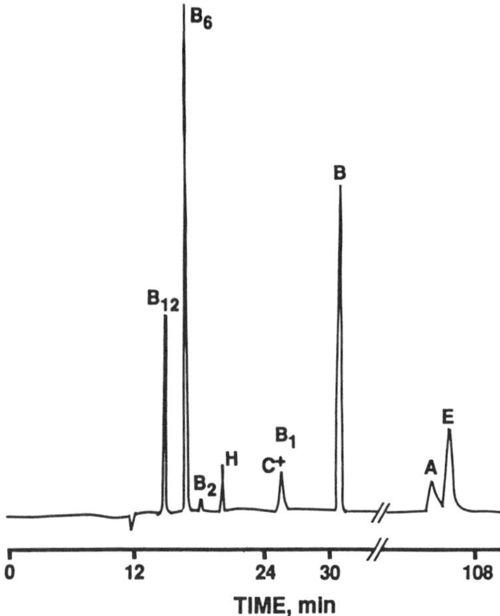

FIGURE 1. Separation of water- and fat-soluble vitamins in a single analysis. Separation buffer, 30 m*M* SDS in 0.1 *M* borate–0.05 *M* phosphate with 3% IPA, pH 7.6; capillary dimensions, 50 μm I.D. × 50 cm; separation voltage, 15 kV; amount injected, 0.75 nl. (From Ong, C. P., Ng, C. L., Lee, H. K., and Li, S. F. Y., *J. Chromatogr.*, 547, 419, 1991. With permission.)

Both fat-soluble vitamins (A and E) and water-soluble vitamins (B, B_2, B_6, B_{12}, B_1; C and H) were separated by MECC in the same analysis[13] (Figure 1). As discussed in Chapter 3, the combination of γ-cyclodextrin with SDS (mixed micelles) in the electrophoretic medium provided the selectivity needed for separating the water- and fat-soluble vitamins by MECC. The separation of mixtures of fat- and water-soluble vitamins has not been reported in the HPLC literature.

Corticosteroids and aromatic hydrocarbons were separated by cyclodextrin-modified MECC.[15] The addition of cyclodextrin to the SDS solution improved the resolution and provided the lipophilic compounds with smaller migration times than if no cyclodextrin was present. γ-Cyclodextrin showed the largest reduction in migration times compared to β-cyclodextrin. A more stable or effective inclusion complex formation between the solute and γ-cyclodextrin was offered as an explanation.

Micellar electrokinetic capillary chromatography with SDS as the surfactant has been used for analysis of eight water-insoluble drugs without ionic charge.[16] The pseudoeffective mobility, a retention parameter, instead of the capacity factor, was used as a parameter for evaluating the best conditions for the separation of these water-insoluble drugs. The composition of the sample solution, especially with organic solvents such as methanol, was shown to have an enormous effect on the migration time and the separation efficiency. Quantitation of one of the water-insoluble drugs, dapsone, in a tablet formulation was obtained by utilizing a calibration curve.

2. Micellar Electrokinetic Capillary Chromatography with Surfactants Other than Sodium Dodecyl Sulfate

The most frequently used surfactant in MECC has been SDS. Other surfactants have been evaluated and their separation efficiencies were compared with SDS for several types of compounds. Peptides with similar net charge but different hydrophobicity were used as model

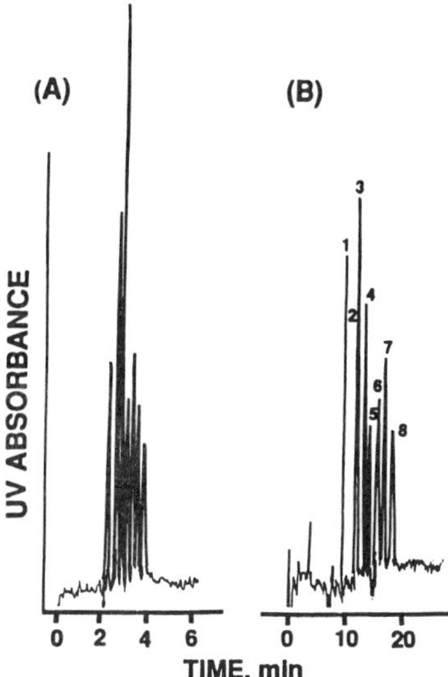

FIGURE 2. Separation of underivatized angiotensin analogs, using the surfactants dodecyltrimethylammonium bromide (DTAB) and hexadecyltrimethylammonium bromide (HTAB). (A) Separation buffer, 10 m*M* Tris–10 m*M* Na_2HPO_4, pH 7.05, and 0.05 *M* DTAB; separation voltage, –15 kV; detection, 220 nm. (B) Separation buffer, 25 m*M* Tris–25 m*M* Na_2HPO_4, pH 7.05, and 0.05 *M* HTAB; separation voltage, –20 kV; detection, 220 nm. (From Liu, J., Cobb, K. A., and Novotny, M., *J. Chromatogr.*, 585, 123, 1991. With permission.)

compounds for evaluating the micelle-forming compounds dodecyltrimethylammonium bromide (DTAB) and hexadecyltrimethylammonium bromide (HTAB)[17] (Figure 2). Cyclodextrins were investigated as additives for the separation of these peptides. Both UV and fluorescence detection, using *o*-phthalaldehyde and 3-(4-carboxybenzoyl)-2-quinolinecarboxaldehyde (CBQCA) as fluorescence derivatization agents were utilized.

Penicillin and cephalosporin antibiotics have also been studied with MECC. Mixtures of seven penicillin antibiotics and nine cephalosporin antibiotics were used as model systems to determine the optimum separation conditions in the presence of two different anionic surfactants, SDS and sodium *N*-lauroyl-*N*-methyltaurate (LMT).[18] The effect of surfactant concentration, analyte functional groups, and pH on migration time was determined. The optimum separation conditions involved a 20 m*M* phosphate–borate buffer, pH 9, with surfactant concentration ranging from 0.1 to 0.3 *M*. Separation under these conditions produced baseline resolution of all the molecules. Addition of tetraalkylammonium salts to a buffer containing SDS improved the resolution for cephalosporin antibiotics.[19] Addition of 20 m*M* tetraalkylammonium salt to a buffer containing 50 m*M* SDS was then applied to closely related antibiotics having both anionic and cationic groups on the same molecule. A combination of SDS and sodium octyl sulfate surfactants led to a greater selectivity for the separation of borate-complexed catechols.[20]

In a systematic approach to control ionic mobilities of similar analytes, Salomon et al.[21] used CE to separate seven tricyclic antidepressants that have similar structure and mass. Resolution was achieved by adjusting pH, buffer concentration, and concentration of methanol added to the buffer. Selectivity enhancement for the tricyclic antidepressants desipramine and nortripytyline was accomplished by Swedberg[22] with the addition of nonionic or zwitterionic surfactants. The advantages of these surfactants for the separation of the peptides bradykinin, luteinizing hormone-releasing hormone, and angiotensin were: less impact on the electroosmotic flow compared to ionic surfactants, and less impact on protein structure of the analyte.

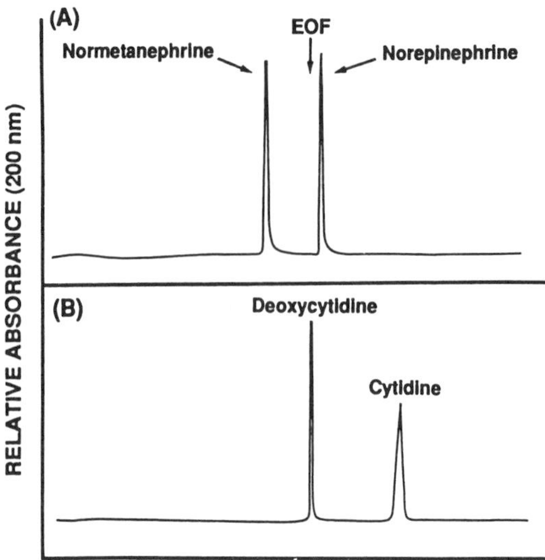

FIGURE 3. Separation of *cis*-diol compounds differing by a single hydroxyl group. (A) Separation of normetanephrine and norepinephrine. (B) Separation of deoxycytidine and cytidine. Separation buffer, 100 m*M* borate buffer, pH 8.4; separation voltage, 25 kV; capillary dimensions, 57 cm × 50 μm; current, 16 μA. (From Landers, J. P., Oda, R. P., and Schuchard, M. D., *Anal. Chem.*, 64, 2846, 1992. With permission.)

3. Borate Complexation

An alternative approach for the separation of molecules that differ only slightly in structure and charge is to obtain the desired selectivity through complexation. Landers et al.[23] found a borate buffer system to be essential in the high-performance capillary electrophoretic separation of several biologically active compounds containing *cis*-diols and differing by only a single hydroxyl group (Figure 3). This complexation of borate with compounds containing a *cis*-diol results in a significant change in the charge-to-mass ratio of the complexed compound due to the freely ionized hydroxyl group of borate. Borate complexation with the *cis*-diol-containing compounds results in a significantly different migration (electrophoretic mobility), allowing separation from the non-*cis*-diol-containing molecules differing only slightly in structure and having identical net charge.[23]

4. Biological Fluids

Micellar electrokinetic capillary chromatography has been used for analyses of drugs and their metabolites in body fluids such as serum and urine. Cefpiramide in human plasma was separated and quantitated with SDS as the surfactant and antipyrine as the internal standard.[24] The method was optimized for SDS concentration and pH. The calibration curve was linear from 10 to 300 μg/ml, with a correlation coefficient of 0.998. The relative peak area reproducibility of cefpiramide and antipyrine in buffer and in plasma gave a CV of ≤4.3% for seven injections.

Micellar electrokinetic capillary chromatography was used to identify seven barbiturates in urine and serum samples from patients undergoing pharmacotherapy.[25] The barbiturates were characterized by their migration times and UV spectra (between 195 and 320 nm). They were identified by comparison with computer-stored data from the analysis of standards.

5. Impurity Profiling/Fingerprinting

In analyzing complex mixtures, CE has been used to establish a "fingerprint" for the mixture without absolute identification of each peak in the electropherogram. A mixture of

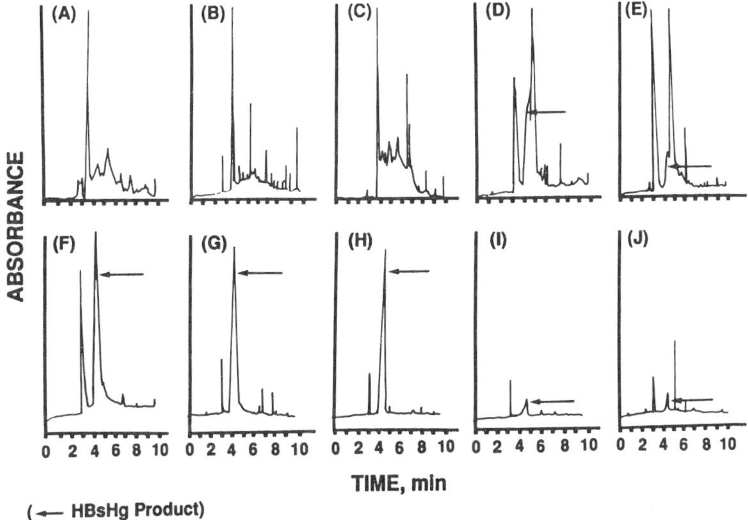

FIGURE 4. Electropherograms of each step in the purification process of Recombivax HB hepatitis B vaccine. (A) Cell lysate; (B) filtered lysate; (C) concentrated lysate; (D) silica product; (E) HIC product; (F) Amberlite product; (G) diafiltered Amberlite product; (H) thiocyanate product; (I) sterile product; (J) formaldehyde product. (A) and (B) are plotted at attenuation 16, all others plotted at attenuation 128. (From Kenndler, E., Schwer, C., and Kaniansky, K., *J. Chromatogr.*, 508, 203, 1990. With permission.)

quercetin, kaempferol, and isorhammetin, three flavonal-3-*O*-glycosides from medicinal plants, gave baseline-resolved peaks using the MECC technique.[26] The MECC separation was faster than reversed-phase HPLC and the peaks were better resolved. The sensitivity level for this separation was reported to be higher with MECC. This is not a typical CE observation. Future plans are to extend this method to other flavonoids and to establish a "fingerprint" for medicinal plants.[26]

Using a similar approach, a "fingerprint" for each step in the lot-to-lot production process of the licensed Recombivax HB hepatitis B vaccine was determined using MECC.[27] Unique "fingerprints" were found at each step of the purification process as illustrated in the electropherograms presented in Figure 4. Another specific application of CE for "fingerprinting" is to establish impurity profiles for different lots of bulk drug. Capillary zone electrophoresis was used for impurity profiling of riboflavin-5′-phosphate.[28] Different electrophoretic "fingerprints" were produced when the pH of the buffering electrolyte was varied. An improved separation of riboflavin-5′-phosphate from its main impurities, riboflavin and different riboflavin phosphates, was achieved by lowering the pH of the buffer from 9.1 to 8.2. However, at pH 7, two additional peaks were resolved from the riboflavin-5′-phosphate, making pH 7 the buffer of choice for quantitation.[28]

Similarly, Altria and Smith illustrated the separation of the anti-depressant GR50360A from a potential manufacturing impurity, the desfluoro analog.[29] The authors indicate that this separation was difficult to achieve by HPLC.

6. Detection

To date, most CE separations of pharmaceutical compounds have used UV detection; however, alternative detection methods have been investigated. The use of laser-induced fluorescence (LIF) for detection provided a very sensitive and selective method for the CE analysis of anthracyclines[30] (Figure 5). Analyte interaction with the capillary wall, and thereby peak shape distortion, was prevented with a high concentration (70%) of acetonitrile in the buffer.

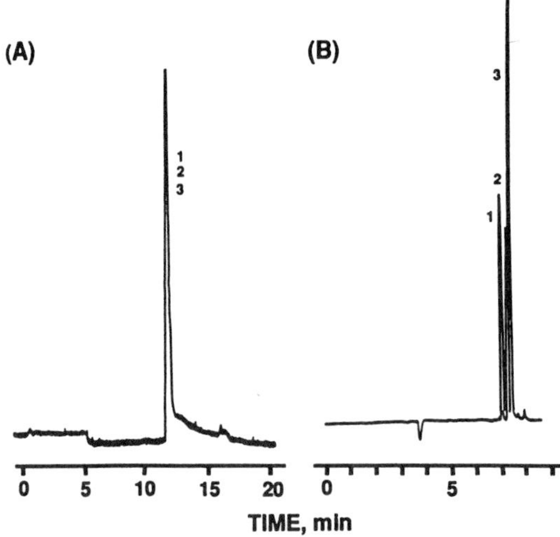

FIGURE 5. Effect of the addition of acetonitrile to the electrophoresis buffer on the separation of anthracyclines with LIF detection. (A) A 1-μg/ml mixture of (1) doxorubicin, (2) epirubicin, and (3) doxorubicin in 20 m*M* sodium phosphate buffer, pH 4.2; injection, 10 kV for 5 s. (B) A 100-ng/ml mixture of (1) doxorubicin, (2) epirubicin, and (3) doxorubicin in 100 m*M* sodium phosphate buffer, pH 4.2, diluted 3:7 with acetonitrile; injection at 12 kV for 5 s. Separation voltage, 20 kV; capillary dimensions, 75 μm × 70 cm. (From Reinhoud, N. J., Tjaden, U. R., Irth, H., and van der Greef, J., *J. Chromatogr.*, 574, 327, 1992. With permission.)

Capillary electrophoresis coupled to LIF provides a viable alternative to the more frequently used high-performance liquid chromatographic determination of doxorubicin and its epimer epirubicin.[30] Roach, Gozel, and Zare described a simple, rapid, and sensitive CE method with LIF for monitoring the anticancer drug methotrexate and 7-hydroxymethotrexate.[31] Laser-induced fluorescence allows detection of methotrexate down to 10^{-10} *M* (*S/N* = 3) in serum. Capillary electrophoresis with LIF detection was used to analyze proline and hydroxyproline in urine by derivatizing the analytes with fluorescamine.[32]

7. Consumer Products

Capillary electrophoresis has also been used for separation of the components in several consumer products. Nishi and co-workers[33] studied the retention behavior of 12 cold medicine ingredients as a function of pH and surfactant type with MECC (Figure 6). The 12 different ingredients investigated in each cold medicine were caffeine, trimetoquinol, ethenzamide, noscapine, acetaminophen, guaifenesin, phenacetin, chlorpheniramine, Sulpyrin, naproxen, isopropylantipyrine, and tipepidine. The migration times of the 12 ingredients were studied with 5 different surfactants: SDS, sodium *N*-lauroyl-*N*-methyltaurate (LMT), sodium tetradecene sulfonate (OS-14), sodium dioxyethylene lauryl ether sulfate (SBL), and sodium trioxyethylene alkyl ether acetate (ECT). The migration times of the analytes differed significantly depending on the hydrophilic group of the surfactant. All of the surfactants allowed analyses of the ingredients within 30 min.

Electrophoretic analysis of cations and anions in prenatal vitamins was described by Swartz.[34] Following a simple sample workup, capillary ion analysis (discussed in Chapter 9) was found to be reproducible with an RSD <1% for migration times. Good correlation was obtained between inductively coupled plasma spectroscopy and capillary ion analysis data for calcium, iron, and zinc cations. Some of these applications can be found in Waters ion analysis database (part number 42404).

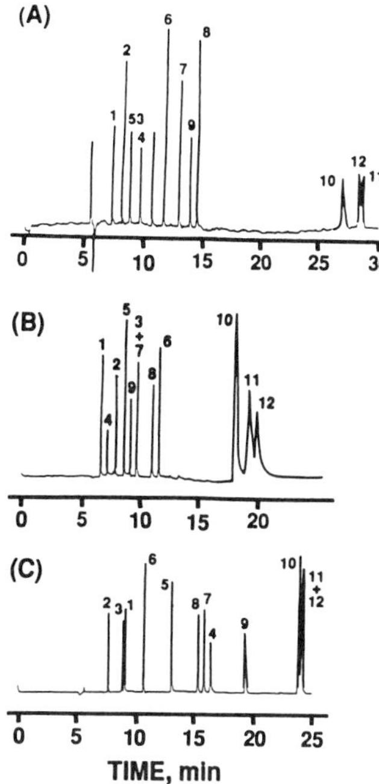

FIGURE 6. Separation of 12 cold medicine ingredients by MECC as a function of pH and surfactant type. Surfactant type and concentration: (A) 0.1 *M* sodium *N*-lauroyl-*N*-methyltaurate (LMT), (B) sodium trioxyethylene alkyl ether acetate (ECT), (C) sodium dodecyl sulfate (SDS). Other conditions: separation buffer, 0.02 *M* phosphate–borate buffer, pH 9.0; separation voltage, 20 kV; detection, 210 nm; temperature, ambient. (From Nishi, H., Fukuyama, T., Matsuo, M., and Terabe, S., *J. Pharm. Sci.*, 79, 519, 1990. With permission.)

8. Experimental Design for Capillary Electrophoresis Methods Development

Capillary electrophoresis methods development requires several parameters to be optimized, as discussed earlier in this chapter and in Chapter 2. Experimental design makes this process a quick and efficient method to study the various types of compounds encountered within the pharmaceutical industry. An experimental design scheme known as overlapping resolution mapping (ORM) has been applied to the separation of eight sulfonamides.[35] This optimization method has been applied previously to HPLC separations. The optimum pH and β-cyclodextrin concentration of the electrophoretic medium were predicted from nine preplanned experiments. Separations at the predicted optimum conditions were then carried out to demonstrate the usefulness of the optimization scheme. Using the conditions obtained from ORM, baseline separation of the eight sulfonamides was obtained within 8 min.

B. QUANTITATIVE

Although CE methods have been developed for the analysis of several different compounds, these methods have not always been suitable for quantitation as a result of poor peak area reproducibility. Methods development for quantitation and characterization of pure drug substance seems to have a higher degree of success than for formulated material due to the absence of matrix effects, which are discussed in detail in Chapter 20. Several attempts have been made to evaluate the factors that influence reproducibility in CE separations, and some of these are discussed below.

1. System Evaluation and Optimization

As mentioned earlier, consistent and reproducible injections are necessary for quantitative reproducibility in CE. In a study by Honda et al.,[36] benzyl alcohol was used as a model compound for evaluating a CE automated hydrostatic injector, using peak area reproducibility measured by percent coefficient of variation (% CV). The reproducibility was found to be less than 4.4% for manual introduction and less than 0.84% with automatic sampling for 12 injections. Similar peak area reproducibility (1 to 8%) was found when chlorinated phenols were quantitated with an internal standard, using MECC and a manual hydrostatic injection system.[37]

As discussed in Chapter 2, when electrokinetic injections are used, a sampling bias is introduced based on the ionic charge of the molecule. Buffer composition, pH, and applied voltage influence the electroosmotic flow and, consequently, the electrophoretic injections. Consistent and reproducible electroosmotic flow is necessary to obtain rugged CE methods. For example, separation of anti-inflammatory drugs was evaluated with uncoated and polyacrylamide-coated capillaries.[38] The longer uncoated capillaries gave better efficiencies than the shorter coated capillaries under the same field strength conditions, even though comparable resolution and analysis time were obtained in both cases. In the coated capillaries, the analytes are separable with a lower field strength, due to their net mobility because the electroosmotic flow is low. This may explain the lower efficiencies observed with the coated capillaries; however, no explanation was given by the authors. Using the uncoated capillaries, CZE and MECC were optimized for the following compounds: antibiotics, sulfonamides, cephalosporins, penicillins, peptides, barbiturates, and pain, cold, and allergy medicines. Standards were analyzed containing the following pain, cold, and allergy medicines: phenylpropanolamine, chlorpheniramine, pseudoephedrine, caffeine, acetaminophen, aspirin, and salicylic acid. Peak area RSD values of less than or equal to 3.5% for eight electrokinetic injections were obtained. The two active ingredients in Congest-Away, phenylpropanolamine HCl and acetominophen, had peak area RSD values of 2.9%.

Buffer composition, capillary dimensions, ionic strength and pH, applied voltage, and buffer additives affect the electroosmotic flow in the capillary and indirectly affect the selectivity of the separation. The selectivity in CZE and MECC was evaluated for several analgesics by variations in pH, buffer ionic strength, applied voltage, SDS concentration, and percent organic modifier in SDS solution.[39] The optimum separation conditions were validated for salicylamide, one of the analgesics, with peak area RSD values of less than 2.31% at the nominal concentration. The limit of quantitation for caffeine was determined to be 0.1% of the salicylamide level, with a 9.3% peak area RSD.

When MECC is used, factors influencing electroosmotic flow, reproducibility, and quantitation must be controlled. The separation mechanism becomes more complicated as a result of the added surfactants (see Chapter 3). Factors related to run buffer composition and instrument design in MECC were studied with a modular CE instrument and the reproducibility of several quantitative parameters was evaluated.[40] The parameters investigated were absolute and relative peak area measurements, RSD of migration times, efficiencies, and resolution. The influence of temperature, pH, and percent organic modifiers on analysis time and resolution of p-hydroxybenzoic acid and benzoic acid was investigated. At higher temperatures, the analysis time and resolution decreased due to lower viscosity and a more rapid solute diffusion. However, a linear relationship between individual migration times and temperature was not observed. The optimum temperature range for conducting electrophoretic separations was determined to be 5 to 67°C depending on whether organic modifiers or surfactants were used. Addition of organic modifiers to SDS buffer solutions was generally found to improve the resolution. Peak area and migration time RSD values, which were obtained by variation of temperature, pH, and percent organic modifiers, ranged from 0.5 to 3%.

The choice of sample diluent can have a significant effect on separation in CZE and MECC. The influence of sample type and concentration on the linear dynamic range of MECC was investigated with a series of nonsteroidal anti-inflammatory drugs (NSAIDs) serving as model compounds.[41] Up to 10% methanol in the injection solution and did not affect peak shape, peak

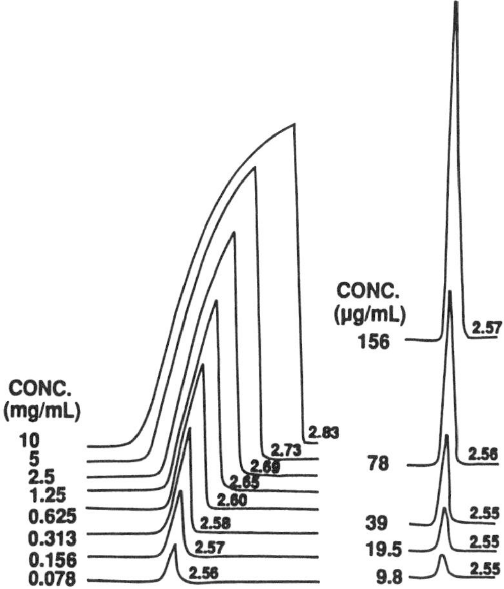

FIGURE 7. Band profile dependence on solute concentration in MECC. Solute, naproxen; capillary dimensions, 50 μm × 38 cm; separation buffer 20 m*M* borate, pH 9.2; temperature, 50°C; separation voltage, 25 kV; injection, 1-s vacuum; detection, UV 230 nm. (From Weinberger, R. and Albin, M., *J. Liq. Chromatogr.*, 14, 953, 1991. With permission.)

width, or the retention characteristics. Peak shape in CZE is dependent on the relative mobilities of the solute and the carrier in free solution. A similar behavior is observed in MECC. At high solute concentrations, where the solute ion has a slower mobility than the carrier ion, the leading edge of the zone appears diffuse whereas the tailing edge is sharp (Figure 7).

An improved peak shape was observed at solute concentrations below 100 μg/ml. At this concentration, both peak area and peak height are linear with concentration for solutes such as naproxen. At solute concentrations of 100 μg/ml, the RSD of peak height and peak area is below 2% with an external standard. The linear dynamic range can be extended by increasing the ionic strength of the separation buffer. However, this requires further dissipation of the Joule heat generated at the higher buffer concentrations.

The effect of sample amount on theoretical plates, peak shape, and resolution was studied using an optimized separation for cinnamic acid and its analogs.[42] Increasing amounts of sample introduced (between 30 and 300 ng) and the sampling time resulted in a significant decrease in the number of theoretical plates and resolution for dimethoxycinnamic acid. Peak area is linear as a function of concentration for both dimethoxycinnamic acid and ferulic acid when this concentration level is maintained. Above 30 ng, a convex relationship is obtained. When the lower sample amounts are introduced into the capillary, poor reproducibility is found. The precision of six consecutive 5-s injections of a 10^{-4} *M* cinnamic acid solution was 4.0% for peak height and 4.4% for peak area. If dimethoxycinnamic acid was used as an internal standard, the precision was improved and ranged between 0.9 to 2.8% for peak height and 1.2 to 2.0% for peak area for 4-methoxycinnamic acid, cinnamic acid, ferulic acid, and 4-hydroxycinnamic acid. This method, which had a detection limit of 10 ng/ml at an *S/N* ratio of 2, was successfully used to monitor fuerlic acid in dog plasma.

2. Biological Fluids

This section covers analyses of the parent drug molecule in a biological matrix. For a more detailed discussion, the reader is referred to Chapters 17 and 20.

The small sample volume requirement makes CE very suitable for analyses of biological fluids. However, the inadequate detection sensitivity in CE, in addition to possible matrix effects from biological fluids, makes quantitative analysis in biological fluids a challenge. In spite of the possible matrix interferences, the following examples show that analyses with acceptable RSD values for peak area are possible.

A thoroughly validated CE assay of cytosine-β-D-arabinoside, an antileukemic agent, in human plasma was reported.[43] The validation included both intra- and interday variability, as well as the determination of unknowns. Migration time reproducibility was best at low pH with a resultant coefficient of variation of 4%. On-capillary peak stacking, an on-column concentration procedure, was used to achieve adequate detection limits. The method was shown to be linear from 1 to 10 μg/ml, with a correlation coefficient of 0.996. The variability study was based on peak height, since peak area gave unacceptable results. Migration times longer than 5 min were used for a system suitability test, and if the requirement was not met, the run was ignored. To date, this is the only detailed report describing a CE validation procedure.

Thiopental and its metabolite, pentobarbital, were quantitated in human serum and plasma using MECC and HPLC with UV detection at 290 nm.[44] Thiopental samples were quantitated down to 2 μg/ml in human serum when a calibration curve was used. When the internal standard, carbamazepine, was added, the best peak area RSD for thiopental was 3.73% for ten injections and 2.04% for carbamazepine. The MECC separation was carried out in a phosphate–borate buffer with 50 m*M* SDS in about 8 min, while the HPLC separation took 12 min on a C_8 column with an acetonitrile and phosphate buffer mobile phase. Micellar electrokinetic capillary chromatography and HPLC data from 66 patient samples compared well with one another based on linear regression analysis.

Derivatization of the analytes is often used to improve the detection sensitivity. Glyphosate and its major metabolite, aminomethylphosphonic acid (AMPA), were separated and quantitated in serum by capillary electrophoresis, using *p*-toluenesulfonyl chloride for derivatization.[45] The within-run precision for peak area was ≤2.9% RSD, whereas the precision from day to day was ≤5.7% for both compounds. The detection limit for both compounds was 0.1 μg/ml in serum at $S/N = 2$. Recoveries from serum were <90% for both compounds.

3. Dosage Formulation and Bulk Drug Characterization

Analyses of pharmaceutical compounds in formulation matrices are of great importance. The formulation excipients can be as simple as a saline solution, to a more complex matrix like a tablet, cream, or gel. Depending on the complexity of the formulation matrix, assay interference can be more or less significant. Removing the matrix by selective extraction is a possibility, but it sometimes involves a sample preparation that is not compatible with the carrier buffer. Bulk drug characterization avoids matrix interferences but generally involves a more rigorous investigation of the drug purity. In spite of these challenges, CE has been applied to each and its application is described in the following examples.

Quantitation of the *d*/*l*-epinephrine ratio in a sterile ophthalmic solution by capillary electrophoresis has been described.[46] The compound *l*-pseudoephedrine was used as an internal standard. Analysis of ten separate standard preparations of *l*-epinephrine over 2 days gave peak area RSD values of 1.4 and 1.8%. The recovery of *l*-epinephrine from the vehicle standard solution was 99 and 101% on the two consecutive days, with an RSD ≤2.0%. The method was linear from 12.5 to 37.5 ppm for both *d*- and *l*-epinephrine. The use of an internal standard improved the quantitation.

Quantitation of cimetidine, a first-generation antiulcer drug developed by SmithKline and French, was investigated by capillary electrophoresis.[47] Cimetidine tablets (Tagamet) were dissolved, diluted, and extracted with petroleum ether before separation in phosphate buffer.

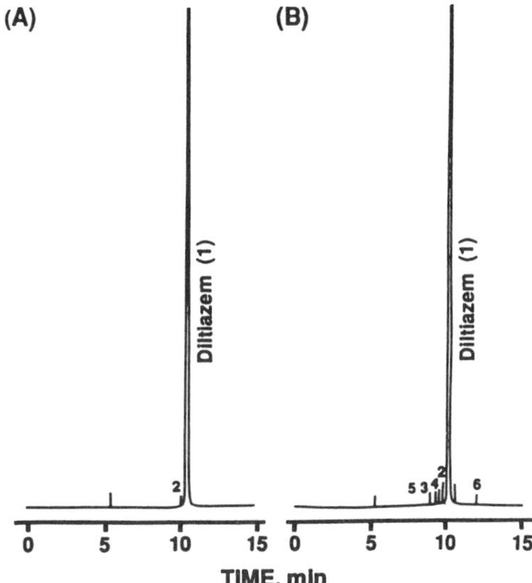

FIGURE 8. Impurity profiling of diltiazem hydrochloride. (A) Diltiazem hydrochloride and (B) authentic diltiazem with 0.3 to 0.5% related substances added. Separation buffer, 0.2 *M* phosphate–borate, pH 8.0, containing 0.1 *M* sodium cholate; separation voltage, 20 kV; temperature, ambient; detection, 210 nm. (From Nishi, H., Fukuyama, T., Matsuo, M., and Terabe, S., *J. Chromatogr.*, 513, 279, 1990. With permission.)

Quantitation of cimetidine in over 60 samples from commercially available formulations (tablets, liquids, and injectables) gave assay RSD values ranging from 1.9 to 6.4%. When 20 duplicate injections from the same sample of cimetidine were analyzed, peak height ratios RSD values were 3.3% compared to 1.8% for peak area ratios. The linearity of the method reached a maximum at approximately 0.18 mg/ml. Direct proportionality was lost above this concentration.

An MECC method for drug substance characterization and quantitation of drug in a tablet formulation was developed for corticosteroids and benzothiazepin analogs by evaluating SDS and bile salts as surfactants.[48] Separation was obtained only with the bile salts as the micellar phase, since SDS promotes almost complete solubilization of the very lipophilic solutes into the micelle and, therefore, yielding no difference in migration time. Quantitative impurity profiling of diltiazem hydrochloride, a benzothiazepin analog, at the 0.2% level was obtained using an internal standard (Figure 8). Quantitation of diltiazem hydrochloride in tablets gave an approximate recovery of 100% with acceptable reproducibility. However, quantitation of 0.5% fluocinonide, a corticosteroid, in a nonionic surfactant cream formulation was associated with sample separation reproducibility of 9.32%. This was speculated to result from formulation matrix effect.

Capillary electrophoresis was used to quantitate the active ingredients in a commercial antipyretic analgesic tablet.[14] Ethyl-*p*-aminobenzoate was used as an internal standard for the five active ingredients: acetylsalicylic acid, anhydrous caffeine, *p*-acetaminophenol, α-ethoxybenzamide, and salicylamide. Optimization of the separation was achieved by varying the pH and SDS concentration. Theoretical plates ranging from 70,000 to 130,000 were obtained. The RSD for the determination of the five different species ranged from 0.8 to 1.8%, using the internal standard. Recovery of these five species from an excipient sample ranged from 98.7 to 100.4%.

Penicillin G (benzylpenicillin) was quantitated in tablets and in injectables by CE with an internal standard, phenol.[49] Quantitation was obtained using a calibration curve ranging from 0.2 to 1 mg/ml. The variation for tablets and parenterals was ±20%. The RSD value for injectable penicillin G, based on peak height ratio, was 5.3% for eight injections. Baseline separation of penicillin G from structurally similar compounds, such as potassium penicillin G, sodium ampicillin, and potassium penicillin V, was accomplished by adding 50 m*M* SDS to the separation buffer. This buffer composition enabled separation of a possible degradation product, which was hypothesized to be penicilloate.

The quantitation of authentic samples and injectable dosage forms of insulin by HPLC and by CZE was compared.[11] The quantitation parameters considered were accuracy, precision, and ease of use. Electrokinetic sample introduction gave superior peak shape and more reproducible peak response compared to vacuum injection. It was pointed out that this mode of sample introduction is biased as a result of analyte-dependent differences in electrophoretic mobility. Peak area response for the insulin standards and authentic samples analyzed gave RSD values between 1.72 and 2.41%. Capillary zone electrophoresis analysis of an injectable insulin formulation had recovery ranging from 85.8 to 108.1% combined with poor precision. The introduction of an internal standard did not improve the results. High-performance liquid chromatography analysis exhibited a clear advantage for both accuracy (98.5 to 101.4%) and precision (0.53 to 0.71%).

Pharmaceutical formulation of proteins and peptides can consist of the active substance in a buffer solution containing human serum albumin. Capillary electrophoresis analysis can be complicated by the added serum in the formulation, which can intefere with the separation by adsorption to the capillary wall. For a more detailed discussion about analysis of proteins with CE see Chapters 13, 18, and 19.

Formulated leukocyte A interferon and interleukin 1α, two recombinant human cytokines, were separated and quantified by capillary electrophoresis.[50] Leukocyte A interferon, which usually assays greater than 95% with HPLC, showed two peaks in CE. The nature of the second peak is unknown. An increased selectivity was observed with a sodium tetraborate buffer that was supplemented with 0.025 *M* lithium chloride. The migration time precision and peak area precision for six consecutive injections of human serum albumin and one of the excipients were 0.43 and 3.52%, respectively. Controlling the temperature of the capillary column during the separation was found to be essential for reproducible and reliable results (see Chapter 21). Quantitative information could not be obtained from the data.

Quantitation of a monoclonal antibody, anti-TAC, in bulk and in therapeutic form, was conducted using CE.[51] The excipients in the parental formulation were L-arginine and *N*-acetyltryptophan. To prevent adsorption of protein onto the capillary wall, and thereby maximizing the reproducibility, tetramethylammonium chloride (TMAC), trimethylammonium propylsulfonate (TMAPS), and lithium chloride (LiCl) were added to the carrier buffer. Lithium chloride has been found to stabilize tertiary structures of some proteins in solution. Peak area RSD values for L-arginine, *N*-acetyltryptophan, and anti-TAC in the sample formulation were obtained using an internal standard. The values ranged from 1.94 to 3.67 for six injections. The most reproducible separations for all three compounds were obtained with TMAPS as the buffer additive.

Berberine chloride in pharmaceutical preparations has also been determined by CZE.[52] A phosphate buffer at pH 5.0 gave complete separation within 20 min. Injection was by the hydrostatic method with detection at 214 nm. The berberine chloride method was linear from 10 to 100 μg/ml, with a correlation coefficient of 0.999. The peak height precision over five injections had an RSD value of 1%.

4. Detection

As discussed in detail in Chapter 7, detection in capillary electrophoresis can either be on-column or postcolumn. Ultraviolet is an example of on-column detection, and is the most commonly used detection mode to date. Postcolumn detection is highly selective and could be very useful for capillary electrophoresis applications in the medical and pharmaceutical laboratory. A postcolumn fluorescence detection method for capillary electrophoresis that is based on three syringe pumps and two mixing stations has been proposed for detection of putrescine.[53] A calibration curve for putrescine showed nonlinear behavior and the detection limit was at the picomole level.

A highly purified monoclonal antibody was characterized using capillary electrophoresis with UV and fluorescence detection.[54] Multiple parallel capillaries were used to increase detection sensitivity and sample load.

Ultraviolet photodiode array detection allows peak purity checks and facilitates identification and quantitation. This detection method was used to detect antibiotics separated by capillary electrophoresis.[55] Samples were prepared in methanol and an injection time of 8 s was used. The migration time RSD values were 0.78%.

C. ENANTIOMERIC SEPARATIONS

1. Separation Mechanism

Interest in enantiomeric composition is based on the pharmacodynamic and pharmacokinetic differences between drug enantiomers. Traditional approaches for measuring enantiomeric excess involve direct and indirect methods. The derivatization of a mixture of enantiomers with an optically active reagent into diastereoisomers is the indirect mode. The diastereoisomers are separated using conventional HPLC or gas chromatography. The direct mode depends on the formation of labile diastereomers (performed via hydrogen bonding, dipole–dipole, π–π, and/or hydrophobic interactions) between the enantiomers and the chiral environment, usually a chiral stationary phase, with which they interact. The utility of HPLC using chiral stationary phases has been the subject of several articles.[56-61]

The selectivity, sensitivity, reproducibility, and high efficiency obtained with CE make it a viable technique for enantiomeric separations of a variety of different types of compounds. The separation method in which CE and chromatography are combined is known as electrokinetic chromatography (EKC). Micellar electrokinetic capillary chromatography involves the use of a pseudostationary phase (in most applications, SDS) and permits the separation of neutral or nonionic compounds. Separations using SDS micelles exhibit migration characteristics similar to conventional reversed-phase HPLC. Optically active bile salts, cyclodextrins (CDs), crown ethers, and chiral molecular micelles are added to the mobile phase and enantiomeric resolution by stereoselective interactions is obtained with the solute. With CDs, this interaction occurs within the molecular cavity by formation of an inclusion complex. Through a mechanism similar to conventional MECC, interaction of the solute with the micelle or chiral additive leads to a slowing of its migration velocity relative to the bulk phase. The relationship between migration time, selectivity, and the number of theoretical plates has been discussed by Cole and Sepaniak.[62] In general, enantiomers are separated on the basis of formation constants of the host–guest complexes. The enantiomer that forms the more stable complex has a longer migration time because of this effect. Chiral additive concentration, micelle concentration, pH, and temperature all affect the enantiomeric resolution. For a more detailed discussion of the separation mechanism for optical resolution by micellar EKC the reader is referred to several review articles on this subject.[62-66] Applications of the direct and indirect enantioseparations using MECC are described in this chapter.

2. Examples

Sepaniak et al.[67] discussed the use of native and chemically modified cyclodextrins for the capillary electrophoretic separation of selected enantiomers. Experimental parameters important in attaining short analysis times are examined. Two sets of binaphthyl enantiomers, 1,1′-binaphthyl diyl hydrogen phosphate and 1,1′-binaphthyl-2,2′-dicarboxylic acid, and dansylated phenylalanine enantiomers are separated in less than 6 min (Figure 9).

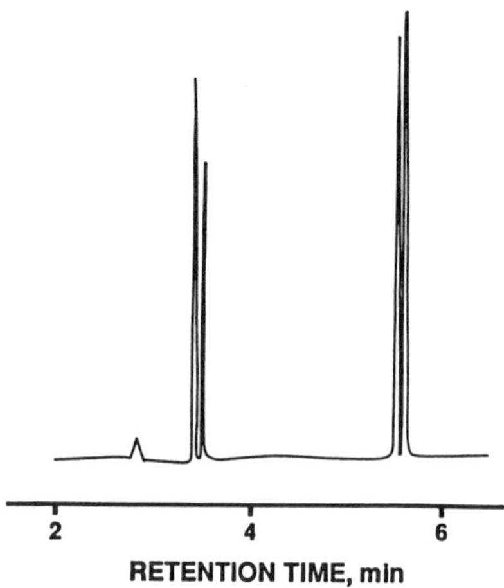

FIGURE 9. Separation of charged binaphthyl enantiomers: first set, 1,1 binaphthyl diyl hydrogen phosphate; second set, 1,1′ binaphthyl 2,2′ dicarboxylic acid. Separation buffer, 0.01 *M* disodium phosphate, 0.006 *M* disodium borate (pH 9) with 0.01 *M* G1α-cyclodextrin; separation voltage, 20 kV; detection, 210 nm. (From Sepaniak, M. J., Cole, R. O., and Clark, B.K., *J. Liq. Chromatogr.,* 15(6–7), 1023, 1992. With permission.)

Differences in the separation mechanisms, and effect of buffer composition and temperature on enantiomeric resolution with cyclodextrins and crown ethers, were studied by Kuhn et al.[68] Separations included the racemate of quinagolide, a potent dopamine agonist, a precursor of drug ENX 792, and several α-amino acids. A detection limit of 0.2% compared to a 100% peak of an antipode could be achieved for ENX 792, using the full dynamic range of the detector.

Separation of amino acid racemates using the chiral crown ether, [18]-crown-6, was demonstrated.[68] While the separation of optical isomers with cyclodextrin is based on a host–guest complex and/or hydrophobic interaction, the crown ether forms a complex with primary alkly amines by dipole–ionic interactions. The examples with [18]-crown-6 expand the application range of enantiomeric separations using CE. Kuhn et al.[69] used a chiral crown ether as a pseudostationary phase in capillary electrophoresis to separate optically active amines. Two recognition mechanisms were proposed on the basis of their separation of more than 20 amines. The results were confirmed by thermodynamic studies on the host–guest complexes. Optimal resolution was achieved when the chiral center was adjacent to the amine functionality.

The quantitation of *l*-epinephrine and the determination of the *d*/*l*-epinephrine enantiomer ratio in pharmaceutical formulations is described by Peterson and Trowbridge.[46] The optical isomers of epinephrine were resolved by capillary electrophoresis with a buffer containing heptakis-(2,6-di-*o*-methyl)-β-cyclodextrin. An internal standard correction with *l*-pseudoephedrine improved the reproduciblity and precision of the quantitative method. The RSD of the standard solutions ranged from 1.4 to 1.8%. Tryptophan enantiomers and the enantiomeric composition in commercial preparations of epinephrine were also investigated by Fanali and Bocek.[70] The quantitative method used (–) isoproterenol as an internal standard and the separations were fast and reproducible. Nishi et al. resolved diltiazem hydrochloride, trimetoquinol hydrochloride, and seven related compounds, using bile salts as chiral surfactants.[64,71] Application to the optical purity testing of trimetoquinol hydrochloride, which has a minor (R) and a major (S) enantiometric form, showed that 1% of the minor enantiomer of trimetoquinol hydrochloride could be detected at an *S*/*N* ratio of 3. The authors concluded that the chiral separation by micellar electrokinetic chromatography with anionic bile salts is successful when the solutes have a relatively rigid conformation, due to the rigid structure of

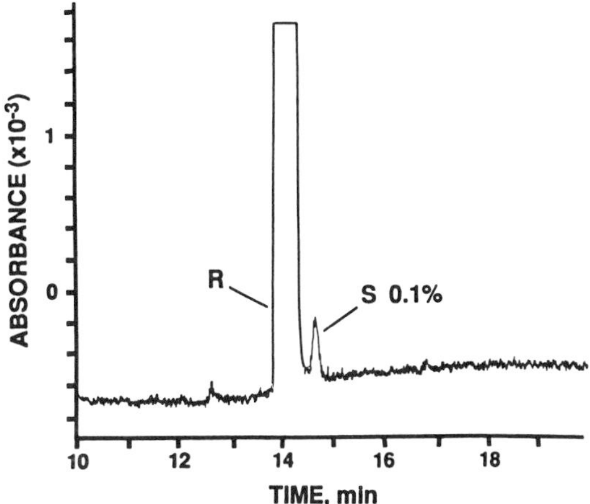

FIGURE 10. Separation of *R*- and *S*-bupivacaine (333 and 0.34 μg/ml, respectively). Separation buffer, 10 m*M* heptakis(2,6-di-*O*-methyl)-β-cyclodextrin, 18 m*M* Tris (Trizma base), pH 2.7, 0.1% methylhydroxyethylcellulose 4000, 0.03 *M* hexadecyltrimethylammonium bromide; pressure injection, 5 s; separation voltage, 18 kV; temperature, 25°C; detection, 254 nm. (From Soini, H., Snopek, J., and Novotny, M., Beckman Technical Bulletin, Capillary Electrophoresis, DS-836, 1992. With permission.)

the bile salt. Enantiomeric compounds that are positively charged or basic will be more effectively resolved due to increasing ionic interaction between the solute and the anionic bile salt micelle.[71] Since small differences in the solute structure affect the chiral recognition, it is difficult to predict enantiomeric resolution.

Applications of stereospecific cyclodextrins in capillary electrophoresis using model isomeric compounds, including enantiomers, is reviewed by Snopek et al.[72] Soluble alkylhydroxyalkylcellulose derivatives were added to the cyclodextrin-modified background electrolytes under study. Their presence enhanced enantioselectivity and separation efficiency. The results obtained in both CZE and isotachophoresis are compared and discussed by the authors.

A comprehensive study by Altria, et al.[73] showed the resolution effects on clenbuterol due to experimental manipulation of cyclodextrin concentration. Effects of temperature, voltage, modifiers to increase cyclodextrin solubility, pH, non-CD chiral additives, and other parameters were shown. The use of anionically modified β-cyclodextrin and sulfobutyl ether derivatives for the resolution of small drug molecules was investigated by Tait et al.[74]

Chiral resolution of five basic drugs was accomplished with run buffers consisting of CD modified with cationic detergents and cellulose derivatives.[75] Electropherograms and experimental conditions for the separation of the optical isomers of carvediol, bupivacaine, mepivacaine, verapamil, and fluoxetine are given. It was possible to quantitate the enantiomers of bupivacaine at the 0.1% level (Figure 10). The authors are developing similar methods for acidic drugs.

Terbutaline and propranolol were resolved using CE.[76] The effects of the cyclodextrin cavity size and concentration added to the electrolyte on the resolution and migration time were studied. Separations were achieved within 10 min and the minimum sample concentration that could be injected to obtain an *S/N* ratio of 2:1 was 1.1×10^{-7} *M* for terbutaline. Enantiomeric separtion of dansyl amino acids and barbituates was obtained by CE based on complexation with CD.[77] The magnitude of enantiomeric resolution was presented for unmodified and methylated CDs.

The resolution of six phenylthiohydantoin derivatives of DL-amino acids was studied using MECC.[78] Digitonin, which is a nonionic compound, was used as a chiral surfactant with SDS to form mixed micelles. Tryptophan, norleucine, norvaline, valine, α-aminobutyric acid, and alanine were separated from each other and their enantiomers resolved; however, long separation times (90 min), were required.

A number of dansyl amino acids were resolved by capillary zone electrophoresis with rapid (10 min) analysis times.[79] The derivatized amino acids were resolved by diastereomeric interaction between the DL-amino acid and the copper(II) complex of L-histidine present in the electrolyte and detected by LIF. In a subsequent study, the same investigators reported subfemtomole amounts of resolved racemic mixtures of 14 different amino acids in less than 12 min.[80] The separation was based on the diastereomeric interaction between the DL-amino acids and a chiral Cu(II)–aspartame complex. Experimental conditions investigated included electrolyte composition, temperature, and pH. Linearity test, response index (*n*), and correlation coefficents for arginine, glutamine, and valine were reported.

Resolution of enantiomeric N-acylated amino acid isopropyl esters was studied with the micellar solution *N*-dodecanoyl-L-valinate.[81] The 3,5-dinitrobenzoyl derivatives were the most effectively resolved and the D-enantiomers eluted faster than the corresponding L-enantiomers in all cases. The migration times of each pair of amino acid derivatives were dependent on the hydrophobicity of the amino acid side chain in the solute. The addition of SDS, urea, and methanol to sodium *N*-dodecanoyl-L-valinate improved peak shape and resolution for phenylthiohydantoin (PTH) DL-amino acids, warfarin, and benzoin.[82] Capacity factors (k'), separation factors (α), and resolution of the PTH derivatives of the amino acids are given.

Birnbaum and Nilsson[83] extended chiral CE applications to separation methods based on affinity interactions for the optical resolution of D- and L-tryptophan on immobilized bovine serum albumin (BSA) as the separation phase. They named this method capillary affinity gel electrophoresis (CAGE). Using this methodology, it was possible to obtain a resolution value of 6 (baseline resolution, 1.5) with a theoretical plate value of 91,000. One of the limitations of this method is the amount of material that can be injected onto the capillary. It was possible to detect 1 μ*M* at an *S*/*N* ratio of 2. The range of concentrations separated and analyzed was 1 to 25 μ*M*. In another study the (6R)- and (6S)-stereoisomers of leucovorin were resolved with BSA as an additive.[84] Resolution was obtained in approximately 12 min with efficiencies greater than 200,000 theoretical plates. More than 100 runs could be made with the capillary before it had to be replaced. A general equation was derived to calculate the free energy of interaction between leucovorin isomers and the BSA molecule.

The use of human protein chiral selectors in CE was described for the separation of benzoin, promethazine, thioridizine, and propriomazine enantiomers. The methods use human serum albumin and α-1-acid glycoprotein immobilized onto silica particles and packed into a capillary. Initial results and electropherograms were presented.[85]

Analytical MECC and CZE can be used successfully for the enantiomeric purity control of drugs. A review article, by Snopek et al.,[63] provides an excellent starting place for the neophyte. The pseudophases discussed include micellar, inclusion (both cyclodextrins and crown ethers), and enantioselective metal complexes and applications in high-performance electromigration methods are given. A more recent article by McLaughlin et al.[86] provides guidelines and limitations of CZE in the separation of pharmaceutical compounds with a major area of discussion being chiral drugs.

III. CONCLUSION

In spite of the pharmaceutical applications presented in this chapter, CE has several parameters that need to be optimized before it can be considered a true alternative to HPLC. Reproducibility, compatibility of sample preparation with separation buffer, detection sensitivity, and relevant system suitability parameters need to be improved if CE is to be equivalent to HPLC. Capillary electrophoresis will remain the preferred technique when sample volumes are limited, i.e., in the analysis of biological fluids with the condition that the detection sensitivity is adequate. There are advantages and disadvantages to the respective techniques.

Enantiomeric separations with HPLC are challenging. Capillary electrophoresis offers a viable alternative for resolution of enantiomers due to its greater separation capacity, small chiral selector requirement, and generally faster analysis times. Because CE may be more economical than HPLC, it will continue to play a complementary role in the resolution of enantiomers.

While the number of CE publications has grown with the commercial availability of instruments, its full potential has not been reached. The final decision in the acceptance of CE as an analytical technique in the pharmaceutical industry is dependent on the requirements or limitations regulatory agencies like the FDA places on this new separation technique.

REFERENCES

1. **Steuer, W., Grant, I., and Erni, F.,** Comparison of high-performance liquid chromatography, supercritical fluid chromatography and capillary zone electrophoresis in drug analysis, *J. Chromatogr.,* 507, 125, 1990.
2. **Fazio, S., Vivilecchia, R., Lesueur, L., and Sheridan, J.,** Capillary zone electrophoresis: some promising pharmaceutical applications, *J. Am. Biotechnol. Lab.,* 8(1), 10, 1990.
3. **Jorgenson, J. W. and Lukacs, K. D.,** Capillary zone electrophoresis, *Anal. Chem.,* 53, 1298,1981.
4. **Cheng, Y. F. and Dovichi, N. J.,** Subattomole amino acid analysis by capillary zone electrophoresis and laser induced flourescence, *Science,* 242, 562, 1988.
5. **Amanakwa, L. N., Scholl, J., and Kuhr, W. G.,** Characterization of the oligomer dispersion of poly(oxyalkylene) diamine polymers by precolumn derivatization and capillary electrophoresis with fluorescence detection, *Anal. Chem.,* 62, 2189, 1990.
6. **Sweedler, J. V., Shear J. B., Fishman H. A., Zare R. N., and Scheller R. H.,** Fluorescence detection in capillary zone electrophoresis using a charge-coupled device with time delayed integration, *Anal. Chem.,* 63, 496, 1991.
7. **Kuhr, W. G. and Yeung, E. S.,** Indirect fluorescence detection of native amino acids in capillary zone electrophoresis, *Anal. Chem.,* 60, 1832, 1988.
8. **Kuhr, W. G. and Yeung, E. S.,** Optimization of sensitivity and separation in capillary zone electrophoresis with indirect fluorescence detection, *Anal. Chem.,* 60, 2642, 1988.
9. **Garner, T. W. and Yeung, E. S.,** Indirect fluorescence detection of sugars separated by capillary zone electrophoresis with visible laser excitation, *J. Chromatogr.,* 515, 639, 1990.
10. **Odake, T., Kitamori, T., and Sawada, T.,** Direct detection of laser-induced capillary vibration by a piezoelectric transducer, *Anal. Chem,* 64, 2870, 1992.
11. **Lookabaugh, M., Biswas, M., and Krull, I. S.,** Quantitation of insulin injection by high-performance liquid chromatography and high-performance capillary electrophoresis, *J. Chromatogr.,* 549(1–2), 357, 1991.
12. **Fuijwara, S., Iwase, S., and Honda, S.,** Analysis of water-soluble vitamins by micellar electrokinetic capillary chromatography, *J. Chromatogr.,* 447, 133, 1988.
13. **Ong, C. P., Ng, C. L., Lee, H. K., and Li, S. F. Y,** Separation of water- and fat-soluble vitamins by micellar electrokinetic chromatography, *J. Chromatogr.,* 547, 419, 1991.
14. **Fujiwara, S. and Honda, S.,** Determination of ingredients of antipyretic analgesic preparations by micellar electrokinetic capillary chromatography, *Anal. Chem.,* 59, 2773, 1987.
15. **Nishi, H. and Matsuo, M.,** Separation of corticosteroids and aromatic hydrocarbons by cyclodextrin-modified micellar electrokinetic chromatography, *J. Liq. Chromatogr.,* 14(5), 973, 1991.
16. **Akermans, M. T., Everaerts, F. M., and Beckers, J. L.,** Determination of some drugs by micellar electrokinetic capillary chromatography, *J. Chromatogr.,* 585, 123, 1991.
17. **Liu, J., Cobb, K. A., and Novotny, M.,** Capillary electrophoretic separations of peptides using micelle-forming compounds and cyclodextrins as additives, *J. Chromatogr.,* 519, 189, 1990.
18. **Nishi, H., Tsumagari, N., Kakimoto, T., and Terabe, S.,** Separation of β-lactam antibiotics by micellar electrokinetic chromatography, *J. Chromatogr.,* 477, 259, 1989.
19. **Nishi, H., Tsumagari, N., and Terabe, S.,** Effect of tetraalkylammonium salts on micellar electrokinetic chromatography of ionic substances, *Anal. Chem.,* 61, 2434, 1989.
20. **Wallingford, R. A., Curry, P. D., and Ewing, A. G.,** Retention of catechols in capillary electrophoresis with micellar and mixed micellar solutions, *J. Microcolumn Sep.,* 1, 23, 1989.

21. **Salomon, K., Burgi, D. S., and Helmer, J. C.,** Separation of seven tricyclic antidepressants using capillary electrophoresis, *J. Chromatogr.,* 549, 375, 1991.
22. **Swedberg, S. A.,** Use of non-ionic and zwitterionic surfactants to enhance selectivity in high-performance capillary electrophoresis. An apparent micellar electrokinetic capillary chromatography mechanism, *J. Chromatogr.,* 503, 449, 1990.
23. **Landers, J. P., Oda, R. P., and Schuchard, M. D.,** Separation of boron-complexed diol compounds using high-performance capillary electrophoresis, *Anal. Chem.,* 64, 2846, 1992.
24. **Nakagawa, T., Oda, Y., Shibukawa, A., Fukuda, H., and Tanaka, H.,** Electrokinetic chromatography for drug analysis. Separation and determination of cefpiramide in human plasma, *Chem. Pharm. Bull.,* 37, 707, 1989.
25. **Thormann, W., Meier, P., Marcolli, C., and Binder, F.,** Analysis of barbiturates in human serum and urine by high-performance capillary electrophoresis-micellar electrokinetic capillary chromatography with on-column multi-wavelength detection, *J. Chromatogr.,* 545, 445, 1991.
26. **Pietta, P. G., Mauri, P. L., Rava, A., and Sabbatini, G.,** Application of micellar electrokinetic capillary chromatography to the determination of flavonoid drugs, *J. Chromatogr.,* 549, 367, 1991.
27. **Hurni, W. M. and Miller, W. J.,** Analysis of a vaccine purification process by capillary electrophoresis, *J. Chromatogr.,* 559, 337, 1991.
28. **Kenndler, E., Schwer, C., and Kaniansky, K.,** Purity control of riboflavin-5′-phosphate (vitamin B_2 phosphate) by capillary electrophoresis, *J. Chromatogr.,* 508, 203, 1990.
29. **Altria, K. D. and Smith, N. W.,** Pharmaceutical analysis by capillary zone electrophoresis and micellar electrokinetic capillary chromatography, *J. Chromatogr.,* 538, 506, 1991.
30. **Reinhoud, N. J., Tjaden, U. R., Irth, H., and van der Greef, J.,** Bioanalysis of some anthracyclines in human plasma by capillary electrophoresis with laser-induced fluorescence detection, *J. Chromatogr.,* 574, 327, 1992.
31. **Roach, M. C., Gozel, P., and Zare, R. N.,** Determination of methotrexate and its major metabolite, 7-hydroxymethotrexate, using capillary zone electrophoresis and laser-induced fluorescence detection, *J. Chromatogr.,* 426, 129, 1988.
32. **Guzman, N. A., Moschera, J., Iqbal, K., and Malick, A. W.,** A quantitative assay for the determination of proline and hydroxyproline by capillary electrophoresis, *J. Liq. Chromatogr.,* 15, 1163, 1992.
33. **Nishi, H., Fukuyama, T., Matsuo, M., and Terabe, S.,** Effect of surfactant structures on the separation of cold medicine ingredients by micellar electrokinetic chromatography, *J. Pharm. Sci.,* 79, 519, 1990.
34. **Swartz, M. E.,** Capillary electrophoretic analysis of inorganic ions in pharmaceutical formulations, 7th Annu. AAPS Meeting, San Antonio, Texas, November 16–19, 1992.,
35. **Ng, C. L., Lee, H. K., and Li, S. F. Y.,** Systematic optimization of capillary electrophoretic separation of sulphonamides, *J. Chromatogr.,* 598, 133, 1992.
36. **Honda, S., Iwase, S., and Fujiwara, S.,** Evaluation of an automatic siphonic sampler for capillary zone electrophoresis, *J. Chromatogr.,* 404, 313, 1987.
37. **Otsuka, K., Terabe, S., and Ando, T.,** Quantitation and reproducibility in electrokinetic chromatography with micellar solutions, *J. Chromatogr.,* 396, 350, 1987.
38. **Wainright, A.,** Capillary electrophoresis applied to the analysis of pharmaceutical compounds, *J. Microcolumn Sep.,* 2, 166, 1990.
39. **Swartz, M. E.,** Method development and selectivity control for small molecule pharmaceutical separations by capillary electrophoresis, *J. Liq. Chromatogr.,* 14(5), 923, 1991.
40. **Lux, J. A., Yin, H. F., and Schomburg, G.,** Construction, evaluation and analytical operation of a modular capillary electrophoresis instrument, *Chromatographia,* 30, 7, 1990.
41. **Weinberger, R. and Albin, M.,** Quantitative micellar electrokinetic capillary chromatography: linear dynamic range, *J. Liq. Chromatogr.,* 14, 953, 1991.
42. **Fujiwara, S. and Honda, S.,** Determination of cinnamic acid and its analogues by electrophoresis in a fused silica capillary tube, *Anal. Chem.,* 58, 1811, 1986.
43. **Lloyd, D. K., Cypess, A. M., and Wainer, I. W.,** Determination of cytosine-β-arabinoside in plasma using capillary electrophoresis, *J. Chromatogr.,* 568, 117, 1991.
44. **Meier, P. and Thormann, W.,** Determination of thiopental in human serum and plasma by high-performance capillary electrophoresis-micellar electrokinetic chromatography, *J. Chromatogr.,* 559, 505, 1991.
45. **Tomita, M. and Okuyama, T.,** Determination of glyphosate and its metabolite, (aminomethyl)phosphonic acid, in serum using capillary electrophoresis, *J. Chromatogr.,* 571, 324, 1991.
46. **Peterson, T. E. and Trowbridge, D.,** Quantitation of *l*-epinephrine and determination of the *d*-/*l*-epinephrine enantiomer ratio in a pharmaceutical formulation by capillary electrophoresis, *J. Chromatogr.,* 603, 298, 1992.
47. **Arrowood, S. and Hoyt, A. M.,** Determination of cimetidine in pharmaceutical preparations by capillary zone electrophoresis, *J. Chromatogr.,* 586, 177, 1991.

48. **Nishi, H., Fukuyama, T., Matsuo, M., and Terabe, S.,** Separation and determination of lipophilic corticosteroids and benzothiazepin analogues by micellar electrokinetic chromatography using bile salts, *J. Chromatogr.,* 513, 279, 1990.
49. **Hoyt, A. M. and Sepaniak, M. J.,** Determination of benzylpenicillin in pharmaceuticals by capillary zone electrophoresis, *Anal. Lett.,* 22(4), 861, 1989.
50. **Guzman, N. A., Ali, H., Moschera, J., Iqbal, K., and Malick, A. W.,** Assessment of capillary electrophoresis in pharmaceutical applications. Analysis and quantification of a recombinant cytokine in an injectable dosage form, *J. Chromatogr.,* 559, 307, 1991.
51. **Guzman, N. A., Moschera, J., Iqbal, K., and Malick, A. W.,** Effect of buffer constituents on the determination of therapeutic proteins by capillary electrophoresis, *J. Chromatogr.,* 608, 197, 1992.
52. **Akada, Y, Ishii, N., Nobuyukia, N., and Nakane, Y.,** Determination of berberine chloride in pharmaceutical preparations by capillary electrophoresis, *Bunseki Kagaku,* 41, 349, 1992.
53. **Tsuda, T., Kobayashi, Y., Hori, A., Matsumoto, T., and Suzuki, O.,** Postcolumn detection for capillary zone electrophoresis, *J. Chromatogr.,* 456, 375, 1988.
54. **Guzman, N. A., Trebilcock, M. A., and Advis, J. P.,** Capillary electrophoresis for the analytical separation and semi-preparative collection of monoclonal antibodies, *Anal. Chim. Acta,* 249, 247, 1991.
55. **Yeo, S. K., Lee, H. K., and Li, S. F. Y.,** Separation of antibiotics by high-performance capillary electrophoresis with photodiode-array detection, *J. Chromatogr.,* 585, 133, 1991.
56. **Pirkle, W. H., Finn, J. M., Schreiner, J. L., and Hamper, B. C.,** A widely useful chiral stationary phase for the high-performance liquid chromatography separation of enantiomers, *J. Am. Chem. Soc.,* 103, 3964, 1981.
57. **Hinze, W. L.,** Applications of cyclodextrins in chromatographic separations and purification methods, *Sep. Purif. Methods,* 10, 159, 1981.
58. **Hermansson, J.,** Liquid chromatographic resolution of racemic drugs using a chiral α_1-acid glycoprotein column, *J. Chromatogr.,* 298, 67, 1984.
59. **Armstrong, D. W., Ward, T. J., Armstrong, R. D., and Beesley, T. E.,** Separation of drug stereoisomers by the formation of β-cyclodextrin inclusion complexes, *Science,* 232, 1132, 1986.
60. **Bethod, A. B., Jin, H. L., Beesley, T. E., Duncan, J. D., and Armstrong, D. W.,** Cyclodextrin chiral stationary phases for liquid chromatographic separations of drug stereoisomers, *J. Pharm. Biomed. Anal.,* 8(2), 123, 1990.
61. **Allenmark, S.,** *Chromatographic Enantioseparation Methods and Applications,* Ellis Horwood, Chichester, England, 1988.,
62. **Cole, R. O. and Sepaniak, M. J.,** The use of bile salt surfactants in micellar electrokinetic capillary chromatography, *LC-GC,* 10, 380, 1992.
63. **Snopek, J., Jelinek, I., and Smolkova-Keulemansova, E.,** Micellar, inclusion and metal-complex enantioselective pseudophases in high-performance electromigration methods, *J. Chromatogr.,* 452, 571, 1988.
64. **Nishi, H., Fukuyama, T., and Matsuo, M.,** Chiral separation of optical isomeric drugs using micellar electrokinetic chromatography and bile salts, *J. Microcolumn Sep.,* 1(5), 234, 1989.
65. **Nishi, H., Terabe, S., and Seiyaku, T.,** Application of micellar electrokinetic chromatography to pharmaceutical analysis, *Electrophoresis,* 11, 691, 1990.
66. **Wren, S. A. C.,** Theory of chiral separations in capillary electrophoresis, *Electrophoresis,* 1993, in press.
67. **Sepaniak, M. J., Cole, R. O., and Clark, B. K.,** Use of native and chemically modified cyclodextrins for the capillary electrophoretic separation of enantiomers, *J. Liq. Chromatogr.,* 15(6–7), 1023, 1992.
68. **Kuhn, R., Stoecklin, F., and Erni, F.,** Chiral separations by host-guest complexation with cyclodextrin and crown ether in capillary zone electrophoresis, *Chromatographia,* 33(1–2), 32, 1992.
69. **Kuhn, R., Erni, F., Bereuter, T., and Hausler, J.,** Chiral recognition and enantiomeric resolution based on host-guest complexation with crown ethers in capillary zone electrophoresis, *Anal. Chem.,* 64, 2815, 1992.
70. **Fanali, S. and Bocek, P.,** Enantiomer resolution by using capillary zone electrophoresis: resolution of racemic tryptophan and determination to the enantiomer composition of commercial pharmaceutical epinephrine, *Electrophoresis,* 11, 757, 1990.
71. **Nishi, H., Fukuyama, T., Matsuo, M., and Terabe, S.,** Chiral separation of diltiazem, trimetoquinol and related compounds by micellar electrokinetic chromatography with bile salts, *J. Chromatogr.,* 515, 233, 1990.
72. **Snopek, J., Soini, H., Novotny, M., Smolkova-Keulemansova, E., and Jelinek, I.,** Selected applications of cyclodextrin selectors in capillary electrophoresis, *J. Chromatogr.,* 559, 215, 1991.
73. **Altria, K. D., Goodall, D. M., and Rogan, M. M.,** Chiral separation of β-amino alcohols by capillary electrophoresis using cyclodextrins as buffer additives. Effect of varying operating parameters, *Chromatographia,* 34, 19, 1992.
74. **Tait, R. J., Thompson, D. O., Stobaugh, J. F., and Stella, V. J.,** Chiral displacement chromatographic separations in the normal-phase mode, 7th Annu. AAPS Meeting, San Antonio, Texas, November 16–19, 1992.

75. **Soini, H., Snopek, J., and Novotny, M.,** Chiral separations of basic drugs with modified cyclodextrin-containing buffers by P/ACE capillary electrophoresis, Beckman Technical Bulletin Information, Capillary Electrophoresis DS-836, 1992.,
76. **Fanali, S.,** Use of cyclodextrins in capillary zone electrophoresis. Resolution to terbutaline and propranolol enantiomers, *J. Chromatogr.,* 545, 437, 1991.
77. **Tanaka, M., Asano, S., Yoshinago, M., Kawaguchi, Y., Tetsumi, T., and Shono, T.,** Separation of racemates by capillary zone electrophoresis based on complexation with cyclodextrins, *Fresenius J. Anal. Chem.,* 339, 63, 1991.
78. **Otsuka, K. and Terabe, S.,** Enantiomeric resolution by micellar electrokinetic chromatography with chiral surfactants, *J. Chromatogr.,* 515, 221, 1990.
79. **Gassman, E., Kuo, J. E., and Zare, R. N.,** Electrokinetic separations of chiral compounds, *Science,* 230, 813, 1985.
80. **Gozel, P., Gassmann, E., Michelsen, H., and Zare, R. N.,** Electrokinetic resolution of amino acid enantiomers with copper(II)-aspartame support electrolyte, *Anal. Chem.,* 59, 44, 1987.
81. **Dobashi, A., Ono, T., Hara, S., and Yamaguchi, J.,** Enantioselective hydrophobic entanglement of enantiomeric solutes with chiral functionalized micelles by electrokinetic chromatography, *J. Chromatogr.,* 480, 413, 1989.
82. **Otsuka, K., Kawahara, J., Tatekawa, K., and Terabe, S.,** Chiral separations by micellar electrokinetic chromatography, *J. Chromatogr.,* 559, 209, 1991.
83. **Birnbaum, S. and Nilsson, S.,** Protein-based capillary affinity gel electrophoresis for the separation of optical isomers, *Anal. Chem.,* 64, 2872, 1992.
84. **Barker, G. E., Russo, P., and Hartwick, R. A.,** Chiral separation of leucovorin with bovine serum albumin using affinity capillary electrophoresis, *Anal. Chem.,* 64, 3024, 1992.
85. **Li, S., Ryan, P., and Lloyd, D. K.,** The use of proteins as chiral selectors in capillary electroseparations: initial results, HPLC 92, Abstract No. 262, Baltimore, Maryland, 1992.,
86. **McLaughlin, G. M., Nolan, J. A., Lindahl, J. L., Palmieri, R. H., Anderson, K. W., Morris, S. C., Morrison, J. A., and Bronzert, T. J.,** Pharmaceutical drug separations by HPCE: practical guidelines, *J. Liq. Chromatogr.,* 15(6–7), 961, 1992.

Chapter 11

CARBOHYDRATE ANALYSIS BY CAPILLARY ELECTROPHORESIS

Joseph D. Olechno and Kathi J. Ulfelder

TABLE OF CONTENTS

0-8493-8690-X/94/$0.00+$.50

I. INTRODUCTION

Not only are carbohydrates ubiquitous, they may be the most common organic compounds on the face of the earth. Our view of the world is colored and flavored by carbohydrates. All vascular plants owe their size to the carbohydrate, cellulose. Without this polysaccharide to provide structure, all plants would be a green film on the surface of earth and water. The colors of many fruits and flowers are derived from the glycosides of numerous phenolic compounds in plants. The exoskeletons of insects and the shells of lobsters and crabs are polymers of *N*-acetylglucosamine. Much of our food is modified with polysaccharides (e.g., carageenan, derivatized celluloses) to provide a creamy texture. The color of cola drinks and the odor of baking bread are due, in part, to carbohydrates. Carbohydrates provide the fuel for the internal engines of life on the planet. The oxygen we breathe is a fortuitous result of the formation of carbohydrates from carbon dioxide and water. Carbohydrates are essential not only in supplying energy but in allowing cells to communicate with one another. Internal to the cells, carbohydrates (inositol phosphates) may act as secondary messengers. The carbohydrates of a glycoprotein affect the folding of the protein (and, therefore, may affect biological activity). They can also affect the solubility of the protein, its ability to be translocated (from one cellular compartment to another), its antigenicity, its resistance to proteolysis, and its clearance from the body.[89] Carbohydrates help in the detoxification of chemicals not only in humans (through the formation of urine soluble glucuronides) but in plants (with the transfer of toxic materials to vacuoles). Carbohydrates (proteoglycans) act as the lubricants in our joints. They even act as antifreeze in Antarctic fish (as special *O*-linked glycoproteins) or as protectants against both freezing and dehydration in the small animals known as tardigrades. It has become apparent that carbohydrates are important from the very beginning of life (they are essential in the fertilization of the egg cell) through its end (a number of life-threatening situations are mediated through carbohydrates).

Carbohydrates may be generally defined as polyhydroxylated compounds. Most, but not all, carbohydrates contain a carbonyl functionality in the form of an aldehyde or ketone group. The large number of hydroxyl groups as well as the chemistry of the carbonyl group allow these molecules to form complex macromolecules through chemical bond formation. These complex materials may be hydrolyzed (usually with acid, although under some circumstances alkali or enzymes may be used) to release carbohydrate building blocks referred to as monosaccharides. While the hydroxyl group is the most characteristic attribute of the carbohydrate molecule, some hydroxyl groups may be replaced with other functional groups, including carboxylic acid groups (uronic, aldonic, and saccharinic acids), halides, amines (glycosamines), sulfur (thio sugars), hydrogen (deoxy sugars, including deoxyribose), double bonds (glycals), etc. Typical monosaccharides include glucose, fructose, sorbitol, inositol, glucosamine, and deoxyribose.

Oligosaccharides (*oligo*, a few) are covalent structures composed of a number, usually 2 to 25, of usually different monosaccharides. The oligosaccharides are composed of monosaccharide residues joined by glycosidic linkages formed by the elimination of water between a hemiacetal (or, rarely, a hemiketal) group of one monosaccharide and the hydroxyl group of another. The resultant structures may be linear, branched, or cyclic. While the study of oligosaccharides has recently focused on those derived from glycoproteins and glycolipids, they are also important components of secondary plant metabolites (e.g., digitonin, quercitin, etc.) and may exist as free, nonbound materials (e.g., antibiotics such as kanamycin). In general, the structure of oligosaccharides may be very complex. Polysaccharides, also known as glycans, are polymers of repeating mono-, di-, or tetrasaccharides. These may be as simple as cellulose (repeating polymer of $\beta(1 \rightarrow 4)$-linked glucose) or as complicated as amylopectin or the sulfated glycosaminoglycan, heparin. Naturally occurring polysaccharides may range in molecular weight from 5000 to several million daltons. In general, three basic structures of polysaccharides exist: linear structures, substituted linear structures in which small side chains are attached, and complexed branched structures in which long monosaccharide side chains are attached to the main side structure and to its side chains.

Glycoconjugates are molecules that covalently link carbohydrates to another moiety. Among the more common glycoconjugates are proteoglycans and glycoproteins (where the carbohydrate is linked to a protein), glycolipids (where the carbohydrate is linked to an aliphatic, fat-soluble structure), and phenolic glycosides of both plants and animals.

Proteoglycans are macromolecules composed of a number of highly charged polysaccharide chains attached to a central protein core. The polysaccharide portions, referred to as glycosaminoglycans, consist of heterorepeating dissacharides, which are composed of an aminosaccharide and another monosaccharide. These disaccharide blocks are highly substituted with sulfate esters or carboxylic acids. Five classes of glycosaminoglycans may be found in connective tissue: chondroitin sulfate, keratan sulfate, heparin (and heparan sulfates), dermatan sulfate, and hyaluronic acid.

Glycoproteins are macromolecules composed of oligosaccharide chains linked to the amide nitrogen (N-linked) of asparagine or the oxygen (O-linked) of serine, threonine, and occasionally hydroxyproline (plant glycoproteins), tyrosine (glycogenin), or hydroxylysine (collagen). Rather than the linear structures found in proteoglycans, these structures, while smaller, tend to be composed of a greater number of different monosaccharides and may be highly branched. The N-linked oligosaccharides may be further classified as "high mannose" (those containing only mannose and *N*-acetylglucosamine), "complex" (possibly containing fucose, galactose, and neuraminic acid as well as mannose and glucosamine), and "hybrid" (containing both high mannose and complex elements). O-Linked oligosaccharides, while generally having fewer branches than N-linked structures, offer more types of attachment to the protein and a greater number of different monosaccharide building blocks, including *N*-acetylgalactosamine, xylose, glucose, and arabinose as well as the sugars found in N-linked structures. Those wishing greater depth in carbohydrates and their analyses are directed to the literature.[1-3]

Carbohydrates and glycoconjugates provide a unique challenge to the analytical chemist. In general, carbohydrates have very low extinction coefficients (Table 1) and are, therefore, not amenable to detection by either absorbance or fluorescence. In addition, the chemical structures of both simple and complex carbohydrates tend to be very similar, consisting primarily of saturated polyhydroxylated hydrocarbons. Carbohydrates tend to be both extremely polar and nonvolatile. As strongly hydrophilic compounds, they are not directly amenable to separation by either reversed-phase high-performance liquid chromatography (HPLC) or supercritical fluid chromatography (SFC). These techniques, however, have been found useful after derivatization of the carbohydrate. As nonvolatile compounds, carbohydrates must be thoroughly derivatized before analysis by gas chromatography (GC). While derivatized monosaccharides are easily analyzed by GC, larger oligosaccharides resist analysis even after peralkylation.

The complexity of carbohydrate structures is illustrated by Table 2. Where 2 identical amino acids can link into a single dipeptide, 2 molecules of glucose can link into 11 disaccharides. Where 4 different amino acids can link into 24 tetrapeptides, 4 different hexopyranoses (e.g., glucose, galactose, mannose, allose) can produce more than 36,000 different tetrasaccharides.

Analytical methodologies for carbohydrate separations have included HPLC,[4-7] high-pH anion-exchange chromatography (HPAEC),[8-16] GC,[17,18] and SFC.[19-22] Detection schemes have included electrochemical,[8-16] refractive index,[13] fluorescence,[23,24] absorbance,[25,26] and others.

Capillary electrophoresis has been used as an alternative to HPLC for a number of different

TABLE 1
Extinction Coefficients of Carbohydrates Dissolved in Water and Measured at 195 nm

Carbohydrate	ε_{195}
Fructose	12
Galactose	5
Glucose	2
Sorbitol	2
Sucrose	2

Note: The higher extinction coefficient from fructose, a ketose, is due to a higher percentage of carbohydrate in the open-chain "keto" form.

TABLE 2
Complexity of Carbohydrate Oligomers vs. that of Peptides

Number of monomers	Amino acids	Carbohydrates (NR)[a]
2 identical	1	11 (3)
2 different	2	20 (4)
3 identical	1	120 (32)
3 different	6	720 (192)
4 identical	1	1,548 (460)
4 different	24	36,672 (14,592)

[a] NR, Number of nonreducing forms of the oligosaccharides that are possible. Reducing sugars that may anomerize between two forms are counted singly. Potential structures are based on all monomers remaining in the hexopyranose form. Allowing the possibility of furanose ring structures dramatically increases the possible number of structures.

separations, as elucidated in other chapters of this book. Carbohydrates, at first, seem poor candidates for analysis by capillary electrophoresis. They are generally very polar, neutral compounds and appear difficult to separate by traditional CE techniques. As previously described, they have low extinction coefficients and are difficult to detect by standard spectrophotometric means. Various researchers have addressed these problems with a variety of approaches.

II. SEPARATIONS

A. UNDERIVATIZED CARBOHYDRATES

1. Native Charge Separations

A number of carbohydrates have an intrinsic charge due to carboxylic acid groups, phosphates, or sulfates. These may be metabolic intermediates in sugar metabolism (e.g., glucose phosphates), cellular secondary messengers (inositol phosphates), breakdown products of proteoglycans (e.g., Δdi-6S from chondroitin), charged sugars found in glycoproteins (e.g., sialic acid, mannose phosphate) or monomer components of polysaccharides (e.g., galacturonic acid from pectin).

Carney and Osborne[27] showed that a change in position of sulfation on a chondroitin-derived disaccharide (Δdi-4S vs. Δdi-6S) affected the electrophoretic mobility of the analytes strongly enough that they could be baseline resolved. They found that the concentration of phosphoric acid had a profound effect on peak efficiency, with a maximum efficiency found when 150 m*M* orthophosphate (pH 3) was used. This acidic buffer system could not be used if the disaccharide was not sulfated since this leads to full protonation of the carboxylic acid and loss of charge. They further developed a second, alkaline buffer system (pH 9) that allowed the analysis of nonsulfated carboxylic acid containing disaccharides. They reported that sodium dodecyl sulfate (SDS) had a positive effect on peak efficiency but not, apparently, due to a mechanism involving micellar electrokinetic chromatography (MECC). Interestingly, they showed that the mobility of Δdi-4S and Δdi-6S reversed depending on the pH of analysis.

Using a buffer similar to that of Carney and Osborne, Damm et al.[28] resolved larger heparin fragments (up to the decasaccharide) containing as many as ten negative charges per molecule. They found that they obtained the greatest resolution (and the longest analysis times) with the buffers of greatest acidity (pH 2). They found this technique to be an attractive alternative to HPAEC. Motsch et al.[29] achieved resolution of Δdi-4S and Δdi-6S with both agarose (2.5% agarose, 20 m*M* phosphate) and polyacrylamide gel (8% T, 100 m*M* Tris)-filled capillaries. Analysis time with agarose was significantly longer (40 min) than with polyacrylamide gel (20 min).

Hermentin et al.[12] were able to partially resolve as many as 40 sialylated oligosaccharides derived from human plasma α_1-acid glycoprotein. The negative charge imparted by the sialic acid residue(s) was of primary importance in the separation but it is clear that there was at least partial resolution of oligosaccharides based on complex structure.

Oefner et al.[30] resolved a number of aldonic and uronic acids at pH 12.1. They showed that small changes in pH had dramatic effects on time of analysis. The migration time of mannuronic acid increases from 19 min at pH 11.9 to 40 min at pH 12.3. They attribute the increase in analysis time to a decrease in the electroosmotic flow in the capillary caused by an increase in the thickness of the diffusion double layer at the capillary wall.

Neutral carbohydrates, as illustrated in Table 3, are very weak acids.[14,31] The ionization at high pH of these molecules has been exploited in high-pH anion-exchange chromatography, HPAEC.[8-16] A number of researchers have utilized this property of carbohydrates and performed capillary electrophoresis at elevated pH. Under these highly alkaline conditions, the anomeric hydroxyl (the most acidic position) ionizes and the molecules may be resolved by their differing pK_a values. Carbohydrates may isomerize and oxidize under strongly alkaline conditions.[32,33] Hardy[34] and others have reported that at room temperature and with sodium hydroxide low in divalent cations, most carbohydrates are stable under highly alkaline conditions for hours.

Vorndran et al.[35] showed increasing resolution of neutral monosaccharides as the pH was increased over the range of 11.9 to 12.3 but noted a drastic drop in sensitivity as pH increased past 12.2. At pH 12, the nonreducing carbohydrates, raffinose and sucrose, are only slightly ionized and show low electrophoretic mobility. They were poorly resolved and eluted just after the water peak. Similarly, Garner and Yeung[36] used a solution of pH 11.5 to resolve neutral monosaccharides. At this pH it was not possible to resolve raffinose and sucrose. While Garner and Yeung determined that the optimum pH for the separation of their chosen carbohydrates would be 11.65, they found significant problems in detection as they increased the pH toward this value.

In contrast, Colón et al.,[37] using a different detection technique, were able to resolve the three nonreducing carbohydrates, sucrose, raffinose, and stachyose, by increasing the pH to 13 (100 m*M* NaOH). Colón et al. determined that they had best resolution at 100 m*M* alkali metal hydroxide. They compared LiOH, NaOH, and KOH for resolution and found that the lithium and sodium hydroxide solutions gave equivalent resolution but that the sodium hydroxide solution reduced the analysis time slightly. The use of potassium hydroxide led to markedly shorter migration times but much poorer resolution. They postulated that the

TABLE 3
pK_a Values of Carbohydrates

Carbohydrate	pK_a
α-Methyl glucoside	13.71
Sorbitol	13.60
Raffinose	12.74
2-Deoxyglucose	12.52
Galactose	12.39
Arabinose	12.34
1-Deuteroglucose	12.31
2-Deuteroglucose	12.29
Glucose	12.28
Xylose	12.15
Ribose	12.11
Mannose	12.08
Fructose	12.03
Sorbose	11.55

Note: Carbohydrates are weak acids. The hydrogen of the anomeric hydroxyl group is the most acidic. Nonreducing carbohydrates, e.g., methyl glucoside, sorbitol, and raffinose, lacking an anomeric hydroxyl group, are considerably less acidic. All values from Rendleman,[31] except for dueteroglucoses from Rohrer and Olechno.[14]

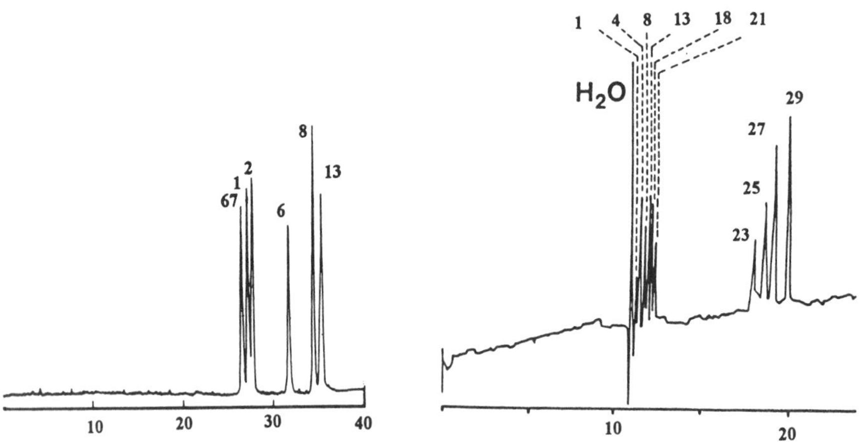

FIGURE 1. Resolution of nonderivatized carbohydrates at high pH. Numbers refer to structures in Table 4. Electropherogram on left shows that when pH is increased to 13.1 (100 m*M* NaOH, fused silica capillary 360 µm O.D. × 50 µm I.D., no internal coating, 70 cm length, 10 kV, gravity injection for 10 s at 10 cm, electrochemical detection), stachyose (67), raffinose (1) , and sucrose (2), unresolved at pH 12.1, are separated. (Reprinted with permission from Colón et al.[37] Copyright 1993 American Chemical Society.) On right, at pH 12.1 (6m*M* sorbic acid adjusted to pH 12.1, fused silica capillary, 50 µm I.D., no internal coating, 122 cm total length, 100 cm effective length, 28 kV, vacuum injection for 2 s at 16.9 kPa, UV detection at 256 nm), carbohydrates are partially resolved. Electrophoretic mobility of carbohydrates is a result of partial ionization with elevated pH. (From Oefner et al., *Chromatographia*, 34, 310, 1992. With permission.) All time in minutes.

difference in separation is due to the size of the sphere of hydration of the alkali metal. Figure compares the separation of neutral carbohydrates at pH 12.1 and 13.1. The increase in pl

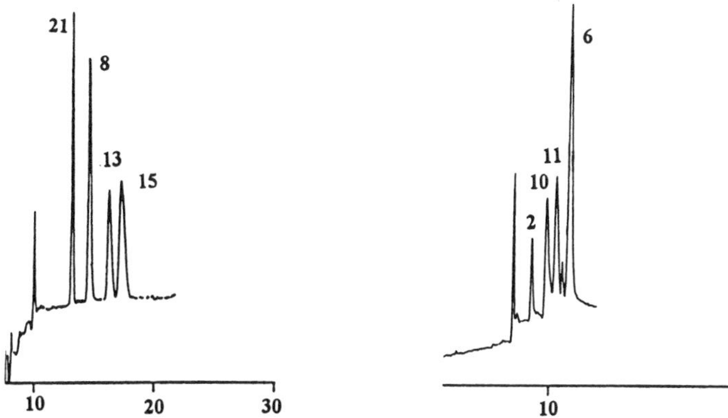

FIGURE 2. Resolution of nonderivatized carbohydrates as borate complexes. These separations were acheived at elevated temperature (60°C). On left, monosaccharides mannose (21), galactose (8), glucose (13), and xylose (15) are resolved with 60 m*M* borate buffer, pH 9.3. Disaccharides sucrose (2), cellobiose (10), maltose (11), and lactose (6) on right resolved with 50 m*M* borate buffer, pH 9.3. In both cases with polyimide outer coating, no internal coating, fused silica capillaries 75 μm I.D. × 360 μm O.D., 94 cm total length, 87 cm effective length. Injections were hydrodynamic for 1 s. Applied voltage of 20 kV. Numbers refer to carbohydrates listed in Table 4. All time in minutes. (Reprinted with permission from Hoffstetter-Kuhn et al.[42] Copyright 1991 American Chemical Society.)

leads to a significant improvement in peak resolution, especially for the early eluting, nonreducing carbohydrates sucrose, raffinose, and stachyose.

O'Shea et al.,[38] using a detection scheme similar to that of Colón et al., were able to resolve a mixture of glucosamine, glucosaminic acid, glucosamine sulfate, and glucosamine phosphate with an electrolytic solution of 10 m*M* NaOH, 8 m*M* Na_2CO_3. They were unable to resolve "neutral" carbohydrates and suggested that the large diameter of their capillaries (75-μm I.D.) vs. those of Garner and Yeung (20-μm I.D.) coupled with the high conductivity of their buffer may have led to excessive Joule heating and loss of resolution.

A number of researchers have used capillary isotachophoretic (ITP) techniques to resolve acidic carbohydrates, including carboxymethyl-D-glucoses,[39] acidic algal polysaccharides,[40] and hyaluronic acid-derived compounds.[41] In this technique, the analytes elute not as individual peaks but as adjacent bands with no intervening space. Chapter 5 of this book describes ITP at length and provides suitable examples.

2. Borate-Assisted Separations

In moderately alkaline pH solutions, borate will combine with polyols to form relatively stable adducts with reduced pK_a values, i.e., they become more anionic and are attracted to the positively charged anode. Borate-assisted CE has been used for the resolution of both free sugars and of derivatized carbohydrates. The tendency of borate to form more stable complexes with cis-oriented hydroxyl groups provides the mechanism by which glucosides (C3–C4 hydroxyls, trans oriented) can be separated from galactosides (C3–C4 hydroxyls, cis oriented). The galactoside gains a greater negative charge and is attracted more strongly to the positively charged anode and elutes later.

Hoffstetter-Kuhn et al.[42] showed that free sugars (as opposed to glycosides) elute in the opposite order (Figure 2); in a 50 m*M* borate solution (pH 9.3), galactose elutes earlier than glucose. They attribute the stronger apparent negative charge on glucose to be due to a borate complexation taking place with glucose (cis configuration of C2–C4) in the open chain form. This is supported by the observation that the extinction coefficient of glucose increases 20-fold in the presence of borate, presumably due to the increased amount of open chain aldehyde form of glucose in solution. They went on to show that efficiency, resolution, and analysis time were all improved when the analysis was carried out at elevated temperatures. At 60°C, baseline resolution of glucose and xylose was obtained in less than 20 min. At 20°C, xylose and glucose coeluted in a broad peak after 40 min. Arentoft et al.[43] resolved the low molecular weight oligosaccharides of the raffinose family (raffinose, stachyose, verbascose, and ajugose) with a 100 m*M* borate buffer, pH 9.9. Larger oligosaccharides (verbascose and ajugose) eluted later than smaller carbohydrates, indicating that electrophoretic mobility increased with increasing size.

Henshall et al.[44] were able to resolve the derivatized cyclic alditols, inositol 1-monophosphates and inositol 2-monophosphate, with two different borate buffer–cationic detergent systems in less than 10 min. They showed the utility of CE for the separation of other inositol phosphates, including inositol 1,4,5-triphosphate, as well. It was unclear whether the resolution of these analytes was due to intrinsic differences in their pK_a values and resultant differences in their electrophoretic mobilities or to the differential formation of borate complexes between the analytes and the borate in the buffer. Tait et al.[45] were able to resolve strongly charged sulfoalkyl ethers of β-cyclodextrin, with as many as seven anionic sites, by taking advantage of the native charge.

Taverna et al.[46] compared the separations of sialylated oligosaccharides and glycopeptides obtained with phosphate buffers to those obtained with borate buffers and concluded that borate complexation did not participate significantly in their resolution. Neither the phosphate nor Tricine buffers these researchers developed were able to resolve neutral oligosaccharides. They noted that while the resolution of the neutral oligosaccharides was poor with borate buffers, there was a tendency for the elution to be in the order of increasing peripheral branches. The presence of unsubstituted mannose residues led to an increase in retention times due to a greater degree of borate complexation.

A number of researchers have described the separations of free carbohydrates or glycosides as borate based. Landers et al.[47] have suggested an equilibrium dissociation constant for the formation of the borate–cytidine complex of 15 m*M* and that this is a reasonable estimate for other *cis*-diols. They noted that complete complexation of the analytes was not reached until the concentration of borate in the buffer had reached a concentration of approximately 100 m*M*. Likewise, Steiner et al.[48] showed that while some resolution could be obtained with concentrations of borate as low as 10 m*M*, complete resolution of analytes was not achieved until the concentration of borate had reached 105 m*M*. It is clear from the work of Taverna et al. that merely plotting resolution or efficiency vs. concentration of borate cannot determine if borate is acting as a complexing agent. They found that peak separation was maximized with a concentration of approximately 160 m*M* borate, pH 9.0. However, they were able to obtain comparable separation in a borate-free buffer (100 m*M* phosphate, pH 6.6) and showed that neither the peak separation factor nor the order of elution was changed, indicating that the borate, even at these high concentrations, was not acting as a complexing agent. They suggested that the lack of complexation was due to the presence of sialic acid residues at the periphery of the molecules, reducing the amount of potential complexation.

B. DERIVATIZED CARBOHYDRATES

Carbohydrates can be derivatized through a number of schemes. Most of these techniques have been developed for the RP-HPLC analysis of carbohydrates and recently adapted to CE. In many cases the derivatizing reagent lends a charge to the carbohydrate, which assists in the resolution of the carbohydrate in CE. Honda's laboratory has made extensive use of the reductive amination of carbohydrates with 2-aminopyridine developed by Hase to form 2-pyridylamines. This formed the basis of the first paper specifically addressing CE analysis of carbohydrates. Honda et al.[49] showed the resolution of 12 carbohydrates after derivatization with 2-aminopyridine. A 200 m*M* borate buffer provided optimal separation of carbohydrates at pH 10.5. As described above, the elution order of cis vs. trans pairs of carbohydrates when derivatized to glycosides is the opposite of the elution order of the free carbohydrates. While galactose elutes before glucose as the free sugars, the 2-aminopyridine derivative of galactose

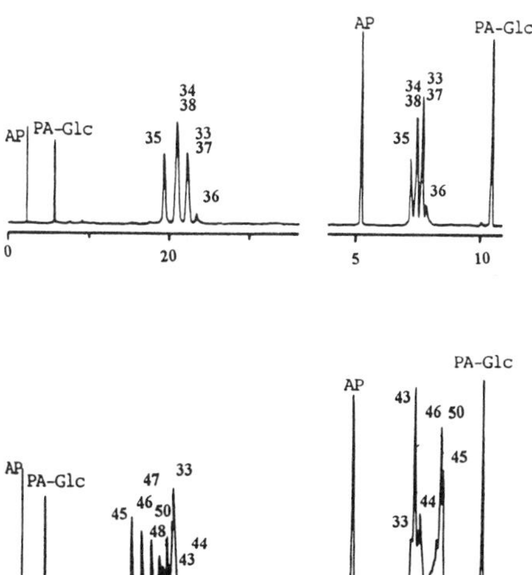

FIGURE 3. Resolution of 2-aminopyridine-derivatized oligosaccharides. *Top*: Resolution of oligosaccharides derived from human immunoglobulin G, using two different buffer systems. *Bottom*: Resolution of porcine thyroglobulin, using two different buffer systems. Numbers correspond to the structures listed in Table 4. Electropherograms on left obtained with 100 m*M* phosphate buffer with 0.1% hydroxypropylcellulose, pH 2.5, 20 kV applied voltage. Electropherograms on right show the separation of the borate complexes of the same oligosaccharide mixtures obtained with a 200 m*M* borate buffer, pH 10.5, 12 kV applied voltage. All CE done with polyimide outer coating, no internal coating, fused silica capillaries 50 μm I.D., 60 cm total length, 30 cm effective length. Fluorescence detection with excitation wavelength of 320 nm and emission of 390 nm. All injections with electromigration for 5 s at 10 kV. All time in minutes. (From Suzuki et al., *Anal. Biochem.*, 205, 232, 1992. With permission.)

elutes after that of glucose. Honda et al. were able to identify the monosaccharides released by the acid hydrolysis of various compounds, including gum arabic and digitonin.

In a second paper, Honda et al.[50] utilized the 2-aminopyridine scheme to separate, at pH 2.5, both the oligomeric series of isomaltooligosaccharides with degree of polymerization 1 through 20 as well as to resolve five classes of oligosaccharides derived from ovalbumin. The five peaks derived from ovalbumin corresponded to the oligosaccharides resolved on a Dowex 50W X-2 column and were identified as the hepta- through undecasaccharides. Using a borate buffer (200 m*M*, pH 10.5) they were able to further resolve the five peaks seen at low pH into at least nine peaks. They were able to tentatively assign structures to all of the peaks although there was some coelution.

In a third related paper, Suzuki et al.[51] extended the concept of both low-pH and borate-assisted, high-pH resolution of 2-aminopyridine-derivatized carbohydrates to develop a 2-dimensional map of 32 oligosaccharides, including complex, high-mannose, and hybrid types (Figure 3). This two-dimensional map defined three different domains, one for each specific type of glycoprotein-derived oligosaccharide.

Nashabeh and El Rassi[52] have demonstrated the potential of 2-aminopyridyl derivatives in the separation of a number of larger carbohydrates, including maltooligosaccharides, both high-mannose and complex-type oligosaccharides derived from glycoproteins, xyloglucan oligosaccharides from cotton cell walls, and *N*-acetyl-chitooligosaccharides. They used not only 2-aminopyridine for derivatization but also 6-aminoquinoline, which gave better sensitivity (see below). They found that the addition of 50 m*M* tetrabutylammonium bromide to the running buffer (100 m*M* phosphate, pH 4.75) significantly improved efficiency and resolution. Smith and El Rassi[53] have resolved 2-aminopyridine-derivatized homooligosaccharides of galacturonic acid (prepared from the partial hydrolysis of pectin) at pH 6.5 (100 m*M* phosphate), using a specially derivatized capillary. The derivatization of the wall allowed the direction and velocity of electroosmotic flow to be controlled by pH.

One limitation of 2-aminopyridine (and, presumably, 6-aminoquinoline) as a derivatization reagent is that it is nonreactive toward ketoses. Vorndran et al.[54] proposed the use of ethyl *p*-aminobenzoate as a derivatizing reagent that will couple to both aldoses and ketoses. They obtained electrophoretic mobilities using borate buffers for 26 different carbohydrates, including the ketoses sorbose and fructose. As anticipated, fructose with cis-oriented hydroxyl groups at C4–C5 had a higher mobility than sorbose, which has the trans orientation. This method was used to determine the monosaccharides found in two different plant polysaccharides and could be used to determine both neutral and acidic sugars (galacturonic and glucuronic acids) in a single analysis. Oefner et al.,[30] in the same laboratory, expanded this work with derivatizations using *p*-aminobenzoic acid as well as the ethyl ester. They found that the esterification of the acid group led to differences in selectivity when compared to the free acid. For example, glucose and mannose were resolved when derivatized with ethyl *p*-aminobenzoate but not when derivatized with the free acid. On the other hand, rhamnose, fructose, mannose, and arabinose could not be resolved after derivatization with the ester but were resolved with the free acid.

The electrophoretic mobilities (or relative migration) of both derivatized and underivatized carbohydrates are compared in Table 4. The electrophoretic mobilities of borate complexes are significantly greater than that of the carbohydrates ionized at high pH. The mobilities of pyridylamine derivatives are compared in both phosphate buffers at low pH and borate buffers at elevated pH. It is clear that high-mannose oligomers are poorly resolved in borate buffers but reasonably well resolved at low pH. Other glycoprotein oligosaccharides showed reasonable resolution with both buffer systems.

Carbohydrates reductively aminated with 8-amino-naphthalene-1,3,6-trisulfonate were separated by Jackson, using slab gel electrophoresis.[55] Chiesa and Horvath[56] reported the use of this derivatizing reagent for the resolution of maltooligosaccharides by CE. The disulfonic analog was used by Hirtzer[57] in the resolution of glycoprotein-derived oligosaccharides. The strong negative charge of this derivatizing reagent remains at all pH values, allowing positively charged impurities in the sample to be easily eliminated with a reduction in electropherogram complexity. Lee et al.[58] used 7-amino-naphthalene-1,3-disulfonic acid as the derivatizing reagent for carbohydrates. These labeled molecules were used as acceptors for $\beta(1 \rightarrow 4)$-galactosyltransferase and $\alpha(1 \rightarrow 3/4)$-fucosyltransferase. Resolution of the acceptor molecules and the enzyme products of the fucosyltransferase was achieved with a borate buffer (10 m*M* sodium borate, 50 m*M* boric acid, pH 8.8). The authors suggest that CE may be a suitable alternative to radiolabeled analyses with paper electrophoresis, HPLC, or thin-layer chromatography.

Liu et al., in a series of four publications,[59-62] used 3-(4-carboxybenzoyl)-2-quinoline carboxaldehyde (CBQCA) and 3-benzoyl-2-naphthaldehyde (BNA) in the presence of cyanide as derivatizing reagents. CBQCA and BNA react with primary amines to form fluorescent isoindole products. CBQCA-derivatized products of the amino sugars 1-aminoglucose, 1-aminogalactose, 2-aminoglucose, 2-aminogalactose, 6-aminoglucose, and galactosaminic acid could be resolved in less than 25 minutes with good resolution between the glucose and galactose analogs. They were also able to resolve the components of a partially digested sample of chitosan (polyglucosamine) after derivatization with CBQCA. To analyze carbohydrates that do not contain a primary amine, these researchers reacted the reducing terminus of the carbohydrates with ammonia in the presence of sodium cyanoborohydride to produce the 1-amino-1-deoxyalditol derivatives. These were then reacted with CBQCA to form the isoindole products. The CBQCA product of a partially hydrolyzed Dextrin 15 sample (maltooligosaccharides) could be resolved up to the 15th oligomer in an open tubular format using a borate buffer (10 m*M* Na_2HPO_4–10 m*M* $Na_2B_4O_7$, pH 9.4). The larger the oligomer the poorer the resolution and the smaller the difference between the migration of the analyte and a potential neutral marker. They found no improvement in resolution with the addition of detergents or cyclodextrin modifiers to the buffer.

TABLE 4
Mobility of Carbohydrates in Different Buffers

No.	Name[a]	Mobility b	c	d	e	RM_b[f]	RM_p[g]	h
1	Raffinose	1.312	nd[i]	nd	nd			0.13
2	Saccharose	1.356	nd	nd	nd			0.06
3	2-Deoxy-D-galactose	2.419	1.540	1.486	2.912			
4	2-Deoxy-D-ribose	2.798	1.230	1.240	2.704			
5	D-Fucose	3.096	2.210	2.138	3.393			0.36
6	Lactose	4.175	1.830	1.855	2.849			0.17
7	Maltotriose	4.313	1.310	1.363	2.178			0.12
8	D-Galactose	4.358	2.350	2.257	3.509			0.39
9	Melibiose	4.623	1.710	1.705	2.746			
10	Cellobiose	4.794	1.560	1.602	2.662			0.12
11	Maltose	4.813	1.490	1.540	2.619			0.17
12	L-Arabinose	5.121	2.000	1.913	3.274			0.40
13	D-Glucose	5.135	1.920	1.871	3.204			0.44
14	Lactulose	5.403	nd	nd	nd			
15	D-Xylose	6.268	1.740	1.664	3.159			0.47
16	D-Lyxose	6.440	1.610	1.522	3.041			
17	L-Sorbose	6.468	nd	1.731	3.121			
18	L-Rhamnose	6.599	1.570	1.490	nd			
19	D-Fructose	7.140	nd	1.898	3.218			0.37
20	D-Ribose	7.419	1.770	1.753	2.949			0.38
21	D-Mannose	7.462	1.970	1.908	3.204			0.36
22	D-Galactonic acid	24.950	nd	nd	nd			
23	D-Gluconic acid	25.518	nd	nd	nd			
24	D-Mannonic acid	25.677	nd	nd	nd			
25	D-Galacturonic acid	26.789	2.990	2.841	4.034			0.58
26	D-Arabonic acid	27.593	nd	nd	nd			
27	D-Glucuronic acid	27.796	2.820	2.626	3.989			0.61
28	D-Ribonic acid	28.139	nd	nd	nd			
29	D-Mannuronic acid	29.315	2.800	2.585	3.835			
30	TRF-9C(Di)($M_3G_2GN_4$)					0.64	0.24	
31	TRF-11C(Tri)($M_3G_3GN_5$)					0.67	0.21	
32	FET-11C(Tri)($M_3G_3GN_5$)					0.69	0.21	
33	IGG-10C(Di)($M_3G_2GN_4F_1$)					0.63	0.22	
34	IGG-9C(Di)($M_3G_1GN_4F_1$)					0.58	0.24	
35	IGG-8C(Di)($M_3GN_4F_1$)					0.54	0.26	
36	IGG-11C(Di)($M_3G_2GN_5F_1$)					0.66	0.21	
37	IGG-10C(Di)($M_3G_1GN_5F_1$)					0.62	0.22	
38	IGG-9C(Di)($M_3GN_5F_1$)					0.58	0.24	
39	AGP-12C(Tri)($M_3G_3GN_5F_1$)					0.62	0.20	
40	AGP-13C(Tetra)($M_3G_4GN_6$)					0.68	0.18	
41	AGP-14C(Tetra)($M_3G_4GN_6F_1$)					0.66	0.18	
42	AGP-15C(Tetra)($M_3G_5GN_7$)					0.77	0.16	
43	TGB-11C(Di)($M_3G_3GN_4F_1$)					0.67	0.22	
44	TGB-12C(Tri)($M_3G_3GN_5F_1$)					0.63	0.19	
45	RNB-7M(M_5GN_2)					0.81	0.32	
46	RNB-8M(M_6GN_2)					0.81	0.29	
47	RNB-9M(M_7GN_2)					0.81	0.26	
48	RNB-10M(M_8GN_2)					0.81	0.24	
49	RNB-10M(M_9GN_2)					0.81	.024	
50	RNB-11M(M_9GN_2)					0.81	0.22	
51	INV-11M(M_9GN_2)					0.80	0.23	
52	INV-12M($M_{10}GN_2$)					0.80	0.21	
53	INV-13M($M_{11}GN_2$)					0.80	0.20	
54	INV-15M($M_{13}GN_2$)					0.80	0.18	

TABLE 4 (Continued)
Mobility of Carbohydrates in Different Buffers

No.	Name[a]	Mobility						
		b	c	d	e	RMb[f]	RMp[g]	h
55	INV-16M($M_{14}GN_2$)					0.80	0.17	
56	OVA-8H(M_4GN_4)					0.56	0.29	
57	OVA-9H(M_4GN_5)					0.63	0.26	
58	OVA-9H(M_5GN_4)					0.69	0.26	
59	OVA-10H($M_4G_1GN_5$)					0.66	0.24	
60	OVA-10H(M_5GN_5)					0.73	0.24	
61	OVA-11H($M_5G_1GN_5$)					0.73	0.23	
62	Sialic acid							0.43
63	*N*-Acetylgalactosamine							0.17
64	*N*-Acetylglucosamine							0.13
65	Gentiobiose							0.29
66	Trehalose							0.06
67	Stachyose							0.16

[a] Common names are used for monosaccharides and small oligosaccharides. Glycoprotein-derived oligosaccharides terminology from Suzuki et al.[51] Three-letter designation refers to protein source: TRF, IGG, AGP, TGB, RNB, INV, and OVA refer to transferrin, immunoglobulin, α_1-acid glycoprotein, thyroglobulin, ribonuclease B, invertase, and ovalbumin, respectively. The first number refers to the total number of carbohydrates in the oligosaccharide. The designation of C, M, or H refers to complex, high mannose, or hybrid, respectively. Di, Tri, and Tetra refer to the form of antennary structure. Monosaccharide species residues are in parentheses. M, G, F, and GN refer to mannose, galactose, fucose, and *N*-acetylglucosamine, respectively. Subscripts refer to number of residues in oligosaccharide.

[b] Value × 10^{-5} $cm^2V^{-1}s^{-1}$, underivatized carbohydrates, pH 12.1; from Oefner et al.[30]

[c] Value × 10^{-4} $cm^2V^{-1}s^{-1}$, pyridylglycamines, 150 m*M* borate, pH 10.0; from Oefner et al.[30]

[d] Value × 10^{-4} $cm^2V^{-1}s^{-1}$, ethyl *p*-aminobenzoates, 175 m*M* borate, pH 10.5; from Oefner et al.[30]

[e] Value × 10^{-5} $cm^2V^{-1}s^{-1}$, *p*-aminobenzoates, 150 m*M* borate, pH 10.0; from Oefner et al.[30]

[f] Relative electrophoretic mobility, $(RM)_b = (t - t_o)/t$, where t is the migration time of the analyte and t_o is the migration time of a neutral marker, pyridylglycamines, 200 m*M* borate, pH 10.5; Suzuki et al.[51]

[g] Relative electrophoretic mobility, $(RM)_p = t_{PAGlc}/t$, where t_{PAGlc} is the migration time of the glucose derivative of 2-aminopyridine and t is the migration time of the analyte, pyridylglycamines, 200 m*M* borate, pH 10.5; Suzuki et al.[51] Value × 10^{-5} $cm^2V^{-1}s^{-1}$, pyridylglycamines, 100 m*M* phosphate, 0.1% hydroxypropylcellulose, pH 2.5; Suzuki et al.[51]

[h] Value × 10^{-3} $cm^2V^{-1}s^{-1}$, underivatized, 60 m*M* borax, 60°C; Hoffstetter-Kuhn et al.[42]

[i] nd, Not determined.

When a polyacrylamide gel-filled capillary was used for the resolution of the CBQCA-derivatized oligosaccharides, the elution pattern was, as expected, reversed with the smallest molecule, derivatized glucose, eluting first. Oligomers to a least degree of polymerization (dp) of 18 could be baseline resolved. To resolve the fragments produced by enzymatic digestion of chondroitin or hyaluronic acid, the authors found it necessary to increase the concentration of the gel from 10% T and 3% C (adequate for resolution of the maltooligomers) to ≥15% T and 3% C. They noted peak splitting in the analysis of digested chondroitin and suggested that it might be due to structural isomers of sulfation, consistent with the analysis of these compounds described earlier.

Honda et al.[63] reported the use of 3-methyl-1-phenyl-2-pyrazolin-5-one (MPP), as a derivatizing reagent for aldoses. In other papers this same reagent is designated 1-phenyl-3-methyl-5-pyrazolone, or PMP. They reported that derivatization with this reagent was both simpler and easier than derivatization with 2-aminopyridine. The product of derivatization includes two molecules of the derivatizing reagent and should lead to a higher extinction coefficient for the product than is obtained with 2-aminopyridine. The authors were able to resolve all four D-aldopentoses and all eight aldohexoses in a single run requiring less than 30 min. These separations required the use of borate buffers (200 m*M*, pH 9.5). Kakehi et al.[26] have shown that the *p*-methoxyphenyl pyrazolinone (PMPMP) derivative may undergo a reversible loss of a single molecule of PMPMP at elevated pH. Presumably, the elimination of pyrazolinone may occur with any chemical analog.

In a separate paper, Honda et al.[64] were able to resolve PMP derivatives without the use of borate by taking advantage of complexation between the pyrazolone group and divalent cations. Where monovalent sodium acetate buffers were unable to resolve a mixture of PMP-pentoses, a 20 m*M* calcium acetate buffer gave full resolution of arabinose, ribose, galactose, glucose, and mannose. The authors noted that the electroosmotic flow of the capillary decreased and eventually reversed in the presence of calcium acetate, leading to a situation where the results of the separation depended on the length of conditioning time for the capillary. For most of their analyses, the capillary was conditioned so that electroosmotic flow was from the cathode to the anode, to reverse the flow obtained with monovalent cationic solutions. Excess reagent eluted first while a neutral marker, mesityl oxide, eluted after all of the derivatized carbohydrates. In the presence of the reverse flow, this indicated that the derivatized carbohydrates had gained a negative charge due to the PMP groups. The choice of divalent cation affected the time required for the analysis, peak efficiency, and the resolution of the analytes. Calcium acetate analyses were the fastest but barium acetate produced peaks of higher efficiency. Strontium acetate provided the best resolution but analysis times were twice that observed with barium acetate. Only poor resolution of some disaccharides (maltose, cellobiose, melibiose, gentiobiose, lactose) could be obtained with this system. In a related paper, Honda et al.[65] were able to resolve the oligosaccharides produced by enzymatic digestion of glycosaminoglycans. They showed that the O-sulfate groups were stable to the derivatization conditions (0.25 *M* NaOH in methanol at 70°C). The order of elution of PMP-derivatized Δdi-4S and Δdi-6S in borate buffer is reversed compared to their elution as borate-complexed, underivatized carbohydrates. This analysis worked well on enzyme-digested urine samples and allowed the researchers to determine that the analyzed samples had no appreciable concentration of chondroitin sulfate B.

Steffansson and Westerlund[66] separated phenyl and substituted phenyl glycosides, 1-thiopyranosides, and 5-thiopyranosides at high pH in both the presence and absence of borate. While the pK_a values of glycosides are generally one pH unit higher than that of the free reducing sugars, the running buffer (100 m*M* NaOH, 50 m*M* Na_3PO_4, pH 13.9) was sufficiently alkaline to ionize some of the remaining hydroxyl groups of the glycosides to allow resolution of the analytes. They reported that the order of elution could be changed by the addition of the cationic detergent, hexadecytrimethylammonium hydroxide (CTAOH). Use of the commercially available bromide salt of CTA rather than the prepared hydroxide form led to a greater than 30% increase in migration time. The use, at high pH, of a polymer analog of CTA, polybrene ($[-(CH_2)_6N^+(CH_3)_2(CH_2)_3N^+-(CH_3)_2-]_n$), provided an elution order different from that seen with borate, cationic detergent at high pH, or high pH alone. Steffansson and Westerlund, in agreement with other researchers, reported that there were no indications of instability of the fused silica capillary or of the glycosides in elevated pH solutions.

Cyclodextrins were resolved by taking advantage of their tendency to bind aromatic compounds into the central hydrophobic core of the molecule. Nardi et al.[67] reported that α-,β-, and δ-cyclodextrin could be resolved using a buffer composed of 30 m*M* benzoic acid titrated with Tris to pH 6.2. They also showed how the degree of complexation of the benzoate with the cyclodextrin can be quantitated. This technique allowed the quick analysis of a β-cyclodextrin-containing drug formulation.

In an investigation of coordination chemistry and biodistribution of radioactive complexes, Steinmetz and Schwochau[68] showed that α-amine oximes of sugars formed stable complexes with transition metals (nickel and technetium) and could be easily resolved from side reaction materials.

III. DETECTION

Detection in capillary electrophoresis is often excellent on a mass basis but, because of the small volume of material that can be loaded, of intermediate sensitivity on a concentration basis. Table 5 lists sensitivity of carbohydrate detection methods on both a mass and a concentration basis.

A. UNDERIVATIZED CARBOHYDRATES

1. Detection by Ultraviolet Absorbance

While most carbohydrates have extremely low extinction coefficients and make poor candidates for direct detection by UV absorbance, a few carbohydrates have significant absorbance in the low UV. Al-Hakim and Linhardt[69] have shown that disaccharides enzymatically released from chondroitin sulfate and dermatan sulfate can be directly detected at 232 nm. The relatively long wavelength for and high value of λ_{max} is due to the presence of an unsaturated carboxylic acid produced by the effects of the enzyme on the polymer. In a related paper, Ampofo et al.[70] showed the resolution of disaccharides obtained from heparin and heparan sulfate. Similarly, Damm et al.[28] showed the detection at 214 nm of unsaturated uronic acid oligosaccharides derived from enzymatic digests of heparin.

Likewise, direct detection of glycosaminoglycan-derived unsaturated oligosaccharides has been reported by Carney and Osborne[26] at both 232 and 200 nm. Detection at 200 nm was especially useful for Carney and Osborne when an acidic buffer was used to resolve the components. The extremely clean electropherograms were attributed to the possibility that most UV-absorbing contaminants were positively charged under the separation conditions and migrated in the opposite direction of the negatively charged analytes (away from the detector). When alkaline conditions were used for separation, the contaminants comigrated with the analytes. Both an increase in wavelength for detection and sample pretreatment were recommended.

Motsch et al.,[29] using gel-filled capillaries for resolution, found that they were required to detect at 250 nm when a polyacrylamide gel was used but were able to detect at 232 nm with greater sensitivity when an agarose gel, which exhibited lower background absorbance, was used. Taverna et al.[46] detected both oligosaccharides and glycopeptides at 200 nm. They noted that the sensitivity increased as the number of sialic acid residues increased, presumably due to the absorbance of the amide bond found in the sialic acid.

Borate buffers exhibit very low background in the UV. Hoffstetter-Kuhn et al.[42] showed that this low background coupled with an unexpected increase in extinction coefficient for borate-complexed sugars allowed the direct determination of carbohydrates at 200 nm. The researchers suggested that the increase in absorbance is due to an increase in the concentration of open chain aldoses in solution. While there is evidence that there is a greater amount of the open chain form of reducing carbohydrates present when borate is added to the buffer, sorbitol also showed a significant increase in absorbance despite its lack of carbonyl group. Limits of sensitivity were reported in the nanogram range. Arentoft et al.[43] found that increasing the concentration of borate in the buffer significantly increased the sensitivity of analysis of oligosaccharides of the raffinose family. They found the best compromise buffer for sensitivity, speed of analysis, efficiency, and baseline stability to be 100 mM $Na_2B_4O_7$, pH 9.9.

Steinmetz and Schwochau[68] found that the α-amine oximes of reducing sugars formed colored complexes with certain transition metals (nickel, technetium) and could be easily and selectively detected at 405 nm.

Vorndran et al.[35] showed that indirect UV absorbance is a viable means of analyzing underivatized carbohydrates. With high alkalinity of analysis (pH 12), the carbohydrates become ionized. The negatively charged carbohydrate molecules displace the strongly UV-absorbing buffer ions (sorbic acid), producing inverse peaks, i.e., a reduction in absorbance rather than the normal increase in UV. These researchers were able to resolve 11 different carbohydrates in 20 min. These researchers found a significant loss of sensitivity when the pH was greater than 12.1. They found a mass limit for sensitivity of approximately 2 pmol and a concentration sensitivity of about 0.5 m*M*. In a related paper, Oefner et al.[30] optimized the concentration of UV-absorbing buffer (sorbic acid) at 6 m*M*. At lower concentrations the concentration of hydroxide ion adversely affected the transfer ratio (see Equation 1 below). At higher concentrations, the signal deteriorated because less light was able to reach the detector.

2. Detection by Indirect Fluorescence

Garner and Yeung[36] used the same ability of a negatively charged carbohydrate molecule to displace background ions from the separating buffer for detection by indirect fluorescence. They listed four essentials for the background fluorophore. First, it should have a high molar absorptivity for the excitation wavelengths available. This is very important if excitation is with a laser rather than a broad source coupled with a monochromator. Second, the quantum efficiency should be high. Third, the fluorophore must be compatible with the separation system and not react with the capillary or the analytes. Finally, the fluorophore should have a charge of 1 for optimum transfer ratio. Transfer ratio was defined as the number of fluorophore molecules displaced by a single analyte molecule. The authors began their work with fluorescein but found that the baseline became extremely unstable as the pH increased from 7 to 9. This may have been due to wall interactions. The researchers switched to Coumarin 343, a fluorophore with a high molar absorptivity at a wavelength (442 nm) available from the helium-cadmium laser. This fluorophore was also relatively stable under the highly alkaline conditions required for ionization of the carbohydrates but did show some degradation above pH 11.5. The researchers reported that working sensitivity levels for carbohydrates were in the single millimolar range. They did demonstrate high mass sensitivity by detecting fructose at the 2-fmol mass level. This was done with a 5-μm I.D. capillary and illustrated the possibility of analyzing the contents of a single cell.

Garner and Yeung describe why all indirect analyses of carbohydrates, by both fluorescence or absorbance, are limited by Equation 1. At constant concentration of fluorophore $[FL^-]$ (or chromophore) and carbohydrate, the transfer ratio, TR, is affected by the value of α, the percent of carbohydrate in the ionized form and the concentration of hydroxide in the solution. Since the percentage of ionization increases as pH increases, these affect the transfer ratio in competing directions.

$$TR = \alpha[\text{sugar}]/([FL^-] + [OH^-]) \tag{1}$$

Carbohydrates that have a strong negative charge at lower pH (e.g., uronic acids, phosphates, sulfates, etc.) can be easily detected by indirect means. At lower pH, the effect of the hydroxide ion in Equation 1 is minimized.[44,45]

3. Other Detection Techniques

Colón et al.[37] were able to detect underivatized carbohydrates by oxidizing them electrochemically. Using a 25-μm copper wire as the working electrode, these researchers were able to detect both reducing and nonreducing carbohydrates in the femtomole range. The calibration plot showed that the detection scheme was linear over three orders of magnitude. In HPLC analyses of carbohydrates, pulsed amperometry at high pH (PAD) has been used successfully with a gold electrode. Work by Baldwin has indicated that it is also possible to detect carbohydrates at constant potential if a copper electrode is used. Colón et al. adapted this to

TABLE 5
Sensitivity of Various Analytical Modes

Detection mode		Mass sensitivity[a]	Concentration sensitivity[b]	NR sugars[c]	Ketoses[d]	λ^{e}	Ref.
Absorbance[f]							
	Ethyl *p*-aminobenzoic acid	7×10^{-15}	2×10^{-6}	No	Yes	305	54
	p-Aminobenzoic acid	15×10^{-15}	4×10^{-6}	No	Yes	285	30
	2-Aminopyridine	$1–10 \times 10^{-12}$	$1–10 \times 10^{-3}$	No	No	240	49,50
	Direct as N-acetylated oligosaccharides	1×10^{-8}	2 g/l	Yes	NR[g]	190	12
	Direct as borate complexes	Nanogram range	5×10^{-2}	Yes	Yes	195	42
	Direct as unsaturated sugars	$10–20 \times 10^{-12}$	2×10^{-3} g/l	NR	NR	232	27
	Transition metal complexes	NR	NR	No	NR	405	68
	Indirect	2×10^{-12} 2×10^{-4} g/l	5×10^{-4}	Yes	Yes	254	35 30
Fluorescence							
	CBQCA	$0.5–0.3 \times 10^{-18}$	$0.24–1.1 \times 10^{-9}$	No	NR	457 (552)	61
	2-Aminopyridine	5×10^{-9}	1×10^{-4}	No	NR		50
	7-Aminonaphthalene-1,3-disulfonic acid	8×10^{-14}	NR	No	NR	250 (420)	57
	Indirect Coumarin 343	2×10^{-15}	8×10^{-10}	Yes	Yes	442	36

Electrochemical						
ADCP	5×10^{-14}	Micromolar	Yes	Yes	Constant potential copper electrode	37
PAD	9×10^{-7}	2.25×10^{-14}	Yes	Yes	Triple-pulsed gold electrode	38
HPLC-RI	2×10^{-9}	1×10^{-4}	Yes	Yes		124
HPLC-PAD	4×10^{-10}	2×10^{-6}	Yes	Yes	Triple-pulsed	124
	4×10^{-5}	2 g/l			gold electrode	12

Note: Different authors report dramatically different limits of sensitivity. This table is not exhaustive but is meant to give an expected range of sensitivity for the various methods of detection. Two common HPLC detection methods are added for comparison.

[a] Moles unless otherwise noted.
[b] Molar unless otherwise noted.
[c] Signifies whether nonreducing sugars can be detected by this technique.
[d] Signifies whether ketoses can be determined by this technique.
[e] Lambda maximum of absorbance or lambda excitation of fluorescence. Lambda emission in parentheses.
[f] Carbohydrate derivatized with reagent listed.
[g] NR, Not reported or unknown.

CE with very good results. Unlike the indirect analyses cited above, where high pH tended to decrease sensitivity due to the competing effect of the hydroxide ions on the transfer ratio, high pH actually improved sensitivity, presumably because the ionized carbohydrate is more easily oxidized. They found that the limit of detection was below 50 fmol for all of the 15 sugars studied. They did not observe any deterioration of the electrode or the capillary even after extended use. This mode of detection was also useful for simple alcohols, including ethanol.

O'Shea et al.[38] adapted pulsed amperometry with a gold electrode to CE. Using a running buffer of 10 m*M* NaOH, 8 m*M* Na_2CO_3, they were able to estimate a limit of sensitivity for glucose of 9×10^{-7} *M*, which, for a 25-nl injection, corresponded to a mass detection limit of 22.5 fmol, close to that reported by Colón et al. for a nonpulsed copper electrode. These researchers showed that this technique could be used to determine blood glucose in a sample with minimal pretreatment (dilution, centrifugation, filtration), indicating a high degree of detector selectivity.

Mass spectrometry has also been recognized as a potential but important tool in the analysis of glycoproteins and carbohydrates derived from glycoproteins.[71] For instance, Duffin et al.[72] characterized the N-linked oligosaccharides of glycoproteins with both electrospray and tandem mass spectrometry. Rudd et al.[73] used both CE and mass spectrometry in their study of ribonuclease but not in conjunction. Tsuji and Little[74] used CE-mass spectrometry for the study of recombinant proteins that were not glycosylated. Leroy et al.[22] reported the direct coupling of a capillary resolution technique, SFC, and mass spectrometry in the analysis of high-mannose oligosaccharides. It appears that all of the components are in order for the direct coupling of CE and mass spectrometry for the analysis of glycoproteins. While this technique may not be available to all laboratories, it promises the researcher an enormous amount of information concerning the structure of the glycoprotein of interest, including significant understanding of carbohydrate microheterogeneity.

When Huang et al.[75] published their description of a postcapillary conductivity detector, they did not choose to use it for the analysis of any sugar acids. However, conductivity measurements, both direct and indirect, have been made in isotachophoretic (ITP) separations. In many cases, the conductivity in ITP is monitored by a change in temperature of the buffer in the detector. Later eluting materials have lower conductivity (i.e., greater resistance to carrying an electrical current) and, at constant current, the resulting Joule heat can be measured. Nebinger[76] used this technique for the detection of oligosaccharides derived from hyaluronic acid, while Hiraoka et al.[40] found it useful in the analysis of acidic polysaccharides. Using an ITP system with a conductimetric detector, Pospîsilík[39] analyzed the carboxymethyl-D-glucoses obtained from the depolymerization of carboxymethylcellulose.

B. DERIVATIZED CARBOHYDRATES

By far the most popular way to detect carbohydrates by CE has been to derivatize the carbohydrates before analysis. The earliest CE separations with derivatized carbohydrates were done by Honda with 2-aminopyridine derivatives. The reductive amination of reducing carbohydrates was first shown by Hase et al.[24] in the analysis of carbohydrates by HPLC. The most popular carbohydrate derivatizations fall into a few categories:

1. Reductive amination of reducing sugars with arylamines (e.g., 2-aminopyridine, 4-aminobenzoate)
2. Reaction of amino sugars with 1,2-dioxo aromatic compounds to form isoindoles (e.g., CBQCA)
3. Reductive amination with ammonia to form amino alditols followed by reaction with dioxo aromatic compounds
4. Reaction of aldoses with pyrazalones (e.g., PMP)

TABLE 6
Derivatization Schemes

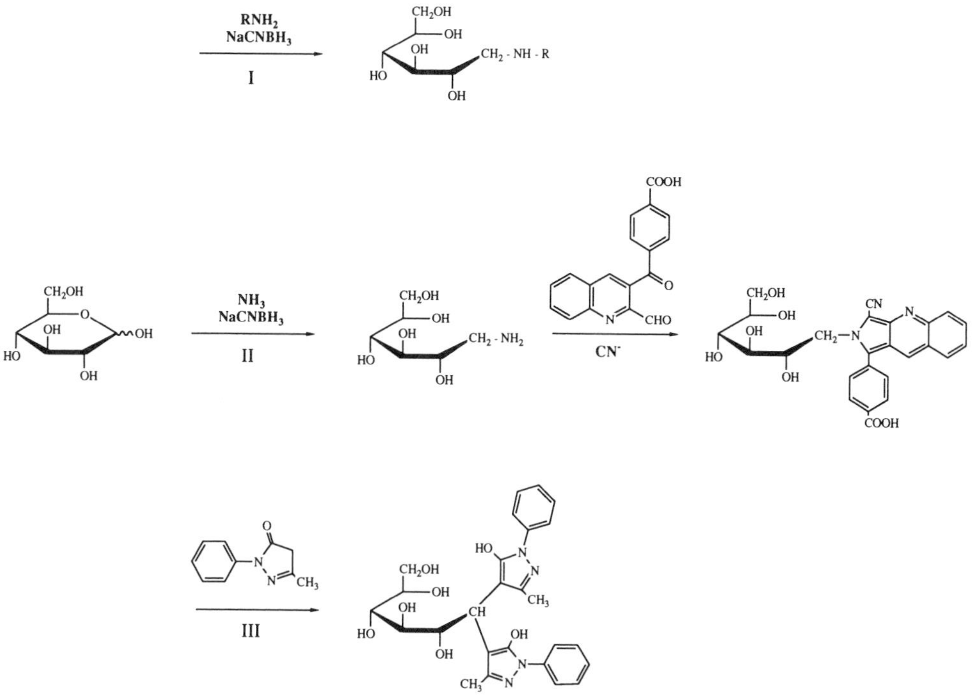

Derivatizing reagent	Scheme	Reacts with ketoses	Fluorescent product/ fluorescent reagent	Charged product pH 3/7/10
2-Aminopyridine	I	No	Yes/yes	+/+/0
4-Aminobenzoic acid	I	Yes	No/no	+/0/–
Ethyl 4-aminobenzoic acid	I	Yes	No/no	+/+/0
8-Aminonaphthalene-1, 6-disulfonic acid	I	—	Yes/yes	–/–/–
BNA	II	—	Yes/no	0/0/0
CBQCA	II	—	Yes/no	0/–/–
PMP	III	—	No/no	0/0/0
PMPMP	III	—	No/no	0/0/0

Note: Scheme I has been applied to the largest number of carbohydrates and with the greatest number of different derivatizing groups. Scheme II has been favored because, while the product is fluorescent, the reagents are not. The isoindole group may become positively charged at low pH. Scheme III uses mild conditions for derivatization. Honda et al. indicate that there is some charge on the PMP derivative at elevated pH.

The choice of derivatizing reagent is determined by the carbohydrate of interest, the available equipment for analysis, the matrix in which the analyte is found, and the simplicity of the derivatization. There are no convenient techniques available for the derivatization of nonreducing carbohydrates that contain no amine groups. This eliminates methyl glycosides, alditols (e.g., sorbitol and inositol), and such sugars as sucrose, trehalose, and raffinose. Ketoses (e.g., fructose, lactulose, sorbose) cannot be derivatized by 2-aminopyridine (see Table 6). Carbohydrates that contain multiple free primary amines may be derivatized with reagents such as CBQCA but the potential for incomplete derivatization exists as does the possibility of self-quenching if the derivatized product is analyzed by fluorescence. While

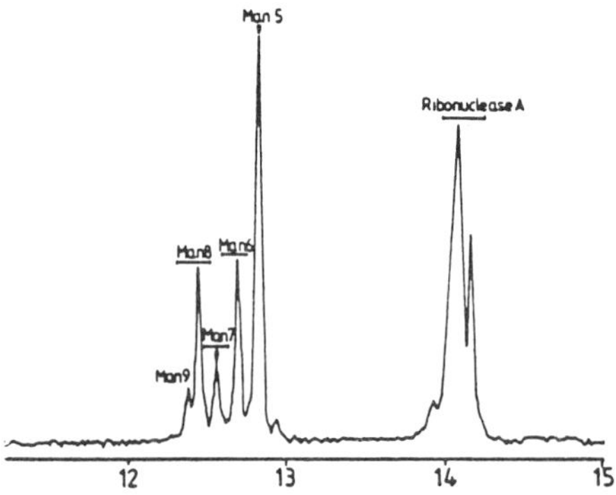

FIGURE 4. Capillary electrophoretic profile of ribonuclease A (nonglycosylated) and ribonuclease B (glycosylated), showing resolution of ribonuclease B into five glycoforms corresponding to the different amounts of D-mannose in the N-linked oligosaccharide side chain. Results obtained with a 75 mm I.D., 72 cm fused silica capillary. Voltage 1 kV for 1 min, 20 kV for 19 min, 30°C, wavelength 200 nm. Injection with vacuum for 1.5 s. Equilibrium buffer of 20 m*M* sodium phosphate, 50 m*M* sodium dodecylsulfate, 5 m*M* sodium tetraborate, pH 7.2. All time in minutes. (From Rudd et al., *Glycoconjugate J.*, 9, 86, 1992, Chapman and Hall, UK. With permission.)

fluorescence detection using lasers for excitation of the fluorescent tag can be extremely sensitive, lasers are both expensive and provide a limited choice of available wavelengths. Fluorescence detection can have monochromator-based excitation with a broad source but this tends to be significantly less sensitive than laser-induced fluorescence. There are no reports in the literature of fluorescence analysis of derivatized carbohydrates with a nonlaser system. With laser-induced detection, it is also important to reduce the possibility of photobleaching of the dye.

Table 5 catalogs the various modes of carbohydrate detection, including sensitivity, reported limits of detection by both mass and concentration, and reaction limitations inherent to each.

While there are no reports of the electrochemical detection of derivatized carbohydrates in CE, this technique has been employed in HPLC, notably with the PMP derivatives. Future reports will undoubtedly compare sensitivity of the derivatized and underivatized carbohydrates by amperometric detection.

IV. DIRECT ANALYSIS OF GLYCOCONJUGATES

A. GLYCOPROTEINS

The study of glycoproteins has expanded rapidly over the last few years. A few decades ago, carbohydrates attached to proteins were considered to be biological artifacts. Now we know that the carbohydrates conjugated to a protein or peptide may have numerous biological effects. Carbohydrates on the surface of a protein may affect the folding of the protein, its biological activity, its solubility, its antigenicity, its transport, and clearance. Because of the increased interest, there has been a concerted effort to develop techniques capable of resolving the glycoforms of proteins.

Grossman et al.[77] noted in 1989 that a highly alkaline running buffer (20 m*M* cyclohexylaminopropanesulfonic acid, pH 11.0) was able to resolve ribonuclease A, a nonglycosylated protein, from ribonuclease B_1 and B_2, two glycosylated forms of the same protein. Ribonuclease B_2 has a chitobiose core with a single mannose residue. It, like ribonuclease A, exhibited a single sharp peak. The peak identified as ribonuclease B_1 shows a great deal of heterogeneity, presumably due to partial resolution of the glycoforms that differ in the number of mannose residues.

Figure 4 illustrates how Rudd et al.,[73] using a borate–phosphate–SDS buffer at the physiologically compatible pH of 7.2, were able to resolve ribonuclease B into five components attributed to the Man_5 through Man_9 forms of the glycoprotein. Presumably, complexation between the oligosaccharide portion of the glycoprotein and the borate buffer provided the mechanism of separation. Further, they were able to run a time course of the digestion of the glycoprotein with a mannosidase and show the gradual conversion of the five-component mixture to a single glycoform, the Man_5 ribonuclease.

A number of other researchers have resolved glycoforms of various glycoproteins both with and without the aid of borate buffers. Tsuji and Little[74] demonstrated high efficiency and good resolution for a chimeric glycoprotein with a citrate–acetate buffer at pH 5.2 but did not show that the resolution was based on carbohydrate content or structure. Wu et al.,[78] using a coated capillary, were able to resolve recombinant T4 receptor protein (rCD4) from a single peak at pH 2.5 to five well-resolved peaks at pH 5.5. Treatment of the rCD4 with neuraminidase resulted in a loss of complexity of the electropherogram as the negative charge-bearing sialic acids were excised from the glycoprotein. Similarly, Tran et al.,[79] using an uncoated capillary, resolved the glycoforms of recombinant human erythropoietin into four peaks at low pH. They ran numerous optimizations to determine the best pH, ionic strength, modifiers, etc., for the analysis. They found that a buffer of 100 m*M* sodium acetate titrated to pH 4.0 with 100 m*M* phosphoric acid gave optimal results. They also showed that an unrealistic preanalysis equilibration of 10 h gave significantly better resolution but that a 30-min equilibration could be used with a phosphate buffer at pH 4.0. The poor buffering capacity of this solution led to a lack of reproducibility.

Nashabeh and El Rassi[80] used a capillary coated with a hydrophilic material characterized as a hydroxy polyether. They found that they could resolve the glycopeptides released by tryptic digestion of α_1-acid glycoprotein with a 100 m*M* phosphate buffer at pH 5.0. Essential to the resolution was the addition of 50 m*M* tetramethylammonium bromide. Glycopeptides were resolved from nonglycosylated peptides with a home-made concanavalin A chromatography column that retained the glycopeptides. They also treated the glycopeptide fragments with PNGase F, an endoglycosidase that cleaves N-linked oligosaccharides from the glycopeptides at the asparagine residue. The oligosaccharides were reductively aminated with 2-aminopyridine (see above) and run under conditions identical to those used for glycopeptides. The oligosaccharides resolved into at least six peaks without the requirement for borate. They attribute the selectivity to ion pairing between the quaternary ammonium salt and the sialic acid residues. Finally, these researchers tried to resolve the compositional monosaccharides (*N*-acetylgalactosamine, *N*-acetylglucosamine, galactose, mannose, and *N*-acetylneuraminic acid) under the same conditions but without success. To resolve these, they exploited the system developed by Honda et al.[49]

On the other hand, a number of researchers have had success with borate buffers for the resolution of glycoproteins. Steiner et al.[48] resolved a leech-derived O-linked glycopeptide from its aglycone form with a borate buffer at pH 8.3. Likewise, Watson and Yao[81] were able to resolve two forms of sialic acid containing recombinant granulocyte colony-stimulating factor from each other. They were able to show that the two forms had different electrophoretic properties than either the aglycone form of the protein expressed from *Escherichia coli* or from the asialo glycoprotein. They used 100 m*M* borate, pH 9.0, in an uncoated capillary to achieve resolution. Landers et al.[82] used a borate buffer (100 m*M* with 1 m*M* putrescine, pH 8.5) to resolve ovalbumin into 5 major peaks and 10 to 12 minor components. Digestion with phosphatase resulted in a shift in migration times without any apparent change in pattern, leading to the conclusion that the amount of phosphorylation was constant for all glycoforms. Finally, Kilár and Hjertén[83] resolved iron-free human transferrin (18 m*M* boric acid, 18 m*M* Tris, 0.3 m*M* ethylenediaminetetraacetic acid [EDTA], coated capillary) into the disialo- through hexasialoglycoforms. Treatment with neuraminidase converted almost all of the transferrin to the asialo form, which was also resolved from all of the other forms.

Kilár and Hjertén also showed the use of capillary isoelectric focusing (IEF) for the resolution of iron-free transferrin. Although resolution was slightly better, analysis time was longer than the results obtained with CZE, probably a result of the capillary coating required to eliminate the electroosmotic flow. In a second paper,[84] these researchers used capillary IEF to resolve iron-containing human transferrin. There are two different forms containing a single iron atom, a single form containing two iron atoms and a single iron-free form. Since all 4 forms can exist in 8 forms of sialylation (asialo through heptasialo), there are 32 different forms of transferrin that can be expected. They showed excellent resolution of the various forms and showed how the profile changed as the percent iron saturation increased.

B. GLYCOLIPIDS

Liu and Chan[85] reported that ganglioside micelles including G_{M1}, G_{D1b}, and G_{T1b} were resolved by CE shortly after mixing. Detection limits were as low as 24 fmol (37 pg for G_{M1}). This concentration, approximately 8 μM, is two to three orders of magnitude greater than the critical micelle concentration. Therefore, the analytes always existed as micelles. The authors showed that many ganglioside peaks would coalesce into a single peak. The time required for the formation of mixed micelles depended on the temperature of incubation and on the presence of calcium in the incubation mixture (EGTA, a calcium chelator, decreased the time required for mixed micelle formation). Some monosialylated gangliosides never formed mixed micelles (e.g., G_{M1} and G_{M2}), while polysialylated gangliosides fused rapidly.

The authors were unsuccessful in their attempts to resolve the gangliosides as free molecules rather than micelles. They tried adding tetrahydrofuran to the buffer and found that it caused irreversible damage to the uncoated silica capillary. The addition of methanol–chloroform to the running buffer did not eliminate the formation of micelles nor did the addition of methanol (5 to 90%), propanol (10 to 50%), acetonitrile (10 to 50%), dimethylsulfoxide (50%), dimethylformamide (50%), tetramethylammonium chloride (10 to 30 m*M*), hydroxylamine (2.5 to 5.0 m*M*), or sodium dodecyl sulfate (10 to 30 m*M*). However, the benefits of CE for ganglioside studies are significant. Analysis is rapid (less than 10 min), mass sensitivity is excellent (10^4 to 10^5 times more sensitive than resorcinol–hydrochloric acid assay), and the micelles can be studied under physiological pH conditions.

Capillary electrophoresis presents the possibility of determining whether the formation of ganglioside mixed micelles is a random process. Capillary electrophoresis also can be used to study ganglioside–phospholipid and ganglioside–protein interactions.

C. OTHER GLYCOCONJUGATES

Carbohydrates are found as glycosides in a vast number of secondary plant products. Danish researchers, in a series of papers,[86-88] used CE to resolve a number of glucosinolates and desulfoglucosinolates, detecting them at 235 nm. These compounds, found in rapeseed oil and members of the mustard family (including cabbage and kale), are involved in off-flavors, can be antinutritive, and have toxic effects. They are usually analyzed by HPLC. The HPLC method is prone to problems that arise from the naturally occurring nitrate in the samples. Capillary electrophoresis eliminates this problem and can be used with crude samples, although the authors suggest that some pretreatment be retained. Using a buffer with cetyltrimethylammonium bromide (CTAB) and an uncoated fused silica capillary, the authors were able to resolve a number of neutral desulfoglucosinolates as well as negatively charged *N*-sulfate containing glucosinolates in the presence of a number of naturally occurring phenolic acids. All glucosinolates eluted after the desulfoglucosinolates but retained the same elution pattern as the desulfo analogs. They showed that both CTAB and the ionic strength of the buffer had a marked negative effect on detection sensitivity, especially on the quantitation of glucotropaeolin and gluconasturtlin. There was a marked positive effect for these two analytes as the pH of the running buffer was increased.

French researchers, in two papers, used CE for the analysis of flavonoid glycosides. Flavonoids are one of the largest groups of secondary plant metabolites. They account for many of the yellow, orange, red, and blue colors in plants. Flavonoids may differ in the structure of the phenolic portion of the molecule (the aglycone) or in the number, position, and type of carbohydrates bound to the aglycone. In the first paper, Morin et al.[90] show that flavonoids that have an identical carbohydrate, rutinose (6-*O*-α-rhamnosyl-D-glucose), attached at the 7-position of the aglycone could be resolved with a borate buffer (200 m*M* borate, pH 10.5). With a phosphate buffer at the same pH, full resolution could not be obtained. Identity of each flavonoid was determined by recording its spectrum on-line with a multiple-wavelength detector. This study also showed that the aglycones were resolved from one another as well as from the glycosides with an SDS-based MECC separation.

Five different 3-*O*-glycosides of quercetin (the arabinoside, glucoside, galactoside, rhamnoside, and arabinoglucoside) were resolved by Morin et al.[91] They found optimum analysis with 200 m*M* borate at pH 10.5. They were able to rationalize the order of elution based on the complexation of borate with *cis*-diols of the carbohydrates.

A number of different flavonoid glycosides were resolved under MECC conditions by Pietta et al.[92] They, too, resolved the 3-*O*-glycosides of quercetin but with much lower borate concentration (20 m*M*, pH 8.3) and with a significant concentration of detergent (50 m*M* SDS). Using this procedure, the researchers were able to resolve the naturally occurring flavonoids from *Ginko biloba* with resolution comparable to that seen with RP-HPLC and requiring only half the time for analysis.

Brunner et al.[93] used an isotachophoretic separation with UV detection at 254 nm to resolve the glucuronide of enterodiol, one of the first identified mammalian lignans, from other products of a synthesizing enzyme mixture. They reported that mammalian lignans have characteristics similar to steroids in many chromatographic systems but can be clearly separated by isotachophoresis. Seitz et al.[94] used isotachophoresis to resolve plant flavonoids with a leading electrolyte of 15 m*M* HCl–ammediol, 0.2% hydroxypropylmethylcellulose, 30% methanol, and a terminating electrolyte of 10 m*M* glycine–barium hydroxide, 0.2% hydroxypropylmethylcellulose, 30% methanol.

V. USE OF CARBOHYDRATES IN ANALYSIS OF OTHER COMPOUNDS

A. BINDING STUDIES

Carbohydrates may function not only as analytes for CE but as part of the resolving system. Heegaard and Robey[95] analyzed the binding of peptides derived from the heparin-binding domain of human serum amyloid P component. Using mannose-1-phosphate in the buffer (1.5 m*M* NaCl, 0.05 m*M* Tris-HCl, 3.5 m*M* mannose phosphate, pH 7.4), they were able to show different binding affinities for the different peptides.

Honda et al.[96] determined the association constants of agglutins from peanut, soy bean, and *Ricinus communis* with the carbohydrate lactobionic acid (β-galactosylgluconic acid) as a component of the electrophoretic buffer. Increasing the concentration of the lactobioic acid retarded the migration of the lectins. For *R. communis* agglutin, saturation was achieved at approximately 5 m*M* lactobionic acid at pH 6.8. The authors point out that the method requires only very small amounts of either the protein or the binding carbohydrate and has better reproducibility than standard techniques. They note that while the carbohydrate used was acidic, in principle the technique could be expanded to analyses with neutral carbohydrates, although the accuracy may not be as good due to a smaller change in migration time.

B. CHIRAL RESOLUTION

Carbohydrates are intrinsically optically active compounds. Addition of these molecules to the buffer should allow resolution based, at least in part, on the chirality of the analytes. As discussed in Chapter 10, the cyclodextrins (CDs), ring-shaped oligosaccharides composed of six, seven, or eight (α, β, and γ, respectively) glucose molecules linked $\alpha(1 \rightarrow 4)$, have been used extensively in HPLC for the resolution of optically active components.[97-105] Snopek et al.[106] have reviewed the use of various cyclodextrins in CE, but since that publication many other papers have been published showing their utility. Soini et al.[107] used modified cyclodextrins in the analysis of optically active drugs in serum samples. They examined a number of parameters of the system, including different chemically modified cyclodextrins, cationic detergents, methylhydroxyethylcellulose, methanol, ethylene glycol, heptanesulfonic acid, and polyacrylamide-coated capillaries. They suggest that the use of micellar agents may decrease the need for chemically modified cyclodextrins. This, in turn with the suppression of biological interference, may make CD-based CE a feasible technique for quantifying drugs at their therapeutic levels. Yamashoji et al.[108] showed that minor changes in the CD structures could invert the elution order of analytes. Yik et al.[109] showed that with CD modifiers in the running electrolyte, compounds that would normally seem inappropriate for analysis by CE (e.g., polyaromatic hydrocarbons) can be quickly and fully resolved.

Cruzado and Vigh[110] reported the use of gel-filled capillaries where the gel was a copolymer of acrylamide, bis-acrylamide, and allyl carbamoylated β-cyclodextrin. This gel allowed for the resolution of dansylated amino acid enantiomers, as well as the enantiomers of homatropine, but was not able to resolve the enantiomers of atropine. The authors suggested that further chemical modifications of the CD molecule might be required. With a similar approach, Guttman et al.[111] used noncovalently bound cyclodextrin in polyacrylamide gels to resolve optically active components.

Cyclodextrins have received a great deal of attention for their ability to resolve enantiomers. A paper by D'Hulst and Verbeke[112] indicates that much simpler and much less expensive compounds, maltodextrins, may be used in CE to resolve some racemic mixtures. Maltodextrins are prepared by the partial acid or enzyme hydrolysis of corn starch. Using three 2-arylpropionic acid, nonsteroidal antiinflammatory drugs (flurbiprofen, ibuprofen, and ketoprofen) as test compounds, they found that resolution of the racemic pairs was improved as the concentration of the oligosaccharides in the buffer increased or as the DE* value of the oligosaccharide decreased. Interestingly, oligosaccharides that had an odd number of glucose residues tended to perform poorer than those that contained an even number. Also, they noted that the isomaltooligosaccharides were unable to resolve any of the enantiomeric pairs. Various surfactants tended to have deleterious effects on resolution.

Ishihama and Terabe[113] reported the use of saponins for the enantiomeric resolution of dansylated and phenylthiohydantoin (PTH) DL-amino acids. The saponins used, glycyrrhizic acid and β-escin, are both triterpene glycosides. These naturally occurring surfactants are optically active in both their hydrophobic and hydrophilic portions. Mixed micelles of 30 m*M* glycyrrhizic acid, 50 m*M* octyl glucoside, 10 m*M* sodium dodecyl sulfate, 20 m*M* borate, 10 m*M* phosphate, pH 7.0 successfully resolved a number of dansylated amino acid enantiomeric pairs. A mixed micelle of β-escin and SDS was able to effect the enantiomeric separation of PTH DL-amino acids. While resolution was often very good, they report that reproducibility was occasionally poor and that the saponin-based buffers had a tendency to form gels.

* DE, or dextrose equivalent, is defined as the percent reducing sugars calculated as glucose on a dry-substance basis.

C. NONCHIRAL USES

Kaneta et al.[114] reported that the addition of glucose to the running buffer in the MECC separation of nucleosides extended the elution range and improved selectivity. In a buffer containing only SDS and phosphate, it was not possible to resolve the cytidine–deoxyuridine pair, or the adenosine–deoxyadenosine pair. While the addition of 10% methanol to the buffer system would bring about the resolution of adenosine and deoxyadenosine, the other pair remained unresolved. With a buffer of 1.0 *M* glucose, 150 m*M* SDS, both pairs (as well as seven other nucleosides) were resolved and the elution order of the adenosine–deoxyadenosine pair was reversed compared to the separation achieved with the added methanol. All analyses were run with a constant current of 35 μA. Micellar electrokinetic capillary chromatography buffer without glucose exhibited a voltage of 12.2 kV; with glucose the voltage was 17.8 kV. While it was apparent that the addition of glucose allowed higher voltages to be run without excessive current, the mechanism of the improved resolution was not clear.

Bruin et al.[115] derivatized the internal walls of capillaries with maltose in order to reduce the binding of proteins to the fused silica. The maltose-modified wall appeared to reduce protein binding up to pH 7 as opposed to a diol coating, also described in the article, which was stable only to pH 5. While the efficiency was good, it was not as high as the diol coating. Antimicrobial agents had to be added to the buffers to maintain stability of the maltose coating.

A number of papers have reported the use of carbohydrate polymers as sieving material for the resolution of proteins and nucleic acids.[116-122] Guttman et al.[123] compared the effects of temperature on the resolution of proteins in capillaries filled with either polyethylene oxide or dextran (an $\alpha(1 \rightarrow 6)$-glucose polymer) based buffer systems. In both cases, the time required for analysis decreased as the temperature increased. However, with polyethylene oxide, the resolution deteriorated as the temperature increased from 20 to 40°C and the efficiency decreased sixfold. In contrast, as the temperature increased, the efficiency of the dextran-based system more than tripled over the same temperature range. They attribute the improvement in the dextran system to the formation of channel-like structures due to polymer–polymer interactions and the orientation force of the applied electric field. The effect of temperature on other carbohydrate sieving buffers has not been reported.

VI. CONCLUSIONS

In the CE analysis of carbohydrates the researcher must first answer a number of important questions: what carbohydrates are of interest, what are their concentrations, how much sample is available, what is the matrix in which they are found, are they reducing carbohydrates (if so, are they aldoses or ketoses), do they have free primary amine groups, are they acidic, are there time or equipment limitations on the analysis?

Free carbohydrates may be resolved by their intrinsic charge at moderate pH if they have acidic functionalities (e.g., glycosaminoglycan breakdown products, sugar acids, and phosphates, etc.). They may be resolved under moderately alkaline conditions as their borate complexes. Some separations, reportedly of borate complexes but with very low concentrations of borate in the electrophoretic buffer, may actually be resolved because of intrinsic charge or other, non-borate-related differences. They can be resolved at high pH (~12) near the pK_a values of neutral carbohydrates. Isotachophoretic techniques may be used to resolve acidic carbohydrates as well as carbohydrates complexed with borate.

Carbohydrates that exist as glycosides, either in their original condition (e.g., glycopeptides, secondary plant metabolites, etc.) or after derivatization in the laboratory, may be resolved by intrinsic charge of either the carbohydrate or the aglycone portion of the analyte.

They have been well resolved with borate-assisted CE. At very high pH, following the ionization of nonanomeric hydroxyl groups, they may also be resolved.

While the molar absorbtivity of most carbohydrates is very low, they may be detected directly especially if they are unsaturated (e.g., enzymatic products of chondroitinase). The presence of N-acetyl groups in a mono- or oligosaccharide significantly increases the direct detection sensitivity. Surprisingly, borate complexes of carbohydrates absorb significantly more strongly than uncomplexed sugars.

Indirect detection of carbohydrates, both fluorescence and absorbance based, is a viable option but sensitivity decreases at high pH due to the effect of hydroxide ion on the transfer ratio. Acidic, but nonabsorbing carbohydrates (e.g., sugar acid and phosphates) are good candidates for this technique. This technique is one of the few generally available for the analysis of nonreducing carbohydrates (e.g., alditols, glycosides, etc.).

Conductimetric detection is necessarily limited to charged molecules but has been used successfully in ITP separations. New advances in amperometric detection offer significant possibilities but there are no commercially available instruments for this technique. Mass spectrometry, while of limited availability, offers significant advantages, including absolute identification of carbohydrates.

The detection of both naturally occurring glycosides, and those synthesized to facilitate both separation and detection, is determined by the aglycone. All fluorescent tags may also be used with absorbance detection while the reverse is not true. While aldoses are reactive to many derivatizing groups, ketoses may be far less reactive. Fluorogenic labeling groups (e.g., CBQCA) are generally preferable to fluorescent labeling groups for high-sensitivity analyses because there is little or no background fluorescence. On the other hand, these reagents require the presence of a primary amine in the carbohydrate. Reductive amination with ammonia to produce an aminoalditol is an added complication and requires extra analysis time. When employing laser-based detection, it is essential that the absorbance of the labeling group, under conditions employed for the separation, match the emission wavelength of the laser.

Carbohydrates themselves can be used in CE to effect chiral resolution of enantiomers, prevent or reduce binding of proteins to the capillary wall, and may act as sieving buffers for the size analysis of macromolecules. In general, the advance of CE technologies significantly increases the opportunities for the carbohydrate researcher.

VII. FUTURE DIRECTIONS

The field of carbohydrate analysis is rapidly expanding on all fronts. As the biological importance of carbohydrates becomes more clear and as technology advances to provide tools for the analysis of these difficult compounds, new researchers with new ideas are entering the field. The most significant anticipated advances in technology are the development of new detectors to improve the sensitivity or selectivity of those currently available. Absorbance-based detection improvements, based on increased path length, have been developed for a number of commercially available instruments and are expected on others. The pioneering work of Cólon et al.[29] and O'Shea et al.[30] holds out the promise that derivatization of carbohydrates may become a drudgery of the past and eliminates the questions concerning the analyses of ketoses or nonreducing carbohydrates. Radioisotope detection, especially important in metabolism studies, may become available to the researcher in the near future. But by far the most interesting detector is the direct coupling of CE to mass spectrometry. Mass spectrometry detection offers the possibility of absolute determination of complex carbohydrate structure, using only microquantities of starting material.

Capillary electrophoresis offers more than improvements in quantitative and qualitative analyses of carbohydrates. Because physiological buffers can be used as the resolving buffers, capillary electrophoresis offers the possibility of following the reaction rates of glycosylation, carbohydrate degradation, cell metabolism, and micelle formation under conditions approaching those found *in vivo* and under conditions that constantly remove product while maintaining a constant concentration of substrate. It offers a direct means for determination of binding constants of carbohydrates to carbohydrate-binding proteins. Interactions with hydrophobic surfaces or strongly absorbing materials, as in RP-HPLC or normal-phase HPLC, can be eliminated. Extraordinarily small samples, including the contents of single cells, may be analyzed. The inherent simplicity of capillary electrophoresis suggests its use in clinical or repetitive analysis laboratories as specially designed, dedicated equipment for the analysis of materials ranging from foodstuffs to biological matrices, including blood and urine. Dedicated instrumentation may be able to quickly and easily inform the biotechnology quality control manager that the glycosylation pattern of a recombinant glycoprotein is changing, alert the neonate physician to abnormal disorder-related carbohydrates in the urine of newborns, signal the food analyst of adulterants in honey or fruit juices, or track the degradation of wood pulp in the paper industry.

The expansion of CE technology couples well with the concurrent increased need for information in carbohydrate analyses. The dedicated commitment of researchers in both fields will have a synergistic effect that may make CE the method of choice in carbohydrate analyses.

ACKNOWLEDGMENTS

The authors would like to thank Drs. J. Nolan, M. Hardy, J. Rohrer, Ms. P. Hirtzer and the librarians of the Beckman Research Library for their help, information and suggestions for this chapter.

REFERENCES

1. **Pigman, W. and Horton, D.,** *The Carbohydrates*, Vols. IIA and IIB, 2nd ed., Academic Press, London and New York, 1970, p. 375 (Vol. IIA) and p. 471 (Vol. IIB).
2. **Chaplin, M. F. and Kennedy, J. F.,** *Carbohydrate Analysis: A Practical Approach*, IRL Press, Oxford, 1975.
3. **Harrison, R. and Lunt, G. G.,** *Biological Membranes*, Halsted Press, New York, 1975, pp. 59–90.
4. **Takemoto, N., Hase, S., and Ikenaka, T.,** Microquantitative analysis of neutral and amino sugars as fluorescent pyridylamino derivatives by high-performance liquid chromatography, *Anal. Biochem.*, 145, 245, 1985.
5. **Jentoft, N.,** Analysis of sugars in glycoproteins by high-pressure liquid chromatography, *Anal. Biochem.*, 148, 424, 1985.
6. **Tomiya, N., Kurono, M., Ishihara, H., Tejima, S., Endo, S., Arata, Y., and Takahashi, N.,** Structural analysis of N-linked oligosaccharides by a combination of glycopeptidase, exoglycosidases, and high-performance liquid chromatography, *Anal. Biochem.*, 163, 489, 1987.
7. **Kakehi, K. and Honda, S.,** Analysis of carbohydrates in glycoproteins by high-performance liquid chromatography and high-performance capillary electrophoresis, in *Methods in Molecular Biology*, Vol. 14, *Glycoprotein Analysis in Biomedicine*, Hounsel, E. F., Ed., Humana Press, Totowa, NJ, 1993.
8. **Basa, L. J. and Spellman, M. W.,** Analysis of glycoprotein-derived oligosaccharides by high-pH anion-exchange chromatography, *J. Chromatogr.*, 499, 205, 1990.
9. **Hardy, M. R., Townsend, R. R., and Lee, Y. C.,** Monosaccharide analysis of glycoconjugates by anion exchange chromatography with pulsed amperometric detection, *Anal. Biochem.*, 170, 54, 1988.
10. **Townsend, R. R., Hardy, M. R., Cumming, D. A., Carver, J. P., and Bendiak, B.,** Separation of branched sialylated oligosaccharides using high-pH anion-exchange chromatography with pulsed amperometric detection, *Anal. Biochem.*, 182, 1, 1989.
11. **Anumula, K. R. and Taylor, P. B.,** Rapid characterization of asparagine-linked oligosaccharides isolated from glycoproteins using a carbohydrate analyzer, *Eur. J. Biochem.*, 195, 269, 1991.

12. **Hermentin, P., Witzel, R., Doenges, R., Bauer, R., Haupt, H., Patel, T., Parekh, R. B., and Brazel, D.,** The mapping by high-pH anion-exchange chromatography with pulsed amperometric detection and capillary electrophoresis of the carbohydrate moieties of human plasma α_1-acid glycoprotein, *Anal. Biochem.*, 206, 419, 1992.
13. **Paskach, T. J., Lieker, H.-P., Reilly, P. J., and Thielecke, K.,** High-performance anion-exchange chromatography of sugars and sugar alcohols on quaternary ammonium resins under alkaline conditions, *Carbohydr. Res.*, 215, 1, 1991.
14. **Rohrer, J. S. and Olechno, J. D.,** Secondary isotope effect: the resolution of deuterated glucoses by anion-exchange chromatography, *Anal. Chem.*, 64, 914, 1992.
15. **Lee, D. P. and Bunker, M. T.,** Carbohydrate analysis by ion chromatography, *J. Chromatogr. Sci.*, 27, 496, 1989.
16. **Hermentin, P., Witzel, R., Vliegenthart, J. F. G., Kamerling, J. P., Nimtz, M., and Conradt, H. S.,** A strategy for the mapping of N-glycans by high-pH anion-exchange chromatography with pulsed amperometric detection, *Anal. Biochem.*, 203, 281, 1992.
17. **Chaplin, M. F.,** A rapid and sensitive method for the analysis of carbohydrate components in glycoproteins using gas-liquid chromatography, *Anal. Biochem.*, 123, 336, 1982.
18. **Elwood, P. C., Reid, W. K., Marcell, P. D., Allen, R. H., and Kolhouse, J. F.,** Determination of the carbohydrate composition of mammalian glycoproteins by capillary gas chromatography mass spectrometry, *Anal. Biochem.*, 175, 202, 1988.
19. **Sheeley, D. M. and Reinhold, V. N.,** Characterization of N-linked glycans by supercritical fluid chromatography-mass spectrometry, *Anal. Biochem.*, 193, 240, 1991.
20. **Leroy, Y., Lemoine, J., Ricart, G., Michalski, J.-C., Montreuil, J., and Fournet, B.,** Separation of oligosaccharides by capillary supercritical fluid chromatography and analysis by direct coupling to high resolution mass spectrometer: application to analysis of oligomannosidic N-glycans, *Anal. Biochem.*, 184, 235, 1990.
21. **Reinhold, V. N., Sheeley, D. M., Kuei, J., and Her, G.-R.,** Analysis of high molecular weight samples on a double-focussing magnetic sector instrument by supercritical fluid chromatography/mass spectrometry, *Anal. Chem.*, 60, 2719, 1988.
22. **Chester, T. L. and Innis, D. P.,** Separation of oligo- and polysaccharides by capillary supercritical fluid chromatography, *J. High Resolut. Chromatogr. Chromatogr. Commun.*, 9, 209, 1986.
23. **Suzuki, J.,** Methods for analysis of component sugars by fluorescence labeling, *Trends Glycosci. Glycobiol.*, 3, 48, 1991.
24. **Hase, S., Hara, S., and Matsushima, Y.,** Tagging of sugars with a fluorescent compound, 2-aminopyridine, *J. Biochem.*, 85, 217, 1979.
25. **Honda, S., Akao, E., Suzuki, S., Okuda, M., Kakehi, K., and Nakamura, J.,** High-performance liquid chromatography of reducing carbohydrates as strongly ultraviolet-absorbing and electrochemically sensitive 1-phenyl-3-methyl-5-pyrazolone derivatives, *Anal. Biochem.*, 180, 351, 1989.
26. **Kakehi, K., Suzuki, S., Honda, S., and Lee, Y. C.,** Precolumn labeling of reducing carbohydrates with 1-(*p*-methoxy)phenyl-3-methyl-5-pyrazolone: analysis of neutral and sialic acid-containing oligosaccharides found in glycoproteins, *Anal. Biochem.*, 199, 256, 1991.
27. **Carney, S. L. and Osborne, D. J.,** The separation of chondroitin sulfate disaccharides and hyaluronan oligosaccharides by capillary electrophoresis, *Anal. Biochem.*, 195, 132, 1991.
28. **Damm, J. B. L., Overklift, G. T., Vermeulen, B. W. M., Fluitsma, C. F., and van Dedem, G. W. K.,** Separation of natural and synthetic heparin fragments by high-performance capillary electrophoresis, *J. Chromatogr.*, 608, 297, 1992.
29. **Motsch, S. R., Kleemiß, M.-H., and Schomberg, G.,** Production and application of capillaries filled with agarose gel for electrophoresis, *J. High Resolut. Chromatogr.*, 14, 629, 1991.
30. **Oefner, P. J., Vorndran, A. E., Grill, E., Huber, C., and Bonn, G. K.,** Capillary zone electrophoretic analysis of carbohydrates by direct and indirect UV detection, *Chromatographia*, 34(5-8), 308, 1992.
31. **Rendleman, J. A., Jr.,** Carbohydrates in solution, in *Advances in Chemistry Series*, Vol. 117, American Chemical Society, Washington, D.C., 1973, pp. 51–69.
32. **Lowbry de Bruyn, M. M. C. A. and van Ekenstein, W. A.,** Action des alcalis sur les sucres. II. Transformation réciproque des uns dans les autres des sucres glucose, fructose et mannose, *Rec. Trav. Chim.*, 14, 203, 1895.
33. **Lowbry de Bruyn, M. M. C. A. and van Ekenstein, W. A.,** Action des alcalis sur les sucres. V. Transformation de la galactose. Les tagatoses, et la galtose, *Rec. Trav. Chim.*, 16, 262, 1897.
34. **Hardy, M. (Immunogen),** personal communication, 1993.
35. **Vorndran, A. E., Oefner, P. J., Scherz, H., and Bonn, G. K.,** Indirect UV detection of carbohydrates in capillary zone electrophoresis, *Chromatographia*, 33(3/4), 163, 1992.
36. **Garner, T. W. and Yeung, E. S.,** Indirect fluorescence detection of sugars separated by capillary zone electrophoresis with visible laser excitation, *J. Chromatogr.*, 515, 639, 1990.

37. **Colón, L. A., Dadoo, R., and Zare, R. N.,** Determination of carbohydrates by capillary zone electrophoresis with amperometric detection at a copper microelectrode, *Anal. Chem.*, 65, 476, 1993.
38. **O'Shea, T. J., Lunte, S. M., and LaCourse, W. R.,** Detection of carbohydrates by capillary electrophoresis with pulsed amperometric detection, *Anal. Chem.*, 65, 948, 1993.
39. **Pospîsilík, K.,** Determination of carboxymethyl-*d*-glucoses in the product of hydrolytic depolymerization of carboxymethylcellulose by isotachophoresis, *Electrophoresis*, 10, 20, 1989.
40. **Hiraoka, A., Harada, N., Uehara, T., Sekiguchi, M., and Maeda, M.,** Capillary-isotachophoretic analyses of algal acidic polysaccharides and their applications to a survey of heparinoid active sulfated polysaccharides in Chlorophyta, *Chem. Pharm. Bull.*, 40, 783, 1992.
41. **Yamamoto, S., Ohta, T., and Morikawa, Y.,** Electrophoretic behavior of sodium hyaluronate by capillary tube isotachophoresis, *Bunseki Kagaku,* 31, 557, 1982.
42. **Hoffstetter-Kuhn, S., Paulus, A., Gassmann, E., and Widmer, H. M.,** Influence of borate complexation on the electrophoretic behavior of carbohydrates in capillary electrophoresis, *Anal. Chem.*, 63, 1541, 1991.
43. **Arentoft, A. M., Michaelsen, S., and Sorensen, H.,** Quantitative determination of oligosaccharides by capillary zone electrophoresis, in press.
44. **Henshall, A., Harrold, M. P., and Tso, J. M. Y.,** Separation of inositol phosphates by capillary electrophoresis, *J. Chromatogr.*, 608, 413, 1992.
45. **Tait, R. J., Skanchy, D. J., Thompson, D. P., Chetwyn, N. C., Dunshee, D. A., Rajewski, R. A., Stella, V. J., and Stobaugh, J. F.,** Characterization of sulfoalkyl ether derivatives on β-cyclodextrin by capillary electrophoresis with indirect UV detection. *J. Pharm. Biomed. Anal.* 10, 615, 1992.
46. **Taverna, M., Baillet, A., Biou, D., Schlüter, Werner, R., and Ferrier, D.,** Analysis of carbohydrate-mediated heterogeneity and characterization of N-linked oligosaccharides of glycoproteins by high performance capillary electrophoresis, *Electrophoresis*, 13, 359, 1992.
47. **Landers, J. P., Oda, R. P., and Schuchard, M. D.,** Separation of boron-complexed diol compounds using high-performance capillary electrophoresis, *Anal. Chem.*, 64, 2846, 1992.
48. **Steiner, V., Knecht, R., Börnsen, O., Gassmann, E., Stone, S. R., Raschdorf, F., Schlaeppi, J.-M., and Maschler, R.,** Primary structure and function of novel O-glycosylated hirudins from the leech *Hirudinaria manillensis*, *Biochemistry*, 31, 2294, 1992.
49. **Honda, S., Iwase, S., Makino, A., and Fujiwara, S.,** Simultaneous determination of reducing monosaccharides by capillary zone electrophoresis as the borate complexes of N-2-pyridylglycamines, *Anal. Biochem.*, 176, 72, 1989.
50. **Honda, S., Makino, A., Suzuki, S., and Kakehi, K.,** Analysis of the oligosaccharides in ovalbumin by high-performance capillary electrophoresis, *Anal. Biochem.*, 191, 228, 1990.
51. **Suzuki, S., Kakehi, K., and Honda, S.,** Two-dimensional mapping of N-glycosidically linked asialo-oligosaccharides from glycoproteins as reductively pyridylaminated derivatives using dual separation modes of high-performance capillary electrophoresis, *Anal. Biochem.*, 205, 227, 1992.
52. **Nashabeh, W. and El Rassi, Z.,** Capillary zone electrophoresis of pyridylamino derivatives of maltooligosaccharides, *J. Chromatogr.*, 514, 57, 1990.
53. **Smith, J. T. and El Rassi, Z.,** Capillary zone electrophoresis of biological substances with surface-modified fused silica capillaries with switchable electroosmotic flow, *J. High Resolut. Chromatogr.*, 15, 573, 1992.
54. **Vorndran, A. E., Grill, E., Huber, C., Oefner, P. J., and Bonn, G. K.,** Capillary zone electrophoresis of aldoses, ketoses and uronic acids derivatized with ethyl *p*-aminobenzoate, *Chromatographia*, 34(3/4), 109, 1992.
55. **Jackson, P.,** The use of polyacrylamide-gel electrophoresis for the high-resolution separation of reducing saccharides labeled with the fluorophore 8-aminonaphthalene-1,3,6-trisulfonic acid, *Biochem. J.*, 270, 705, 1990.
56. **Chiesa, C. and Horváth, C.,** Capillary zone electrophoresis of malto-oligosaccharides derivatized with 8-aminonaphthalene-1,3,6-trisulfonic acid, *J. Chromatogr.*, 1993, in press.
57. **Hirtzer, P. (Glycomed),** personal communication, 1993.
58. **Lee, K. B., Desai, U. R., Palcic, M. M., Hindsgaul, O., and Linhardt, R. J.,** An electrophoresis-based assay for glycosyltransferase activity, *Anal. Biochem.*, 205, 108, 1992.
59. **Liu, J., Shirota, O., and Novotny, M.,** Capillary electrophoresis of amino sugars with laser-induced fluorescence detection, *Anal. Chem.*, 63, 413, 1991.
60. **Liu, J., Shirota, O., and Novotny, M.,** Separation of fluorescent oligosaccharide derivatives by microcolumn techniques based on electrophoresis and liquid chromatography, *J. Chromatogr.*, 559, 223, 1991.
61. **Liu, J., Shirota, O., Wiesler, D., and Novotny, M.,** Ultrasensitive fluorometric detection of carbohydrates as derivatives in mixtures separated by capillary electrophoresis, *Proc. Natl. Acad. Sci. U.S.A.*, 88, 2302, 1991.
62. **Liu, J., Shirota, O., and Novotny, M. V.,** Sensitive, laser-assisted determination of complex oligosaccharide mixtures separated by capillary gel electrophoresis at high resolution, *Anal. Chem.*, 64, 973, 1992.

63. **Honda, S., Suzuki, S., Nose, A., Yamamoto, K., and Kakehi, K.,** Capillary zone electrophoresis of reducing mono- and oligo-saccharides as the borate complexes of their 3-methyl-1-phenyl-2-pyrazolin-5-one derivatives, *Carbohydr. Res.*, 215, 193, 1991.
64. **Honda, S., Yamamoto, K., Suzuki, S., Ueda, M., and Kakehi, K.,** High-performance capillary zone electrophoresis of carbohydrates in the presence of alkaline earth metal cations, *J. Chromatogr.*, 588, 327, 1991.
65. **Honda, S., Ueno, T., and Kakehi, K.,** High-performance capillary electrophoresis of unsaturated oligosaccharides derived from glycosaminoglycans by digestion with chondroitinase ABC as 1-phenyl-3-methyl-5-pyrazolone derivatives, *J. Chromatogr.*, 608, 289, 1992.
66. **Steffansson, M. and Westerlund, D.,** Capillary electrophoresis of glycoconjugates in alkaline media, *J. Chromatogr.*, 632, 195, 1993.
67. **Nardi, A., Fanali, S., and Foret, F.,** Capillary zone electrophoretic separation of cyclodextrins with indirect UV photometric detection, *Electrophoresis*, 11, 774, 1990.
68. **Steinmetz, H. J. and Schwochau, K.,** Preparation, isolation and characterization of technetium complexes containing α-amine oxime sugars, publication of Institut für Chemie 1 (Nuklearchemie), Forschungszentrum Jülich GmbH (KFA), D-5170 Jülich, Germany, 1991.
69. **Al-Hakim, A. and Linhardt, R. J.,** Capillary electrophoresis for the analysis of chondroitin sulfate- and dermatan sulfate-derived disaccharides, *Anal. Biochem.*, 195, 68, 1991.
70. **Ampofo, S. A., Wang, H. M., and Linhardt, R. J.,** Disaccharide compositional analysis of heparin and heparan sulfate using capillary zone electrophoresis, *Anal. Biochem.*, 199, 249, 1991.
71. **Harvey, D. J.,** The role of mass spectrometry in glycobiology, *Glycoconjugate J.*, 9, 1, 1992.
72. **Duffin, K. L., Welply, J. K., Huang, E., and Henion, J. D.,** Characterization of N-linked oligosaccharides by electrospray and tandem mass spectrometry, *Anal. Chem.*, 64, 1440, 1992.
73. **Rudd, P. M., Scragg, I. G., Coghill, E., and Dwek, R. A.,** Separation and analysis of the glycoform populations of ribonuclease B using capillary electrophoresis, *Glycoconjugate J.*, 9, 86, 1992.
74. **Tsuji, K. and Little, R. J.,** Charge-reversed, polymer-coated capillary column for the analysis of a recombinant chimeric glycoprotein, *J. Chromatogr.*, 594, 317, 1992.
75. **Huang, X., Zare, R. N., Sloss, S., and Ewing, A. G.,** End-column detection for capillary zone electrophoresis, *Anal. Chem.*, 63, 189, 1991.
76. **Nebinger, P.,** Isotachophoretic determination of hyaluronate oligosaccharide-degrading enzyme activities, *J. Chromatogr.*, 354, 530, 1986.
77. **Grossman, P. D., Colburn, J. C., Lauer, H. H., Nielsen, R. G., Riggin, R. M., Sittampalam, G. S., and Rickard, E. C.,** Application of free-solution capillary electrophoresis to the analytical scale separation of proteins and peptides, *Anal. Chem.*, 61, 1186, 1989.
78. **Wu, S.-L., Teshima, G., Cacia, J., and Hancock, W. S.,** Use of high-performance capillary electrophoresis to monitor charge heterogeneity in recombinant-DNA derived proteins, *J. Chromatogr.*, 516, 115, 1990.
79. **Tran, A. D., Park, S., Lisi, P. J., Huynh, O. T., Ryall, R. R., and Lane, P. A.,** Separation of carbohydrate-mediated microheterogeneity of recombinant human erythropoietin by free solution capillary electrophoresis — effects of pH, buffer type and organic additives, *J. Chromatogr.*, 542, 459, 1991.
80. **Nashabeh, W. and El Rassi, Z.,** Capillary zone electrophoresis of α_1-acid glycoprotein fragments from trypsin and endoglycosidase digestions, *J. Chromatogr.*, 536, 31, 1991.
81. **Watson, E. and Yao, F.,** Capillary electrophoretic separation of recombinant granulocyte-colony-stimulating factor glycoforms, *J. Chromatogr.*, 630, 442, 1993.
82. **Landers, J. P., Oda, R. P., Madden, B. J., and Spelsberg, T. C.,** High-performance capillary electrophoresis of glycoproteins: the use of modifiers of electroosmotic flow for the analysis of microheterogeneity, *Anal. Biochem.*, 205, 115, 1992.
83. **Kilár, F. and Hjertén, S.,** Separation of the human transferrin isoforms by carrier-free high-performance zone electrophoresis and isoelectric focusing, *J. Chromatogr.*, 480, 351, 1989.
84. **Kilár, F. and Hjertén, S.,** Fast and high resolution analysis of human serum transferrin by high performance isoelectric focusing in capillaries, *Electrophoresis*, 10, 23, 1989.
85. **Liu, Y. and Chan, K.-F. J.,** High performance capillary electrophoresis of gangliosides, *Electrophoresis*, 12, 402, 1991.
86. **Bjergegaard, C., Michaelsen, S., Moller, P., and Sorensen, H.,** High performance capillary electrophoresis: determination of individual anions, carboxylates, intact- and desulfoglucosinolates, paper presented at Proc. 8th Int. Rapeseed Congress, July 9–11, Saskatoon, Saskatchewan, Canada, 3, 822, 1991.
87. **Michaelsen, S., Moller, P., and Sorensen, H.,** Factors influencing the separation and quantitation of intact glucosinolates and desulphoglucosinolates by micellar electrokinetic capillary chromatography, *J. Chromatogr.*, 608, 363, 1992.
88. **Michaelsen, S., Moller, P., and Sorensen, H.,** Factors influencing the separation of intact glucosinolates and desulfoglucosinolates by micellar electrokinetic capillary chromatography, *J. Chromatogr.*, 608, 363, 1992.

89. **Varki, A.,** Biological roles of oligosaccharides: all of the theories are correct, *Glycobiology,* 3, 97, 1993.
90. **Morin, Ph., Villard, F., and Dreux, M.,** Borate complexation of flavonoid-*O*-glycosides in capillary electrophoresis. I. Separation of flavonoid-7-*O*-glycosides differing in their flavonoid aglycone, *J. Chromatogr.,* 628, 153, 1993.
91. **Morin, Ph., Villard, F., Dreux, M., and André, P.,** Borate complexation of flavonoid-*O*-glycosides in capillary electrophoresis. II. Separation of flavonoid-3-*O*-glycosides differing in their sugar moiety, *J. Chromatogr.,* 628, 161, 1993.
92. **Pietta, P. G., Mauri, P. L., Rava, A., and Sabbatini, G.,** Application of micellar electrokinetic capillary chromatography to the determination of flavonoid drugs, *J. Chromatogr.,* 549, 367, 1991.
93. **Brunner, G., Tegtmeier, F., Kirk, D. N., Wynn, S., and Setchell, K. D. R.,** Enzymatic synthesis and chromatographic purification of lignan glucuronides, *Biomed. Chromatogr.,* 1, 89, 1986.
94. **Seitz, U., Bonn, G., Oefner, P., and Popp, M.,** Isotachophoretic analysis of flavanoids and phenolcarboxylic acids of relevance to phytopharmaceutical industry, *J. Chromatogr.,* 559, 499, 1991.
95. **Heegaard, N. H. H., and Robey, F. A.,** Use of capillary zone electrophoresis to evaluate the binding of anionic carbohydrates to synthetic peptides derived from human serum amyloid P component, *Anal. Chem.,* 64, 2479, 1992.
96. **Honda, S., Taga, A., Suzuki, K., Suzuki, S., and Kakehi, K.,** Determination of the association constant of monovalent mode protein-sugar interaction by capillary zone electrophoresis, *J. Chromatogr.,* 597, 377, 1992.
97. **Terabe, S., Miyashita, Y., Shibata, O., Barnhart, E. R., Alexander, L. R., Patterson, D. G., Karger, B. L., Hosoya, K., and Tanaka, N.,** Separation of highly hydrophobic compounds by cyclodextrin-modified micellar electrokinetic chromatography, *J. Chromatogr.,* 516, 23, 1990.
98. **Fanali, S.,** Separation of optical isomers by capillary zone electrophoresis based on host-guest complexation with cyclodextrins, *J. Chromatogr.,* 474, 441, 1989.
99. **Fanali, S.,** Use of cyclodextrins in capillary zone electrophoresis — resolution of terbutaline and propranolol enantiomers, *J. Chromatogr.,* 545, 437, 1991.
100. **Kuhn, R. and Hoffstetter, S.,** Chiral separations by capillary electrophoresis, *Chromatographia,* 34, 505, 1992.
101. **Liu, J., Cobb, K. A., and Novotny, M.,** Capillary electrophoretic separations of peptides using micelle-forming compounds and cyclodextrins as additives, *J. Chromatogr.,* 519, 189, 1990.
102. **Nishi, H. and Matsuo, M.,** Separation of corticosteroids and aromatic hydrocarbons by cyclodextrin-modified micellar electrokinetic chromatography, *J. Liq. Chromatogr.,* 14, 973, 1991.
103. **Rogan, M. M., Drake, C., Goodall, D. M., and Altria, K. D.,** Enantioselective enzymic biotransformation of 2′-deoxy-3′-thiacytidine (BCH 189) monitored by capillary electrophoresis, *Anal. Biochem.,* 208, 343, 1993.
104. **Schutzner, W. and Fanali, S.,** Enantiomers resolution in capillary zone electrophoresis by using cyclodextrins, *Electrophoresis,* 13, 687, 1992.
105. **Snopek, J., Jelinek, I., and Smolkova-Keulemansova, E.,** Micellar, inclusion and metal-complex enantioselective pseudophases in high-performance electromigration methods, *J. Chromatogr.,* 452, 571, 1988.
106. **Snopek, J., Soini, H., Novotny, M., Smolkova-Keulemansova, E., and Jelinek, I.,** Selected applications of cyclodextrin selectors in capillary electrophoresis, *J. Chromatogr.,* 559, 215, 1991.
107. **Soini, H., Riekkola, M.-L., and Novotny, M. V.,** Chiral separations of basic drugs and quantitation of bupivacaine enantiomers in serum by capillary electrophoresis with modified cyclodextrin buffers, *J. Chromatogr.,* 608, 265, 1992.
108. **Yamashoji, Y., Ariga, T., Asano, S., and Tanaka, M.,** Chiral recognition and enantiomeric separation of alanine β-naphthylamide by cyclodextrins, *Anal. Chim. Acta,* 268, 39, 1992.
109. **Yik, Y. F., Ong, C. P., Khoo, S. B., Lee, H. K., and Li, S. F. Y.,** Separation of selected PAHs by using high performance capillary electrophoresis with modifiers, *Environ. Monit. Assess.,* 19, 73, 1991.
110. **Cruzado, I. D. and Vigh, G.,** Chiral separations by capillary electrophoresis using cyclodextrin-containing gels, *J. Chromatogr.,* 608, 421, 1992.
111. **Guttman, A., Paulus, A., Cohen, A. S., Grinberg, N., and Karger, B. L.,** Use of complexing agents for selective separation in high-performance capillary electrophoresis — chiral resolution via cyclodextrins incorporated within polyacrylamide gel columns, *J. Chromatogr.,* 448, 41, 1988.
112. **D'Hulst, A. and Verbeke, N.,** Chiral separation by capillary electrophoresis with oligosaccharides, *J. Chromatogr.,* 608, 275, 1992.
113. **Ishihama, Y. and Terabe, S.,** Enantiomeric separation by micellar electrokinetic chromatography using saponins, *J. Liq. Chromatogr.,* 16, 933, 1993.
114. **Kaneta, T., Tanaka, S., Taga, M., and Yoshida, H.,** Effect of addition of glucose on micellar electrokinetic capillary chromatography with sodium dodecyl sulphate, *J. Chromatogr.,* 609, 369, 1992.
115. **Bruin, G. J. M., Huisden, R., Kraak, J. C., and Poppe, H.,** Performance of carbohydrate-modified fused-silica capillaries for the separation of proteins by zone electrophoresis, *J. Chromatogr.,* 480, 339, 1989.

116. **Chin, A. M. and Colburn, J. C.,** Counter-migration capillary electrophoresis (CMCE) in DNA restriction fragment analysis, *Am. Biotech. Lab. News Ed.*, 7, 16, 1989.
117. **Grossman, P. D., Hino, T., and Soane, D. S.,** Dynamic light-scattering studies of hydroxyethyl cellulose solutions used as sieving media for electrophoretic separations, *J. Chromatogr.*, 608, 79, 1992.
118. **Guszczynski, T. and Chrambach, A.,** Electrophoretic separation of *S. pompe* chromosomes in polyacrylamide solutions using a constant field, *Biochem. Biophys. Res. Commun.*, 179, 482, 1991.
119. **Schwartz, H. E., Ulfelder, K., Sunzeri, F. J., Busch, M. P., and Brownlee, R. G.,** Analysis of DNA restriction fragments and polymerase chain reaction products towards detection of the AIDS (HIV-1) virus in blood, *J. Chromatogr.*, 559, 267, 1991.
120. **Strege, M. A. and Lagu, A. L.,** Capillary electrophoretic protein separations in polyacrylamide-coated silica capillaries and buffers containing ionic surfactants, *J. Chromatogr.*, 630, 337, 1993.
121. **Zhu, M., Hansen, D. L., Burd, S., and Gannon, F.,** Factors affecting free zone electrophoresis and isoelectric focusing in capillary electrophoresis, *J. Chromatogr.*, 480, 311, 1989.
122. **Grossman, P. D. and Soane, D. S.,** Experimental and theoretical studies of DNA separations by capillary electrophoresis in entangled polymer solution, *Biopolymers*, 31, 1221, 1991.
123. **Guttman, A., Horváth, J., and Cooke, N.,** Influence of temperature on the sieving effect of different polymer matrices in capillary SDS gel electrophoresis of proteins, *Anal. Chem.*, 65, 199, 1993.
124. **Martens, D. A. and Frankenberger, W. T., Jr.,** Determination of saccharides by high performance anion-exchange chromatography with pulsed amperometric detection, *Chromatographia*, 29, 7, 1990.

Chapter 12

CAPILLARY ZONE ELECTROPHORESIS OF PEPTIDES

Randy M. McCormick

TABLE OF CONTENTS

0-8493-8690-X/94/$0.00+$.50

I. HISTORICAL BACKGROUND

Separation of a simple mixture of several synthetic dipeptides was one of the first demonstrated applications in the initial reports on capillary zone electrophoresis (CZE) by Jorgenson et al. in 1981.[1] Soon to follow was the separation of the peptides in a fluorescamine-labeled tryptic digest of lysozyme.[2] Resolution of the 50 or so fragments in this digest (Figure 1) in less than 20 min clearly demonstrated the potential of this emerging technology to separate a large number of peptide fragments in a short time period with high efficiency.

These early results laid the groundwork for development of a number of areas where CZE would later have applicability. Due to inherently high separation efficiency, CZE has found use for detection of single amino acid substitutions in polypeptides as well as for discerning subtle differences in the degree of glycosylation, phosphorylation, sulfonation, etc., of peptides and proteins arising during posttranslational processing. In addition, separation of the degradation products of peptides due to deamidation, proteolysis, methionine oxidation, or covalent attachment between peptides has become a common use for the technique. This application is of particular value for quality control analysis of recombinant peptide drugs, which are often accompanied by variants formed by hydrolysis, oxidation, deamidation, etc. Capillary zone electrophoresis has also been used for peptide mapping, in which a protein is chemically and/or enzymatically cleaved into numerous, smaller peptides and these fragments are electrophoretically separated to yield a characteristic map or "fingerprint" of the protein. This qualitative, comparative technique can also be used to discern subtle differences between proteins such as single amino acid substitutions or posttranslational modifications as well as serving as the initial step in protein sequencing. All of these applications can be easily achieved since, unlike with large proteins where optimal separation conditions cannot be readily predicted, a pseudoquantitative theory has been developed for peptides, allowing the bench chemist to optimize CZE separations in a systematic manner.

II. THEORY DESCRIBING SEPARATIONS

Nyberg et al.[3] first demonstrated that relative migration of small peptides (synthetic substance P peptides) in CZE exhibited a linear relationship with the parameter $M^{2/3}/Z$, where M is the molecular mass of the peptide and Z is its charge or valency. The basis of this relationship derived from the pioneering work on peptides in paper electrophoresis by Offord,[4] who observed a linear correlation between the logarithm of electrophoretic mobilities and the logarithm of molecular weights for over 100 peptides (with the exception of peptides containing histidine [excluded because at the buffer pH used, imidazole residues were only partially ionized] and cysteic acid [cystine and cysteine residues were converted to cysteic acid and thus excluded]). A plot of these two parameters by Offord yielded a set of lines, each corresponding to a subset of peptides with identical charge and each line having a slope of –2/3. To describe the relationship of the electrophoretic mobility μ to peptide charge and molecular mass, Offord proposed the relationship

$$\mu = kZM^{-x} \tag{1}$$

with $x = 2/3$ as derived from the slopes of the plots. This relationship assumes that, during electrophoresis, an ion moving through a conducting medium experiences a retarding shear force proportional to the surface area of the analyte ion, i.e., the electrophoretic mobility is inversely proportional to the square of the molecular radius r (rather than proportional to $M^{-1/3}$ [$\propto r^{-1}$]) as described by Stoke's law.

The validity of this relationship for CZE has been substantiated by a number of workers. Deyl et al.[5,6] found that the cyanogen bromide fragments of collagen strictly followed the

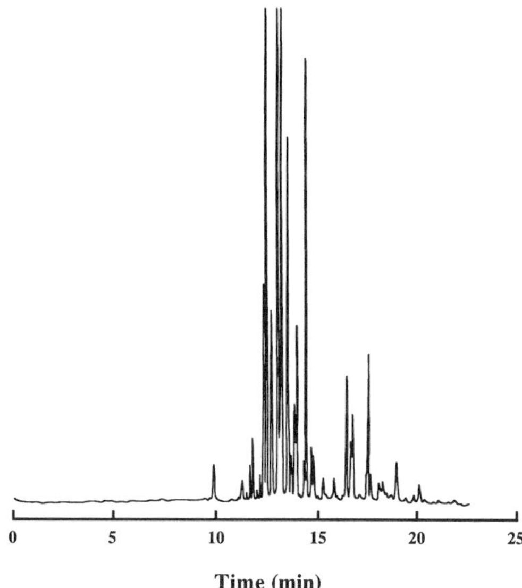

FIGURE 1. Separation of fluorescamine-labeled peptides obtained from a tryptic digest of reduced and carboxymethylated egg white lysozyme. (From Jorgenson, J. W. and Lukacs, K. D., *J. High Resolut. Chromatogr. Chromatogr. Commun.*, 4(5), 230, 1981. With permission.)

linear relationship of Offord (Equation 1). In addition, a linear relationship between the relative migration time and the isoelectric point (p*I*) of low molecular weight proteins over a broad pH range (6.9 to 10.5) was observed, indicating that the effect of molecular size is apparently smaller than the effect of charge for small and moderate-sized proteins.

Using migration data on a diverse set of 33 peptides derived from enzymatic digests as well as from 10 intact proteins in both acidic (pH 2.35) and moderately basic (pH 8.15) buffers, Rickard et al.[7] evaluated the correlation between electrophoretic mobility and a variety of charge-to-size functions (specifically $q/M^{1/3}$ [based on Stoke's law], $q/M^{1/2}$ (size parameter based on radius of gyration of the molecule, which is proportional to the square root of peptide chain length [$\propto M^{-1/2}$]), and $q/M^{2/3}$ [Offord's parameter]). For a subset of the data on the tryptic digest fragments of human growth hormone (hGH), the strongest correlation (Figure 2c) as measured by linear regression fit was found between CZE mobilities and the quantity $q/M^{2/3}$ (q is equivalent to Z), in agreement with the results of Nyberg and Offord; poorest correlation was found with the quantity $q/M^{1/3}$ (Figure 2a). Issaq et al.[8] tested the correlation of electrophoretic mobility not only with the parameters listed above but also with the quantity $\ln(Z + 1)/n^{0.43}$ (from Grossman et al., see below); best correlation was obtained between mobility and $Z/M^{2/3}$. Kurosu et al.,[9] using pH 2.0, 4.0, 7.0, and 10.0 buffers, investigated the migration behavior of 35 peptides (ranging in size from 2 to 42 amino acids in length and with charges from –1.44 to 6.93) and 20 proteins (with molecular masses of 5 to 480 kDa and p*I* values of 4.1 to 11.0), and attempted correlation of electrophoretic mobility with a variety of molecular mass functions M^x (where x = –1/3, –1/2, –2/3, and –1). At pH 2.0, they found the best correlation with $M^{-2/3}$. Landers et al.[10] also established good correlation between the parameter $M^{2/3}/Z$ and peptide migration time (proportional to μ^{-1}) for a group of peptides as well as for a peptide dimer formed by disulfide linkage. Finally, Florance et al.[11] found that the best correlation of migration of 24 motilin peptide fragments was with the parameter log $Z/M^{2/3}$, although they plotted that parameter against peptide migration time.

As alluded to above, a second and more empirical relationship between mobility and a peptide charge/mass parameter was proposed by Grossman et al.[12] On the basis of the

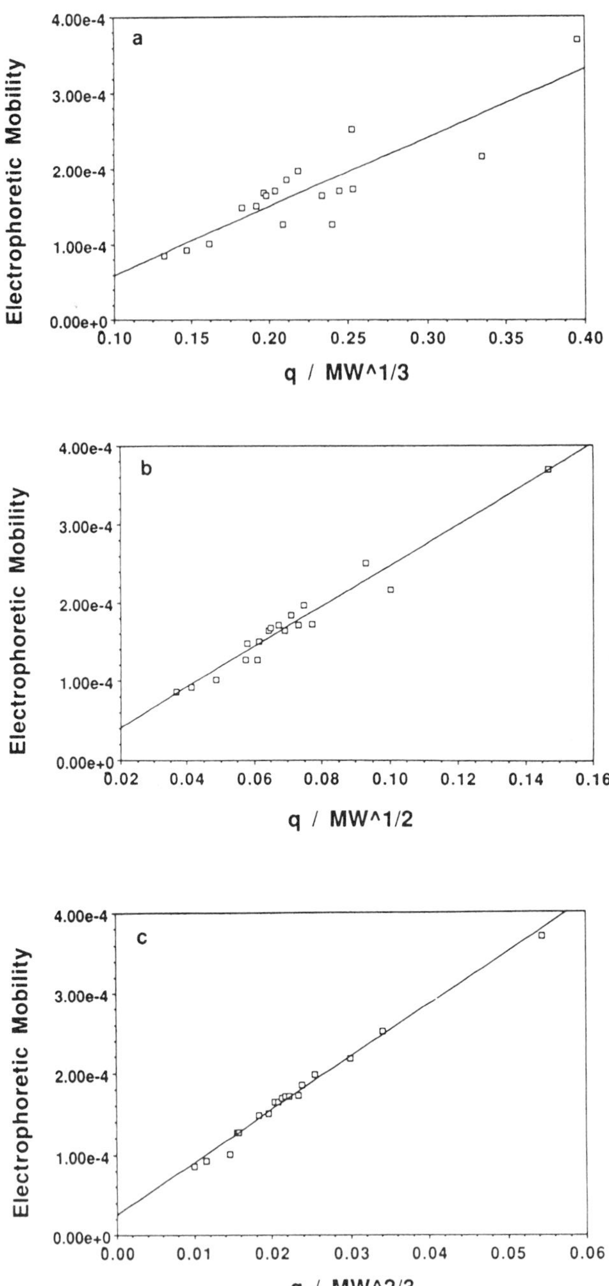

FIGURE 2. Fit of electrophoretic mobility ($cm^2/V{\cdot}s$) vs. charge-to-size parameter for peptides from a hGH digest separated in pH 2.35, 0.1 *M* glycine buffer. (From Rickard, E. C., Strohl, M. M., and Nielsen, R. G., *Anal. Biochem.*, 197(1), 197, 1991. With permission.)

electrophoretic mobilities of a set of 40 peptides varying in size from 3 to 39 amino acids and ranging in charge from 0.33 to 13.0, they fitted their data to the empirical relationship

$$\mu = \frac{k_1 \times \ln(Z+1) + k_2}{n^{0.43}} \qquad (2)$$

where k_1 and k_2 are constants characteristic of the buffer system used and n is the number of amino acids in the polypeptide chain of charge Z. To account for the nonlinear increase in mobility with respect to charge observed in their data, the authors arrived at the above relationship by modification of the Stoke's equation for electrophoretic mobility in combination with use of the radius of gyration of the molecule as an approximate measure of peptide size. Also, by plotting mobility vs. a hydrophobicity index, the authors further concluded that the hydrophobicity of the residues in the peptide chain is a determinant in electrophoretic mobility; however, this latter relationship may arise solely from an increase in molecular weight of the peptide due to incorporation of larger hydrophobic residues into the polypeptide. The authors could not directly account for an increase in peptide mass in their empirical relationship (Equation 2) since they used the number of amino acid residues (and not the molecular weight) as the size parameter.

A relationship similar to that in Equation 2 has also been proposed by Castagnola and coworkers,[13] who evaluated their experimental data using the model of Grossman et al. Differences between the coefficients (k_1 and k_2 of Equation 2) derived in their work and those proposed by Grossman et al. were explained as arising from differences in buffer ionic strength and capillary dimensions. On the basis of their data, a relationship between mobility and $\ln(Z + 1)/n^{0.5}$ was proposed, with the coefficient of the size parameter derived from the radius of gyration model.

Application of these relationships for prediction of mobilities of a given set of peptides and subsequent optimization of separation by CZE is relatively straightforward. Computation of the charge on a peptide at a given pH can be accomplished with the Henderson–Hasselbalch equation, which uses solution pH and the ionization constants of various functionalities in the amino acid residues in the peptide to calculate the charge on the molecule at that pH. Accurate knowledge of the ionization constants of amino acid residues is required to implement this calculation. Use of the p*K* values of free amino acids for this calculation is common, although this practice assumes ionization of each amino acid residue in the peptide will not be affected by neighboring residues. However, formation of the peptide bonds induces an electrostatic change in the charge on amino and carboxy groups in neighboring amino acids, and ionization constants can shift when amino acids are joined to form peptides. The ionization of the C-terminus carboxy group can become tenfold weaker and the terminal amino moieties can be affected even more. In addition, microenvironments within the peptide, as well as changes in the local dielectric constant arising from changes in ionic strength of the surrounding medium, can significantly alter ionization constants of amino acids upon incorporation into peptides. Tabulated values of adjusted as well as free amino acid ionization constants are given by Rickard et al.[7] and Skoog and Wichman[14] and are reproduced in Table 1.

Several computer programs permit calculation of the charge on a peptide at a given pH using the above amino acid ionization constants. Skoog and Wichman[14] developed a Pascal program to calculate the p*I* of a polypeptide based on contributions to the net charge of the terminal amino and carboxylic moieties as well as charged amino acid side chains within the peptide sequence; the program employs the p*K* values of the free amino acids in an iterative process to calculate the p*I* of a polypeptide and can, with minor modification, generate titration curves (charge as a function of solution pH) of polypeptides. The program uses only the primary structure (amino acid sequence) of the peptide, so discrepancies can arise between measured and calculated values due to the presence of carbohydrate moieties or modifications of amino acid residues, neither of which can be taken into account in the program. CHARGPRO™,* available as part of the PC/Gene package from IntelliGenetics (Mountain View, CA) is a more sophisticated program that finds the charge of a specified peptide (or protein) sequence as a function of pH and indicates the isoelectric point. It can take posttranslational modifications (number of deamidated asparagine and glutamine residues, blocked arginine and lysine residues, cysteine residues blocked via disulfide bonds, blocked N- and/

* CHARGPRO™ is a registered trademark of IntelliGenetics, Inc. (Mountain View, CA).

TABLE 1
Values of Amino Acid Ionization Constants Used for the Calculation of Charge

	Adjusted pK_a values[7]			Isolated amino acid pK_a values[7]			Skoog and Wichman pK_a values[14]		
Amino acid	C Terminal	N Terminal	Side chain	C Terminal	N Terminal	Side chain	C Terminal	N Terminal	Side chain
Ala (A)	3.20	8.20		2.34	9.87		2.4	9.9	
Arg (R)	3.20	8.20	12.50	1.91	9.02	12.48	2.2	9.0	12.5
Asn (N)	2.75	7.30		2.06	8.82		2.1	8.8	
Asp (D)	2.75	8.60	3.50	2.02	9.85	3.82	2.1	9.8	3.9
Cys (C)	2.75	7.30	10.30	1.93	10.40	8.26	1.7	10.8	8.3
Gln (Q)	3.20	7.70		2.17	9.13		2.2	9.1	
Glu (E)	3.20	8.20	4.50	2.15	9.57	4.18	2.2	9.7	4.3
Gly (G)	3.20	8.20		2.35	9.78		2.4	9.8	
His (H)	3.20	8.20	6.20	1.79	9.18	6.08	1.8	9.2	6.0
Ile (I)	3.20	8.20		2.34	9.72		2.3	9.8	
Leu (L)	3.20	8.20		2.35	9.67		2.3	9.7	
Lys (K)	3.20	7.70	10.30	2.17	9.06	10.66	2.2	9.0	10.5
Met (M)	3.20	9.20		2.28	9.24		2.1	9.3	
Phe (F)	3.20	7.70		2.37	9.21		2.2	9.2	
Pro (P)	3.20	9.00		1.98	10.62		2.0	10.6	
Ser (S)	3.20	7.30		2.20	9.18		2.1	9.2	
Thr (T)	3.20	8.20		2.09	9.10		2.1	9.1	
Trp (W)	3.20	8.20		2.40	9.42		2.4	9.4	
Tyr (Y)	3.20	7.70	10.30	2.20	9.11	10.11	2.2	9.1	10.0
Val (V)	3.20	8.20		2.30	9.68		2.3	9.7	

TABLE 2
Predicted and Measured Migration Time Order for Peptides

Peptide sequence[a]	Calculated charge	$Z/M^{2/3}$	Measured migration time (min)[a]
KWK	2.75	0.04611	10.17
RKDVY	2.62	0.03389	13.04
MEHFRWG	2.74	0.02811	13.99
RPPGFSPFR	2.75	0.02587	14.36
GIVEQCCASVCSLYQLENYCN	—	—	15.75
DRVYIGPF	2.62	0.02540	16.95
pyro-EHP-amide	—	—	17.38
DSDPR	1.48	0.02027	19.82
VAAF	0.75	0.01367	24.70
EAE	0.65	0.01315	25.54
YGGFL	0.75	0.01110	25.96
YGGFM	0.75	0.01086	27.53
YAGFM	0.75	0.01069	28.48
CYIQNCPLG-amide	0.75	0.00749	28.98

[a] Reproduced from Strickland, M. and Strickland, N., *Am. Lab. (Fairfield, Conn.)*, November, 60, 1990.

or C-termini that suppress positive/negative charge on the terminal groups) into account. The program also allows editing of the default (free amino acid) p*K* values. Finally, Pennino[15] developed a computer program for prediction of the migration times of peptides based on the semiempirical model of Grossman et al. The program calculates the charge on peptides using the Henderson–Hasselbalch equation, and subsequently will calculate peptide migration times, given the amino acid sequence, buffer pH, etc.

Application of Equation 1 for predicting CZE peptide separations can be demonstrated with published data of Strickland and Strickland.[16] Using peptide sequence data (Table 2, column 1), calculation of the charge (column 2) and the charge/size parameter $Z/M^{2/3}$ (column 3) were accomplished using the CHARGPRO™* and PHYSCHEM™* programs in the PC/Gene package. These charge/size parameters can be used to predict the relative migration order of the peptide fragments, since peptide mobility is directly proportional to $Z/M^{2/3}$. The migration order of the fragments (based on the published migration time data of Strickland, column 4 of Table 2) correlates exactly in the expected inverse fashion with the parameter $Z/M^{2/3}$; use of the calculated charge/size parameters ($Z/M^{2/3}$) for a group of peptide fragments would be an aid for peak identity in a CZE separation of peptides.

A further example illustrative of the value of Equation 1 for optimizing CZE separations of peptides can be found in the work of Bongers et al.[17] Using a modification of the computer program and p*K* data of Skoog and Wichman[14] to calculate charge vs. pH on a set of three synthetic undecapeptides that differed by substitution of Asn, Asp, or β-Asp at a single residue in the polypeptide chains, they obtained plots of charge (q) vs. pH for the three peptides shown in Figure 3a. On the basis of these titration curves, one can predict optimum separation of the three peptides should occur in a buffer pH range of 3 to 5, with maximum differences in charge (and concomitantly maximum differences in electrophoretic mobilities, since the masses of the three peptides are essentially equivalent) between the three peptides occurring around pH 4. Indeed, as illustrated in Figure 3b, the authors found an optimum separation of the three peptides at pH 4.3.

As illustrated above, use of peptide sequence information to calculate the charge and mass and subsequent use of these data to predict the electrophoretic mobilities of the peptides afford a straightforward approach to achieving separation of a mixture of peptides by estimating the optimum separation buffer pH. Factors important in the optimization of peptide separations

* CHARGPRO™ and PHYSCHEM™ — registered trademarks of IntelliGenetics, Inc. (Mountain View, CA).

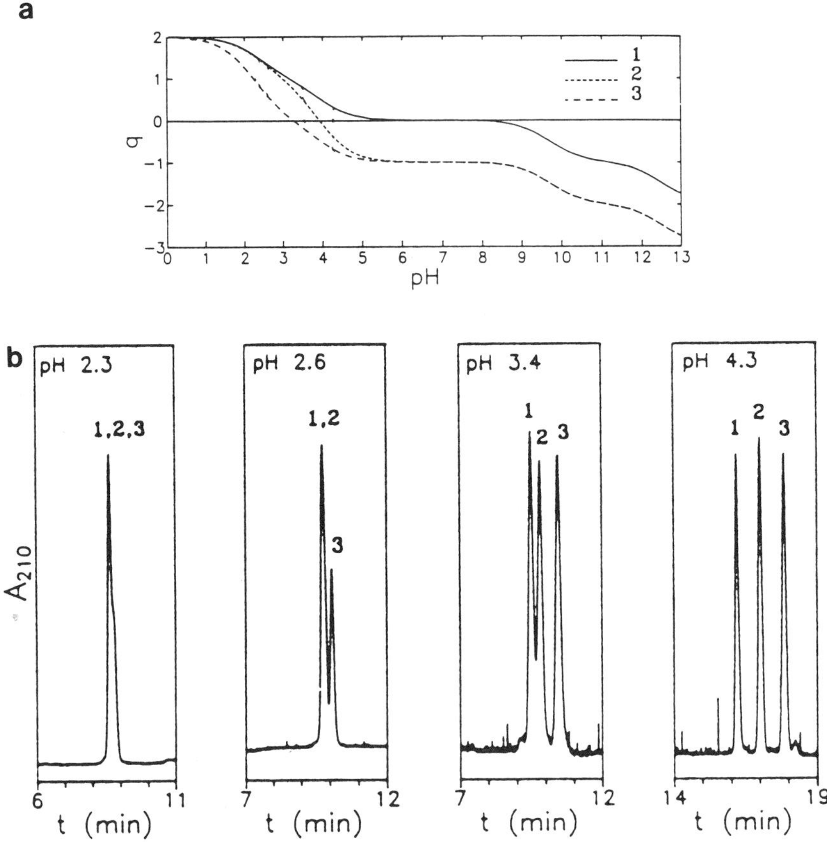

FIGURE 3. (a) Calculated charge (q) vs. pH profiles and (b) electropherograms at various pH for synthetic model peptides:1, [Leu[27], Asn[28]]-GRF(22–32)-OH; 2, [Leu[27], Asp[28]]-GRF(22–32)-OH; and 3, [Leu[27], β-Asp[28]]-GRF(22–32)-OH. The pH of the 25 m*M* Na_2HPO_4/H_3PO_4 running buffer is shown at upper left in each panel. (From Bongers, J., Lambros, T., Felix, A. M., and Heimer, E. P., *J. Liq. Chromatogr.*, 15(6–7), 1115, 1992. With permission.)

by CZE include not only buffer pH, as discussed above, but also a number of buffer additives can be employed to achieve special selectivities in the separation buffer.

III. OPTIMIZATION OF CAPILLARY ZONE ELECTROPHORETIC SEPARATIONS OF PEPTIDES

In CZE, separation selectivity is determined by differences in the electrophoretic mobilities of the various analyte ions. Parameters such as buffer ionic strength, electrical field strength, capillary temperature, diameter, length, etc., can affect the separation performance obtained for peptides; however, none of these parameters directly impacts on the selectivity achievable in the separation. Of the aforementioned parameters, only the buffer pH influences the selectivity and variation of this parameter is thus the most direct route for optimization of the separation of peptides (and proteins). In addition to changes in buffer pH, enhancement of selectivity can be achieved through use of several types of buffer additives. Unlike chromatography, where both the stationary and mobile phases can be varied to enhance separations, the only route to selectivity variation in CZE is manipulation of the separation buffer (mobile phase) composition.

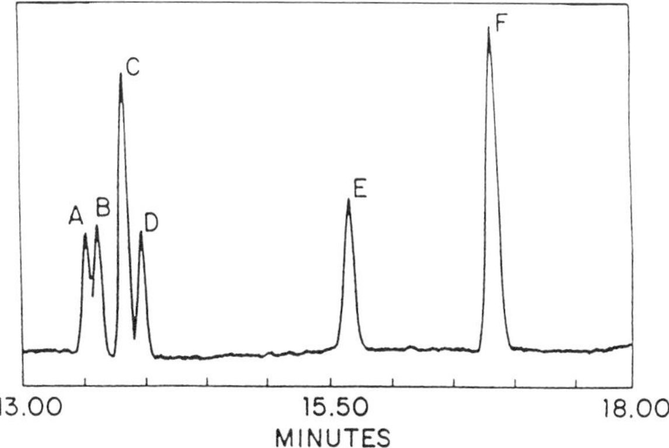

FIGURE 4. Separation of angiotensin II octapeptides. Capillary, 110 cm of 53-μm I.D. fused silica, 75-cm separation distance; buffer, 150 m*M* NaH_2PO_4, pH 3.0; detection, 190 nm; injection, 5 s at 2.5 kV; separation voltage, 2.5- to 30-kV linear ramp in 300 s; Sample A, Sar Arg Val Tyr Ile His Pro Gly; B, Sar Arg Val Tyr Ile His Pro Leu; C, Sar Arg Val Tyr Ile His Pro Thr; D, Sar Arg Val Tyr Ile His Pro Phe; E, Asp Arg Val Tyr Val His Pro Phe; F, Asp Arg Val Tyr Ile His Pro Phe. (From McCormick, R. M., *Anal. Chem.*, 60(21), 2322, 1988. With permission.)

A. BUFFER pH

On the basis of previous discussions, two physicochemical molecular properties, the charge on the molecule and the molecular weight of the molecule, play a major role in determining the mobility of peptides. To alter selectivity in peptide separations, changes in one of these parameters must occur. Since molecular weight is essentially invariant (with the negligible exception of gain or loss of protons), the parameter that can alter selectivity to the greatest extent is a change in the charge on the peptide. Peptide charge is changed most readily by altering the pH of the separation buffer, thereby directly affecting the extent of ionization of the amino and carboxylic acid moieties in the peptide. The importance of buffer pH was illustrated by Kirkland and McCormick,[18] who demonstrated substantial changes in the relative mobilities of a group of five small proteins as buffer pH varied from 1.5 to 4.0. Manipulation of the mobilities of peptides to optimize separation using this strategy was subsequently demonstrated by Grossman et al.[19] and McCormick.[20] The selectivity achievable in low-pH buffers was illustrated by McCormick[20] for a mixture of six angiotensin octapeptides separated in less than 20 min in a pH 3.0 phosphate buffer. Separation of two angiotensin II analogs (Figure 4) that differed at residue 5 by substitution of isoleucine for valine (difference of one methylene group) was achieved with resolution greater than 10, illustrating the dramatic effect substitution of different amino acids can have on changes in the degree of ionization of adjacent amino acids. Grossman et al.[19] further elaborated on the importance of buffer pH, utilizing both acidic and basic buffers for optimizing separations of mixtures of closely related heptapeptides; resolution of two heptapeptides that differ by substitution of alanine for isoleucine (difference of C_3H_6) at residue 4 in the peptide chains was demonstrated. The ability to resolve these two heptapeptides was attributed to changes in the local hydrophobic environment around adjacent charged amino acid side chains in the peptide molecules due to a change in the neighboring neutral amino acid.

In buffers of low pH (e.g., pH ~2), mobility differences between peptides will be determined by the total number of protonated basic residues in the chain and separation will be based on the positive charge density of the peptides.[16] Conversely, above a buffer pH of ~10, peptides will be completely deprotonated, and the number of negatively charged acidic residues in the chain will determine the charge and mobility.[21] At these extremes of pH, changes in selectivity are difficult to achieve, since all residues in the peptides are either fully protonated or deprotonated and the peptides have maximum electrophoretic mobilities (charges).

In buffers of intermediate pH, selectivity can be enhanced by varying the buffer pH to alter the ionization of the side chain residues. For acidic residues, greatest changes in selectivity can be achieved in the range of pH 3 to 6, whereas for basic residues greatest changes in selectivity are realized around pH 10. These observations are readily adaptable to the optimization of separations of peptides derived from various enzymatic digests of proteins. Since trypsin cuts proteins at arginine and lysine residues, most peptides derived from tryptic digestion of a protein will have a single basic residue (at the N terminus), and greatest selectivity changes would thus be realized in buffers of acidic pH.[21] In contrast, *Staphylococcus aureus* strain V8 protease cleaves at the carboxy linkage of glutamic acid residues and thus optimization of the separation of peptides in such digests could best be accomplished in buffers of basic pH. Endoproteinases such as chymotrypsin, thermolysin, and pepsin, which preferentially cut at neutral amino acids (i.e., those with aliphatic or aromatic side chains), yield fragments with both acidic and basic residues and thus the full pH range of 2 to 10 can be utilized to optimize the separation.

B. SPECIAL SELECTIVITY MECHANISMS

Although adjustment of buffer pH is the most direct and the most general route to optimize separations of peptides by CZE, a number of special selectivities can be achieved through use of buffer additives. These additives impart unique selectivities to the buffers by exploiting special interaction mechanisms with select chemical moieties in the peptide analyte molecules. Most significant among these are addition of ion pair agents to form micellar buffers and addition of organic modifiers, with changes in buffer ionic strength playing only a minor role in altering selectivity.[22]

1. Micellar Buffers

One of the methods of broadest applicability for imparting special selectivity to CZE separations is the use of micellar buffer modifiers (see Chapter 3 for more details on this technique). Surfactants added to the aqueous CZE buffer at concentrations above critical micelle concentration (CMC) form a pseudostationary or "mobile" stationary phase. Interaction between analyte molecules and the micellar pseudostationary phase is based on hydrophobic forces, with more lipophilic analyte molecules partitioning into the micelles to a greater extent relative to their more hydrophilic counterparts. During residence in the micelles, the partitioned molecules possess a different electrophoretic mobility than when free in the bulk aqueous buffer phase. More hydrophobic molecules experience longer times in the micellar phase relative to more hydrophilic counterparts and thus molecules with essentially identical free zone electrophoretic mobilities but differing hydrophobicities can be readily separated because of differing degrees of partitioning into the micellar phase.

Micellar CZE buffers have been found to be beneficial in the separation of peptides with similar charge-to-size ratios but different hydrophobicities.[23] Using either sodium dodecyl sulfate (SDS), hexadecyltrimethylammonium bromide (HTAB), or dodecyltrimethylammonium bromide (DTAB) to form the micellar phase, separation of angiotensin peptides with virtually identical p*I* values can be readily accomplished in neutral aqueous micellar buffers. Figure 5 illustrates the improvement in separation of these peptides by CZE that can be achieved by addition of DTAB to the running buffer. When run in a buffer with the surfactant DTAB below the CMC, the seven angiotensin analogs were poorly resolved into four overlapping bands (Figure 5A); however, when the surfactant concentration was increased above the CMC (Figure 5B), a dramatic improvement in resolution of all of the components was achieved. Use of micellar selectivity might thus be most useful for enhancing the separation of large hydrophobic peptides as well as for separating peptides that differ by single amino acid

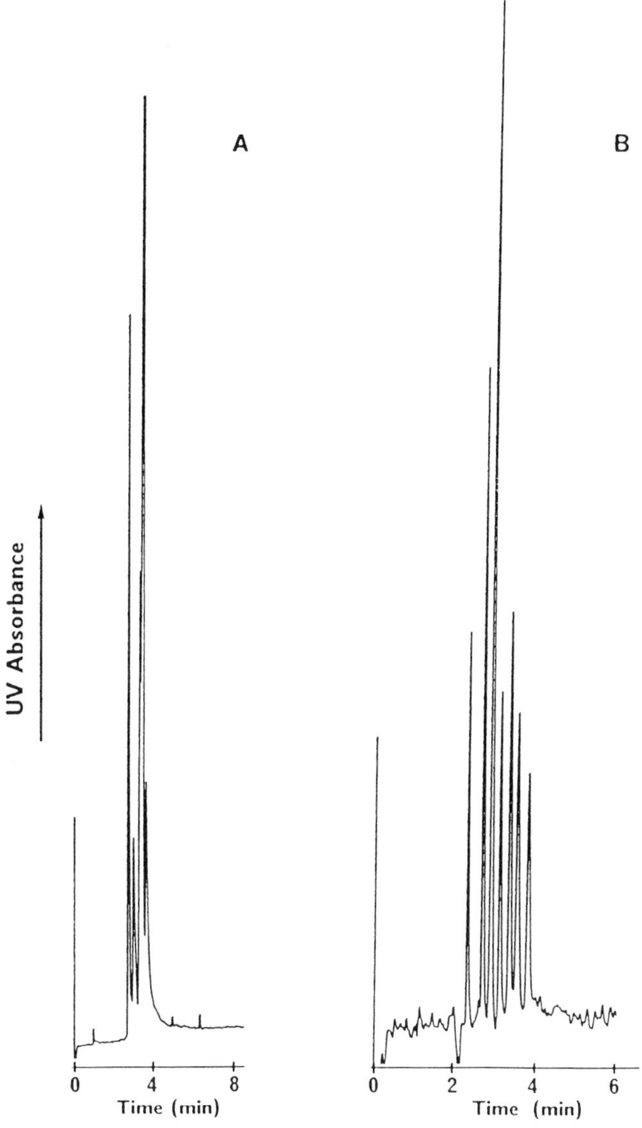

FIGURE 5. Comparison of separations of angiotensin analogs, using cationic detergent above and below the critical micelle concentration. Capillary: 45 cm in length (25 cm to detector), 50-μm I.D. Separation buffer: 10 m*M* Tris–10 m*M* Na_2HPO_4 (pH 7.05) with (A) 0.002 M and (B) 0.05 *M* DTAB. Operating voltage: –15 kV. UV detection at 220 nm. (From Liu, J., Cobb, K. A., and Novotny, M., *J. Chromatogr.*, 519(1), 189, 1990. With permission.)

substitutions involving alanine, valine, leucine, isoleucine, phenylalanine, etc. Table 3 lists CMC values for some common surfactants useful for capillary electrophoretic separations.

2. Metal Additives

Mosher[26] developed a special selectivity for histidine-containing peptides through the use of metal ions in the CZE buffer. With the addition of either Zn^{2+} or Cu^{2+} ions to acidic (pH 2.5) buffers in concentrations up to 30 m*M*, Mosher demonstrated the ability to alter significantly the mobility of histidine-containing dipeptides. These metal ions are known to complex specifically with this amino acid residue. Two comigrating dipeptides containing histidine (L-His L-His and D-His L-His) in pH 2.8 buffer were baseline resolved by addition of up to 30 m*M*

TABLE 3
Critical Micelle Concentrations for Select Surfactants

Surfactant	CMC (m*M*)	Ref.
Dodecyltrimethylammonium bromide (DTAB)	14	24
Sodium dodecyl sulfate (SDS)	~8	25
Hexadecyltrimethylammonium bromide (HTAB)	50	23
Decanesulfonic acid	32.6	24
Cetylpyridinium chloride	0.9	24
Tetradecyltrimethylammonium bromide	0.28	24
3-[(3-Cholamidopropyl)-dimethylammonio]-1-propane-sulfonate (CHAPS)	8	24
Taurocholic acid, Na^+	10–15	24

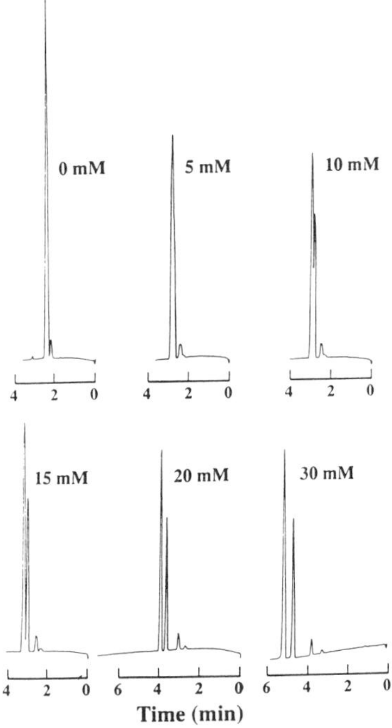

FIGURE 6. The impact of the concentration of $ZnSO_4$ on the resolution of a sample of DL-His-DL-His in 0.1 *M* phosphate buffer, pH 2.5. Sample concentrations were 2 m*M* each in the running buffer. The Zn(II) concentration is indicated in each electropherogram. Electrophoretic injection for 8 s and 8 kV; running voltage was 8 kV. Detection by absorbance at 200 nm. (From Mosher, R. A., *Electrophoresis*, 11(9), 765, 1990. With permission.)

Zn^{2+} to the buffer, as illustrated in Figure 6. Stover et al.[27] also observed improved resolution of histidine-containing heptapeptides by addition of 1 m*M* Zn^{2+} to the separation buffer. In addition, Issaq et al.[8] observed that the addition of Zn^{2+} to the buffer at low pH affected the migration time and enhanced the peak shape of peptides in general; a significant improvement in the resolution of two histidine-containing heptapeptides was demonstrated as well.

3. Borate Complexation

Complexation of carbohydrates with borate ions has been utilized to improve separations of sugars in several forms of chromatography (high-performance liquid chromatography [HPLC] and thin-layer chromatography [TLC]) and it is not surprising that this well-known selective interaction would be of use in electrophoretic separations as well (see Chapter 11 for more details). Hoffstetter-Kuhn et al.[28] demonstrated the utility of this selectivity mechanism in the separation of a glycosylated hexapeptide from its nonglycosylated counterpart; variation of the borate concentration between 50 and 200 m*M* at pH 10 resulted in reversal of the elution order of the two analyte species. Complexation of the sugar moiety in the glycopeptide imparted greater negative charge (arising from the borate anion) to the glycosylated peptide relative to its nonglycosylated counterpart.

4. ^2H_2O-Based Separations

Okafo and Camilleri,[29] Camilleri and Okafo,[30] and Camilleri and co-workers[31,32] have advocated the use of deuterium oxide rather than water in CZE buffers as a means of improving separations. Higher resolution was achieved in buffers where ^2H_2O replaced H_2O and this improvement was thought to arise from reduction of the electroosmotic flow in buffers containing the former because of the higher viscosity of ^2H_2O (1.23X) relative to water, as well as from differences in the ionization constants of the two solvents. Differences in separations in the two buffer systems are observed to be more significant at higher buffer pH (e.g., 7.8) than at lower pH (2.95). The inherently low electroosmotic flow in silica capillaries at low pH[20,33] would not be affected as dramatically as at higher buffer pH and thus the magnitude of changes in electroosmotic flow by addition of ^2H_2O at lower pH is reduced. Electroosmotic mobility in ^2H_2O was found to be ~25% lower than in water,[30] in agreement with the measured differences in viscosity of the two solvents.

Although not a direct means of altering the selectivity in a CZE separation, use of ^2H_2O-based buffers should afford some value by increasing the migration times of all analyte molecules, thereby allowing a longer "on-capillary" time for the separation to take place, thus improving resolution. Use of ^2H_2O to slow the electroosmotic flow offers another independent variable to control the separation. Alteration of electroosmotic flow can also be accomplished with buffer pH, but variation in this parameter also changes the charge (and selectivity) of the peptide molecules. Changes in viscosity of the electrophoresis buffer can be accomplished by other means such as addition of organic modifiers or addition of inert, neutral polymers and these additives are probably more cost effective than ^2H_2O to affect the increase in buffer viscosity.

5. Organic Modifiers

Pessi et al.[34] found addition of methanol, ethanol, or acetonitrile to the separation buffer useful for improving separations of peptides with single neutral amino acid residue substitutions. Addition of 40% acetonitrile to a 50 m*M* phosphate buffer at pH 3.5 permitted separation of peptides that differed by substitution of serine and alanine residues in the peptide. Possible explanations of the effect are differences in solvation of the peptide chains or differences in ionization of the amino acid side chains or terminus moieties due to the presence of the organic modifier. Addition of an organic modifier to the CZE buffer has also been observed to substantially decrease the magnitude of the electroosmotic flow,[35] either by altering the electrical double-layer thickness at the capillary wall and/or by changing the viscosity of the electrophoresis buffer. For example, addition of 5 to 25% (v/v) methanol to a 20 m*M* phosphate buffer has been found to reduce the electroosmotic flow to less than 50% of the flow in aqueous buffer alone[35] and this reduction of flow might also explain the observed improvement in resolution.

6. Miscellaneous

Stover et al.[27] discussed the benefits of addition of putrescine to the buffer for separation of histidine-containing compounds. For the separation of mono-, di-, and trihistidine peptides, they found improvement in efficiency and peak symmetry by addition of up to 2 m*M* putrescine to the buffer, which reduced unwanted interaction of these basic peptides with acidic silanol moieties on the wall of the silica capillary.

Tran et al.[36] described derivatization of peptides with Marfey's reagent to affect separation of chiral isomers of di- and tripeptides using micellar buffers. L-L-L, L-D-L, and D-D-D isomers of alyl-alyl-alanine were resolved in less than 30 min using this approach. Monitoring racemization of amino acids in proteins via deamidation during posttranslation processing as well as during peptide synthesis was suggested as possible uses of this approach.

IV. IMPLEMENTATION OF CAPILLARY ZONE ELECTROPHORETIC SEPARATIONS OF PEPTIDES

A. CAPILLARIES

The most widely used type of capillary for CZE separation of peptides is unmodified fused silica (50- to 100-µm I.D.) which has been pretreated for 10 to 30 min with caustic hydroxide solution ranging in concentration between 0.05 and 1 *M* (typically 0.1 *M*). This treatment cleans the inner surface of the silica capillary of contaminants and hydrolyzes any siloxane bonds formed during the capillary drawing process to yield free, ionizable silanol groups on the inner wall of the capillary. Pretreatment of the capillary is recommended to ensure a reproducible starting point if the capillary is to be used either in unmodified form or modified with some derivatization agent. Because silica behaves as a weak acid (pK_a reported to be between 6 and 10), for separation of basic peptides, modification of the capillary wall to yield a hydrophilic surface other than native silica is sometimes desired. Detailed descriptions of modification of capillaries for CZE and capillary isoelectric focusing separations can be found in Chapters 18 and 19, and only a few particularly pertinent modification procedures will be discussed here.

Hjertén[37] has described a widely applicable capillary modification procedure that used a bifunctional vinyl-containing silane to derivatize the silica surface; this treatment is followed by a polymerization reaction in which a covalently attached hydrophilic polymer is synthesized on the surface of the capillary, using an appropriate monomer (acrylamide) in conjunction with a free-radical initiator/catalyst (ammonium persulfate/*N,N,N′,N′*-tetramethylethylenediamine). Hjertén demonstrated the utility of the polyacrylamide coating for the separation of a tryptic digest of horse myoglobin at pH 2.5.

A second potentially useful series of coatings, in which the capillary wall is covered with "fuzzy" or interlocked polyether coatings based on γ-glycidoxypropyltrimethoxysilane as an anchor for polyethylene glycol (PEG) moieties, has been developed by Nashabeh and El Rassi.[38,39] Hydrophilic coatings of this type are likely to be more stable than earlier reported coatings of PEG, which were based on single anchor points of polymer onto the silica surface of the capillary inner wall.

A final type of coating useful for separation of peptides is based on derivatization of the capillary inner surface with aminopropylsilane.[40] When used in conjunction with acidic buffers, these derivatized capillaries possess a net positive charge on the interior capillary wall and thus the normal cathodal electroosmotic flow characteristic of fused silica is reversed and electroosmotic flow in these derivatized capillaries is toward the anode. Reversal of the polarity of the power supply restores the direction of electroosmotic flow toward the detector, and at acidic buffer pH (3.4), basic peptides, although positively charged, do not possess sufficiently high electrophoretic mobilities to overcome the high anodal electroosmotic flow and are thus carried to the detector. Because both the peptides and the aminopropyl-derivatized silica capillary wall possess net positive charge, adsorption of the peptides onto the capillary is negligible, permitting successful CZE separation followed by mass spectrometry (MS) analysis of the positively charged polypeptides.

TABLE 4
Useful Electrolytes for Capillary Zone Electrophoresis of Peptides

Electrolyte	Useful pH[a]	Approximate minimum useful λ (nm)[b]
Phosphate	2.12, 7.21, 12.32	195
Citrate	3.06, 4.74, 5.40	260
CHES	9.50	<190
Acetate	4.75	250
Borate	9.24	<190
CAPS	10.40	~210
BisTris propane	6.8, 9.0	~250
Tris	8.2	230
Tricine	8.1	230
PIPES	6.8	215

[a] Useful buffer pH range is ±1 pH unit above and below specified value.
[b] For UV absorbance.

B. BUFFERS

A wide variety of electrolytes can be used to prepare buffers for CZE separations of peptides. For absorbance detection, a major requirement of any component used in the buffer system is a low-ultraviolet (UV) absorbance at the wavelength used for detection. This restriction thus substantially limits the choices to a few non-UV-absorbing electrolytes. For low-pH buffers, phosphate[16,20,23,26,29,30,32,41,42] and citrate[11,39,43-45] are routinely used, whereas for basic pH buffers, Tris/Tricine,[46-49] borate,[45,50-52] and 3-(cyclohexylamino)–1-propane-sulfonic acid (CAPS)[52,53] are acceptable buffer salts. Borate, although limited to an effective pH buffering range of 7.5 to 9.5, has proved to be a particularly useful CE buffer because of its inherently low conductance, allowing separations to be run at high voltages in this buffer without excessive heating of the capillary. For detection modes other than UV absorbance, a number of other electrolytes can be utilized. Table 4 presents a partial listing of useful electrolytes for preparing running buffers for CZE separation of peptides; an indication of the minimum usable wavelength (for absorbance detection) is given, as well as the pH at which these electrolytes yield maximum buffering capacity.

Buffer concentration is restricted by the capillary internal diameter and length, the applied electrical field strength, and the efficiency of capillary cooling. Generally, buffer concentrations in the range of 10 to 100 m*M* are usable, although with high-mobility electrolytes (e.g., chloride, sulfate, citrate), use of salt concentrations at the high end of this range can overheat an uncooled capillary as a result of excess current at higher applied voltages. Use of higher ionic strength buffers is generally desirable for suppression of ion-exchange effects between charged analyte ions and the oppositely charged silica capillary wall. Higher buffer ionic strength also increases the sample loading capacity of the capillary[42] and gives more symmetrical analyte bands.

C. SAMPLE INJECTION AND PREPARATION

As discussed in Chapter 2 by Oda and Landers, introduction of sample into the CZE capillary is accomplished by either hydrostatic injection or electromigration. Because of problems with low sample volume capacity, precautions are required to ensure efficient sample injection and to avoid nonlinearity in peak height response. Direct injection of sample volumes larger than 5 to 10 nl[54] can lead to a significant loss of resolution in the separation. Also, high concentrations of salt must be removed from peptide samples (by HPLC or sample-desalting cartridges) prior to CZE separation to achieve efficient injection of samples without excessive peak broadening (see Chapter 20 for greater detail on this topic). Samples should also be concentrated by removing solvent via evaporation or lyophilization prior to injection. Alternatively, a number of on-capillary sample concentration techniques have been developed that permit direct injection of dilute samples and accomplish the concentration step in the separation capillary prior to and/or during the initial stage of the separation process.

1. Isotachophoretic Sample Enrichment

Dolnik et al.[55] implemented a unique sample enrichment technique by coupling a sampling capillary, a preconcentration capillary, and a CZE separation capillary with Teflon™* sleeves, and used these in conjunction with leading and trailing buffer ions to accomplish preconcentration of dilute samples within the preconcentration capillary before the analytical separation began. By proper choice of the separation electrolyte as well as leading and trailing ions (see Chapter 5 for more detailed explanations on choice of leading/trailing ions as well as on isotachophoresis [ITP] in general), which possessed respective electrophoretic mobilities greater than and less than the fastest and slowest analyte ions in the sample, they demonstrated preseparation enrichment of the cyanogen bromide peptide fragments of cytochrome *c*. Sample enrichment was accomplished in a 10-cm preconcentration capillary that was loaded with the dilute sample from a 6-cm length of sampling capillary. Subsequent separation occurred in a 40-cm length of the analytical separation capillary.

Substantially better results were obtained, however, when the peptide sample was simply diluted 20-fold with the leading electrolyte buffer (same electrolyte as in the separation capillary) and this diluted sample was injected from the sampling capillary directly into the separation capillary with no preconcentration capillary. In this implementation, sample enrichment occurred within the sampling capillary during the initial phase of the analysis, as the trailing edge of the leading ion in the sample swept through the length of the original sample plug, compressing the dilute sample into a sharp starting band. Although effective at enriching dilute samples, the ITP concentration technique has the disadvantages that, in addition to being cumbersome to implement as several lengths of connected capillary and numerous electrolyte solutions are used, only one type of ion (either cations or anions but not both) can be determined in a given separation.

2. Electrostacking Sample Enrichment

Compared to the ITP sample concentration method described above, a more broadly applicable and easily implemented technique for on-capillary sample concentration is sample stacking using electrical field amplification. A comprehensive review of on-column sample concentration (sample stacking) techniques in CZE can be found in the literature[56] and will be only briefly described here.

Sample stacking techniques utilize the consequences of matrix differences between the plug of injected sample and the surrounding buffer (elevated local electrical field strength within the sample zone relative to the surrounding buffer because of the increased electrical resistance of the sample zone) to cause ions of the sample to accelerate toward the adjacent buffer zone when the separation voltage is applied. As the fast-moving sample ions cross the boundary between sample and buffer zones, a lower local electrical field strength is encountered, causing a decrease in velocity and the slowed sample ions compress into a buffer zone substantially smaller than the original sample zone volume.[57,58]

Matrix differences that can generate the different electrical fields required for implementing sample stacking can be a change in solution pH and/or a change in concentration of buffer ions in the separation buffer relative to the sample buffer. In the simplest implementation of sample stacking, a large plug of sample dissolved in water is introduced into the CZE capillary

* Teflon™ is a registered trademark of E. I. Du Pont de Nemours and Co. (Wilmington, DE).

by hydrostatic injection or by electroinjection. Alternatively, a long plug of low-concentration buffer containing the sample is introduced into the separation capillary filled with buffer of the same chemical composition but of higher concentration than that in the sample. Satow et al.[25] demonstrated the impact of injection time on peak height linearity for peptides injected (hydrostatic injection) in either water or a low ionic strength buffer (0.01% trifluoroacetic acid) relative to a high-concentration phosphate buffer; maximum injection times in the water or the low ionic strength buffer were ~20 s, whereas significant deviations from linearity in peak height response and peak efficiency were encountered in the phosphate buffer for injection times above a few seconds.

A second approach to sample enrichment by electrostacking is based on a difference in pH of the sample zone relative to the separation buffer. In one implementation of this mode of on-capillary sample concentration for the separation of peptides, the pH of the sample was raised above the p*I* values of the peptides to a value above pH 10 by addition of ammonia.[54] The zone of high-pH sample was interspaced between low-pH separation buffer. On application of the separation voltage, the negatively charged peptides in the initial sample zone migrated backward toward the anode. On migration into the lower pH separation buffer moving into the capillary from the anode, the peptides became positively charged and the direction of mobility was suddenly reversed, compressing the sample ions into a narrow zone at the interface of the low-pH/high-pH buffers. The pH discontinuity rapidly dissipated and the peptides began moving from the sample concentration zone toward the cathode. When using this method of sample stacking for peptides, a fivefold improvement in sensitivity and a dramatic sharpening of peaks, even for injection times up to 30 s, was reported.

Using these sample electrostacking techniques, the final sample zone width theoretically is proportional to the ratio of buffer concentration in the original sample to that in the separation capillary. However, electroosmotic pressure gradients at the concentration discontinuities between the sample and the separation buffers originating from differences in the electrolyte concentrations in the two zones, give rise to a laminar backflow of buffer within the separation capillary, limiting the maximum concentration factor achievable with these techniques. About an order of magnitude improvement in concentration detection limit can be routinely realized with no loss of resolution with these electrostacking sample enrichment techniques.[58]

3. Protein Digestion with Enzymes or Chemical Cleavage

A number of methods are available for preparation of peptide samples from proteins.[46,53,59] The first step usually involves denaturation of the protein to disrupt the native secondary and tertiary structure of the polypeptide chain; heating an aqueous solution of the protein in boiling water for several minutes or trichloroacetic acid (TCA) precipitation (equal volumes of 10% TCA solution and 1% aqueous protein solution) of the sample are simple and effective treatments for preparing protein substrates for further degradation. Denaturation is typically followed by breakage of the disulfide bonds in the protein, which can be done by reduction and S-carboxymethylation with mercaptoethanol and iodoacetic acid;[59] performic acid oxidation is another effective treatment.[59] Once the protein substrate is prepared, the sample is subjected to chemical and/or enzymatic degradation treatments to break down the protein molecule into smaller peptide fragments. Many of these chemical agents and proteolytic enzymes exhibit specificity for one or a limited number of peptidic bonds and selectively attack the protein only at specific cleavage sites. A number of these chemical and enzymatic treatments are described in Table 5, along with an indication of the specificity of the agents. Interested readers are referred to the literature citations in Table 5, which describe in detail the use of the agents to prepare peptide samples for CZE separations.

TABLE 5
Chemical and Enzymatic Methods for Peptide Preparation

Digestion	Primary cleavage[a]	Secondary cleavage[b]	Comments	Ref.
Chemical				
N-Bromosuccinimide	Tyr	Tyr, His	Undesirable side reactions	6, 59
Cyanogen bromide	C-Side of Met	None	Most specific and generally applicable chemical method	53, 55, 57
Partial acid hydrolysis	Nonspecific	Multiple	Peptides near terminal stage of hydrolysis provide most useful information	59
Enzymatic				
Trypsin	C-Side of Lys, Arg		Endopeptidase of highest known degree of specificity	30, 39, 46, 47, 53, 59–63, 70, 84
Chymotrypsin	C-Side of Tyr, Phe, Trp, Leu	C-Side of Met, Gln, Asn, His, Thr	Broad specificity, favors cleavage at aromatic	7, 53
Pepsin	N- or C-side of Phe, Tyr, Glu, Cys, cysteine	His	Favored by hydrophobic side chains on both sides of the sensitive bond; amino and carboxyl side of aromatic residues, Leu	7, 59
Thermolysin	Ile, Leu, Val, Phe, Ala Met, and Tyr	N-Side of hydrophobic amino acids		7, 59
Staphylococcus aureus V8	C-Side of Glu, Asp	None		17, 19, 64
Endoproteinase Arg-C	C-Side of Arg			
Endoproteinase Asp-N	N-Side of Asp			
Endoproteinase Lys-C	C-Side of Lys			
Clostripain	Arg			64
Amillaria protease	N-Side of Lys			44
Peptide-*N*-glycosidase F	N-Linked oligosaccharides	Asn		39
Carboxypeptidase	C-Terminal residue			97

[a] Indicates primary cleavage site(s) of enzyme in peptides. C-side, cleaves at carboxy-terminal side of specified amino acid. N-side, cleaves at amino-terminal side of specified amino acid.

[b] Indicates secondary cleavage sites of enzyme.

D. DETECTION

Several modes of detection have been successfully interfaced with CE, and these are described in detail in Chapter 7. Those modes pertinent to CZE analysis of peptides will be discussed here.

1. Ultraviolet Absorbance

Measurement of absorbance at low-UV wavelengths is probably the most common method of detection used in CZE separations of peptides. Low-UV wavelengths are used to exploit the absorptivity of the carbonyl bonds in the peptides and of the residues containing the UV-active side chains found in phenylalanine, tryptophan, and tyrosine, as well as some absorbance contributions from histidine, arginine, glutamine, and asparagine residues. Typically, detection is done in the wavelength range of 190 to 220 nm, depending on the exact design (fiber optics, refractive, or reflective optics) of the detector. Low mass detection limits (10 to 100 pg for the typical peptide[16]) but poor concentration limits of detection (~1 μg/ml[65]) are achieved with absorbance detection due to the short optical path in the absorbance flow cell, which is a short section of the narrow diameter (50 to 100 μm) separation capillary itself. A concentration limit of detection between 10^{-5} and 10^{-6} *M* analyte can be expected when using a 75-μm I.D. capillary.

2. Fluorescence

Although direct fluorescence detection of peptides separated by CZE is possible,[65,66] this approach is useful only for peptides containing the naturally fluorescent amino acids tryptophan, tyrosine, and phenylalanine. Indeed, selective determination of tryptophan-containing peptides by native fluorescence using a xenon lamp and excitation at 280 nm[65] has been reported. For a more universal approach toward fluorescence detection of peptides, derivatization of the peptides with a fluorescent tag is required. Unlike with proteins, where a homogeneous sample can yield a broad mixture of labeled molecules when derivatized with a fluorogenic agent because of the multitude of label attachment sites in each protein molecule, the relative simplicity of peptides (and amino acids) results in much simpler mixtures after derivatization. Thus, fluorescent labeling is much more applicable to these types of compounds than to proteins.

Protocols for both preseparation[50,61,65] and on-line postseparation[65,67] derivatization of peptides have been developed for capillary electrophoretic separations. Preseparation derivatization permits a prolonged reaction time of labeling agent with analyte, ensuring complete derivatization of the analyte molecules. However, potential selectivity differences between peptides can be lost because the derivatization agent neutralizes some of the amino acids responsible for imparting charge differences to the peptides, primarily through derivatization of the N-terminal and side chain amino moieties of the residues. Postcapillary derivatization obviates these problems by adding the labeling agent to the analyte bands in the capillary effluent, using a low dead volume mixing Tee after the separation has occurred. However, postcapillary derivatization techniques employ a hydrostatic or hydrodynamic flow to introduce the labeling agent and to transport the labeled analyte bands through the reaction chamber and to the detector; this imparts a laminar component to electroosmotic flow, which can significantly increase band broadening and degrade the separation.

Excitation sources used for fluorescence detection in capillary electrophoresis range from lamp-based systems utilizing deuterium, tungsten, or xenon arc lamps (190 to 800 nm) to laser-induced fluorescence (LIF) based on argon (476, 488, 514 nm) or helium-cadmium (325, 442 nm) lasers. A range of fluorescent labeling agents has also been used, the majority of which react with the free amine functionalities in the peptide. Table 6 lists fluorophores used for both preseparation as well as postcapillary derivatization of peptides. In addition, the excitation/emission wavelengths employed, the particular advantages or disadvantages of

TABLE 6
Common Fluorophores for Peptide Derivatization in Capillary Zone Electrophoresis

Fluorogenic agent	ex/em λ[a]	Pre/postseparation[b]	Reacts with	Advantages/ disadvantages	Sensitivity[c]	Ref.
o-Phthaldialdehyde (OPA)	350/455	Pre, post	Primary amines	Excess reagent not fluorescent Derivatives are unstable	60 ng/ml; 3.3 fmol	23, 65, 67
Fluorescamine	390/450	Pre, post	Primary amines	Reacts quickly Excess reagent hydrolyzed to nonfluorescent product	20 fmol; 440 ng/ml	1, 2, 23, 65, 68–71
3 -(4-Carboxybenzoyl) -2-quinoline- carboxaldehyde (CBQCA)	442/550, He/Cd	Pre	Primary amines		low amol	23, 50
Benzoin	325/440	Pre	Arg residues	Arg-selective label	270 amol	53, 61
Naphthalene-2,3-dicarbox- aldehyde (NDA)	440/490	Post	Primary amines	Reacts slower than OPA		60, 67
9-Fluorenylmethyl chloroformate (FMOC)	260/305	Pre	Primary and secondary amines	Reaction time < 1 min Excess fluorescent reagent extracted with pentane	10 ng/ml	65
Fluorescein isothiocyanate (FITC)	488/525	Pre	Primary and secondary amines	Reaction relatively slow Artifact signal from free label	10–80 zmol	65, 71, 72
4-Methoxy-1,2- phenylenediamine	325/438, He/Cd	Pre	Tyr residues	Tyr-selective label	390 amol	61
2-Hydroxy-3*H*-phenoxa- zin-3-one	476, argon	Pre			1 pg; 10^{-9} *M*	72
Native fluoresence	257–325 280–305	None	Phe, Tyr, Trp; native fluorescence	No labeling reaction required	14–25 n*M*	65, 66

[a] Excitation/emission wavelengths.
[b] Pre- or postseparation derivatization.
[c] Attomole (amol) 10^{-18} mol; Zeptomole (zmol) 10^{-21} mol.

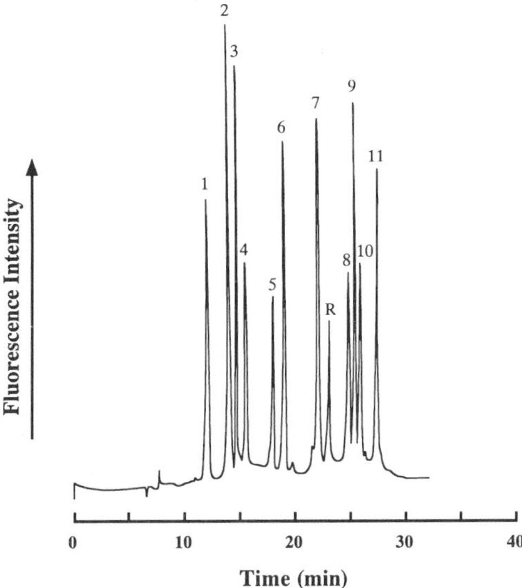

FIGURE 7. Capillary electrophoretic separation of tryptic digest of chicken egg white lysozyme derivatized with benzoin. Numbered peaks indicate signals due to derivatized peptides and the peak labeled R is due to the reagent. Separation buffer, 0.05 *M* 2-(*N*-cyclohexylamino)ethane sulfonic acid (pH 9.0), 0.07 *M* SDS, 10% CH_3CN; applied voltage, 20 kV; current, 22 μA. (From Cobb, K. A. and Novotny, M. V., *Anal. Biochem.*, 200, 149, 1992. With permission.)

each tagging agent, and an indication of the reported sensitivity achieved in CZE peptide separations are also given. Interested readers are directed to the references listed in Table 6 (e.g., see Refs. 68 to 72 and Chapter 7) for greater detail on use of specific reagents for labeling peptide samples for fluorescence detection.

Figure 7[61] serves as an exemplary illustration of the sensitivity achievable with laser-induced fluorescence detection of peptides from a real-world sample separated by CZE. Detection of the benzoin-labeled tryptic peptides in an enzymatically digested sample of chicken egg white lysozyme is shown. Only 8 fmol of digested lysozyme was injected into the capillary and the authors report a detection limit of 360 attomoles (10^{-18} mol) ($S/N = 3$).

In addition to native fluorescence and derivatization of peptides with a fluorogenic agent, a third approach based on indirect fluorescence detection of peptides has been described.[52,73,74] Indirect detection is based on charge displacement of a comigrating, fluorescent ion present in the separation buffer by nonfluorescent, nonabsorbing analyte ions. The signal is derived from analyte displacement of the buffer additive from the analyte band because, to maintain charge neutrality, the concentration of fluorescent additive is reduced in the analyte bands relative to the steady state concentration of fluorescent buffer ion that exists when no charged analyte is present. With indirect fluorescence detection, no derivatization step is required and thus the method provides a general means of detection of all ions. To date, two fluorescent detection ions have been employed; for visualization of anionic peptide molecules (run in a high-pH electrophoresis buffer), salicylate anion was used, whereas for indirect fluorescence detection of cationic peptides, cationic quinine was employed. Mass determination limits 180 times lower than those for the UV absorbance detectors commonly used in CZE systems and 3000 times lower than for commercial HPLC absorbance detectors were realized. Figure 8 illustrates the type of separation achievable on a tryptic digest of bovine serum albumin, utilizing salicylate ion as the indirect visualization agent; detection of subfemtomole quantities in individual peptide bands was reported.

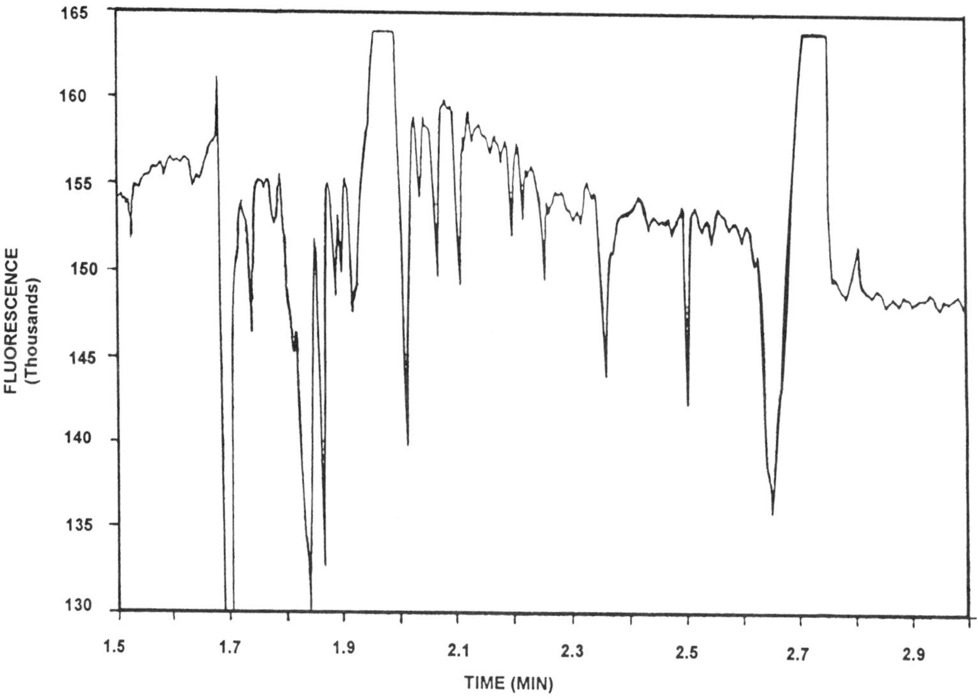

FIGURE 8. Tryptic digest of bovine serum albumin detected by indirect fluorescence. Total albumin concentration, 1.98 mg/ml before injection (~3 × 10^{-5} *M*), ~6.4 fmol of peptide injected. Electrolyte buffer, pH 10.9: 500 μ*M* salicylate, 3-(cyclohexylamino)–1-propane sulfonic acid running buffer. Capillary, 10-μm I.D. fused silica, 70-cm total length, 60 cm from injection to detector window; electromigration injection, 1 s, +10 kV. Running voltage, +40 kV. (From Hogan, B. L. and Yeung, E. S., *J. Chromatogr. Sci.*, 28(1), 15, 1990. With permission.)

3. Mass Spectrometry

The coupling of a mass spectrometer (MS) to the effluent of a CZE capillary is considered by many to be the optimum detection mode for capillary electrophoresis. As a detector, MS offers both high sensitivity as well as provides information (molecular weight, chemical structure) about peptide molecules. Several characteristics of CZE make the technique inherently compatible with MS. A total volumetric flow rate of 1 to 5 μl/min through the separation capillary is easily accommodated by the vacuum system of the mass spectrometer. Use of aqueous buffers with moderate concentrations of volatile electrolytes further facilitates the interface. Finally, development of ionization techniques such as electrospray, ion spray, and continuous flow-fast atom bombardment (CF-FAB), each of which extends the mass range of MS to accommodate large biomolecules such as peptides and small proteins, has allowed successful interfacing of CZE and MS in a relatively straightforward manner. Electrospray, ion spray, and CF-FAB all exhibit "soft ionization" characteristics for analyzing polar high molecular weight compounds; "soft" ionization typically produces only protonated or deprotonated molecular ions with little or no fragmentation products. To date, CZE has been successfully coupled to both triple quadrupole and magnetic sector mass spectrometers, with the former utilizing predominantly the electrospray interface whereas the latter is based on the CF-FAB approach. A more detailed description of CZE-MS and its uses can be found in Chapter 8.

a. Capillary Zone Electrophoresis and Electrospray/Ion Spray Ionization Mass Spectrometry

Interface of CZE to a quadrupole mass spectrometer via the electrospray ionization (ESI) interface by Olivares and co-workers[75] was the first demonstration of the combination of the two techniques. Electrospray ionization directly from the end of CZE separation capillary[76] as well as interface through a coaxial liquid sheath electrode[77] have been reported. The sheath flow interface affords greater versatility by allowing a makeup fluid (consisting typically of a volatile acid and an organic solvent such as methanol, acetonitrile, acetone, etc.) to supplement the electroosmotically pumped flow of effluent from the CZE capillary to achieve the desired volumetric flow rates of 3 to 5 μl/min. This makeup fluid serves to increase the resistivity of the electrospray liquid and to provide an adequate volume of fluid to the electrospray electrode, both facilitating formation of a stable electrospray as well as protonating analyte molecules that normally are ionized as cationic species. A second advantage of the sheath flow interface is the ability to control the composition of electrospray fluid independently of CZE buffer. This permits variation of the CZE buffer to achieve optimum separation while allowing manipulation (dilution of high ionic strength buffers, neutralization of buffer pH) of the composition of the fluid fed to the electrospray needle to yield optimal electrospray conditions.

A number of features make the electrospray or ion spray interface suitable for connecting a CZE to a mass spectrometer for analysis of peptides and proteins. Electrospray ionization and ion spray ionization of nonvolatile biomolecules produce multiply charged molecular ions, allowing a mass spectrometer with a limited mass range to analyze species with molecular weights in excess of 10^5 kDa. Biomolecules with molecular masses up to 200 kDa have been successfully analyzed with ESI, because multiprotonated molecular ions of the form $[M + nH]^{n+}$ are produced during the ionization process, bringing the m/z ratio of the ionized analyte molecules within the range of the MS. Precision and accuracy of molecular mass measurements (typically, ±0.005 to 0.05%) are reportedly superior to gel electrophoresis.[78,79]

Using the ESI interface, Smith and co-workers obtained mass spectra on ~5 pmol of various peptides such as bradykinin and angiotensin[80] and on ~1 pmol of Leu-enkephalin.[81] Moseley et al.,[82] in a comprehensive study of the conditions for optimization of CZE/ESI-MS, found a major improvement in MS detection sensitivity could be achieved using acidic buffers of low ionic strength for the CZE separation; these buffers not only maximized the signal-to-noise (S/N) ratio by increasing the total ion current but also fully protonated the peptides, minimizing the distribution of ion current across the ensemble of possible charge states that would normally occur in neutral or basic buffers. Formation of multiply charged species effectively divides the analyte signal into numerous mass/charge ratio signals. Combination of acidic CZE buffers with an aminopropyl-derivatized CZE capillary (see Section IV.A above) allowed acquisition of full-scan mass spectral data on as little as 160 fmol of peptide and collection of fragmentation data using CZE/MS/MS on less than 1 pmol of peptide. Thibault et al.,[79] using a similar strategy based on acidic buffers and derivatized capillaries, coupled CZE to an ionspray (nebulizer-assisted electrospray) coaxial sheath flow interface and were able to obtain full-scan MS analysis of glucagon and associated tryptic peptides at the picomole level. Similar picomole sensitivities for dynorphin[83] and enkephalin[84] peptides have been reported by Henion and co-workers, who employed an atmospheric pressure ionization source with ion spray interface and a liquid junction coupling (see below) to add the makeup fluid.

b. Capillary Zone Electrophoresis and Continuous Flow-Fast Atom Bombardment Mass Spectrometry

A second type of CZE-MS interface has evolved around the use of a double-focusing magnetic sector MS instrument and a continuous flow-fast atom bombardment ionization source.[85,86] Coaxial capillary columns deliver the effluent from the CZE separation capillary and the FAB matrix from an independent reservoir to the tip of the FAB probe;[85] alternatively, the effluent from the CZE capillary can be mixed with the FAB matrix through a 50-μm gap between the CZE capillary and the capillary inlet probe to the mass spectrometer via the liquid junction interface.[86]

Compatibility of CZE with mass spectrometry via the CF-FAB ionization interface is evidenced in a number of areas. The CF-FAB ionization technique provides a powerful desorption/ionization source for analysis of the polar, nonvolatile and/or thermally labile biopolymers that are typically separated by CZE. Depending on the pH of the FAB matrix, either positive or negative molecular ions can be formed in the ionization process, permitting analysis of peptides in either anion or cation form. Unlike electrospray/ion spray ionization, CF-FAB does not readily form multiply charged ions and thus the technique is more limited in mass range; peptides up to 3.20 kDa have been successfully separated by CZE and analyzed by on-line CF-FAB mass spectrometry. Also, unlike the electrospray techniques, which yield a distribution of multiply charged molecular ions that can diminish the sensitivity attainable, CF-FAB ionization, which is also a soft ionization process, yields predominantly the $[M^+]$ ion and should ultimately exhibit better detection limits relative to ESI.

Using the on-line coaxial CF-FAB interface, Moseley et al.[87] have achieved low femtomole (picogram) detection levels of small chemotactic and neuropeptides separated by CZE. Furthermore, they demonstrated acquisition of full mass spectral data up to $m/z = 650$ for three pentapeptides from a CZE separation of the mixture; to acquire full-scan mass spectra required ~500 fmol of each of the analytes in the mixture.[40] For more complex mixtures and/or to acquire mass spectral data on higher molecular weight analytes, concomitantly larger quantities of sample are needed and thus detection limits are not as sensitive as for the smaller peptides. Caprioli et al.[86] demonstrated acquisition of full-scan mass spectral data and identification of nine of the ten tryptic fragments of cytochrome *c* by CZE/CF-FAB MS; for this rather complex sample, only 60 pmol of protein was required for the analysis.

V. SELECT APPLICATIONS

A. PEPTIDE CHARACTERIZATION

The utility of CZE has evolved from the initial separations on mixtures of small, synthetic peptides and enzymatic digests of common low molecular weight proteins to high-resolution separations of large recombinant peptides and complex mixtures derived from enzymatic digests of real samples. At present, CZE is widely applied to the characterization and analysis of hormonal and therapeutic peptides and proteins derived from recombinant DNA (r-DNA) technology;[88-90] characterization of these materials with regard to degradation, deamidation, glycosylation, and methionine oxidation as well as sialic acid isoforms of peptides due to carbohydrate microheterogeneity is of interest. Figure 9 illustrates application of CZE to

FIGURE 9. Example of application of CZE to analysis of various types of peptide samples. (A) Human growth hormone (hGH) was separated from deamidation derivatives by CZE. The net charge difference of the derivatives under these experimental conditions with respect to hGH is –1 and –2 for [desamido-149]hGH and [didesamido-149–152]hGH, respectively. Peaks: (1) hGH; (2) [desamido-149]hGH; and (3) [didesamido-149–152]hGH. (From Nielsen, R. G., Sittampalam, G. S., and Rickard, E. C., *Anal. Biochem.*, 177, 20, 1989, With permission.) (B) Disulfide bond formation in IGF. Electropherogram: (1) r-IGF by-product (0.1%); (2) r-IGF I (0.1%). The irregularities in the baseline between 14 and 20 min are due to the buffer used for the stock solutions of the peptides (20 m*M* Tricine–10 m*M* $Na_2B_4O_7$–1 m*M* EDTA, pH 8.2). Electrolyte buffer, pH 11.1: 10 m*M* CAPS–5 m*M* $Na_2B_4O_7$–1 m*M* EDTA. Injection, 10 kV for 10 s; length of capillary, 120 cm (105 cm to the detector); voltage for electrophoresis, 250 V/cm at 25 μA; detection at 215 nm. (From Lüdi, H., Gassmann, E., Grossenbacher, H., and Märki, W., *Anal. Chim. Acta,* 213, 215, 1988. With permission.) (C) Deletion products. Electropherogram: (65 aa) r-hirudin (1–65 aa; 0.15%); (64 aa) r-hirudin (1–64 aa; 0.15%); (63 aa) r-hirudin (1–63 aa; 0.15%); (TRP) tryptophan (0.01%). Other by-products observed between 32 and 43 min are of unknown structure. Electrolyte buffer, pH 6.7: 16.7 m*M* PIPES–12 m*M* $Na_2B_4O_7$–1 m*M* EDTA. Injection, 10 kV for 10 s; length of capillary, 100 cm (75 cm to detector); voltage for electrophoresis, 300 V/cm at 25 μA; detection at 215 nm. (From Lüdi, H., Gassmann, E., Grossenbacher, H., and Märki, W., *Anal. Chim. Acta,* 213, 215, 1988. With permission.)

A.

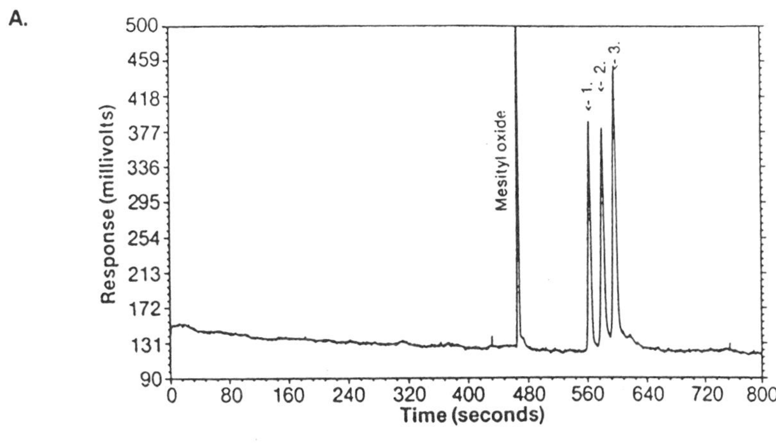
Response (millivolts)
500
459
418
377
336
295
254
213
172
131
90
Mesityl oxide
<- 1.
<- 2.
<- 3.
0
80
160
240
320
400
480
560
640
720
800
Time (seconds)

B.

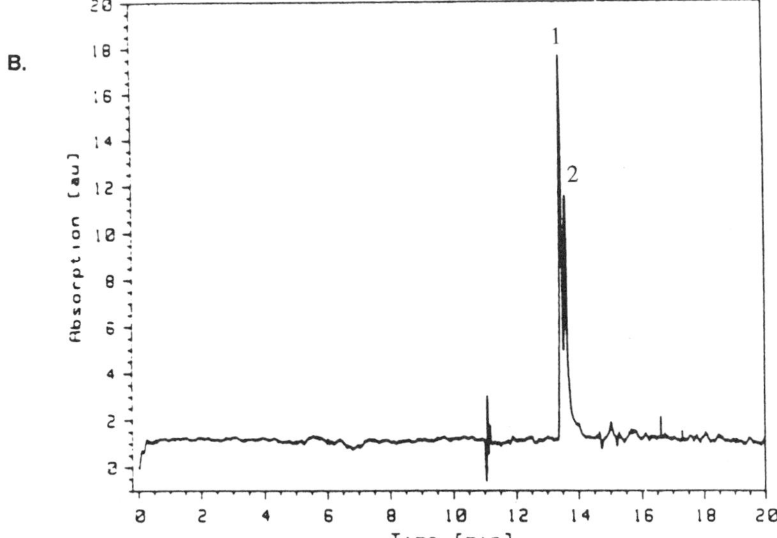
Absorption [au]
20
18
16
14
12
10
8
6
4
2
0
1
2
0
2
4
6
8
10
12
14
16
18
20
Time [min]

C.

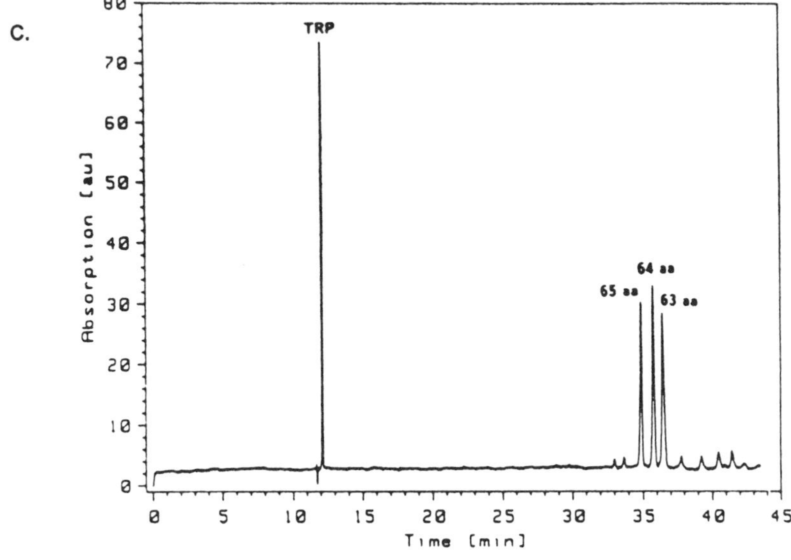
Absorption [au]
80
70
60
50
40
30
20
10
0
TRP
64 aa
65 aa
63 aa
0
5
10
15
20
25
30
35
40
45
Time [min]

FIGURE 9.

analysis of several such samples, where separation of degradation products varying from the parent molecule via subtle changes in structure has been demonstrated.

A primary degradative mechanism of peptide drugs is deamidation at asparagine and/or glutamine residues. Separation of intact recombinant human growth hormone (r-hGH) from the mono- and didesamido degradation products due to deamidation of asparagine residues at positions 149 and 152 in this 191-residue acid residue peptide is shown in Figure 9A.[48] Figure 9B[91] shows the separation of native recombinant insulin-like growth factor (r-IGF I) from a variant that differs from the native molecule in that the two disulfide bonds in the variant molecule are between cysteines 6–47 and 48–52 instead of between cysteines 6–48 and 47–52 as in r-IGF I. Incorrect pairing of these two disulfide bonds in this 70-residue peptide can be discerned easily via CZE in a 15-min separation. Another common degradative mechanism of peptides is loss of amino acids to yield deletion products. Figure 9C,[91] which illustrates the application of CZE to this type of analysis, shows the separation of recombinant hirudin, a 65-amino acid peptide, from two deletion products produced by loss of one and two amino acids from the C terminus of the native peptide. Each of the above examples serves to illustrate the capability of CZE to resolve biomolecules that differ subtly in structure and/or chemical composition, which is one of the reasons the technique has become widely employed for analysis of recombinant DNA-derived peptides and proteins.[92-95]

Capillary zone electrophoretic has also been used to monitor interaction of certain types of molecules such as antibiotics with peptides and small proteins[49,96] as well as peptide–peptide interactions.[10] A comprehensive listing of reported applications of CZE to peptide analyses and studies is given in Table 7. This listing gives an overview of the types of peptides that have been found to be amenable to CZE analysis and provides an initiation point for more in-depth reading about CZE separations of peptides of particular interest to individual readers.

B. PROTEIN MAPPING/SEQUENCING

In addition to quality control of genetically engineered peptides as discussed above, peptide mapping by CZE is also useful for detection of posttranslational amino acid modifications and for identification and localization of genetic variants in peptides and proteins. Peptide mapping is the first step in any protein-sequencing strategy;[13] the technique yields both the peptides fragments to be used in subsequent sequencing protocols as well as information on protein variants and posttranslation protein modification by comparative analyses. Several novel applications have been developed from the use of CZE in this area.

Application of CZE for peptide mapping and for use in protein sequencing[46] derives from the minute quantities of samples that are required. Development of microreactors for cleavage of minute quantities of proteins into peptide fragments has led to the use of CZE for peptide mapping on very small quantities of proteins. Trypsin, chymotrypsin, and cyanogen bromide were employed in a strategy that required only ~5 pmol (nanograms) of protein to obtain a peptide map.[46,53] Use of tributylphosphine as both protein denaturant and reducing agent to cleave disulfide bonds and simultaneous alkylation of reduced disulfide bonds via 2-methyl-aziridine precluded substantial protein losses encountered by employing the traditional denaturation steps with urea or guanidine chloride followed by dialysis to remove unwanted salt. Chemical cleavage of the protein with cyanogen bromide or site-specific enzymatic cleavage by passage through a microcolumn digestion reactor packed with immobilized enzyme on agarose beads were used to produce the peptide fragment samples. Finally, separation and detection of the resulting peptides was performed by CZE with either conventional UV absorbance or LIF detection to yield peptide maps on low picomole quantities of model proteins such as trypsinogen and human serum albumin. A similar approach[60] based on a microreactor with trypsin immobilized on the inner surface of a 50-μm I.D capillary via biotin–avidin linkages for on-line digestion of minute amounts of protein has also been reported; the tryptic map of β-casein was shown.

TABLE 7
Peptide Applications in Capillary Zone Electrophoresis

Peptide	Size	Buffer pH	Application	Ref.
Recombinant and synthetic peptides				
Human growth hormone (r-hGH)	191 res; 22 kDa	8.1	Tryptic map	47, 62
		2.56, 8.1	Separation of native from deamidated variant	63
			Tryptic map	
		8.0	Separation of desamido-149 hGH, didesamido-149–152 hGH, and hGH	48
		6.5	Separation of deamidated from native	88
		2.4, 6.1, 8.1, 10.4; pH 8.1 optimum	19 peptide fragments of hGH	
Insulin	51 res	8.2, 2.2	Separation of derivatives of biosynthetic human insulin	48
Anticoagulant peptide (MDL 28,050)	10 res	5.5	Separation of native peptide from deletion by-products	89
Interferon		8.3	Tryptic map; quality control	92
T4 receptor protein (r-CD4)	40 kDa	4.5	Separation of variants	88
Tissue plasminogen activator (r-tPA)	66 kDa	4.5	Assess charge heterogeneity due to variable sialic acid content	88, 93
Multiple antigen peptides	44 res	2.5–4.5	Assess purity of synthetic branched peptides	34
Insulin-like growth factor (r-IGF)	70 res	11.1	Separation of by-products with small variations in structure; differences in location of two disulfide bonds	91
Insulin-like growth factor II (r-IGF-II)	67 res	2.35, 8.15		7
Bovine aprotinin (pancreatic trypsin inhibitor)	58 res		Separated from deletion products	44
Interleukin 1α		8.3	Separation of native from deaminated	92
Natural peptides				
Motilin peptides	22 res	2.5, 9.4	Effect of charge, hydrophobicity, secondary structure, and length. Deamidation of the Gln residues	11, 45
Luteinizing hormone-releasing hormone		8.3		50
Calcitonin gene-related peptides	37 res	3.5–4.5	Separation of α, β species variants from rat amygdala	93
Calcitonin, elcatonin	32 res	2.93, 7.93	Tryptic digest; micropreparative CZE for sequencing	30
Hirudin	65 res; 12 kDa		Deletion products	91
		6.5	Sulfate and nonsulfate forms	41
7.5-kDa peptide			Monitor purity of material prepared by RPLC	94
Growth hormone-releasing factor	Fragments, 1 to 29 res	2.3–4.3	Separation of degradative and cyclic lactam derivatives. Deamidation of Asn residues	17
Marine mollusk neuropeptides			Detection of subattomole quantities by LIF	71

TABLE 7 (Continued)
Peptide Applications in Capillary Zone Electrophoresis

Peptide	Size	Buffer pH	Application	Ref.
Bovine somatostatin (bST)	199 res	2.35, 8.15		7
Human proinsulin	86 res	2.35, 8.15		7
Substance P peptides		2.6	Degradation products	3
Cholecystokinin octapeptide	8 res	6.5	Sulfate and nonsulfate form	41
Enkephalin pentapeptide	5 res	6.5	Sulfate and nonsulfate form	41
Bradykinin	9 res	7.8	Preparative CE for sequencing	31
Vasoactive intestinal peptide	28 res		Detection in rat cerebral cortex	95
Peptides from chemical and enzymatic digests of proteins				
Cytochrome *c* proteins			Identify single residues from rabbit, bovine	43
			Substitutions between homologous peptides	
Chicken ovalbumin		7.0	Tryptic digest	68
α_1-Acid glycoprotein fragments	181 res	5.0	Trypsin and endoglycosidase digestions; mapping of glycosylated and nonglycosylated forms	39
Collagen type I, type III		9.2	Separation of chains, chain polymers, and CNBr fragments	6
Glucagon		7.8	Tryptic digests	29, 31
Miscellaneous peptide applications				
Alyl-alanine analogs	2 res	7.1, 7.5	Affinity measurements with vancomycin	96
	2 res	8.28	Binding constants with vancomycin	49

Following peptide mapping, the second step in a protein–sequencing strategy is determination of both the total amino acid composition of the protein as well as the actual sequence information; pure fractions of individual peptides are required for each of these operations. In this respect, CZE has been used in a micropreparative mode, where pure individual peptide fractions have been collected and analyzed for total amino acid composition as well as peptide sequence. Bergman et al.[97] described the use of CZE for preparation of peptides for sequencing; four or five consecutive CZE runs yielded sufficient pooled material for either gas- or solid-phase sequence analysis and a single CZE separation produced enough material for determination of total amino acid composition via phenylthiocarbamyl amino acid analysis. Using only 200 fmol of human insulin B-chain, they demonstrated separation of the three Glu-C endoproteinase fragments of this model protein in 10 min. Two of the three peptide fragments were collected from four or five sequential CZE runs, pooled, and subjected to sequencer analysis. The authors also demonstrated the use of CZE for C-terminal characterization of a porcine intestinal peptide by digestion with carboxypeptidase followed by separation and identification of the liberated amino acid.

Camilleri and co-workers[31,32] also demonstrated the preparative capabilities of CZE to yield sufficient peptide for amino acid sequencing. Using a ^2H_2O-based electrophoresis buffer to diminish the magnitude of the electroosmotic flow and thus improve resolution of adjacent peptide bands (see Section III.B.4), CZE separations of the fragments of calcitonin as a model peptide were accomplished. A high-pressure syringe pump was used to deliver buffer through the CZE separation capillary at flow rates of ~1 to 2 μl/min (in a process they termed "dynamic elution") to drive the separated peptides past the detector and allow for fraction collection.

Using the effluent from a single injection (35 pmol or 120 ng of total protein) of the tryptic digest of calcitonin, the complete sequence of three of the four tryptic fragments was determined by Edman degradation. Chen et al.[42] also demonstrated micropreparative separations of tryptic peptide fragments for sequencing, using a large inner diameter (200-μm) capillary and a high ionic strength buffer (280 m*M* borate). Use of the larger inner diameter capillary increased sample loading capacity and thus allowed sequencing to be accomplished on fractions collected from a single injection of 90 pmol of protein.

C. COMPLEMENTARITY TO REVERSED-PHASE LIQUID CHROMATOGRAPHY

Use of CZE as a second-dimension separation technique following HPLC has been explored as a method of expanding the information attainable from separation methods. A first-dimension separation by reversed-phase liquid chromatography (RPLC) followed by a second-dimension separation by CZE produces a two-dimensional separation strategy, yielding much more information than either of the individual techniques alone. As with any multidimensional separation strategy, complementarity of the combined techniques is essential for maximizing the peak capacity and information content of the combined methods. Capillary zone electrophoresis selectivity, based primarily on analyte charge (and to a much lesser degree on mass or diameter), coupled with RPLC, which separates on the basis of relative hydrophobicity of the analyte molecules, yields the desired combination.

Castagnola et al.[13] demonstrated the appropriateness of combining RPLC and CZE by performing separations of tryptic digests of horse myoglobin in the same buffers by both methods. Good correlation between the peak areas measured by both methods was observed; however, correlation between migration/elution times as measured by the two methods was poor. This is illustrated in Figure 10, which shows separations of the peptide mixture by both CZE (Figure 10A) and RPLC (Figure 10B) as well as a plot of the migration time of the peptide fragments measured by CZE against the elution time of the fragments measured by RPLC (Figure 10C). Correlation between the data (which would be indicated by data points falling on a diagonal line) would have indicated similarity between the two separation techniques. However, the large degree of scatter in the data in Figure 10C, along with no obvious relationship between the elution order of the peptide fragments (Figure 10A and B) indicates orthogonality of the separation mechanisms for the two methods. Combination of the two techniques will likely result in few if any unresolved components in a complex mixture of peptides.

Bushey and Jorgenson[70,98] have described a computer-automated two-dimensional separation system using RPLC followed by CZE. Effluent containing analyte bands from a narrow bore reversed-phase column is collected in the loop of a computer-controlled valve and the collected effluent fraction is periodically sampled into a CZE capillary by electromigration and subsequently separated while the valve loop is filling with the next effluent fraction from the chromatograph. Because of the inherent speed of separation by CZE, the second-dimension electrophoretic separation could be completed in about 1 min (in more recent investigations, the CZE analysis time was reduced to several seconds[99]) and thus in excess of 100 CZE analyses could be performed during the RPLC separation. Automated data collection and processing yielded three-dimensional contour plots termed "chromatoelectropherograms" of the migration elution times and peak intensities of the multitude of bands in tryptic digests.

Finally, Strickland[64] has advocated the use of CZE both for verification of the purity of synthetic peptides produced on a synthesizer and subsequently purified by RPLC as well as for ascertaining the purity of RPLC-purified peptide fractions to be subsequently used for amino acid analysis. The speed of the method permits rapid verification of the purity of a

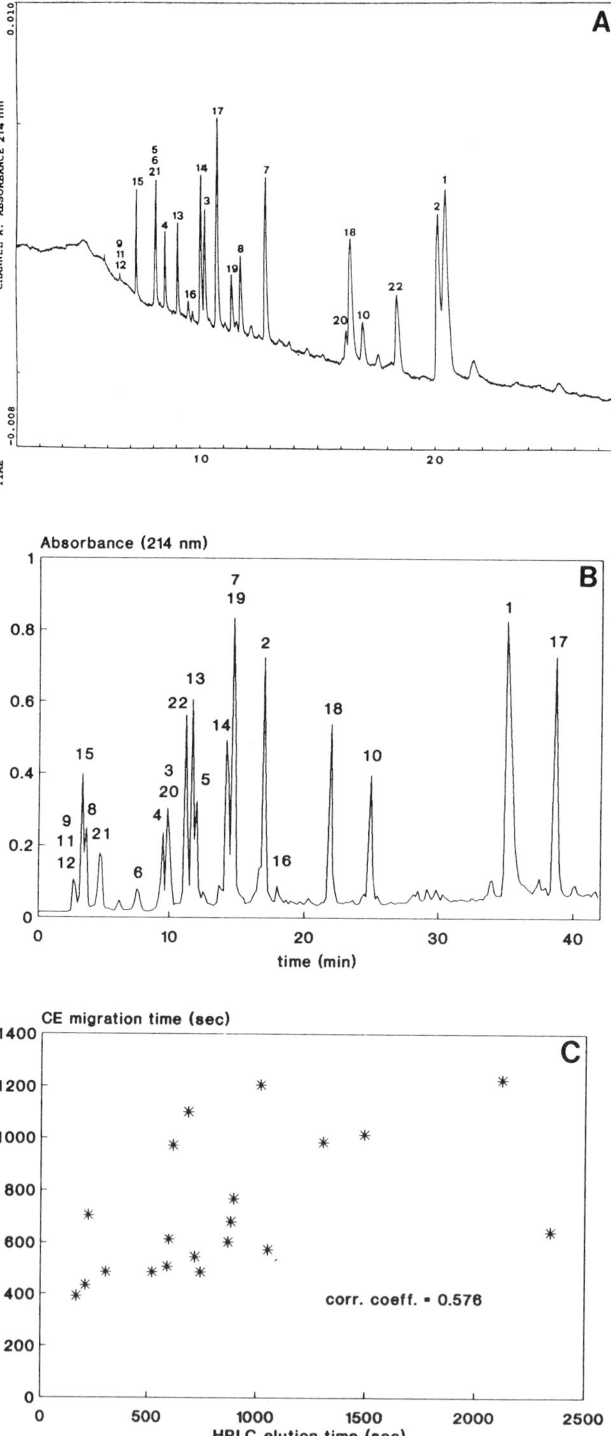

FIGURE 10. Horse myoglobin tryptic mapping obtained by (A) CZE on a modified capillary and (B) reversed-phase HPLC. (C) Correlation between HPLC elution times and CE migration times of the peptides identified in the tryptic digest of horse myoglobin. (Reproduced from Castagnola, M., Cassiano, L., Rabino, R., Rossetti, D. V., and Bassi, F. A., *J. Chromatogr.*, 572(1–2), 51, 1991. With permission.)

sample isolated by a different separation mechanism (in this case, microbore and narrowbore RPLC) and can ultimately save sequencer time as well as ensuring that amino acid analyses are run only on pure samples of peptides.

VI. FUTURE DIRECTIONS

Capillary electrophoretic separations of peptides have progressed significantly during the past decade as evidenced by the wide range of applications discussed above. Several exciting developments in the implementation and application of the technique are on the horizon. As discussed above, manipulation of the electroosmotic flow to improve CZE separations can be accomplished by changes in buffer pH, ionic strength, viscosity, and chemical composition, although these can also simultaneously cause unwanted changes in selectivity in a CZE separation. Control of the electroosmotic flow by means other than changes in the chemical or physical properties of the separation buffer could result in significant improvement in the reproducibility and implementation of CZE separations. Promising approaches[100-102] to control electroosmotic flow based on externally applied radial electrical fields across the capillary wall to supplement or to reduce the magnitude of the ζ potential, which is responsible for the magnitude of the electroosmotic flow through the capillary, are currently under investigation (see Chapter 22 for more details). The impact of control of electroosmotic flow on the quality of separation of peptides is illustrated in Figure 11.[102] The four panels in this figure represent separations of the tryptic fragments of r-hGH in a capillary in which the electroosmotic flow progressively slows from that in native silica (Figure 11A) to a flow of approximately zero (Figure 11D) in response to a change in electric (offset) voltage across the capillary wall from –23 to +1 kV. Migration times become progressively longer as the flow is reduced, providing a longer time during which the separation can occur. The quality of the separation improves dramatically during the longer separation period, with substantially better resolution of the tryptic fragments. Since this method of flow control can be implemented independent of changes in buffer composition, the technique could offer a significant enhancement in the ability to optimize separation of peptides (and other types of molecules as well).

Use of CZE for sequencing smaller quantities of peptides and proteins also appears to offer great promise. Collection of protein and peptide bands on elution from a CZE capillary has been demonstrated[103-105] using moving disks of filter paper or filtration membrane and nitrocellulose-coated aluminum foils used for plasma desorption mass spectrometry. Direct sequencing of the collected bands of separated protein in the 1- to 10-pmol range via Edman degradation and subsequent separation of the PTH-amino acids by HPLC has been accomplished.[104] Peptide sequencing using plasma desorption mass spectrometry on 300 to 500 fmol of peptide fractions collected on elution from the CZE capillary has also been reported.[105] Application of CZE for the sequencing separations coupled with laser-induced fluorescence detection and improvements in on-capillary concentration techniques could reduce the protein sample requirements for sequencing by several orders of magnitude.

Another exciting approach to sequencing derives from the use of CZE coupled to a mass spectrometer for direct sequencing of peptides subsequent to separation in the capillary. Hunt et al.[106] have demonstrated on-line sequencing of proteins and peptides at the 1-pmol level, using CZE coupled to an electrospray mass spectrometer. Sequence data was derived from the collision-activated dissociation spectra, using both the fragmentation pattern and differences in the *m/z* values for adjacent fragments in the spectra. Subpicomole-level sequencing of peptides in mixtures, using the electrospray interface and the ion trap mass spectrometer, appears possible in the near future.[106]

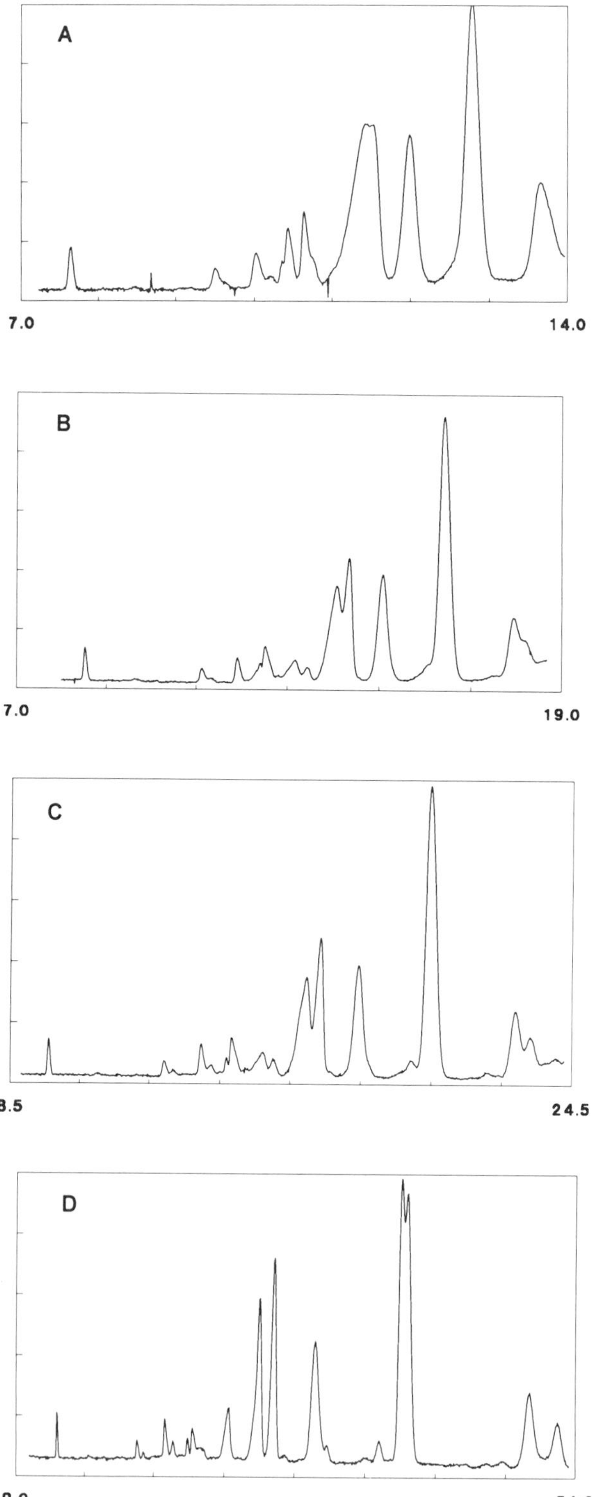

FIGURE 11. Variation of bulk flow by variation of offset voltage. The electropherograms shown illustrate the sharpening of the bands with progressively slower bulk electroosmotic flow. Effect of external radial electrical field on separation of tryptic fragments of recombinant human growth hormone. External voltage: (A) –23 kV, (B) –18 kV, (C) –10 kV, and (D) +1 kV. The electroosmotic flow in (D) is very close to zero. (Holloway, R. R., Keely, C. A., McManigill, D., and Young, J. E., paper presented at HPCE '93, Orlando, FL, January 25–28, 1993. With permission.)

ACKNOWLEDGMENTS

The author would like to thank Ms. Maria Rey for conducting a thorough search on the background literature for this chapter. The author also acknowledges appreciation to Dionex Corporation for facilities and time used in preparation of this chapter.

REFERENCES

1. **Jorgenson, J. W. and Lukacs, K. D.,** Zone electrophoresis in open tubular capillaries, *Anal. Chem.*, 53(18), 1298, 1981.
2. **Jorgenson, J. W. and Lukacs, K. D.,** Zone electrophoresis in open-tubular glass capillaries: preliminary data on performance, *J. High Resolut. Chromatogr. Chromatogr. Commun.*, 4(5), 230, 1981.
3. **Nyberg, F., Zhu, M.-D., Liao, J.-L., and Hjertén, S.,** High-performance electrophoresis in studies of substance P degradation, in *Electrophoresis '88*, Schafer-Nielsen, C., Ed., VCH, Weinheim, Germany, 1988, p. 141.
4. **Offord, R. E.,** Electrophoretic mobilities of peptides on paper and their use in the determination of amide groups, *Nature (London)*, 211, 591, 1966.
5. **Deyl, Z., Rohlicek, V., and Adam, M.,** Separation of collagens by capillary zone electrophoresis, *J. Chromatogr.*, 480, 371, 1989.
6. **Deyl, Z., Rohlicek, V., and Struzinsky, R.,** Some rules applicable to capillary zone electrophoresis of peptides and proteins, *J. Liq. Chromatogr.*, 12(13), 2515, 1989.
7. **Rickard, E. C., Strohl, M. M., and Nielsen, R. G.,** Correlation of electrophoretic mobilities from capillary electrophoresis with physicochemical properties of proteins and peptides, *Anal. Biochem.*, 197(1), 197, 1991.
8. **Issaq, H. J., Janini, G. M., Atamna, I. Z., Muschik, G. M., and Lukszo, J.,** Capillary electrophoresis separation of small peptides: effect of pH, buffer additives and temperature, *J. Liq. Chromatogr.*, 15(6–7), 1129, 1992.
9. **Kurosu, Y., Sato, Y., and Senda, M.,** Investigation of migration behaviors of peptides and proteins by capillary zone electrophoresis, *Anal. Sci.*, 7 (Suppl., Proc. Int. Congr. Anal. Sci.), 273, 1991.
10. **Landers, J. P., Oda, R. P., Liebenow, J. A., and Spelsberg, T. C.,** The utility of high resolution capillary electrophoresis for monitoring peptide homo- and hetero-dimer formation, *J. Chromatogr.*, 1993, in press.
11. **Florance, J. R., Konteatis, Z. D., Macielag, M. J., Lessor, R. A., and Galdes, A.,** Capillary zone electrophoresis studies of motilin peptides. Effects of charge, hydrophobicity, secondary structure and length, *J. Chromatogr.*, 559(1–2), 391, 1991.
12. **Grossman, P. D., Colburn, J. C., and Lauer, H. H.,** A semiempirical model for the electrophoretic mobilities of peptides in free-solution capillary electrophoresis, *Anal. Biochem.*, 179(1), 28, 1989.
13. **Castagnola, M., Cassiano, L., Rabino, R., Rossetti, D. V., and Bassi, F. A.,** Peptide mapping through the coupling of capillary electrophoresis and high-performance liquid chromatography: map prediction of the tryptic digest of myoglobin, *J. Chromatogr.*, 572(1–2), 51, 1991.
14. **Skoog, B. and Wichman, A.,** Calculation of the isoelectric points of polypeptides from the amino acid composition, *Trends Anal. Chem.*, 5(4), 82, 1986.
15. **Pennino, D. J.,** Free-zone capillary electrophoresis. A simple computer program for predicting peptide migration times, *BioPharm*, Sept 2(8), 41, 1989.
16. **Strickland, M. and Strickland, N.,** Free-solution capillary electrophoresis using phosphate buffer and acidic pH, *Am. Lab. (Fairfield, Conn.)*, November, 60, 1990.
17. **Bongers, J., Lambros, T., Felix, A. M., and Heimer, E. P.,** Capillary zone electrophoresis of degradative and cyclic lactam derivatives of the growth hormone-releasing factor peptide, *J. Liq. Chromatogr.*, 15(6–7), 1115, 1992.
18. **Kirkland, J. J. and McCormick, R. M.,** Liquid phase separation methods: HPLC, FFF, electrophoresis, *Chromatographia*, 24, 58, 1987.
19. **Grossman, P. D., Wilson, K. J., Petrie, G., and Lauer, H. H.,** Effect of buffer pH and peptide composition on the selectivity of peptide separations by capillary zone electrophoresis, *Anal. Biochem.*, 173(2), 265, 1988.
20. **McCormick, R. M.,** Capillary zone electrophoretic separation of peptides and proteins using low pH buffers in modified silica capillaries, *Anal. Chem.*, 60(21), 2322, 1988.
21. **Wheat, T. E., Young, P. M., and Astephen, N. E.,** Use of capillary electrophoresis for the detection of single-residue substitutions in peptide mapping, *J. Liq. Chromatogr.*, 14(5), 987, 1991.
22. **Steuer, W., Grant, I., and Erni, F.,** Comparison of high-performance liquid chromatography, supercritical fluid chromatography, and capillary zone electrophoresis in drug analysis, *J. Chromatogr.*, 507, 125, 1990.

23. **Liu, J., Cobb, K. A., and Novotny, M.,** Capillary electrophoretic separations of peptides using micelle-forming compounds and cyclodextrins as additives, *J. Chromatogr.*, 519(1), 189, 1990.
24. **Sigma Catalog,** Sigma Chemical Company, St. Louis, Missouri, 1993, p. 1561.
25. **Satow, T., Machida, A., Funakushi, K., and Palmieri, R.,** Effects of the sample matrix on the separation of peptides by high performance capillary electrophoresis, *J. High Resolut. Chromatogr. Commun.*, 14, 276, 1991.
26. **Mosher, R. A.,** The use of metal ion-supplemented buffers to enhance the resolution of peptides in capillary zone electrophoresis, *Electrophoresis*, 11(9), 765, 1990.
27. **Stover, F. S., Haymore, B. L., and McBeath, R. J.,** Capillary zone electrophoresis of histidine-containing compounds, *J. Chromatogr.*, 479, 241, 1989.
28. **Hoffstetter-Kuhn, S., Paulus, A., Gassmann, E., and Widmer, H. M.,** Influence of borate complexation on the electrophoretic behavior of carbohydrates in capillary electrophoresis, *Anal. Chem.*, 63(15), 1541, 1991.
29. **Okafo, G. N. and Camilleri, P.,** Capillary electrophoretic separation in both H_2O- and ^2H_2O-based electrolytes can provide more information on tryptic digests, *J. Chromatogr.*, 547(1–2), 551, 1991.
30. **Camilleri, P. and Okafo, G. N.,** Replacement of H_2O by D_2O in capillary zone electrophoresis can increase resolution of peptides and proteins, *J. Chem. Soc. Chem. Commun.*, 3, 196, 1991.
31. **Camilleri, P., Okafo, G. N., Southan, C., and Brown, R.,** Analytical and micropreparative capillary electrophoresis of the peptides from calcitonin, *Anal. Biochem.*, 198(1), 36, 1991.
32. **Camilleri, P., Okafo, G. N., and Southan, C.,** Separation by capillary electrophoresis followed by dynamic elution, *Anal. Biochem.*, 196(1), 178, 1991.
33. **Lukacs, K. D. and Jorgenson, J. W.,** Capillary zone electrophoresis: effect of physical parameters on separation efficiency and quantitation, *J. High Resolut. Chromatogr. Chromatogr. Commun.*, 8, 407, 1985.
34. **Pessi, A., Bianchi, E., Chiappinelli, L., Nardi, A., and Fanali, S.,** Application of capillary zone electrophoresis to the characterization of multiple antigen peptides, *J. Chromatogr.*, 557(1–2), 307, 1991.
35. **Altria, K. D. and Simpson, C. F.,** Measurement of electroendosmotic flows in high-voltage capillary electrophoresis, *Anal. Proc.*, 23, 453, 1986.
36. **Tran, A. D., Blanc, T., and Leopold, E. J.,** Free solution capillary electrophoresis and micellar electrokinetic resolution of amino acid enantiomers and peptide isomers with L- and D-Marfey's reagents, *J. Chromatogr.*, 516, 241, 1990.
37. **Hjertén, S.,** High-performance electrophoresis: elimination of electroendosmosis and solute adsorption, *J. Chromatogr.*, 347, 191, 1985; U.S. Patent 4,680,201, 1987.
38. **Nashabeh, W. and El Rassi, Z.,** Capillary zone electrophoresis of proteins with hydrophilic fused-silica capillaries, *J. Chromatogr.*, 559(1–2), 367, 1991.
39. **Nashabeh, W. and El Rassi, Z.,** Capillary zone electrophoresis of α_1-acid glycoprotein fragments from trypsin and endoglycosidase digestions, *J. Chromatogr.*, 536(1–2), 31, 1991.
40. **Moseley, M. A., Deterding, L. J., Tomer, K. B., and Jorgenson, J. W.,** Determination of bioactive peptides using capillary zone electrophoresis/mass spectrometry, *Anal. Chem.*, 63, 109, 1991.
41. **Hortin, G. L., Griest, T., and Benutto, B. M.,** Separation of sulfated and nonsulfated forms of peptides by capillary electrophoresis: comparison with reversed-phase HPLC, *BioChromatography*, 5(3), 118, 1990.
42. **Chen, F. A., Kelly, L., Palmieri, R., Biehler, R., and Schwartz, H.,** Use of high ionic strength buffers for the separation of proteins and peptides with capillary electrophoresis, *J. Liq. Chromatogr.*, 15(6–7), 1143, 1992.
43. **Young, P. M., Astephen, N. E., and Wheat, T. E.,** Effects of pH and buffer composition on peptide separations by high performance liquid chromatography and capillary electrophoresis, *LC-GC*, 10(1), 26, 1992.
44. **Vinther, A., Bjørn, S. E., Sørensen, H. H., and Søeberg, H.,** Identification of aprotinin degradation products by the use of high-performance capillary electrophoresis, high-pressure liquid chromatography and mass spectrometry, *J. Chromatogr.*, 516, 175, 1990.
45. **Florance, J.,** CZE studies of motilin peptides, *Am. Lab. (Fairfield, Conn.)*, May, 32L, 1991.
46. **Cobb, K. A. and Novotny, M.,** High-sensitivity peptide mapping by capillary zone electrophoresis and microcolumn liquid chromatography, using immobilized trypsin for protein digestion, *Anal. Chem.*, 61(20), 2226, 1989.
47. **Nielsen, R. G. and Rickard, E. C.,** Method optimization in capillary zone electrophoretic analysis of hGH tryptic digest fragments, *J. Chromatogr.*, 516(1), 99, 1990.
48. **Nielsen, R. G., Sittampalam, G. S., and Rickard, E. C.,** Capillary zone electrophoresis of insulin and growth hormone, *Anal. Biochem.*, 177, 20, 1989.

49. **Carpenter, J. L., Camilleri, P., Dhanak, D., and Goodall, D.,** A study of the binding of vancomycin to dipeptides using capillary electrophoresis, *J. Chem. Soc. Chem. Commun.*, 11, 804, 1992.
50. **Liu, J., Hsieh, Y.-Z., Wiesler, D., and Novotny, M.,** Design of 3-(4-carboxybenzoyl)-2-quinolinecarboxaldehyde as a reagent for ultrasensitive determination of primary amines by capillary electrophoresis using laser fluorescence detection, *Anal. Chem.*, 63(5), 408, 1991.
51. **Advis, J. P., Hernandez, L., and Guzman, N. A.,** Analysis of brain neuropeptides by capillary electrophoresis: determination of luteinizing hormone-releasing hormone from ovine hypothalamus, *Pept. Res.*, 2, 389, 1989.
52. **Hogan, B. L. and Yeung, E. S.,** Indirect fluorometric detection of tryptic digests separated by capillary zone electrophoresis, *J. Chromatogr. Sci.*, 28(1) 15, 1990.
53. **Cobb, K. A. and Novotny, M. V.,** Peptide mapping of complex proteins at the low-picomole level with capillary electrophoretic separations, *Anal. Chem.*, 64(8), 879, 1992.
54. **Aebersold, R. and Morrison, H. D.,** Analysis of dilute peptide samples by capillary zone electrophoresis, *J. Chromatogr.*, 516(1), 79, 1990.
55. **Dolnik, V., Cobb, K. A., and Novotny, M.,** Capillary zone electrophoresis of dilute samples with isotachophoretic preconcentration, *J. Microcolumn Sep.*, 2, 127, 1990.
56. **Chien, R.-L. and Burgi, D. S.,** On-column sample concentration using field amplification in CZE, *Anal. Chem.*, 64(8), 489A, 1992.
57. **Burgi, D. S. and Chien, R.-L.,** Optimization in sample stacking for high performance capillary electrophoresis, *Anal. Chem.*, 63, 2042, 1991.
58. **Burgi, D. S. and Chien, R.-L.,** Application of sample stacking to gravity injection in capillary electrophoresis, *J. Microcolumn Sep.*, 3, 199, 1991.
59. **Kasper, C. B.,** Fragmentation of proteins for sequence studies and separation of peptide mixtures, in *Protein Sequence Determination*, Needleman, S. B., Ed., Springer-Verlag, Berlin, 1975, Chapter 5, pp. 141–161.
60. **Amankwa, L. N. and Kuhr, W. G.,** Trypsin-modified fused-silica capillary microreactor for peptide mapping by capillary zone electrophoresis, *Anal. Chem.*, 64(14), 1610, 1992.
61. **Cobb, K. A. and Novotny, M. V.,** Selective determination of arginine-containing and tyrosine -containing peptides using capillary electrophoresis and laser-induced fluorescence detection, *Anal. Biochem.*, 200(10), 149, 1992.
62. **Nielsen, R. G., Riggin, R. M., and Rickard, E. C.,** Capillary zone electrophoresis of peptide fragments from trypsin digestion of biosynthetic human growth hormone, *J. Chromatogr.*, 480, 393, 1989.
63. **Frenz, J., Wu, S.-L., and Hancock, W. S.,** Characterization of human growth hormone by capillary electrophoresis, *J. Chromatogr.*, 480, 379, 1989.
64. **Strickland, M.,** Use of capillary electrophoresis to screen peptide fractions from micro- and narrow-bore reversed-phase columns, *Am. Lab. (Fairfield, Conn.)*, March, 70, 1991.
65. **Albin, M., Weinberger, R., Sapp, E., and Moring, S.,** Fluorescence detection in capillary electrophoresis: evaluation of derivatizing reagents and techniques, *Anal. Chem.*, 63, 417, 1991.
66. **Swaile, D. F. and Sepaniak, M. J.,** Laser-based fluorimetric detection schemes for the analysis of proteins by capillary zone electrophoresis, *J. Liq. Chromatogr.*, 14(5), 869, 1991.
67. **Nickerson, B. and Jorgenson, J. W.,** Characterization of a post-column reaction-laser-induced fluorescence detector for capillary zone electrophoresis, *J. Chromatogr.*, 480, 157, 1989.
68. **Green, J. S. and Jorgenson, J. W.,** High speed zone electrophoresis in open-tubular fused silica capillaries, *J. High Resolut. Chromatogr. Commun.*, 7, 529, 1984.
69. **Jorgenson, J. W. and Lukacs, K. D.,** Free zone electrophoresis in glass capillaries, *Clin. Chem., (Winston Salem, N.C.)*, 27(9), 1551, 1981.
70. **Bushey, M. M. and Jorgenson, J. W.,** Automated instrumentation for comprehensive two-dimensional high-performance liquid chromatography/capillary zone electrophoresis, *Anal. Chem.*, 62(15), 978, 1990.
71. **Sweedler, J. V., Shear, J. B., Fishman, H. A., Zare, R. N., and Scheller, R. H.,** Analysis of neuropeptides using capillary zone electrophoresis with multichannel fluorescence detection, *Proc. SPIE-Int. Soc. Opt. Eng.*, 1439, 37, 1990.
72. **Toulas, C. and Hernandez, L.,** Applications of a laser-induced fluorescence detector for capillary electrophoresis to measure attomolar and zeptomolar amounts of compounds, *LC-GC*, 10(6), 471, 1992.
73. **Gross, L. and Yeung, E. S.,** Indirect fluorometric detection of cations in capillary zone electrophoresis, *Anal. Chem.*, 62, 427, 1990.
74. **Yeung, E. S. and Kuhr, W. G.,** Indirect detection methods for capillary separations, *Anal. Chem.*, 63(5), 275A, 1991.

75. **Olivares, J. A., Nguyen, N. T., Yonker, C. R., and Smith, R. D.,** On-line mass spectrometric detection for capillary zone electrophoresis, *Anal. Chem.*, 59(15), 1230, 1987.
76. **Smith, R. D., Olivares, J. A., Nguyen, N. T., and Udseth, H. R.,** Capillary zone electrophoresis-mass spectrometry using an electrospray ionization interface, *Anal. Chem.*, 60(5), 436, 1988.
77. **Smith, R. D., Barinaga, C. J., and Udseth, H. R.,** Improved electrospray ionization interface for capillary zone electrophoresis-mass spectrometry, *Anal. Chem.*, 60(18), 1948, 1988.
78. **Edmonds, C. G., Loo, J. A., Loo, R. R., Udseth, H. R., Barinaga, C. J., and Smith, R. D.,** Application of electrospray ionization mass spectrometry and tandem mass spectrometry in combination with capillary electrophoresis for biochemical investigations, *Biochem. Soc. Trans.*, 19(4), 943, 1991.
79. **Thibault, P., Paris, C., and Pleasance, S.,** Analysis of peptides and proteins by capillary electrophoresis/mass spectrometry using acidic buffers and coated capillaries, *Rapid Commun. Mass Spectrom.*, 5(10), 484, 1991.
80. **Loo, J. A., Udseth, H. R., and Smith, R. D.,** Peptide and protein analysis by electrospray ionization-mass spectrometry and capillary electrophoresis-mass spectrometry, *Anal. Biochem.*, 179(2), 404, 1989.
81. **Loo, J. A., Jones, H. K., Udseth, H. R., and Smith, R. D.,** Capillary zone electrophoresis-mass spectrometry with electrospray ionization of peptides and proteins, *J. Microcolumn Sep.*, 1(5), 223, 1989.
82. **Moseley, M. A., Jorgenson, J. W., Shabanowitz, J., Hunt, D. F., and Tomer, K. B.,** Optimization of capillary zone electrophoresis/electrospray ionization parameters for the mass spectrometry and tandem mass spectrometry analysis of peptides, *J. Am. Soc. Mass Spectrom.*, 3(4), 289, 1992.
83. **Lee, E. D., Mück, W., Henion, J. D., and Covey, T. R.,** On-line capillary zone electrophoresis-ion spray tandem mass spectrometry for the determination of dynorphins, *J. Chromatogr.*, 458, 313, 1988.
84. **Johansson, I. M., Huang, E. C., Henion, J. D., and Zweigenbaum, J.,** Capillary electrophoresis-atmospheric pressure ionization mass spectrometry for the characterization of peptides. Instrumental considerations for mass spectrometric detection, *J. Chromatogr.*, 554(1–2), 311, 1991.
85. **Moseley, M. A., Deterding, L. J., Tomer, K. B., and Jorgenson, J. W.,** Coupling of capillary zone electrophoresis and capillary liquid chromatography with coaxial continuous-flow fast atom bombardment tandem sector mass spectrometry, *J. Chromatogr.*, 480, 197, 1989.
86. **Caprioli, R. M., Moore, W. T., Martin, M., DaGue, B. B., Wilson, K., and Moring, S.,** Coupling capillary zone electrophoresis and continuous-flow fast atom bombardment mass spectrometry for the analysis of peptide mixtures, *J. Chromatogr.*, 480, 247, 1989.
87. **Moseley, M. A., Deterding, L. J., Tomer, K. B., and Jorgenson, J. W.,** Capillary zone electrophoresis-mass spectrometry using a coaxial continuous-flow fast atom bombardment interface, *J. Chromatogr.*, 516(1), 167, 1990.
88. **Wu, S.-L., Teshima, G., Cacia, J., and Hancock, W. S.,** Use of high-performance capillary electrophoresis to monitor charge heterogeneity in recombinant-DNA derived proteins, *J. Chromatogr.*, 516, 115, 1990.
89. **Yim, K. W.,** Fractionation of the human recombinant tissue plasminogen activator (rtPA) glycoforms by high performance capillary zone electrophoresis and capillary isoelectric focusing, *J. Chromatogr.*, 559, 401, 1991.
90. **Chen, T.-M., George, R. C., and Payne, M. H.,** Separation of anticoagulant peptide MDL 28,050 from its deletion by-products by capillary zone electrophoresis, *J. High Resolut. Chromatogr. Chromatogr. Commun.*, 13(11), 782, 1990.
91. **Lüdi, H., Gassmann, E., Grossenbacher, H., and Märki, W.,** Capillary zone electrophoresis for the analysis of peptides synthesized by recombinant DNA technology, *Anal. Chim. Acta*, 213, 215, 1988.
92. **Guzman, N. A., Ali, H., Moschera, J., Iqbal, K., and Malick, A. W.,** Assessment of capillary electrophoresis in pharmaceutical applications; analysis and quantification of a recombinant cytokine in an injectable dosage form, *J. Chromatogr.*, 559, 307, 1991.
93. **Saria, A.,** Identification of α- and β-species of calcitonin gene-related peptide in the rat amygdala after separation with capillary zone electrophoresis, *J. Chromatogr.*, 573(2), 219, 1992.
94. **Guarino, B. C. and Phillips, D.,** High-performance ion exchange chromatography as a preparative adjunct to capillary electrophoresis, *Am. Lab.* (*Fairfield, Conn.*), March, 68, 1991.
95. **Soucheleau, J. and Denoroy, L.,** Determination of vasoactive intestinal peptide in rat brain by high-performance capillary electrophoresis, *J. Chromatogr.*, 608, 181, 1992.
96. **Chu, Y.-H. and Whitesides, G. M.,** Affinity capillary electrophoresis can simultaneously measure binding constants of multiple peptides to vancomycin, *J. Org. Chem.*, 57(13), 3524, 1992.
97. **Bergman, T., Agerberth, B., and Jörnvall, H.,** Direct analysis of peptides and amino acids from capillary electrophoresis, *FEBS Lett.*, 283(1), 100, 1991.
98. **Bushey, M. M. and Jorgenson, J. W.,** A comparison of tryptic digests of bovine and equine cytochrome c by comprehensive reversed-phase HPLC-CE, *J. Microcolumn Sep.*, 2, 293, 1990.
99. **Jorgenson, J. W., Moore, A. W., Larmann, J. P., and Lemmo, A. W.,** *Advances in Two-Dimensional Separations by LC/CE*, presented at HPCE '93, Orlando, FL, January 25–28, 1993.
100. **Ghowsi, K. and Gale, R. J.,** Field effect electroosmosis, *J. Chromatogr.*, 559, 95, 1991; U.S. Patent 5,092,972.

101. **Wu, C.-T., Lopes, T., Patel, B., and Lee, C. S.,** Effect of direct control of electroosmosis on peptide and protein separations in capillary electrophoresis, *Anal. Chem.*, 64(8), 886, 1992.
102. **Holloway, R. R., Keely, C. A., McManigill, D., and Young, J. E.,** Enhancement of Performance in Free Solution Capillary Electrophoresis by Bulk Flow Control, presented at HPCE '93, Orlando, FL, January 25–28, 1993.
103. **Huang, X. and Zare, R. N.,** Continuous sample collection in capillary zone electrophoresis by coupling the outlet of a capillary to a moving surface, *J. Chromatogr.,* 516, 185, 1990.
104. **Cheng, H.-F., Fuchs, M., Andrews, D., and Carson, W.,** Membrane fraction collection for capillary electrophoresis, *J. Chromatogr.,* 608, 109, 1992.
105. **Takigiku, R., Keough, T., Lacey, M. P., and Schneider, R. E.,** Capillary zone electrophoresis with fraction collection for desorption mass spectrometry, *Rapid Commun. Mass Spectrom.*, 4(1), 24, 1990.
106. **Hunt, D. F., Shabanowitz, J., Moseley, M. A., McCormack, A. L., Michel, H., Martino, P. A., Tomer, K. B., and Jorgenson, J. W.,** Protein and peptide sequence analysis by tandem mass spectrometry in combination with either capillary electrophoresis or micro-capillary HPLC, *Methods Protein Sequence Anal. Proc. Int. Conf., 10th,* 257, 1991.

Chapter 13

PROTEIN CAPILLARY ELECTROPHORESIS: THEORETICAL AND EXPERIMENTAL CONSIDERATIONS FOR METHODS DEVELOPMENT

Richard Palmieri and Judith A. Nolan

TABLE OF CONTENTS

0-8493-8690-X/94/$0.00+$.50

I. PROPERTIES OF PROTEINS AND PEPTIDES

A. INTRODUCTION

Capillary electrophoresis is a new and emerging technology that has generated considerable attention because of its ease of use, high flexibility, high resolution, and its ability to provide information complementary to classic separation techniques. While earlier applications focused on small organic molecules and pharmaceuticals, advances in column technology, e.g., static and dynamic coatings and packings, have allowed this technique to be applied to the separation of proteins and other macromolecules.[1]

The extension of this technique to protein macromolecules was faced with a plethora of new problems. These involved the nature of the analyte (e.g., stability, net charge, matrix effects, association behavior) and the nature of the silica surface (e.g., properties, adsorption, osmotic flow). However, some of these issues were not unique to capillary electrophoresis (CE). Not only do other transport processes (i.e., free boundary electrophoresis, sedimentation analysis, and gel filtration) share common concerns for the efficacy of the analyte, but other separation/material science techniques (i.e., gas chromatography [GC], reversed/normal-phase liquid chromatography [HPLC], size-exclusion chromatography [SEC], biocompatible materials, and biosensors) share a common concern for the interaction of biomolecules with solid surfaces.[4,5]

The purpose of this section is to develop a more effective strategy for protein separation and to provide a better rationale for the interpretation of experimental results than is currently available. Selected references in the areas of physical/chemical properties, electrophoretic mobility, protein interactions, adsorption behavior, and protein stability have been reviewed. It is anticipated that these topics will also become an integral part of the ongoing discussion about capillary electrophoresis.

B. PHYSICAL PROPERTIES

1. Shape and Size

a. Globular vs. Fibrous Proteins

Proteins are usually divided into two general classes (i.e., globular and fibrous) that are differentiated on the basis of shape or axial ratios. It is convenient to represent the overall shape as ellipsoid (prolate or oblate) with typical axial ratio (a:b) values of the 3:1 or 4:1 (prolate) for globular proteins. By contrast, rodlike molecules can have axial ratios >20:1.[2] (See Table 1.) It is important to note that the actual shape of the molecule may be irregular and the surface may be rough and/or creviced.

Differences in globular and fibrous shape are readily apparent during transport in aqueous media. As the axial ratio increases, solution viscosity increases because the macromolecule crosses a number of "flow lines" and thus distorts the pattern of flow. Viscosity may increase from 10- to 100-fold in changing from a globular to a fibrous protein.[2] As expected, this parameter will be found in the electrophoretic transport equation (see Section I.D below) and is defined by the frictional coefficient.

b. Frictional Coefficient

The proportionality between the force applied to a particle and the resulting velocity can be expressed as

$$\text{Force} = f \times \text{velocity} \tag{1}$$

where f is defined as the frictional coefficient. Often, however, it is convenient to determine the frictional ratio (f/f_0) rather than the frictional coefficient. This latter ratio reflects deviation

TABLE 1
Selected Hydrodynamic Properties of Globular and Fibrous Proteins[a]

Protein	Molecular weight	Viscosity[b] η (cm³/g)	Diffusion coefficient D (10^{-7} cm²/s)	Frictional coefficient f/f_0	Dimensions (Å)
Globular					
Ribonuclease A (b)	13,690	3.30 (13.9)	10.7	1.290	38 × 28 × 22
Chymotrypsinogen (b)	23,660		9.48	1.261	50 × 40 × 40
Carbonic anhydrase (h)	27,020		10.7	1.053	47 × 41 × 41
Carboxypeptidase A (b)	35,040		9.2	1.063	50 × 42 × 38
Hemoglobin, oxy (e)	67,980	3.6	6.02	1.263	70 × 55 × 55
Alcohol dehydrogenase (e)	79,070		6.23	1.208	45 × 55 × 110
					Axial ratio (*a*/*b*)
Fibrous[c]					
Tropomyosin	93,000	52	2.24	3.22	29
Fibrinogen	330,000	27	2.02	2.34	20
Collagen	345,000	1150	0.69	6.8	175
Myosin	493,000	217 (93)	1.16	3.53	68

[a] Adapted from Creighton, T. E., *Proteins,* W. H. Freeman, New York, 1983; and from Squire, P. E. and Himmel, M. E., *Arch. Biochem. Biophys.*, 196, 165, 1979. *Note:* The parenthetical letters following proteins refer to the source, i.e. (b) bovine, (h) human, and (e) equine.

[b] Viscosity data from Schachman, H. K., *Cold Spring Harbor Symp. Quant. Biol.*, 28, 409, 1963. Values for the denatured state are given in parentheses. *Note*: The viscosity of ovalbumin changes from 4.0 to 34 and serum albumin changes from 3.7 to 53 after reduction in urea.

[c] The data for fibrous proteins was taken from C. Tanford, *Physical Chemistry of Macromolecules,* 1961, chap. 6, p. 395.

in behavior from that of a spherical particle of the same mass and degree of hydration. The value of f_0 can be found by applying Stokes' law coupled with the relationship between volume and radius *r* of a sphere to give the following:[3]

$$f_0 = 6\pi\eta r = 6\pi\eta(3M\bar{\upsilon}/4\pi N)^{1/3} \tag{2}$$

Note: Stokes' law assumes a solid, rigid sphere of equivalent volume and mass. As is discussed below, the frictional coefficient is also an integral part of the electrophoretic mobility equation, where it (*f*) is inversely related to mobility (see Section I.D).

The frictional coefficient is functionally dependent on two factors: the molecular asymmetry and the degree of solvation.[2] Thus, identical frictional ratios are not an indication of identical axial ratios. X-ray crystallography shows that alcohol dehydrogenase (ADH) has an axial ratio of 45 × 55 × 110 Å, while the elastase axial ratio is 55 × 40 × 38 Å — this represents nearly a threefold increase in the *z* axis.[6] However, the frictional ratios are f/f_0 = 1.208 (ADH) and 1.214 (elastase), respectively. The frictional coefficient is actually smaller for the longer *z* axis. It is apparent that shape factors alone do not account for the observed difference. The extent of solvation and the nature of the surface (rough, smooth, and/or creviced) are additional factors.

As the three-dimensional structures of more proteins are completed, it becomes possible to obtain reliable estimates of the bound water. Frictional ratios can then be calculated that agree with the measured values. In this case, it is expected that the tightly bound solvent increases the mass of the molecule; and irregular or rough surfaces result in increased resistance to flow.[6]

Although it does not seem intuitively obvious, the frictional ratio does not change appreciably with molecular mass, other factors being equal. (This would suggest that it is the ratio of the displaced solvent to mass that is critical and not the mass itself.) The following two proteins have more than a tenfold difference in molecular weight yet still have the same frictional ratios, e.g., β-lactoglobulin (Mw = 35,000 Da) and urease (Mw = 480,000 Da); f/f_m = 1.20. While this may be the case for molecular mass, other changes readily alter this ratio and affect the results, e.g., molecular weight, mobility, sedimentation or diffusion coefficients, etc. It is important to note that the value f/f_0 often changes with alterations in temperature or electrolyte concentration.[7] This would suggest that for any comparison of mobility measurements to be valid they should be made under identical conditions.

c. Mass and Hydration

In solution, water may be bound to the macromolecular ion either through interaction with ionic or polar groups on the surface or through being trapped in crevices or holes in the interior. In either case, the bound solvent travels with the same velocity as the macromolecule and modifies the molecular mass and volume in transport processes.[2] This solvation must be taken into account when the mass and volume of the hydrodynamic particle formed by the macromolecule are computed. Thus, in a two-component system the mass of each hydrodynamic particle can be represented as:

$$m_h = M(1+\delta_1)/N \tag{3}$$

In this equation, m_h is the hydrodynamic mass of molecular weight M, δ_1 is grams of solvent per gram of dry protein, and N is Avogadro's number. Typical values for a globular protein are about 0.2, although values as high as 0.38 g/g protein have also been reported.[8,9] Bound water can be difficult to remove completely. Even after lyophilization, 10 to 20% moisture can remain in "dry" proteins. Thus, accurate mass measurements require extensive drying to obtain a constant weight or the use of such independent methods as UV measurements, amino acid analysis, or colorimetric methods.

d. Mass and Denaturation

During denaturation, it is possible to observe both a change in conformation and/or a change in the extent of hydration.[10] In contrast to the native state, the denatured state of a protein can adopt a variety of conformations depending on the mode of denaturation (see Section I.D.5 below). Strong denaturants such as 8 *M* urea or 6 *M* guanidine hydrochloride will cause the protein chains to unfold to give a new conformation that appears to be larger in size. However, denaturation is not synonymous with linearization of the chain. True, the protein chain appears random and disordered but it is not rodlike in shape. These denaturants can also affect the extent of hydration compared to the native state.[2]

2. Electrostatic Properties

a. Ionizable Amino Acids

Of the 20 naturally occurring amino acids, only 5 (side chain groups) are usually charged at physiological pH, i.e., Asp, Glu, His, Lys, and Arg—the N- and C-terminal groups are also ionized under these conditions. In addition, both Cys and Tyr residues also ionize as the pH is increased to 9 to 10. For many proteins, between 20 and 30% of the total residues are present as ionizable groups and most of these are located on the surface of the molecule (Table 2).

TABLE 2
Selected Properties of Amino Acid Residues[a,b]

Amino acid residue	Mass (Da)	Relative hydrophobicity (kcal/mol)[c]	pK_a of amino acid side chains[d]	pK_a of protein side chains[e]	Percent occurrence/ protein[f]
Alanine	71.08	0.5	—		9.0
Arginine	**156.20**	**–11.2**	**12**	**>13**	**(14.8) 4.7**
Asparagine	114.11	–0.2	—		4.4
Aspartic acid	**115.09**	**–7.4**	**4.5**	**4.0–4.8**	**(17.3) 5.5**
Cysteine	**103.14**	**–2.8**	**9.1–9.5**	**9.5**	**(8.8) 2.8**
Glutamine	128.14	–0.3	—		3.9
Glutamic acid	**129.12**	**–9.9**	**4.6**	**4.0–4.8**	**(19.5) 6.2**
Glycine	57.06	0	—		7.5
Histidine	**137.15**	**0.5**	**6.2**	**6.3–7.8**	**(6.6) 2.1**
Isoleucine	113.17	2.5	—		4.6
Leucine	113.17	1.8	—		7.5
Lysine	**128.18**	**–4.2**	**10.4**	**9.6–10.4**	**(22.0) 7.0**
Methionine	131.21	1.3	—		1.7
Phenylalanine	147.18	2.5	—		3.5
Proline	97.12	–3.3	—		4.6
Serine	87.08	–0.3	—		7.1
Threonine	101.11	0.4	—		6.0
Tryptophan	186.21	3.4	—		1.1
Tyrosine	**163.18**	**2.3**	**9.7**	**9.5–10.5**	**(11.0) 3.5**
Valine	99.14	1.5	—		6.9
(α-Amino)			**6.8–7.9**	**7.5–8.5**	
(α-Carboxyl)			**3.5–4.3**	**3.6–3.8**	

[a] Adapted from Creighton, T. E., *Proteins*, W. H. Freeman, New York, 1983.
[b] Amino acids in boldface usually charged at pH 7.2; Cys and Tyr ionize at pH 9 to 10.
[c] Values from Zamyatnin, A. A., *Prog. Biophys. Mol. Biol.*, 24, 107, 1972.
[d] Estimated from the pK_a values of small model compounds, from Tanford, C., *Adv. Protein Chem.*, 17, 69, 1962.
[e] Adapted from Timasheff, S. N., *Biol. Macromol.*, Ser. 3, 1, 1970.
[f] Values from Klapper, M. H., *Biochem. Biophys. Res. Commun.*, 78, 1018, 1977. The numbers in parentheses represent the percentage of total ionizable side chains.

b. *Charge vs. pH*

Unlike individual amino acids, side chain residues in proteins are represented as having a range of possible pK_a values, rather than a unique value (Table 2). This is indicative of the local environmental differences, i.e., dielectric constants, steric factors, folding constraints, intramolecular bonding, and degree of solvent accessibility. In fact, much wider ranges have been reported for buried residues.[11]

Taken as a group, the pK_a value of the ionizable groups tend to overlap a large portion of the pH range (see Table 2). However, the change in net charge with pH produces a titration curve that is a combination of several sigmoidal curves. This can be attributed to the nonequivalent distribution of the ionizable side chains in a given protein. Typically, histidine groups may constitute only a few percent of the total residues—an average of 2.1% histidine has been reported in 207 unrelated proteins.[12] The first leg of the titration curve is steep and represents the titration of the C-terminal and side chain aspartic and glutamic acid groups (about 36.8% of the total ionizable residues). By contrast, the slope in the middle portion is much shallower due to the presence of significantly fewer basic histidine side chains. The final segment contains a mixture of basic (lysine, 22% and arginine, about 14.8%) residues and the

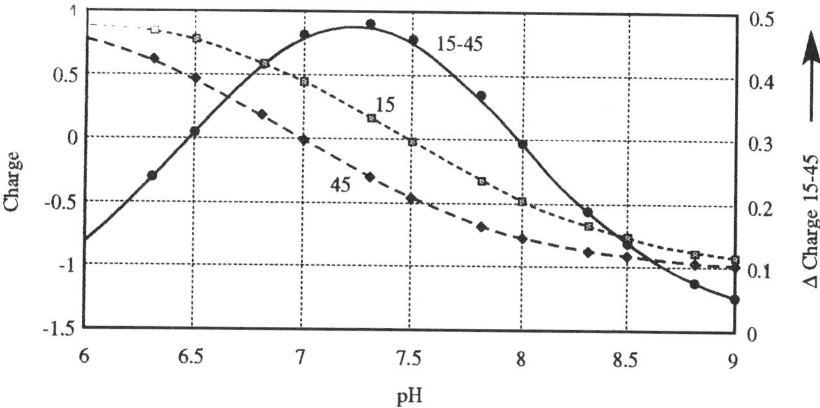

FIGURE 1. The effect of temperature on net charge. In each case the net charge was calculated for the synthetic peptide Asp-His-Ala-Tyr-Leu-Leu-Ala-Ala-Arg after correction for the effect of temperature. For this calculation, the enthalpy of ionization was taken as 1.0 kcal/mol for carboxyl, 10 kcal/mol for imidazole, and 12 kcal/mol for amino groups. The following pK_a values were assumed: C terminal (3.0), side-chain carboxyl (4.5), imidazole (7.0), α-amino (8.0) at 25°C. Note that the net charge is plotted along the left-hand *y* axis while the difference in net charge at 15 and 45°C is plotted along the right-hand *y* axis (R. Palmieri, M. Field, and L. Holaday, unpublished results). These results indicate that a maximum difference in mobility would be observed at pH ~7.3 when the temperature is changed from 15 to 45°C. ●, 15°C; □, 45°C; ◇, 15 to 45°C.

acidic residues of cysteine (8.8%) and tyrosine (11%) side chains. Again, this portion of the curve is much steeper than the middle section. Note that this latter segment of the curve contains both acidic (tyrosine and cysteine) and basic (lysine) side chains which tend to ionize in the same pH region. (The majority of proteins have been reported to be acidic with isoelectric points between pH 4.5 and 7.0.[13a]) Thus, it is reasonable to expect rather slight changes in mobility with pH changes in the neutral region compared to the midsegment pH values of the other two sections, i.e., pH < 6 or pH > 9. Unfortunately, these latter conditions can also be the regions of potential protein instability.

c. pK_a and Temperature

While the influence of temperature changes is not particularly significant for carboxyl groups, it is significant for amino groups. The enthalpy of ionization for amino groups is about 12 kcal/mol and about 10 kcal/mol for an imidazole side chain.[13b] (See Figure 1.) Changes of 10 or 30°C above the set temperature are not uncommon in capillary electrophoresis; however, this may result in several tenths change in the overall net charge. Depending on the pH, this difference is capable of producing changes in mobility. Temperature fluctuation can also perturb native protein structure and produce anomalous results during capillary electrophoresis (see Section I.E.5 below).

d. Isoelectric Point

The isolectric point is defined as the pH at which the protein has zero net charge. Below this value the protein will have a net positive charge and above this value it will have a net negative charge (Figure 2). It is important to note that this measured value may differ from the literature depending on the conditions of measurement i.e., buffer or metal ions may also bind (specific/nonspecific) to the protein and alter the net charge and the expected isoelectric point.[14]

In most cases, proteins usually display minimum solubility at or near their isoelectric points so that in practice it is desirable to operate under conditions that are at least several tenths of a pH unit above/below this value.[15]

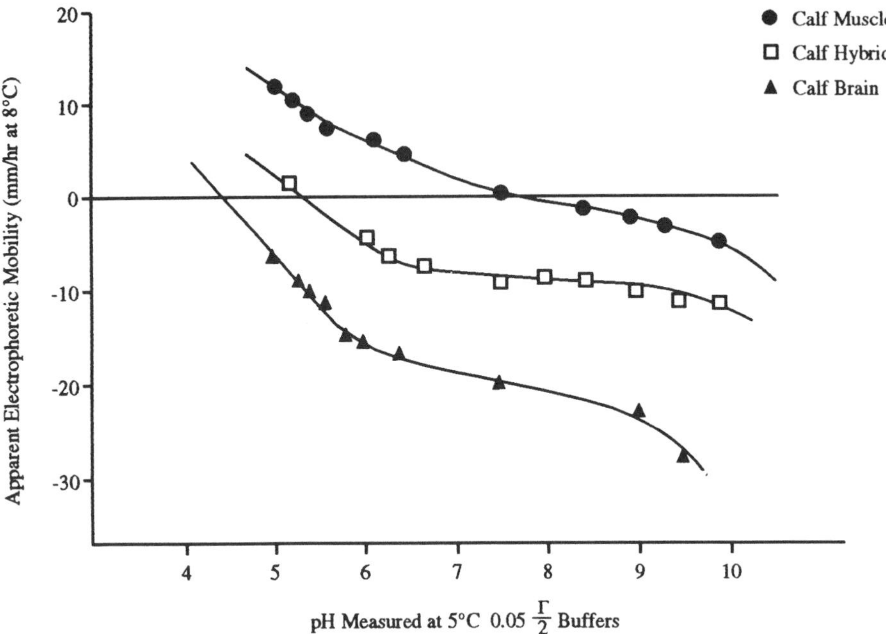

FIGURE 2. Determination of p*I* values for the isoenzymes of calf muscle and brain ATP-creatine transphosphorylase (MM, MB, and BB). p*I* values were measured by cellulose acetate strip electrophoresis at an average temperature of 8°C. The following buffers were all 0.05 *M* in total buffer species adjusted to a total ionic strength of 0.05 with KCl: for pH 5 to 6.0, sodium acetate (acetic acid); for pH 6.5 to 8.0, imidazole (HCl); and for pH values 8.5 to 9.8, Tris (HCl). Apparent mobilities were measured from densitometric tracings or from the positions of the enzymatically stained bands and expressed in arbitrary units of millimeters per hour. It was necessary to obtain the p*I* value for the brain-type enzymes by extrapolation since this isoenzyme was unstable at the lower pH values. The measured values for the calf muscle and brain enzymes were 7.3 and 5.6, respectively. (From Kuby, S. A., et al., *J. Protein Chem.*, 2, 269, 1983. With permission.)

In some cases, it is possible to obtain a reasonable estimate of the p*I* from the amino acid composition alone if one knows the amide content.[16] In Figure 3, a theoretical titration curve was constructed from the amino acid composition of the creatine kinase isoenzymes, assuming no electrostatic factor correction to the Henderson–Hasselbach equation and average population pK_a values for protein side chains at room temperature.[17,18] Surprisingly, the calculated p*I* values for the calf muscle- and brain-type creatine kinases seem to agree quite well with those obtained by electrophoretic measurements (compare p*I* values from Figures 2 and 3). While it is tempting to assume that all charged groups are completely exposed and titrate normally, in actuality the results could also be attributed to compensating factors rather than a statement of unique physical arrangement (i.e., groups of a specific type are mostly exposed).

e. *Charge Distribution*

The distribution of charged side chains and their effect on function continues to be part of ongoing investigations. This is especially true as more complete structures based on X-ray crystallography data are published. However, it is possible to draw certain conclusions from results to date. The following summary given below addresses this issue and has been taken from Alber:[19]

1. Ionizable groups are not distributed uniformly over the protein surfaces. (This ordering of charge groups suggest a structural and/or functional role.)
2. Charged groups tend to be surrounded by charges of the *opposite* sign.[20,21]
3. The asymmetric charge distributions facilitate the binding of charged ligands.[22]

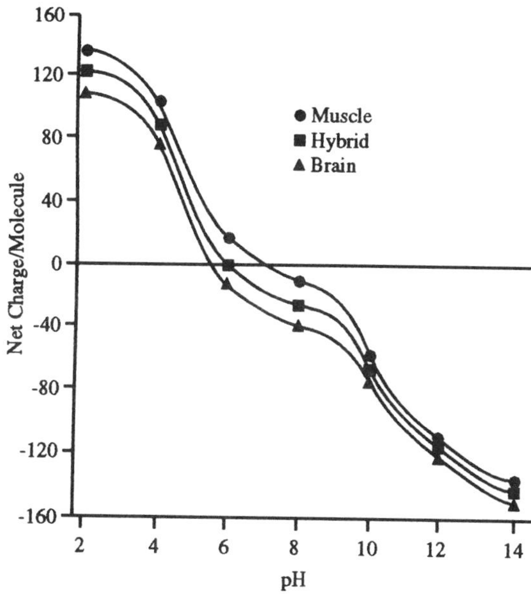

FIGURE 3. Theoretical titration curves for the calf isoenzymes of ATP-creatine transphosphorylase [MM, MB, and BB refer to the various dimeric species from muscle (MM), heart (MB), or brain (BB)]. Calculations were made with the assumption that there is no electrostatic interaction between ionizable groups and that each member of each species is identical. The p*K* values for the individual groups were taken from Edsall and Wyman.[17] Both glutamic and aspartic acid were corrected for amide content. This was determined by extrapolating the ammonia values from 20, 40, 70, and 120 h of hydrolysis to zero time. For the sake of calculation, the amide was assumed to be evenly distributed between glutamic and aspartic acid. The isoelectric points for the calf muscle and brain were p*I* 6.8 (MM) and p*I* 5.4 (BB). Compare these values to the measured values of p*I* 7.3 (MM) and p*I* 5.6 (BB) as shown in Figure 2. (From Kuby, S. A., et al., *J. Protein Chem.*, 2, 269, 1983. With permission.)

4. Approximately one third of the charged residues in proteins are involved in ion pair formation where, for the most part, they are involved in stabilizing different elements of secondary structure.[20]
5. Approximately 17% of charged groups are buried; they are stabilized either by ion pair formation or in some cases by hydrogen bonding.[23]

Taken together, the conclusions from points 1 and 3 can provide some interesting scenarios. First, point 1 seems intuitively obvious, whereas point 3, if not obvious, is an understandable functional application. Consider the case in which two proteins have the same charge-to-mass ratios but different charge distributions. This condition could portend differences in isoelectric point, electrophoretic mobility, and association and/or aggregation states. This could be a "good news/bad news" situation. While it is possible to obtain separation, it is not possible to accurately predict mobility; or monomer may be isolated in one case and dimer in another from two proteins with very similar general physical properties.

The conclusion from point 2 provides several explanations of the electrophoretic phenomenon. It suggests a rationale for the effectiveness of zwitterion additives and an expected decreased effectiveness of classic HPLC ion-pairing agents.[24,25] It may provide some insight into the existence of the mixed ionic characteristics of the biological membranes. Here, membrane lipids contain a variety of head groups that can be negative, neutral, or zwitterionic in net charge. These characteristics could facilitate "stable" interactions with intracellular proteins (see Section I.E.2 below).

Finally, the results from points 4 and 5 would tend to argue against the possibility of determining the net charge from the amino acid composition alone since ion-pair formation and buried groups would probably not share a common pK_a. However, other studies suggest that the number of buried residues is much lower than 17%, so that these residues are not likely to account for the differences between calculated and measured mobilities.[26] The existence of salt bridges and the asymmetric distribution would be expected to have significant impacts on the expected pK_a values.

f. Summary Charge Effects

To a first approximation it is possible to calculate the net charge on a protein at a given pH if one assumes: (1) a common pK_a value for all ionizable groups of a given set; (2) all groups are completely exposed to solvent; and (3) electrostatic interactions can be ignored. However, the actual net charge and hence the mobility may differ due to: (1) buried charged groups not accessible to solvent; (2) differences in intrinsic pK_a values; (3) specific/nonspecific ion binding; and (4) uncorrected temperature and ionic strength effects.

C. ELECTROPHORETIC PROPERTIES

1. Mobility

a. Equation of State

Because of the assumptions involved, it is not possible to derive an exact equation of state for electrophoretic mobility (of macromolecules) and its molecular parameters, which parallel those that describe either the determination of molecular weight (sedimentation analysis) or the determination of diffusion coefficient.[27] At best, the equation for electrophoresis may provide only semiquantitative information about the molecular charge, size, or shape, so that it does not have the same validity as other molecular parameters. Difficulties arise in attempting to assign the net charge, the charge distribution, and the interaction of the charged particle with the field. It is, however, still informative to consider the parameters that are involved in the determination of electrophoretic mobility. (The electrophoretic mobility, μ, depends only on molecular parameters and corresponds to the sedimentation coefficient for macromolecules.) Mobility is defined as:

$$\mu = v / E(\text{velocity/potential gradient}) \tag{4}$$

The following relationship was derived by Booth and represents the first term of the mobility equation:[27,28]

$$\mu = (Ze / f)\left[\Phi_1(\kappa r) / (1 + \kappa r)\right] \tag{5}$$

for a sphere of radius r, $f = 6\pi\eta r$; so that the equation can be written as

$$\mu = (Ze / 6\pi\eta r)\left[\Phi_1(\kappa r) / (1 + \kappa r)\right] \tag{6}$$

Definitions for the various components in the equation are as follows: μ = mobility (cm^2/[V·s]); I = ionic strength; E = potential gradient (V/cm); r = radius of sphere; v = velocity (cm/s); D = dielectric constant; Ze = net charge; e = electron charge; f = frictional coefficient; k = Boltzmann constant; κ = Debye–Huckel constant = $(8\pi Ne^2/1000\ DkT)^{1/2}\ I^{1/2}$; and the function Φ_1 (κr) has a value between 1.0 and 1.5. *Note*: This equation is somewhat similar to the Smoluchowski equation for large particles but it lacks the ζ potential term.[27]

Additional information can be obtained if equations for the frictional coefficient and the Debye–Huckel constant are substituted in the Booth equation. Besides the net charge (Ze) they include the effects of shape and degree of solvation (f/f_0) and the effects of temperature (T) and ionic strength contribution (κ). All of these parameters, however, do not carry the same weight. The fractional powers for several of these parameters tend to dilute their effect so that net charge appears as a dominant factor. This can be seen since f/f_0 is between 1.0 to 1.7 for globular proteins, κr has a value between 1.0 and 1.5, and viscosity increases are not expected to have major influences on the mobility since the frictional coefficient has little enhancement on resolution.[29] It is readily obvious from the above equation, that the mobility has a significant dependence on the net charge as compared to the asymmetry of the molecule (f). However, ionic strength can play an important part in discriminating between different molecular weights.

b. Mobility and Molecular Weight

It is possible to start with the first term of the Booth equation, and with certain simplifying assumptions about the shape, porosity, net charge, and field effects on a macromolecule, to arrive at a relationship between the electrophoretic mobility and the molecular weight. This relationship was derived by Compton, and a portion of his argument is reconstructed here.[29] For a solid sphere of radius r, it is possible to substitute the following expression for the radius:

$$r = (3M/4\pi N)^{1/3}(f/f_0) \tag{7}$$

into the first term of the Booth equation; and after rearrangement and consolidation into constants K_1, K_2, and K_3, the mobility can be expressed as the following:

$$\mu = K_1 Z\Phi(\kappa r)/(K_2 M^{1/3} + K_3 M^{2/3}) \tag{8}$$

From Equation 8 it is apparent that mobility does not have a simple dependency on molecular weight. Previously, Offord had determined an empirical relationship between net charge and migration distance for peptides on paper electrophoresis, i.e., migration distance $\propto Z/M^{2/3}$.[30] Instead, mobility appears to be a complex function of the protein molecular weight and the ionic strength, which is contained in K_3. For small molecules at low molecular weight the $M^{-1/3}$ term dominates; while at high ionic strength and high molecular weight the molecular weight has an $M^{-2/3}$ dependency. Although some effects of ionic strength have been reported for small molecules, further work is necessary to clarify the impact of this parameter on the overall mobility. It may be possible to provide greater differentiation between analytes of different molecular weight by alteration of this parameter.

2. Mobility Determinants

a. Influence of Shape

While both empirical and theoretical relationships have demonstrated a well-defined relationship between the mass:charge ratio and the mobility for small molecules, an equivalent parallel does not exist for macromolecules. First, as noted in Section I.B, proteins possess unique rather than random structures in solution. Spheres migrate faster than rods of the same molecular weight; smooth surfaces have less resistance than rough surfaces. Perturbations in the shape during the transport process can alter mobility. Second, net charge becomes harder to estimate as the molecular mass increases, and the potential for ion pairing and matrix effects increases. Finally, it can be difficult to determine the extent of hydration since the amount of water adsorbed can depend on the conditions of analysis.[2] Thus it is important to view mobility as a function involving the effective charge as well as the effective mass.

b. Net Charge

Given the dependence of the pK_a on the local environment it is reasonable to expect that the net charge will also be influenced by the introduction of groups that alter the charge through steric or electrostatic influences. Thus, the addition of bulky substituents, introduction of charged groups, or alterations in the folding or contact surface interactions are potentially capable of altering the local charge.[22,23,31] These changes are particularly evident at pH values near the pK_a, where differences in ionization would be observed for the modified and unmodified species. Strategically, one titrates across a common pK_a region or a region where either charge or side chain modifications have occurred in order to exaggerate differences in electrophoretic mobility.[32] Splitting a pK_a in this fashion does not produce multiple electrophoretic peaks since the protonation/deprotonation step is rapid relative to the electrophoretic transfer.

c. Resolution and Electroosmotic Flow

Jorgenson and Lukacs[33] developed an equation that relates resolution, mobility, and osmotic flow:

$$\text{Res} = C_1\left(\mu_{\text{ave}} + \mu_{\text{osm}}\right)^{-1/2} \tag{9}$$

where Res is the resolution, C_1 is a constant, μ_{ave} is the average mobility, and μ_{osm} is the osmotic mobility. In the ideal case, resolution can be optimized when $\mu_{ave} = -\mu_{osm}$. Practically speaking, however, it is advantageous for either μ_{osm} or μ_{ave} mobility to be greater so that results can be obtained in a timely manner.

These results argue in favor of controlling rather than eliminating electroosmotic flow (EOF) for the separation of proteins, e.g., through column coatings. If EOF has been eliminated, resolution is dependent on the charge-to-mass ratio and manipulation of net charge alone might be insufficient in achieving the desired resolution. The choice of pH can be further restricted by stability considerations. Experimental analysis time could be unreasonably long for proteins near their isoelectric points. Bidirectional separation conditions could exist so that only the net positively charged proteins would be observed. Complete analysis would require a reversal in the electric field so that analytes with a net negative charge could be observed. Finally, it will not be possible to quantitate analytes with low to very low charge-to-mass ratios since these compounds will not be eluted from the column during the experimental allowed time. Strategically, it is preferable either to retain some EOF or to control the existing EOF in protein separations. With this approach it is possible to exploit the full options in resolution by manipulating either the net charge, the EOF, or both parameters.

II. PROTEIN–PROTEIN INTERACTIONS

A. CHARACTERISTICS

Protein interactions can be of several types and must be considered in any transport process. They include both association–dissociation and aggregation processes with homogeneous proteins; the association/aggregation process between heterogeneous proteins or between other macromolecules, e.g., lipids, oligonucleotides, etc.; the site-specific binding of substrates and/or cofactors; association-dependent molecule binding; and the specific/nonspecific binding with sample matrix components.[34] Interactions can be further classified as specific and ordered or nonspecific and random. In any case, these effects must be taken into account not only in experimental design but also in the interpretation of analytical results.

Charged groups are often implicated in driving the association/aggregation process (hydrogen bonds and hydrophobic forces are involved in the stabilization of the assembly).[35a,35b] The presence of unpaired charge groups is often implicated in stabilizing the quaternary as well as the tertiary structures — electrostatic binding is not directionally controlled and possesses a considerable dependence on molecular weight. From the discussion above (Section I.B.2e), it is apparent that about one third of the polar side chains are involved in stabilizing the tertiary structure of biopolymers through intramolecular interactions. Considering the viability of this mechanism, it is possible that some of the remaining groups could be involved in intermolecular interactions that promote association. Indeed, it is reasonable to expect that this type of interaction would become more prominent as the concentration of protein increases.[3,36]

These interactions are not restricted to like species but may involve heterogeneous macromolecules, small molecules, or other matrix elements. The essential criterion needed to observe these processes is that the rate of interaction must be on the same time scale as the transfer process.[37]

Protein interactions can be either ideal or nonideal. While ideal behavior is described as the formation of well-ordered, well-defined interactions, nonideal behavior is associated with random aggregrate or cluster formation and complex binding behavior. The latter behavior is most likely to be observed with (1) highly asymmetric molecules, (2) highly charged proteins, and (3) strong denaturing solvents.[38] In each of these cases, either the structure, charge, or elements dominate the formation of the final product.

B. ASSOCIATING AND AGGREGATING SYSTEMS

One way of looking at associating–dissociating and aggregating systems is to compare protein behavior under a variety of conditions. To do this, Klotz et al. compiled a list of nearly 300 proteins.[39] This set included proteins that were normally monomeric under physiological conditions (e.g., ribonuclease, lysozyme, and serum albumin), proteins that displayed well-ordered, stable quaternary structures (e.g., hemoglobin and hemerythrin), and proteins that displayed association–dissociation behavior (e.g., β-lactoglobulin, insulin) and/or aggregation behavior under different conditions (e.g., chymotrypsinogen A and glutamate dehydrogenase).

The results from this study were revealing. By far the majority of proteins (76%) possessed quaternary structure that consisted of even-numbered subunits, i.e., either two (36%), four (31%), or six (9%) subunits. Although about 15% of the entries had subunits that were nonidentical, the tendency to form even-numbered associating species was still observed. The thermodynamic analysis did reveal some surprising results. *There did not seem to be any significant difference between the normal associating–dissociating system and the somewhat artificially induced aggregating systems.*[39] Nonideal aggregation behavior (i.e., chymotrypsinogen A) does not seem to be any less favored over the ideal protein association–dissociation behavior (i.e., hemoglobin). Thermodynamic parameters for the particular proteins cited above are shown in Table 3.

It is apparent from the results listed in Table 3 that association–dissociation/aggregation phenomena are not restricted to a few isolated examples. The process may occur with monomeric or oligomeric species;[40,41] the final form seems to depend on the initial protein concentration and/or conditions of analysis. For example, the aggregation state of chymotrypsin A is dependent on the initial protein concentration,[42] while the association behavior of β-lactoglobulin depends on the pH of the system. Over the pH range of 7 ± 1.5, β-lactoglobulin exists primarily as a dimer [M_1 (monomer) = 18,500 Da].[43,44] However, if the pH is lowered to about 4.5 and the temperature to about 5°C, the dimer associates to form an octamer.[45] If the pH is lowered still further (pH<3) the monomeric form of the protein predominates. Since it is not obvious what effect different concentrations or analysis conditions may have on the protein behavior, it will be important to investigate a variety of conditions prior to deriving any final conclusions.

TABLE 3
Thermodynamic Parameters of Subunit Association

System	K_i (molar scale)	$-\Delta G_m$ (kcal/mol) subunit	Ref.
Hemoglobin			
pH 7.0, 20°C			
$2\alpha\beta \rightleftharpoons (\alpha\beta)_2$	8.0×10^4	–3.2	40,40a
Hemerythrin			
pH 7.0, 5°C			
$8A_i \rightleftharpoons A_8$	3.4×10^{36}	–5.8	41
Chymotrypsinogen A			
pH 7.9, 25°C			
$A_i + A_{i-1} \rightleftharpoons A_i$	1.3×10^3	–2.1	42
β-Lactoglobulin			
pH 4.55, 4.6°C			
$4A_2 \rightleftharpoons A_8$	2.4×10^{11}	–3.8	43
pH 5.6, 25°C			
$2A_1 \rightleftharpoons A_2$	4.3×10^4	–4.0	44
pH 8.8, 20°C			
$2A_1 \rightleftharpoons A_2$	4.1×10^3	–2.4	45

Note: α and β refer to the heterodimer of hemoglobin, while subscript "i" indicates the degree of association.

Adapted from Klotz, I. M., Darnall, D. W., and Langerman, N. R., *The Proteins,* Neurath, H. and Hill, R. L., Eds., Academic Press, New York, 1975, Vol. 1, chap. 5, pp. 358–359. With permission.

From the preceding examples, it is apparent that experimental conditions can promote different association states that can impact data analysis. The extent by which experimental results are altered does not depend on the equilibrium constant but on the forward and reverse rate constants (see Table 4). Thus, it is possible for a simple monomer–dimer equilibrium to produce not one or two but three peaks (see Figure 4). This kind of behavior is to be expected when the reactions are so slow that chemical equilibrium cannot keep pace with the transport process; yet they are fast enough that appreciable changes take place during the course of transport.[38] The cautionary note described by van Holde for sedimentation analysis is also appropriate here: "If chemical equilibria are suspected, the number of peaks in the...pattern should not be taken as necessarily equal to the number of species present."[38]

Protein–protein interactions are not restricted to the cases cited above. Similar alterations in properties may also be produced by conformation alterations (e.g., disulfide bond rearrangements, pH perturbations, ligand-mediated association).

The theoretical work originally provided by Gilbert and Jenkins[46a,46b] for the behavior of rapid monomer-*n*-mer reactions during sedimentation analysis[38] has also been applied to other association processes in free-boundary electrophoresis,[47,48] and in gel filtration.[49] It is expected that at least to a semiquantitative extent, the same theory is applicable to capillary electrophoresis.

C. PROTEIN INTERACTIONS AND CAPILLARY ZONE ELECTROPHORESIS

It is reasonable to assume that association/aggregation phenomena can also be expected in capillary electrophoresis, due to the presence of a number of contributing factors: (1) the presence of high initial protein concentrations, (2) the existence of secondary sample equilibria involving sodium dodecyl sulfate (SDS), substrates, selective ions, or other macromolecules, (3) the sample-focusing mechanism, which tends to use low ionic strength buffers, and

TABLE 4
Classification of Interacting Systems

$$A + B \underset{k_r}{\overset{k_f}{\rightleftharpoons}} C$$

Class	Magnitude of rate constants	Results (system initially at equilibrium)
I	$k_f \simeq k_r$, both small	All species present
II	$k_f << k_r$	Reactants present
III	$k_r << k_f$	Complex and excess reactants present
IV	$k_f \simeq k_r$, both large	Variable
V	k_f, k_r, intermediate	Variable

From Nichol, L. W., Bethune, J. L., Kegeles, G., and Hess, E. L., *The Proteins*, Neurath, H., Ed., Academic Press, New York, 1964, Vol. 2, p. 305. With permission.

(4) the voltage-ramping technique, which generates high potential gradients across the sample zone. The problem is that one or more of these factors may be operative during a given electrophoresis run.

High protein sample concentrations are often required because of problems with sensitivity. This requirement combined with zone-sharpening techniques that employ low ionic strength and high potential gradients (sample zone) can result in unusually high localized concentrations that can promote association. The resulting zone broadening, or peak splitting, may be misinterpreted as adsorption or as heterogeneity. Using existing techniques, initial protein concentrations of about 0.1 mg/ml may exceed 10 mg/ml after focusing.[50] To control these effects and obtain unbiased estimates of the mobility, it is important to (1) keep the sample concentration as low as possible,[38] (2) increase the ionic strength in the sample to 0.2,[38] (3) decrease the voltage ramp during injection,[51,52] and (4) employ buffer additives such as ethylene glycol to improve peak shape.[53]

A further note on analyte properties should also be considered. High protein net charge tends to promote nonideal behavior during transfer processes. This primary charge effect can be removed for most proteins studied during sedimentation analysis by using a buffer with an ionic strength of 0.2 or greater and protein concentrations below a few milligrams per milliliter.[38]

Above all, it is important to keep in mind that a single determination by any technique is not a necessary and sufficient criterion to establish homogeneity. Indeed, anomalous results can be observed with an experimental design that contains too few conditions or that is too limited in scope, i.e., the proposed model is too restrictive.

III. PROTEIN ADSORPTION

A. SCOPE OF THE PROBLEM

The adsorption of proteins to silica surfaces has been an ongoing problem in many different applications, e.g., "biocompatible" materials, general analytical methods, pharmaceutical formulations, and biosensor technology.[54] It comes as no surprise that the need to control this process has been the focus of research for many years. The successful application of capillary electrophoresis to the field of protein separation requires a better understanding of the factors involved in this process. Adsorption dynamics are usually created by the interaction of at least

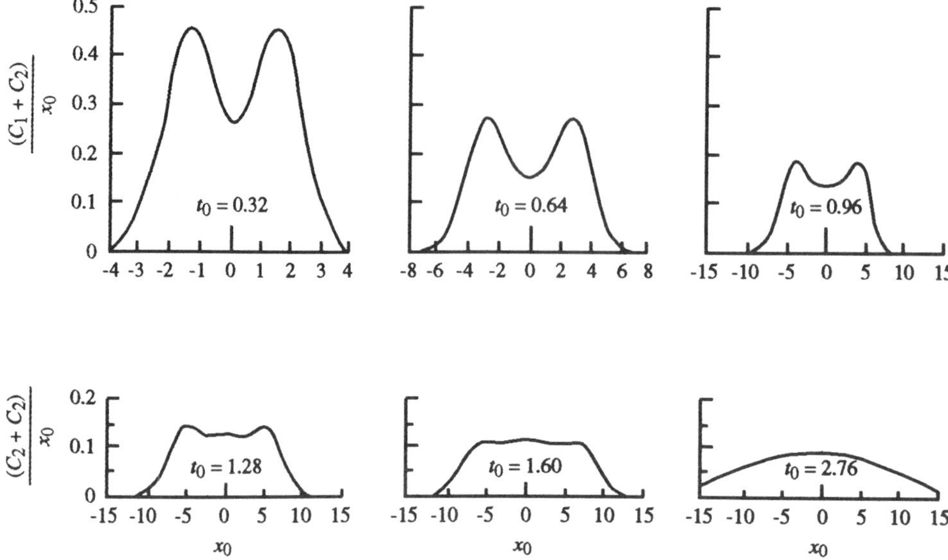

FIGURE 4. Time course of development of theoretical moving boundary electrophoretic patterns for an isomerizing system A $\rightleftharpoons$ B, in which $D_1 = D_2$, $k_1 = k_2$, and $\alpha = 0.01$. $\alpha = Dk_1/(\mu_2-\mu_1)^2 E^2$ and t_o is a reduced time. The same predictions apply to zone electrophoresis and chromatography. (From Giddings, J. C., *J. Chromatogr.*, 3, 443, 1960; and Keller, R. A. and Giddings, J. C., *J. Chromatogr.*, 3, 205, 1960.)

three factors: the silica surface (e.g., net charge, ionic strength, charge density), the protein physical properties (e.g., isoelectric point, stability, charge distribution), and the sample/buffer matrix (e.g., ionic strength, buffer type, salt properties). It is most important to consider the influence of net charge and analyte properties on adsorption. Initially, we will compare the electrostatic properties of biomembranes and fused silica.

B. BIOMEMBRANE–SOLUTION INTERFACE

Under physiological conditions, the biological membrane carries a negative charge density that is not significantly different from that of the fused silica. Yet adsorption is not a problem with biomembranes. The influence of the membrane charge has been discussed by McLaughlin,[55] who used the Gouy–Chapman theory of the diffuse double layer to predict the impact of surface charge and ionic strength on the potential in solution. The magnitude of the surface potential can be estimated from the average distribution of charged groups within membranes. Assume that about 20% of the membrane lipid bears a single net negative charge originating from the ionization of phosphatidylserine, -glycerol, or -inositol. The remaining lipids are either zwitterions (e.g., phosphatidylcholine; phosphatidylethanolamine) or neutral (e.g., cholesterol). This distribution of charged groups corresponds to an average charge density (S) = 1 (charge)/300 Å^2 (or 0.335/nm^2, or 100 Å^2). If the concentration of salt in the bulk solution is 10^{-1} M, the theory predicts that the potential at the aqueous side of the membrane–solution interface $\phi = -60$ mV.

The magnitude of the potential decreases in an exponential manner with distance from the membrane.[55] The Debye length, $1/k$, is defined as the distance from the wall at which the potential falls to $1/e$ of its value at the surface of the membrane. This length is about 10 Å when the concentration of monovalent ions is 10^{-1} M (at $10^{-3}M$ the Debye distance increases to 100 Å). Of special interest is the fact that the concentration of monovalent cations at the surface of the bilayer will be an order of magnitude higher than the concentration of these ions in the bulk aqueous phase. In this case, the pH at the wall should also be about 1 pH unit lower than the bulk solution. While this phenomenon may not be significant in membranes under

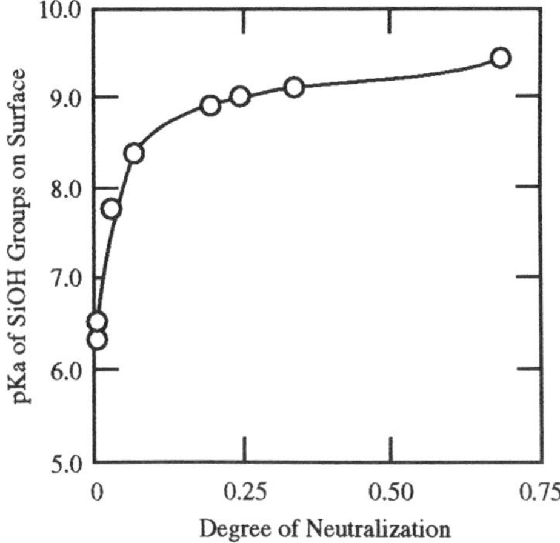

FIGURE 5. Relation between pK_a of silanol groups on the surface of amorphous silica and the extent of SiOH titration. Since silica acts as a polyprotic acid, the functional pK_a increases as more groups are titrated and the surface becomes more negatively charged. Above pH 9 silica can be expected to dissolve, so that further titration would be unsuccessful. (From Strazhesko, D. N., et al., *J. Chromatogr.*, 102, 191, 1974. With permission.)

physiological conditions (pH 7.3 to 7.5), it might be significant in fused silica where a broader range of pH values is employed either to modulate the osmotic flow or to control adsorption during capillary electrophoresis (see the next section). The lower than expected pH even near the wall may be disadvantageous to analyte stability and actually facilitate adsorption.

C. THE SILICA SURFACE

1. General Properties

In contrast to biomembranes, silica has a higher potential density of ionizable groups, a variety of pK_a values, and potentially lower stability constants.[56,57] (The properties of surface silanol groups discussed in this section have come from a study of colloidal powders and silica gels; it is assumed that these conclusions can be extended in whole or in part to fused silica.) The silica surface can usually be considered to be irregular with repeating units of -SiOH and a certain percentage of -$Si(OH)_2$ when chain termination occurs. Using a variety of techniques, it can be shown that there are about 4.6 -SiOH/nm^2 (100 $Å^2$) on the surface of silica gel.[57,58] Of these, 1.4 were free and 3.2 were hydrogen-bonded pairs. When fully ionized, the charge density of silica would be ten times that of the biomembrane. However, as will be shown below, these conditions prevail only under extremes of pH, e.g., conditions where silica dissolves.

2. Hydroxyl Groups per Square Nanometer

Adsorption and/or ion-exchange both depend on the degree of ionization of the -SiOH groups on the surface. Since the surface silanols actually originate from a polyprotic acid, these groups cannot be described with a single pK_a.[59] Once ionization begins, the apparent acidity of the remaining SiOH groups decreases as the degree of ionization of the surface increases. Complete ionization is effected only at the pH just short of the point where silica dissolves, and then only in the presence of a strong salt solution.[57] Although it is possible to describe silica as having a pK_a of about 6.5,[33] the point of zero net charge is actually at pH 2, which is 4.5 units below the initial pK_a value. This is especially apparent with silica, where the pK_a varies from 6.5 initially to about 9.2 at 50% ionization as shown in Figure 5.[59] In fact, the first silica groups ionize at a much lower pK_a (more strongly acidic) than monosilic acid, which has a pK_a of 9.8. The degree of ionization is further impacted by the salt concentration, as shown in Figure 6. Note: above pH 8 to 9, silica begins to dissolve and forms silicate ions, $HSiO_3^-$, which is also present in increasing amounts as the pH is raised.

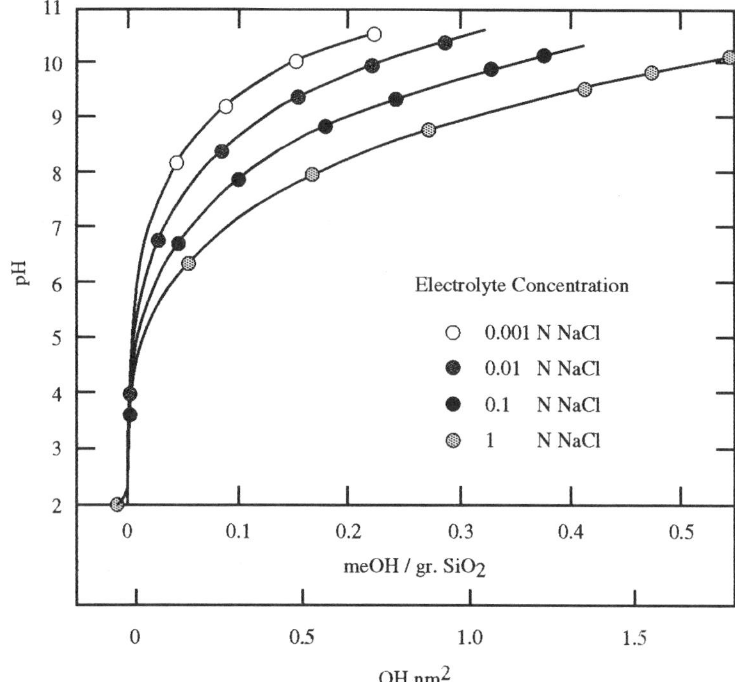

FIGURE 6. Charge density expressed as OH^- ions on the surface of particles of colloidal silica (surface area, 180 $m^2\ g^{-1}$) in the presence of different concentrations of sodium chloride. The charge density increases for a given pH as the salt concentration increases. At low ionic strengths, it is not possible to reach the maximum surface charge even at extremely high pH values, i.e., pH >10, where silica dissolves. Sodium ion binding to the ionized silica reduces charge–charge repulsion and enables more SiOH groups to be titrated. (Adapted from Blot, G. H., *J. Phys. Chem.*, 61, 1166, 1957. With permission.)

3. Distance between Charges

It can also be informative to interpret the distance between charges in terms of -Si-O-Si or -C-C bond lengths. The number of charges per square nanometer or 100 Å^2 as a function of pH are shown in Tables 5A and 5B. The area in angstroms squared that contains a single charge can be expressed either as a square of a specific length. In an idealized case with uniform distribution of charge groups, the charge would be located at the center of each square. In terms of bond lengths between charges, at pH 8 there would be about 4 potentially charged SiOH groups (3.1 Å between OH groups), or about 10 carbons (2.51 Å between C1 and C3) would be needed to bridge this distance if the charge resided at the center of each square. However, at pH 10, the respective values change so that about 6 carbon atoms are needed to bridge this distance. Since the distances are larger than one might expect, based on the total possible ionizable groups (4.5/nm^2), questions might reasonably be raised about the effectiveness of various additives, e.g., diamino compounds and primary alcohols, polymeric celluloses, etc. It is important to remember that a dynamic rather than static state exists in the capillary, so that actual silica-binding sites can be closer than the above calculations would indicate.These results also explain why hydrocarbon length is an important determining factor in additive effectiveness (see Table 6).

D. PROTEIN STRUCTURE AND ADSORPTION

1. General Considerations

It is commonly assumed that adsorption is primarily related either to the p*I* or the net charge on the molecule and the net charge on the silica surface, i.e., basic proteins stick to acidic silica

TABLE 5A
Effect of pH on Net Charge on the Silica Surface

pH	Charge per nm²	Å² per charge	Dimensions of square[a] (Å)	Distance between charges[b] (Å)
8	0.76	132	11.5	11.5
9	1.45	69	8.3	8.3
10	2.52	40	6.3	6.3

Adapted from Iler, R. K., *The Chemistry of Silica,* John Wiley & Sons, New York, 1979, chap. 6. With permission.

[a] Dimensions of a square to give required area for a single charge group.
[b] Average distance between charged groups assuming that charges are located at the center of each square.

through electrostatic interactions. However, this tends to focus on only part of the mechanism since it is possible that two proteins with the same p*I* can have different adsorptive characteristics. This result would suggest that adsorption entails more than charge-charge interaction. An acidic surface and a basic protein are not a necessary and sufficient condition for adsorption.

2. Functional Groups and Adsorption

In fact, no single mode of binding adequately describes protein adsorption (see Figure 7).[60] Compared to smaller molecules, the potential for multiple binding modes within a single molecule lend themselves to adsorption. While the stability constant may be rather small for any single functional group, the wide distribution of different charges actually promotes this process. In addition to electrostatic binding, both hydrogen binding with OH, NH, or CO groups, and hydrophobic binding modes have also been described.[61] Surprisingly, infrared analysis demonstrated that about 70 amide bonds were involved in the adsorption of each molecule of bovine albumin. This calculation assumed the formation of a monolayer that held about four amide segments/nm^2 and an average amide segment molecular weight of 100.[57] Second, and also somewhat unpredictable, is the fact that proteins tend to display maximum adsorption at or near their isoelectric points.[61] This would seem to be more significant considering that charged groups tend to be surrounded by charges of the opposite sign (see Section I.B). In this case also, the bonding involves both the electrostatic as well as hydrogen bonds. In many respects, it is not unreasonable to assume that electrostatic forces would predominate (see Table 5B). Attractive force between two atoms contains both direction as well as distance components.[62,63] Electrostatic interactions have no restriction due to direction, and a $1/r$ dependence on distance; hydrogen bonds, however, display both a directional limitation as well as a distance limitation restriction.

3. Protein Structure vs. Surface Activity

One way of evaluating the efficacy of different organic compounds in competing with proteins for the adsorption sites on silica is to test the resolubilization of silica–protein precipitates.[64] Two factors seem to be of prominent importance in deciding the activity between organic molecules and the silica surface, i.e., multiple bonding and surface activity. It is obvious that when a molecule is adsorbed from solution onto a silica surface through hydrogen bonds that are being continuously formed and broken at ordinary temperature, if there are several such points of attachment the probability of all the bonds being broken simultaneously becomes very low. This was shown by Iler[65] in the case of the hydrogen-bonding effectiveness of polyethylene glycols. It is beneficial to compare the relative effectiveness of different classes of organic compounds to form hydrogen bonds (see Table 6). A somewhat unexpected result was obtained for all types of additives studied. Both the presence

TABLE 5B
Relative Strength of Bonds in Biological Systems

Bond type[a]	ΔH (kcal/mol)	Distance dependence[b]	Average bond distance (Å)
Electrostatic	3–7	$1/r$	
Hydrogen[c]	3–6	—	2.7–3.1[d]
van der Waals[e]	1.0	$1/r^6$	
Thermal	0.6		
Covalent (C1–C2)	83		1.54[d]
Covalent (C1–C2–C3)	—		2.51
Covalent Si-O	89		1.91[d]
Covalent Si-O-Si[f]	—		3.1 (2.4–2.8)
Covalent (general)	50–100		

[a] Stryer, L., *Biochemistry*, 3rd ed., W. H. Freeman, New York, 1988, p. 6.
[b] Mathews C. K. and van Holde, K. E., *Biochemistry*, Benjamin Cummings, Redwood City, 1990, p. 31; r represents the distance between the two bonding atoms.
[c] March, J., *Advanced Organic Chemistry*, McGraw Hill, New York, 1977, pp. 75–78.
[d] Gould, E. S., *Mechanism and Structure in Organic Chemistry*, Holt, Rinehart and Winston, New York, 1959, pp. 34–52.
[e] Creighton, T. E., *Proteins, Structure and Molecular Function,* W. H. Freeman, New York, 1983, p. 136.
[f] Xu, B. and Vermeulen, N. P. E., *J. Chromatogr.*, 445, 1, 1988; distance between surface hydroxyl groups. Numbers in parentheses give distance between surface hydroxyl groups when hydrogen bonded.

and the length of the attached hydrocarbon chains were important factors in the determination of additive effectiveness. Not only must a particular molecule find a binding site, but it must be stabilized at the site to be really effective. With respect to protein adsorption, notice the relative comparison of the amide additives.

E. INFLUENCE OF ADSORPTION ON CAPILLARY ZONE ELECTROPHORESIS

From the discussion above, it is apparent that a model employing electrostatic interactions alone is inadequate to describe silica adsorption of proteins. Undoubtedly, electrostatic interactions do play a role, and assist in getting the protein to the surface. Once there, both hydrogen bonding and hydrophobic forces can join in anchoring the analyte to the surface.

IV. PROTEIN STABILITY

A. THE NATIVE STRUCTURE

Biopolymers, and proteins in particular, differ from other polymers in that they possess a unique structure in solution. Surprisingly, considering the importance of their functions, most proteins possess a rather fragile structure that imposes certain constraints when working not only with the native forms but also the denatured forms of these molecules.[66]

B. PROTEIN DENATURATION

1. General Features

Typically, proteins are held together by relatively weak forces.[67] It is the combination of these forces that provides the molecule with its stability in solution. However, relatively small changes in charge, solution pH, denaturant concentration, ionic strength, or temperature can initiate the unfolding process (see Figure 8). While there are exceptions, most proteins tend to unfold in a

TABLE 6
Relative Effectiveness of Hydrogen-Bonding Agents

Compounds	Relative molar effectiveness
Reference STD	
Dimethoxytetraethylene glycol	100
Alcohols	
Ethanol	6
t-Butyl alcohol	16
Glycols	
Ethylene glycol	0
3-Methyl-1,2-butanediol	18
Ketones	
Acetone	17
Methyl ethyl ketone	25
Amides	
Acetamide	11
N,N-Diethylacetamide	54
Urea	7
Tetramethylurea	44
Primary amines	
2-Ethylhexylamine	32
Cyclohexylamine	25
Secondary amines	
Diethylamine	19
Dibutylamine	65
Tertiary amines	
Trimethylamine	14
Cyclohexyldiethylamine	117

Adapted from Iler, R. K., *The Chemistry of Silica*, John Wiley & Sons, New York, 1979, chap. 6.With permission.

simple, first-order process going from N (native) → U (unfolded) state without going through intermediate states.[68] This first-order process seems to be independent of the mode of denaturation.

Denaturation is usually accompanied by an increase in molecular dimensions, viscosity, and aggregation, and a decrease in ordered structures (α helix and β sheets), solubility, and solvation.[67] The definition of this latter state needs further clarification; likewise, the influence of different modes of denaturation needs to be elucidated (e.g., pH, temperature, surfactant, or denaturant concentration). Proteins can adopt a variety of conformations from native-like states,[69] such as the so-called "molten globule" state,[70a] to the highly unfolded states that are apparent at high concentrations of denaturants.[70b]

2. Size and Denaturation

In contrast to the native state, the denatured state of a protein can adopt a variety of conformations depending on the mode of denaturation — the degree of hydration can also be affected by these processes. Gast and colleagues[72a] studied the influence of different denaturing conditions on the compactness of various proteins and the dependence of the dimensions of a disordered protein chain on the solvent conditions.[72a] The diffusion coefficient D was measured for the native and "denatured" states, using differential light scattering (DLS). With this value the Stokes' radius was calculated from the Stokes–Einstein equation:

$$R_s = kT/6\pi\eta D$$

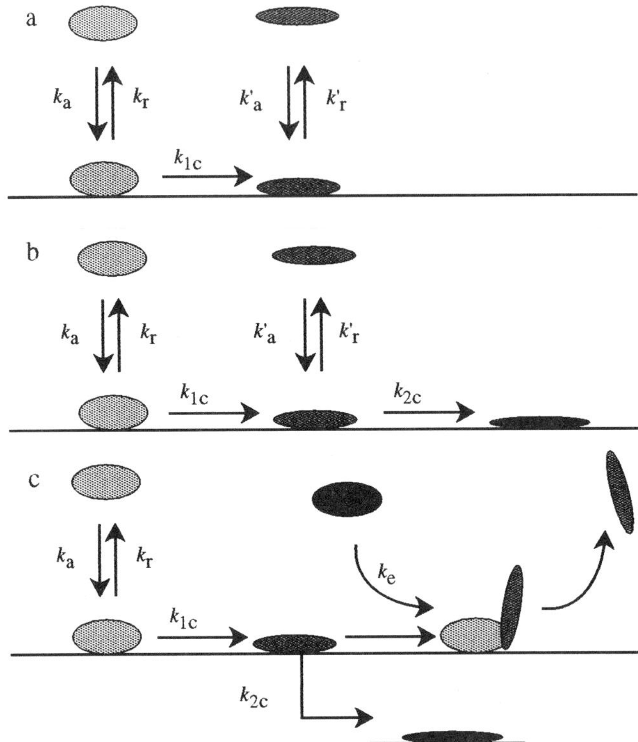

FIGURE 7. Aggregation of proteins by interactions in an aqueous–organic solvent mixture.

where k is Boltzmann's constant, T is the temperature in Kelvin, and η is the solvent viscosity. The increase in size is defined as $S = R_{s,d}/R_{s,n}$. In addition, the radius of gyration (R_g) can be calculated for a particular model or from the relationship $R_g = 1.51R_s$. This ratio R_g/R_s is in the range of 0.8 to 1 for most globular proteins on moderate axial ratios and on the order of 1.5 for a randomly disordered chain[72a] — the radius of gyration, R_g, is the square root of the weight average of R_i^2 for all mass elements measured from the center of mass.[70b]

In the several examples discussed by Gast et al. (lysozyme, α-lactalbumin, streptokinase, and apocytochrome c) the relative increase in Stokes' radius [$S = R_{s,d}/R_{s,n}$] was approximately the same for all proteins denatured in guanidine hydrochloride (GuHCl). By contrast, this was not the case for denaturation by acid or heat, where changes in size could be smaller or larger than those obtained with GuHCl. Unless special precautions are taken, denaturation can produce a range of different conformer and aggregation states.

While the results from denaturation by GuHCl showed comparable increases in size ($R_{s,d}/R_{s,n} = 1.5$), the results from acid and thermal denaturation were variable, i.e., 1.2 to 1.8. With the latter modes of denaturation (acid and temperature), it is possible to observe relative small increases in size, the presence of randomly ordered chains, or alternatively no significant change in conformation. How is it possible for a thermally "denatured" protein to regain its ordered structure? Gast et al. argue that the presence of intramolecular disulfide bonds seems to be an additional stabilizing factor that allows the "denatured" form to revert back to its native state.

C. APPLICATION TO CAPILLARY ZONE ELECTROPHORESIS

These results are particularly applicable with respect to capillary electrophoresis. Since denatured proteins can adopt a variety of conformations from native-like states, such as the so-called "molten globe" state, to highly unfolded states at high concentrations of strong

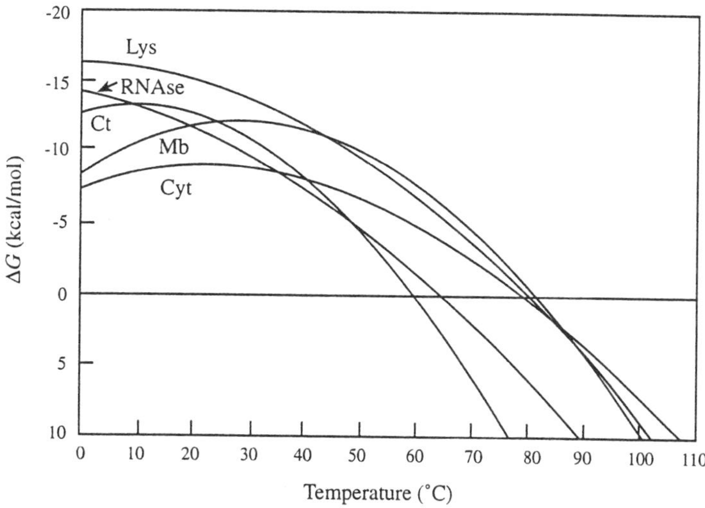

FIGURE 8. Temperature dependence of the difference in free energy (per mole of protein) between the folded and unfolded state. The five different proteins are designated as follows: lysozyme, Lys; ribonuclease A, RNase; metmyoglobin, Mb; α-chymotrypsin, Ct; and cytochrome *c*, Cyt. In each case, the solution pH was chosen for maximum stability. Although a number of samples appear to have maximum stability near room temperature, two samples appear to decrease in stability as the temperature is lowered from room temperature to 0°C. (From Privalov, P. L. and Khechinashvili, N. N., *J. Mol. Biol.*, 86, 665, 1974. With permission.)

denaturants,[72a] special precautions need to be taken. Successful electrophoretic separation of these proteins depends on the elimination of these states and the formation of a "normal" distribution or randomized structures. As discussed above, neither pH nor temperature denaturation can be expected to give completely randomized structures. For capillary electrophoresis, complete denaturation is usually accomplished by the addition of high concentrations of a selective denaturant such as urea (6 to 8 *M*). While guanidine hydrochloride is equally effective in unfolding the native structure, it introduces extremely high ionic strength conditions into the sample, which should be avoided.[72b] By following this protocol, it should be possible to eliminate the apparent peak broadening due to the presence of multiple forms of the same species.

V. EXPERIMENTAL APPROACHES TO METHODS DEVELOPMENT

A. INTRODUCTION

The appearance of CE as a commercially available technique was predicted to capture the interest of protein chemists concerned with separation sciences. Electrophoresis was already a prevalent technique in multiple formats for protein separations and analysis.[73-75] CE initially appeared to be simply a microformat of established procedures, which might benefit small sample handling issues. In spite of the initial apparent promise of this new technique, the protein community did not find immediate applications for CE. There were several difficulties discovered that were not theoretically predicted. Sensitivity assessments of the technique were misleading; the initial papers reported femtomole detection limits, but these referred to the sample amount on column.[76-78] A more relevant value is the amount of material needed to apply to the capillary. Since the columns are usually on the order of 75 to 50 μm I.D., with loading volumes between 10 and 100 nl, it is necessary to start with 10 and 100-μg/ml quantities of protein. Very early reports on home-made CE units without optimized detection systems had required 1-mg/ml starting concentrations.[79,80] Capillary columns presented a second difficulty. The bare fused silica walls exhibit an increasingly negative charge from pH 2.0 upward.[81] Proteins frequently would stick to the walls,[82-87] either preventing them from traveling through the capillary at all, or severely affecting quantitation determinations. Finally, the use of a buffer as a separation matrix, instead of a gel, induced separation on the basis of charge-to-mass ratios,[88] rather than size separation, so that CE results would not be comparable to SDS-PAGE analysis. Attempts to formulate gel capillaries to mimic these separations were initially unsuccessful.

Fortunately, despite these difficulties, the potential benefits of protein analysis by CE were realized. Significant gains have been made in the last 3 years, permitting many protein samples to be analyzed successfully by CE. These gains include: prevention of protein adsorption onto capillary walls by judicious choices of buffer systems, coated and gel-filled capillaries, varied detection systems to improve sensitivity, and development of interface techniques to provide additional information. Since CE is still a new technology for most protein chemists, the following discussion of protein CE separations will be provided as a guide to areas of consideration in developing a separation strategy. General starting conditions will be provided, with a rationale for improved separations and data interpretation.

B. CHOICE OF BUFFERS

1. Analyte Charge

The most important consideration in methods development for CE is the choice of run buffer and sample matrix buffer. This is also the first variable that should be considered in designing a separation strategy. In free solution mode, analytes are separated based on differences in their charge-to-mass ratios. Therefore, a buffer should be chosen that will maximize these differences. Any mixture of proteins is likely to contain a diversity of p*I* values. While the protein may be soluble in the sample buffer this may not be the case for the run buffer. It may occur that the sample is not known to be soluble in the selected run buffer. It is worth attempting one CE run under these conditions, since the solubility characteristics may be different in sample concentrations on capillary as opposed to the much higher concentration of the starting sample. For example, this author (J. N.) had an instance where a protein was known to precipitate in ionic concentrations of sodium borate above 100 m*M*, but was successfully analyzed by using a 500 m*M* sodium borate as a run buffer. The additional ion binding that was evident at the higher buffer concentration was capable of altering the net charge sufficiently to reduce protein–protein interactions and prevent precipitation.

2. Capillary Wall Charge

Buffers will affect not only the analyte, but also the capillary wall. The silanol groups of the bare fused silica will demonstrate a weakly negative charge at low pH. This negative characteristic dramatically increases with increasingly basic buffer systems.[89] This outcome of the induced charge on the capillary wall is to increase the electroosmotic flow (EOF).[90,91] Increased bulk fluid movement through the capillary may be detrimental to separations, by pushing a mixture of proteins through the capillary faster than a separation would occur based on their differences in charge to mass. Conversely, EOF is essential in analyzing proteins that have opposite net charges. Positively and negatively charged proteins would move in the capillary in opposite directions. Regardless of the selected polarity, one would never move past the detector. EOF of sufficient magnitude will not only increase the mobility of the compound moving toward the detector, but also overcome the electrophoretic attraction of the protein that is moving away from the detector and subsequently moves it past the detection point (Figure 9). Therefore, buffer pH should be selected to control the rate of EOF, and thus optimize the speed and resolution of a separation.

A second consideration of buffer effects on the wall charge is the prevention of protein adsorption. Low-pH buffers will generally keep a protein below its p*I* point, creating a net

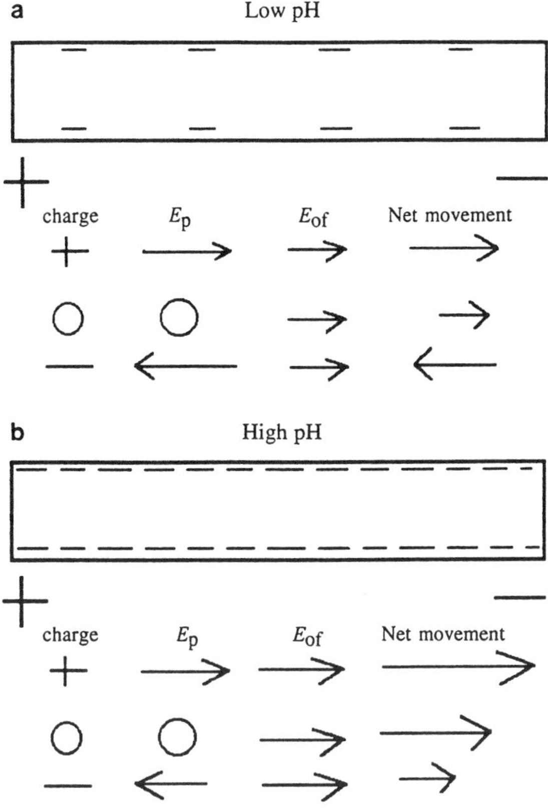

FIGURE 9. Net movement of particle for a given charge, based on the sum of electrophoretic (E_p) and electroosmotic (E_{of}) forces in (a) low-pH and (b) high-pH run buffers.

positive charge on the protein.[92] Under these conditions, the wall will be only weakly negatively charged, so that electrostatic attraction between the protein and capillary surface should be minimized. Although there has been some success with using extremely low-pH buffers (pH 1.5) to inhibit any charge of the silanol groups,[82] proteins can still adsorb to the capillary wall under these conditions. An alternate approach has proved more generally applicable. Working 1 to 2 pH units above the p*I* of a protein will induce a net negative charge on the protein. The charge on the capillary wall will also be negative, and the repulsion of the protein from the wall will permit analysis. It is important to remember that the p*I* is a *net* charge for the protein.[93] There is still the possibility that basic amino acids presenting on the surface of the protein will cause protein–wall interactions even with high-pH run buffers. In this instance, buffer changes other than pH need to be considered.

3. Buffer Additives

a. General Classification

Initial concerns in methods development for CE concentrated on the optimal pH that would adjust the capillary wall pH to prevent protein adsorption while not increasing EOF such that resolution would be destroyed. Buffers were originally chosen on the basis of their buffering capabilities at the pH of interest. Subsequent research efforts have shown that the type of buffer salt selected may significantly affect a separation. Tran et al.[94] demonstrated a glycoprotein separation using a mixed acetate–phosphate system in place of a single buffer salt. An improvement in separation was achieved in the mixed buffer, when compared to an individual acetate or phosphate system, keeping the pH constant for all three variations. There have been many papers discussing the use of additives to buffer systems to inhibit protein–wall interactions. The general, principle mechanism is that of a buffer additive that either masks or competes for the charge on the wall, so that it is not available to the protein interaction. A variety of different agents have been used, including divalent amines,[95,96] high ionic strength buffers,[87,95,97-99] detergents,[100-104] organic solvents,[105] and zwitterionic salts.[80,102,106-108] The definitive mechanism for these reagents is unknown. It is presumed that they combine with and mask charges on the capillary wall, the protein, or both. Buffer additives are frequently claimed to affect wall charge, but it must be remembered that buffer is in contact with both the capillary and the analytes. Therefore, the effect of any buffer component must be considered from a sample, as well as a capillary, perspective. In addition, there are protein–protein interactions that must also be controlled, such as hydrophobic forces, which are general protein–handling issues.[109] These can decrease resolution and efficiency to the same, or even greater, extent as wall interaction phenomena.

b. Anionic and Neutral Surfactants

Several examples exist that demonstrate the use of both ionic and nonionic detergents as buffer additives. As a general rule, proteins are too large to partition into a true micellar system.[110] Therefore, SDS is used in free solution to coat proteins with negative charge. One innovative method was developed by Nakagawa et al.[111] to remove proteins from a crude serum matrix. In the absence of detergent, drug presence in the serum could not be detected due to protein peaks coeluting with the compounds of interest. By adding a low concentration of SDS (10 m*M)* which is slightly higher than the critical micelle concentration (CMC), Nakagawa et al. were able to increase the negative charge of the proteins and shift their mobility away from the drug components. Another use of SDS for protein separations was developed by Nolan and Palmieri.[102] In general, most proteins bind SDS in a constant ratio (1.4 g of SDS:1 g of protein),[112] negating any charge-to-mass differences. In SDS-PAGE this permits differential migration on the basis of mass alone, and the subsequent estimation of molecular weight.[113] However, it is well known that the weights determined by this method can be in error since they are based on the assumption of a constant SDS binding ratio. For example, proteins with posttranslational modifications such as glycosylation may bind SDS in significantly different amounts than the predicted ratios, thereby providing aberrant molecular weight determinations.[114] This fact, combined with the idea of coating proteins with a negative charge to create a repulsion from the negatively charged capillary wall, was used to develop a separation system of proteins based on their differential binding of SDS. Figure 10 demonstrates the application of this method for a glycosylated and nonglycosylated protein. Because of the net negative charge of the samples, the analysis is performed in reverse polarity, in a low-pH buffer to decrease EOF. Streptavidin, the nonglycosylated protein, migrates first, presumably due to its ability to bind more SDS. This should increase its negative charge over avidin, causing it to migrate toward the anode ahead of its glycosylated counterpart.

Nonionic detergents have also been employed for protein CE analysis. In this instance, the detergent may be used to supplement a hydrophobically coated capillary to increase the coverage of the coating.[115] While an octadecylsilane coating of the capillary reduced electrostatic interactions, it was found that hydrophobic forces were still detrimental to the separation. Addition of nonionic detergents created a hydrophilic layer, and were also thought to mask any uncovered silanol groups. Improvements in resolution and efficiency for protein samples were achieved when the detergent selected had the appropriate size and branching pattern to enhance the initial capillary wall coating.

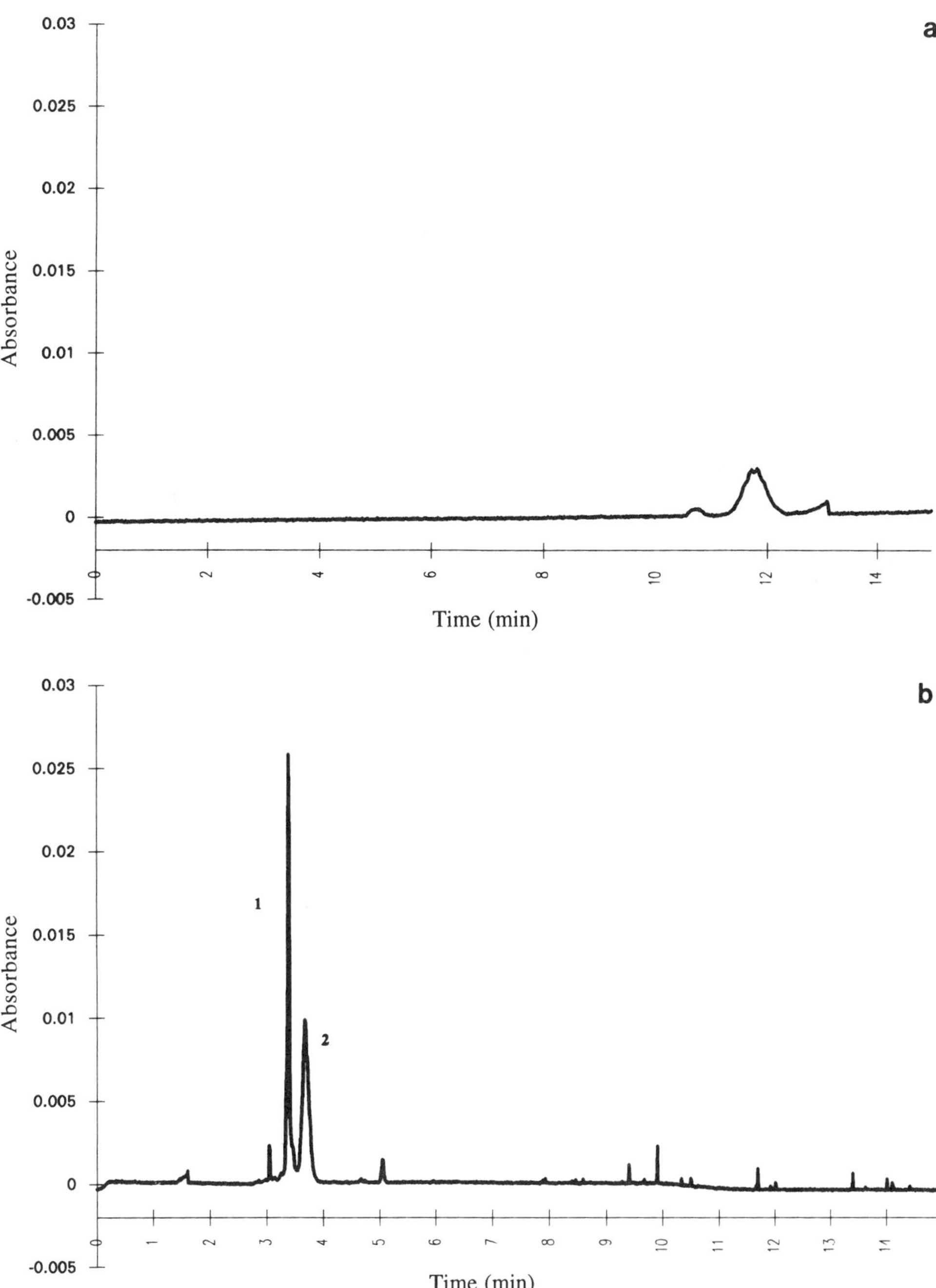

FIGURE 10. Sodium dodecyl sulfate (SDS)-containing run buffer for the separation of glycosylated and nonglycosylated proteins. Capillary, 50-μm I.D. × 27-cm length; wavelength, 200 nm; proteins, streptavidin (peak 1) and avidin (peak 2) stocks made 1 mg/ml each in 25% phosphate-buffered saline (PBS). (a) Run buffer, 500 m*M* sodium borate, pH 8.5; sample preparation, diluted 1:3 in water for analysis; voltage, 0 to 10-kV linear ramp over 20 min, normal polarity (toward cathode). (b) Run buffer, 100 m*M* sodium phosphate, pH 2.0, 0.1% SDS; sample preparation, stocks diluted 1:3 in 1% SDS, 5% 2-mercaptoethanol, 2.5 m*M* Tris-glycine (pH 8.9), 10% glycerol, and boiled for 5 min; voltage, 10 kV, constant, reverse polarity (toward anode).

4. Analyte Matrix

As discussed in Chapter 9, the sample environment that is used to load the sample onto the column will dramatically affect subsequent resolution and efficiency. A low ionic strength sample matrix, relative to the run buffer ionic strength, is preferred, inducing stacking[116-118] to improve peak efficiency and detectability. The high field strength experienced by the sample plug induces the analyte to traverse this region rapidly until it is slowed at the interface of the run buffer, which experiences a lower field strength. On-column concentration of the sample occurs as the analyte is pushed into a smaller volume than originally used in initial loading of the sample. There are two concerns with this technique. Chien and Burgi[119-121] have discussed the possibility that large differences in ionic strength between sample and run buffer will create a very large EOF over the sample plug, which may be detrimental to the resolution. The speed at which the sample is introduced into the capillary may overcome the separation predicted to occur on the basis of sample size and charge. It has also been suggested that an extremely high field strength over the sample region, which is induced by ionic stacking, can cause protein aggregation and denaturation, due to excessive heat generation.[122,123] In addition, certain samples may not be amenable to very dilute sample buffers, thus the ability to stack on the basis of differential ionic strength sample and run buffers is decreased. To alleviate the latter difficulty, it is possible to use high ionic strength run buffers (500 m*M*)[87] in conjunction with sample buffers that are one tenth the ionic strength to induce stacking of the sample. A different consideration of sample environment was examined by Gordon et al.[124] In an attempt to prevent protein adsorption to uncoated capillaries, ethylene glycol was added to the sample buffer prior to injection for analysis by CE. It is postulated that the additive decreased wall interaction, as evidenced by improved reproducibility when compared to sample in the absence of ethylene glycol. However, it was also considered that protein–protein interaction may have been reduced by this technique as well.

C. INSTRUMENTAL PARAMETERS

1. Capillary

a. Dimensions

Capillary dimensions may be altered in an approach similar to HPLC columns. The first consideration is the placement of the detection system. Currently, the most widely used capillary format is coated with an opaque polyimide to protect it from breakage. Therefore, it is necessary to remove part of that coating to provide a detection window. The detector is preferably placed near the end of the capillary, since once detection occurs it is unnecessary to continue migration through a separation matrix. If fraction collection is to be performed, the short capillary length after the detector will help to diminish any band spreading, particularly important for closely resolved peaks.

The lengths of capillaries used range from 20 to 100 cm. A longer capillary may be used to increase resolution, but for some protein separations there seems to be a decrease in resolution with use of longer capillaries. It is hypothesized that the longer capillary may provide a greater opportunity for protein–wall interaction.[87] It must be remembered that increased capillary length will create a decreased voltage and pressure drop. In transferring a method to a longer capillary, increased voltages will be needed to maintain shorter run times. Furthermore, in loading a sample either hydrodynamically or electrokinetically, a longer time period with the same pressure/voltage, or increased pressure/voltage, must be used to apply the same amount of sample.

Varying the capillary diameter is usually considered for two purposes. For fraction collection, it is desirable to load as much sample as possible on column to reduce the number of runs that is necessary to collect a useful amount of material. There has been confusion as to the usefulness of fraction collection by CE. Initially, CE was viewed as an analytical tool, dealing with sample amounts too minute for collection after separation. However, if CE is the only technique that will afford a separation of certain components, this emphasizes the need to develop fraction collection techniques. Early work in fraction collection[125] recommended multiple runs using standard capillary diameters to obtain sufficient material for analysis. Most commonly used capillaries for analytical purposes are 20 to 75 μm in internal diameter. An increase to 150 μm improves the sample capacity for preparative applications.[126] While there are still difficulties in routine fraction collection by CE, this will undoubtedly be an area of further research.

b. Coatings

In contrast, small-diameter capillaries have been employed for high current-generating buffer conditions. Charged buffer additives and high ionic strength buffers, frequently used for resolution enhancement, may require the use of 50- or 20-μm I.D. capillaries. The smaller diameters permit more efficient heat dissipation,[127] required to prevent Joule heating and subsequent convection in the capillary, which would diminish resolution.

There has been tremendous debate concerning the use of coated vs. uncoated capillaries, and the types of compounds that would be appropriate to use for developing CE coated columns. Initially, there was the proposal that the use of uncoated, bare fused silica columns would actually be advantageous. It was believed that the lack of surface modification would alleviate any batch-to-batch reproducibility problems. However, there may be significant differences in uncoated capillary lots to which we have not been attributing separation differences.[128] It has also been argued that since bare silica columns in HPLC have not been successful for protein analysis due to interaction problems, this same concern applies to CE.[97,99,129-132] Initial attempts at coating capillaries, taken largely from GC column technology, were unsuccessful with regard to homogeneity and longevity. The requirement for stability over a large range of pH values was more demanding than for an HPLC column, and short column lengths were less forgiving with regard to the percentage of the surface that may not be fully treated. Despite these problems, great improvements have been made. The future availability of capillaries with a range of covalently treated surfaces will improve the success of protein analysis by CE. Dynamic coatings are gaining in popularity,[133,134] permitting a replenishment of capillary wall treatment prior to the start of each run, by simply rinsing the capillary with a solution containing a coating agent. One misconception of coated capillaries is that they will completely prevent any protein adsorption to the wall. In certain cases, proteins will still adsorb to a coating, but to a lesser extent compared to an uncoated surface. Furthermore, the coating may permit an easier and more thorough removal of residual protein than from bare silica.[135] This topic is thoroughly discussed in Chapters 14 and 15.

2. Detectors

a. UV Detectors

Ultraviolet detection has been the most universal system adapted to CE. On-column detection creates a flow cell with a 50 to 75-μm path length, which in turn decreases sensitivity when compared to HPLC flow cells.[93,136,137] This provides the ability to work at very low-UV wavelengths (190 to 200 nm) since sensitivity to buffer absorbances is also decreased. Absorbance spectra for proteins show a large increase in this range, so an increase in analyte sensitivity may be achieved. Other components of the sample mixture may also be detected, since most compounds will show absorbance at 200 nm.

b. Other Options

Other detection systems have been discussed in the literature as research instruments, including electrochemical,[138,139] radioisotope,[140] and mass spectrometry interface.[141-150] Fluorescence detectors are the one commercially available detection option. Both traditional monochromator-based[151,152] and laser sources[153-159] have been developed. While the laser systems offer the highest sensitivities, the unique excitation line usually requires the derivatization of a protein with the appropriate fluorophore. Nonreacted probe can be removed by traditional exchange methods such as dialysis or gel-filtration chromatography. The expected gain in sensitivity over UV methods is two to three orders of magnitude. Diode array detectors are beginning to become available[160] and while these systems may not offer great advantages in protein analysis, they will be helpful for samples containing proteins and other analytes with unique spectra.

3. Voltage

Almost all separations to date have been performed under constant voltage conditions. The use of voltage ramping, originally described by McCormick,[82] has been reexamined for protein separations performed under severe sample-stacking conditions.[102] Figure 11 shows the separation of four proteins under constant and ramped voltage conditions. The ramping is proposed to decrease the extreme difference in field strength between a low ionic strength sample matrix and high ionic strength buffer, which can compromise resolution. It is also possible that the high field strength over the sample plug may cause some localized heating effects, which could denature and aggregate the proteins. It has been suggested that a constant current mode is preferable for isoelectric focusing applications.[161]

4. Temperature

The most critical factor regarding temperature is consistency. Most separations are performed between 20 and 30°C, and held constant to within 0.1°C, since there is a 2% mobility change for every 1°C change over the capillary.[127] Temperature changes will affect several parameters in the separation. Buffer pH will change with temperature,[162] and an increase in temperature in one part of the capillary may set up a temperature gradient. Different run temperatures will change the pH of the buffer in the capillary, which may affect migration times. Viscosity of buffers will also change with temperature, and gels in particular will be sensitive to this phenomenon. Both their viscosity, or rigidity,[163] and their pore size will be affected. Finally, a protein may exhibit conformational changes as temperatures vary, presenting a differently charged surface and altering its mobility[164] (Figure 12). Temperature effects in CE are further discussed in Chapter 21.

D. METHODS DEVELOPMENT GUIDE

Table 7 lists a series of protein classifications and applications that are currently of interest in CE analysis of proteins. While this list is by no means comprehensive, it is meant to serve as a general guide for initial conditions in methods development. The following section will elaborate on the considerations for separation of proteins by CE.

1. Starting Conditions

The following conditions are a good starting point for a CE protein separation if the behavior of the protein is unknown:

Run buffer: 500 m*M* sodium borate, pH 8.5 to 9.0
Sample buffer: one-tenth strength run buffer
Capillary: uncoated, 50-μm I.D. × 27-cm length
Wavelength: 200 nm
Temperature: 25°C
Sample concentration: 100 μg/ml
Run time: 30 min

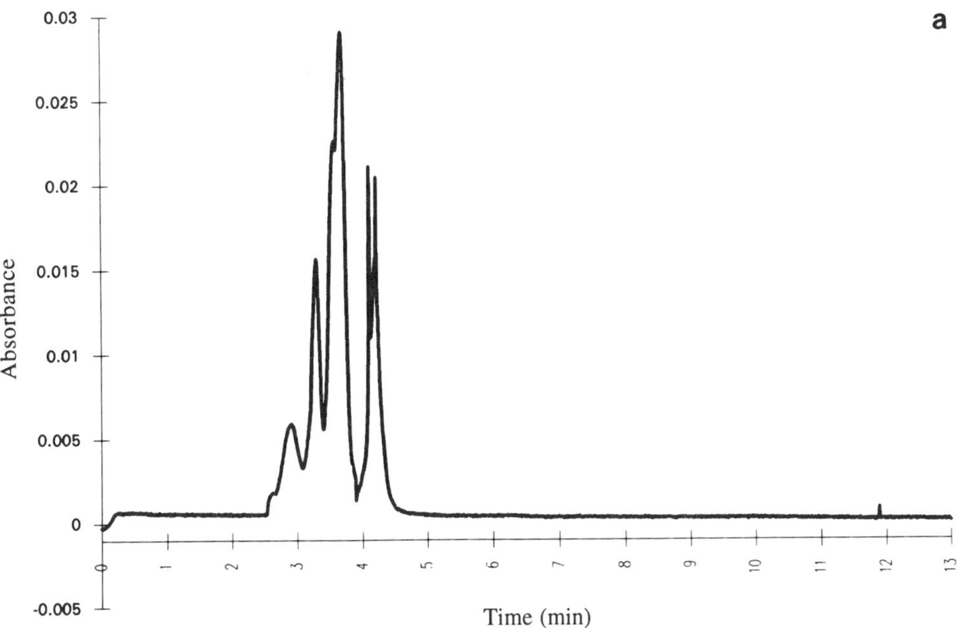

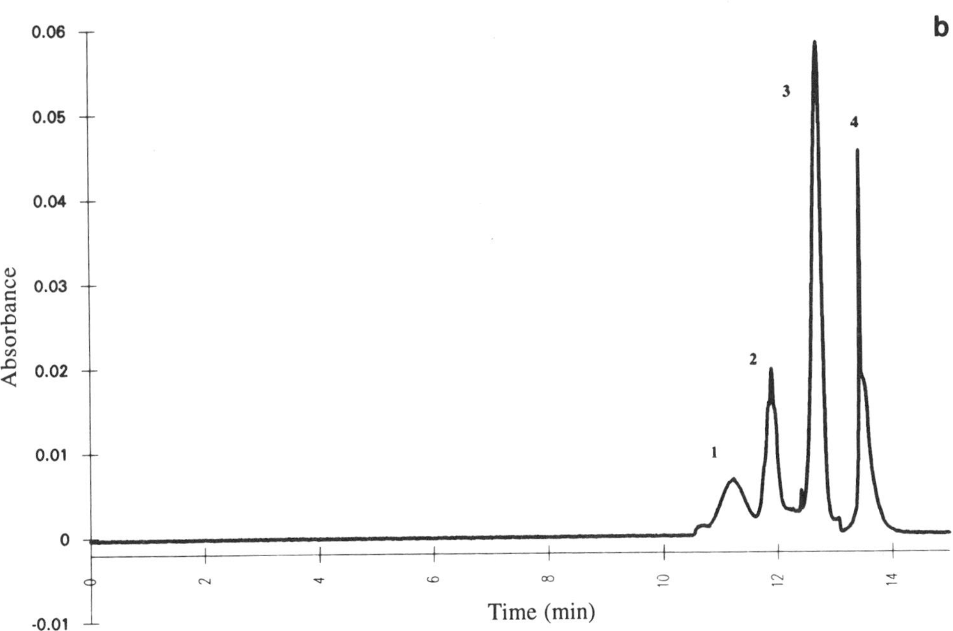

FIGURE 11. Voltage ramping applied to improve resolution for protein mixture. Rabbit immunoglobulin G (peak 1), transferrin (peak 2), α_2-macroglobulin (peak 3), and bovine serum albumin (peak 4) stocks initially made 1 mg/ml each in 25% PBS and combined in equal volumes for a final concentration of 250 µg/ml each. Capillary, 50-µm I.D. × 27-cm length; run buffer, 500 m*M* sodium borate, pH 8.5; wavelength, 200 nm. Voltage: (a) 10 kV, (b) 0 to 10-kV linear ramp over 20 min, normal polarity.

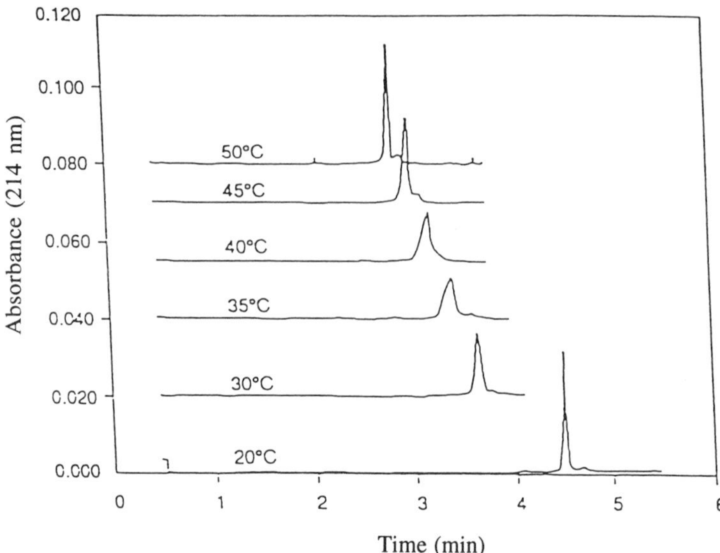

FIGURE 12. Temperature effect on protein conformation. α-Lactalbumin analyzed at increasing run temperatures to demonstrate the change in conformation of the protein. Unfolding of protein at elevated temperatures presents alternate charged surfaces to run buffer, causing a change in migration behavior. (From Rush, R. S., Cohen, A. S., and Karger, B. L., *Anal. Chem.*, 63, 1346, 1991. With permission.)

These parameters are chosen to provide a system in which most proteins will migrate through the capillary. It is analogous to using a relatively high organic solvent concentration to initially develop a reversed-phase method; it is necessary to determine that the protein passes through the column and is detected before optimum resolution can be obtained. This is the rationale for the relatively high sample concentration and detection at 200 nm. If the material is available, it is always easier to develop a separation method without the added difficulty of working at the limit of detection. Another point to note is the relatively long run time.

While CE is considered a high-speed technique, a common mistake is to terminate the first trial run prematurely. Since the mobility characteristics of the sample may be unknown, it is important to wait a sufficient period of time on the first run to make certain that all components of the sample have been given adequate time to migrate through the capillary. After the first electropherogram is obtained, there are four areas to be considered for possible improvement: speed, detection limit, resolution, and reproducibility.

2. Separation Optimization

a. Decrease Analysis Time

If all the other requirements have been met for the separation, the speed of the separation may be increased by the following:

1. Increasing the EOF by
   a. increasing the buffer pH
   b. decreasing buffer ionic strength
2. Increasing the run voltage to decrease migration times
3. Increasing the temperature to decrease buffer viscosity
4. Decreasing the capillary length

b. Increase Detection Limits

Increased sensitivity may be obtained from the following:

TABLE 7
Summary of Protein Categories Analyzed by Capillary Electrophoresis

Classification	Ref.
Basic proteins	85, 133, 134, 191
Single cell proteins	192
Binding assays	190, 193
Glycoproteins	94, 96, 194
Amino acid analysis	185, 195, 196
Sulfation	197
Antibody–antigen complexes	189
Conformation determination	164
Enzyme activity	184
Fraction collection	125, 126, 184, 185
Enzyme-antibody conjugates	104
Thiols	198
Milk proteins	87

1. Loading more sample
2. Increasing the ionic strength difference between sample and run buffers to induce greater sample stacking, without exceeding the heat dissipation capabilities of the CE instrumentation
3. Monitoring at low-UV wavelengths, 190 to 200 nm
4. Using isotachophoretic loading to increase the amount of material on column
5. Using fluorescence detection with sample derivatization
6. Treating samples in biological matrices (urine, serum) with a clean-up step prior to CE analysis to reduce background

c. Improve Resolution/Efficiency

In any separation technique, this is probably the area that will require the most development. CE techniques for improvement include the following:

1. Decreasing EOF by
   a. increasing buffer ionic strength
   b. increasing buffer viscosity with additives
   c. decreasing capillary wall ζ potential with additives
   d. voltage ramping to decrease sample plug EOF
   e. selection of coated capillary
2. Decreasing analyte–wall interactions with
   a. buffer additives such as divalent amine, zwitterions
   b. coated capillaries
   c. different run buffer pH to change the charge on the analyte
   d. gels/entangled polymer networks
   e. ionic detergents
3. Decreasing analyte–analyte interactions with
   a. nonionic detergents in run buffer
   b. ion-pairing reagents in run buffer
   c. adjustment of sample matrix ionic strength, addition of additives
   d. ethylene glycol addition to sample buffer[124]

d. Improve Migration Time Reproducibility

Although earlier investigators experienced initial problems with reproducibility, many of these issues have been resolved. This has been especially true as more information about the dynamics of capillary electrophoresis has emerged. To date, the following factors have been implicated: buffer matrix, capillary regeneration, buffer concentration, length of analysis, temperature fluctuations, run pH, buffer volume, and analyte stability. While no one recommended procedure can insure good reproducibility, experience dictates some common fundamentals. An important consideration is appropriate rinsing conditions of the capillary following each run.[165,166] Some suggestions follow, although the best conditions for any particular sample and capillary type must be determined empirically. Many of the variables can be controlled by the appropriate use of internal reference standards. With appropriate software the relative migration time can then be calculated. Thus, changes in temperature, viscosity, ionic strength, EOF, and dielectric constant can be obviated.[166a]

1. Rinse with 0.1 *N* NaOH, followed by run buffer.
2. Rinse with 1 *N* HCl, followed by run buffer.
3. Rinse with run buffer at 5× ionic strength.
4. Insert a water rinse between base/acid and buffer rinse.
5. For gel capillary electrophoresis, replaceable polymer networks, which may be rinsed fresh into the capillary prior to each run, seem to be preferable over chemical gel systems for both reproducibility and capillary longevity.
6. Use fresh buffer reservoirs for each run; this is particularly important for long runs (30 min) with high current-generating buffers (above 60 μA)

e. Improve Quantitation

Compared to migration time, work is still in progress with regard to quantitation. However, the following procedures can improve the reproducibility and accuracy of quantitation:

1. For electrokinetic injections, use a fresh sample vial for each injection to avoid nonhomogeneous sample and sample matrix depletion.[167]
2. For pressure injections, use at least a 3-s pressure injection time; times around 1 s or less are usually less reproducible.
3. Inject a plug of run buffer behind sample plug; if there is any localized heating over the sample plug due to sudden application of high field strengths, expansion of the sample plug will induce a volume loss by pushing the sample out of the capillary and into the inlet vial.
4. Use a short voltage ramp to reach the run voltage (same rationale as above).

VI. FUTURE DIRECTIONS

Advances in CE have been rapidly appearing over the last 3 years, but the protein analysis segment has probably seen the most dramatic improvements. High-efficiency separations have now been achieved, with a combination of buffer additives and coated capillaries, both in separations on the basis of charge-to-mass ratio and by molecular weight in gel systems. Continued developments for protein analysis in CE will most probably concentrate in the following areas.

A. DETECTION

Increasing the sensitivity of a 50-μm I.D. path length cell, along with nanoliter volumes of loaded sample, has always been an area of interest in CE. Increased loading capacities with the use of isotachophoretic sample preconcentration[168-171] has already shown significant promise. Samples are initially loaded and run with isotachophoretic buffer systems, which are then changed to more traditional CE buffers after focusing has been achieved. On-column concentration occurs, which is a derivative of a stacking procedure, based on differential mobilities of buffer and analyte ions. The run is then completed in a CE mode, providing data output as an electropherogram. Detection limits can be increased 10- to 100-fold with this technique. Derivatization schemes for laser-based fluorescence detectors will continue, possibly starting to develop on-column derivatization procedures.[152,156,159] Secondary detection schemes, such as mass spectrometry interface, diode array detectors for spectral analysis, and the Bushey and Jorgenson LC-CE concepts[172,173] will provide more information than do currently existing single detector options. It is these interface techniques that will truly make CE a complementary technique by developing two-dimensional analyses on line. Another avenue of detection improvement involves changes to the capillary geometry. Increasing the dimensions of the flow-cell portion of the capillary by either monitoring down the axis of the capillary[174,175] or making a bubble in the detection window[176] has met with success. The increased detection limits, however, are at the expense of resolution. There are applications where this loss is minimal and does not impair the resulting analysis.

B. SIZE SEPARATION AND ISOELECTRIC FOCUSING

These are two very well-established techniques that are of interest to the protein chemist in terms of obtaining important physical information concerning a sample. Capillary electrophoresis offers several advantages to slab gel procedures: direct detection provides true quantitation capabilities[135,177] that are currently biased due to dye binding affinities; speed and ease of gel preparation, particularly in the case of IEF; the ability to examine separations by charge-to-mass ratio and mass alone on the same apparatus. Figure 13 shows the use of capillary gel electrophoresis to examine the disulfide bond reduction for an immunoglobulin. Similar degrees of accuracy are obtained for molecular weight determination, permitting transfer of slab gel techniques to the CE format. Although the demand for gel procedures for CE was great, it is only recently that they have been demonstrated to be a commercially available option. In a similar manner, isoelectric focusing was presumed to be a method that would be well suited to the capillary format. The first attempts at IEF for a CE procedure were similar to the slab gel methods: focus the proteins in an ampholyte mixture, then detect. For a slab gel, this is easily accomplished by staining the gel. In CE, it is necessary to mobilize the gradient past the detection point without disturbing the gradient. This originally involved using coated capillaries with no EOF flow for initial focusing, and then using a chemical mobilization of the gradient.[178,179] An alternative method was to focus and mobilize simultaneously in an uncoated capillary, or a coated capillary with residual EOF.[180] Yet a third solution was to focus under conditions of no EOF, and then mobilize using a hydrodynamic method,[181] while voltage is still applied,[182] to inhibit band spreading caused by laminar flow (Figure 14). The latter two techniques have been introduced to the CE community, and it seems that there are now three viable techniques to perform IEF by CE. Research continues for this methodology, with an alternate technique of moving the detector past a fixed gradient.[183]

C. FRACTION COLLECTION

Fraction collection was originally viewed as an analytical technique; when it became apparent that certain separations could only be attained by a CE method, it was necessary to develop preparative techniques to recover an isolated component of a sample. While collection into microliter volumes has proved successful[126,184,185] this methodology will move toward direct collection onto membranes for subsequent immunochemical analysis and protein sequencing.[186,187] Undoubtedly, many scientists are also waiting for an integrated software package that can reproducibly locate preselected peak(s) and place them into a specific vial.

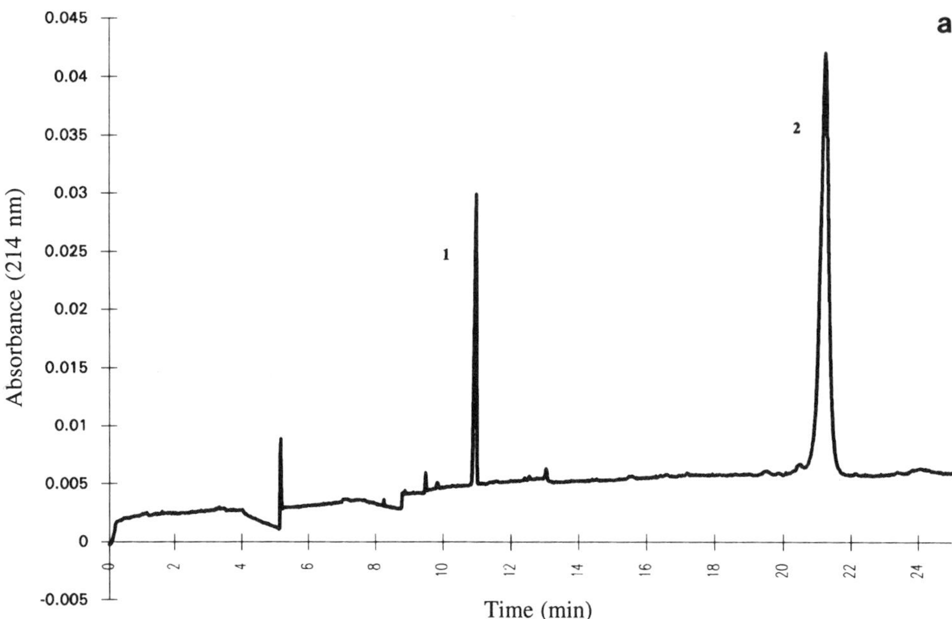

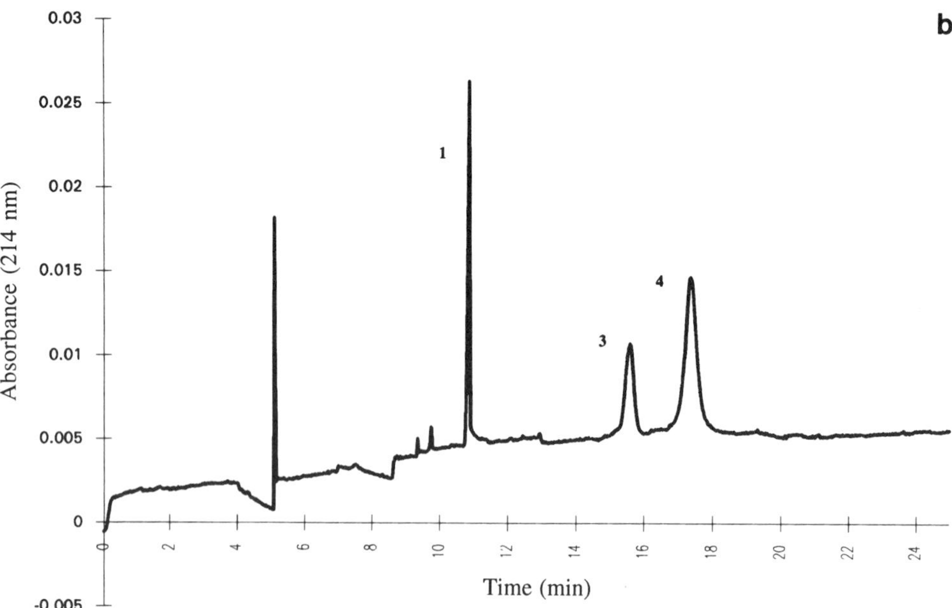

FIGURE 13. Protein gel capillary electrophoresis. A replaceable gel matrix, eCAP™ SDS 200, is rinsed into the capillary and acts as a sieving matrix for protein separation on the basis of molecular weight. Capillary, eCAP™-coated capillary, 100-µm I.D. × 47-cm length; wavelength, 200 nm; temperature, 20°C; polarity, reversed; sample preparation, pierce mouse monoclonal antibody (IgG) to human follicle-stimulating hormone (FSH) at 1 mg/ml, diluted 1:1 in eCAP SDS 200 sample buffer with 10 µl of orange G internal standard, was boiled for 5 min (a) in the absence and (b) in the presence of 5% 2-mercaptoethanol. Peak identification: (1) orange G, (2) intact IgG, (3) light chain IgG, (4) heavy chain IgG.

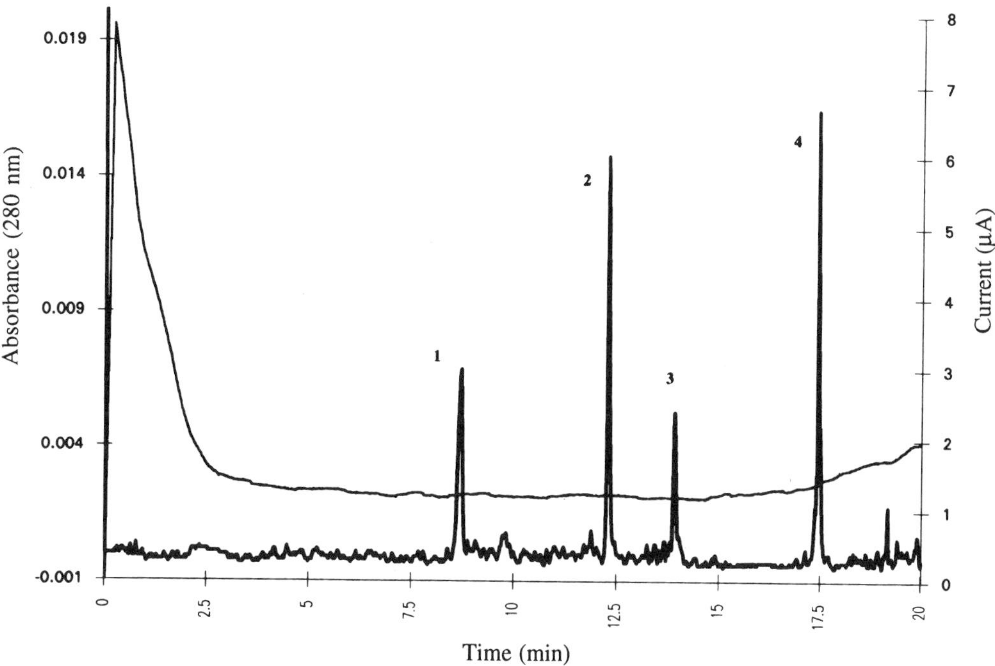

FIGURE 14. Capillary isoelectric focusing. Myoglobin (peak 1, p*I* 7.0), carbonic anhydrase II (peak 2, p*I* 5.9), β-lactoglobulin A (peak 3, p*I* 5.1), and soybean trypsin inhibitor (peak 4, p*I* 4.6) stock solutions were made 2 mg/ml in 25% PBS. They were mixed in equal volumes and the mix was diluted 1:1 with a 1.5% ampholyte 3-10 in 0.4% methyl cellulose. The capillary, DB-1 (J&W Scientific), 50-μm I.D. × 27 cm, was filled up to the detector window with 1.5% ampholyte in 0.4% methyl cellulose, using a 20-psi pressure rinse. Sodium hydroxide (20 m*M* in 0.4% methyl cellulose) fills the remainder of the capillary and the outlet vial. Phosphoric acid (100 m*M*) fills the inlet vial. Sample is pressure loaded and run at 20-kV constant voltage for 3 min. At this time, the current drops (as seen by the overlay trace), indicating completion of focusing. The gradient is mobilized by application of 0.5 psi while maintaining a 20-kV constant voltage to move the focused zones past the detection point.

Alternatively, an entirely new approach that more easily accommodates the needs of capillary electrophoresis should be forthcoming. This would obviate the need for a more sophisticated software package.

D. BINDING AND KINETICS STUDIES

The ability to closely control and change the capillary temperature will permit the continued investigation of enzyme kinetics,[188] antibody–antigen binding,[189] and perhaps the development of CE protein–DNA gel retardation assays.[190]

Capillary electrophoresis has started to become an electrophoretic analysis tool for protein purification, and will continue to develop for the purposes of gaining more information concerning protein physical parameters and biological activities.

REFERENCES

1. **Deyl, Z. and Struzinsky, R.,** Capillary zone electrophoresis: its applicability and potential in biochemical analysis, *J. Chromatogr.*, 569, 63, 1991.
2. **Tanford, C.,** *Physical Chemistry of Macromolecules*, John Wiley & Sons, New York, 1961, chap. 6, pp. 326 & 358.
3. **Bowen, T. J. and Rowe, A. J.,** The determination of molecular monoformation, in *An Introduction to Ultracentrifugation,* Wiley-Interscience, London, 1970, chap. 7.

4. **Baier, R. E., Meyer, A. E., Natiella, J. R., Natiella, R. R., and Carter, J. M.,** *J. Biomed. Mater. Res.*, 18, 337, 1984.
5a. **Klein, J.,** Surface interactions with adsorbed macromolecules, *J. Colloid Interface Sci.*, 111, 305, 1986.
5b. **Andrade, J. D.,** Heparin interaction with adsorbed protein layers, *J. Colloid Interface Sci.*, 111, 314, 1986.
6. **Creighton, T. E.,** *Proteins, Structure and Molecular Function*, W. H. Freeman, New York, 1983, chap. 7, pp. 268, 269.
7. **Budd, P. M.,** Sedimentation analysis of synthetic polyelectrolytes, in *Analytical Ultracentrifugation in Biochemistry and Polymer Science*, Harding, S. E., Rowe, A. J., and Harton, J. C., Eds., Royal Society of Chemistry, Cambridge, 1992, pp. 593–607.
8. **Kuntz, I. D. and Kauzmann, W.,** Hydration of proteins and polypeptides, *Adv. Protein Chem.*, 28, 239, 1974.
9. **Rupley, J. A. and Carer, G.,** Protein hydration and function, *Adv. Protein Chem.*, 41, 38, 1991.
10. **Tanford, C.,** *Physical Chemistry of Macromolecules*, John Wiley & Sons, New York, 1961, chap. 6, p. 383.
11. **Creighton, T. E.,** *Proteins, Structure and Molecular Function*, W. H. Freeman, New York, 1983, chap. 4, p. 135f.
12. **Klapper, M. H.,** Frequency of occurrence of amino acids in proteins, *Biochem. Biophys. Res. Commun.*, 78, 1018, 1977.
13a. **Scopes, R. K.,** *Protein Purification*, Springer-Verlag, New York, 1982, chap. 1.
13b. **Izatt, R. M. and Christensen, J. J.,** Heats of protein ionization, pK_a, and related thermodynamic quantities, in *Handbook of Biochemistry and Molecular Biology, 3rd ed., Vol. 1*, Fasman, G. D., Ed., CRC Press, Cleveland, OH, 1976.
14. **Scatchard, G., Coleman, J. S., and Shen, A. L.,** The attractions of proteins for small molecules and ions, *J. Am. Chem. Soc.*, 79, 12, 1957.
15. **Scopes, R. K.,** *Protein Purification*, Springer-Verlag, New York, 1982, chap. 3.
16. **Kuby, S., Palmieri, R., Okabe, K., Cress, M., and Yue, R.,** A physicochemical comparison of the creatine kinase isoenzymes from man, calf and rabbit, *J. Protein Chem.*, 2, 269, 1983.
17. **Edsall, J. T. and Wyman, J.,** *Biophysical Chemistry, Vol. 1*, Academic Press, New York, 1958, pp. 466–534.
18. **Timasheff, S. N.,** Polyelectrolyte properties of globular proteins, *Biol. Macromol.*, Ser. 3, 1, 1970.
19. **Alber, T.,** Stabilization energies of protein conformation, in *Prediction of Protein Structure and the Principles of Protein Conformation*, Fasman, G. D., Ed., Plenum Press, New York, 1991, chap. 5.
20. **Wada, A. and Nakamura, H.,** Nature of the charge distribution in proteins, *Nature*, 293, 757, 1981.
21. **Barlow, D. J. and Thornton, J. M.,** Ion pairs in proteins, *J. Mol. Biol.*, 168, 867, 1983.
22. **Barlow, D. J. and Thornton, J. M.,** The distribution of charged groups in proteins, *Biopolymers*, 25, 1717, 1986.
23. **Rogers, N. K.,** The role of electrostatic interactions in the structure of globular proteins, in *Prediction of Protein Structure and the Principles of Protein Conformation*, Fasman, G. D., Ed., Plenum Press, New York, 1991, chap. 8.
24. **Chen, F. A., Kelly, L., Palmieri, R., Biehler, R., and Schwartz, H. E.,** Use of high ionic strength buffers for the separation of proteins and peptides with capillary electrophoresis, *J. Liq. Chromatogr.*, 15, 1143, 1992.
25. **McLaughlin, G. M., Nolan, J. A., Lindahl, J. L., Palmieri, R. H., Anderson, K. W., Morris, S. C., Morrison, J. A., and Bronzert, T. J.,** Pharmaceutical drug separations by HPCE: practical guidelines, *J. Liq. Chromatogr.*, 15, 961, 1992.
26. **Rashin, A. A. and Honig, B.,** On the environment of ionizable groups in globular proteins, *J. Mol. Biol.*, 173, 515, 1984.
27. **Tanford, C.,** *Physical Chemistry of Macromolecules*, John Wiley & Sons, New York, 1961, chap. 6, pp. 412–420.
28. **Booth, F.,** Electrophoretic properties of proteins, *Proc. R. Soc.*, A203, 514, 1950.
29. **Compton, B. J.,** Electrophoretic mobility modeling of proteins in free zone capillary electrophoresis and its application to monoclonal antibody microheterogeneity analysis, *J. Chromatogr.*, 559, 357, 1991.
30. **Offord, R. E.,** Electrophoretic mobilities of peptides on paper and their use in the determination of amide groups, *Nature*, 211, 591, 1966.
31. **Tran, A. D., Park, S., Lisi, P. J., Huynh, O. T., Ryall, R. R., and Lane, P. A.,** Separations of carbohydrate-mediated microheterogeneity of human erythropoietin by free solution CE, *J. Chromatogr.*, 542, 459, 1991.
32. **Field, M., Keck, J., O'Connor, J., Palmieri, R., and Ohms, J.,** The effect of amino acid sequence on the electrophoretic mobility and reverse phase retention of peptides, 2nd International Symposium on High Performance Capillary Electrophoresis, San Francisco, CA, 1990, p. 214.
33. **Jorgenson, J. W. and Lukacs, K. D.,** Capillary zone electrophoresis, *Science*, 222, 266, 1983.
34. **Frieden, C.,** The regulation of protein polymerization, in *Proteins: Form and Function*, Bradshaw, R. and Purton, M., Eds., Elsevier Trends Journals, Cambridge, pp. 241–247.
35a. **Janin, J. and Chothia, C.,** The structure of protein-protein recognition sites, *J. Biol. Chem.*, 265, 16027, 1990.
35b. **Duquerroy, S., Cherfils, J., and Janin, J.,** Protein-protein interaction: an analysis by computer simulation, in *Protein Conformation*, Wiley, Chichester (Ciba Foundation Symposium 161), 1991, pp. 237–252.

36. **Winzor, D. J. and Scheraga, H. A.,** Studies of chemically reacting systems on Sephadex. I. Chromatographic demonstration of the Gilbert theory, *Biochemistry,* 2, 1263, 1963.
37. **Nichol, L. W., Bethune, J. L., Kegeles, G., and Hess, E. L.,** Interacting protein systems, *The Proteins,* Neurath, H., Ed., Academic Press, New York, 1964, vol. 2, p. 305.
38. **van Holde, K. E.,** Sedimentation analysis of proteins, in *The Proteins, Vol. I,* 3rd ed., Neurath, H. and Hill, R. L., Eds., Academic Press, New York, 1975, chap. 4.
39. **Klotz, I. M., Darnall, D. W., and Langerman, N. R.,** Quaternary structures of proteins, in *The Proteins, Vol. I,* 3rd ed., Neurath, H. and Hill, R. L., Eds., Academic Press, New York, 1975, chap. 5.
40. **Kellett, G. L.,** Dissociation of hemoglobin into subunits, *J. Mol. Biol.*, 59, 401, 1971.
40a. **Evans, W. J., Forlani, L., Brunori, M., Wyman, J., and Antonini, E.,** *Biochim. Biophys. Acta,* 214, 64, 1970.
41. **Langerman, N. R. and Klotz, I. M.,** Free energy of subunit interactions in hemerythrin, *Biochemistry,* 8, 4746, 1969.
42. **Hancock, D. K. and Williams, J. W.,** Sedimentation equilibrium studies with chymotrypsinogen A in solution at pH 7. 9 and I = 0. 03, *Biochemistry,* 8, 2598, 1969.
43. **Roark, D. E. and Yphantis, D. A.,** *Ann. N.Y. Acad. Sci.*, 164, 245, 1969.
44. **Kelly, M. J. and Reithel, F. J.,** A thermodynamic analysis of monomer–dimer association of β-lactoglobulin A at the isoelectric point, *Biochemistry,* 10, 2639, 1971.
45. **Georges, C., Guinand, S., and Tonnelat, J.,** Thermodynamic studies on the reversible dissociation of β-lactoglobulin at pH 5.5, *Biochem. Biophys. Acta,* 59, 737, 1962.
46a. **Gilbert, G. A. and Jenkins, R. C. L,** *Proc. R. Soc., Ser. A,* 253, 420, 1959.
46b. **Gilbert, G. A.,** *Proc. R. Soc., Ser. A,* 276, 354, 1963.
47. **Kim, H., Deonier, R., and Williams, J. W.,** The investigation of self-association reactions by equilibrium ultracentrifugation, *Chem. Rev.*, 77, 659, 1977.
48. **Cann, J. R.,** in *Physical Principles and Techniques of Protein Chemistry,* Part A, Leach, S., Ed., Academic Press, New York, 1969, chap. 8.
49. **Ackers, G. K.,** Molecular sieve methods of analysis, in *The Proteins, Vol. 1,* 3rd ed., Neurath, H. and Hill, R. L., Eds., Academic Press, New York, 1975, chap. 1.
50. **Burgi, D. S. and Chien, R.-L.,** Optimization in sample stacking for high performance capillary electrophoresis, *Anal. Chem.,* 63, 2042, 1991.
51. **McCormick, R. M.,** Capillary zone electrophoretic separation of peptides and proteins using low pH buffers in modified silica capillaries, *Anal. Chem.*, 60, 2322, 1988.
52. **Nolan, J. A. and Palmieri, R.,** Development of separation strategies for proteins by capillary electrophoresis, *Techniques in Protein Chemistry IV,* Hogue-Angeletti, R. A., Ed., Academic Press, San Diego, 1993.
53. **Gordon, M., Lee, K. -J., and Zare, R.,** Protocol for resolving protein mixtures in CZE, *Anal. Chem.*, 63, 69, 1991.
54. **Wahlgren, M. and Arnebrant, T.,** Protein adsorption to solid surfaces, *TIBTECH,* 9, 201, 1991.
55. **McLaughlin, S.,** Electrostatic potentials at membrane–solution interfaces, *Current Topics Membranes Transport,* 9, 71, 1977.
56. **Martell, A. G. and Smith, R. M.,** *Critical Stability Constants,* Vols. 1 and 2, Plenum Press, New York.
57. **Iler, R. K.,** *The Chemistry of Silica,* John Wiley & Sons, New York, 1979, chap. 6.
58. **Xu, B. and Vermeulen, N. P. E.,** Preparation of wall-coated open-tubular capillary columns for gas chromatography, *J. Chromatogr.*, 445, 1, 1988.
59. **Strazhesko, D. N., Strelko, V. B., Belyakov, V. N., and Rubanik, S. C.,** Mechanism of cation exchange on silica gels, *J. Chromatogr.*, 102, 191, 1974.
60. **Norde, W., Fraaye, J., and Lykema, J.,** Protein adsorption at solid-liquid interfaces: a colloid–chemical approach, in *Protein at Interfaces,* Brash, J. L. and Horbett, T. A., Eds., American Chemical Society, Washington, D. C., 1987, chap. 2.
61. **Horbett, T. A. and Brash, J. L.,** Proteins at interfaces: current issues and future prospects, in *Protein at Interfaces,* Brash, J. L. and Horbett, T. A., Eds., American Chemical Society, Washington, D. C., 1987, chap. 1.
62. **Banaszak, L. J., Birktoft, J. J., and Barry, C. D.,** Protein–protein interactions and protein structures, in *Protein–Protein Interactions,* Frieden, C. and Nichol, L. W., John Wiley and Sons, New York, 1981, chap. 2.
63. **Mathews, C. and van Holde, K. E.,** *Biochemistry,* Benjamin Cummings, Redwood City, CA., 1990, chap. 4.
64. **Iler, R. K.,** Polymerization of silica, in *The Chemistry of Silica,* Iler, R. K., Ed., John Wiley & Sons, New York, 1979, chap. 3.
65. **Iler, R. K.,** Association between polysilicic acid and polar organic compounds, *J. Phys. Chem.*, 56, 673, 1952.
66. **Burchard, W.,** Static and dynamic light scattering approaches to structure determination of bio-polymers, in *Laser Light Scattering in Biochemistry,* Harding, S. E., Sattelle, D. B., and Bloomfield, V. A., Eds., Royal Society of Chemistry, 1991, chap. 1.

67. **Alber, T.**, Stabilization energies of protein conformation, in *Prediction of Protein Structure and the Principles of Protein Conformation*, Fasman, G. D., Ed., Plenum Press, New York, 1991, chap. 5.
68. **Pfeil, W.**, The problem of the stability of globular proteins, *Mol. Cell. Biochem.*, 40, 3, 1981.
69. **Privalov, P. L.**, Stability of proteins, *Adv. Protein Chem.*, 33, 176, 1979.
70a. **Ptitsyn, O. B.**, Protein folding: hypothesis and experiments, *J. Protein Chem.*, 6, 273, 1987.
70b. **Tanford, C.**, Protein denaturation, *Adv. Protein Chem.*, 24, 1, 1970.
71. **Kuwajima, K.**, The molten globule state as a clue for understanding the folding and cooperativity of globular-protein structure, *Proteins*, 6, 87, 1989.
72a. **Gast, K., Damaschum, G., Damaschum, R., Misselwitz, D., and Bychkova, V. A.**, in *Laser Light Scattering in Biochemistry*, Harding, S. E., Sattelle, D. B., and Bloomfield, V. A., Eds., Royal Society of Chemistry, 1991, chap. 14.
72b. **Satow, T., et al.**, The effect of salts on the separation of bioactive peptides by capillary electrophoresis, *J. High Resolut. Chromatogr.*, 14, 276, 1991.
73. **Hames, B. and Rickwood, D.**, *Gel Electrophoresis of Proteins: A Practical Approach*, IRL Press, Oxford, 1981.
74. **Andrews, A. T.**, *Electrophoresis: Theory, Techniques and Biochemical and Clinical Applications*, 2nd ed., Oxford Science Publications, New York, 1986.
75. **Gordon, A. H.**, *Electrophoresis of Proteins in Polyacrylamide and Starch Gels*, revised, Elsevier Science Publishers B. V., Amsterdam, 1975.
76. **Guzman, N. A., Hernandez, L., and Hoebel, B. G.**, Capillary electrophoresis: a new era in microseparations, *BioPharm*, January 22, 1989.
77. **Gordon, M. J., Huang, X., Pentoney, S. L., Jr., and Zare, R. N.**, Capillary electrophoresis, *Science*, 242, 224, 1988.
78. **Ewing, A. G., Wallingford, R. A., and Olefirowicz, T. M.**, Capillary electrophoresis, *Anal. Chem.*, 61, 292A, 1989.
79. **Green, J. S. and Jorgenson, J. W.**, Design of a variable wavelength uv absorbance detector for on-column detector for on-column detection in comparison of its performance to a fixed wavelength uv absorption detector, *J. Liq. Chromatogr.*, 12, 2527, 1989.
80. **Bushey, M. M. and Jorgenson, J. W.**, Capillary electrophoresis of proteins in buffers containing high concentration of zwitterionic salts, *J. Chromatogr.*, 480, 301, 1989.
81. **Ginn, M. E.**, Cationic surfactants, in *Surfactant Science Series, Vol. 4*, Jungermann, E., Ed., Marcel Dekker, New York, 1970, 343.
82. **McCormick, R. M.**, Capillary zone electrophoretic separation of peptides and proteins using low pH buffers in modified silica capillaries, *Anal. Chem.*, 60, 2322, 1988.
83. **Novotny, M. V., Cobb, K. A., and Liu, J.**, Recent advances in capillary electrophoresis of proteins, peptides and amino acids, *Electrophoresis*, 11, 735, 1990.
84. **Towns, J. K. and Regnier, F. E.**, Capillary electrophoretic separations of proteins using nonionic surfactant coatings, *Anal. Chem.*, 63, 1126, 1991.
85. **Gurley, L. R., London, J. E., and Valdez, J. G.**, High-performance capillary electrophoresis of histones, *J. Chromatogr.*, 559, 431, 1991.
86. **Wu, C.-T., Lopes, T., Patel, B., and Lee, C. S.**, Effect of direct control of electroosmosis on peptide and protein separations in capillary electrophoresis, *Anal. Chem.*, 64, 886, 1992.
87. **Chen, F. A., Kelly, L., Palmieri, R., Biehler, R., and Schwartz, H.**, Use of high ionic strength buffers for the separation of proteins and peptides with capillary electrophoresis, *J. Liq. Chromatogr.*, 15, 1143, 1992.
88. **Hjertén, S.**, Free zone electrophoresis, *Chromatogr. Rev.*, 9, 122, 1967.
89. **Lukacs, K. D. and Jorgenson, J. W.**, Capillary zone electrophoresis: effect of physical parameters on separation efficiency and quantitation, *J. High Resolut. Chromatogr. Chromatogr. Commun.*, 8, 407, 1985.
90. **Tsuda, T., Nomura, K., and Nakagawa, G.**, Separation of organic and metal ions by high voltage capillary electrophoresis, *J. Chromatogr.*, 264, 385, 1983.
91. **Fujiwara, S. and Honda, S.**, Determination of cinnamic acid and its analogues by electrophoresis in a fused silica capillary tube, *Anal. Chem.*, 58, 1811, 1986.
92. **Lehninger, A. L.**, *Biochemistry*, 2nd ed, Worth Publishers, New York, 1975, chaps. 4 and 5.
93. **Landers, J. P., Oda, R. P., Spelsberg, T. C., Nolan, J. A., and Ulfelder, K. J.**, Capillary electrophoresis: a powerful microanalytical technique for biologically active molecules, *BioTechniques*, 14, 98, 1993.
94. **Tran, A. D., Park, S., Lisi, P. J., Huynh, O. T., Ryall, R. R., and Lane, P. A.**, Separations of carbohydrate-mediated microheterogeneity of human erythropoietin by free solution CE, *J. Chromatogr.*, 542, 459, 1991.
95. **Lauer, H. H. and McManigall, D.**, Capillary zone electrophoresis of proteins in untreated fused silica tubing, *Anal. Chem.*, 58, 166, 1986.
96. **Landers, J. P., Madden, B. J., Oda, R. P., and Spelsberg, T. C.**, The use of modifiers of endoosmotic flow for analysis of microheterogeneity, *Anal. Biochem.*, 205, 115, 1992.
97. **Jorgenson, J. W. and Lukacs, K. D.**, Capillary electrophoresis, *Science*, 222, 266, 1983.

98. **Green, J. S. and Jorgenson, J. W.,** Minimizing adsorption of proteins on fused silica in capillary zone electrophoresis by the addition of alkali metal salts to the buffer, *J. Chromatogr.*, 478, 63, 1989.
99. **Swedberg, S. A.,** Characterization of protein behavior in high performance capillary electrophoresis using a novel capillary system, *Anal. Biochem.*, 185, 51, 1990.
100. **Kenndler, E. and Schmidt-Beiwl, K.,** Effect of sodium dodecyl sulfate in protein samples on separation with free capillary zone electrophoresis, *J. Chromatogr.*, 545, 397, 1991.
101. **Hjertén, S., Vatcheva, L., Elenbring, K., and Eaker, D.,** High performance electrophoresis of acidic and basic low-molecular weight compounds and of proteins in the presence of polymers and neutral surfactants, *J. Liq. Chromatogr.* 12, 2471, 1989.
102. **Nolan, J. A. and Palmieri, R.,** Development of separation strategies for proteins by capillary electrophoresis, in *Techniques in Protein Chemistry IV*, Hogue-Angeletti, R. A., Ed., Academic Press, San Diego, 1993, in press.
103. **Bietz, J. A. and Simpson, D. G.,** Electrophoresis and chromatography of wheat proteins: available methods and procedures for statistical evaluation of the data, *J. Chromatogr.*, 624, 53, 1992.
104. **Harrington, S. J., Varro, R., and Li, T. M.,** High-performance capillary electrophoresis as a fast in-process control method for enzyme-labelled monoclonal antibody conjugates, *J. Chromatogr.*, 559, 385, 1991.
105. **Cobb, K. A. and Novotny, M. V.,** Peptide mapping of complex proteins at the low-picomole level with capillary electrophoretic separations, *Anal. Chem.*, 64, 879, 1992.
106. **Kristensen, H. K. and Hansen, S. H.,** Open-tubular capillary electrokinetic chromatography using a zwitterionic surfactant for the separation of peptides with equal mass/charge ratios, poster M-21, presented at 4th Int. Symp. on High Performance Capillary Electrophoresis, Amsterdam, 1992.
107. **Astephen, N. and Wheat, T.,** Separations of proteins by capillary electrophoresis, poster M-18, presented at Fourth Int. Symp. on High Performance Capillary Electrophoresis, Amsterdam, 1992.
108. **Merion, M. and Petersen, J. S.,** U.S. Patent 5,192,405, 1991.
109. **Deutscher, M. P., Ed.,** *Guide to Protein Purification, Methods in Enzymology*, Vol. 182, Academic Press, San Diego, 1990.
110. **Terabe, S., Otsuka, K., Ichikawa, K., Tsuchiya, A., and Ando, T.,** Electrokinetic separations with micellar solutions and open-tubular capillaries, *Anal. Chem.*, 56, 111, 1984.
111. **Nakagawa, T., Oda, Y., Shibukawa, A., Fukuda, H., and Tanaka, H.,** Separation and determination of Cefpiramide in human plasma by electrokinetic chromatography with a micellar solution and an open tubular-fused silica capillary, *Chem. Pharm. Bull.*, 36, 1622, 1988.
112. **Reynolds, J. and Tanford, C.,** The gross conformation of protein-sodium dodecyl sulfate complexes, *J. Biol. Chem.*, 245, 5161, 1970.
113. **Laemmli, U. R.,** Cleavage of structural proteins during the assembly of the head of bacteriophage T4, *Nature (London)*, 227, 680, 1970.
114. **Andrews, A. T.,** *Electrophoresis: Theory, Techniques, and Biochemical and Clinical Applications*, 2nd ed., Oxford University Press, New York, 1986, chap. 5.
115. **Towns, J. K. and Regnier, F. E.,** Capillary electrophoretic separations of proteins using nonionic surfactant coatings, *Anal. Chem.*, 63, 1126, 1991.
116. **Mikkers, F. E. P., Everaerts, F. M., and Verheggen, T. P. E. M.,** Concentration distributions in free zone electrophoresis, *J. Chromatogr.*, 169, 1, 1979.
117. **Beckers, J. L. and Everaerts, F. M.,** Isotachophoresis with two leading ions and migration behavior in capillary zone electrophoresis. II. Migration behaviour in capillary zone electrophoresis, *J. Chromatogr.*, 508, 19, 1990.
118. **McLaughlin, G. M., Nolan, J. A., Lindahl, J. L., Palmieri, R. H., Anderson, K. W., Morris, S. C., Morrison, J. A., and Bronzert, T. J.,** Pharmaceutical drug separations by HPCE: practical guidelines, *J. Liq. Chromatogr.*, 15, 961, 1992.
119. **Burgi, D. S. and Chien, R.-L.,** Optimization in sample stacking for high performance capillary electrophoresis, *Anal. Chem.*, 63, 2042, 1991.
120. **Chien, R.-L. and Burgi, D. S.,** Field amplified sample injection in high performance capillary electrophoresis, *J. Chromatogr.*, 559, 141, 1991.
121. **Chien, R.-L. and Burgi, D. S.,** Field-amplified polarity switching sample injection in high performance capillary electrophoresis, *J. Chromatogr.*, 559, 153, 1991.
122. **Vinther, A. and Soeberg, H.,** Temperature elevations of the sample zone in free solution capillary electrophoresis under stacking conditions, *J. Chromatogr.*, 559, 27, 1991.
123. **Vinther, A., Soeberg, H., Nielsen, L., Pedersen, J., and Biedermann, K.,** Thermal degradation of a thermolabile *Serratia marcescens* nuclease using capillary elelctrophoresis with stacking conditions, *Anal. Chem.*, 64, 187, 1992.
124. **Gordon, M. J., Lee, K.-J., Arias, A. A., and Zare, R. N.,** Protocol for resolving protein mixtures in capillary zone electrophoresis, *Anal. Chem.*, 63, 69, 1991.

125. **Biehler, R., Anderson, K., and Schwartz, H.,** Micropreparative application of capillary electrophoresis, poster S123, presented at the Protein Society 5th Symp., Baltimore, MD, 1991.
126. **Ohms, J. and Smith, A.,** Micropreparative capillary electrophoresis (MPCE) and micropreparative high performance liquid chromatography of protein digests, in *Techniques in Protein Chemistry IV*, Hogue-Angeletti, R. A., Ed., Academic Press, San Diego, 1993.
127. **Wieme, R. J.,** *Chromatography: A Laboratory Handbook of Chromatographic and Electrophoretic Methods*, 3rd ed., Heftmann, E., Ed., Van Nostrand Reinhold, New York, 1975, chap. 10.
128. **Engelhardt, H., Kohr, J., Beck, W., and Schmitt, T.,** Characterization and application of coated capillaries in capillary electrophoresis, poster M301, presented at the 5th Int. Symp. on High Performance Capillary Electrophoresis, Orlando, Florida, 1993.
129. **Hjertén, J.,** High performance electrophoresis: elimination of electroendosmosis and solute adsorption, *J. Chromatogr.*, 347, 191, 1985.
130. **Cobb, K. A., Dolnik, V., and Novotny, M.,** Electrophoretic separation of proteins in capillaries with hydrolytically stable surfaces, *Anal. Chem.*, 62, 2478, 1990.
131. **Bruin, C. J. M., Huidsen, R., Kraak, J. C., and Poppe, H.,** Performance of carbohydrate-modified fused silica capillaries for the separation of proteins by zone electrophoresis, *J. Chromatogr.*, 480, 339, 1989.
132. **Bruin, C. J. M., Chang, J. P., Kuhlman, R. H., Zegers, K., Kraak, J. C., and Poppe, H.,** Capillary zone electrophoretic separations of proteins in polyethylene glycol-modified capillaries, *J. Chromatogr.*, 471, 429, 1988.
133. **Wiktorowicz, J. E. and Colburn, J. C.,** Separation of cationic proteins via charge reversal in capillary electrophoresis, *Electrophoresis*, 11, 769, 1990.
134. **Bullock, J. A. and Yuan, L.-C.,** Free solution capillary electrophoresis of basic proteins in uncoated fused silica capillary tubing, *J. Microcolumn Sep.*, 3, 241, 1991.
135. **Guttman, A., Nolan, J. A., and Cooke, N.,** Capillary sodium dodecyl sulfate gel electrophoresis of proteins, *J. Chromatogr.*, 632, 171, 1993.
136. **Tsuda, T., Sweedler, J. V., and Zare, R. N.,** Rectangular capillaries for capillary zone electrophoresis, *Anal. Chem.*, 62, 2149, 1990.
137. **Wang, T., Aiken, J. H., Huie, C. W., and Hartwick, R. A.,** Nanoliter-scale multireflection cell for absorption detection in capillary electrophoresis, *Anal. Chem.*, 63, 1372, 1991.
138. **Olefirowicz, T. M. and Ewing, A. G.,** Capillary electrophoresis in 2 and 5 μm diameter capillary: application to cytoplasmic analysis, *Anal. Chem.*, 62, 1872, 1990.
139. **Engstrom-Silverman, C. E. and Ewing, A. G.,** Copper wire amperometric detector for capillary electrophoresis, *J. Microcolumn Sep.*, 3, 141, 1991.
140. **Pentoney, S., Zare, R., and Quint, J.,** On-line radioisotope detection for capillary electrophoresis, *Anal. Chem.*, 61, 1642, 1989.
141. **Smith, R. A., Olivares, J. A., Nguyen, N. T., and Usdeth, H. R.,** Capillary zone electrophoresis-mass spectrometry using an electrospray ionization interface, *Anal. Chem.*, 60, 436, 1988.
142. **Loo, J. A., Jones, H. K., Udseth, H. R., and Smith, R. D.,** Capillary zone electrophoresis-mass spectrometry with electrospray ionization of peptides and proteins, *J. Microcolumn Sep.*, 1, 223, 1989.
143. **Moseley, M., Deterding, L., Tomer, K., and Jorgensen, J.,** Coupling of capillary zone electrophoresis and capillary liquid chromatography with coaxial continuous-flow fast atom bombardment tandem sector mass spectrometry, *J. Chromatogr.*, 480, 197, 1989.
144. **Caprioli, R., Moore, W., Martin, M., DaGue, B., Wilson, K., and Moring, S.,** Coupling capillary zone electrophoresis and continuous-flow fast atom bombardment mass spectrometry for the analysis of peptide mixtures, *J. Chromatogr.*, 480, 247, 1989.
145. **Edmonds, C. G., Loo, J. A., Barinaga, C. J., Udseth, H. R., and Smith, R. D.,** Capillary electrophoresis-electrospray ionization mass spectrometry, *J. Chromatogr.*, 474, 21, 1989.
146. **Johannson, I. M., Huang, E. C., Henion, J. D., and Zweigenbaum, J.,** Capillary electrophoresis-atmospheric pressure ionization mass spectrometry for the characterization of peptides: instrumental considerations for mass spectrometric detection, *J. Chromatogr.*, 554, 311, 1991.
147. **Garcia, F. and Henion, J. D.,** Gel-filled capillary electrophoresis/mass spectrometry using a liquid junction-ion spray interface, *Anal. Chem.*, 64, 985, 1992.
148. **Wahl, J. H., Goodlett, D. R., Udseth, H. R., and Smith, R. D.,** Attomole level capillary electrophoresis-mass spectrometric protein analysis using 5-μm-i.d. capillaries, *Anal. Chem.*, 64, 3194, 1992.
149. **Keough, T., Takigidu, R., Lacey, M. P., and Purdon, M.,** Matrix-assisted laser desorption mass spectrometry of proteins isolated by capillary zone electrophoresis, *Anal. Chem.*, 64, 1594, 1992.
150. **Tsuji, K., Baczynskyj, L., and Bronson, G. E.,** Capillary electrophoresis-electrospray mass spectrometry for the analysis of recombinant bovine and porcine somatotropins, *Anal. Chem.*, 64, 1864, 1992.
151. **Olechno, J. D., Tso, J. M. Y., and Thayer, J.,** Capillary electrophoresis: a multifaceted technique for analytical chemistry. 3. Detection, *Am. Lab. (Fairfield, Conn.)*, 223, 59, 1991.
152. **Rose, D. R. J. and Jorgensen, J. W.,** Post-capillary fluorescence detection in capillary zone electrophoresis using *o*-phthaldialdehyde, *J. Chromatogr.*, 447, 117, 1988.

153. **Kuhr, W. G. and Yeung, E. S.,** Optimization of sensitivity and separation in capillary zone electrophoresis with indirect fluorescence detection, *Anal. Chem.,* 60, 2642, 1988.
154. **Gassman, E., Kuo, J. E., and Zare, R. N.,** Electrokinetic separation of chiral compounds, *Science,* 230, 813, 1985.
155. **Gozel, P., Gassman, E., Michelsen, H., and Zare, R. N.,** Electrokinetic resolution of amino acid enantiomers with copper(II)-Aspartame support electrolyte, *Anal. Chem.,* 59, 44, 1987.
156. **Nickerson, B. and Jorgensen, J. W.,** High speed capillary zone electrophoresis with laser induced fluorescence detection, *J. High Resolut. Chromatogr.,* 11, 533, 1988.
157. **Wu, S. and Dovichi, N. J.,** High-sensitivity fluorescence detection for fluorescein isothiocyanate derivatives of amino acids separated by capillary zone electrophoresis, *J. Chromatogr.,* 480, 141, 1989.
158. **Nickerson, B. and Jorgensen, J. W.,** Characterization of a post-column reaction-laser-induced fluorescence detector for capillary zone electrophoresis, *J. Chromatogr.,* 480, 157, 1989.
159. **Pentoney, S. L., Huang, X., Burgi, D. S., and Zare, R. N.,** On-line connector for microcolumns: application to the on-column *o*-phthaldialdehyde derivatization of amino acids separated by capillary zone electrophoresis, *Anal. Chem.,* 60, 2625, 1988.
160. **Kobayashi, S., Ueda, T., and Kikumoto, M.,** Photodiode array detection in high performance capillary electrophoresis, *J. Chromatogr.,* 480, 179, 1989.
161. **Yowell, G. G., Fazio, S. D., and Vivilechhia, R. V.,** The analysis of a recombinant granulocyte macrophage colony stimulating factor (GM-CSF) dosage by capillary electrocapillary isoelctric focusing and high performance liquid chromatography, *J. Chromatogr.,* 1993, in press.
162. **Cooper, T. G.,** *The Tools of Biochemistry,* John Wiley & Sons, New York, 1977, chap. 1.
163. **Guttman, A., Horvath, J., and Cooke, N.,** Influence of temperature on the sieving effect of different polymer matrices in capillary SDS gel electrophoresis of proteins, *Anal. Chem.,* 65, 199, 1993.
164. **Rush, R. S., Cohen, A. S., and Karger, B. L.,** Influence of column temperature on the electrophoretic behavior of myoglobin and alpha-lactalbumin in high performance capillary electrophoresis, *Anal. Chem.,* 63, 1346, 1991.
165. **Strickland, M. and Strickland, N.,** Free-solution capillary electrophoresis using phosphate buffer and acidic pH, *Am. Lab. (Fairfield, Conn.),* 22, 60, 1990.
166. **Smith, S. C., Strasters, J. K., and Khaledi, M. G.,** Influence of operating parameter on reproducibility in capillary electrophoresis, *J. Chromatogr.,* 559, 57, 1991.
166a. **McLaughlin, G., Palmieri, R., and Anderson, K.,** Benefits of automation in the separation of biomolecules by high performance capillary electrophoresis, in *Techniques in Protein Chemistry II,* Villafranca, J. J., Ed., Academic Press, Inc., San Diego, 1991, section 1, p. 3.
167. **Moring, S. E., Colburn, J. C., Grossman, P. D., and Lauer, H. H.,** Analytical aspects of an automated capillary electrophoresis system, *LC-GC,* 8, 34, 1990.
168. **Foret, F., Sustacek, V., and Bocek, P.,** On-line isotachophoretic sample preconcentration for enhancement of zone detectability in capillary zone electrophoresis, *J. Microcolumn Sep.,* 2, 229, 1990.
169. **Dolnik, V., Cobb, K. A., and Novotny, M.,** Capillary zone electrophoresis of dilute samples with isotachophoretic preconcentration, *J. Microcolumn Sep.,* 2, 127, 1990.
170. **Stegehuis, D. S., Tjaden, U. R., and van der Greef, J.,** Analyte focusing in capillary electrophoresis using on-line isotachophoresis, *J. Chromatogr.,* 591, 341, 1992.
171. **Schwer, C. and Lottspeich, F.,** Analytical and micropreparative separation of peptides by capillary zone electrophoresis using discontinuous buffer systems, *J. Chromatogr.,* 623, 345, 1992.
172. **Bushey, M. M. and Jorgenson, J. W.,** Automated instrumentation for comprehensive two-dimensional high performance liquid chromatography/capillary zone electrophoresis, *Anal. Chem.,* 62, 978, 1990.
173. **Bushey, M. M. and Jorgenson, J. W.,** A comparison of tryptic digests of bovine and equine cytochrome C by comprehensive reversed-phase HPLC-CE, *J. Microcolumn Sep.,* 2, 293, 1990.
174. **Chervet, J. P., van Soest, R. E. J., and Ursem, M.,** Z-shaped flow cell for UV detection in capillary electrophoresis, *J. Chromatogr.,* 543, 439, 1991.
175. **Albin, M. and Moring, S.,** Sensitivity enhancements for capillary electrophoresis, poster M 203, presented at the 5th Int. Symp. on High Performance Capillary Electrophoresis, Orlando, Florida, 1993.
176. **Heiger, D. N.,** Sensitivity improvements in HPCE, Wednesday plenary session, presented at the 5th Int. Symp. on High Performance Capillary Electrophoresis, Orlando, Florida, 1993.
177. **Werner, W. E., Demorest, D. M., Dubrow, R., Stevens, J., and Wiktorowicz, J.,** Determination of protein molecular weights by capillary electrophoresis, poster T79, presented at The Protein Society 6th Symp., San Diego, California, 1992.
178. **Hjertén, S., Liao, J.-L., and Yao, K.,** Theoretical and experimental study of high-performance electrophoretic mobilization of isoelectrically focused protein zones, *J. Chromatogr.,* 387, 127, 1987.
179. **Zhu, M., Rodriguez, R., and Wehr, T.,** Optimizing separation parameters in capillary isoelectric focusing, *J. Chromatogr.,* 559, 479, 1991.
180. **Mazzeo, J. R. and Krull, I. S.,** Coated capillaries and additives for the separation of proteins by capillary zone electrophoresis and capillary isoelectric focusing, *BioTechniques,* 10, 638, 1991.

181. **Nelson, T. J.,** Rapid isoelectric focusing of proteins in hydrolytically stable capillaries, *J. Chromatogr.,* 623, 357, 1992.
182. **Chen, S.-M. and Wiktorowicz, J.,** High resolution full-range (pI 2. 5-10. 0) isoelectric focusing of proteins and peptides in capillary electrophoresis, poster T80, presented at the Protein Society 6th Symp., San Diego, California, 1992.
183. **Wang, T. and Hartwick, R. A.,** Whole column absorbance detection in capillary isoelectric focusing, *Anal. Chem.,* 64, 1745, 1992.
184. **Banke, N., Hansen, K., and Diers, I.,** Detection of enzyme activity in fractions collected from free solution capillary electrophoresis of complex samples, *J. Chromatogr.,* 559, 325, 1991.
185. **Bergman, T., Agerberth, B., and Jornvall H.,** Direct analysis of peptides and amino acids from capillary electrophoresis, *FEBS Lett.,* 283, 100, 1991.
186. **Eriksson, K.-O., Palm, A., and Hjertén, S.,** Preparative capillary electrophoresis base on adsorption of the solutes (proteins) onto a moving blotting membrane as they migrate out of the capillary, *Anal. Biochem.,* 201, 211, 1992.
187. **Cheng, Y.-F., Fuchs, M., Andrews, D., and Carson, W.,** Membrane fraction collection for capillary electrophoresis, *J. Chromatogr.,* 608, 109, 1992.
188. **Wu, D. and Regnier, F. E.,** Native protein separations and enzyme assays by capillary gel electrophoresis, Thursday plenary session, presented at the 5th Int. Symp. on High Performance Capillary Electrophoresis, Orlando, Florida, 1993.
189. **Nielsen, R. G., Rickard, E. C., Santa, P. F., Sharknas, D. A., and Sittampalam, G. S.,** Separation of antibody-antigen complexes by capillary zone electrophoresis, isoelectric focusing and high performance size-exclusion chromatography, *J. Chromatogr.,* 539, 177, 1991.
190. **Maschke, H. E., Frenz, J., Williams, M., and Hancock, W. S.,** Investigation of protein-DNA-interaction by mobility shift assays in capillary electrophoresis, poster T121, presented at the 5th Int. Symp. on High Performance Capillary Electrophoresis, Orlando, Florida, 1993.
191. **Emmer, A., Jansson, M., and Roeraade, J.,** Improved capillary zone electrophoretic separations of basic proteins, using a fluorsurfactant buffer additive, *J. Chromatogr.,* 547, 544, 1991.
192. **Lee, T. T. and Yeung, E. S.,** Quantitative determination of native proteins in individual human erythrocytes by capillary zone electrophoresis with laser-induced fluorescence detection, *Anal. Chem.,* 64, 3045, 1992.
193. **Heegaard, N. H. H. and Robey, F. A.,** Use of capillary zone electrophoresis to evaluate the binding of anionic carbohydrates to synthetic peptides derived from human serum amyloid P component, *Anal. Chem.,* 64, 2479, 1992.
194. **Tsuji, K. and Little, R. J.,** Charge-reversed, polymer-coated capillary column for the analysis of a recombinant chimeric glycoprotein, *J. Chromatogr.,* 594, 317, 1992.
195. **Liu, J., Cobb, K. A., and Novotny, M.,** Separation of pre-column ortho-phthalaldehyde-derivatized amino acids by capillary zone electrophoresis with normal and micellar solutions in the presence of organic modifiers, *J. Chromatogr.,* 468, 55, 1988.
196. **Higashijima, T., Fuchigami, T., Imasaka, T., and Ishibashi, N.,** Determination of amino acids by capillary zone electrophoresis based on semiconductor laser fluorescence detection, *Anal. Chem.,* 64, 711, 1992.
197. **Hortin, G. L., Griest, T., and Benutto, B. M.,** Separations of sulfated and non-sulfated forms of peptides by capillary electrophoresis: comparison with reversed-phase HPLC, *BioChromatography,* 5, 118, 1990.
198. **Stamler, J. S. and Loscalzo, J.,** Capillary zone electrophoretic detection of biological thiols and their S-nitrosated derivatives, *Anal. Chem.,* 64, 779, 1992.

Chapter 14

CAPILLARY GEL ELECTROPHORESIS FOR DNA SEQUENCING: SEPARATION AND DETECTION

Norman J. Dovichi

TABLE OF CONTENTS

0-8493-8690-X/94/$0.00+$.50

I. INTRODUCTION

The Human Genome Initiative is an international program with the goal of mapping and sequencing the human genome. A genome is the complete complement of genetic material in a single cell; in humans, it contains ~3×10^9 base pairs (bp) of DNA organized into 23 pairs of chromosomes.

The process of DNA sequencing involves several steps. Because chromosomes are far too large to be sequenced directly, the first steps in the sequencing process involve breaking the DNA into successively smaller pieces. These smaller pieces are then subjected to sequencing reactions. The reaction products are separated by gel electrophoresis and detected by either autoradiography or laser-induced fluorescence. Last, the raw sequence data are compiled into a complete sequence. This chapter deals primarily with the use of capillary gel electrophoresis for the separation and identification of the sequencing reaction products. However, other steps in the sequencing process are described briefly.

II. DNA BASICS

DNA consists of a long, linear polymer made from four monomeric components called nucleotides: deoxyadenosine (dA, sometimes shortened to A), deoxycytidine (dC or C), deoxyguanosine (dG or G), and deoxythymidine (dT or T). The nucleotides consist of three parts, a base that varies for the four nucleotides, the sugar deoxyribose, and a phosphate group.

The phosphate is attached to the 5′ end of the sugar and a single hydroxyl group is attached to the 3′ sugar position. The monomeric units are joined together by phosphodiester bonds between the 5′ phosphate and the 3′ hydroxyl group of adjacent sugars. As a result, a DNA strand has orientation: there is a 5′ end with a phosphate group and a 3′ end with a hydroxyl group (Figure 1).

In most organisms, DNA exists in a double helix. The two strands are complementary to each other; a deoxyadenosine is always paired with a deoxythymidine and a deoxycytidine is always paired with a deoxyguanosine. If the sequence of one strand is known, the sequence of the complementary strand is also known.

A. RESTRICTION NUCLEASES

Restriction nucleases are powerful tools for manipulation of DNA. These enzymes are produced by bacteria, apparently as a defense against viral infection. The enzymes recognize certain sequences, and cut DNA near the recognition site. For example, the enzyme *Hin*dIII recognizes the sequence AAGCTT, and cuts the DNA between the two A's:

5′...A^ A G C T T...3′		5′...A		A G C T T...3′
	→		+	
3′... T T C G A^ A...5		3′...T T C G A		A...5′

where ^ denotes the cutting site. Note that the recognition sequence is identical when read from the 5′ end of either strand; this type of sequence is called a palindrome.

The average size of a restriction fragment is related to the number of bases in the recognition sequence. An enzyme with a 4-base recognition site will cut DNA into pieces that are roughly 4^4 or 256 bases long. An 8-base recognition site leads to pieces that are roughly 4^8 or 65,000 bases long. The distribution of fragment length is characteristic of the particular piece of DNA that was digested by the restriction nuclease.

In the case of *Hin*dIII, each fragment has a 4-base overhang of single-stranded DNA. These short pieces of single-stranded DNA are called *sticky ends* because they can hybridize to their

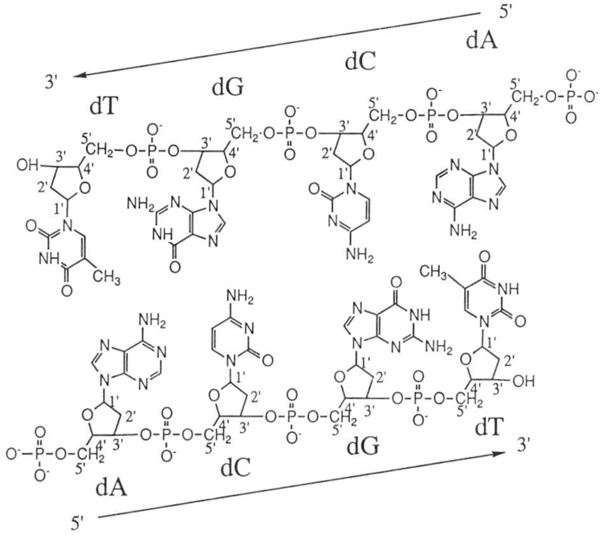

FIGURE 1. Double-stranded DNA. A short segment of DNA with the sequence 5′ dA dC dG dT 3′. Note the opposite orientation of the two strands. One strand has a 5′ phosphate group on the left and a 3′ hydroxyl group on the right. The complementary strand is oriented with the 3′ hydroxyl group on the left and the 5′ phosphate group on the right. In addition, the strands are complementary. A dT on one strand is always coupled with a dA on the complementary strand; a dC on one strand is always coupled with a dG on the complementary strand.

complementary pair. Hybridization refers to the formation of hydrogen bonds between the two strands. Another enzyme, called ligase, is used to connect the cut strands.

B. GENERATION OF SEQUENCING TEMPLATE

Preparation of DNA for sequencing takes advantage of several molecular biology tools. These tools include restriction enzymes, cloning to introduce pieces of DNA into organisms for propagation, and specific vectors, which are organisms that are genetically engineered to be cloned easily.

Rather than dealing with whole chromosomes, which are fragile and awkward to work with, the chromosome is randomly broken into smaller pieces, which are then inserted into vectors. The vectors can be grown to produce relatively large amounts of the chromosomal fragments. The largest vectors are YACs, or yeast artificial chromosomes. Yeast artificial chromosomes can hold inserts of up to $\sim 2 \times 10^6$ bp of DNA. A few hundred different YACs can contain the DNA of an entire chromosome. This collection of YACs is called a library.

The YACs contain randomly generated pieces of chromosomal DNA. There is a large amount of overlap between the DNA insterted into the YACs. To organize the library, the overlaps must be identified. Often, restriction nucleases are used to cut the YACs; the distribution of fragment lengths is characteristic of that particular region of DNA. The restriction fragments are separated by agarose electrophoresis. By noting the presence of identical-length restriction fragments in different YACs, the library of YACs can be organized. Determining the set of YACs that spans the chromosome with minimal overlap (this set is called a *contig*) is an important step in sequencing the human genome.

While YACs are much more convenient to work with than human chromosomes, they are still too large to be sequenced directly. Instead, they must be cut into smaller pieces, which are grown in a different vector. Often cosmids, engineered from λ phage, are used with DNA inserts of up to 50,000 bp. As with YACs, a mapping effort is required to construct a contig of cosmids for each YAC.

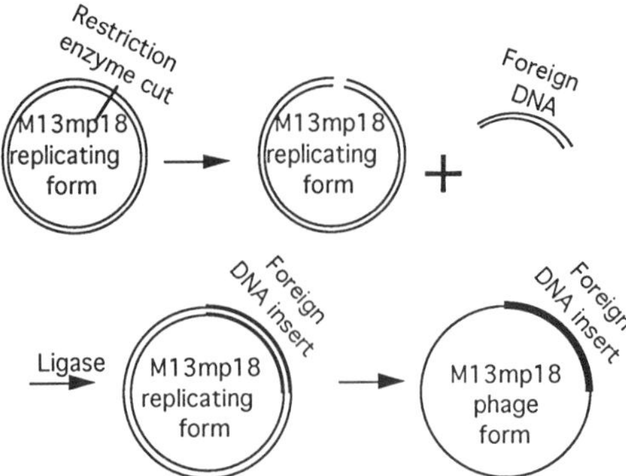

FIGURE 2. Cloning of foreign DNA into M13mp18. First, a restriction nuclease is used to cut the double-stranded form of the phage. A piece of foreign DNA, which was cut with the same nuclease, hybridizes to the phage. Ligase is used to permanently join the DNA insert to the phage. The single-stranded form of the phage is expressed with the foreign insert. Last, the phage is used to infect *E. coli*, producing large amounts of the single-stranded M13mp18 DNA.

Some libraries are huge, with over 100,000 clones. Manual analysis of the clones is a very tedious task. Instead, robots have been constructed to automate many steps in the process.[1]

C. M13mp18: A SEQUENCING VECTOR

At the present time, cosmids are too large to be sequenced directly. Instead, the cosmids are cut into smaller pieces, and these pieces are inserted into another vector. Most commonly, this vector is M13mp18, a genetically engineered bacteriophage (also called a phage), or virus that infects bacteria. The M13mp18 genome is a circular piece of DNA with 7249 nucleotides.[2]

The wild-type phage, M13, infects *Escherichia coli.* The virus has been the subject of much study, and in 1983, Norrander et al. described a genetically engineered version of the phage for DNA sequencing.[3] The authors mutated M13 by introduction of recognition sites for specific restriction enzymes. The phage was also engineered to produce β-galactosidase; in the presence of specific reagents, colonies of these mutated phages will produce a blue color. However, on insertion of foreign DNA, the gene coding β-galactosidase is disrupted, and the colonies incorporating the foreign DNA are white.

Foreign DNA is introduced into the phage in a series of steps (Figure 2). First, the foreign DNA is cut with a specific restriction nuclease. Next, the same nuclease is used to cut the phage vector. The foreign DNA is mixed with the phage, and a small fraction of the phage and foreign DNA pieces will hybridize. The hybridized pieces are permanently joined together with ligase.

An M13mp18 vector that contains an insert is called a sequencing template; it is this randomly generated template that is subjected to the sequencing reaction. Usually, these templates will contain ~2000 bases of inserted DNA. Most laboratories will sequence a large number of these templates, eventually producing enough sequence to piece together the sequence of the cosmid. These "shot-gun" approaches require that each piece of DNA be sequenced an average of six to eight times. Because of statistics, certain parts of the cosmid will be sequenced many more times while other regions will be sequenced rarely. Alternatively, approaches to sequencing can be performed in which an ordered set of templates is sequenced; these templates may be analyzed by restriction mapping to produce a contig of the original cosmid.

D. DNA SEQUENCING BY DIDEOXY-CHAIN TERMINATION

In 1977, Sanger and colleagues reported an enzymatic method for DNA sequencing.[4] This work, for which Sanger received his second Nobel Prize, is the main method used for DNA

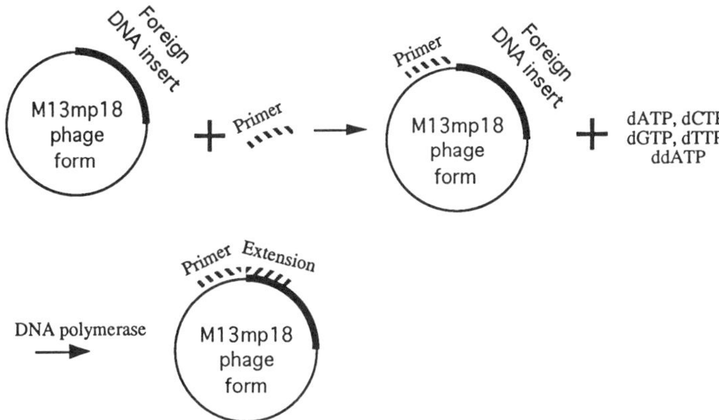

FIGURE 3. Sequencing reaction. The primer is designed to hybridize to the template adjacent to the foreign DNA insert. In the presence of the four deoxynucleotides, DNA polymerase will start from the primer, synthesizing the strand that is complementary to the DNA insert.

sequencing today. The starting point for this sequencing technology is an M13mp18 clone that contains an insert of foreign DNA. In the phage form, the M13mp18 clone consists of single-stranded DNA; this form is produced in high concentration in infected *E. coli* bacteria.

The sequencing reaction is used to generate a set of DNA fragments that are complementary to the DNA insert. First, a short (~21 bases) piece of DNA is synthesized that is complementary to the single-stranded M13mp18 clone. This piece of DNA is called a primer because it acts as the starting site of the enzymatic synthesis. The primer is synthesized to be complementary to the M13mp18 sequence that is adjacent to the insert. A commonly used sequencing primer for M13mp18 is chosen to be adjacent to the only *Hin*dIII restriction nuclease site in M13mp18:

3′ ··· ACATTTTGCTGCCGGTCACGGTTCGAA ··· 5′

5′ <u>TGTAAAACCGACGGCCAGTGCC</u> 3′

The 21-base synthetic primer is shown underlined and hybridized to its complementary sequence on the template. The insert begins after the recognition site. Since this priming site can be used with all M13mp18 clones, a large amount of the primer can be synthesized, making the cost per sequencing reaction relatively low.

The primer is placed in solution with the M13mp18 template (Figure 3). The primer anneals, or binds, to the complementary portion of the template. In the presence of an enzyme called DNA polymerase, DNA will be synthesized to extend from the 3′ end of the primer and to be complementary to the sequence of the foreign DNA insert. The chain extension requires the presence of all four deoxynucleotide triphosphates. If there were no other components added to the mixture, the DNA polymerase would extend the primer to completely encircle the template.

Sanger employed modified nucleotides in the sequencing reaction. These nucleotides are called dideoxynucleotides, which are identical to normal nucleotides except that the sugar is missing a 3′ hydroxyl group. The polymerase can add a dideoxynucleotide to a growing chain. However, because there is no hydroxyl group on the sugar, the polymerase enzyme cannot form a phosphodiester bond between the dideoxynucleotide and a subsequent nucleotide. Chain extension is terminated on incorporation of a dideoxynucleotide (Figure 4). If there were only dideoxynucleosides present, then all chains would terminate on addition of the first nucleotide. Instead, the dideoxynucleotides are added in low concentration with

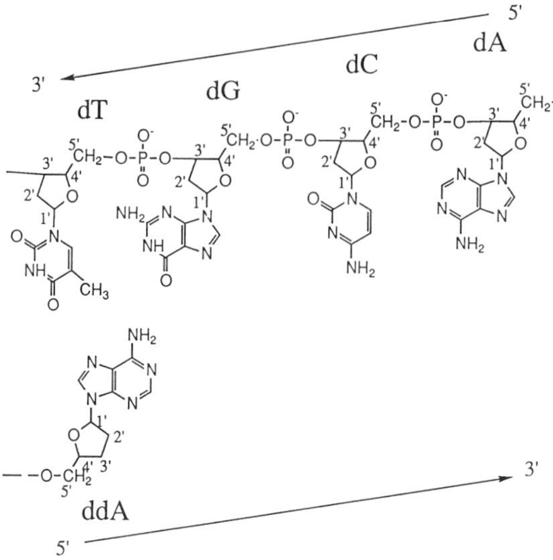

FIGURE 4. Incorporation of a dideoxyadenosine terminates the chain extension by DNA polymerase. Because the dideoxynucleoside has no 3′ hydroxyl group, a phosphodiester bond cannot be formed with a subsequent nucleotide.

deoxynucleotides. Chains will occasionally be terminated by enzymatic addition of the dideoxynucleotide; the rate of termination depends on the ratio of deoxynucleotide and dideoxynucleotide.

In Sanger's sequencing method, four reactions are carried out simultaneously (Figure 5). In the first reaction, the template, the four deoxynucleotides, and dideoxyadenosine triphosphate (ddATP) are mixed with the polymerase enzyme. The enzyme will extend the primer, synthesizing a strand of DNA that is complementary to the template. Occasionally, a dideoxyadenosine will terminate the growth of the strand by incorporation at a location that is opposite from a deoxythymidine in the template. The reaction product consists of fragments with the primer at the 5′ end and a dideoxyadenosine at the 3′ end. The reaction conditions are manipulated so that every A site in the growing chain will incorporate some dideoxyadenosine.

The other three reactions are identical except that they employ ddCTP, ddGTP, and ddTTP as chain terminators. These sequencing fragments are separated based on their size by use of polyacrylamide gel electrophoresis. In conventional sequencing, the samples are made radioactive through incorporation of either a phosphorus or sulfur isotope. The isotope may be added through addition of a radioactive nucleotide to the sequencing reaction.

III. SEPARATION BY GEL ELECTROPHORESIS

DNA sequencing fragments are separated based on size by polyacrylamide gel electrophoresis. The slab gels are ~0.5 mm thick, 30 cm wide, and 50 cm long. Conventionally, the products of the sequencing reaction are loaded onto 4 adjacent lanes in a slab gel; up to 36 lanes may be run simultaneously on the gel. Under the influence of the electric field, the DNA fragments migrate from the negative toward the positive electrode. The DNA fragments are separated by size: small fragments elute first and large fragments elute last.

The total acrylamide plus bisacrylamide concentration is typically 6%. Bisacrylamide is added as a cross-linker, typically at 5% of the acrylamide concentration. In shorthand notation, the gel formula would be given as 6%T and 5%C, where%T refers to the total acrylamide plus cross-linker concentration and where %C refers to the ratio of bisacrylamide to acrylamide plus bisacrylamide. The gel is typically prepared in 1× TBE buffer (89 m*M* Tris, 89 m*M*

```
3'...ACATTTTGCTGCCGGTCACGGTTCGAA....... + dNTP + ddATP→
   5'TGTAAAACGACGGCCAGTGCC3'

3'...ACATTTTGCTGCCGGTCACGGTTCGAA....... +
   5'TGTAAAACGACGGCCAGTGCA3'

3'...ACATTTTGCTGCCGGTCACGGTTCGAA....... + .....
   5'TGTAAAACGACGGCCAGTGCAA3'


3'...ACATTTTGCTGCCGGTCACGGTTCGAA....... + dNTP + ddCTP→
   5'TGTAAAACGACGGCCAGTGCC3'

3'...ACATTTTGCTGCCGGTCACGGTTCGAA....... + .....
   5'TGTAAAACGACGGCCAGTGCAAGC3'


3'...ACATTTTGCTGCCGGTCACGGTTCGAA....... + dNTP + ddGTP→
   5'TGTAAAACGACGGCCAGTGC3'

3'...ACATTTTGCTGCCGGTCACGGTTCGAA....... + .....
   5'TGTAAAACGACGGCCAGTGCAAG3'


3'...ACATTTTGCTGCCGGTCACGGTTCGAA....... + dNTP + ddTTP→
   5'TGTAAAACGACGGCCAGTGC3'

3'...ACATTTTGCTGCCGGTCACGGTTCGAA....... +
   5'TGTAAAACGACGGCCAGTGCAAGCT3'

3'...ACATTTTGCTGCCGGTCACGGTTCGAA....... + .....
   5'TGTAAAACGACGGCCAGTGCAAGCTT3'
```

FIGURE 5. Four sequencing reactions. In the first reaction, ddATP is used to terminate some of the fragments at A. In the second reaction, ddCTP is used to terminate some of the fragments at C. In the third reaction, ddGTP is used to terminate some of the fragments at G. In the last reaction, ddTTP is used to terminate some of the fragments at T.

borate, and 2 m*M* ethylenediaminetetraacetic acid [EDTA], pH 8.3) and 7 *M* urea. Polymerization is initiated by addition of *N,N,N′,N′*-tetramethylethylenediamine (TEMED) and ammonium persulfate; the solution is mixed and carefully poured between two glass plates.

The glass plates that form the walls of the slab are usually treated with two bifunctional silane reagents. The first reagent, γ-methacryloxypropyltrimethoxysilane, is used to coat one of the glass plates. The reagent cross-links with the polyacrylamide gel, holding the gel tightly to the plate. The other glass plate is treated with a second silane regent such as octadecyl silane to provide a coating that does not react with acrylamide.

The gels are operated at an electric field of ~40 V/cm, and the separation is complete in about 8 h. To provide uniform temperature across the gel, a water bath is placed in contact with one of the electrophoresis plates. The separation is typically performed at 40°C.

On completion of the separation, the top plate of the gel is removed and the gel is dried. A piece of film is put in contact with the gel for typically 8 h, during which time the radioactive decay of the labeled nucleotides creates bands on the film (Figure 6). The film is developed to reveal the migration distance of the sequencing fragments. The film is scanned by eye, and the sequence is recorded manually.

Sanger's method, as conventionally performed, suffers from serious limitations when applied to large-scale sequencing projects. The method is slow: roughly 8 to 16 h for the electrophoretic separation, 8 to 16 h for autoradiography, and perhaps another 4 h for visual inspection of the autoradiogram. Large amounts of acrylamide are handled; the material is toxic and must be handled with care. Moreover, visual inspection of sequencing autoradiograms is a painfully tedious and error prone process. In addition, Sanger's method requires use

FIGURE 6. Slab gel electropherogram of DNA sequencing reaction products. The fragments are loaded at the top of the gel, electric field is applied, and the smallest fragments move fastest. In this case, the sequence would be read as AAGCTTGCATGCCTGC.

```
      3'...ACATTTTGCTGCCGGTCACGGTTCGAA........
    FAM-5'TGTAAAACGACGGCCAGTGC 3'

      3'...ACATTTTGCTGCCGGTCACGGTTCGAA........
    JOE-5'TGTAAAACGACGGCCAGTGC 3'

      3'...ACATTTTGCTGCCGGTCACGGTTCGAA........
  TAMRA-5'TGTAAAACGACGGCCAGTGC 3'

      3'...ACATTTTGCTGCCGGTCACGGTTCGAA........
    ROX-5'TGTAAAACGACGGCCAGTGC 3'
```

FIGURE 7. Primer labeled with four different fluorescent dyes. FAM, Fluorescein; JOE, modified fluorescein; TAMRA, tetramethyl rhodamine; ROX, Texas red.

of radionucleotides; the expense of disposal of low-level waste is rapidly becoming a significant item in the sequencing budget.

A. FLUORESCENCE DETECTION

Conventional slab gel electrophoresis separation of sequencing fragments relies on autoradiography for detection. Ansorge et al. reported the replacement of radioactive label with a fluorescent label. In this technique, the 5′ end of the sequencing primer was labeled with rhodamine.[5] The sequencing reactions proceeded normally, and the products of the four sequencing reactions were separated in adjacent lanes of a slab gel. Fluorescence is excited by a single laser, and fluorescence is detected from each lane. Identification of the terminal nucleotide is based on the particular lane in which the fragment appears.

A much more sophisticated approach had been described in the same year by L. Smith and co-workers in Hood's laboratory.[6] In their method, the primer must be labeled with four different fluorescent labels (Figure 7). Each labeled primer is used for one of the four sequencing reactions; the products of the reaction are combined and separated in a single lane of a slab gel. Because a single lane is used for the separation, sample density can be four times larger than in autoradiography, where each sample requires separation in four lanes (Figure 7). Fluorescein (FAM) and modified fluorescein (JOE) are trade names; they are excited by the 488-nm line of an argon ion laser. The rhodamine-based dyes TAMRA (tetramethyl rhodamine) and ROX (Texas red) are excited by the 514.5-nm line of the argon ion laser. The fluorescence instrumentation is rather complex. Each lane is probed in succession with the blue and green argon ion laser beams. Fluorescence is monitored in four spectral bands through a rotating filter wheel. The four dyes are identified based on their spectral characteristics. Identification of the terminal nucleotide is determined by the fluorescence label attached to the primer.

In 1987, Prober and co-workers at Du Pont (Wilmington, DL) reported the use of fluorescently

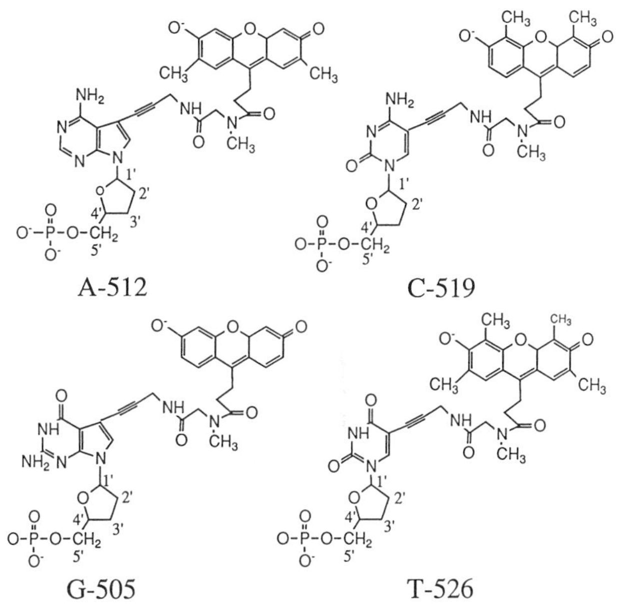

A-512 C-519

G-505 T-526

FIGURE 8. Du Pont dye-labeled dideoxynucleotides. The dyes used to label the dideoxynucleotides have a common structure. Their emission maximum shifts to the red with addition of substituents. The emission maximum, in nanometers, is appended to the name of each nucleotide.

labeled dideoxynucleotides in DNA sequencing.[7] They synthesized four different modified fluorescein dyes and attached them to the deoxynucleotides (Figure 8). A single chain termination reaction is employed with all four dyes present. Identification of the terminal nucleotide is encoded in the fluorescence spectrum of the attached dye. The dyes are excited by the 488-nm line of the argon ion laser; emission is recorded in two spectral bands. The ratio of fluorescence intensity in the two bands is used to identify the terminal nucleotide. However, because of the poor spectral resolution between the dyes, sequencing accuracy suffers unless the signal-to-noise ratio is high. Du Pont no longer manufactures this sequencer.

Two groups, Ansorge's group in Germany and Tabor and Richardson at Harvard, independently reported DNA sequencing based on a single dye and a single lane.[8,9] A single fluorescently labeled primer is used in a single sequencing reaction. T7 DNA polymerase, in the presence of manganese, uniformly incorporates dideoxynucleotides, generating reaction peaks of nearly equal height. If a single sequencing reaction is performed with ddA:ddC:ddG:ddT present in the ratio of 8:4:2:1, then termination will proceed in the same proportion. As a result, sequence is identified by peak amplitude. The biggest peaks will be A, the next biggest will be C, the next biggest will be G, and the smallest will be T. Unfortunately, it is difficult to control peak height precisely, even when using manganese in the sequencing reaction. As a result, the simple single-color peak height-encoded sequencing protocol produces poor accuracy.

A two-color peak height-encoded sequencing protocol was developed by this group for application in capillary gel electrophoresis.[10] In this case, fluorescein-labeled primer is used with ddA and ddC in a 3:1 ratio for one sequencing reaction while rhodamine-labeled primer is used with ddG and ddT in a 3:1 ratio for a second reaction. Large peaks associated with fluorescein represent A, small fluorescein peaks represent C, big rhodamine peaks represent G, and small rhodamine peaks represent T. Sequence information is based on both spectral and amplitude information. The accuracy of the two-color peak height-encoded sequencer is similar to that produced by commercial sequencers.[11]

B. CAPILLARY GEL ELECTROPHORESIS

Increased sequencing rates are produced by operating the gels at high electric field strength. Unfortunately, the finite resistance of the separation buffer leads to unacceptable heating of conventional 500-μm thick gels at electric field strengths much greater than 50 V/cm. Gel-filled capillaries have attracted interest because their high surface-to-volume ratio provides excellent heat transport properties, allowing use of very high electric field strength. Typical capillaries are 20 to 75 μm I.D. and 25 to 100 cm long.

A number of investigators have reported the use of gel-filled capillaries for separation of oligonucleotide standards with detection by ultraviolet absorbance.[12-16] Early reports made remarkable claims on the performance of these gels; in particular, over 150 separations were reported with the same gel-filled capillary.[13]

While surprising performance has been reported for separation of oligonucleotide standards, much poorer performance is observed for separation of sequencing fragments. Separation efficiency is observed to drop after a few separations; bubbles often form near the injection tip of the capillary. These artifacts appear to be associated with the presence of template in the sequencing mixture. The high molecular weight, and high charge, associated with the template produces a band of high conductivity in the gel. The slow-moving template generates a hot region in the capillary, which can damage the gel. A number of approaches can be envisioned to remove the template. It can be digested with a nuclease; it can be separated from the smaller sequencing products through use of an appropriate filter; it can be attached to a solid support; or a high-sensitivity detector can be used with low loading levels.[18]

To maximize the separation rate, high electric fields are used to separate the DNA sequencing fragments. Several factors limit the maximum separation field. At electric fields greater than ~200 V/cm, polyacrylamide gels tend to extrude from the capillary, destroying the separation. Temperature rise in the capillary is insufficient to lead to expansion of the matrix.[19] Instead, extrusion of the gel probably is due to the presence of residual charge on the gel, which drives the gel from the capillary. Alternatively, residual charge on the capillary wall leads to electroosmotic force that can move the gel.

Several patents have been issued to prevent migration of the gel.[20,21] The walls of the capillary are treated with the same bifunctional silane reagent used in slab gel electrophoresis, γ-methacryloxypropyltrimethoxysilane. During acrylamide polymerization, the gel will cross-link to the wall of the capillary. However, during the polymerization step, the gel tends to shrink by a few percent. Because the gel is rigidly bound to the walls of the capillary, the gel cannot relax during this shrinkage, and a series of bubbles are formed in the center of the capillary. A number of exotic methods have been used to eliminate the bubbles, including addition of polymer additives and polymerization under high pressure. We found that the use of additive leads to poor resolution of sequencing fragments more than 100 bases in length. Polymerization under high pressures seems to be too cumbersome for routine use.

We developed a simple method for production of bubble-free gels that are stable at high electric field.[22] Rather than treating the entire length of the capillary, we treat only the last few centimeters of the gel with the bifunctional silane reagent. After a few minutes of reaction time, the capillaries are then filled with the acrylamide mixture for polymerization. By binding the gel to a portion of the capillary wall, the gel cannot migrate. Because only a short piece of capillary is treated by the reagent, stress associated with shrinkage is easily accommodated by the gel. These gels are produced with a success rate exceeding 95%, and the polyacrylamide gels are stable at electric fields up to 450 V/cm.

Alternatively, if residual charge on polyacrylamide leads to extrusion of the gel under high electric fields, then a neutral gel should be stable at high fields. We have found that Long-Ranger, a modified acrylamide distributed by J. T. Baker (Phillipsburg, NJ) is stable at electric fields of at least 800 V/cm.[23] Less than 7 min is required for the separation of fragments up

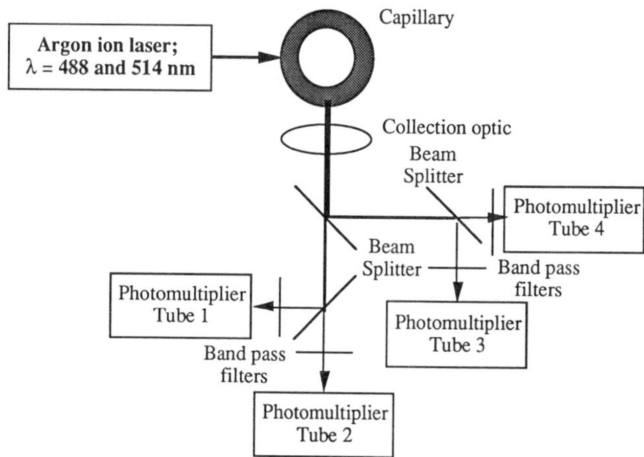

FIGURE 9. Four-color DNA sequencer. This instrument uses a set of beam splitters to divert fluorescence to four different photomultiplier tubes. The beam splitters are aluminum-coated plates, and they reflect and transmit about 50% of the incident light. Interference filters determine the spectral band detected by each photomultiplier tube.

to 250 bases in length. Larger fragments suffer from the phenomenon of biased reptation; they are not well resolved at high electric fields.

C. DNA SEQUENCING BY CAPILLARY GEL ELECTROPHORESIS

Luckey and co-workers reported DNA sequence by capillary gel electrophoresis in 1990.[24] They used the ABI Biotechnology (Winnipeg, Manitoba, Canada) fluorescent labeled primers in four sequencing reactions. The products were pooled and separated in a single capillary. A multiline argon ion laser was used to illuminate the capillary simultaneously with light at 488 and 514 nm. Fluorescence was collected at right angles from the capillary (Figure 9). A set of beam splitters was used to direct the fluorescence to a set of four photomultiplier tubes, each equipped with a bandpass spectral filter. This system records fluorescence simultaneously in the four spectral channels. At an electric field of 300 V/cm, 80 min was required to separate fragments up to 360 bases in length. A 3%T, 3%C gel was used for the separation. No sequencing errors were observed for fragments ranging from 80 to 356 bases in length.

In 1991, a second sequencer was reported for the ABI fluorescently labeled primer system. A. Karger and co-workers at the University of Utah used a spectrograph and charge-coupled device (CCD) camera to resolve fluorescence from the four dyes (Figure 10).[25] At an electric field of 190 V/cm, roughly 60 min was required to separate fragments up to 340 bases in length. Sequencing accuracy of 96.5% was observed for fragments ranging from 60 to 350 bases in length.

This research group demonstrated three different sequencing techniques by capillary gel electrophoresis.[22] In the first method, ABI primers were used. Fluorescence was excited with two lasers, an argon ion laser operating at 488 nm and a helium-neon laser operating at 543.5 nm (Figure 11). The two beams were sequentially focused onto the capillary, with the 488-nm beam exciting fluorescence from FAM and JOE and the 543.5-nm beam exciting fluorescence from TAMRA and ROX. The fluorescence was collected with a single microscope objective and fluorescence was discriminated with a filter wheel. A second detector was built for the Du Pont fluorescently labeled dideoxynucleotides. A 488-nm argon ion laser excited fluorescence, and a two-channel direct-reading fluorescence spectrometer was used to discriminate fluorescence between the dyes (Figure 12). The poor incorporation rate of the highly modified dye-labeled deoxynucleotides, along with the severe spectral overlap, led to poor

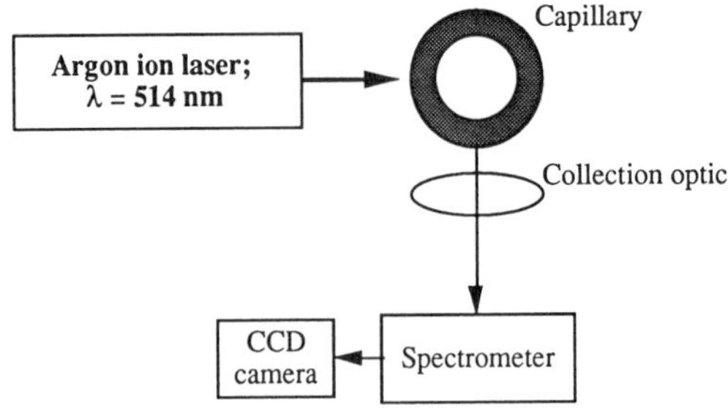

FIGURE 10. Charge-coupled device-based sequencer. The capillary is imaged through a spectrometer onto a charge-coupled device (CCD); spectral information is dispersed along one axis of the CCD and spatial information is dispersed along the orthogonal axis. Charge generated in selected spectral and spatial bands is integrated and used to identify the eluting nucleotide.

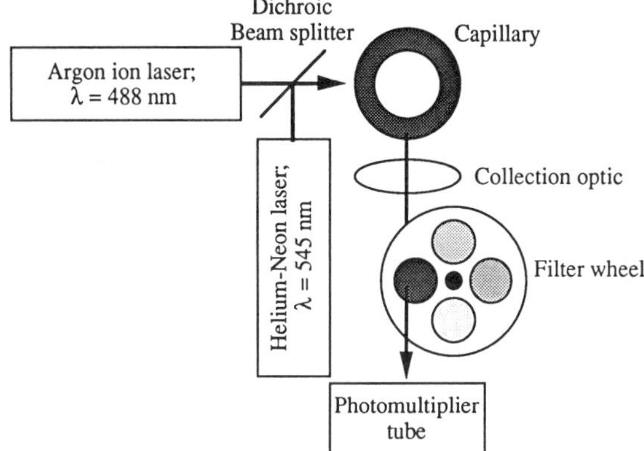

FIGURE 11. Filter wheel-based four-color capillary electrophoresis system. A rotating filter wheel isolates, in turn, the emission spectrum of the four dyes used to label the sequencing fragments. To minimize Raman scatter effects, a sector wheel, not shown, alternately blocks each laser beam. The two filter wheels are synchronized through use of stepper motors with a common controller.

sequencing performance. The Du Pont technology has subsequently been taken off of the market, and is not easily available today. A third instrument was built for the four-level peak height-encoded sequencing technique. A helium-neon laser operating at 543.5 nm was the excitation source for a single-channel fluorescence spectrometer. TAMRA-labeled primers were used for the sequencing reaction. The four-level sequencing protocol requires excellent electrophoretic separation between adjacent peaks. Any overlap between the peaks leads will degrade the estimation of peak height, leading to poor sequence accuracy.

The two-color sequencer based on the Du Pont dideoxynucleotide chemistry provided mediocre sequencing accuracy. In that system, the terminal nucleotide was identified based on a single parameter: the ratio of fluorescence signal in the two emission channels. Sequencing accuracy was degraded by two phenomena. First, low-amplitude peaks suffered from poor signal-to-noise ratio; accurate determination of peak amplitude was difficult. Second, ghost peaks sometimes appear, primarily associated with false priming events. On the other hand, the instrumentation required for the two-emission channel sequencer is robust and powerful. Detection limits in the low zeptomole (zmol; 10^{-21} mol) to high yoctomole (ymol; 10^{-24} mol) range are routinely produced with the instrument. The elimination of moving parts produces a rugged instrument. Last, the use of a dichroic filter produces high optical throughput. We feel that the two-color instrumentation is ideally suited for capillary electrophoresis.

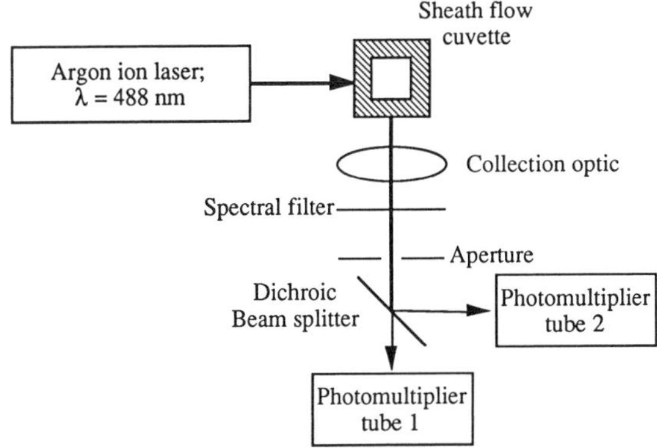

FIGURE 12. Postcolumn fluorescence detector for two-channel spectral emission. A dichroic filter reflects green light and transmits yellow light. As a result, fluorescence in two spectral channels can be monitored simultaneously. In a sense, the instrument is a simple two-channel direct reading spectrometer.

We find that the two-color peak height-encoded instrumentation produces excellent performance in capillary electrophoresis.[10] The instrumentation is identical to that shown in Figure 12. Figure 13 presents a sequencing run generated at 200 V/cm from a template generated from the malaria genome. In this 4%T, 5%C gel, fragments 570 bases in length are separated in less than 100 min. No errors were observed for fragments ranging from bases 53 to 520; sequencing accuracy exceeds 99% for the first 541 bases.

The high accuracy of the two-color sequencer arises from use of two pieces of information in sequence determination: spectral information and amplitude information. In Figure 13, big peaks in both channels represent T, big peaks only in the dashed channel represent A, small peaks in both channels represent G, and small peaks only in the dashed channel represent C. Figure 13B presents a closer look at a region corresponding to the elution of fragments around 425 bases in length. The resolution in this region is about 0.8 to 0.9; identification of adjacent peaks is simple. Determination of sequence is trivial in this portion of the electropherogram. Figure 13 does point out a limitation of the peak height-encoded technology. While a resolution of 0.5 is sufficient to discriminate between peaks of equal height, a higher resolution is required to discriminate between peaks of dissimilar amplitude. It has become clear that the high resolution of capillary electrophoresis is required to produce accurate sequencing data. Near the end of the run, the resolution degrades precipitously due to the phenomenon of biased reptation. Sequence information cannot be extracted near the end of the separation, irrespective of the fluorescence labeling scheme.

D. COMPARISON OF SLAB GEL AND CAPILLARY GEL ELECTROPHORESIS

There has been some discussion in the literature comparing the resolution produced by commercial slab gel electrophoresis instruments and capillary gel electrophoresis instruments.[26] This comparison is often difficult, because most commercial sequencers display highly processed data. However, the raw data can be extracted from the instrument. Figure 14 presents a comparison of electropherograms obtained with a Pharmacia (Piscataway, NJ) ALF sequencer and our capillary system for fragments ranging from 359 to 388 bases in length. The capillary gel data have a resolution of 1, measured by regression analysis for the doublet A bases 363–364. The resolution of the ALF system is 0.5, measured for bases 360–361. Similar results are produced by the ABI instrument.

Resolution can be dominated by injection volume, thermal gradients, and mass diffusion. In stacking buffers, injection volume effects are negligible. Thermal gradients are minimal in the narrow diameter capillaries at these electric fields. If mass diffusion dominates the peak width, then resolution (Res) is given by

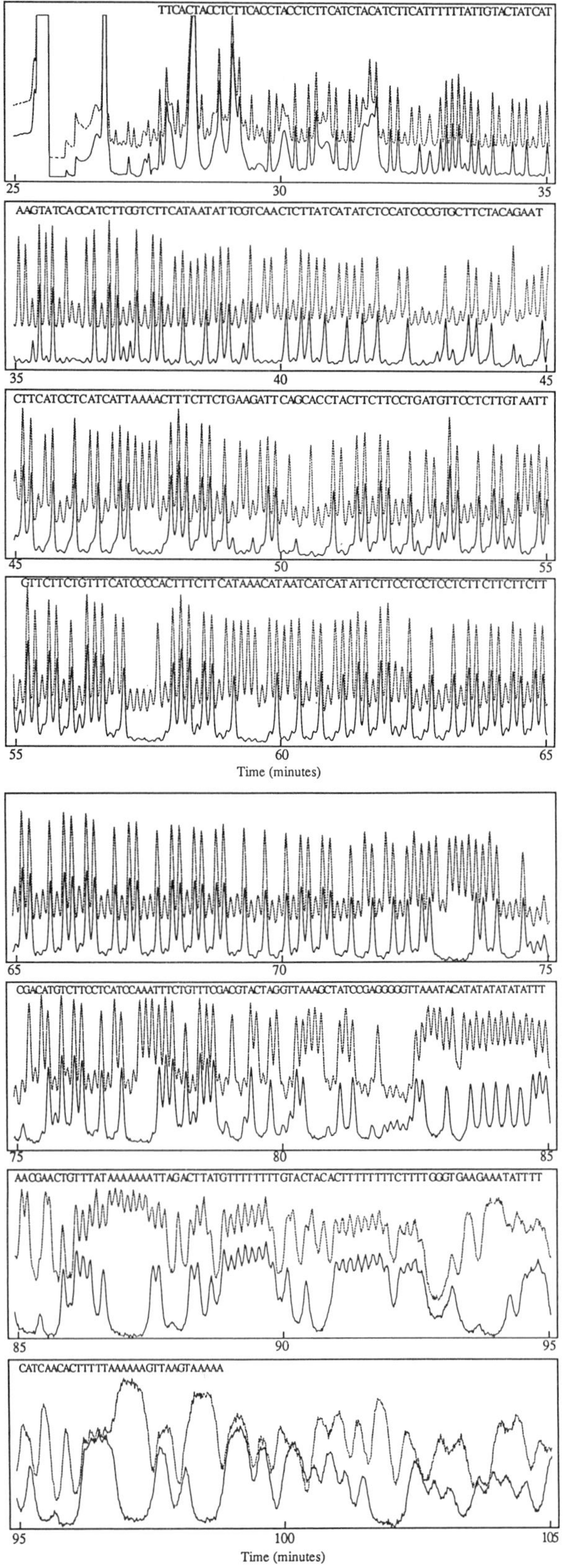

FIGURE 13A.

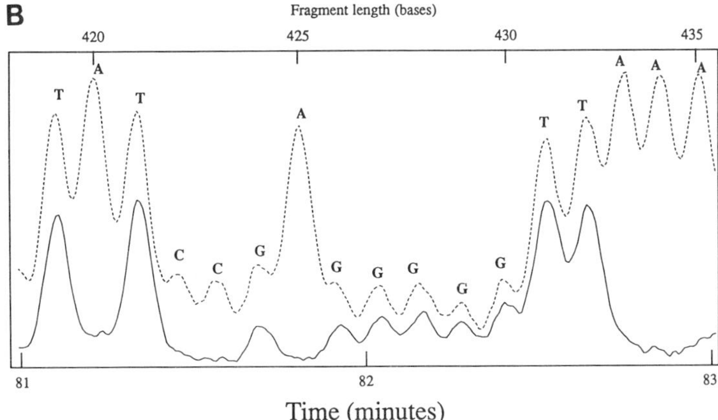

FIGURE 13. Capillary gel electrophoresis separation of two-color peak height-encoded sequencing product. The sample was a clone taken from the malaria genome. The separation took place at an electric field of 200 V/cm in a 30-cm long, 50-μm I.D. capillary. A 4%T, 5%C gel was used for the separation. (A) Entire separation; (B) close-up of bases 419 to 435.

$$\text{Res} = \frac{\sqrt{\mu V / 2D}}{4} \frac{\Delta\mu}{\bar{\mu}}$$

where μ is the average mobility of two components, Δμ is the difference in mobility between the two components, *V* is the separation voltage, and *D* is the analyte diffusion coefficient. Resolution is proportional to the square root of the separation voltage, not electric field.

The ALF system operates at an electric field of about 50 V/cm over a separation distance of about 40 cm; the separation voltage is about 2000 V. The capillary data are obtained at an electric field of 200 V/cm over a distance of 30 cm; the separation voltage is about 6000 V. The capillary electrophoresis system should have a resolution that is $(6000/2000)^{1/2}$ or 1.7 times larger than the slab gel instrument, assuming that mass diffusion dominates the separation. This value is consistent with the twofold higher resolution observed for the capillary system compared with the slab gel system. Capillaries produce higher resolution than slab gels because the capillaries operate at a higher voltage.

E. COMPRESSIONS

Peak overlap occurs late in the sequence as resolution degrades. However, peak overlap is occasionally observed as an artifact relatively early in the DNA sequence. This overlap is associated with formation of secondary structure within the DNA. For example, a compression is often noted near base 65 in the M13mp18 sequence. The sequence in this region reads ...GGTACC. A hairpin structure can form, where GGT is paired with ACC (Figure 15). This compact structure migrates at a faster rate compared to migration in the absence of secondary structure. As a result, the fragments that terminate near this sequence will have smaller peak spacing than normal regions. The electropherogram is said to suffer from a compression at this point in the sequence; sequencing accuracy suffers in compressions because resolution is degraded.

In Figure 16A, a single sequencing reaction was performed with ddCTP; only fragments ending in C produce a peak. Peaks 56 to 59 are well resolved, with addition of one nucleotide to the sequence producing a significant retardation. On the other hand, peaks 65 to 66 are compressed. Addition of a single C produces only a small change in mobility. In Figure 16B, 10% formamide has been added to the sequencing gel to disrupt secondary structure. Peaks 56 to 59 appear unchanged by the addition, whereas the compression associated with peaks 65 to 66 has been decreased.

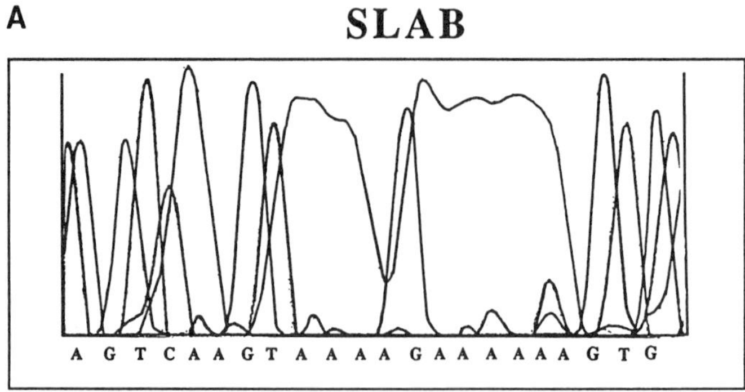

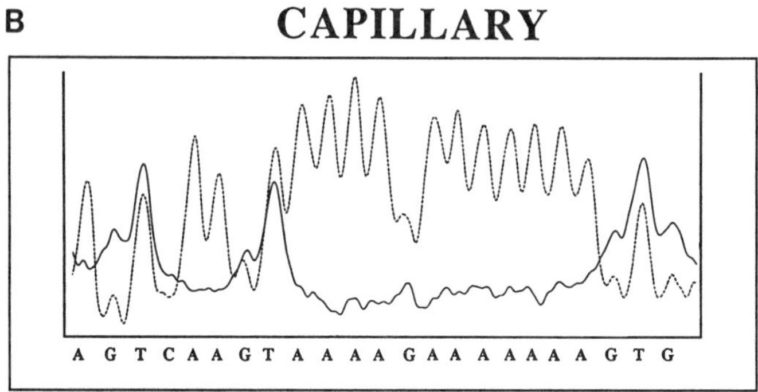

FIGURE 14. Comparison of a separation with the Pharmacia ALF sequencer (A) and a capillary gel system (B). The sample was taken from the malaria genome; it is a different sample than was used in Figure 14. The capillary separation used the same electrophoretic conditions as in Figure 13.

...GGA
ƆƆ⅄

FIGURE 15. Schematic of a compression. Because of the self-complementary nature of the sequence, the bases are able to pair with each other, forming a compact secondary structure. Addition of a single base to this compact structure does not change the electrophoretic mobility; as a result, sequence fragments near a region with secondary structure are poorly resolved.

To eliminate secondary structure, DNA sequencing gels contain a high concentration of urea, typically 7 *M*. The urea breaks up secondary structure. Conventional slab gel electrophoresis plates operate at a temperature above ambient, typically 40°C. This increased temperature also helps to eliminate compressions. Rather than operating the capillary at high temperature, additional denaturing reagent can be added to the sequencing gel. We have reported the use of formamide as a denaturing agent; roughly 10% formamide, in addition to 7 *M* urea, provides enhanced resolution of regions that normally suffer from compressions.[24] Higher concentrations of formamide lead to very slow separations, while lower concentrations do not relax the secondary structure effectively.

IV. FUTURE PROSPECTS

Capillary electrophoresis produces higher resolution separation of sequencing data compared with slab gel electrophoresis. This improvement in resolution is associated with the

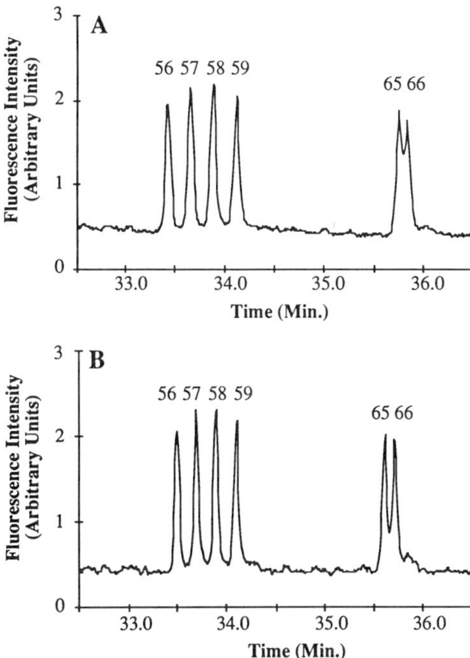

FIGURE 16. Elimination of compressions by addition of formamide. Peaks 56 to 59 migrate normally whereas peaks 65 to 66 form a compression. Addition of denaturant acts to break up the secondary structure, allowing more normal migration. (A) Urea (7 *M*), 6%T, 5%C polyacrylamide gel operating at 200 V/cm; (B) 30% formamide, 7 *M* urea, 6%T, 5%C polyacrylamide gel operating at 200 V/cm.

higher separation potential employed in the capillary system. As a result, capillary electrophoresis systems should inherently produce higher accuracy data compared with slab gel systems. Capillary electrophoresis systems produce a factor of five faster separation of sequencing fragments compared with slab gel electrophoresis. The increased separation speed is due to the high electric field employed in capillary electrophoresis.

The high-resolution and high-speed separation of sequencing fragments suggests that capillary electrophoresis is a very attractive alternative to conventional slab gel electrophoresis for DNA sequencing. However, there are two issues that must be addressed before the capillary system can replace slab gel electrophoresis. First, multiple capillaries must be operated in parallel. Conventional slab gel electrophoresis typically separates 24 to 32 samples simultaneously; to be competitive, a capillary gel electrophoresis system must operate many capillaries in parallel. The first multiple capillary gel electrophoresis system was reported in 1990 by Zagursky and McCormick.[27] They modified a standard Du Pont Genesis 2000 sequencer to run with 500-μm I.D. capillaries; a single laser beam was scanned across the capillaries and fluorescence was detected in a two-color spectrometer. The instrument operated at 50 V/cm; 9.5 h was required to separate fragments 500 bases in length. Sequencing accuracy was less than 97% for fragments ranging from 29 to 512 bases in length. In 1992, Mathies and co-workers used a more sophisticated scanning instrument to image 100-μm I.D. capillaries.[28] In this system, an array of capillaries is scanned under a confocal microscope, which is a two-color version of an instrument manufactured by Europhore.[29] The instrument is currently operating with relatively large inner diameter capillaries, 100 μm, and has produced sequence information for fragments of 320 bases in length. Early this year, a group from Hitachi presented preliminary results for an instrument based on a sheath flow cuvette fluorescence detector.[30] The instrument is currently operating with relatively large inner diameter capillaries, 100 μm, and only the products of a single base reaction have been separated; no sequence information has been published.

A second issue deals with the temperature of the gel used during the separation. While formamide may be added to reduce the formation of secondary structure in DNA, it would be desirable to operate gels above ambient temperatures. Higher temperatures break up the compressions. Higher temperatures also increase the separation rate. The mobility of DNA fragments increases by about 2% per degree of temperature rise. If the sequencing gel is operated at 40°C, then the separation time will decrease by about 40%; capillary systems will then produce separations that are almost an order of magnitude faster than slab gel systems. We have performed some preliminary separations at elevated temperatures. Sequencing rate increases as predicted. Compressions are also minimized. Last, longer pieces of DNA can be resolved compared with room temperature operation of the capillary.

The combination of multiple capillary systems operating at 40°C produces separations that are an order of magnitude faster and a factor of two higher in resolution compared with conventional electrophoretic instrumentation. There is a bright future ahead for this technology.

REFERENCES

1. **Meier-Ewert, S., Maier, E., Ahmadi, A., Curtis, J., and Lehrach, H.,** An automated approach to generating expressed sequence catalogs, *Nature (London)*, 361, 375, 1993.
2. **Yanisch-Peron, C., Vieira, J., and Messing, J.,** Improved M13 phage cloning vectors and host strain: nucleotide sequences of the M13mp18 and pUC19 vectors, *Gene*, 33, 103, 1985.
3. **Norrander, J., Kempe, T., and Messing, J.,** Construction of improved M13 vectors using oligodeoxynucleotide-directed mutagenesis, *Gene*, 26, 101, 1983.
4. **Sanger, F., Nicklen, S., and Coulson, A. R.,** DNA sequencing with chain-terminating inhibitors, *Proc. Natl. Acad. Sci. U.S.A.*, 74, 5463, 1977.
5. **Ansorge, W., Sproat, B., Stegemann, J., and Schwager, C.,** A non-radioactive automated method for DNA sequence determinations, *J. Biochem. Biophys. Methods*, 13, 315, 1986.
6. **Smith, L. M., Sanders, J. Z., Kaiser, R. J., Hughes, P., Dodd, C., Connell, C. R., Heiner, C., Kent, S. B. H., and Hood, L. E.,** Fluorescence detection in automated DNA sequence analysis, *Nature (London)*, 321, 674, 1986.
7. **Prober, J. M., Trainor, G. L., Dam, R. J., Hobbs, F. W., Robertson, C. W., Zagursky, R. J., Cocuzza, A. J., Jensen, M. A., and Baumeister, K.,** A system for rapid DNA sequencing with fluorescent chain-terminating dideoxynucleotides, *Science*, 238, 336, 1987.
8. **Ansorge, W., Zimmermann, J., Stegemann, J., Erfle, H., and Voss, H.,** One label, one tube, Sanger DNA sequencing in one and two lanes on a gel, *Nucleic Acids Res.*, 18, 3419, 1990.
9. **Tabor, S. and Richardson, C. C.,** DNA sequence analysis with a modified bacteriophage T7 DNA polymerase, *J. Biol. Chem.*, 265, 8322, 1990.
10. **Chen, D. Y., Harke, H. R., and Dovichi, N. J.,** Two-label peak-height encoded DNA sequencing by capillary gel electrophoresis: three examples, *Nucleic Acids Res.*, 20, 4873, 1992.
11. **Bay, S., Harke, H. R., Elliott, J., and Dovichi, N. J.,** Accuracy of two-color peak height encoded DNA sequencing by capillary gel electrophoresis and laser-induced fluorescence, *Proc. Soc. Photo-Opt. Instrum. Eng.*, 891, 8, 1993.
12. **Cohen, A. S., Najarian, D. R., Paulus, A., Guttman, A., Smith, J. A., and Karger, B. L.,** Rapid separation and purification of oligonucleotides by high-performance capillary gel electrophoresis, *Proc. Natl. Acad. Sci. U.S.A.*, 85, 9660, 1988.
13. **Guttman, A., Cohen, A. S., Heiger, D. N., and Karger, B. L.,** Analytical and micropreparative ultrahigh resolution of oligonucleotides by polyacrylamide gel high-performance capillary electrophoresis, *Anal. Chem.*, 62, 137, 1990.
14. **Yin, H. F., Lux, J., and Schomburg, G.,** Production of polyacrylamide gel filled capillaries for capillary gel electrophoresis (CGE): Influences of capillary surface pretreatment on performance and stability, *J. High Resolut. Chromatogr.*, 13, 624, 1990.
15. **Wang, T., Bruin, G. J., Kraak, J. C., and Poope, H.,** Preparation of polyacrylamide gel-filled fused silica capillaries by photopolymerization with riboflavin as the initiator, *Anal. Chem.*, 63, 2207, 1991.

16. **Baba, Y., Matsuura, T., Wakamoto, K., and Stuhako, M.,** A simple method for the preparation of polyacrylamide gel filled capillaries for high performance separation of polynucleotides by using capillary gel electrophoresis, *Chem. Lett.*, p. 371, 1991.
17. **Guttman, A.,** Chapter 5 of this book.
18. **Swerdlow, H., Dew-Jager, K. E., Brady, K., Gesteland, R., Grey, R., and Dovichi, N. J.,** Stability of capillary gels for automated sequencing of DNA, *Electrophoresis*, 13, 475, 1992.
19. **Harke, H. R., Bay, S., Zhang, J. Z., Rocheleau, M. J., and Dovichi, N. J.,** The effect of total percent polyacrylamide in capillary gel electrophoresis for sequencing of short DNA fragments: a phenomenological model, *J. Chromatogr.*, 608, 143, 1992.
20. **Bente, P. F. and Myerson, J.,** U.S. Patent 4,810,456, 1989.
21. **Karger, B. L. and Cohen, A. S.,** U.S. Patent 4,865,706, 1989.
22. **Swerdlow, H., Zhang, J. Z., Chen, D. Y., Harke, H. R., Grey, R., Wu, S., Fuller, C., and Dovichi, N. J.,** Three DNA sequencing methods based on capillary gel electrophoresis and laser-induced fluorescence, *Anal. Chem.*, 63, 2835, 1991.
23. **Rocheleau, M. J. and Dovichi, N. J.,** Separation of DNA sequencing fragments at 53 bases/minute by capillary gel electrophoresis, *J. Microcolumn Sep.*, 4, 449, 1992.
24. **Luckey, J. A., Drossman, H., Kostichka, A. J., Mead, D. A., D'Cunha, J., Norris, T. B., and Smith, L. M.,** High speed DNA sequencing by capillary electrophoresis, *Nucleic Acids Res.*, 18, 4417, 1990.
25. **Karger, A. E., Harris, J. M., and Gesteland, R. F.,** Multiwavelength fluorescence detection for DNA sequencing using capillary gel electrophoresis, *Nucleic Acids Res.*, 19, 4955, 1991.
26. **Rocheleau, M. J., Grey, R. J., Chen, D. Y., Harke, H. R., and Dovichi, N. J.,** Formamide modified polyacrylamide gels for DNA sequencing by capillary gel electrophoresis, *Electrophoresis*, 13, 484, 1992.
27. **Zagursky, R. J. and McCormick, R. M.,** DNA sequencing separations in capillary gels on a modified commercial DNA sequencing instrument, *BioTechniques*, 9, 74, 1990.
28. **Huang, X. C., Quesada, M. A., and Mathies, R. A.,** DNA sequencing using capillary array electrophoresis, *Anal. Chem.*, 64, 2149, 1992.
29. **Hernandez, L., Secalona, J., Joshi, N., and Guzman, N.,** Laser induced fluorescence and fluorescence microscopy for capillary electrophoresis zone detection, *J. Chromatogr.*, 559, 183, 1991.
30. **Kambara, H. and Takahashi, S.,** Multiple-sheathflow capillary array DNA analyser, *Nature (London)*, 361, 565, 1993.

PART IV
SPECIALIZED CAPILLARY ELECTROPHORESIS APPLICATIONS

Chapter 15

CAPILLARY ELECTROPHORESIS FOR THE ANALYSIS OF SINGLE CELLS

Sandra Sloss and Andrew G. Ewing

TABLE OF CONTENTS

0-8493-8690-X/94/$0.00+$.50

I. INTRODUCTION

Chemical analysis of single cells is an area of great interest in the biological and medical sciences. Knowledge of the chemical composition of individual cells should lead to a better understanding of how specific cells function in a heterogeneous population of cells. Information of this type promises to advance our knowledge of such diverse processes as cellular differentiation, intracellular communication, neurotransmission, and the physiological effects of external stimuli such as drugs and toxins. Advances in these areas require the development of analytical methods for studying single whole cells, subcellular components, and extracellular fluids. The ability to handle small volumes, to analyze a variety of compounds simultaneously, and to give good qualitative and quantitative information are all key requirements for development of methods for the analysis of single cells.

The considerable interest in studying single-cell chemistry has resulted in the development of a number of analytical techniques. These include enzyme activity measurements,[1] immunoassay,[2,3] microgel electrophoresis,[4] fluorescence imaging techniques,[5] microscale ion-selective electrodes,[6,7] voltammetric microelectrodes,[8-15] optical and electron microscopic techniques,[16,17] and secondary ion mass spectrometry.[18] Although these methods have provided good information, they have important limitations. Most suffer from inadequate sensitivity, poor quantitative capabilities, or an inability to determine multicomponent samples.

Microcolumn liquid chromatography (LC) separation methods with their small volume requirements and high-resolution capabilities offer several advantages, especially in the areas of quantitation and multicomponent analysis. To date, microcolumn techniques coupled to electrochemical detection or laser-induced fluorescence (LIF) detection offer the best performance for single-cell analysis.[19-28] Results obtained through microcolumn high-performance liquid chromatography (HPLC) in the open tubular and packed column formats have clearly demonstrated quantitative determination of multicomponents in individual cells.[19-23]

Capillary electrophoresis (CE) in narrow bore (2- to 10-μm I.D.) capillaries provides a means to obtain the high-efficiency separations promised by microcolumns without the need for a bonded or coated phase.[19,24,29,30] Narrow bore CE is ideally suited for single-cell analysis.[25] It provides rapid, highly efficient separations of ionic species in extremely small volume samples. One of the major advantages of CE is that it provides enhanced heat dissipation through the capillary wall, thus reducing zone broadening due to convection.

Capillary electrophoresis offers other advantages in addition to enhanced heat dissipation. One of the most important characteristics of CE is electroosmotic flow. Electroosmotic flow is the bulk movement of liquid through the capillary due to the effect of the high electric field on the double layer next to the negatively charged capillary wall. Electroosmotic flow provides three key advantages for the separation of small biological samples. First, this flow is strong enough to cause elution of all species, cations, neutrals, and anions at the detector end of the column. Second, electroosmosis produces pluglike flow that has a flat velocity distribution across the capillary diameter. This flat flow profile results in high-efficiency separations. Finally, electroosmosis results in very small volume flow rates, which permits sampling from single cells. Injection volumes as low as 270 fmol have been demonstrated using electroosmotic flow.[25] Thus, CE in narrow bore capillaries possesses unique characteristics that make it especially suitable as a technique for single-cell analysis.

This chapter describes the development of CE for application to the analysis of single cells. The extremely small sample volumes involved in these analyses place stringent requirements on sample-handling and injection techniques, and on detection methods. Developments in these areas, as well as specific applications of CE to single-cell experiments, are discussed.

II. METHODOLOGY

A. CAPILLARY ELECTROPHORESIS IN NARROW BORE CAPILLARIES

There are several reasons for developing CE in smaller diameter capillaries. First, solution heating from ion migration limits the potential field that can be employed in CE. Decreasing the bore of the capillary maximizes the surface area to liquid volume ratio and provides for very efficient dissipation of heat via the capillary wall. Therefore, very high separation potentials can be employed in CE to achieve rapid, high-resolution analyses. Second, small inner diameter capillaries require smaller sample volumes. Total column volumes for a 1-m long capillary with an internal diameter of 2 to 10 μm range from 13 to 300 nl, with injection volumes even smaller. In addition, adaptation of concentration-sensitive detectors to smaller capillaries should improve mass detection limits. This is important when demands are placed on sample size rather than sample concentration as is the case in biological and cellular analyses. Finally, the major goal in developing CE in very small capillaries is its compatibility with direct analysis of single cells.

The miniaturization of CE for application to single-cell analysis requires the use of very small capillaries. Amperometric detection is one of the few methods compatible with small diameter capillaries. Off-column amperometric detection has been applied to capillaries with internal diameters of 12.7 μm.[31] This work was significant for several reasons. It was one of the first demonstrations of the small volume capabilities of CE in narrow bore capillaries. Calculated injection volumes as low as 430 pl (picoliters) were achieved. In addition, it showed the advantage of lower mass detection limits possible with small capillaries. A detection limit of 6 amol (attomoles) was obtained for serotonin. This corresponded to approximately 4×10^6 molecules injected onto the column. Finally, this work represented the first application of CE to single-cell analysis. Use of a 12.7-μm capillary attached to an 8-μm diameter microinjector permitted sampling from the giant dopamine neuron of the pond snail, *Planorbis corneus.*

In an effort to acquire small volume samples from cells, CE with amperometric detection was later extended to 5- and 2-μm I.D. capillaries.[25] Injection volumes as low as 270 fl were achieved in 5-μm I.D. capillaries with typical injection volumes in the 50- to 60-pl range. With a 2-μm I.D. capillary, injection volumes as low as 4 pl were achieved. In one experiment, 2 amol of serotonin and 1 amol of catechol were injected onto a 5-μm I.D. column. The detection limits obtained in these experiments were 0.5 and 0.6 amol, respectively, and represent the best detection limits obtained for this system. The use of narrow bore capillaries in CE has been an extremely important development. In addition to providing high-efficiency separations, small capillaries can accommodate very small sample volumes and lead to increased mass sensitivity. These are key advantages in the development of CE for single-cell analysis.

B. SAMPLING TECHNIQUES

To preserve the high separation efficiencies available with narrow bore CE, it is necessary to introduce samples onto the column with minimum volume and without significant zone broadening. Capillary electrophoresis systems are easily overloaded by large sample volumes. Therefore, an important requirement of any injection technique in CE is the ability to introduce small volumes of sample (<100 nl) onto the column efficiently and reproducibly. It is also desirable to minimize sample handling and dilution. Development of CE in narrow bore capillaries has resulted in improved sample-handling techniques and in the development of microinjectors. This section reviews general injection techniques and describes specific techniques for acquiring samples from single cells.

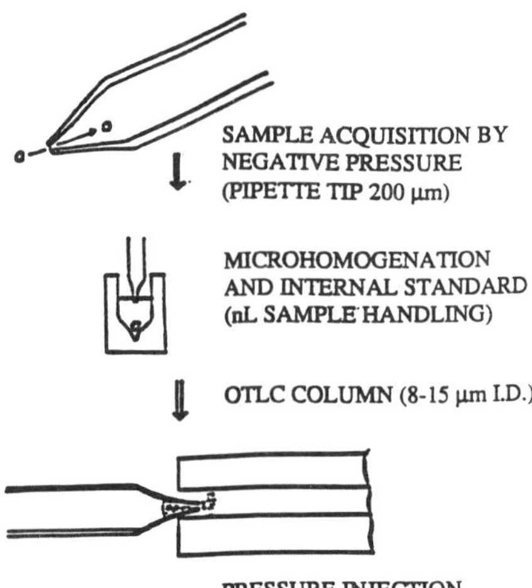

FIGURE 1. Procedure for whole-cell analysis by open tubular liquid chromatography. (From Ewing, A. G., Strein, T. G., and Lau, Y. Y., *Acc. Chem. Res.*, 25, 440, 1992. With permission.)

1. *In Vitro* Techniques

Traditional *in vitro* injection techniques include injection by electromigration[32,33] and hydrodynamic injection.[34-36] Electromigration is a popular method of injection in CE, particularly for samples that have been derivatized or purified from biological samples. Injection by electromigration involves placement of the injection end of the capillary into the sample and application of voltage for a brief period of time (5 to 30 s). This causes electromigration of the sample into the capillary. After injection, the capillary is returned to the separation buffer and the separation voltage is applied. Hydrodynamic injection delivers small volumes of sample directly onto the column without the charge bias associated with electromigration injection. Hydrodynamic injection forces a small plug of liquid into the sampling end of the capillary by applying a pressure differential across the capillary. This pressure differential may be created by elevating the sample and capillary with respect to its opposite end, by pressure control of the chamber holding the sample, or by application of vacuum at the detector end of the capillary.

2. *In Vivo* Techniques

Electrokinetic and hydrodynamic injection techniques have both been employed in the application of CE to single-cell analysis. Sample preparation and injection methods differ considerably for whole-cell analyses and cytoplasmic analyses. Several techniques exist for sample acquisition in the analysis of whole cells while electromigration appears to be the predominant injection method for cytoplasmic analyses.

a. *Whole-Cell Sampling*

Acquisition of whole cells for analysis by CE has been accomplished in a variety of ways. Jorgenson and co-workers analyzed single neurons of the land snail, *Helix aspersa,* by CE with LIF.[19] An overview of their procedure for sample preparation and injection is shown in Figure 1. First, a cell is isolated by microdissection techniques. Then it is transferred to a 200-nl microvial with a calibrated, pneumatic micropipette (120-μm I.D.) made of borosilicate glass. This is followed by homogenation and centrifugation. After centrifugation, the supernatant is transferred to a clean microvial where it is derivatized with the fluorophore, naphthalene-2,3-dicarboxyaldehyde (NDA). Following derivatization, the NDA-tagged supernatant is injected onto the capillary column, using a microinjector.

The details of the microinjection system are shown in Figure 2. The microinjector consists of a glass micropipette, drawn down to 10-μm O.D. by a commercial pipette puller, connected to a mercury-filled, 250-μl syringe. The syringe is driven by a micrometer and allows samples between 1 and 100 nl to be drawn up or expelled by adjustment of the micrometer. Once the micropipette is loaded with sample, it is mounted into a micromanipulator and its tip is inserted into the separation capillary. The sample is then forced into the column by another adjustment of the micrometer. Once the injection is made, the capillary is connected to a source of separation buffer to begin electrophoresis. Since only a small portion of the cell is used in this analysis it should be possible to use this method of sample preparation and injection on even smaller cells or to perform multiple analyses on the same cell.

A very simple scheme for sampling whole cells has been developed using CE.[26] In this system, the separation capillary is used as a microinjector to pull the cell onto the column by electroosmosis. The outer diameter of the tip of the capillary is etched down from 150 μm to approximately 50 μm O.D. with HF and cut to a blunt tip, using a scalpel (Figure 3). This results in an injector tip small enough to be visualized with a microscope while sampling a single cell.

The procedure for direct sampling of a whole cell by electroosmosis is shown in Figure 4. With the aid of a microscope, the microinjector is placed in direct contact with a cell and an injection potential of 1 to 10 kV is applied. This results in electroosmotic transport of the cell into the capillary. Cells as large as 75 μm in diameter have been drawn into a 25-μm I.D. electrophoresis capillary. During injection, the cell may elongate but does not appear to lyse. After the cell is drawn into the capillary, a 10-s injection of a lysing agent is carried out and the voltage is turned off for 60 s. This delay allows time for the cell to lyse. Following lysing, the cell components are separated and detected in the CE system. This system allows for easy acquisition of a whole cell because the capillary itself is used as a microinjector. Because 100% of the cell compartments is sampled, better detection limits for low-level species can be obtained.

Yeung and co-workers have also utilized the capillary as an injector in their analyses of single whole erythrocytes by CE with both LIF and indirect fluorescence detection.[27,37] In their system, a 20-μl drop of cell suspension is placed on a clean microscope slide. An individual erythrocyte is chosen for injection by examination of the droplet under a microscope with a magnification of ×100 (Figure 5). With a 5-mm section of the injector end of the capillary cleared of its polyimide coating, the tip of the capillary was immersed in the droplet of cell suspension and positioned near a cell with the aid of a micromanipulator. The cell is introduced into the capillary by a pulse of vacuum from the opposite end of the capillary. After injection, the cell is allowed to settle and adhere to the capillary wall. The capillary is then manipulated out of the cell suspension and into a CE separation buffer. On initiation of electrophoresis, the cell lyses as a result of osmotic shock. Once the cell is lysed, intracellular components migrate toward the optical detector window. Although electromigration could also be employed to introduce the cell onto the column, this method suffers from the inability to rapidly adjust the magnitude of the injection velocity or to reverse the flow.

b. Cytoplasmic Sampling

The extremely low volume flow capabilities of CE also allow manipulation of subcellular samples. Injection of cytoplasm from single cells was first accomplished for samples from the giant dopamine neuron of the pond snail *P. corneus* with a system using a small glass pipette placed over the injection end of the capillary.[31] A microinjector was constructed by pulling down a standard glass capillary to a tip of approximately 7.5 μm with a microelectrode puller. The microinjector was then filled with buffer and placed on the sampling end of the separation capillary. The tip of the microinjector was placed into the cell, using a micromanipulator. A

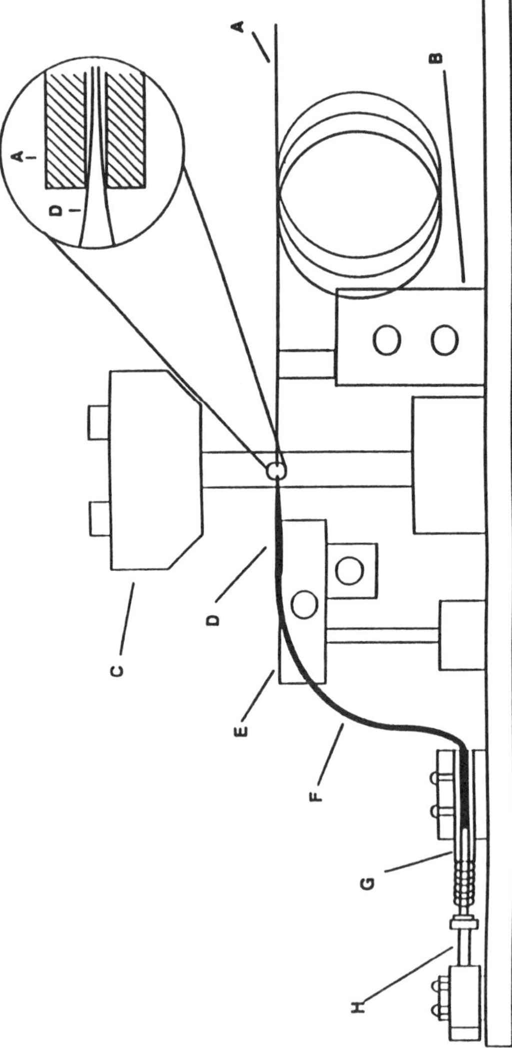

FIGURE 2. Schematic diagram of microinjector. A, Column; B, column-positioning stage; C, microscope; D, micropipette; E, micromanipulator; F, connector tubing; G, syringe; H, micrometer. The inset shows the pipette inserted into the column for injection. (From Kennedy, R. T., St. Claire, R. L., White, J. G., and Jorgenson, J. W., *Mikrochim. Acta*, 2, 37, 1987. With permission.)

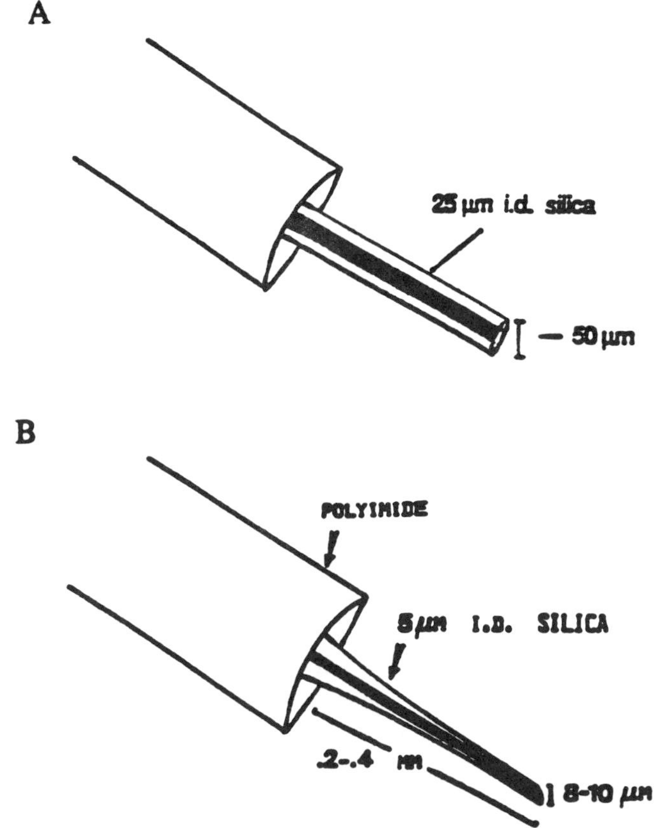

FIGURE 3. Capillary electrophoresis microinjectors. (A) Injector for acquiring and injecting whole cells, constructed at the high-voltage end of a 25-µm I.D. capillary. (B) Injector for acquiring and injecting cytoplasmic samples at the high-voltage end of a 5-µm I.D. capillary. (From Olefirowicz, T. M. and Ewing, A. G., *Chimia*, 45, 106, 1991. With permission.)

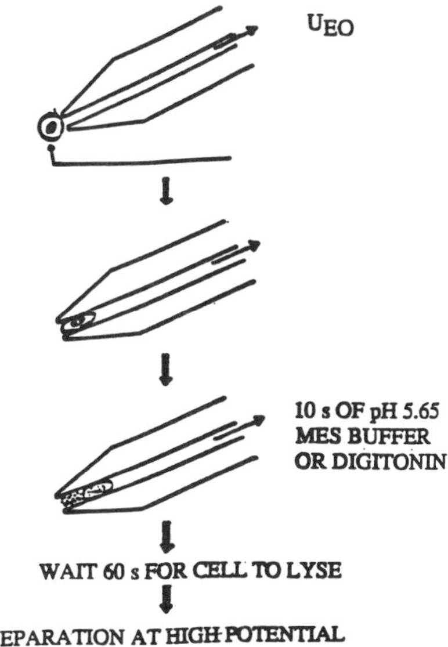

FIGURE 4. Procedure for whole-cell injection and separation, utilizing the separation capillary as a microinjector. (From Ewing, A. G., Strein, T. G., and Lau, Y. Y., *Acc. Chem. Res.*, 25, 440, 1992. With premission.)

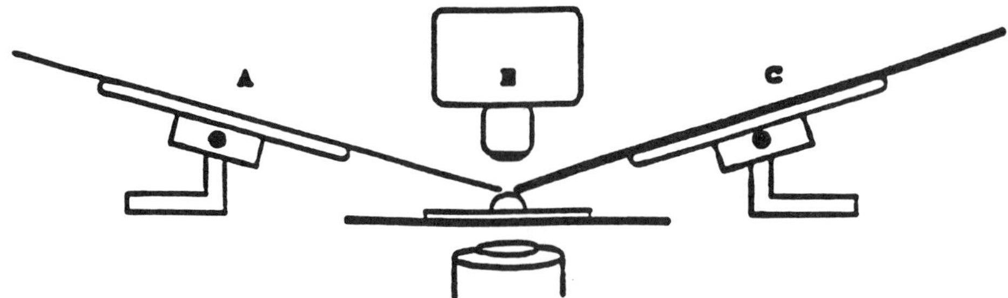

FIGURE 5. Close-up of cell injection region of the apparatus used for acquiring and injecting red blood cells. (A) Ground electrode; (B) ×100 microscope; (C) separation capillary. A cell is selected for injection by manually positioning the orifice of the capillary close to the cell of interest and then applying a pulse of vacuum to the capillary. (From Hogan, B. L. and Yeung, E. S., *Anal. Chem.*, 62, 2841, 1992. With permission.)

Pt wire in contact with the Ringer solution covering the snail preparation served as the electrophoresis injection anode. Injection was carried out by electromigration. After injection, the tip of the capillary was removed from the microinjector and placed in the buffer reservoir separation by electrophoresis.

To avoid problems dealing with a separate pipette as a microinjector, a method has been devised to directly sample single-cell cytoplasm, using the separation capillary as a microinjector.[25,26,28] These microinjectors were constructed using the same HF etching procedure described earlier for whole-cell sampling, except that a 5-μm I.D. capillary is used (Figure 3). Tip diameters as small as 6 μm have been obtained by etching the capillary tip in HF.

Use of HF-etched capillaries for direct sample injection minimizes zone-broadening effects due to large dead volumes and it eliminates laminar flow associated with microsyringe injection. Furthermore, it minimizes sample preparation. Injection volumes as small as 270 fl have been injected with this method. The use of microinjectors fashioned from 2-μm I.D. capillaries should permit even smaller injection volumes to be attained.

C. DETECTION METHODS

One of the key factors limiting any CE experiment is detection. This is especially true when extremely small cellular samples are examined. Capillaries 1 m in length with inner diameters ranging from 100 μm down to 2 μm have total volumes from 80 μl to 3 nl, respectively. Green and Jorgenson have reported that for a 1-m long, 75-μm I.D. capillary, a detection volume of 1 nl is required.[38] Even smaller detection volumes are required for smaller capillaries. Therefore, capillary size, sample injection volumes, and detector volumes place great demand on detection schemes.

Several detection methods have been developed for CE. These include absorption, fluorescence, LIF, mass spectrometric, electrochemical, refractive index and other optical spectroscopic detection schemes (for an overview, see Chapters 7, 8 and refs. 39, 40). At present, only amperometric and LIF detection methods are sensitive enough to permit detection of solutes following CE analysis of single cells.

1. Amperometric Detection

Although electrochemical detectors offer both sensitivity and selectivity for detection in CE, they are not routinely used. This is primarily due to the difficulty of performing electrochemistry in the presence of a high-voltage electric field. Wallingford and Ewing overcame this difficulty by making a crack near the detector end of the capillary (Figure 6).[41] The crack divides the capillary into two segments, a separation capillary and a detection capillary. The crack is covered by small piece of porous glass such that all of the applied separation potential

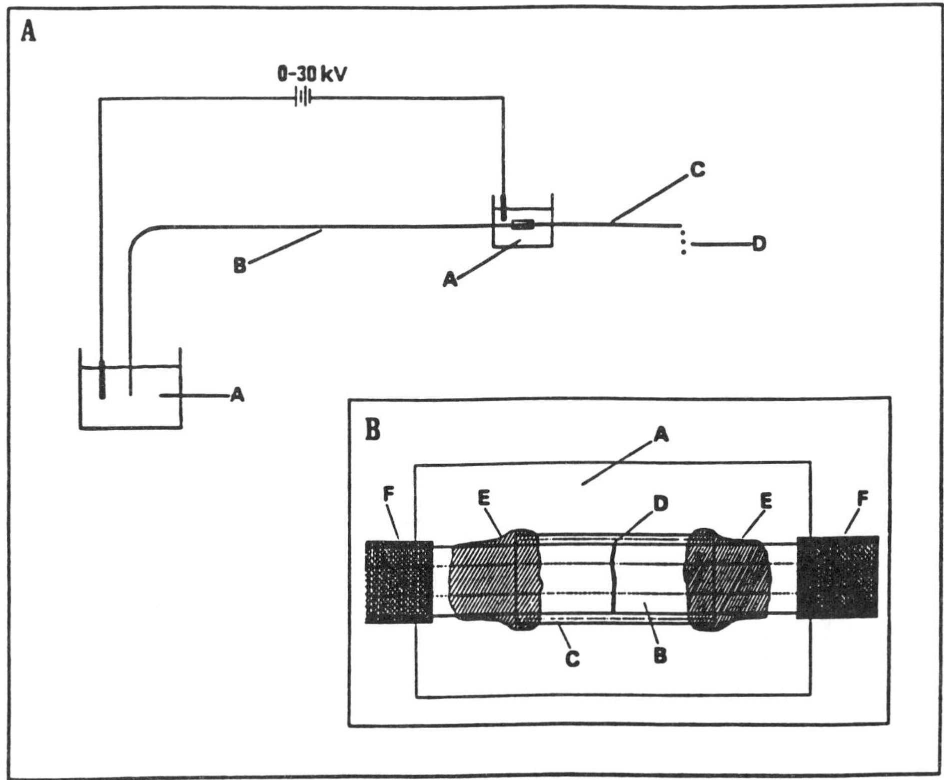

FIGURE 6. (A) Schematic of coupled CE system: A, buffer reservoir; B, separation capillary; C, detection capillary; D, eluent. (B) Detailed schematic of porous glass joint: A, microscope slide; B, fused silica capillary; C, porous glass capillary; D, joint; E, epoxy; F, polymer coating. (From Wallingford, R. A. and Ewing A. G., *Anal. Chem.*, 59 1762, 1987. With permission.)

is dropped across the first segment of capillary. The joint in the capillary is placed at ground so that the second section of capillary is isolated from the high potential field generated in the first segment of capillary. Electroosmotic flow generated in the separation capillary acts as a pump to transport analytes past the crack and into the detection capillary.

Detection is accomplished in this system with a cylindrically shaped carbon fiber electrode (5- to 10-μm diameter) inserted into the end of the detection capillary using a micromanipulator and a microscope (Figure 7). A mercury battery and a variable resistor are connected to the sodium-saturated calomel reference electrode to control the applied potential. Amperometric detection has been used in capillaries with inner diameters ranging from 75 to 2 μm with little loss in separation efficiency.

To perform amperometric detection in smaller capillaries (2- to 5-μm I.D.), electrochemically and flame-etched carbon fiber electrodes have been used.[25] Cylindrically shaped electrodes can be etched from 5 to 2 μm in diameter by cycling the electric potential between ±1.75 V in 3 *M* KOH solution. Separation and amperometric detection of several catecholamines has been achieved in 5-μm I.D. capillaries using electrochemically etched electrodes (Figure 8). Typical detection limits are 3 amol for catechol ($S/N = 2$, peak-to-peak noise). The best detection limits obtained with this system have been on the order of 0.5 amol. Electrodes with diameters ≤2 μm can be made by placing a 5-μm diameter carbon fiber electrode in the flame of a Bunsen burner for a few seconds. This technology has been important for the development of CE in 2-μm I.D. capillaries. Separation of several amines in a 2-μm I.D. capillary with a flame-etched electrode is shown in Figure 9.

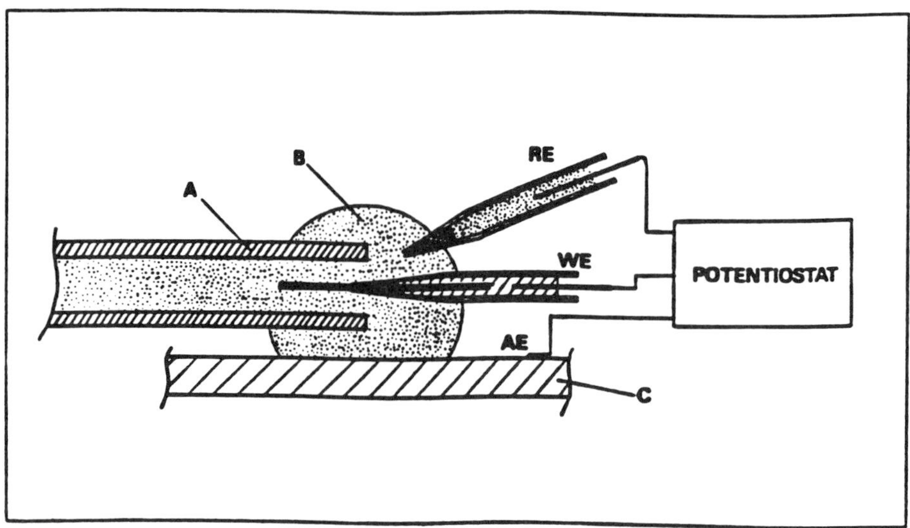

FIGURE 7. Schematic of carbon fiber electrochemical detector. (A) Fused silica capillary; (B) eluent drop; (C) stainless steel plate; RE, reference electrode; WE, working electrode; AE, auxiliary electrode. (From Wallingford R. A. and Ewing, A. G., *Anal. Chem.*, 59, 1762, 1987. With permission.)

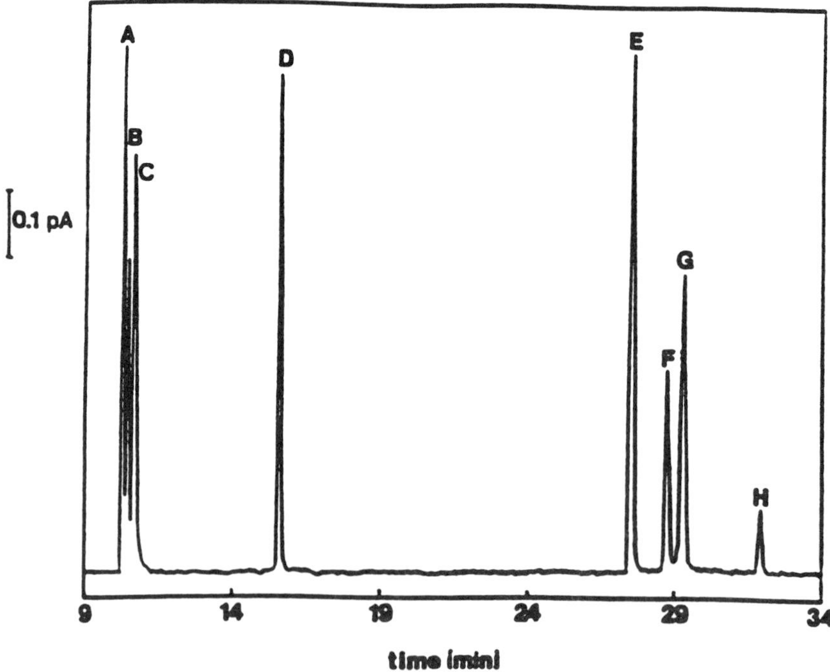

FIGURE 8. Electropherogram obtained in a 5-µm I.D. capillary, using an electrochemically etched carbon fiber microelectrode. Capillary length, 85 cm; buffer, 25 m*M* MES (pH 5.65) modified with 10% (v/v) 2-propanol; injection, 5 s at 25 kV. Peaks: (A) serotonin; (B) norepinephrine; (C) epinephrine; (D) 1-dihydroxyphenylalanine; (E) 5-hydroxyindoleacetic acid; (F) homovanillic acid; (G) dihydroxyphenylacetic acid; (H) ascorbic acid. (From Olefirowicz, T. M. and Ewing, A. G., *Anal. Chem.*, 62, 1872, 1990. With permission.)

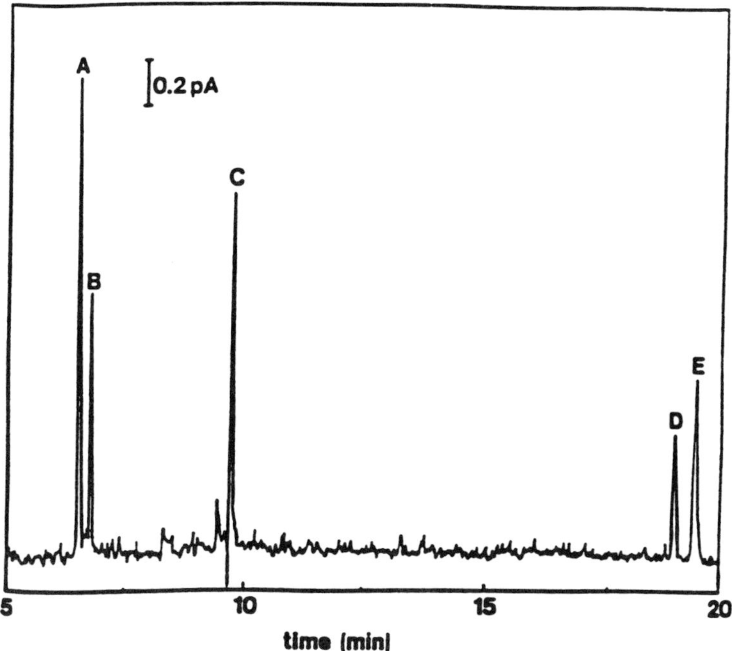

FIGURE 9. Capillary electrophoretic separation of several amines in a 2-μm I.D. capillary. Capillary length, 81 cm; buffer, 25 m*M* MES (pH 5.65); injection, 5 s at 25 kV. Peaks: (A) dopamine; (B) 3-methoxytyramine; (C) 1-dihydroxyphenylalanine; (D) homovanillic acid; (E) dihydroxyphenylacetic acid. (From Olefirowicz, T. M. and Ewing, A. G., *Anal. Chem.*, 62, 1872, 1990. With permission.)

Several other designs exist for carrying out amperometric detection in CE that are similar to the off-column detection scheme developed by Wallingford and Ewing. These designs all employ an on-column joint in the capillary to isolate the working electrode. Differences arise in the material used to make the joint. Yik et al.[42] have employed a porous graphite tube while O'Shea and co-workers[43] have used Nafion tubing to cover a scratch in the capillary and isolate the electrical circuit from the detection end of the capillary. To date, the only off-column amperometric detection system used in the analysis of single cells is that using porous glass tubing.

More recent cell experiments using CE with amperometric detection have involved the use of an end-column amperometric detection mode.[44,45] This method of amperometric detection eliminates the need for the porous glass coupler. In end-column detection, the working electrode is placed in the buffer solution outside of the capillary but positioned up against the bore of the capillary (Figure 10). Amperometric detection is performed without further isolation of the working electrode because virtually all of the high-voltage separation potential is dropped across a high-resistance capillary (10^{12} to 10^{13} Ω). A detection limit of 56 amol has been obtained for catechol in a 5-μm I.D. capillary with a 10-μm diameter carbon fiber microelectrode.

These detection limits have been improved by the development of a conical end-column detector.[46] This new design involves etching the inside wall of the detector end of the separation capillary with HF to form a cone at the tip. Capillaries are etched using a procedure similar to that described for making etched microinjectors; however, the use of lower gas flow

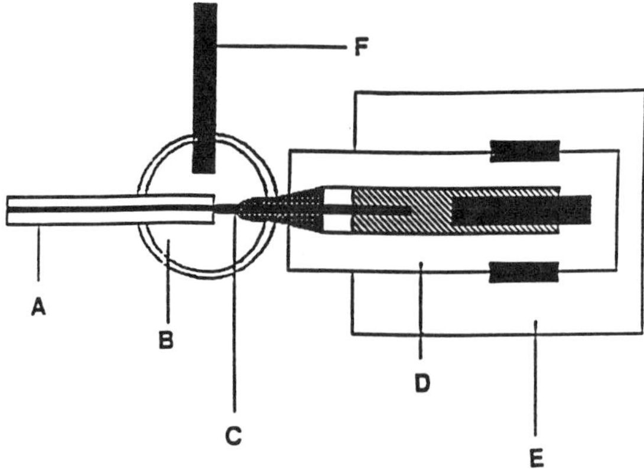

FIGURE 10. Schematic of end-column amperometric detector. (A) Separation capillary; (B) buffer reservoir; (C) carbon fiber microelectrode; (D) microscope slide; (E) micromanipulator; (F) reference electrode.

rates through the capillary results in the formation of a cone-shaped bore at the end of the capillary. Detection is carried out by manipulating the working electrode into the etched portion of the detector end of the capillary. This etched sheath provides a way to achieve and maintain alignment of the electrode with the bore of the capillary. Detection limits as low as 19 amol have been obtained for catechol in 2-μm I.D. capillaries with 10-μm diameter electrodes. Therefore, this improved end-column detection method combines the ease of end-column detection with nearly the same detection limits as off-column detection. More importantly, it allows for the routine use of 2-μm I.D. capillaries, which could be very important for the application of CE to single-cell analysis.

2. Fluorescence Detection

Fluorescence detection is easily adapted to CE since its sensitivity is not path length dependent and the capillary itself may be used for an on-column detection cell. When coupled with derivatization, it provides for analyte specificity. Fluorescence detection also provides extremely low detection limits. Detection limits range from 10^{-15} to 10^{-21} mol depending on the excitation source, the derivatization procedure, and the fluorescent tag.[47-50]

Laser light is monochromatic and well collimated, making it easy to focus on very small capillaries. Focusing of the laser light source is accomplished with imaging optics or fiber optics. Stray light is rejected by collecting the emitted light at right angles to the excitation source and through the use of spectral and spatial filtering. On-column detection is easily accomplished by imaging the laser source onto the column and collecting the emission at an angle perpendicular to the incident light. A typical experimental apparatus is shown in Figure 11. Helium-cadmium lasers seem to be the most popular since they are inexpensive and emit in the UV region (325 nm). Argon ion lasers have also been used extensively. They often have much higher power (1 to 20 W) at longer wavelengths (488 nm). Although still at the development stage, means to detect cellular samples have been explored with both of these lasers.

Lee and Yeung accomplished detection of proteins in single erythrocytes following separation by CE by detecting the native fluorescence of the proteins with a modified argon ion laser.[37] Laser-based detection of proteins can be accomplished by exciting the aromatic amino acid residues of the protein. The amino acids tryptophan, tyrosine, and phenylalanine have excitation maxima between 260 and 280 nm and fluoresce weakly between 280 and 350 nm. The emission spectrum of the protein is a combination of the fluorescence of three fluorophores

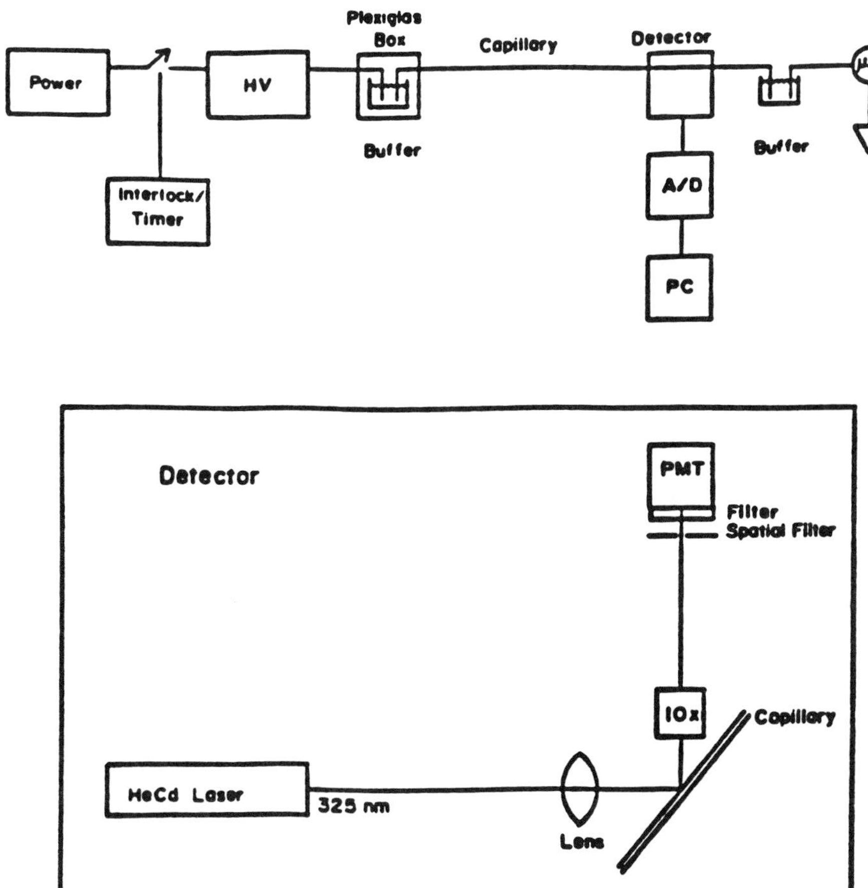

FIGURE 11. Instrumental arrangement for CE with LIF detection. (From Kuhr, W. G. and Yeung, E. S., *Anal. Chem.*, 60, 1832, 1988. With permission.)

and sensitivity is dependent on the number of aromatic amino acids in the protein. This is an extremely powerful technique that holds great promise for cellular analysis by CE.

Swaile and Sepaniak first demonstrated laser-induced fluorescence detection of native proteins in CE.[52] They used a frequency-doubled 514-nm output of an argon ion laser to produce an excitation wavelength of 257 nm. This was sufficient to excite the aromatic amino acids present in the protein. The minimum detectable concentration obtained for conalbumin (61 amino acids) was 14 n*M*. Lee and Yeung improved on this sensitivity by two orders of magnitude by using the 275.4-nm line of a modified argon ion laser. With this system, they achieved a limit of detection of 10^{-10} *M* for conalbumin. *In vitro* experiments with hemoglobin and carbonic anhydrase yielded limits of detection of 0.2 and 8 amol, respectively.

Although the native fluorescence technique is very sensitive, many species of interest do not fluoresce with sufficient quantum yield at easily attained wavelengths. Thus, detection typically involves the use of fluorescent tags covalently attached to the molecules of interest through a derivatization reaction. The samples most frequently analyzed in this manner have been amino acids and peptides, although phenols, amines, carboxylic acids, aldehydes, and ketones can also be converted to fluorescent derivatives.[53,54] Typical derivatizing reagents include 5-dimethylaminonaphthalene-1-sulfonyl (DANSYL) chloride, fluorescein isothiocyanate (FITC), fluorescamine, *o*-phthalaldehyde (OPA), and napthalene-2,3-dicarboxaldehyde (NDA).[39]

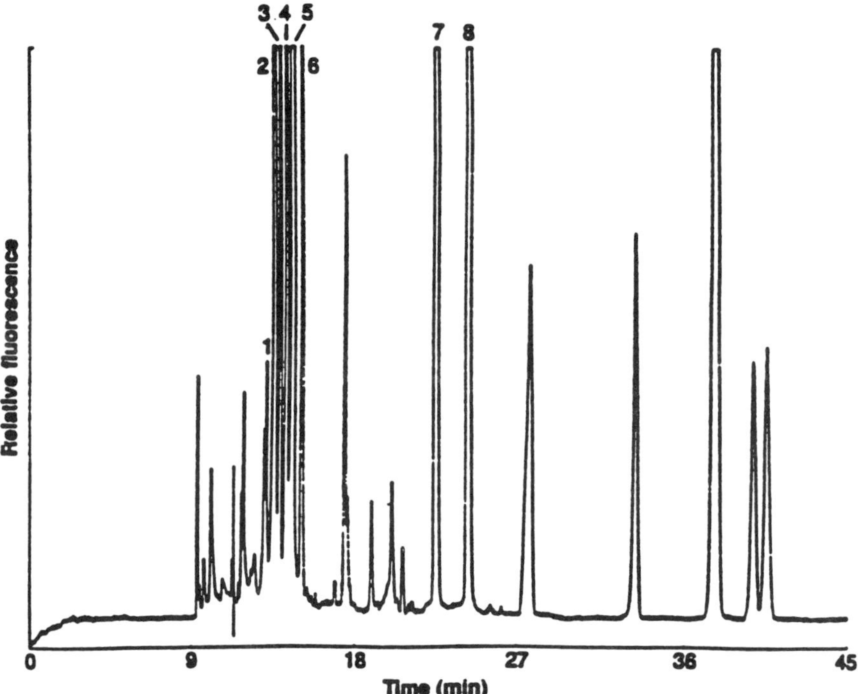

FIGURE 12. CE-LIF separation of NDA derivatives of an E4 cell. The numbered peaks correspond to NDA-labeled amino acids as follows: 1, Trp; 2, Gln, His, Ile, Leu, Met, Phe; 3, Asn, Thr, Tyr, Val; 4, Ser; 5, Ala; 6, Gly; 7, Glu; and 8, Asp. Conditions are as follows: 25-μm I.D. capillary; length, 104.5 cm; buffer, 0.01 *M* borate, 0.04 *M* KCl, pH 9.5; applied potential, −25 kV. (From Kennedy, R. T., Oates, M. D., Cooper, B. R., Nickerson, B., and Jorgenson, J. W., *Science*, 246, 57, 1989. With permission. Copyright 1989 by the AAAS.)

There are basically two ways to derivatize samples for detection with CE: precolumn and postcolumn. Kennedy and co-workers utilized a precolumn derivatization scheme to analyze the contents of single *H. aspersa* neurons by CE with LIF detection.[19] An example of an electropherogram for NDA-derivatized contents of a cell is shown in Figure 12. Peaks were tentatively identified on the basis of migration times of standard NDA-labeled amino acids. No attempts were made to quantitate these data; however, most of the NDA-labeled peaks have been allowed to go off scale to demonstrate that there is ample signal-to-noise ratio for lower level species when precolumn derivatization is used.

Precolumn derivatization possesses several limitations. First, sufficient volumes (>25 nl) must be available for derivatization. Second, multiple derivatizations for the same molecule can occur, yielding several peaks for one analyte. Third, migration times are for the derivatized product and not the analyte. As a result, identification and quantitation can be difficult. Finally, degradation products may be detected if the derivatized samples are not stable throughout the preparation and analysis time.

Postcolumn derivatization avoids many of the problems associated with precolumn derivatization. Its predominant advantage is minimal sample preparation. In addition, because the sample is unaltered during the separation, migration times are those of the native analyte. This can lead to improved identification and quantitation in postcolumn derivatization. Furthermore, detection of degradation products is less likely to occur because the time between the derivatization reaction and detection is minimized. This can be a drawback; however, since the signal is a function of the postcolumn reaction rate, the reactor must be optimized for maximum signal and carefully calibrated for quantitative work.

Other limitations associated with postcolumn derivatization include band broadening in the postcolumn reactor and the necessity for exact timing of detection of derivatized analytes. In addition, optimization of instrument technology can be quite difficult. The postcolumn detection schemes reported to date have been quite elaborate in their design and construction. Although several postcolumn detectors have been reported for CE,[55-58] none have been employed in single-cell analysis. They are mentioned here since their use in single-cell analysis is inevitable.

Indirect fluorescence is an attractive alternative to derivatization of nonfluorescent compounds in CE.[59] Indirect fluorescence detection records the passage of a nonfluorescent analyte through the detection zone by charge displacement effects of the analyte on a fluorophore comigrating through the buffer solution. It is a detection scheme for species that have no inherently useful detection properties and it provides a general detection method for charged species.

III. APPLICATION TO SINGLE CELLS

Capillary electrophoresis at the cellular level should provide highly useful methodology to several fields of cellular chemistry and biology. Two critical limitations to routine analyses of this type have been the development of sampling methods for microenvironments and the development of sensitive detection methods for molecules of biological interest. Advances in these areas have already been discussed. These advances have led to many interesting experiments that have increased our knowledge of chemistry at the single-cell level.

The potential advantage of using CE for analysis of single cells was first demonstrated for cytoplasmic samples removed from the giant dopamine neuron of the pond snail *P. corneus*.[24] This nerve cell contains the easily oxidized neurotransmitter, dopamine, which can be detected amperometrically. Dopamine stores in this cell are believed to be at least partially bound inside vesicles. Independent voltammetric studies using microvoltammetric electrodes have shown that an easily oxidized substance increases in concentration following exposure to an ethanol solution. The identification of this substance as dopamine is difficult by voltammetry, so a sample of cytoplasm was acquired via a microinjector fashioned from a pulled glass capillary and separated by CE.

In these experiments, the cell was exposed to ethanol prior to separation in order to lyse the dopamine-containing vesicles. Then 100 to 300 pl of cytoplasm was drawn into the microinjector by electromigration. The resulting electropherogram is compared with one for a standard sample of dopamine in Figure 13. The migration times are nearly identical and provide further evidence that the substance increasing in concentration is dopamine. The peak area in this experiment corresponds to approximately 14 fmol of dopamine. Estimates of cytoplasmic concentration of dopamine following exposure to ethanol are 4.6×10^{-5} to 1.4×10^{-4} *M*.[10] These correspond closely to the 97 μM change in concentration observed by intracellular voltammetry.

Olefirowicz and Ewing determined the cytoplasmic concentration of dopamine in a single *P. corneus* neuron.[28] Sampling was accomplished by using the capillary as a microinjector. The microinjector was constructed by etching the capillary with HF as previously described. After separation, easily oxidized neurotransmitters were detected with off-column amperometric detection. Figure 14 compares an electrophoretic separation of a cytoplasmic sample from the dopamine neuron to a standard electropherogram of dopamine, dihydroxyphenylalanine, and dihydroxyphenylacetic acid. The peaks were identified by calculating the electrophoretic mobility of each peak and comparing them to those of standards. The small increase in migration time observed for the cytoplasmic sample is thought to be due to an increase in viscosity with the cellular experiments. The injected volume for this experiment was 66 pl. These results indicate that the cytoplasmic levels of dopamine for this cell are 2.2 ± 0.52 μM.

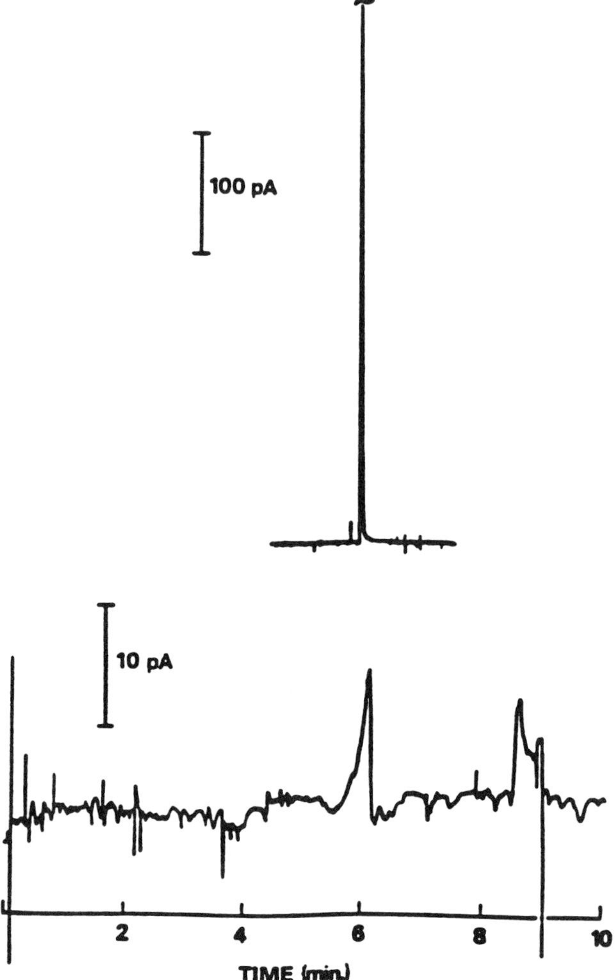

FIGURE 13. Electropherogram of a sample taken from the giant dopamine neuron of an unanesthetized *Planorbis corneus* after treatment with 100 μl of 50% ethanol–Ringer's solution. Conditions: capillary, 14-μm I.D. × 80 cm; detection, amperometric; buffer, 0.025 *M* MES (pH 5.65); injection, 60 s at 15 kV through a 9-μm O.D. microinjector; separation potential, 25 kV. *Top*: Electropherogram of 5×10^{-5} *M* dopamine standard solution assayed under similar conditions except that the injection was for 1 s at 25 kV. (From Chien, J. B., Wallingford, R. A., and Ewing A. G., *J. Neurochem.*, 54, 633, 1990. With permission.)

The same system used to examine the dopamine cell was also used to separate cytoplasmic samples from the large serotonin cell of *P. corneus.*[25] Figure 15 compares an electropherogram of cytoplasm to an electropherogram of serotonin and catechol in a standard solution. In these experiments, apparent injection volumes ranged from 73 to 134 pl. On the basis of these apparent volumes, cytoplasmic levels of serotonin were determined to be 3.1 ± 0.57 *M*.

FIGURE 14. (A) Capillary electrophoretic separation of solutes in a sample of cytoplasm removed from inside the giant dopamine neuron of *Planorbis corneus*. Conditions: capillary length, 66 cm; buffer, 0.025 *M* MES (pH 5.65); injection, 5 s at 10 kV; separation potential, 25 kV; detection, electrochemical, 0.7 V vs. SSCE. Peaks A, B, and C have been identified as dopamine, a nonionic species, and dihydroxyphenylacetic acid, respectively. (B) Capillary electrophoretic separation of a standard solution obtained after the cell experiment. Peaks A, B, and C are dopamine, dihydroxyphenylalanine, and dihydroxyphenylacetic acid, respectively. (From Olefirowicz, T. M. and Ewing, A. G., *J. Neurosci. Methods*, 34, 11, 1990. With permission.)

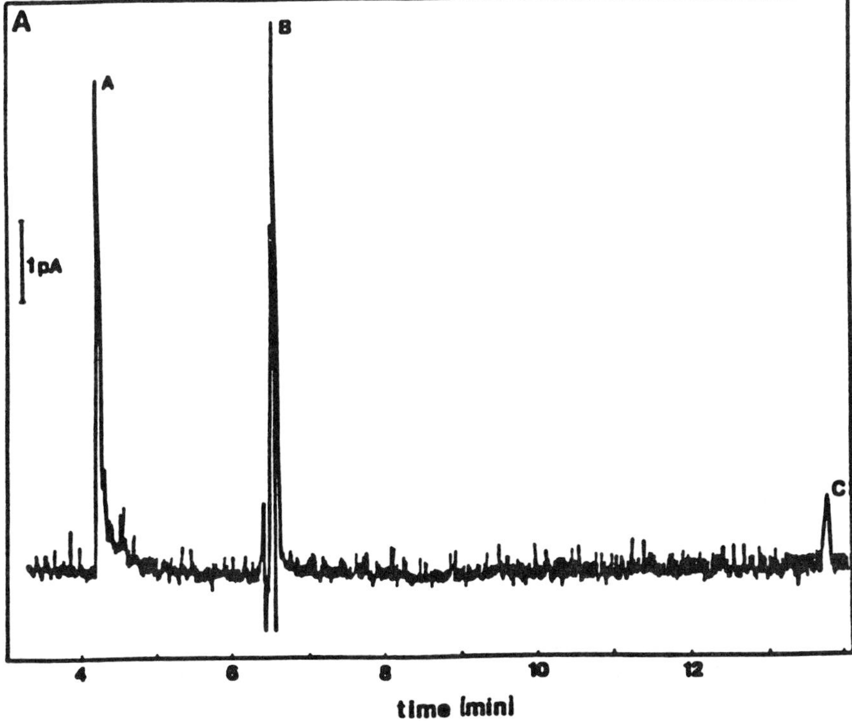

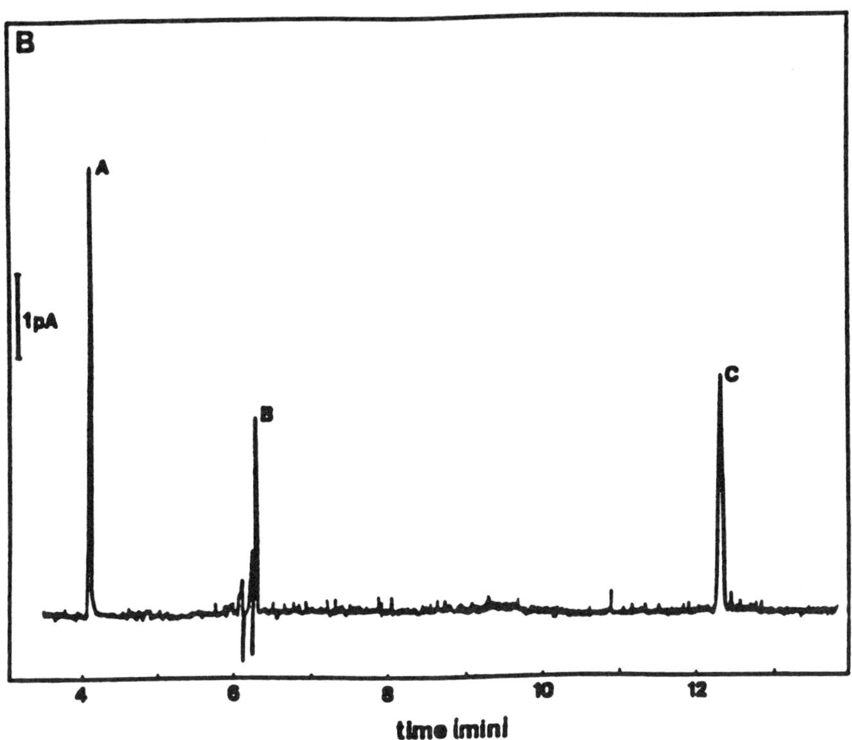

FIGURE 14.

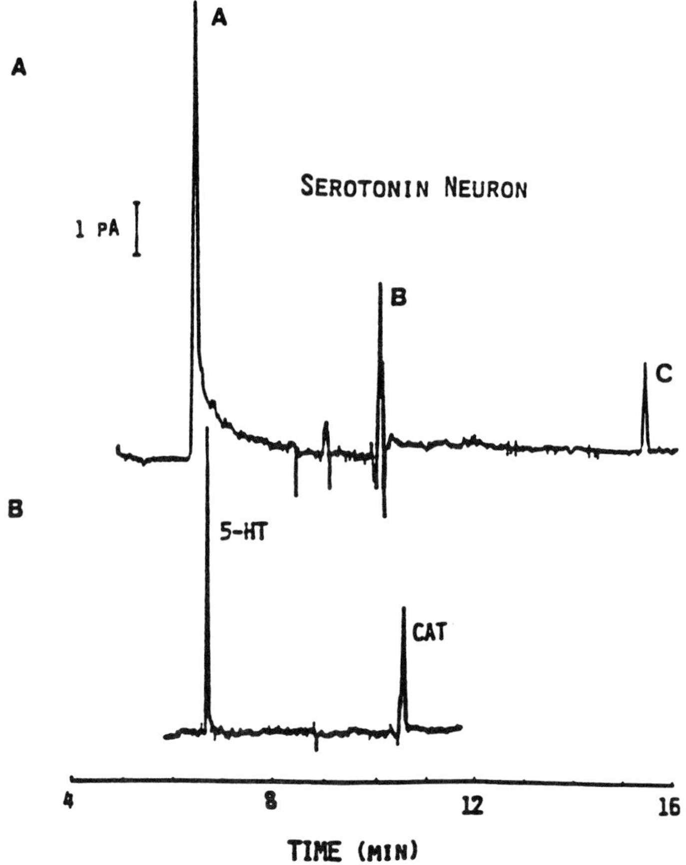

FIGURE 15. Comparison of electropherograms of a cytoplasmic sample obtained from the large serotonin neuron of *Planorbis corneus* (A) to an electropherogram of a standard solution of serotonin (5-HT) and catechol (CAT) (B). Conditions: capillary length, 77 cm; capillary, 5-μm I.D.; buffer, 25 m*M* MES (pH 5.65); injection, 5 s at 10 kV; separation potential, 25 kV. The electrophoretic mobilities of peaks A and C correspond to the calculated electrophoretic mobilities of serotonin and an unidentified anion, respectively. (From Olefirowicz, T. M. and Ewing A. G., *Anal. Chem.*, 62, 1872, 1990. With permission.)

In addition to cytoplasmic analyses, direct sampling of whole cells has led to interesting new data.[26] The analysis of whole cells provides complementary information to the cytoplasmic analyses. The sampling procedure for whole-cell analysis was described in Section II and the system for separation and detection is identical to that for cytoplasmic analysis except that a larger capillary is generally used. In the experiment presented in Figure 16A, a cell with a diameter of approximately 75 μm was analyzed. Three major peaks were observed. The peak labeled A has an eletrophoretic mobility that corresponds to that of dopamine. Peaks B and C have not been identified. However, peak B appears to represent the migration of a cation, whereas peak C appears to be a neutral species.

Kristensen et al. continued the investigation of this system with interesting results.[60] In Figure 17, an electropherogram of a single cell that has been injected into the capillary and then lysed for 60 s (Figure 17B) is compared to a standard injection of dopamine, catechol, uric acid, and dihydroxyphenylacetic acid (Figure 17A). The electropherogram has several peaks that can be tentatively identified by electrophoretic mobility. Peak 1 has a mobility identical to that of dopamine. Peak 3 is a nonionic solute or solutes. Peaks 4 and 5 have mobilities that match uric acid and dihydroxyphenylacetic acid, respectively.

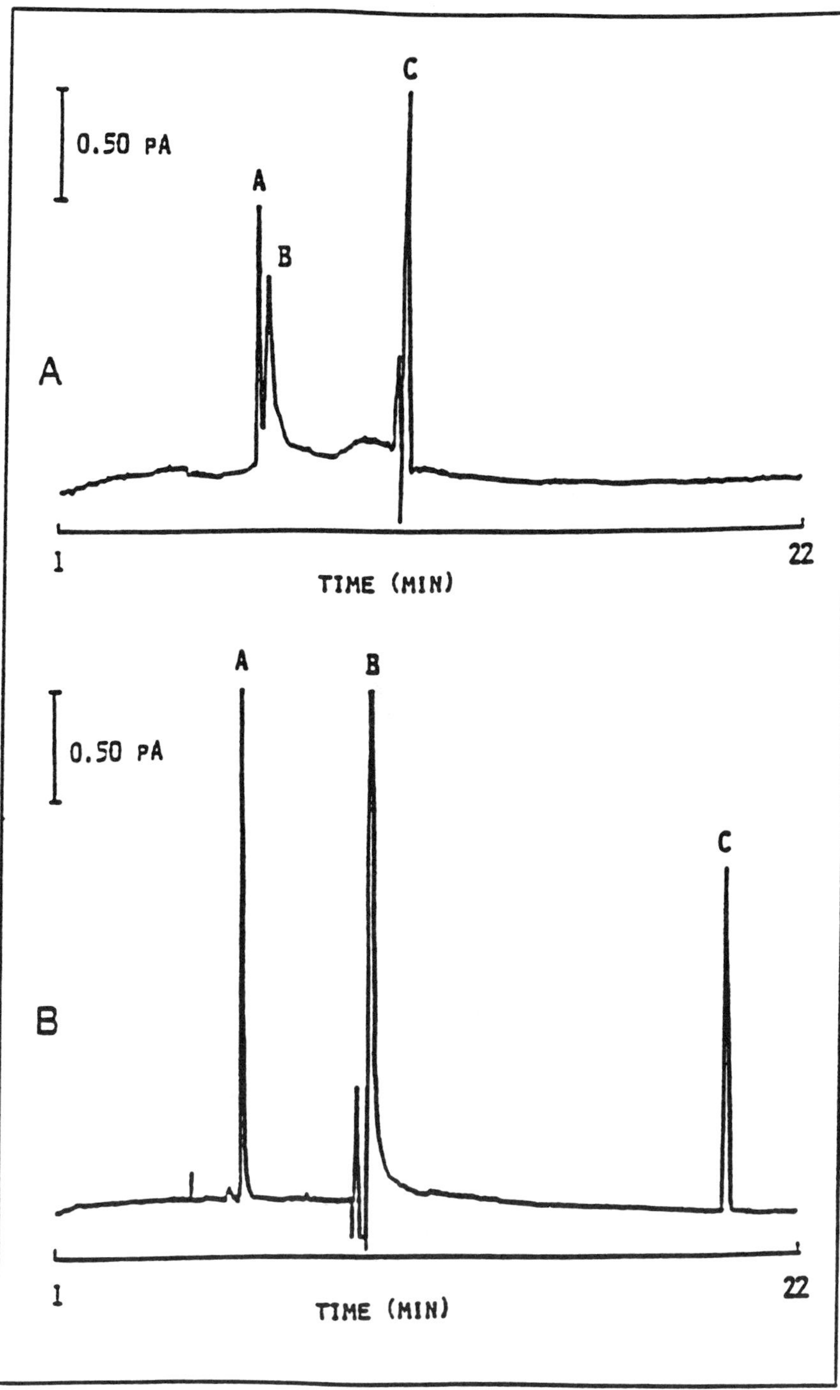

FIGURE 16. Electropherograms of a whole dopamine neuron (A) and a calibration standard (B). (A) Peak A corresponds to dopamine, peak B is unidentified, and peak C corresponds to electroosmotic flow. (B) Calibration standards: peak A is dopamine, peak B is catechol, and peak C is dihyroxyphenylacetic acid. (From Olefirowicz, T. M. and Ewing, A. G., *Chimia*, 45, 106, 1991. With permission.)

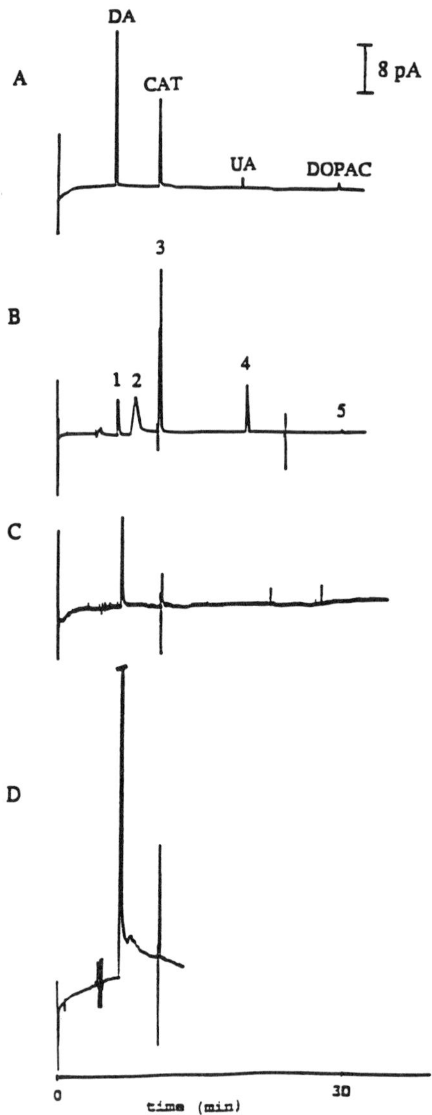

FIGURE 17. (A) Capillary electrophoretic separation of a standard solution containing 10^{-6} *M* dopamine (DA), catechol (CAT), uric acid (UA), and dihydroxyphenylacetic acid (DOPAC); injection volume was 3 nl, based on electroosmotic flow; separation capillary, 75 cm long, 25-μm I.D.; buffer, 25 m*M* MES at pH 5.65; separation potential, 25 kV; detection, carbon fiber electrode at 0.8 V vs. SSCE. (B) Separation of components from an injected dopamine cell after lysing in the capillary tip for 60 s with buffer. (C) Same as (B), except that the lysing time was 5 min. (D) Same as (B), except that the cell was bathed with 10 μ*M* reserpine for 60 min prior to cell injection. (From Kristensen, H., Lau, Y. Y., and Ewing, A. G., *J. Neurosci. Methods*, 1993, in press. With permission.)

Peak 2 is considerably broader than the others and its size can be altered by varying the lysing time of the cell. Increasing the lysing time of the cell from 60 s to 5 min completely removes peak 2 from the electropherogram (Figure 17C). These data appear to indicate a physiological basis for peak 2 that might be explained by two vesicular compartments for the neurotransmitter dopamine.[61,62] In this hypothesis, neurotransmitters in functional compartments are readily available for stimulated release, whereas those in nonfunctional compartments are thought to act as long-term stores. The variation of lyse time in the CE experiment

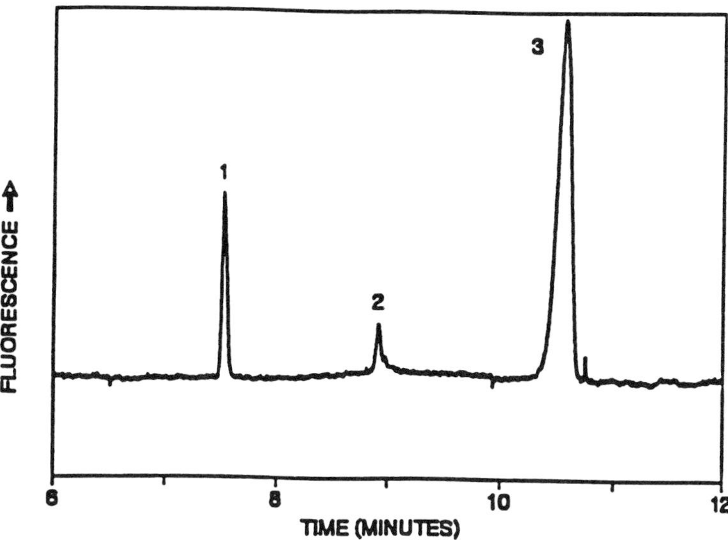

FIGURE 18. Electropherogram of mBBr-derivatized contents of a single human erythrocyte, using direct fluorescence detection. Peak 1 is unreacted mBBr, peak 2 is unidentified intracellular thiol, and peak 3 is glutathione. (From Hogan, B. L. and Yeung, E. S., *Anal. Chem.*, 64, 2841, 1992. With permission.)

apparently leads to selective lysing of vesicles in the functional and nonfunctional compartments and leads to two different peaks for dopamine in a single electropherogram (peaks 1 and 2 in Figure 17B). Treatment of these cells with reserpine, an agent that prevents the vesicular storage of catecholamines, supports this hypothesis. When cells are treated with reserpine prior to sampling with CE, the amount of dopamine observed in the second peak is dramatically reduced (Figure 17D). Thus, CE can be used to directly probe a reserpine-sensitive compartment of dopamine in a single nerve cell. This compartment has been modeled as a storage area for reserve vesicles that are not immediately available for release.

Capillary electrophoresis has been applied not only to the study of single nerve cells, but also to the study of red blood cells, with interesting results. Hogan and Yeung have used a precolumn derivatization scheme to analyze the levels of glutathione (GSH) in single erythrocytes via CE with direct fluorescence detection.[27] Monobromobimane (mBBr) was used to derivatize the glutathione in a pooled population of cells prior to injection into the CE-LIF system. An example of an electropherogram of mBBr-derivatized glutathione from a single cell is shown in Figure 18. The average total GSH content of a single human erythrocyte was determined to be 68 amol ± 48 amol ($n = 6$). Glutathione has been reported to comprise over 95% of the nonprotein thiols in macroscopic cell lysates and glutathione metabolism is an important determinant in the control of cellular response to external stimuli. The high separation efficiency of CE and the specificity of the monobromobimane derivatization reaction allow positive identification of GSH. Improved detection of other thiols present in these cells should be possible with the use of optimized fluorescein-based thiol reagents and the use of specific laser lines.

Hogan and Yeung have also studied the effects of extracellular reagents on intracellular glutathione levels.[27] They used diamide and dithiothreitol as model oxidizing and reducing agents and induced a modulation of GSH between high and low levels. The ability to measure changes in intracellular GSH levels in this way implies that more complex and interesting stimuli could be studied at the single-cell level.

In addition to GSH, sodium and potassium ions have been separated and detected in single human erythrocytes via indirect fluorescence detection.[27] The fluorophore added to the run-

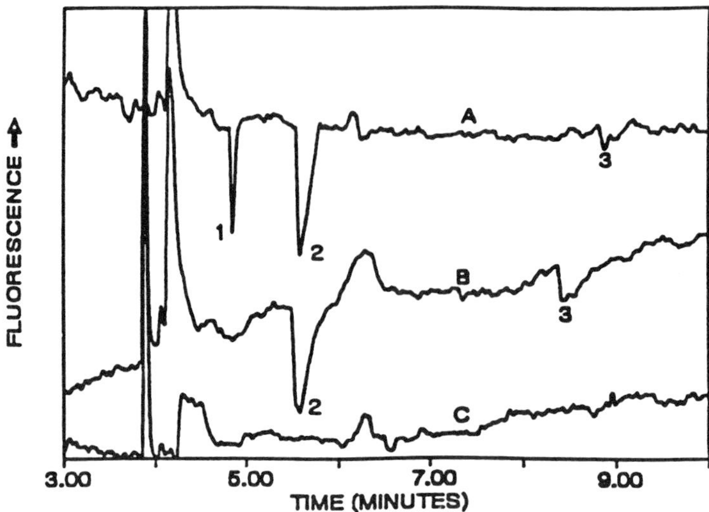

FIGURE 19. Electropherogram, using indirect fluorescence detection. (A) Injection of 45 μ*M* standards. Peak 1 is Li (11.7 fmol injected), peak 2 is Na (10.7 fmol), and peak 3 is K (6.5 fmol). (B) Injection of one human erythrocyte. (C) Blank injection of extracellular matrix. (From Hogan, B. L. and Yeung, E. S., *Anal. Chem.*, 64, 2841, 1992. With permission.)

ning buffer was 6-aminoquinoline. Figure 19 shows an electropherogram obtained with this system. Quantitation of Na and K was not reported for this system.

Single-cell monitoring of this type not only provides information about average cell content but also information about cell-to-cell variation within a group of cells. When multiple injections of single red blood cells are carried out to assess the variability of contents within a population of cells large variations in the cells are found that cannot be explained by variations in sampling or measuring. The distribution appears to be non-Gaussian. It appears that these variations are due to differences in specific cellular properties such as membrane permeability and intracellular thiol content. These studies underscore the potential value of CE for single-cell analysis.

Single red blood cells have also been examined using CE with detection by native fluorescence. Of the more than 100 proteins in the average human red blood cell, hemoglobin A_0 (450 amol/cell), carbonic anhydrase (7 amol/cell), and methemoglobin (a form of hemoglobin in which the Fe^{2+} has been oxidized to Fe^{3+}) are present in sufficient quantities to detect by native fluorescence. Figure 20 compares electropherograms of protein standards (Figure 20A) from commercial sources and a hemolysate of human erythrocytes (Figure 20B). Both hemoglobin A_0 and carbonic anhydrase migrate as sharp peaks, whereas methemoglobin appears broad and exhibits subtle features that indicate that several forms of the protein may be present. The hemolysate separation exhibits several small additional peaks and subtle features, the most prominent of which is an unidentified peak labeled 2. The identified peaks have been quantitated using calibration runs and experimentally measured values of 1.7×10^{-3} and 0.349 g/ml were obtained for carbonic anhydrase and hemoglobin A_0, respectively. These values agree with the 2.5×10^{-3} and 0.335 g/ml for the same analytes obtained by other methods.[63,64]

A total of 39 red blood cells was analyzed consecutively in these experiments. Electropherograms from some of these analyses are shown in Figure 21. These results indicate that individual cells vary in average cell composition. Although the major components, hemoglobin A_0, carbonic anhydrase, methemoglobin, and the unknown peak are clearly recognizable in each cell, there are several minor features that vary from cell to cell. Variations among the

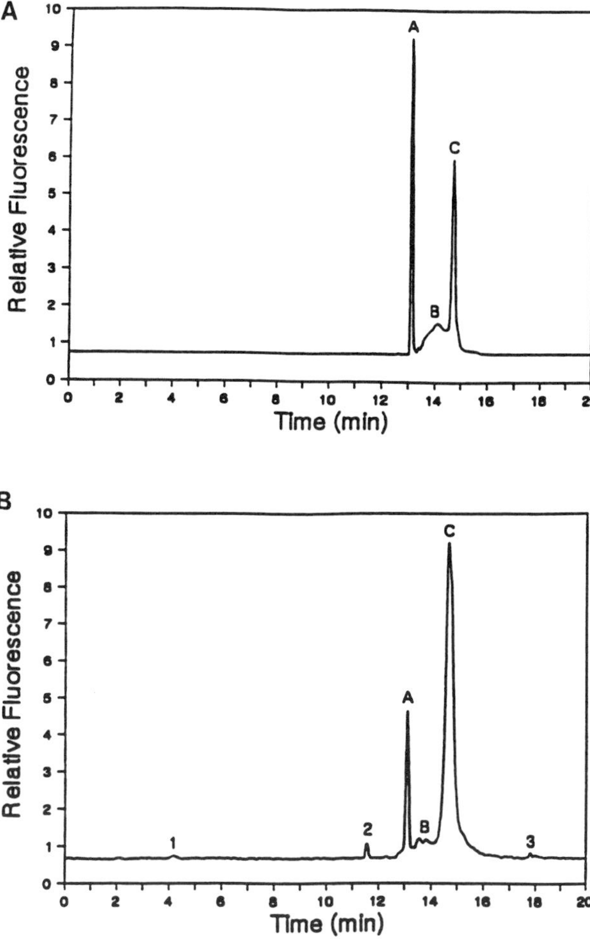

FIGURE 20. (A) Electropherogram of standard proteins. Peaks A, B, and C correspond to carbonic anhydrase, methemoglobin, and hemoglobin A_0, respectively. (B) Electropherogram of hemolysate of human erythrocytes. The lettered peaks are as in (A). The identities of the numbered peaks are unknown. (From Lee, T. T. and Yeung, E. S., *Anal. Chem.*, 64, 3045, 1992. With permission.)

group of 39 cells were as much as 1 order of magnitude even though erythrocytes are known to be fairly homogeneous in size distribution. These variations apparently reflect age differences in the cells. Further studies of the system should elucidate this point. The unexpectedly large variations in individual protein fractions and total protein content obtained in these studies emphasize the importance of single-cell studies.

IV. FUTURE PROSPECTS

Capillary electrophoresis provides a powerful approach to understanding single-cell chemistry. The ability to separate picoliter- and femtoliter-volume samples allows profiling of the chemical composition of single whole cells and subcellular regions. Clearly, the techniques in this area have already led to many intriguing discoveries concerning single-cell composition and chemistry. However, in order to realize the full potential of CE, further improvements must be made in sampling, separation, and, especially, detection methodologies.

Sampling is a key area in the development of CE for analysis of small biological environments. Although progress in this area has been considerable, further development of sampling

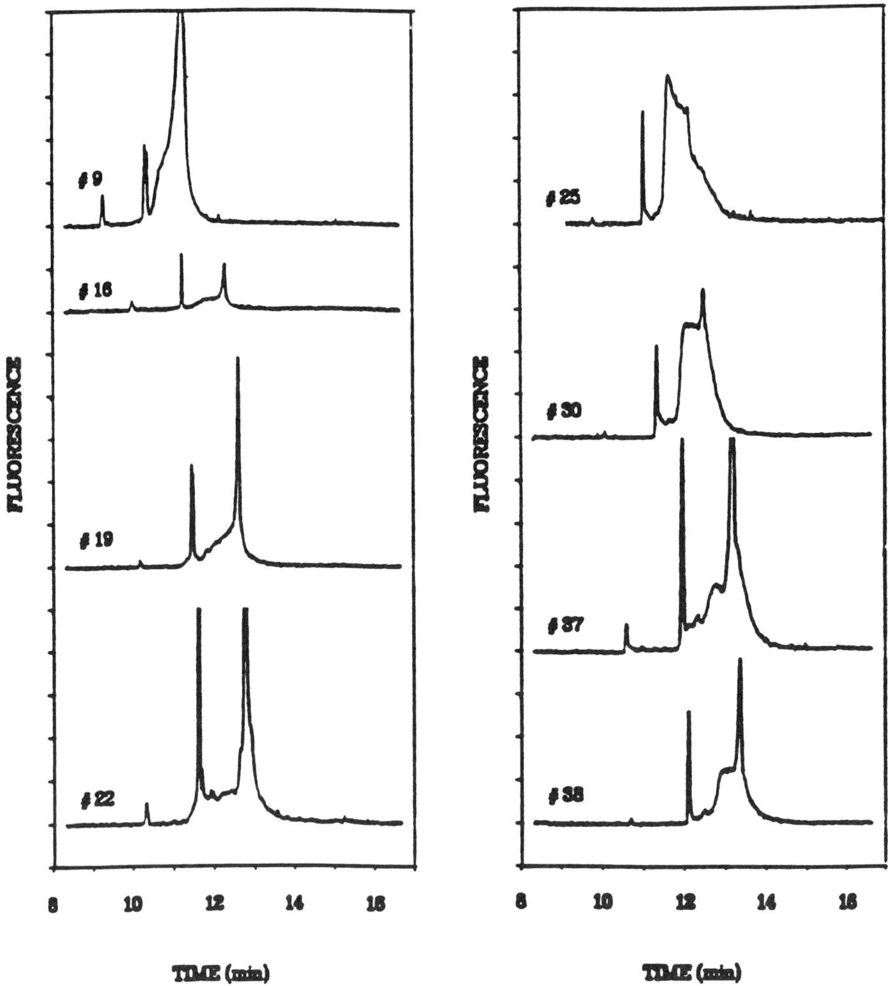

FIGURE 21. Electropherograms of proteins in several individual human erythrocytes. Numbers refer to consecutive run numbers over a series of 39 trials. A common scale factor was used throughout. (From Lee, T. T. and Yeung, E. S., *Anal. Chem.*, 64, 3045, 1992. With permission.)

technology is necessary. For example, further miniaturization is required in order to sample from mammalian cells and smaller structures such as vesicles, axons, and single synapses on a routine basis. The smallest injectors reported have been constructed from 5-μm I.D./150-μm O.D. capillaries. In principle, microinjectors could be made from 2-μm I.D. capillaries, which are now commercially available, or 1-μm I.D. capillaries when they become available. Furthermore, an injection system allowing for continuous small volume sampling would be desirable. Small volume injection from very small structures pushes microscopy and micromanipulator technology to its limits, so improvements in these areas are as important as improvements to the actual sampling devices themselves.

Improvements in CE separation technology will also be important for the application of CE to single-cell analysis. Because many molecules of biological importance are structurally similar, optimization of separation conditions and development of new experimental conditions to improve resolution are necessary. Proteins present a special challenge in CE separations because they tend to adsorb to the capillary wall and cause band broadening. Furthermore, when sampling from biological environments, it is important to develop separation conditions that reduce the effects of adsorption of the molecules in the matrix (i.e., lipids and proteins) on the column and detector. Finally, the area of micellar electrokinetic chromatography will be important because it will allow the sampling and separation advantages of CE to be extended to neutral species (for a review, see Ref. 40).

Improvements in detection technology may prove to be the most challenging, but ultimately the most important, aspect of the further development of CE for single-cell analysis. To detect trace compounds in picoliter or femtoliter volumes, a detector must be capable of attomole or even zeptomole mass detection limits. Although electrochemical, LIF, and radiometric detection methods can all provide subattomole detection limits, each one suffers from specific shortcomings. Electrochemical and LIF detection detect only a few compounds without derivatization while radiometric detection requires radiolabeling of the molecules of interest. Therefore, one area of key importance in detection is development of appropriate derivatization schemes for molecules such as amino acids, peptides, proteins, and cyclic nucleotides. Advances in this area should include schemes for precolumn derivatization, postcolumn derivatization, and on-column derivatization. Development of microscale derivatization reactions will be especially important for single-cell analysis by CE.

In summary, the further development of CE for analysis of single cells depends on the continued development of sampling, separation, and detection methodologies. Immediate goals include the miniaturization of techniques to study cell compartments, organelles, and mammalian cells. Single-cell CE will be useful for correlating relationships between cell function and biochemical composition. Elucidation of the mechanistic effects of new and existing drugs on specific cells is a potentially exciting application in this area. In the area of neuroscience, long-term goals include the identification of transmitters in single vesicles and the analysis of efflux of neurotransmitters from a single synapse.

ACKNOWLEDGMENTS

We would like to acknowledge the help of the many co-workers whose work is referenced in this chapter. The support of the National Science Foundation, the National Institutes of Health, and the Office of Naval Research is gratefully acknowledged.

REFERENCES

1. **Giacobini, E.,** Chemical studies on individual neurons, *Neurosci. Res.*, 1, 1, 1968.
2. **Giacobini, E.,** Neurochemical analysis of single neurons — a mini-review dedicated to Oliver H. Lowry, *J. Neurosci. Res.*, 18, 632, 1987.
3. **Ishikawa, E., Hashida, S., Kohno, T., and Hirota, K.,** Ultrasensitive enzyme immunoassay, *Clin. Chim. Acta*, 194, 51, 1990.
4. **Matioli, G. T. and Niewisch, H. B.,** Electrophoresis of hemoglobin in single erythrocytes, *Science*, 150, 1824, 1965.
5. **Tank, D. W., Sugimori, M., Connor, J., and Llinas, R. R.,** Spatially resolved calcium dynamics of mammalian Purkinje cells in cerebellar slices, *Science*, 242, 773, 1988.
6. **Nicholson, C. and Rice, M. E.,** Use of ion-selective microelectrodes and voltammetric microsensors to study brain cell microenvironment, in *Neuromethods*, Boulton, A. A., Baker, G. B., and Walz, W., Eds., Humana Press, Clifton, NJ, 1988.
7. **Amman, D.,** *Ion-Selective Microelectrodes*, Springer-Verlag, Berlin, 1986.
8. **Meulemans, A., Poulain, B., Baux, G., and Tauc, L.,** Changes in serotonin concentration in a living neurone: a study by on-line intracellular voltammetry, *Brain Res.*, 414, 158, 1987.
9. **Meulemans, A., Poulain, B., Baux, G., Tauc, L., and Henzel, D.,** Micro carbon electrode for intracellular voltammetry, *Anal. Chem.*, 58, 2088, 1987.

10. **Chien, J. B., Wallingford, R. A., and Ewing, A. G.,** Estimation of free dopamine in the cytoplasm of the giant dopamine cell of *Planorbis corneus* by voltammetry and capillary electrophoresis, *J. Neurochem.*, 54, 633, 1990.
11. **Lau, Y. Y., Chien, J. B., Wong, D. K. Y., and Ewing, A. G.,** Characterization of the voltammetric response at intracellular carbon ring electrodes, *Electroanalysis*, 3, 87, 1991.
12. **Ponchon, J.-L., Cespuglio, R., Gonon, F., Jouvet, M., and Pujol, J. F.,** Normal pulse polarography with carbon fiber electrodes for in vitro and in vivo determination of catecholamines, *Anal. Chem.*, 51, 1483, 1979.
13. **Wightman, R. M.,** Microvoltammetric electrodes, *Anal. Chem.*, 53, 1125A, 1981.
14. **Leszczyszyn, D. J., Jankowski, J. A., Viveros, O. H., Diliberto, E. J., Jr., Near, J. A., and Wightman, R. M.,** Nicotinic receptor-mediated catecholamine secretion from individual chromaffin cells, *J. Biol. Chem.*, 265, 14736, 1990.
15. **Kawagoe, K. T., Jankowski, J. A., and Wightman, R. M.,** Etched carbon-fiber electrodes as amperometric detectors of catecholamine secretion from isolated biological cells, *Anal. Chem.*, 63, 1589, 1991.
16. **Sossin, W. S. and Scheller, R. H.,** A bag cell neuron-specific antigen localizes to a subset of dense core vesicles in *Aplysia californica*, *Brain Res.*, 494, 205, 1989.
17. **Betzig, E., Trautman, J. H., Harris, T. D., Weiner, J. S., and Kostelak, R. L.,** Breaking the diffraction barrier: optical microscopy on nanometric scale, *Science*, 251, 1468, 1991.
18. **Mantus, D. S., Valaskovic, G. A., and Morrison, G. H.,** High mass resolution secondary ion mass spectrometry via simultaneous detection with a charge-coupled device, *Anal. Chem.*, 63, 788, 1991.
19. **Kennedy, R. T., Oates, M. D., Cooper, B. R., Nickerson, B., and Jorgenson, J. W.,** Microcolumn separations and the analysis of single cells, *Science*, 246, 57, 1989.
20. **Kennedy, R. T., St. Claire, R. L., White, J. G., and Jorgenson, J. W.,** Chemical analysis of single neurons by open tubular liquid chromatography, *Mikrochim. Acta*, 2, 37, 1987.
21. **Kennedy, R. T. and Jorgenson, J. W.,** Quantitative analysis of individual neurons by open tubular liquid chromatography with voltammetric detection, *Anal. Chem.*, 61, 436, 1989.
22. **Oates, M. D., Cooper, B. R., and Jorgenson, J. W.,** Quantitative amino acid anlaysis of individual snail neurons by open tubular liquid chromatography, *Anal. Chem.*, 62, 1573, 1990.
23. **Cooper, B. R., Jankowski, J. A., Leszyczyszyn, D. J., Wightman, R. M., and Jorgenson, J. W.,** Quantitative determination of catecholamines in individual bovine adrenomedullary cells by reversed-phase microcolumn liquid chromatography with electrochemical detection, *Anal. Chem.*, 64, 691, 1992.
24. **Ewing, A. G., Wallingford, R. A., and Olefirowicz, T. M.,** Capillary electrophoresis, *Anal. Chem.*, 61, 292A, 1989.
25. **Olefirowicz, T. M. and Ewing, A. G.,** Capillary electrophoresis in 2 and 5 μm diameter capillaries: application to cytoplasmic analysis, *Anal. Chem.*, 62, 1872, 1990.
26. **Olefirowicz, T. M. and Ewing, A. G.,** Capillary electrophoresis for sampling single nerve cells, *Chimia*, 45, 106, 1991.
27. **Hogan, B. L. and Yeung, E. S.,** Determination of intracellular species at the level of a single erythrocyte via capillary electrophoresis with direct and indirect fluorescnece detection, *Anal. Chem.*, 64, 2841, 1992.
28. **Olefirowicz, T. M. and Ewing, A. G.,** Dopamine concentration in the cytoplasmic compartment of single neurons determined by capillary electrophoresis, *J. Neurosci. Methods*, 34, 11, 1990.
29. **Jorgenson, J. W. and Lukacs, K. D.,** Capillary zone electrophoresis, *Science*, 222, 266, 1983.
30. **Gordon, M. J., Huang, X., Pentoney, S. L., Jr., and Zare, R. N.,** Capillary electrophoresis, *Science*, 242, 224, 1988.
31. **Wallingford, R. A. and Ewing, A. G.,** Capillary zone electrophoresis with electrochemical detection in 12.7 μm diameter columns, *Anal. Chem.*, 60, 1972, 1988.
32. **Jorgenson, J. W. and Lukacs, K. D.,** High resolution separations based on electrophoresis and electroosmosis, *J. Chromatogr.*, 218, 209, 1981.
33. **Jorgenson, J. W. and Lukacs, K. D.,** Zone electrophoresis in open-tubular glass capillaries, *Anal. Chem.*, 53, 1298, 1981.
34. **Huang, X., Luckey, J. A., Gordon, M. J., and Zare, R. N.,** Quantitative analysis of low molecular weight carboxylic acids by capillary zone electrophoresis/conductivity detection, *Anal. Chem.*, 61, 766, 1989.
35. **Fujiwara, S. and Honda, S.,** Effect of addition of organic solvent on the separation of positional isomers in high voltage capillary zone electrophroresis, *Anal. Chem.*, 59, 487, 1987.
36. **Rose, D. J. and Jorgenson, J. W.,** Characterization and automation of sample introduction methods for capillary zone electrophoresis, *Anal. Chem.*, 60, 642, 1988.
37. **Lee, T. L. and Yeung, E. S.,** Quantitative determination of native proteins in individual human erythrocytes by capillary zone electrophoresis with laser-induced fluorescence detection, *Anal. Chem.*, 64, 3045, 1992.
38. **Green, J. S. and Jorgenson. J. W.,** Variable-wavelength on-column fluorescence detector for open-tubular zone electrophoresis, *J. Chromatogr.*, 352, 337, 1986.

39. **Olefirowicz, T. M. and Ewing, A. G.,** Detection methods in capillary electrophoresis, in *Capillary Electrophoresis*, Grossman, P. D. and Colburn, J. C., Eds., Academic Press, New York, 1992, chap. 2.
40. **Kuhr, W. G. and Monnig, C. A.,** Capillary electrophoresis, *Anal. Chem.*, 64, 389R, 1992.
41. **Wallingford, R. A. and Ewing, A. G.,** Capillary zone electrophoresis with electrochemical detection, *Anal. Chem.*, 59, 1762, 1987.
42. **Yik, Y. F., Lee, H. K., Li, S. Y. O., and Khoo, S. B.,** Micellar electrokinetic capillary chromatography of vitamin B6 with electrochemical detection, *J. Chromatogr.*, 585, 139, 1991.
43. **O'Shea, T. G., Greenhagen, R. D., Lunte, S. M., Lunte, C. E., Smyth, M. R., Radzik, D. M., and Watanabe, N. J.,** Capillary electrophoresis with electrochemical detection employing an on-column Nafion joint, *J. Chromatogr.*, 593, 305, 1992.
44. **Huang, X., Zare, R. N., Sloss, S., and Ewing, A. G.,** End-column detection for capillary zone electrophoresis, *Anal. Chem.*, 63, 189, 1991.
45. **Haber, C., Silvestri, I., Roosli, S., and Simon, W.,** Potentiometric detector for capillary zone electrophoresis, *Chimia*, 45, 117, 1991.
46. **Sloss, S. and Ewing, A. G.,** Improved method for end-column amperometric detection for capillary electrophoresis, *Anal. Chem.*, 65, 577, 1993.
47. **Cheng, Y. F. and Dovichi, N. J.,** Subattomole amino acid analysis by capillary zone electrophoresis and laser-induced fluorescence, *Science*, 242, 562, 1988.
48. **Cheng, Y. F. and Dovichi, N. J.,** Laser techniques in luminescence spectroscopy, *ASTM Spec. Tech. Publ.*, 1066, 151, 1990.
49. **Waldron, K. C., Wu, S., Earle, C. W., Harke, H. K., and Dovichi, N. J.,** Capillary zone electrophoresis separation and laser-based detection of both fluorescein thiohydantoin and dimethylaminoazobenzene thiohydantoin derivatives of amino acids, *Electrophoresis*, 11, 777, 1990.
50. **Chen, D. Y., Swerdlow, H. P., Harke, H. R., Zhang, J. Z., and Dovichi, N. J.,** Single-color laser-induced fluorescence detection and capillary gel electrophoresis for DNA sequencing, *Proc. SPIE Int. Soc. Opt. Eng.*, 1435, 161, 1991.
51. **Swerdlow, H. P., Zhang, J. Z., Chen, D. Y., Harke, H. R., Grey, R., Wu, S., Dovichi, N. J., and Fuller, C.,** Three DNA sequencing methods using capillary gel electrophoresis and laser-induced fluorescence, *Anal. Chem.*, 63, 2835, 1991.
52. **Swaile, D. F. and Sepaniak, M. J.,** Laser-based fluorimetric detection schemes for the analysis of proteins by capillary zone electrophoresis, *J. Liq. Chromatogr.*, 14, 869, 1991.
53. **Lawrence, J. F.,** Fluorimetric derivatization in high performance liquid chromatography, *J. Chromatogr. Sci.*, 17, 147, 1979.
54. **Ohkura, Y. and Nohta, H.,** Flourescence derivatization in high-performance liquid chromatography, in *Advances in Chromatography*, Vol. 29, Giddings, J. C., Grushka, E., and Brown, P. R., Eds., Marcel Dekker, New York, 1989, pp. 221–258.
55. **Pentoney, S., Huang, X., Burgi, D., and Zare, R. N.,** On-line connector for microcolumns: application to the on-column *o*-phthalialdehyde derivative of amino acids separated by capillary zone electrophoresis, *Anal. Chem.*, 60, 2625, 1980.
56. **Nickerson, B. and Jorgenson, J. W.,** Characterization of a post-column reaction-laser-induced fluorescence detector for capillary zone electrophoresis, *J. Chromatogr.*, 480, 157, 1989.
57. **Tsuda, T., Kobayaski, Y., Hori, A., Matsumota, T., and Suzuki, O.,** Post-column detection for capillary zone electrophoresis, *J. Chromatogr.*, 456, 375, 1988.
58. **Rose, D. J. and Jorgenson, J. W.,** Post-capillary fluorescence detection in capillary zone electrophoresis using *o*-phthaldialdehyde, *J. Chromatogr.*, 447, 117, 1988.
59. **Kuhr, W. G. and Yeung, E. S.,** Indirect detection methods for capillary separations, *Anal. Chem.*, 63, 275A, 1991.
60. **Kristensen, H., Lau, Y. Y., and Ewing, A. G.,** Capillary electrophoresis of single cells: observation of two compartments of neurotransmitter vesicles, *J. Neurosci. Meth.*, 1992, submitted.
61. **Besson, M. J., Cheramy, A., Felts, P., and Glowinski, J.,** Release of newly synthesized dopamine from dopamine-containing terminals in the stratium of the rat, *Proc. Natl. Acad. Sci. U.S.A.*, 62, 741, 1969.
62. **Shore, P. A.,** Actions of amfonelic acid and other non-amphetamine stimulants on the dopamine neuron, *J. Pharm. Pharmacol.*, 28, 855, 1976.
63. **Osgood, E. E.,** Normal hematologic standards, *Arch. Intern. Med.*, 56, 849, 1935.
64. **Roughton, F. J. W. and Rupp, J. C.,** Problems concerning the kinetics of the reactions of oxygen, carbon monoxide, and carbon dioxide in the intact red cell, *Ann. N.Y. Acad. Sci.*, 75, 156, 1958.

Chapter 16

CAPILLARY ELECTROPHORESIS FOR THE ROUTINE CLINICAL LABORATORY

Gerald L. Klein and Carl R. Jolliff

TABLE OF CONTENTS

0-8493-8690-X/94/$0.00+$.50

I. INTRODUCTION

Electrophoresis of human serum samples began with the moving boundary experiments of Tiselius in buffer-filled open tubes.[1] These experiments showed that serum separated into five distinct zones. Research with paper strips (Durrum) and microporous acetate membranes (Grunbaum) further established a diagnostic utility for the method.[2,3] Electrophoresis became a standard part of the clinical laboratory through commercialization of these methods. In the 1970s, developments with gel support media further enhanced the clinical utility.

Electrophoretic techniques present a unique method where with a single experiment one can obtain a multitude of information. It can be an effective disease screening tool through qualitative and quantitative assessment of the distribution of the separated components. Widespread usage of the method for screening was realized with partial mechanical automation of the manual procedures for membrane and gel methods. Manual electrophoretic methods entail requisite operator skill and labor to achieve results.

With the introduction of the capillary electrophoresis, a potential new era in diagnostic screening becomes possible due to more rapid separations with a significant reduction in labor requirements and with improvements in the information provided by the separations. In addition, development of fully automated systems is quite achievable. It is ironic that the capillary uses an open tube much like Tiselius, the founder of electrophoresis, with the only difference being the smaller diameter of the tubes. Capillary electrophoresis emerged as a useful analytical tool from the work of Hjertén and Jorgensen in the late 1970s and early 1980s.[4,5] The realization of the method as a valuable clinical laboratory tool has just begun.[6]

II. STANDARD ELECTROPHORETIC METHODS

Capillary electrophoresis (CE) results can be comparable to those of other electrophoretic methods. Understanding the potential for CE warrants a brief discussion of current electrophoretic methods for comparison.

Electrophoresis is an important procedure in the routine clinical laboratory. Before the introduction of automation, it was largely used for secondary and confirmatory testing. Standard clinical electrophoretic methods employ a solid support matrix either as a microporous membrane or as a gel. The procedures for handling these are quite labor intensive and require operator proficiency. Sample application is the most difficult operation required, as the volumes are in the 1- to 5-μl range and require precise placement on the support medim. The technician must handle the support matrix through all the steps of buffering, sample application, separation, staining, destaining, and detecting procedures. Apparatus and accessories provide aid in these operations, but it still requires technician attention. The support matrices can accommodate multiple samples in a single run and, hence, affords some efficiency. Typically a skilled, trained technician can process a routine serum protein gel separation from preparation to final answers in about $1^1/_2$ h. Instruments are on the market that automate most of the wet chemical processes for routine serum protein separations, providing the laboratory with a less labor-intensive means for high-volume electrophoresis screening. These are large instruments that require careful maintenance. They also consume large volumes of reagents for staining and destaining the separations.

In electrophoresis gels, the stained serum protein patterns partition into at least five zones: albumin and the globulins, $\alpha 1$, $\alpha 2$, β, and γ. Methods have been developed that have increased the resolution beyond the five zones, for showing the appearance of monoclonal antibody peaks in the gamma globulin zone and to provide more resolution for the α and β zones. Other, even higher resolution methods such as isoelectric focusing and two-dimensional electrophoresis are used in some reference laboratories and research laboratories. Following is a list of the most common electrophoretic methods available for the clinical laboratory.

1. Serum proteins
2. Cerebral spinal fluid proteins
3. Urine proteins
4. Hemoglobins
5. Glycosylated hemoglobins
6. Isoenzymes (CK, LD, and AP)
7. Lipoproteins

III. CLINICAL CAPILLARY ELECTROPHORESIS

Development of methods using CE for routine clinical laboratory applications has largely been with open tubes. Methods explored to date are discussed in the following sections.

A. SERUM PROTEINS

Since serum proteins were the foundation of clinical electrophoresis, developments for these will lay a new foundation for CE as an important clinical laboratory tool. Immunofixation/subtraction is a new development with CE that provides a method of identifying specific proteins in a separation and, ultimately, further developments of the method hold the promise of quantitation of specific proteins. Isoenzymes presented a particular difficulty with CE. How can the separated isoenzymes be exposed to substrate for detection and measurement? This will be discussed later. Genetic variants of hemoglobins present an important opportunity that can be realized with CE. Diseases associated with certain variants, such as sickle-cell anemia, are important issues in public health. Lipoproteins have been effectively separated with CE with more information than by any current method. This will be important for cardiac risk assessments.

A number of other methods, which appear technically feasible to adapt to the CE format are likely to be explored as the technology becomes more widely used. Some of these are presently being explored and include:

Urines (proteins and metabolites)[7]
Cerebrospinal fluid (for multiple sclerosis confirmation)
Drugs of abuse
Proteolytic enzymes and isoenzymes (cancer markers)
Amino acids (nutritional assessments)
Phospholipids (lecithin and sphingomyelin)
Myoglobin[8]
Glycosylated proteins (diabetic monitoring)
Immune complexes[9]
Carbohydrates
Vitamins[10]
Pharmacology of therapeutic drugs
DNA genetic disease markers
Infectious diseases[11]

These potential applications (and others the authors have not yet imagined) will be an important part of diagnostic and therapeutic medicine amenable to CE analysis.

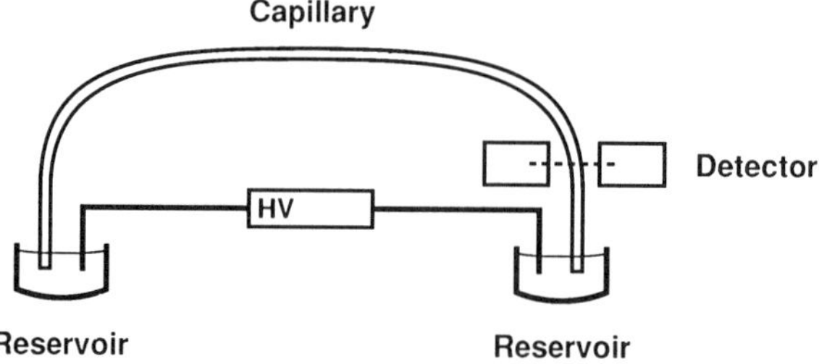

FIGURE 1. Diagram of a capillary electrophoresis (CE) system. HV, High voltage.

IV. INSTRUMENTATION

The development of systems for automating the applications is necessary to the introduction and acceptance of CE as a useful clinical laboratory tool. Instrument design for CE presents certain challenges. The capillaries are very small, requiring accurate dimensional controls. Design considerations for safely handling the high voltages used in the separations place constraints on materials and geometry. The capillaries must be serviced with fluids for separation as well as for washing and rinsing, and for adding buffers. The sensitive detection circuitry must be positioned near high voltages and liquids. Thermal control of the capillary environment is important. Design for easy replacement of the capillaries is essential. Schematically, the basic layout for a capillary electrophoresis system is shown in Figure 1.

To meet the needs of many routine clinical laboratories, a system must offer adequate throughput. For example, some high-volume laboratories may process as many as 500 routine serum protein separations in a single work day. There are two approaches for meeting adequate throughput: (1) reduce the separation time so that high throughput can be achieved (but with a cost in resolution),[12] and (2) run parallel channels at the same time so that more separations can be made without compromising the quality of the separations. Beckman (Fullerton, CA) has developed a six-channel experimental clinical CE system. Preliminary tests have shown that it can provide the equivalent of one separation per minute with a total run time of 6 min/separation. Other design requirements for a routine laboratory CE system include reproducibility, on-board sample preparation (such as dilution), and ease of use for the operator. A block diagram of the six-channel fluid-handling system is shown in Figure 2. This system demonstrates sample-to-answer walkaway operation and has a capacity for up to 60 samples. Samples can be identified either through keyboard entry or from bar codes applied to the sample tubes. The technician then selects the electrophoretic method and attends to other tasks, returning in about 1 h to recover the results.

As with any CE system, there are two principal types of sample injection: electrokinetic and pressure. In the multichannel system, the capillary outlet ends are sealed into a common manifold to which a negative vacuum pressure is applied for both servicing the capillaries and for sample injections. Electrokinetic injection can be used. However, there may be a risk of partitioning the sample, with the consequence that high-mobility components may not be injected or only partially injected. Pressure injection creates a laminar flow profile with a parabolic distribution along the capillary axis, which may result in band broadening. Band broadening may be minimized by using a discontinuous buffer, that is, the sample buffer has a lower ionic strength and may have a different pH than the running buffer. This produces a stacking effect (described in Chapters 2 and 20) in which the sample components become concentrated at the interface between the sample and the running buffer at the start of the separation. Stacking effects are discussed in Chapter 19. These modes of injection are illustrated in Figure 3.

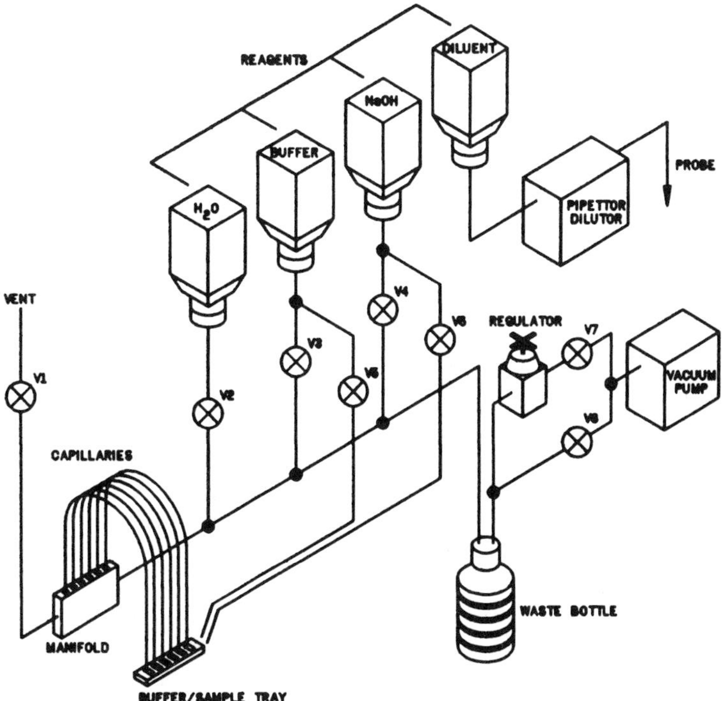

FIGURE 2. Diagram of a six-channel CE system.

A. DETECTION SYSTEM

The detection system shown in Figure 4 illustrates the major elements required to acquire and process signal information from the capillary. A light source, including optical elements, provides illumination across the capillary. The light wavelength is selected according to the absorbance properties of the samples being measured. For proteins, the wavelength used is near 200 nm, corresponding to the high absorptivity of the polypeptide bonds of the protein molecules. Most of the clinical methods developed so far use absorbance detection, although future developments may include methods such as fluorescence and conductivity detection.

For a multichannel clinical CE system, the light from the source lamp must be distributed to all the channels. This can be accomplished effectively with fiber optics. An optical system gathers and relays light from the source lamp onto a bundle of fibers. The fibers are then split to provide an optical channel for each capillary. After passing through the capillary, the light must be directed to the detector. Again, fiber optics can be used to relay the light to individual detectors or to a multiplexing device associated with a single detector. Fiber optics afford a significant advantage by providing for remote noncritical placement of both the light source and the detector components. Remote placement ensures that heat from the source lamp will not affect the capillaries, and the sensitive detection circuitry will not be compromised by the high voltages and liquids associated with the capillaries. A diagram of a six-channel fiber optic system is shown in Figure 5.

B. SERVICING THE CAPILLARY

As described in Chapter 2, Section II.C.4, the capillary must be rigorously maintained. Reproducible and reliable operation is assured with continuous attention to capillary conditioning. Because of the small dimensions of the capillary, plugging is an ever-present problem. Protein buildup on the capillary walls from serum samples is another problem that requires

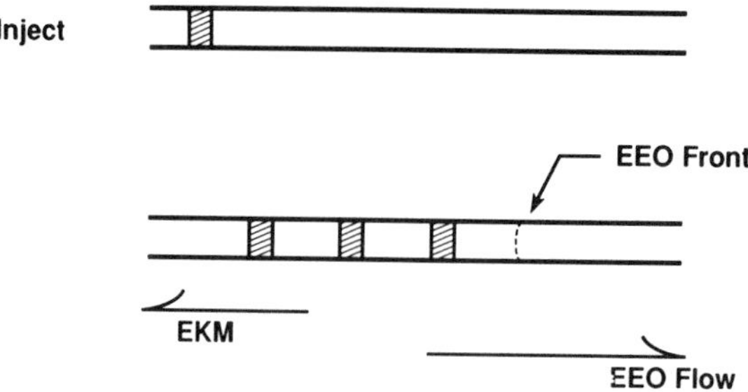

Electrokinetic Injection

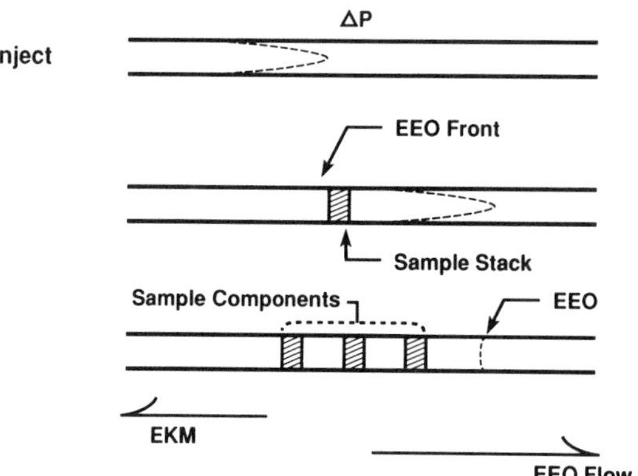

Stacking and Separation
ΔP Injection

FIGURE 3. Modes of injection for CE.

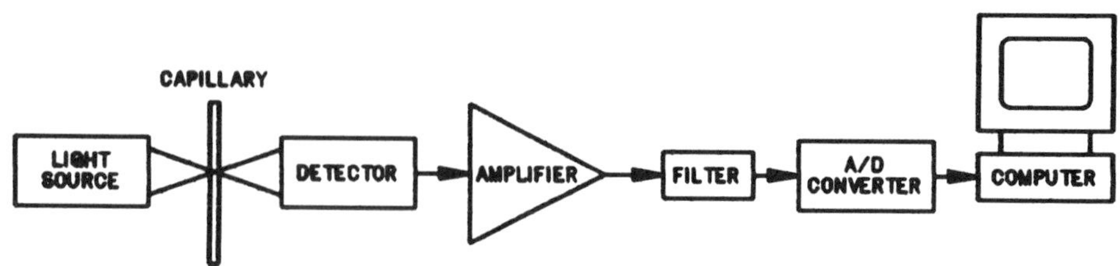

FIGURE 4. Diagram of a CE detection system.

attention between separations. These problems can be avoided by post-run rinsing and keeping both ends immersed in filtered, distilled water during all idle times.

One example of a maintenance routine follows three procedures: First, if the capillary is between scans, a solution of 1 *N* NaOH is pumped through the capillary for a time assuring at least three replacement volumes. This is followed with a reequilibration of the running buffer for an equal time. Second, if the capillary is to remain at rest for longer than 15 min between runs, a wash with NaOH as above is performed after the last scan, followed with a

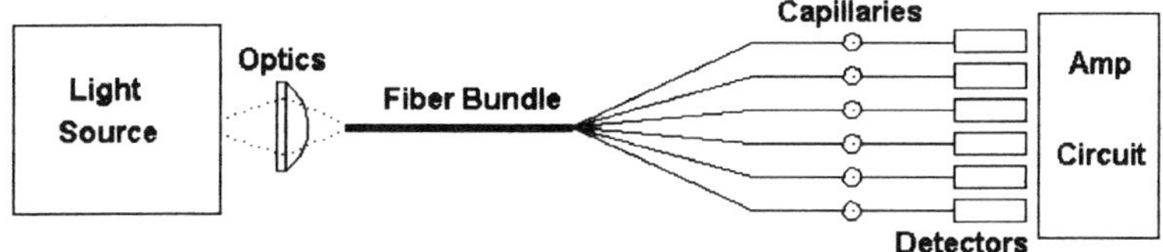

FIGURE 5. Diagram of a fiber optic CE interface.

replacement of at least 5 vol of distilled water. The ends of the capillary are immersed in distilled water during the idle period. Third, if the capillary is to be stored out of use for longer than 15 min, such as daily shut-down, a procedure as in the second condition is performed, followed with either immersion in distilled water or with a drying step, in which the differential pressure is maintained across the capillary for several minutes with neither end immersed. Again, it is preferred to keep the capillary immersed, but if there is a risk of the immersing liquid evaporating during the idle time, the alternative procedure for drying should be performed.

Initial start-up procedure for idled or new capillaries could be as follows: Using differential pressure, wash with at least 5 volumes of 1 *N* NaOH, followed by rinsing with at least 10 volumes of distilled water. Equilibration with the running buffer for 5 min with the high voltage applied should follow. Prior to the first scan, a second injection with running buffer as between scans will prepare the capillary for a scan.

With modern automation in instrumentation, all these procedures can be programmed to perform automatically with a minimum of attention; however, even with the best care, the capillaries will have a finite useful life under the conditions of extreme basic pH (NaOH is known to attack silica, forming a silica gel). If the capillary is allowed to dry without thorough rinsing, the gel can set, forming an impermeable plug.

V. CLINICAL APPLICATIONS

A. SERUM PROTEIN PATTERNS

More than 300 different soluble proteins have been recognized in human plasma, but most of these are below the detection limit of electrophoretic methods, including CE.[13] Two "normal" patterns are shown in Figure 6. Close examination of these show a prealbumin peak distinctly isolated from the balance of the pattern. Most CE separations show a prealbumin peak (to the right of the albumin peak) whereas other electrophoretic methods rarely show the peak. Prealbumin occurs in a relatively narrow range of concentration in all "normal" samples (20 to 40 mg/dl) which represents about 0.3% of the total protein in serum. It can serve as an indicator of the detection sensitivity limit of capillary separations. A list of proteins is given in Table 1 whose reported normal range concentrations are equivalent to or greater than that of prealbumin. It is the authors' opinion that CE has the potential to show all the proteins in Table 1.

The order of the separation shown in the graphic display places the albumin peak toward the right. During separation, the peaks are migrating toward the anode; however, the velocity of the electroosmotic flow (EOF) toward the cathode dominates. Peaks of lower mobility will appear first, with peaks of higher mobility appearing later.

Human serum is a particularly difficult sample to analyze since it is very heterogeneous with many overlapping peaks. To a skilled chromatographer, a serum pattern would seem uninteresting, lacking sharp peak resolution. However, a large body of diagnostic correlation has been established from the electrophoretic separations by comparing the relative concentrations of the distribution of proteins in the five major zones (albumin, $\alpha 1$, $\alpha 2$, β, and γ). It

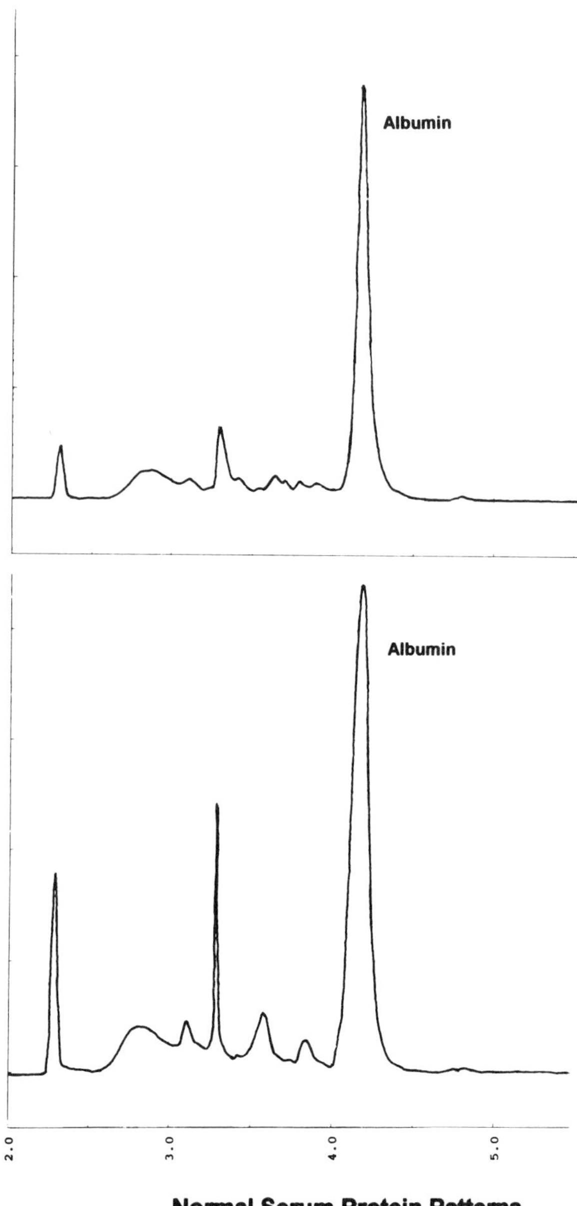

FIGURE 6. Capillary electrophoretic scans of two normal serum protein samples. (*Note:* With all figures in this chapter, the *y* axis will be relative absorbance and the *x* axis will be migration time.)

seems fitting that the first developments for CE should be for serum protein separations, as this was the historical beginning for applying electrophoresis to clinical methods.

An initial goal was to achieve a separation whose pattern resembled the patterns from densitometric scans of acetate membrane and gel separations. This would provide a familiar basis for comparisons with current methodologies. This goal was realized by Chen et al.,[14] using the following separation conditions:

Sample: Human serum diluted 1:10 in phosphate-buffered saline (PBS)
Capillary size: 75 μm × 375 μm × 25 cm

TABLE 1
Proteins in Human Plasma

Protein	Concentration (g/dl)	M_r	Amino acids	Mobility (mm/s/kV)
Prealbumin[a]	0.02–0.036	50,000	508	0.188
Albumin[a]	2.4–4.8	66,500	585	0.160
α_1-Acid-glycoprotein[a]	0.05–0.115	44,000	204	0.149
α_1-Antitrypsin[a]	0.095–0.185	51,000	394	0.143
α-Lipoprotein	0.28–0.38	200,000	243	
α_1-Chymotrypsin	0.03–0.06	58,000	??	
α_1-β-Glycoprotein	0.015–0.03	50,000	400	
Gc-Globulin	0.025–0.035	52,000	~500	
Haptoglobin[a]	0.04 –0.17	100,000	~900	0.131
Ceruloplasmin	0.02–0.055	135,000	1065	
α_2-Macroglobulin[a]	0.18–0.43	725,000	1451	0.109
α_2-HS-Glycoprotein	0.04–0.085	49,000	402	
Hemopexin	0.072	57,000	410	
Transferrin[a]	0.154–0.29	79,600	678	0.105
β_2-Glycoprotein	0.03–0.09	50,000	326	
β-Lipoprotein[a]	0.06–0.155	3,000,000	~250	0.112
Fibrinogen (plasma only)[a]	0.2–0.4	340,000	1482	0.090–0.100
C3[a]	0.07–0.15	180,000	~1000	0.091
IgA[a]	0.05–0.3	324,000	2668	0.090–0.060
IgM[a]	0.015–0.21	935,000	6670	0.090–0.060
IgG[a]	0.67–1.55	150,000	1334	0.090–0.060

Note: Concentration values represent normal ranges reported. Proteins listed are within the sensitivity limit corresponding to prealbumin detection.

[a] Protein fractions have been positively recognized. The clinical significance of the remaining fractions has not been established.

Buffer:	Borate
Injection:	Electrokinetic, 2 s
Temperature:	24°C
Separation:	10 min at 8000 V
Wavelength:	214 nm

Further experimentation showed other operating conditions that could result in equivalent or better separations.[15] Reduction of the capillary bore led not only to improved resolution but also to decreased separation times. Lower ionic strength buffers were found to degrade the resolution, while higher ionic strengths improve resolution. However, with ionic strength above 200 m*M*, the resolution improvements favor the γ region while compressing the remaining zones of the pattern. Other buffer compositions yielded interesting but unfamiliar patterns.[16,17] It was found that pressure injection yields higher relative signals than electrokinetic injection without any apparent loss in resolution when the sample diluent is PBS at pH 7.0. This was probably the result of stacking, which in effect concentrates the diluted sample. Other sample diluents can be used, including distilled water, with equivalent separations.[17] Using PBS, however, avoids the risk of sample denaturation if the delay time between dilution and separation becomes excessive. This was important during our early experiments with serum separations. These were done on a single-channel capillary system with automatic sample queuing, in which the resident time for the later samples could be more than 1 h.

From such experimentation, one can see there are numerous possibilities for modifying the information content obtained from the separations. As the application of capillary electrophoresis technology for routine clinical laboratory testing becomes widely used, modifications of separation conditions may show more sensitive pathophysiological changes in protein-related disease states.

Conditions for serum protein separations during preliminary developments for a screening test were

Sample:	Human serum diluted 1:10 in PBS
Capillary size:	25 μm × 150 μm × 20 cm
Buffer:	Borate
Injection:	Pressure, 10 s
Temperature:	24°C
Separation:	6 min at 10,000 V
Wavelength:	214 nm, 10 nm half band width (HBW)

In a controlled study in a clinical setting, CE has been compared to agarose gel electrophoresis (AGE) to assess its value as a clinical method of examination for serum proteins. Two facets of serum protein electrophoresis are particularly important: (1) Changes in the distribution of specific proteins that reflect pathophysiological conditions, as classified in Table 2 and (2) the ability to detect proteins of restricted mobility at elevated as well as low concentrations.

A series of 100 normal serum samples from adults of both sexes were run to compare the relative percentage values recovered from the 5 major zones used to interpret the serum patterns. This comparison of CE and AGE values is shown in Table 3. Differences in the relative values are attributable to the distribution of specific protein fractions between CE and AGE as well as differences due to measurement of intrinsic absorbance at 214 nm with CE vs. measurements of staining with AGE. Such a study must also identify the specific proteins showing in the separate zones. Additive methods, "spiking," were used to identify specific fractions. Immunosubtractive methods were also used for identification and these proved to be more specific since "pure" proteins purchased from biochemical suppliers proved to be impure. Results of the spiking and immunosubtractive studies are shown in Figures 7 and 8.

With these studies it was possible to assign locations for the major protein constituents in the separation pattern and establish a protein map for the separations. This is illustrated in Figure 9. We were then able to examine the CE patterns in comparison with AGE patterns for the various classification categories as previously published for AGE.[18] Figures 10 through 12 show scans corresponding to certain classes of dysproteinemias listed in Table 2.

It became apparent that excellent visual comparisons could be made between the densitometric scans of AGE separations and the photometric assessments at 214 nm from the CE separations. It is important to note that the CE separations measured at 214 nm record the intrinsic absorbance of the protein polypeptide bonds rather than the staining properties of the proteins as used in AGE. Therefore, quantitation errors that result from staining techniques (time, temperature, stain concentration and age, destaining times, differential staining due to posttranslational modification such as glycosylation, etc.) are avoided with CE. The intrinsic absorbances of the protein fractions are a more accurate measure of their relative concentrations.

With these considerations in mind, the five presently accepted zones for serum patterns may well be better represented by CE separations than by AGE: an important point to consider

TABLE 2
Classification of Serum Patterns with Agarose Gel Electrophoresis

Class	Subclass	Name	Albumin	α1	α2	β	γ
I		Hypoproteinemic					
	A	Malnutrition	↓	N-↑	N-↑	N	N-↑
	B-1	Protein loss — nonselective	N	N	N	N	N
	B-2	Exudative dermatopathies and burns	↓	↑	N	↓	N-↓
	B-3	Exudative pulmonary disease	↓	↑	N-↑	N-↓	N-↓
	B-4	Essential hypoproteinemia	↓	N-↑	N-↑	N-↑	N-↑
	B-5	Gastroenteropathies	↓	↑	N	N	N-↓
	B-6	Blood loss and plasmaphoresis	↓	↑	N-↑	↓↑	N-↑
II		Selective protein loss — nephrotic syndrome	↓	N-↑	↑	N-↓	N-↓
III		Diffuse, acute hepatodegenerative pattern (severe hepatic)	N-↓	N-↓	↓	N-↓	N-↑
IV		Cirrhotic pattern (β–γ bridging)	↓	N-↓	↓	N-↓	↑
V		Acute inflammatory stress	↓	↑	↑	↓	N
VI		Chronic inflammatory	↓	↑	↑	↓	↑
VII		Anemia					
	A	Iron deficiency anemia	N	N	N	↑	N
	B	*In vivo* hemolysis	N	N	↓	N	N
VIII		Polyclonal gammopathy	↓	N-↑	N-↑	↑↓	↑
IX		Monoclonal gammopathy					
	A	With normal heterogeneous immunoglobulins	N	N	N	N	N-↑
	B	With decreased heterogeneous immunoglobulins	N-↓	N	N	N	↑
	C	Hypogammoglobulinemia with urinary Bence–Jones proteins	N	N	N	N	↓
X		Hyperlipidemic pattern	N	N	↑	N	N
	A	Diabetes mellitus pattern	N	↑	↑	N	N
XI		Increased estrogen pregnancy pattern	N-↓	N-↑	N	N-↑	N
XII		Defect dysproteinemias					
	A	Analbuminemia	↓	N-↑	N-↑	N	N
	B	Bisalbuminemia	N	N	N	N	N

TABLE 2 (Continued)
Classification of Serum Patterns with Agarose Gel Electrophoresis

Class	Subclass	Name	Albumin	$\alpha 1$	$\alpha 2$	β	γ
	C	Hypo-α_1-antitrypsin	N	↓	N	N	N
	D	Atransferrinemia	N	N	N	↓	N
	E	An-β_2-betalipoproteinemia	N	N	N-↓	N	N
	F	Afibrinogenemia (plasma samples)	N	N	N	N	N
	G	Presence of fibrinogen in serum	N	N	N	N	N-↑
	H	Hypogammaglobulinemia	N	N-↑	N-↑	N	↓
	I	Hemoconcentration	↑	↑	↑	↑	↑
	J	Decreased complement (C3)	N	N	N	N-↓	N
	K	Ahaptoglobulinemia	N	N	N-↓	N	N

Note: N, zone within normal range; N-↑, zone within normal range or elevated; N-↓, zone within normal range or depressed; ↑, zone above normal range; ↓, zone below normal range; ↑↓, zone varies from below to within or above normal range.

TABLE 3
Paragon SPE-1 (Agarose Gel Electrophoresis) vs. Capillary Electrophoresis: Normal Range Values

Fraction	Paragon values (AGE)	Capillary values (CE)
Albumin	58.1–70.2	51.2–63.2
α1 Globulins	1.9–4.5	2.7–5.1
α2 Globulins	7.8–13.8	6.5–12.7
β Globulins	7.3–11.6	Trf 6.7–9.8
		C3 3.8–8.0
Gamma globulins	7.8–16.9	9.9–20.0

Note: All results are in percent of total protein as determined by the Biuret method; 100 samples from healthy male (50) and female (50) patients were used. Total protein (TP) normal range, 6.1–7.9 g/dl.

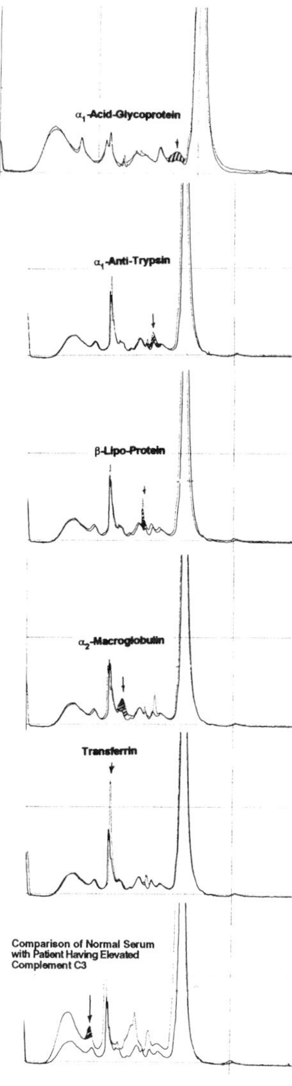

FIGURE 7. Normal serum “spiked” with purified proteins for identification.

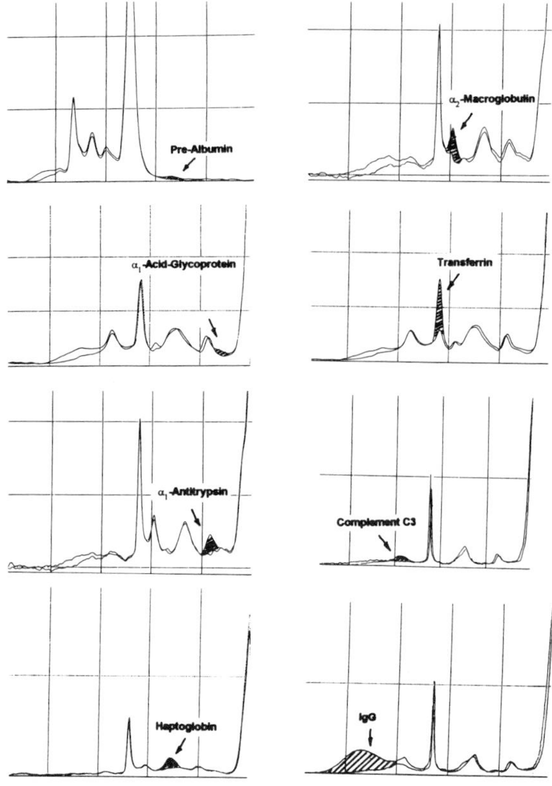

FIGURE 8. Serum peaks identified by immunosubtraction.

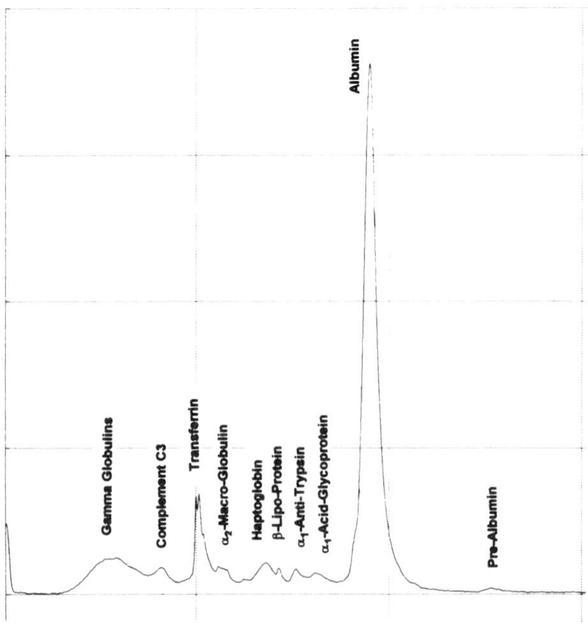

FIGURE 9. Serum protein map for CE scans.

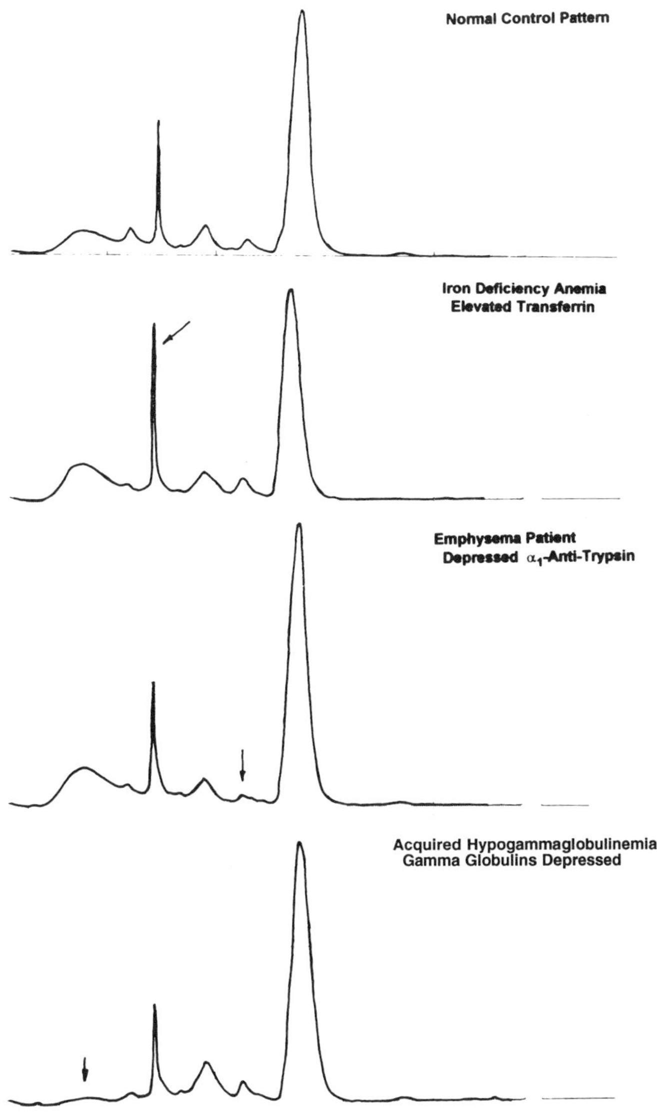

FIGURE 10. Capillary electrophoretic scans of patients with selected diseases.

when determining which method is used to screen for dysproteinemias and paraproteins. The operational parameters for AGE and CE used in this study were as follow.

Agarose gel electrophoresis

Instrument:	Beckman Paragon®* System and Appraise densitometer
Gel:	Beckman Paragon SPE®*
Buffer:	Barbital at pH 8.6
Temperature:	25°C
Run time:	25 min
Stain:	Paragon Blue
Sample:	4 μl of serum diluted 1:5 in buffer
Scan:	Densitometer at 600 nm

* Registered trademark of Beckman Instruments, Inc. (Brea, CA).

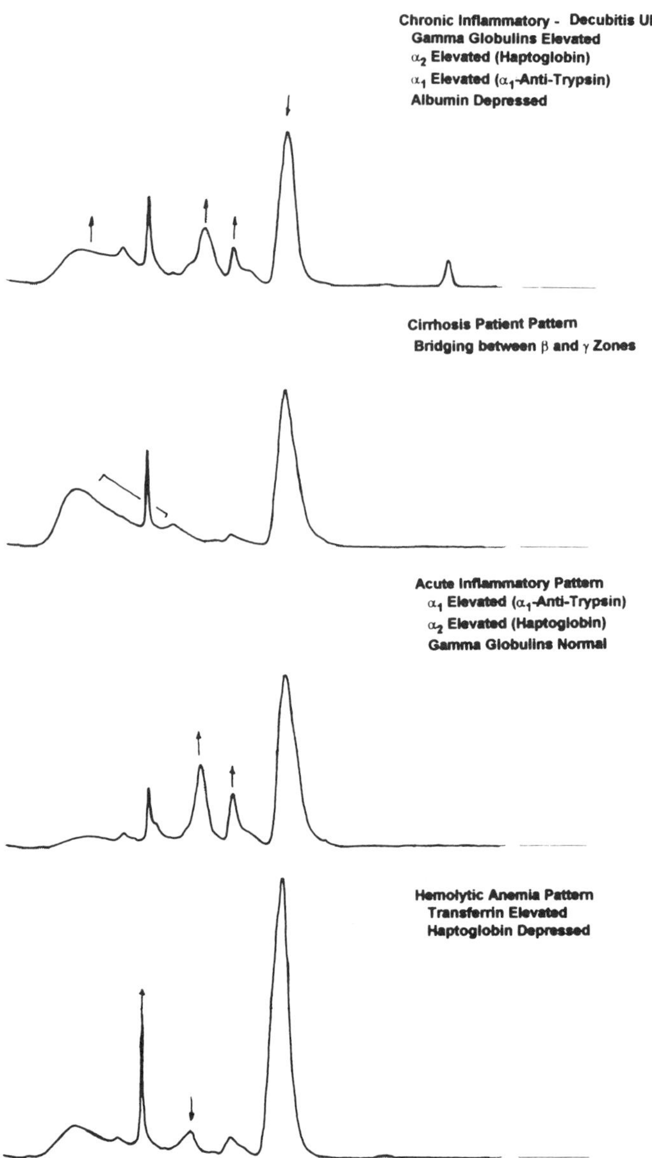

FIGURE 11. Capillary electrophoretic scans of patients with selected diseases.

Capillary electrophoresis

Instrument:	Beckman P/ACE®* capillary electrophoresis system (for research use only)
Capillary:	25-m bore 150-m O.D., 20-cm long
Wavelength:	214 nm, 10 nm HBW
Temperature:	24°C
Injection:	5 psi for 10 s
Run time:	8 min
Voltage:	10,000 V
Buffer:	Borate
Sample:	Serum diluted 1:10 in PBS

* Registered trademark of Beckman Instruments, Inc. (Brea, CA).

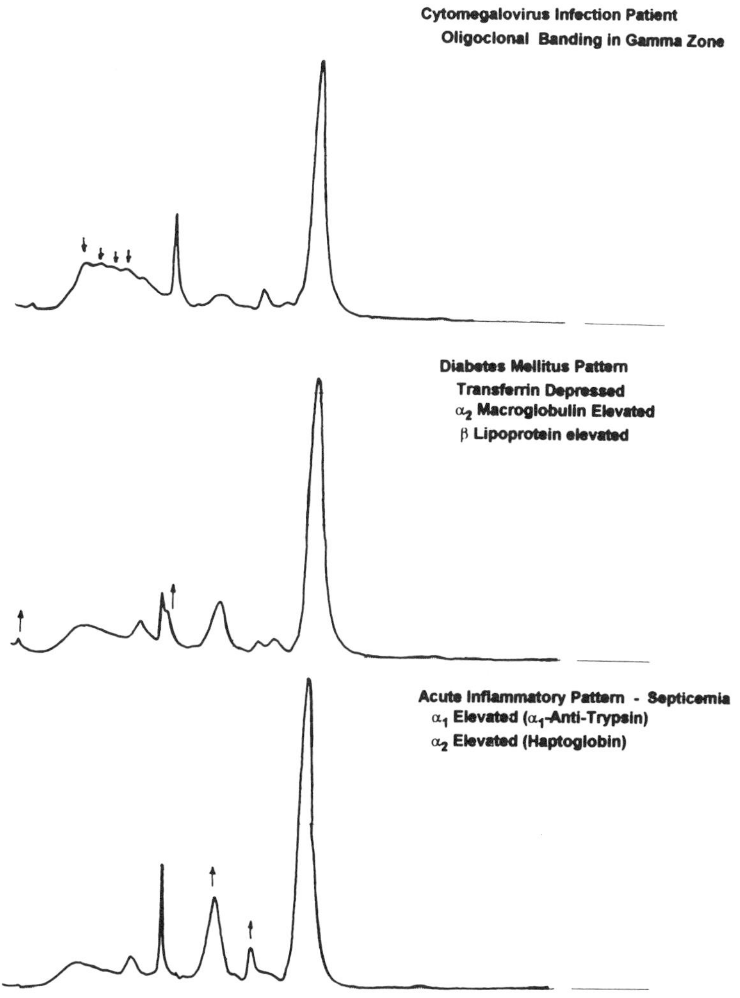

FIGURE 12. Capillary electrophoretic scans of patients with selected diseases.

Agarose gel electrophoresis provides a visual pattern with the dried stained gel. Densitometric scans of the stained gel are used to quantitate the distribution of the proteins in the five accepted zones. It is important to visualize the gel directly and interpret differences between the patient's pattern and the reference control pattern for classifying the dysproteinemia. Using computer graphics, we have been able to generate from the CE amplitude data a "synthetic" pattern that closely resembles the appearance of stained gels. Note that it is the inverse of densitometry. With a densitometer scan of a stained gel, a graphical trace whose relative height is proportional to the intensity of the stained proteins is produced. Photometric scans with a CE detector are, in effect, densitometry at 214 nm of the separated fractions as they pass the detector window.

With CE, the separations occur in a fluid matrix and no direct visual record of the proteins is available; only the graphical trace remains. The relative amplitudes of the data from the trace are "painted" on the computer screen with relative color intensities proportional to the amplitude of the trace. This allows not only quantitation of the graphical data, but qualitative visual comparisons of generated patterns of samples and controls. Figure 13 shows an example

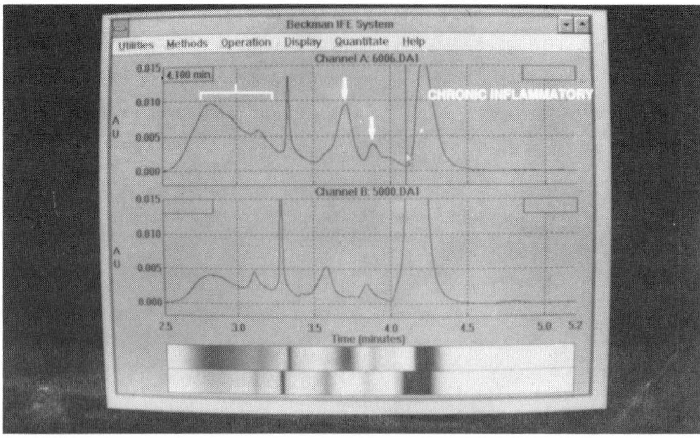

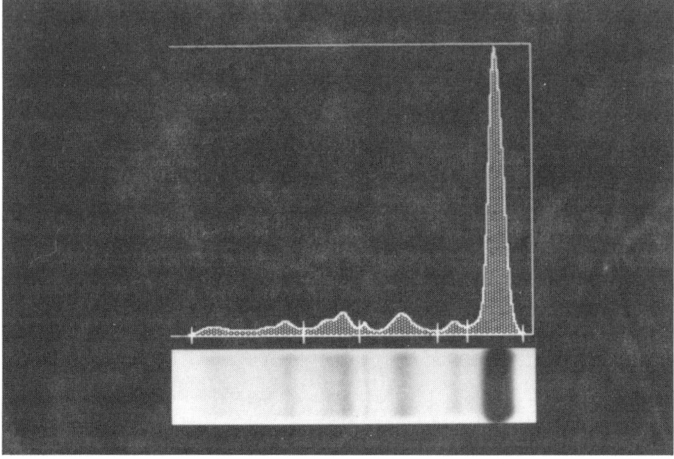

FIGURE 13. Comparison of CE scans with synthetically generated patterns (top) to agarose gel pattern and densitometer scan (bottom).

of the simulation for a normal pattern and a dysproteinemia. For comparison, a typical stained gel pattern with a densitometer scan is also shown. Archiving of the patient's data with AGE requires attaching a densitometer scan to the patient's chart. Since several separations are usually combined on a common matrix (gel or acetate), these are usually labeled and stored separately and must be cross-referenced to the patient's chart. The additional labor and opportunity for error are obvious. Archiving of data with CE will be electronic, requiring less labor. Specific areas of the synthetic pattern can be "enhanced" to visually show subtle distributions of the relative densities. This is equivalent to selectively overstaining a gel, but is not practical with gels.

Detection of proteins of restricted mobility by both gel and acetate methods is important in recognizing possible paraproteinemias. In an unpublished study, 240 known paraprotein samples (elevated gamma globulins) were compared using a multichannel CE system with the standard AGE Paragon method.[19] Diagnostic concordance was shown in the sensitivity of detecting paraproteins of all types.

With the appearance of a sharp immunoglobulin peak in a serum protein profile, it is important to classify the protein for diagnostic interpretation as IgG, IgA, IgM heavy chain proteins, and κ or λ light chain types. With AGE, this is accomplished with either immunoelectrophoresis (IEP) or immunofixation electrophoresis (IFE). Both of these methods are labor intensive and skill intensive and require a relatively long execution time. With CE, since no permanent pattern of proteins on a support is available, it was necessary to develop a procedure with equivalent sensitivity that would supplant the above techniques for classifying the paraproteins. The method of "immunofixation/subtraction" was shown to be effective. With this method, the serum sample is mixed and incubated with a solid phase to which specific antibodies are bound. During the mixing and incubation, the specific proteins in the sample are fixed and held to the solid phase. Subsequently, a series of six separations is performed. One sample has been exposed to the solid phase that has no fixant attached while the remaining five have been exposed to the specific fixants (IgG, IgA, IgM, κ, and λ, as noted above). Comparisons of the subtracted scans with the control scan show deletions of the proteins corresponding to the specific immunofixant, thereby providing a means to classify the paraprotein.

Classification of paraproteins by gel electrophoresis (AGE) methods is accomplished by either IEP or IFE. With IEP, a serum sample is applied to the gel and separated in the usual manner. After separation, a trough is cut into the gel alongside the separation path, and a protein-specific antibody is inserted. After an incubation period, the antibodies and the separated proteins diffuse toward each other through the gel. Where the concentrations of the protein and antibody are equivalent, a complex is formed and precipitates. After staining, these precipitin bands are visible and can be interpreted. While adequate quantitatively this is a lengthy process (usually 18 to 24 h). With IFE, the serum sample is separated in six individual paths, similarly to IEP. After separation, a template is applied on top of the gel with six rectangular slots. Antibodies of known specificities are applied within the slot areas of the template, diffuse into the gel, and the gel is incubated. The control track is exposed to protein fixative such that all the separated proteins are precipitated. As with IEP, where the antigen–antibody concentrations are equivalent, a precipitin complex is formed and becomes fixed. After incubation, the gel is rinsed with saline, washed, and then pressed with a blotter to dispel remaining soluble proteins. The gel is then stained and interpreted by comparing the control track with the specific antibody tracks for the presence of stained bands. With CE a process similar to IFE is employed. The specific proteins are fixed to a solid phase prior to separation through antigen–antibody binding. After separation, identification of the specific proteins is established by their absence from the separation. This procedure is called immunofixation electrophoresis/subtraction (IFE/S).[20] A comparison of CE IFE/S and gel IFE is shown in Figures 14 through 16.

1. Classification of Gammopathies by Capillary Electrophoresis

The classification of gammopathies shown in Table 2 by the gel methods is divided into five distinct classes:

Polyclonal
Monoclonal with normal heterogeneous immunoglobulins
Monoclonal with decreased heterogeneous immunoglobulins
Hypogammoglobulinemias with urinary Bence–Jones proteins
Oligoclonal with multiple monoclonal bands

With CE, these types of antibody separations are characterized by distinct peak shapes. These are illustrated in Figures 17 and 18.

Polyclonals will show a broad, "dome-shaped" peak quite similar to a normal gamma globulin peak but elevated. With a pathological gammopathy, there exists a marked change in the κ-to-λ ratio. This can be determined with IFE/S or with specific κ and λ immunoassays.

Monoclonals with a normal heterogeneous immunoglobulin background are of a relatively low concentration and appear as "mini" monoclonals characteristic of the early stages of a

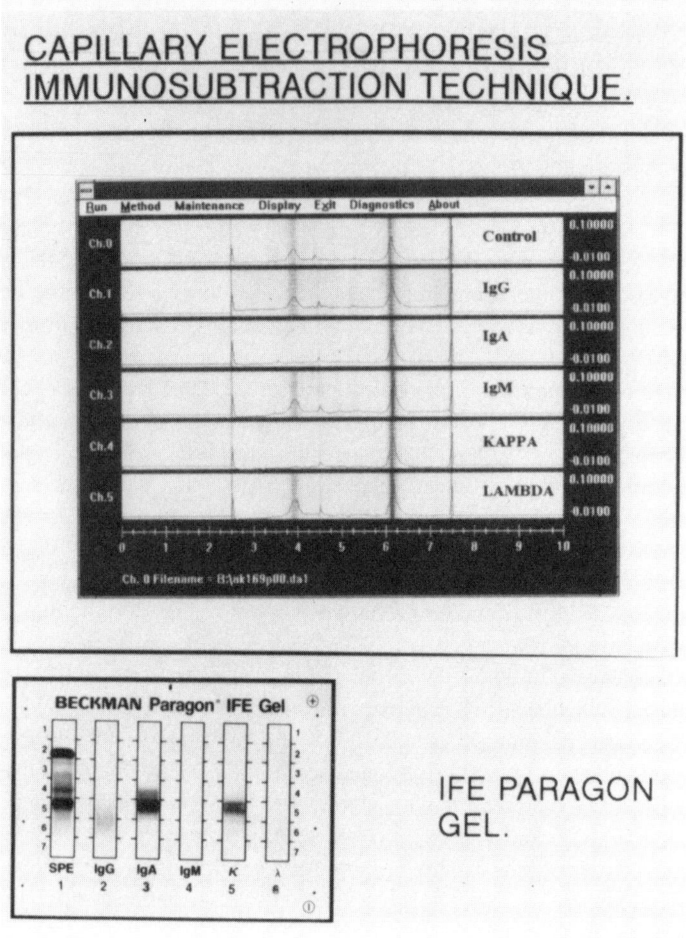

FIGURE 14. Comparison of CE immunosubtraction (top) and agarose gel IFE method (bottom) for an IgA(κ) gammopathy.

disease process. Frequently the mini monoclonals may be an antibody response to epitopes of a specific infectious agent such as a virus and will not be myeloma proteins. Assessment of the κ-to-λ ratio will differentiate such pathologies. They appear as small, sharp peaks on the underlying broad immunoglobulin pattern. These patterns may reflect monoclonal gammopathies of undetermined significance or be an indication of an immunoproliferative process.

Monoclonals with reduced heterogeneous immunoglobulins are more characteristic of a monoclonal gammopathy of a more serious nature. With CE they appear as elevated sharp peaks.

Hypogammoglobulinemias with urinary Bence–Jones proteins are light chain-type gammopathies. A small peak may appear in a reduced immunoglobulin background, which indicates the presence of free light chains in the serum. Full diagnostic interpretation must include a urine analysis for free light chains.

Oligoclonal patterns show more than one sharp peak in the gamma globulin region. Biclonal, triclonal, and multiple peaks can be seen. They may be of the same or different immunoglobulin classes.

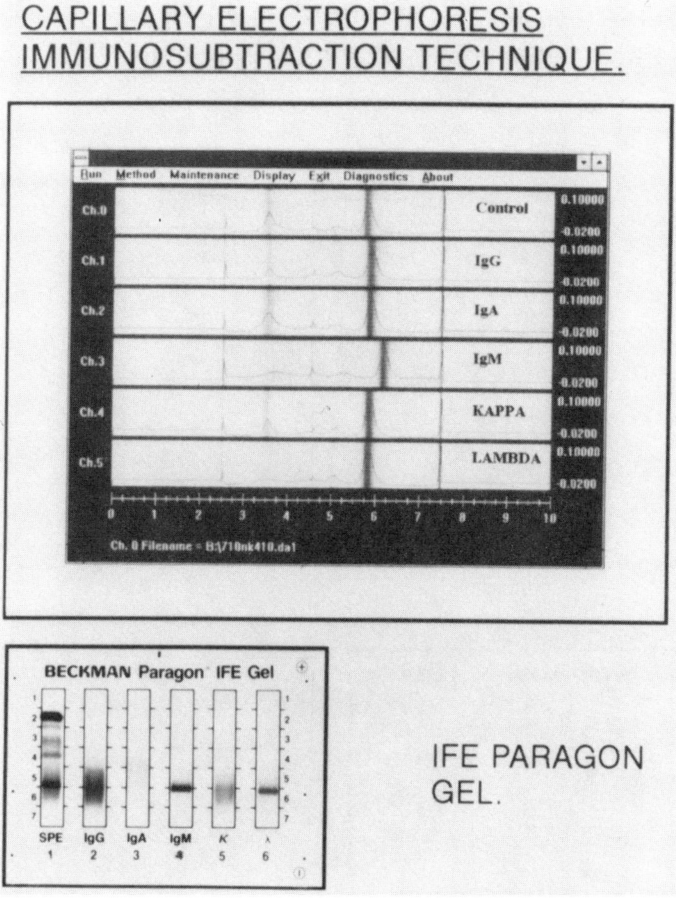

FIGURE 15. Comparison of CE immunosubtraction (top) and agarose gel IFE method (bottom) for an IgM(λ) gammopathy.

In our studies of comparing AGE to CE an interesting pattern was discovered that highlights the advantages of CE from a resolution perspective. This pattern demonstrated a small gammopathy of the IgA κ type. In the gel pattern it appeared as a slightly elevated β fraction that could have been interpreted as an elevated transferrin. In the CE pattern, the small peak is distinctly separated from the transferrin. This is shown in Figure 18 and clearly identifies the potential advantage of CE.

2. Factors Affecting the Serum Pattern Shapes

Joule heating — In the study described previously for routine serum protein screening, Joule heating did not appear to be a problem since no degradation of the patterns was observed. However, some means of dynamic heat removal from the outer wall of the capillary is needed, and can be in the form of free convection if the capillary is not contained in the form of forced air convection or forced liquid conductive transfer if contained. As described in Chapters 2 and 21, monitoring for Joule heating is easily accomplished by measuring the current flowing in the capillary during separation; a continuous increase indicates that severe Joule heating is occurring. Adverse Joule heating will cause band broadening due to thermal turbulence, and may ultimately result in the formation of gas bubbles.

Electroosmotic flow — The EOF is the driving force for the clinical separations with CE. At the high alkaline pH (8.6 or above) used for clinical serum separations the EOF is about

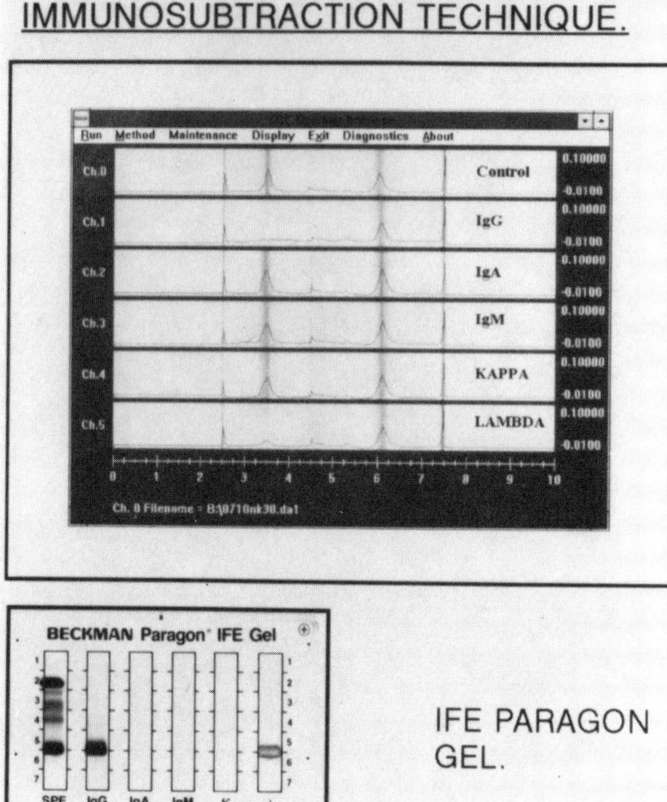

FIGURE 16. Comparison of CE immunosubtraction (top) and agarose gel IFE method (bottom) for an IgG(λ) gammopathy.

2 mm/s toward the cathode. Since the pH of the buffer is selected to be above the p*I* of any serum components, separation occurs with the serum components migrating counter to the EOF. Since the forward velocity of the EOF is greater than the migration velocities of the serum components, all fractions will pass a stationary position along the capillary for detection. There are a number of techniques to reduce or eliminate the EOF by coating the inner surface of the capillary with modifiers.[21] None of these have proved to be sufficiently robust for reproducible performance in high-volume routine applications.[22]

Diffusion — Because of the short scan time and the small bore of the capillaries used for clinical serum separations, diffusion has not been shown to be an important contributor to the pattern shapes.[4]

Buffer composition and pH — This is one of the most important factors in pattern morphology. The ideal buffer should exhibit low absorptivity in the far-ultraviolet (UV) spectral regions. It should be adjusted to a pH value above the p*I* of all serum components (i.e., above 9.10). It should have an adequate ionic strength to bias the equilibrium of dynamic complex formation with the serum constituents toward complexation so that protein adsorption to the capillary wall is minimized. As described in Chapter 18, small changes in pH have a profound effect on the morphology of the patterns, requiring close control of the pH. Figure 19 illustrates the relative migrations of the major components of a serum pattern at differing pH values as measured with a 20-cm capillary length. At different pH values noticeable variations of the patterns occur as compared to the preferred pattern.[19]

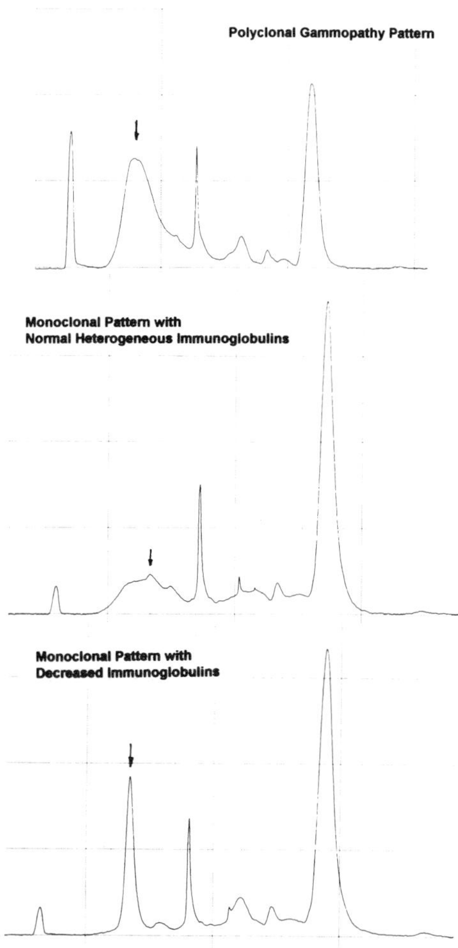

FIGURE 17. Capillary electrophoretic scans of gammopathy patterns.

Ionic strength also has an effect on the pattern. Figure 20 shows the effect of retention times on the major components of the pattern at varying ionic strengths.[19]

Solute concentration effects — Consider a unit volume along the capillary. It contains the solvent (water) plus buffer salts in an ionic equilibrium. Another neighboring unit volume contains additional sample components. The net conductivity of the two volumes will not necessarily be equal due to the contribution of the sample molecules. At the start of a separation the sample volume will contain all the solute molecules at a total maximum concentration. As the separation proceeds the components with higher mobility will partition away from the initial volume. The distribution of solute concentration spreads over the separating pattern. For the serum patterns, the dominant concentration is albumin, which moves rapidly counterflow toward the anode. The remaining proteins migrate progressively slower and interact with each other. The final pattern is a snapshot of a dynamic process captured at a time determined by the capillary length, the buffer characteristics, and the applied field. On close examination, however, one can observe small shifts in the relative positions of recognizable peaks in response to a disease process that has increased or decreased the concentration of the proteins in the peaks relative to a normal pattern. This is a consequence of the solute concentration effect. An example is shown in Figure 21.

Wall–solute interactions — Even though the pH of the running buffer is selected to induce a strong negative net charge on all of the protein molecules in a serum sample, there

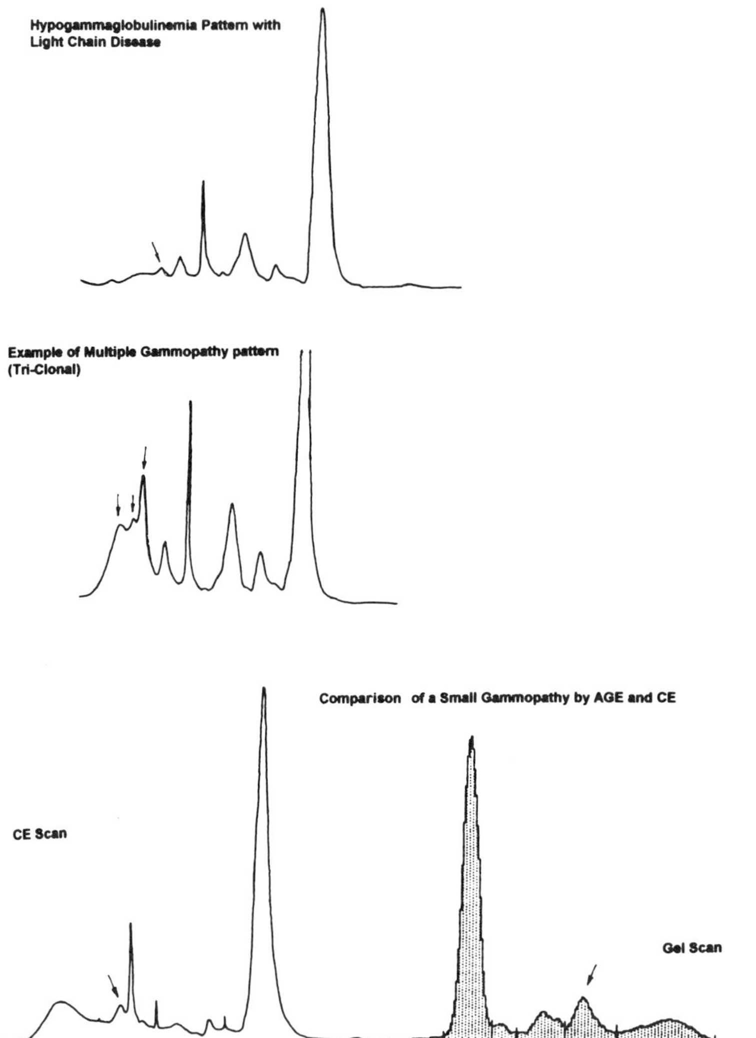

FIGURE 18. Capillary electrophoretic scans of gammopathy patterns.

still remain regions of the protein tertiary structure with net positive charges. If asymmetrically or unevenly distributed, these can replace the singly valent Na^+ buffer ions and achieve a weak equilibrium binding with the wall. Experiments involving successive runs without a NaOH rinse showed progressive lengthening of retention times and a loss in resolution suggestive of a protein buildup on the capillary wall. After a single NaOH rinse, these capillaries returned to normal performance. By examining a serum pattern closely one can observe asymmetric tailing of the albumin peak. Since the albumin is the highest concentration component of serum, it is likely that this is evidence of minor wall interaction. Other fractions of the pattern do not show the asymmetrical tailing; in fact, they closely fit Gaussian models. An example of this is shown in Figure 22. It is an expanded view of a patient sample's haptoglobin peak with a mathematically generated Gaussian peak superimposed.

Temperature — Once a protocol has been established with a given buffer pH, composition, and molarity, it is important that the running temperature be maintained within narrow limits. This is primarily due to the thermal change of the pH of the buffers. Small changes in pH can have a noticeable effect on pattern morphology, as discussed above.

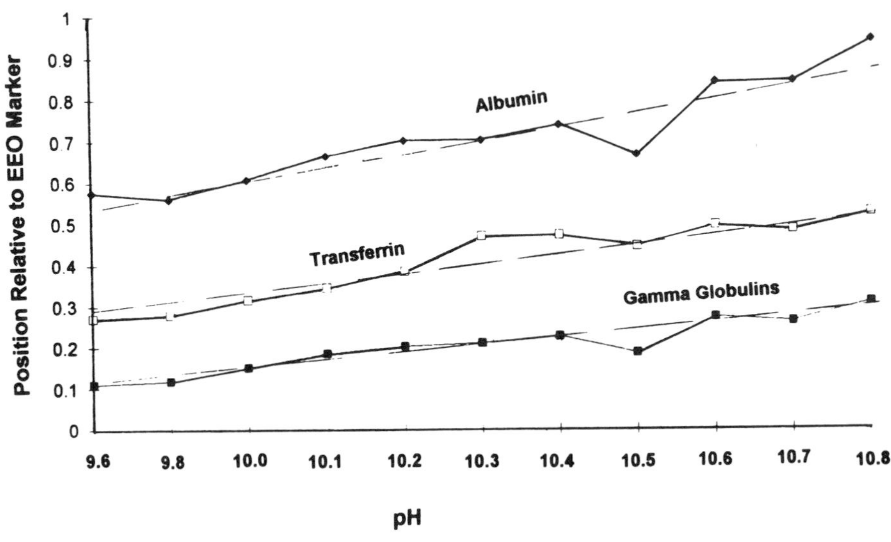

FIGURE 19. Migration times vs. buffer pH (150 m*M* borate).

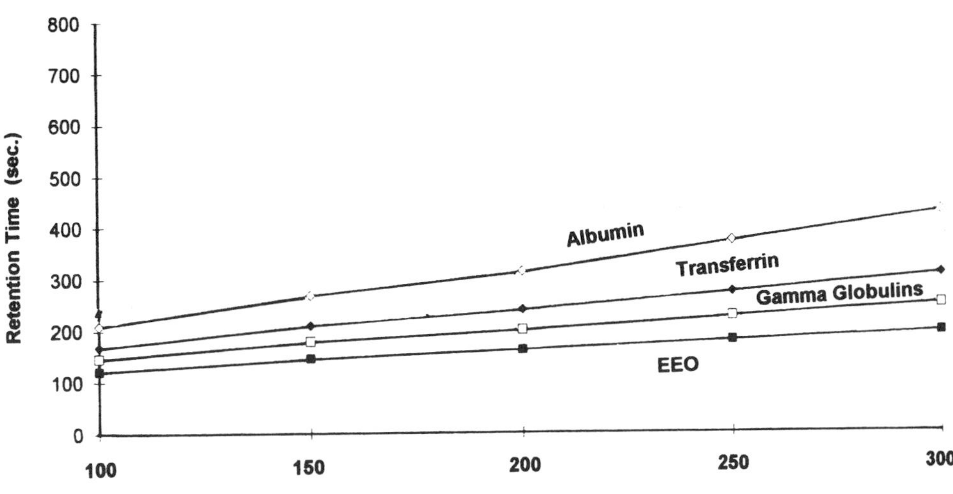

FIGURE 20. Migration times vs. buffer ionic strength (pH 10.2).

Applied field — The applied field seems to affect only the migration times up to the point of adverse Joule heating. As higher fields are used, separation times can be reduced, although apparent resolution may be reduced by a crowding of the separated peaks. This is largely due to the detector sampling rate and the final display format. Ultimately the optimal voltage will be dictated by the needs of the analysis. Shorter separation times (i.e., higher fields) can be an advantage for increased analytical throughput. A plot of the relative separation times at field voltages varying between 10.5 and 15 kV in a 20-cm column is shown in Figure 23.

Capillary diameter — There is a tradeoff, when using larger diameter capillaries for separation, between detection limit and speed of analysis. Higher detection limits can be obtained with larger bore capillaries due to an increase in the absorbing path length. However, Joule heating increases as the square of the capillary diameter. To avoid adverse Joule heating, the applied field must be lowered, resulting in longer separation times. Ultimate selection of the optimal column diameter will depend on the needs of the

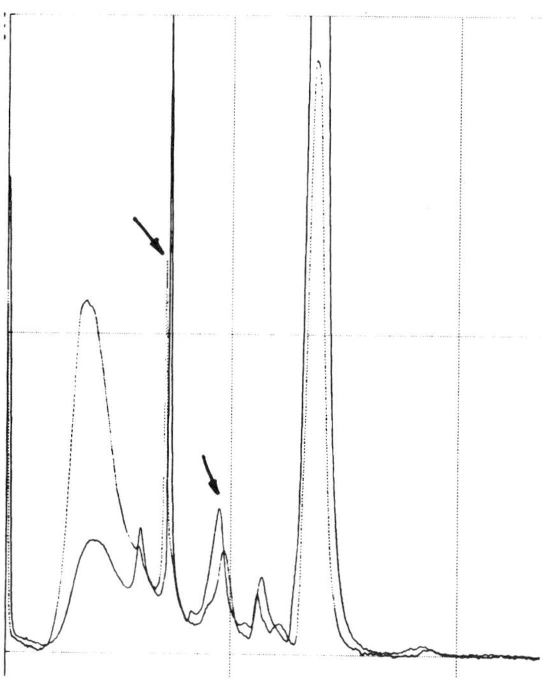

FIGURE 21. Overlay of a normal pattern and hypergammopathy pattern, showing small position shifts of transferrin and haptoglobin due to differences in relative concentration of the proteins.

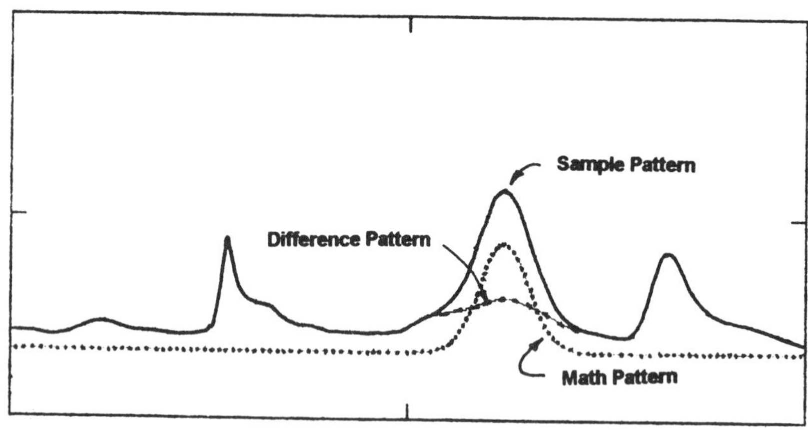

FIGURE 22. Comparison of a patient sample haptoglobin peak with a mathematically generated Gaussian peak of equal area.

analysis. For routine clinical separation of serum proteins a smaller bore capillary is advantageous.

Capillary length — For most clinical applications the capillaries are typically selected to be relatively short (20 cm). Increasing the length appears to offer no immediate benefits. As discussed above, the proteins are very heterogeneous. For example, haptoglobin will consist of many individual molecules that may have slightly different structures due to genetic variations and due to their function as transport proteins. Together these factors create a distribution of molecules, all of which are haptoglobin, but each of which will carry a different net charge and migrate electrophoretically at slightly different velocities. These variations are

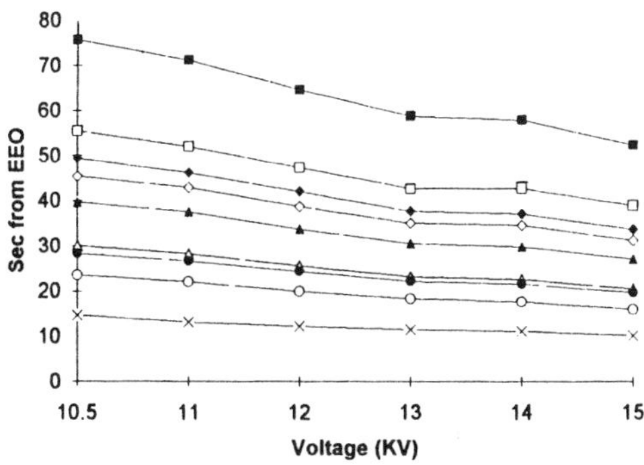

FIGURE 23. Voltage effect on separation time. ■, Prealbumin; □, albumin; ◆, α_1-acid glycoprotein; ◊, α_1-antitrypsin; ▲, haptoglobin; Δ, α_2-macroglobulin; ●, transferrin; ○, complement (C3); ×, gamma globulins.

not resolvable by CE, but a Gaussian distribution does adequately describe the shape of the peak. Experiments with longer columns did not appear to improve the resolution of individual fractions, but simply lead to widening of the Gaussian peaks. Therefore, shorter columns with faster separation times are preferred for routine clinical separations, finding an adequate balance between throughput and information content is desired.

3. Modeling of Serum Protein Patterns

To better understand the information in a serum pattern, models have been generated using mathematical functions. Variables for the functions were derived from the published parameters for specific proteins[13,23] and the basic assumptions for the models were as follow.

1. The profile of the injection volume is preserved during the separation. The theory of plug flow with minimal band broadening is consistent with this assumption.[24]
2. The final pattern represents a superimposition of all the individual components in the sample. The graphical presentation is in absorbance units representing the extinction of the proteins at the measuring wavelength. Each individual fraction will have a peak shape consistent with the sample characteristics and separation parameters. Overlapping peaks will be additive.
3. Position is determined by electrophoretic mobility as a function of net mass-to-charge ratio. Any effects due to sieving will be neglected since serum separations are in open capillaries. A different model will be required for gel separations.
4. Proteins are very heterogeneous. Genetically determined variations in the structure as well as the transport function of many of the molecules will result in spreading of the bands relative to ideal pure compounds. A Gaussian distribution serves as the best model for the distribution of these variations.

A simple model based on these assumptions for the major fractions in a normal serum protein pattern is shown in Figure 24. This model will be used for illustrative purposes to discuss the techniques of peak identification, quantitation, and immunosubtraction. As our understanding of the capillary separations of serums increases with further research, refinements of the mathematical model will be developed. For example, the proteins listed in Table 1 represent those that should be identifiable based on a detection limit evidenced by prealbumin. Not all of these have been recognized. Comparisons of actual serum separations with mathematical models may lead to the recognition of these.

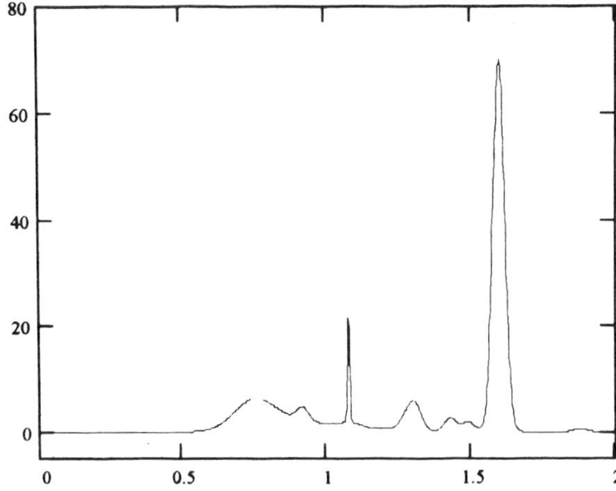

FIGURE 24. Mathematical model of a serum protein pattern.

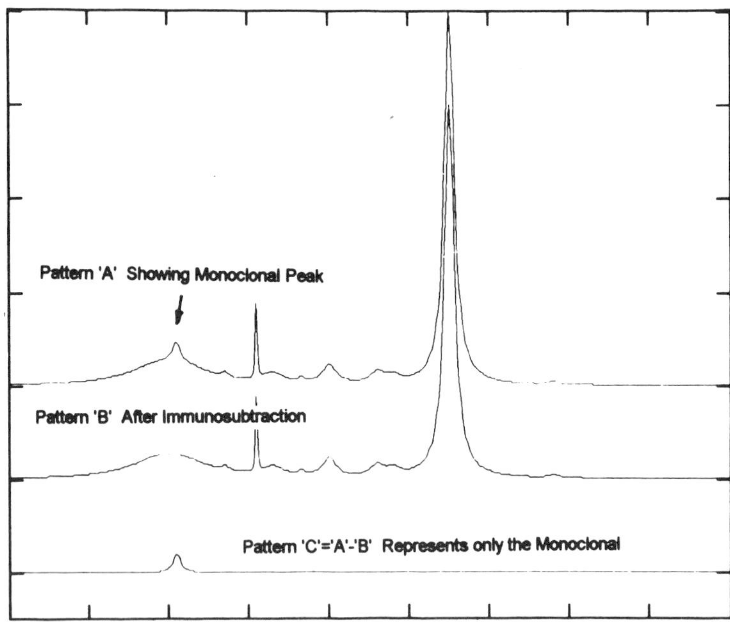

FIGURE 25. Model of immunosubtraction of a small monoconal peak.

4. Immunofixation/Subtraction

Immunofixation electrophoresis/subtraction is a method that allows the identification of specific peaks in a separated pattern. Immunosubtraction was first demonstrated by Aguzzi with acetate membrane electrophoresis.[25] An important application of the method for CE in a clinical setting will be the ability to classify the paraproteins. It has also been used to confirm the location of specific components for producing a CE protein map and has the potential for quantitation of specific proteins.

The method involves incubating the sample with a solid phase to which is bound antibodies or antigens of known specificity. After incubation, the specific components in the serum sample will be fixed by the specific solid-phase ligand and will not inject as part of the serum sample for separation. Comparison of the patterns with and without fixation shows a reduction of the pattern peak corresponding to the fixed moiety. The patterns may be mathematically subtracted from each other, which, under ideal conditions, yields a

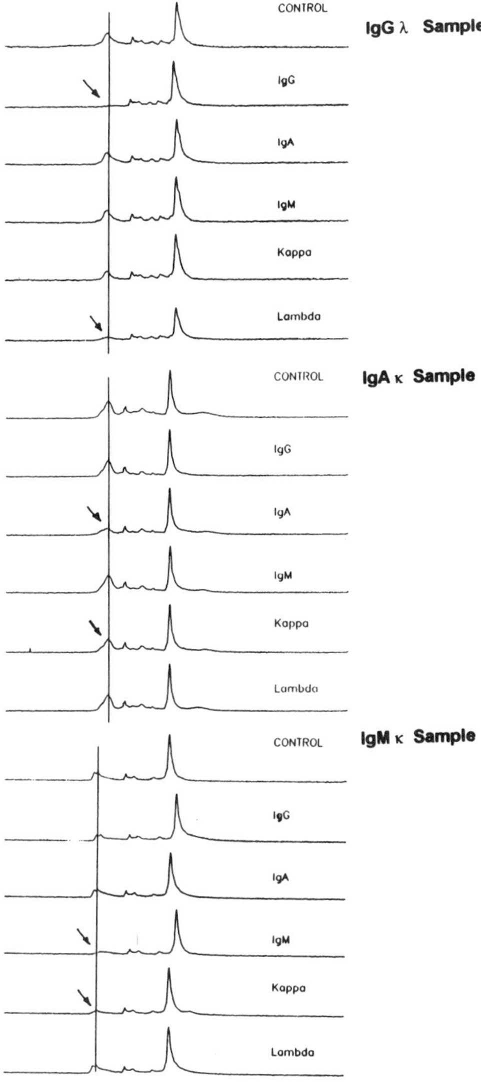

FIGURE 26. Examples of six-channel immunosubtractions patterns used in typing paraproteins.

pattern that represents the fixed peak. This process is shown conceptually with a mathematical model in Figure 25.

The solid phase used in this method involves the covalent attachment of specific fixants (antibodies) to a solid phase corresponding to IgG, IgA, and IgM heavy chains and κ and λ light chains. An additional well is prepared with naked solid phase for a reference control. Six dilutions of a serum sample are added to volumes containing the beads and lightly mixed. The beads are allowed to settle and capillaries are introduced into the wells for sample injection. The separation is performed as with serum separations (described in Section III). The results of the separation are compared with the control separation. If there is a large paraprotein peak it will dominate the gamma globulin region of the separation. In the scans of samples exposed to the immunofixants, there will be a marked reduction of the peak amplitude corresponding to the specific immunofixant of the protein heavy chain type. There will also be a reduction in the amplitude for the corresponding light chain type. From these observations it is easy to classify the gamma fraction as an IgG, IgA, or IgM, or as κ or λ paraprotein.[26] Examples of these patterns are shown in Figures 26 and 27.

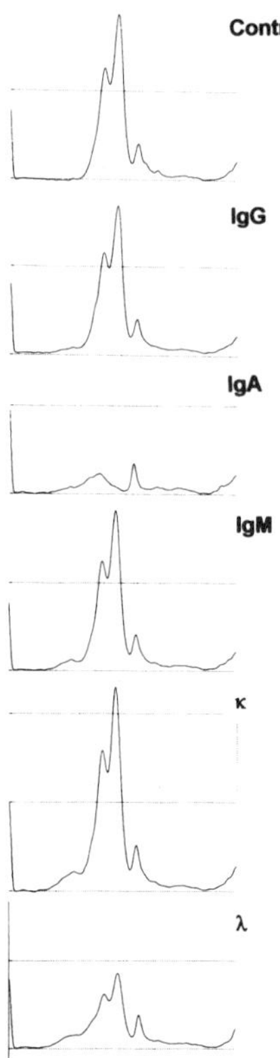

FIGURE 27. Enlarged view of a biclonal IgA — lambda immunosubtraction.

For the mini monoclonals, which will appear as a small discernible peak on top of the broad gamma zone, the interpretation of the subtraction is slightly different. Sufficient concentrations of all the immunoglobulins may be present in the sample, in addition to the sharper monoclonal peak, such that the immunofixants also fix the normal immunoglobulins. The specific immunofixants that eliminate the sharp mini peak or noticeably reduce its amplitude will be the characteristic types for classification of the mini paraprotein. Examples of these patterns are shown in Figure 28.

B. ISOENZYMES

Isoenzymes are tissue-specific forms of an enzyme that, while differing slightly in structure, share the ability to catalyze a common substrate. The structural differences are substantial enough to allow for their separation by electrophoresis. The ability to measure the presence of these molecules in serum is very important from a diagnostic point of view.

Normally the level of enzymes found in the circulatory system is low and due to the natural catabolism of cells. Elevations in the levels of circulating enzymes may show disease correlations that can be diagnostically important. This is evidenced by the routine clinical use of isoenzyme analysis for alkaline phosphatase (AP), lactate dehydrogenase (LD), and creatinine phosphokinase (CK), which began in the 1970s and is widely used today.[27-29]

Can capillary electrophoresis be adapted to assay for isoenzymes? We have developed a model system, using LD, that demonstrates the feasibility of developing CE isoenzyme assays. Lactate dehydrogenase is composed of four polypeptide chains of two types called H (heart) or M (muscle) subunits. There are five combinations of these subunits that give rise to the five electrophoretically separable fractions LD1, LD2, LD3, LD4, and LD5. The LD1 and LD2 types are associated with cardiac muscle, kidney and red blood cells (RBCs). LD4 and LD5

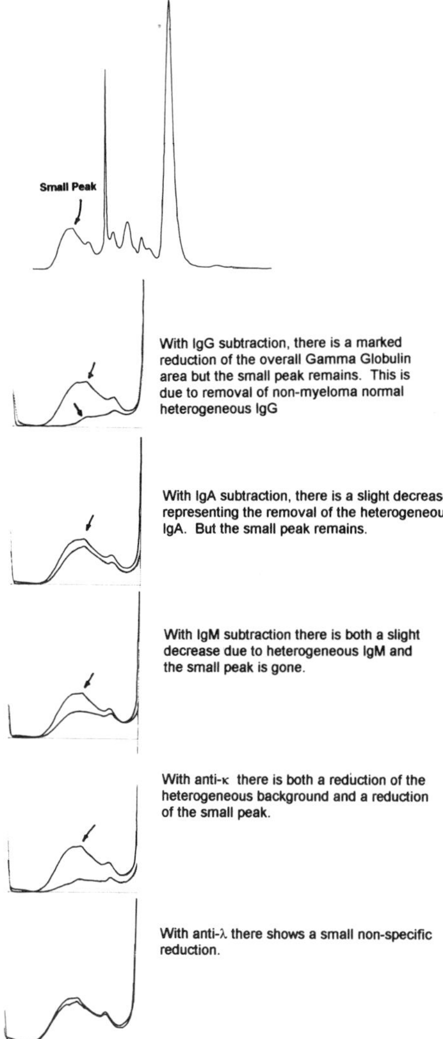

Small Peak
With IgG subtraction, there is a marked reduction of the overall Gamma Globulin area but the small peak remains. This is due to removal of non-myeloma normal heterogeneous IgG
With IgA subtraction, there is a slight decrease representing the removal of the heterogeneous IgA. But the small peak remains.
With IgM subtraction there is both a slight decrease due to heterogeneous IgM and the small peak is gone.
With anti-κ there is both a reduction of the heterogeneous background and a reduction of the small peak.
With anti-λ there shows a small non-specific reduction.

FIGURE 28. Enlarged view of a mini monoclonal IgM — kappa immunosubtraction.

are associated with skeletal muscle and liver. LD3 is associated with endocrine glands, spleen, lung, lymph tissue platelets, and nongravid uterine muscle.

The determination of LD activity is through the measurement of reduced nicotinamide-adenine dinucleotide (NADH). The reaction is reversible and can be biased in either direction by adjusting the pH of the supporting buffer. NAD-to-NADH conversion can be monitored by an increase in absorbance at 340 nm. In gels, a formazan dye is included as a redox indicator to generate a visible indicator.

In standard gel electrophoresis, an isoenzyme method is usually a two-step process. First, a sample is separated in a standard buffered gel. Second, a substrate is overlaid onto the separated gel. As the substrate diffuses into the gel, a measurable product forms at the locations where the enzymes have separated. An example of this is shown in Figure 29.

With capillary electrophoresis it is technically impossible to first separate and then uniformly apply a substrate within the capillary. Therefore, we chose to incorporate the substrate as a component of the running buffer. This results in a complex dynamic process that occurs during separation. The capillary is filled with a buffer containing a suitable substrate with the pH

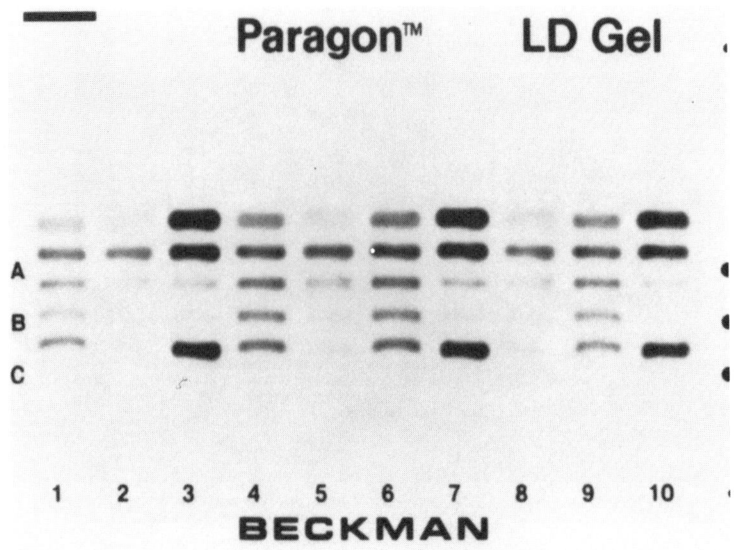

FIGURE 29. Example of a gel LD isoenzyme pattern.

adjusted for optimal enzyme kinetics. An enzyme-containing sample is injected into the capillary. At this time all of the isoenzymes will be exposed to the substrate at the boundary between the sample volume and the buffered substrate. An immediate reaction between the isoenzymes and the substrate occurs and is termed "sample shock." Because of the manipulation times required to place the capillary into the sample solution, inject the sample, move the inlet end to the running buffer, and initiate the separating field, the contact between the sample volume and the substrate in the running buffer in the capillary will be stationary and a diffusion-limited enzyme–substrate reaction will occur, forming a product peak at the boundary. After the separating field is established separation begins. The principal driving force for migration will be EOF, which transports the bulk volume in the capillary toward the cathode end. In this flowing volume, the sample components, including the isoenzymes, will migrate counter to the EOF toward the anode end and will separate according to differences in their net mass/charge ratios. The isoenzymes, isoenzyme–substrate complexes, the substrate, and the catalytic products all exhibit different mobilities. We can ignore the mobility of the substrate as it is homogeneously distributed in the running buffer and always available to the isoenzymes.

During catalysis the formed product, having a different mobility than the isoenzyme, will migrate away from the isoenzyme. Unlike stationary enzyme analyses, the product will not accumulate but will "stream" away from the enzyme. A steady state reaction creates a product stream whose concentration will be a function of the difference in the mobility between the isoenzyme and the product, the concentration of the isoenzyme, and the specific activity of the isoenzyme. At the beginning of the separation, all the isoenzymes are present in the sample volume such that the product stream represents the sum of the contributions of all the isoenzymes. As the separation progresses, the isoenzymes will partition and the product stream will decline to the point where it represents only the contribution of the fastest migrating isoenzyme. Interspersed will be a declining product stream. Assuming EOF toward the cathode represents the fastest velocity of all moving components, the enzymatic reactions can be observed as an increasing signal ending with the sample shock peak. If the boundaries between the individual isoenzymes can be detected, measurement of the relative differences in adjacent parts of the product stream can indicate the concentrations of the individual

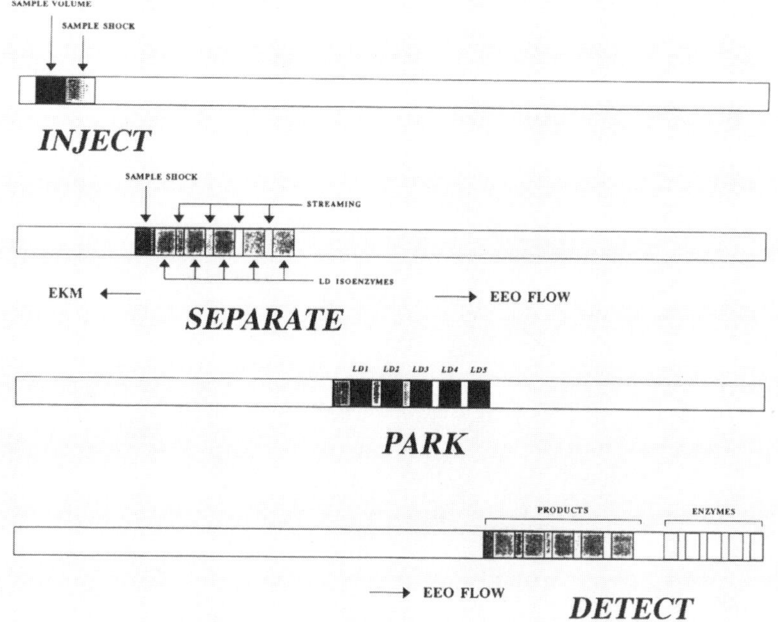

FIGURE 30. Model for CE isoenzyme separations.

isoenzymes. In actual practice, the concentrations of the individual isoenzymes produce a product signal below the detection limit for the capillary system and the boundary between the individual products is often difficult to recognize.

To overcome this difficulty, we add a stop-flow interrupt of the high voltage to "park" the isoenzymes. This is done before any of the separating components pass the detection position, but sufficiently delayed to assure that the individual isoenzymes have been clearly separated. Choice of separation conditions and column length must be determined for each isoenzyme class. With the electric field turned off, the enzymes continue to catalyze product, but the product does not migrate away from the enzyme. An increasing product concentration builds in a diffusion-limited region around the isoenzyme that results in the formation of a detectable localized peak above the signal background created during the dynamic streaming.

After an appropriate incubation "park" time, the separating field is reestablished and the products are driven past the detector window. The detection wavelength is selected to respond only to the product signal. These concepts are illustrated in Figure 30.

Experiments with purified LD isoenzymes demonstrate the above concept. The procedure used was the following:

Sample:	Purified LD extracts (LD1 through LD5)
Matrix:	Normal human serum
Sample buffer:	Tris (20 m*M*) with EDTA (1 m*M*) at pH 8.7
Running buffer:	Beckman Dri-Stat LD with lactic acid and NAD added with pH adjusted to 8.7
Sample injection:	Electrokinetic 1 kV for 1 s
Capillary:	75-μm bore, 375-μm O.D., 37 cm long
Separation:	7.5 kV for 2 min
Park time:	10 min
Detection:	5 kV for 6 min
Wavelength:	340 nm

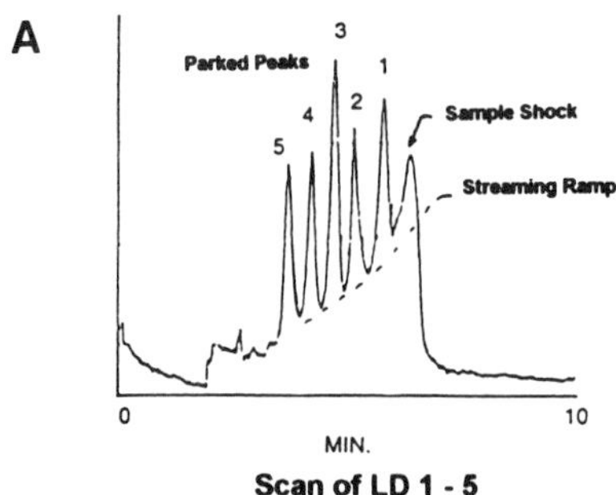

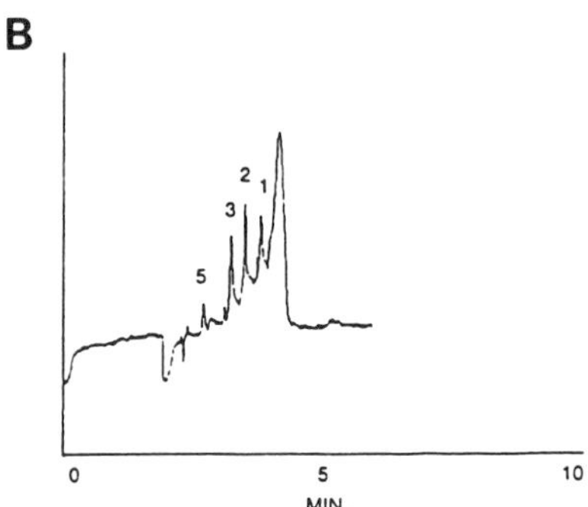

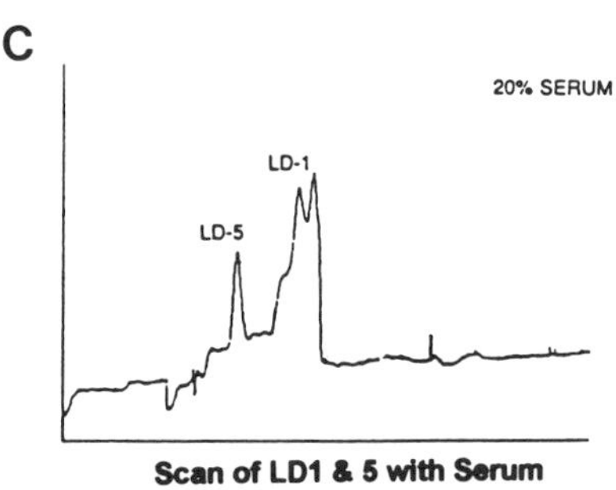

FIGURE 31. Experimental examples of CE scans of LD isoenzymes: (A) scan of LD1–5; (B) scan of LD1–3 and LD5 at low clinical concentration; (C) scan of LD1 and LD5 with serum.

Results of these experiments are shown in Figure 31. Figure 31A represents purified isoenzymes at a concentration of 1000 IU. Figure 31B shows results at 62.5 IU, which is in the low clinical range. Figure 31C shows the results of adding purified isoenzymes to serum that would be equivalent to a patient's sample. There is some broadening of the product peaks, probably due to interactions of the serum components and the capillary wall during separation. By adjusting the initial dilution of the sample and the park time, the matrix effect of the serum can be minimized. Other possible approaches for minimizing broadening may be realized with preconditioning of the capillary wall with coatings.

C. HEMOGLOBINS

Electrophoresis has been an important tool for separating genetic variants of hemoglobins. The amino acid sequence of hemoglobin chains is genetically controlled. At least five types of chains have been identified: α, β, γ, δ, ε. The hemoglobin molecule is assembled from symmetrical pairs of specific chains into a tetramer and the association of the specific chains

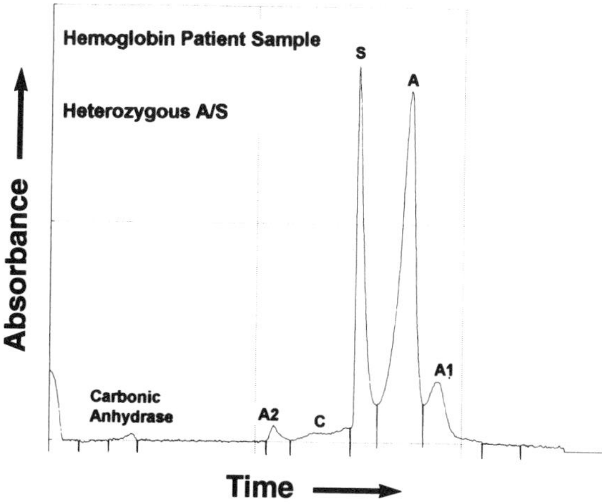

FIGURE 32. Capillary electrophoretic scan of hemoglobin.

defines the genetic type of hemoglobin. During human development, the types of chains present in the hemoglobin change through the embryonic, fetal, and neonatal stages, with the final adult form stabilizing about 6 months after birth.[30]

Normal type A hemoglobin is of the form $\alpha_2\beta_2$ and is genetically dominant with the other types being recessive. Severe hemoglobinopathies can occur with homozygous combinations of other chains. Important among these pathologies are sickle cell anemia, β-thalassemia and α-thalassemia.

Screening for hemoglobin types can aid in predicting disease onset of afflicted individuals and aid in genetic counseling for prospective parents.

Genetic hemoglobin types[31]

A (normal)	J_{Oxford}	Siriraj
F	K	Constant Spring
S	Koln	Seal Rock
$S_{Antilles}$	Lepore	A2
S_{Travis}	$G_{Philadelphia}$	$S/G_{Philadelphia\ hybrid}$
H	D_{Punjab}	$C/G_{Philadelphia\ hybrid}$
I	C	Low O_2 affinity variant
$J_{Baltimore}$	$C_{Georgetown\ (Harlem)}$	

Preliminary experiments with CE show the feasibility of this technique as a rapid tool for hemoglobin screening.

Method

Buffer:	Barbital
Sample:	RBC lysate (lysed with water)
Injection:	Pressure, 5 psi for 10 s
Separation:	12 min at 10,000 V
Temperature:	20°C
Detection:	Absorbance at 415 nm

Results of a series of patients' samples and controls are shown in Figures 32 and 33.

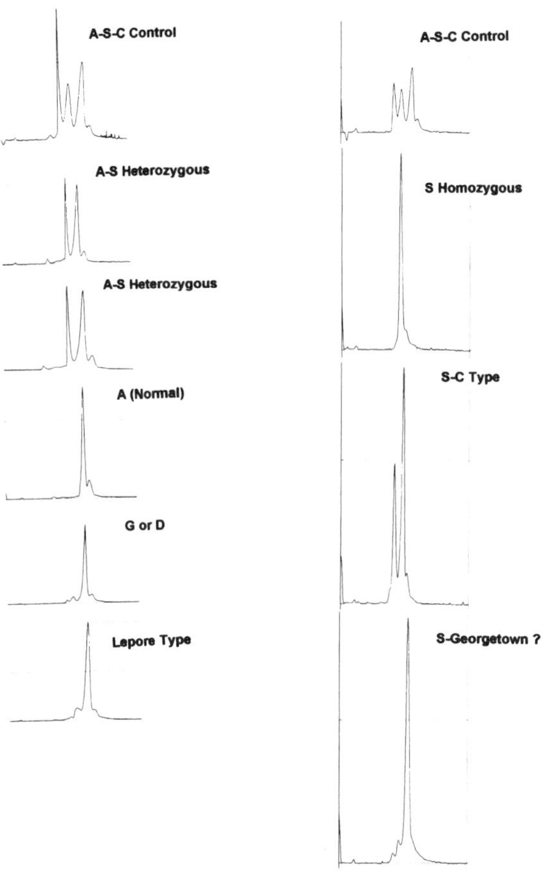

FIGURE 33. Capillary electrophoretic scans of hemoglobins.

D. LIPOPROTEINS

Understanding the role of lipoproteins in the development of atherosclerosis has become important in clinical diagnoses. An effective CE method using isotachophoresis to partition plasma lipoproteins into 14 subfractions has been reported.[32,33] Serum or plasma is prestained with a lipid stain and injected onto the column with a series of spacers. From this procedure an assessment of the quantitative distribution of high-density lipoproteins (HDLs) and low-density lipoproteins (LDLs) can be determined. In addition an assessment of the distribution of the subtypes of LDL as low density, intermediate density, and very low density lipoproteins (LDLs, IDLs, VLDLs) can be determined. Finally, evaluating the ratios of the individual fractions yields information as to the metabolic processes of lipid enzymes in the individual patient. Sample scans of normal and abnormal patients are shown in Figure 34.

The patterns partition into two major groups, with the HDL components on the right and the LDL components to the left. The LDL group partitions further into three subgroups corresponding to LDL, IDL, and VLDL components. The individual peaks represent precursors and postcursors of lipid-active enzymes: lipoprotein lipase (LPL), hepatic lipase (HL), and lecithin cholesterol acyltransferase (LCAT).

Clinically important diagnostic information can be obtained from the patterns by determining lipoprotein distributions and lipid enzyme activities.

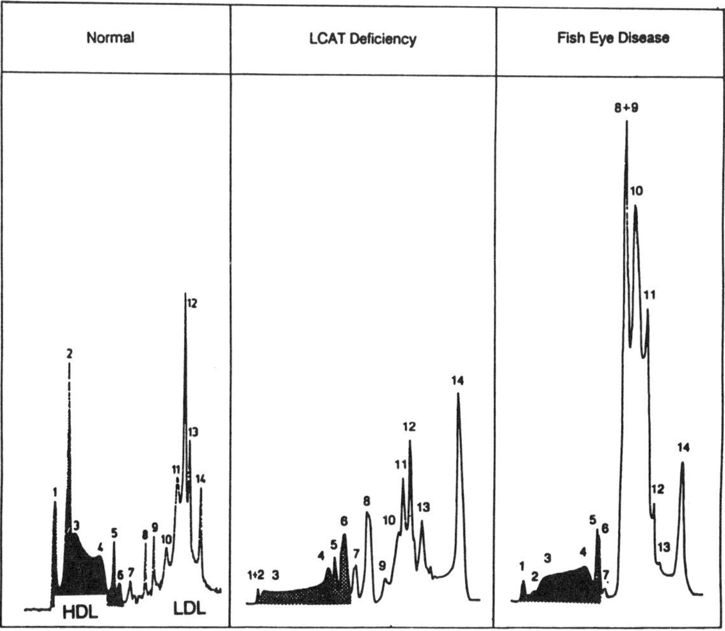

FIGURE 34. Isotachophoresis pattern of serum lipoproteins in LCAT deficiency and fish eye disease.

VI. SUMMARY

From our investigations it is evident that CE, in its present form, is amenable to the separation of human plasma proteins. These preliminary studies clearly demonstrate that recognition of the five currently accepted electrophoretically separated protein zones can be achieved using CE. Further optimizations of the operational parameters will likely lead to more definitive information. Evaluation of the separated proteins at 214 nm is a dramatic change from densitometric evaluation of stained patterns. Value assignments for the five zones and even quantitation of individual protein fractions are achievable.

The ability to automate electrophoresis with a small free-standing system will offer considerable savings in time and labor for the routine laboratory. Quality control of the results will be easier to achieve. The potential for reporting and archiving of the results with electronic media is in keeping with current trends in laboratory management.

The scope of the technology in adapting to other forms of analysis, such as immunosubtraction and those involving isoenzymes, hemoglobins, lipoproteins, and others, suggests CE will be an important tool for the future of the clinical laboratory.

ACKNOWLEDGMENTS

The authors wish to thank Drs. James Sternberg and Fui-Tai Chen for their early work and inspiration in demonstrating that CE can be an important tool for the clinical laboratory, Drs. Cheng-Men Liu and Han-Ping Wang for diligent efforts in optimizing the procedures and development of the methods discussed herein, Cynthia Blessum for providing and coordinating the extensive methods testing, and Beckman Instruments for having the courage to launch a product development project. We wish to thank Phyllis Hergenrader for providing the data for the clinical lab testing. We also wish to thank our early collaborators Dr. Jan Olaf Jeppsson,

Malmo Sweden, Dr. Zacharia Shihabi, Bowman Gray, N. Car., Prof. Gerd Schmitz (Regensberg, Germany) and Prof. Francesco Agguzzi (Stradella, Italy) for their helpful reviews, constructive ideas, and criticisms of the methods.

REFERENCES

1. **Tiselius, A.,** A new apparatus for electrophoretic analysis of colloidal mixtures, *Trans. Faraday Soc.,* 33, 524, 1937.
2. **Durrum, E. L.,** Continuous electrophoresis and ionophoresis on filter paper, *J. Am. Chem. Soc.*, 73, 4875, 1951.
3. **Grunbaum, B. W.,** Microelectrophoresis on cellulose acetate membranes, *Anal. Chem.,* 32(1), 1361, 1960.
4. **Jorgenson, J. W. and Lukacs, K. D.,** Zone electrophoresis in open-tubular glass capillaries, *Anal. Chem.,* 53, 1298, 1981.
5. **Hjertén, S.,** High-performance electrophoresis: the electrophoretic counterpart of high-performance liquid chromatography, *J. Chromatogr.*, 270, 1, 1983.
6. **Spinosa, J. C.,** Challenging the future, *Clin. Lab. Manage. Rev.*, 241, 1992.
7. **Ito, R.,** Analysis of urinary porphyrins by capillary electrophoresis, *AACC Workshop, Beckman Instruments,* 44th Annual Meeting, Chicago, Illinois, 1992.
8. **Shihabi,** personal communication, 1992.
9. **Nielsen, R. G., Rickard, E. C., Santa, P. F., Sharknas, D. A., and Sittampalam, G. S.,** Separation of antibody-antigen complexes by capillary zone electrophoresis, isoelectric focusing and high-performance size exclusion chromatography, *J. Chromatogr.*, 539, 177, 1991.
10. **Koh, E., Bissell, M., and Ito, R.,** Measurement of vitamin C by capillary electrophoresis in biological fluids and fruit beverages using a stereoisomer as an internal standard, *J. Chromatogr.*, 633, 245, 1993.
11. **Schwartz, H. E., Ulfelder, K., Sunzeri, F. J., Busch, M. P., and Brownlee, R. G.,** Analysis of DNA restriction fragments and polymerase chain reaction products towards the detection of the AIDS (HIV-1) virus in blood, *J. Chromatogr.*, 559, 267, 1991.
12. **Chen, F. T.,** Unpublished studies, 1991.
13. **Teitz, N. W., Ed.,** *Fundamentals of Clinical Chemistry*, W. B. Saunders, Philadelphia, 1987.
14. **Chen, F. A., Hsieh, Y., Liu, C., and Sternberg, J. C.,** Capillary electrophoresis — a new clinical tool, *Clin. Chem.*, 37(1), 1991.
15. **Wang, H. P. and Liu, C. M.,** Separation and identification of human serum proteins with capillary electrophoresis, *Clin. Chem.,* 38, 963, 1992. (Abstract)
16. **Jorgenson, J. W. and Lukas, K. D.,** Capillary zone electrophoresis, *Science*, 1883, 222, 1983.
17. **Gordon, M. J., Lee, K. J., Arias, A. A., and Zare, R. N.,** Protocol for resolving protein mixtures in capillary zone electrophoresis, *Anal. Chem.*, 63, 69, 1991.
18. **Jolliff, C. R.,** Classification and Interpretation of Paragon® Serum Protein Electrophoresis Patterns, Beckman Technical Bull. EP1, Beckman Instruments, Fullerton, California.
19. **Blessum, C.,** Unpublished data, 1993.
20. **Liu, C. M. and Wang, H. P.,** Myeloma Protein Classification and Typing by Capillary Electrophoresis (abstract). 5th Int. Symp. on HPCE, January 25–28, 1993, Orlando, Florida.
21. **Hjertén, S. and Kiessling-Johannsson, M.,** High-performance displacement electrophoresis in 0.025 to 0.050 mm capillaries coated with a polymer to suppress adsorption and electroendosmosis, *J. Chromatogr.*, 550, 811, 1991.
22. **Palmeri, R. N.,** Personal communication, 1993.
23. **Lentner C., Ed.,** *Geigey Scientific Tables,* Vol. 3, Ciba Geigey, Basel, 1984.
24. **Grossman, P. D. and Colburn, J. C.,** *Capillary Electrophoresis — Theory and Practice*, Academic Press, San Diego, 1992.
25. **Aguzzi, F. and Poggi, N.,** Immunosubtraction electrophoresis: a simple method for identifying specific proteins producing the cellulose acetate electrophoretogram, *Boll. Inst. Sieroter. Milan*, 56, 212, 1977.
26. **Liu, C. M. and Wang, H. P.,** Unpublished data, 1992.
27. **Moss, D. W.,** Alkaline phosphatase isoenzymes, *Clin. Chem.*, 28, 2007, 1982.

28. **Henderson, A. P.,** Lactate dehydrogenase isoenzymes, in *Methods of Enzymatic Analysis*, 3d ed., Bergmeyer, Bergmeyer, and Grassl, Eds., Verlag-Chemie, Weinhein, 1983.
29. **Lang, H., Ed.,** *Creatine Kinase Isoenzymes*, Springer-Verlag, Berlin, 1981.
30. **Barnhart, M. I., Henry, R. L., and Lusher, J. M.,** Sickle cell, *A Scope Publication*, Upjohn, 1974.
31. **Bunn, H. F., Forget, B. G., and Ranney, H. M.,** *Human Hemoglobin*, W. B. Saunders, Philadelphia, 1977.
32. **Schmitz, G. and Williamson, E.,** High-density lipoprotein metabolism, reverse cholesterol transport and membrane protection, in *Current Opinion in Lipidology*, Vol. 2, Current Science, Philadelphia, 1991.
33. **Schmitz, G., Borgman U., and Assmann, G.,** Analytical capillary electrophoresis: a new technique for the analysis of lipoproteins and lipoprotein subfractions in whole serum, *J. Chromatogr.*, 320, 253, 1985.

Chapter 17

MONITORING DRUG METABOLISM BY CAPILLARY ELECTROPHORESIS

Stephen Naylor, Linda M. Benson, and Andy J. Tomlinson

TABLE OF CONTENTS

0-8493-8690-X/94/$0.00+$.50

I. INTRODUCTION

The structural diversity of the multitude of modern therapeutic drugs available for the treatment of disease is well documented.[1,2] However, after entry into the body, such therapeutic agents are subjected to metabolism. This process involves a series of complex events, including absorption, distribution, metabolic transformation, and, eventually, excretion of the remaining parent drug and/or its metabolites. These events modulate the efficacy of the administered drug in the treatment of the disease, as well as determining its pharmacological and toxicological effects. Therefore, the subsequent isolation and structural characterization of the resulting plethora of metabolites is vitally important to understanding the physical and biological effects of the parent drug.[3-5]

The enzymatic reactions involved in drug metabolism pathways are often classified into three groups (see Table 1). Phase I metabolism consists of a variety of functionalization reactions, including oxidation, reduction, and hydrolysis, which generally produces more polar compounds that are subsequently susceptible to further modification. Phase II metabolism involves conjugation of the drug, or more likely a phase I metabolite, with an endogenous substrate, such as the tripeptide glutathione or glucuronic acid. This normally leads to a water-soluble product, which can then be excreted in bile or urine. The existence of a third phase of metabolism involving breakdown of conjugates by intestinal microflora accompanied by subsequent reabsorption and metabolism (see Table 1) has also been postulated.[6]

Contemporary methods for drug metabolite identification are usually based on the comparison of ultraviolet (UV) spectral data and high-performance liquid chromatography (HPLC) retention times of isolated "unknown" metabolites with synthetic standards.[7] Such methods of detecting drug metabolites and subsequent structural characterization can be a time-consuming process, as well as affording only very limited structural information. Furthermore, since phase I metabolism of a drug often results in only minor structural modification of the parent compound,[3,4] determination of suitable chromatographic conditions to effect HPLC separation of mixtures of drug metabolites is often particularly difficult. Also, since phase II conjugate metabolites are generally thermally unstable and very polar, they are difficult to assay by classic methods, such as HPLC. To this end, the high-efficiency separations of capillary electrophoresis (CE) combined with the ease of rapid analysis and method development presents obvious advantages over the use of HPLC for the detection of new and possibly relatively short-lived reactive drug metabolites. Capillary electrophoresis also offers the unique possibility for simultaneous analysis of both phase I and phase II metabolites.

The use of CE for the analysis of small organic molecules (<600 to 800 Da), such as conventional pharmaceutical agents, natural products, and drug metabolites, has received scant attention.[8,9] Most work has been carried out on mixtures of commercially available pharmaceutical compounds, predominantly by micellar electrokinetic chromatography (MECC), and these studies are summarized in Table 2[10-32] and discussed in detail elsewhere (see Chapter 10).

Investigations of both physiological and drug metabolism by CE are as yet limited. In separate studies, vitamin B_2 and B_6 metabolites have been investigated in blood[33] and urine,[34] respectively, using MECC. Roach and co-workers[35] used free solution CE (CZE), combined

TABLE 1
Classification of Phase I to III Drug Metabolism

Phase I	Phase II	Phase III
Oxidation Aliphatic oxidation Aromatic hydroxylation N-, O-, S-Dealkylation Epoxidation N-, S-Oxidation Dehalogenation Alcohol and aldehyde oxidation Oxidative deamination Reduction Azo-, nitroreduction Hydrolysis Amide, ester, hydrazide and carbamate hydrolysis Epoxide hydration Isomerization	Glucuronidation at –OH, –COOH, $-NH_2$, –SH Acetylation at $-NH_2$, $-SO_2NH_2$, –OH Glycosylation at –OH, –COOH, –SH N-, O-, S-Methylation at $-NH_2$, –OH, –SH Sulfation at $-NH_2$, $-SO_2NH_2$, –OH Glutathione conjugation Fatty acid conjugation	Glutathione-*S*-conjugates of haloalkanes $\xrightarrow{\beta\text{-lyase}}$ Unstable thiols

with laser-induced fluorescence (LIF) detection to monitor the metabolism of the antifolate methotrexate to its major metabolite, 7-hydroxymethotrexate, in serum. Fujiwara and Honda[36] have analyzed the levels of the drug γ-oryzanol (used in the treatment of autonomic neuron diseases) and its metabolite ferulic acid in plasma. Various metabolites of the mucolytic agent *S*-carboxymethylcysteine were detected in urine samples by both CZE and capillary isotachophoresis (ITP) by Tanaka and Thormann.[37] We have reported the investigation of the *in vitro* microsomal metabolism of the neuroleptic drug haloperidol, using both MECC and CZE conditions.[38]

Various metabolic studies utilizing CE have also been carried out to assess the health of individuals (see Chapter 16 for more in-depth discussion). Weinberger et al.[39] investigated the urinary porphyrin content of a patient suffering from porphyria cutanea tarda by MECC, whereas Guzman et al. have assessed the profile levels of urea and creatinine in urine by CZE of healthy, dehydrated, and strenuously exercised individuals.[40] Tomita et al.[41] determined the levels of the toxic herbicide glyphosate and its major metabolite (aminomethyl)phosphonic acid in serum. Both Wernly and Thormann[42,43] and Weinberger and Lurie[44] have reported on the CE analysis of metabolites derived from a variety of different illicit drugs found in the urine of human users.

The development of on-line CE-mass spectrometry (CE-MS), pioneered by Smith[45-49] (see Chapter 8 in this book) and by Henion[50-53] has further enhanced the utility of CE in the analysis of complex drug metabolite mixtures. A considerable amount of literature is already available on the use of MS,[54-56] tandem mass spectrometry (MS/MS),[57-60] and HPLC-MS[56,61-63] to elucidate the structures of both phase I and phase II metabolites. However, to date there have been only three studies reporting the use of CE-MS in metabolism studies. Johansson and co-workers[17] investigated the separation and mass spectral analysis of a series of sulfonamide and benzodiazepines, as well as the *in vivo* metabolic fate of flurazepam. Guzman and co-workers[31] used two different preconcentration steps in CE before analyzing uric acid and methamphetamine in human urine samples by electron ionization mass spectrometry, and we have investigated the *in vitro* metabolic fate of haloperidol, using different animal hepatic microsomes by CE tandem mass spectrometry (CE-MS/MS).[64]

TABLE 2
Types of Pharmaceutical Agents Analyzed by Either Capillary Zone Electrophoresis or Micellar Electrokinetic Chromatography

Compound type	Method	Ref.
Amphetamine	CZE/MS (human urine)	31
Antiinflammatories	CZE	10, 20
	MECC	13, 15, 20
	MECC/cyclodextrin	30
Antibiotics	CZE	27
Cephalosporins	CZE	10
	MECC	12, 13
	MECC (human plasma)	18
Penicillins	CZE	10, 29
	MECC	12, 13
	MECC (human plasma)	13
Sulfonamides	CZE	10
	CZE-MS and CZE-MS-MS	17
Anticancer	CZE	28
Antidepressants	CZE	22
Antileukemic		
Leucovorin	Chiral ECC	16
Cytosine-β-D-arabinoside	CZE (human plasma)	26
Antiepileptic	MECC (human plasma)	32
Barbiturates	MECC	10, 24
	MECC (human plasma and urine)	21
	MECC (human plasma and serum)	24
Benzodiazepines	CZE-MS and CZE-MS-MS	17
Calcium channel blockers	MECC	13, 15
Cold medicines — active ingredients	CZE	10, 29
	MECC	11, 13, 14, 29
Cytokines	CZE	23
H_2-antagonists	MECC (human serum)	25
Vitamins, B series	CZE	29
	MECC	13
Uric acid	CZE	19
	CZE/MS (human urine)	31
Xanthines	MECC	19

II. APPLICATIONS

A. GENERAL CONSIDERATIONS

The parameters affecting the separation of drug metabolites, such as buffer composition and ionic strength, pH, organic additives, voltage and current, capillary dimensions, and temperature, are obviously identical to those for any small organic molecules, and these aspects have been discussed in some detail by Swartz[29] and also in Chapter 10 of this book. However, there are several features of drug metabolism analysis using CE that require special consideration, and they include the following:

1. Advantages and disadvantages of CE vs. HPLC
2. MECC vs. CZE
3. Dynamic range problem associated with the relatively large amount of unmetabolized parent drug in the presence of minor metabolites (<1 to 5% of parent drug)
4. Loading enough sample onto the capillary in order to detect minor metabolites

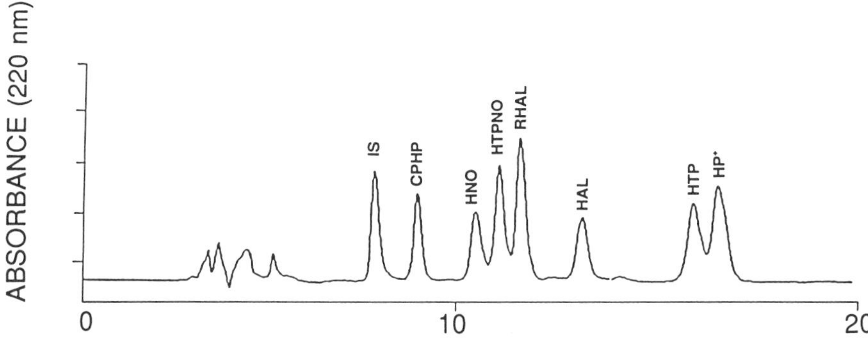

FIGURE 1. Reversed-phase HPLC chromatogram (220 nm) of haloperidol (HAL) and a mixture of six synthetic standards (see Figure 2 for structures), using a 4.6 cm × 250 mm, 5-μm Hypersil CPS-5 column. Mobile phase was 67% CH_3CN with 10 m*M* NH_4OAc and AcOH (pH 5.4) run isocratically at 1 ml/min. IS, internal standard pirenzepine.

5. Loss of metabolites on uncoated capillary walls
6. Stability of metabolites under CE buffer conditions
7. Analysis of metabolite mixtures derived from complex biological matrices, such as blood, urine, and feces

B. HIGH-PERFORMANCE LIQUID CHROMATOGRAPHY AND CAPILLARY ELECTROPHORESIS OF DRUG METABOLITE MIXTURES

The separation of complex mixtures containing drug metabolites, derived from both *in vivo* and *in vitro* sources, has evolved to be specific for a particular drug or class of drugs. Furthermore, because of the structural similarity of phase I metabolites, development of HPLC conditions is often time consuming, and because of the relatively limited resolving capabilities of this technique, also difficult to achieve. This is reflected in the reversed-phase HPLC separation of six synthetic standards/putative metabolites plus the parent drug haloperidol (a neuroleptic agent used in the treatment of schizophrenia and Tourette's syndrome[1]) shown in Figure 1. It was not possible to effect baseline resolution of 5 of the 6 components even after investigating in excess of 50 different solvent systems.[65] In addition, polar molecules, such as FBALD, FBPA, and FAA (see Figure 2 for structures) coelute with the solvent front under the reversed-phase HPLC conditions utilized and, therefore, are difficult to detect. This latter problem is particularly pronounced for the analysis of the much more polar phase II metabolites, such as glutathione and glucuronide conjugates (data not shown).

In contrast, we have reported[38] the CZE separation of a mixture containing haloperidol and ten synthetic standards, using a separation buffer of 50 m*M* ammonium acetate (NH_4OAc) with 10% methanol (MeOH) and 1% acetic acid as a buffer (pH 4.1) at 30 kV (see Figure 3). We were able to baseline resolve 10 of the 11 components.

In other work, it has been noted by Atamna et al.[19] that in situations where high peak capacity is required, such as drug metabolism and biotechnology studies, CE is superior to HPLC. Furthermore, on analyzing cefpiramide in plasma, Nakagawa et al.[18] noted that the advantage of CE over LC is that the former requires only ~10 nl of plasma, which can be analyzed directly without any pretreatment. Fujiwara and Honda[14] have commented that CE can separate a wide range of antipyretic compounds with widely differing polarity and ionic character, which offers an alternative approach to conventional HPLC analysis. In a quanti-

COMPOUND STRUCTURE

CPHP

HP^+

HTP

HAL

HTPNO

RHAL

HNO

FBPOH

FBALD

FBPA

FAA

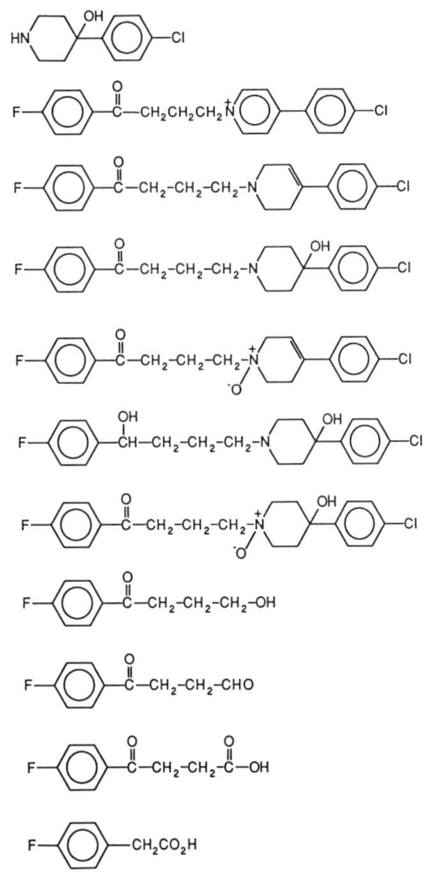

FIGURE 2. Structures of the neuroleptic drug haloperidol (HAL) and ten synthetic standards/putative metabolites.

tative study of the barbiturate, thiopental, in human serum and plasma, Meier and Thormann[24] determined that the linear regression analysis of data from 66 patients was comparable for CE and HPLC. However, they claim that CE has several advantages over HPLC, including "high degree of automation, small sample size, no requirement for large amounts of organic solvents, and rapidity of analysis." Furthermore, they suggest that since capillary conditioning is not a prerequisite, rapid change of buffer composition can be achieved in CE analysis.

It is clear that a disadvantage, at present, of CE compared to HPLC is sample loading capacity. Steuer et al.[66] concluded in a comparative study of CE, HPLC, and supercritical fluid chromatography (SFC) that CE is "constrained in terms of loading capacity and sensitivity." They emphasized that concentration sensitivity, i.e., the amount that can be injected onto the capillary, and not the minimum detectable amount, is most relevant in drug analysis. They concluded that the concentration sensitivity of HPLC is about tenfold greater than for CE. Salomon and co-workers[22] noted that the CE detection limit of seven antidepressants was ~1.0 µg/ml, whereas for HPLC it was 2 to 10 ng/ml. Weinberger et al.,[39] in a study of urinary porphyrins, found that the concentration limits of detection (CLOD) using CE were 80 to 90 times higher, i.e., less sensitive, than HPLC. However, they assert that the mass limit of detection (MLOD) is 50-fold lower than found for HPLC and is brought about by the reduction in volume of the optical window for CE (0.2 nl) vs. HPLC (5 µl).

Sample loading and decreasing CLOD are active areas of research in CE analysis and are discussed further in later sections as they pertain to drug metabolism studies.

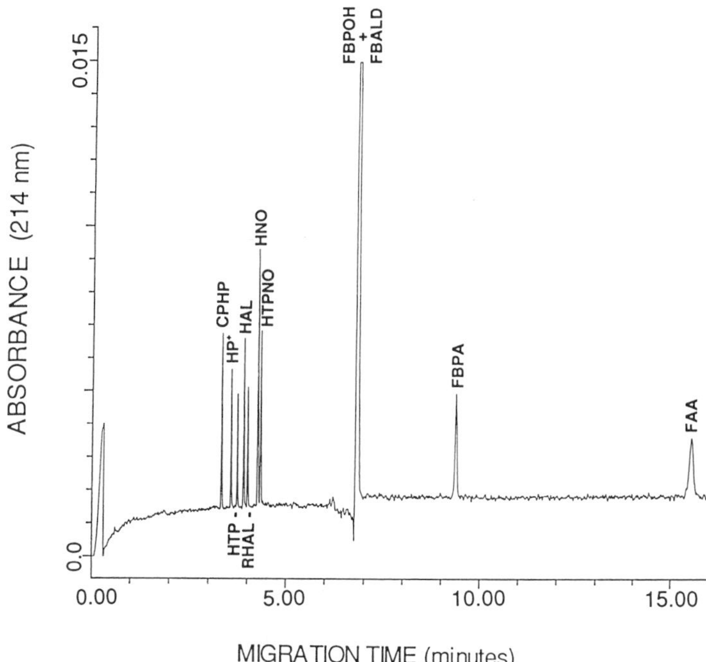

FIGURE 3. Separation of haloperidol (HAL) and ten synthetic standards/putative metabolites by CZE (monitored at 214 nm). Analysis performed on a Beckman (Fullerton, CA) P/ACE 2100. Separation buffer consisted of 50 m*M* NH_4OAc containing 10% MeOH and 1% AcOH (pH 4.1) on an uncoated fused silica capillary (75-μm I.D. × 50 cm [effective length], 57 cm [total length]). Separation voltage, 30 kV at 25°C, with a 1-s pressure injection.

C. MICELLAR ELECTROKINETIC CHROMATOGRAPHY AND CAPILLARY ZONE ELECTROPHORESIS IN DRUG METABOLISM STUDIES

Most of the reported literature on the use of CE in the separation of small molecules, in particular pharmaceuticals, predominantly utilizes MECC (see Chapter 3 for an in-depth discussion of this technique), and this is summarized and highlighted in Table 2. Indeed, comparative studies of MECC and CZE in the separation of small molecules have concluded that the former is superior in resolving such components.[11,15] Furthermore, several workers have reported that MECC is the method of choice to separate neutral compounds, since CZE is not a viable option for such molecules.[15,33]

However, in drug metabolism studies, two pertinent factors must also be considered.

1. The ultimate goal is to determine metabolite structures, and this is readily achievable using CE-MS and CE-MS/MS. Capillary zone electrophoresis is, therefore, the only viable method of choice, since MECC requires the presence of surfactants in the separation buffer. This class of compounds is, at present, incompatible with soft ionization MS techniques, such as electrospray-MS (ESI-MS).[67] Also, the choice of CE buffers is limited to electrolytes that are volatile, e.g., NH_4OAc,[67] when used in conjunction with ESI-MS (also see Chapter 8).
2. The structural diversity of drugs and their metabolites necessitates that a systematic approach to developing optimal CE separation conditions using simple CZE buffers is warranted. This obviates the need for continually developing new CE conditions for each individual drug metabolite mixture that requires separation.

Salomon et al.[22] systematically investigated the effects of buffer pH and concentration on the CZE separation of seven tricyclic antidepressants. They demonstrated that optimal separation was obtained when the pH of the buffer was close to the pK_a values of amino functional groups present in these tricyclic amines. They also determined that by increasing buffer concentration from 10 m*M* 3-(cyclohexylamino)-2-hydroxy-1-propanesulfonic acid (CAPSO) containing 2.7 m*M* NaOH to 50 m*M* CAPSO–14 m*M* NaOH greatly enhanced the resolution of the seven-component mixture. They related this to a decrease in ζ potential and, hence, a slowing of electroosmotic flow (EOF).

We reported[38] that two simple factors are important for selection of initial CZE buffer conditions to effect drug metabolite separations, namely, the hydrophobic character of the drug/putative metabolites and the presence of ionizable functional groups on analyte molecules. The hydrophobic nature of the drug determines the need for addition of organic solvent to the separation buffer to aid in solubilizing components of the mixture. (*Note*: This also serves to enhance resolution and peak shape, and these aspects are discussed in detail in Chapter 2, as well as highlighted in Sections II.D.2 and II.D.3 below.) Since the majority of drug/putative metabolites possess functional groups that are either acidic (-COOH, -SH) or basic (-NR_2), modification of separation buffer pH can effect a change in the charge state of a neutral parent drug and its metabolites. This has been exploited in the separation of ten synthetic standards/putative metabolites of haloperidol,[38] shown in Figure 3. Using 50 m*M* NH_4OAc buffer, 10% MeOH, and 1% acetate (AcOH) (pH 4.1), excellent separation of 10 of the 11 components was achieved. Clearly, at pH 4.1, such structures (see Figure 2) are not fully ionized, but differences in partial positive charge distribution are enough to effect separation, and this demonstrates that by subtle manipulation of buffer pH, compounds that were originally neutral can be separated by CZE.

D. LOADING, RESOLUTION, AND LOSS OF SAMPLE COMPONENTS

As previously discussed in Section II.B, a perceived limitation of CE is the ability to load significant amounts of sample onto the capillary. This is a particularly pronounced problem in drug metabolism studies, for the following reasons:

1. For *in vitro* studies, e.g., microsomal incubations, it is often the case that substantial amounts of unmetabolized parent drug remain in the presence of small amounts of metabolites produced.
2. For *in vivo* studies, it is usual that a sizeable portion of the parent drug is metabolized; therefore dynamic range is not a problem. However, both major and minor metabolites produced are present in only low concentrations in body fluids, such as blood and urine.

We have adopted a simple approach to overcome this problem and have utilized isotachophoresis (ITP) as a stacking step after analyte injection, as suggested by Foret and coworkers[68] and by Gebauer et al.[69]

1. Isotachophoresis Preconcentration

Isotachophoresis has found only very limited use as a separation technique in the analysis of small molecules[37] and, in general, has not been a particularly well-utilized technique for the analysis of all types of compounds (see Chapter 5 for a more in-depth discussion). It has, however, found some utility in increasing the amount of sample loaded onto the capillary, as well as stacking the individual components of the mixture into relatively sharp bands (discussed further in Chapter 5).

We have utilized ITP preconcentration to overcome some of the problems associated with analyzing minor phase I metabolites in the presence of large amounts of unmetabolized parent drug derived from a microsomal incubation. The dramatic difference of a large sample injection followed by ITP preconcentration and then separation, compared to a conventional low-volume pressure injection, is shown in Figure 4. A 1-s pressure injection (sample dissolved in MeOH:H_2O [1:1]) of a guinea pig hepatic microsomal incubate of the H_2-

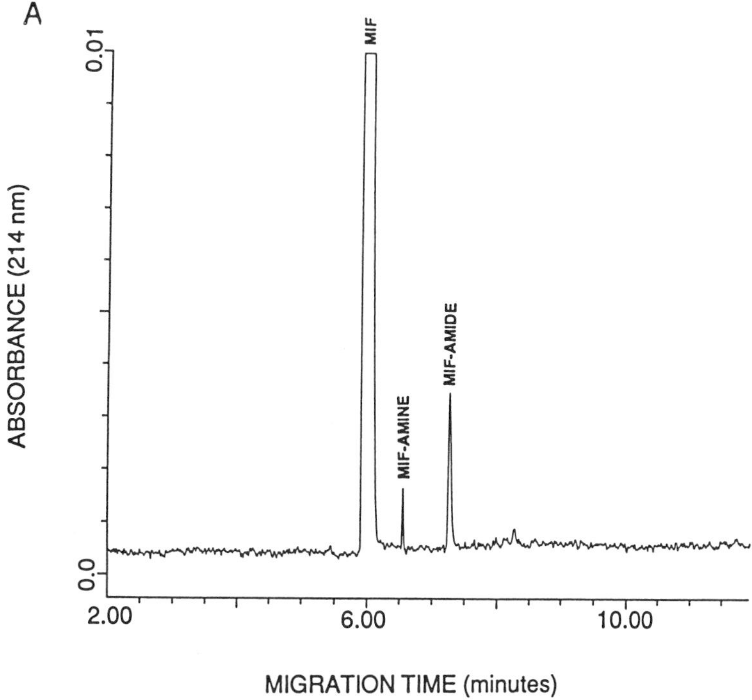

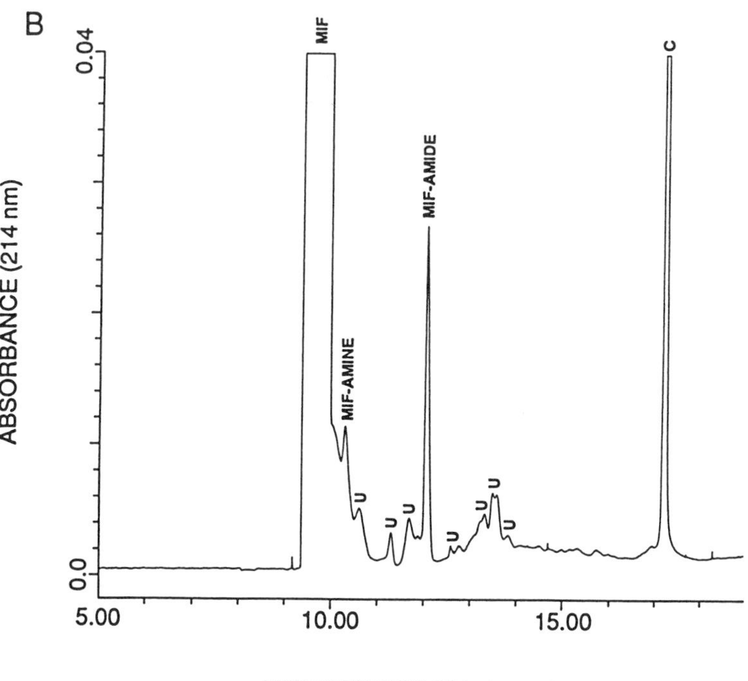

FIGURE 4. Separation of a guinea pig hepatic microsomal incubation of the H_2-antagonist mifentidine containing metabolites (see Figure 5 for structures) by CZE (monitored at 214 nm). Analysis performed on a Beckman P/ACE 2100. Separation buffer consisted of 50 m*M* NH_4OAc containing 30% MeOH and 1% AcOH (pH 4.2) on an uncoated fused silica capillary (75-μm I.D. × 50 cm [effective length], 57 cm [total length]). Separation voltage, 20 kV at 25°C. (A) One-second pressure injection of sample in MeOH:H_2O (1:1) followed by electrophoresis with separation buffer. (B) Isotachophoretic preconcentration: 20-s pressure injection of sample in MeOH:H_2O (1:1), followed by a 1.5-min introduction of 10 m*M* AcOH and then electrophoresis with separation buffer. U, Unknown, not found in control microsomal samples (microsomes deactivated by heating at 100°C for 30 min).

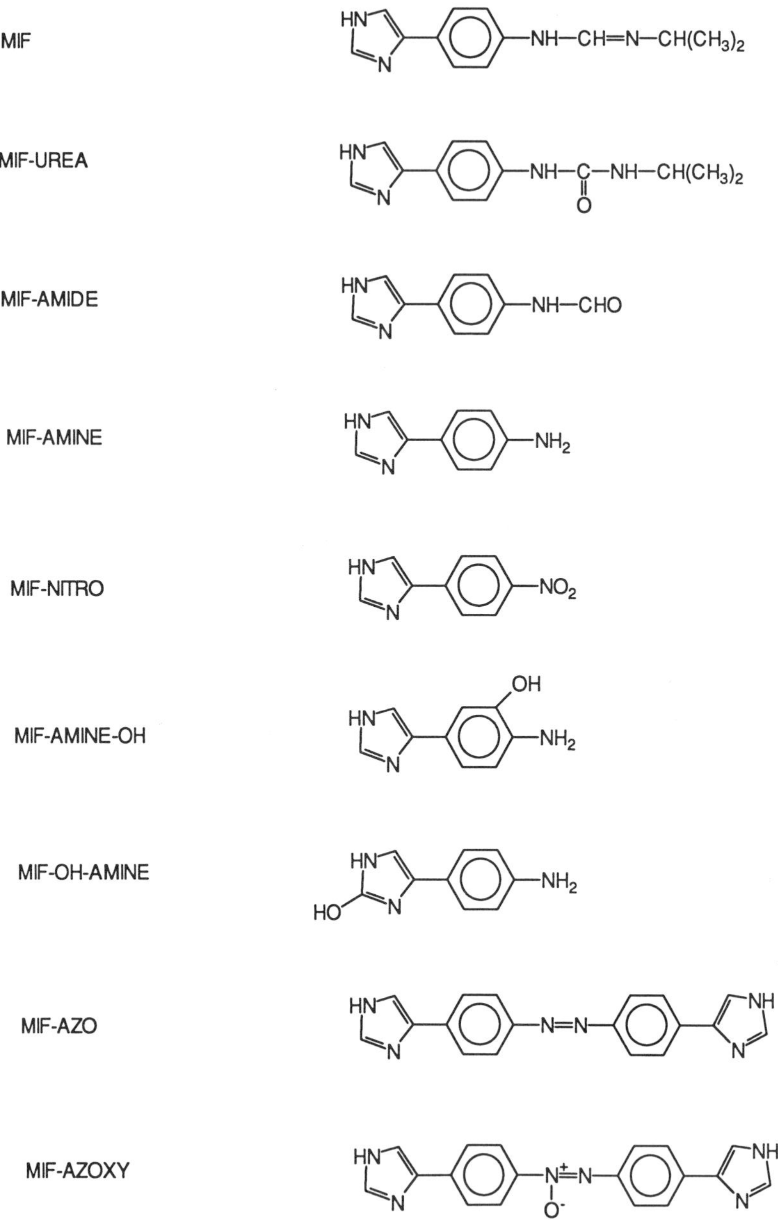

FIGURE 5. Structures of the H_2-antagonist mifentidine (MIF) and eight synthetic standards/putative metabolites.

antagonist, mifentidine (after removal of microsomal protein by $ZnSO_4$ precipitation[70]) is shown in Figure 4A. The separation buffer was 50 m*M* NH_4OAc, containing 30% MeOH and 1% AcOH (pH 4.2). Only two new components were observed in the electropherogram, namely MIF-AMINE and MIF-AMIDE, in addition to unmetabolized parent drug MIF (see Figure 5 for structures). However, an ITP preconcentration consisting of a 20-s pressure injection of the same microsomal mixture, followed by a 1.5-min injection of 10 m*M* AcOH at 10 kV, then electrophoretic separation at 20 kV in the separation buffer used in Figure 4A, resulted in the detection of a number of other unidentified metabolites, which have yet to be

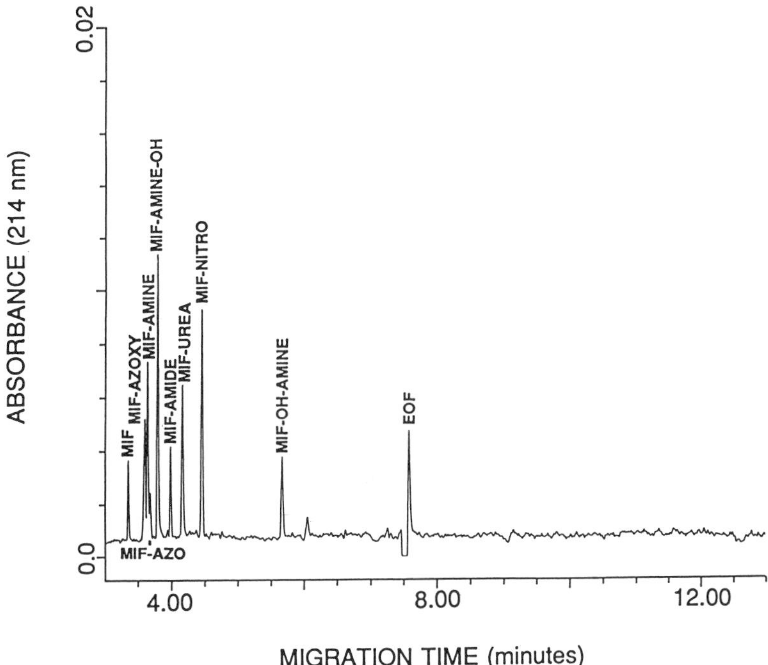

FIGURE 6. Analyte stacking: separation of mifentidine (MIF) and eight synthetic standards/putative metabolites by CZE, demonstrating analyte stacking (monitored at 214 nm). Analysis performed on a Beckman P/ACE 2100. Separation buffer consisted of 50 m*M* NH_4OAc containing 1% AcOH (pH 4.2) on an uncoated fused silica capillary (75 μm × 50 cm [effective length], 57 cm [total length]). Separation voltage, 20 kV at 25°C, with 1-s pressure injection; sample in MeOH:H_2O (1:1).

fully characterized by CE-MS/MS (peaks marked U [unknown] were not detected in control incubations, where mifentidine was incubated with deactivated microsomes). The peak widths of the metabolites in Figure 4B are relatively broad for an ITP preconcentration injection, since focusing of the analytes into narrow bands usually occurs. We believe that at the pH (4.2) of the separation buffer, analyte–wall interactions are responsible for the observed broadening of peaks, and this is still under investigation. However, in general it is possible to substantially reduce peak broadening using the phenomenon of analyte stacking, and this is discussed in the following section.

2. Analyte Stacking

Introduction of analyte mixtures onto the capillary in CZE in a low-conductivity solution relative to the separation buffer leads to a large potential gradient across the injection region. This results in a rapid migration of analyte through the injection medium, which ultimately leads to localized concentration of analytes. Furthermore, since the migration of individual species is likely to be different, enhanced resolution of analyte mixtures can be achieved. This phenomenon has been exploited by Lloyd and co-workers[26] in the analysis of the antileukemic agent cytosine-β-D-arabinoside in human plasma.

We have also utilized this technique in the CZE analysis of eight putative metabolites of the H_2-antagonist mifentidine[71] (see Figure 5 for structures). The nine-component mixture was loaded onto the capillary (1-s pressure injection) in MeOH:H_2O (1:1) and electrophoresed in a separation buffer containing 50 m*M* NH_4OAc, and 1% AcOH (pH 4.2) at 20 kV (see Figure 6). The peak width observed is narrow, and all nine components are resolved, including MIF-AZOXY, MIF-AMINE, and MIF-AZO. Without analyte stacking, it is likely that these components would appear as a single peak.

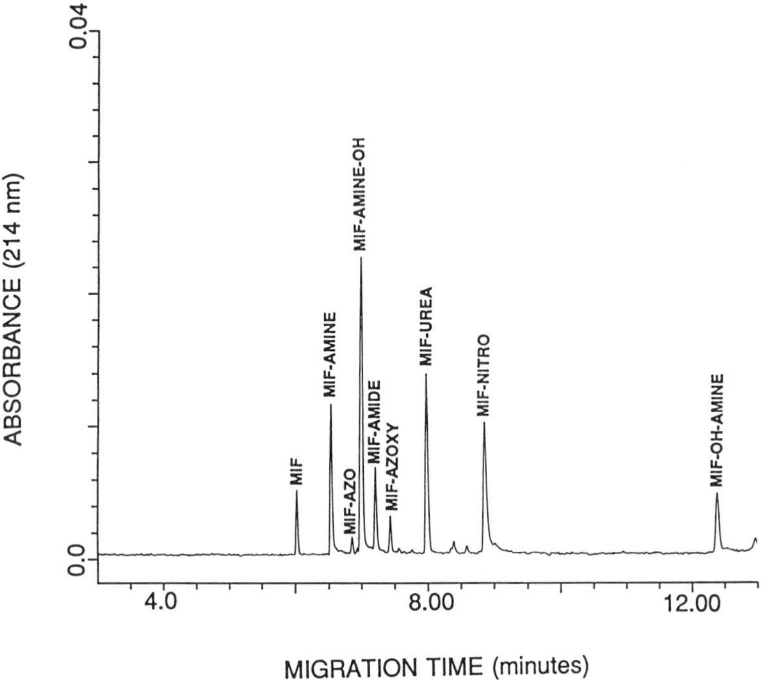

FIGURE 7. Effect of organic modifier: separation of MIF and eight synthetic standards/putative metabolites by CZE. Conditions as for Figure 6, only 30% MeOH added to the separation buffer.

It was not possible to inject the nine-component mixture onto the capillary in separation buffer alone, due to the solubility properties of some of the putative MIF metabolites (in particular, MIF-AZO and MIF-AZOXY). However, a comparative study of injecting the parent compound MIF dissolved in separation buffer alone or in MeOH:H_2O (1:1) (to bring about analyte stacking) revealed a 20% enhancement in peak height with a comparable peak area of the MIF response for the latter injection (data not shown).

3. Organic Modifiers

It has been noted by several investigators[72-74] that organic modifiers, such as acetonitrile, isopropanol, and methanol, can improve the resolution of small molecule mixtures in CZE (see also Chapter 2). This can be related to changes in the viscosity and dielectric constant of the buffer, as well as the ζ potential of the capillary wall. The amounts of organic modifier added to aqueous running buffer have varied, but typically may constitute as high as ~30% by volume. The substantial enhancement in resolution observed on addition of organic modifier to separation buffer is shown by comparison of Figures 6 and 7 for a nine-component mixture containing MIF plus eight putative metabolites. Injection and separation conditions were identical to those described in Figure 6, except that the separation buffer contained 30% MeOH. In separation buffer containing organic modifier, all nine components are baseline resolved, including MIF-AZOXY, MIF-AMINE, and MIF-AZO, which were only partially resolved when no MeOH was present (see Figure 6). It should also be noted that relative migration times of MIF-AZOXY and MIF-AZO have changed on addition of 30% MeOH to the separation buffer.

We and others[20,22,38,39] have made the case that addition of MeOH also enhances the solubility of small molecules in aqueous buffers. This is noteworthy since the solubility of many hydrophobic drug metabolites in aqueous solutions can be problematical and, in numerous cases, a major obstacle to analysis by CZE.

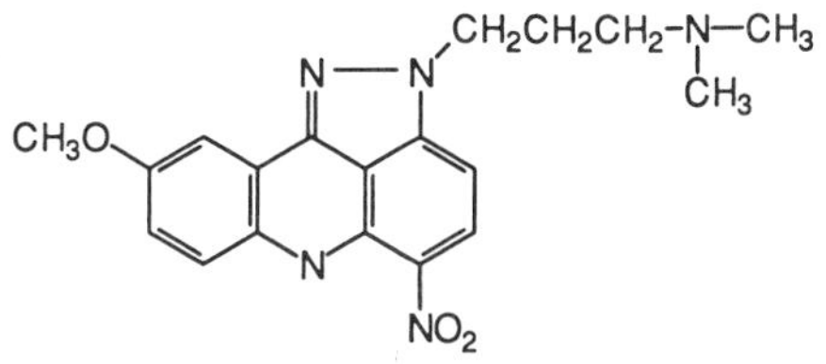

PYRAZOLOACRIDINE (PA)

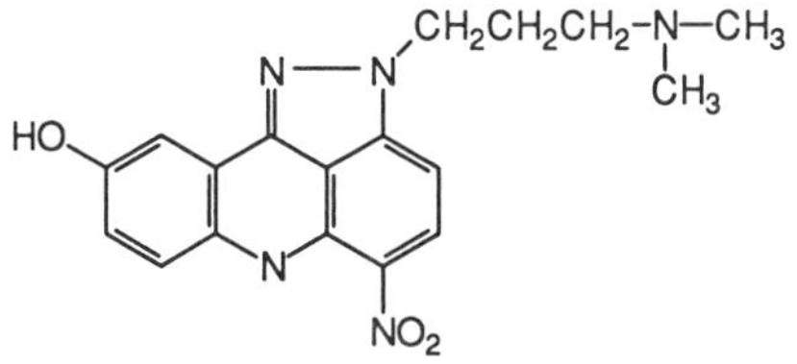

9-DESMETHYL-PYRAZOLOACRIDINE (DE-PA)

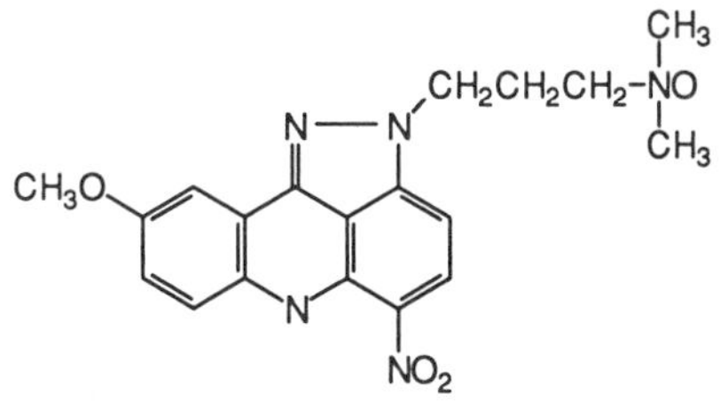

PYRAZOLOACRIDINE-N-OXIDE (PA-NO)

FIGURE 8. Structures of the antitumor agent pyrazoloacridine (PA) and two synthetic standards/putative metabolites, 9-desmethyl-pyrazoloacridine (DE-PA) and pyrazoloacridine-*N*-oxide (PA-NO).

We have investigated the metabolism of the pyrazoloacridines (see Figure 8 for structures) by CZE. Such drugs possess antitumor activity and have been used in the treatment of solid tumors. Their lipophilic character appears to enable them to penetrate tumor masses and intercalate into the DNA.[75] Oda and Landers (detailing our work[76]) have described the improvement in resolution of the three-component mixture consisting of the pyrazoloacridine standards PA, DE-PA, and PA-NO (see Figure 8 for structures) by increasing MeOH concentration from 10 to 30% in a separation buffer of 20 m*M* NH_4OAc in H_2O with 1% AcOH (see Figure 14 in Chapter 2). It was not possible to use 20 m*M* NH_4OAc containing 1% AcOH alone as the separation buffer, due to the lack of solubility of pyrazoloacridines in H_2O.

The lack of solubility of the pyrazoloacridines in aqueous buffers prompted us to investigate the effect of increasing the percentage of organic modifier in the separation buffer. We increased the MeOH content from 30 to 50%, and ultimately to 100%, and in all cases the separation buffer used was 20 m*M* NH_4OAc with 1% AcOH, and these results are shown in

Figure 9.[76] As expected, resolution is considerably improved on going from 30 to 50% to 100% MeOH (see Figures 14 [Chapter 2] and 9A and B, respectively), such that in the nonaqueous buffer, all three components, namely PA, DE-PA, and PA-NO, are baseline resolved (Figure 9B). Furthermore, an impurity (IMP) is now detectable (Figure 9B) and is presumably derived from the synthetic process of making DE-PA and/or PA-NO. However, it is also clear from inspection of Figures 9A and B that much more of the pyrazoloacridine standards are being detected in 100% MeOH. Indeed, the amount of compound detected for all three standards increases approximately threefold on going from 30 to 100% MeOH in the separation buffer. We assume this is due to adsorptive losses of the hydrophobic pyrazoloacridine standards onto the capillary wall as MeOH concentration is decreased, since the solubility of these compounds is much greater in nonaqueous solvents. The use of 100% organic solvents containing small amounts of electrolyte buffer could find widespread use in the analysis of small hydrophobic organic molecules, such as drug metabolites, pharmaceuticals, and natural products.[76]

4. Sample Loss

A substantial problem associated with the CE analysis of drug metabolites is loss of minor components onto the capillary wall. This is particularly pronounced when analyzing metabolites that possess basic functional groups in acidic buffers. This problem could be potentially overcome by using an appropriate coated capillary (see Chapters 18 and 19 for a detailed discussion of the wide variety of coatings that have been used and are available). However, there are three major factors to consider in developing a protocol to overcome this problem of analyte loss in drug metabolism studies.

1. As stated previously (see Section II.C), the ultimate goal is to structurally characterize the newly formed metabolites, using CE-MS and CE-MS/MS. Hence, if a coated capillary is used, it is important that no leaching or bleeding of the capillary occurs, since this would lead to a considerable background contamination and, ultimately, reduced sensitivity.
2. Since parent drugs and, hence, the resulting metabolites possess great structural diversity, any coated capillary used should be applicable to a wide range of compounds.
3. It is also necessary to simultaneously analyze both cationic and anionic metabolites that possess either positive, negative, or neutral character at the pH of the separation buffer.

This latter point is highlighted for separation of the 11 haloperidol compounds shown in Figure 2, where CPHP, HP^+, HTP, HAL, RHAL, HNO, and HTPNO are all cationic in nature; FBPOH and FBALD are neutral, migrating with EOF; and FPBA and FAA are still anionic in character, even at pH 4.2.[38] Wainwright[10] has discussed this problem previously in an analysis of a series of antiinflammatory drugs, using both coated (polyacrylamide) and uncoated capillaries. Wainwright concluded that the uncoated capillary was more useful, since it "allowed separation of both positive and negative species in one run."

In attempting to develop systematic approaches to analyzing drug metabolite mixtures by CE, it is important to keep conditions as simple as possible. Therefore, an approach we have adopted to prevent loss of metabolites onto the capillary wall is to initially condition the column by electrophoresing the parent drug dissolved in the separation buffer,[71] for a period of time, and the effect is shown in Figure 10. A 2-s pressure injection of mifentidine plus seven of the mifentidine standards/putative metabolites (MIF-AMINE, MIF-AZO, MIF-AMINE-OH, MIF-AMIDE, MIF-AZOXY, MIF-UREA, and MIF-NITRO; see Figure 5 for structures) in a separation buffer of 50 m*M* NH_4OAc containing 30% MeOH and 1% AcOH is shown in Figure 10A. The capillary had been high-pressure washed with 1 *M* NaOH (5 min), then H_2O

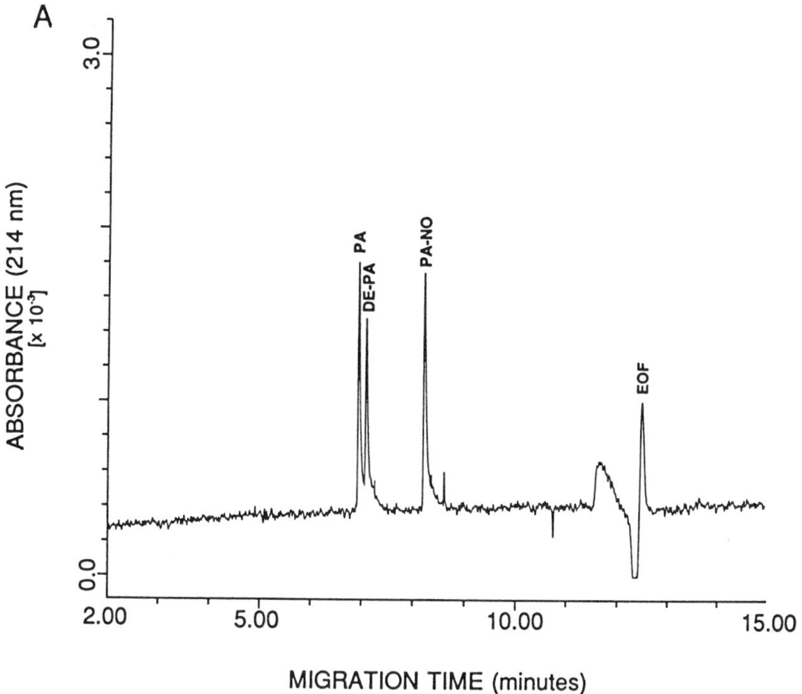

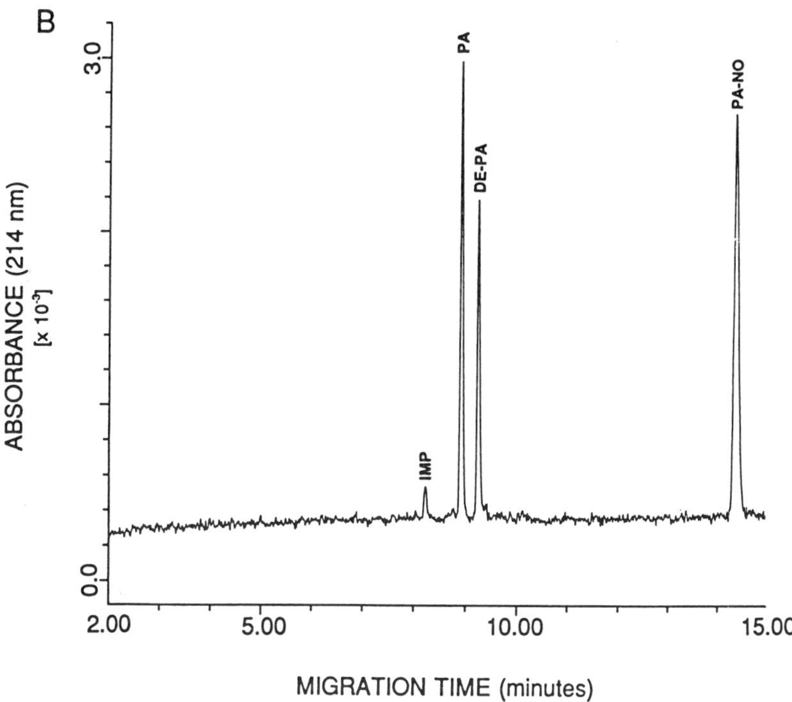

FIGURE 9. Analysis of pyrazoloacridines, using nonaqueous buffer. Analysis performed on a Beckman P/ACE 2100. Separation buffer (monitored at 214 nm) used was (A) 20 m*M* NH_4OAc in 50% MeOH with 1% AcOH and (B) 20 m*M* NH_4OAc in 100% MeOH with 1% AcOH. An uncoated fused silica capillary, 75-µm I.D. × 50 cm (effective length), 57 cm (total length), was used at 20 kV and 40°C with a 1-s pressure injection (sample in 100% MeOH). IMP, Impurity.

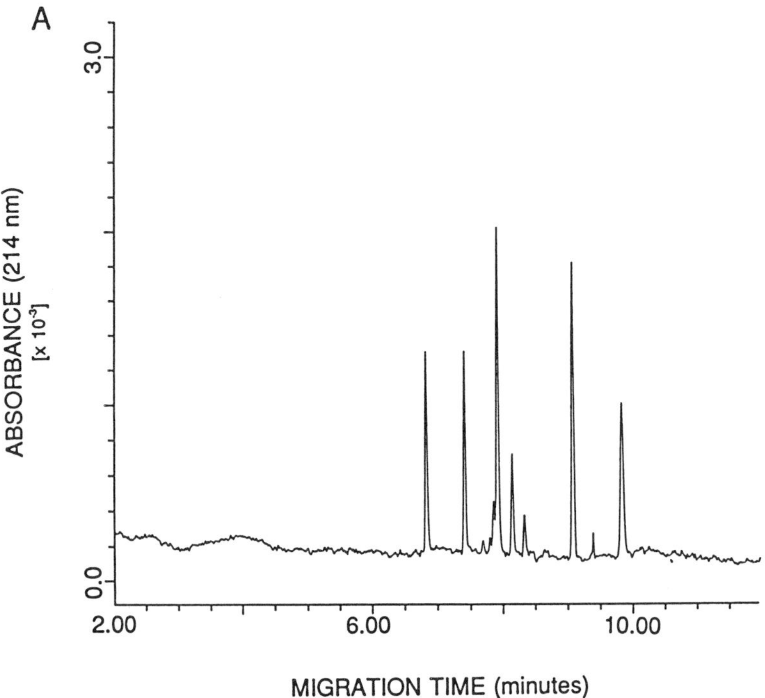
A
ABSORBANCE (214 nm)
[x 10-3]
3.0
0.0
2.00
6.00
10.00
MIGRATION TIME (minutes)

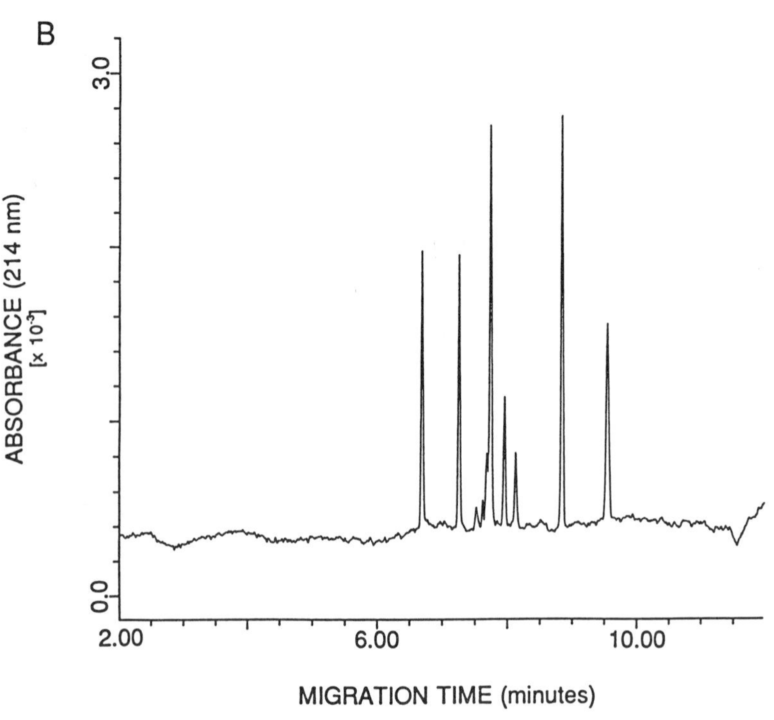
B
ABSORBANCE (214 nm)
[x 10-3]
3.0
0.0
2.00
6.00
10.00
MIGRATION TIME (minutes)

FIGURE 10.

(5 min), and finally separation buffer (5 min) prior to the injection of the mixture. After separation of the mixture (Figure 10A), the capillary was subsequently washed again using the same sequence of solvents, but then "washed" with separation buffer containing a 10 μ*M* concentration of the parent drug mifentidine (5 min). A second injection (2-s pressure) of the eight mifentidine standards was made and electrophoresed in separation buffer as for Figure 10A, but also containing 10 μ*M* mifentidine, and this is shown in Figure 10B. The average increase in peak area is ~30% for all compounds and demonstrates the usefulness of this simple but effective procedure.[71]

This observation prompted us to question the whole process of washing and conditioning capillaries prior to analysis of metabolite mixtures. Initial use of a new capillary often gives rise to

1. Baseline drift during the first two to three analyses
2. Loss of analytes onto the capillary wall
3. Variable migration times

In our experience, it is important to condition a new capillary, as well as desist from frequent severe washing (NaOH) of a used capillary. We suggest that a relatively concentrated sample (10 μ*M*) of parent drug or a mixture of analyte standards should be electrophoresed prior to analysis of an *in vitro-* or *in vivo*-derived metabolite mixture.

E. REACTIVITY OF DRUG METABOLITES

A unique feature of drug metabolites is the variable nature of the functional groups present in such molecules, as well as the unusual reactivity of some of these groups. Therefore, it is of some importance when considering the selection of separation buffer conditions for the analysis of drug metabolite mixtures that extremes of pH and ionic strength be avoided. Furthermore, it is important to consider that organic modifiers also can react with metabolites to afford structural changes.

1. Reactivity with Buffer

N-Methylformamide (NMF) is an experimental antitumor agent,[77] as well as a hepatotoxin in both humans and animals.[78] Threadgill et al.[79] have demonstrated by MS that 5-(*N*-methylcarbamoyl)glutathione (GSH-MC) (see Figure 11 for structure) is a phase II biliary metabolite of NMF in mice. In collaboration with T. A. Baillie (University of Washington, Seattle, WA), we have been interested in developing CZE conditions to separate a variety of similar glutathione conjugates. Three synthetic standards, including GSH-MC (see Figure 11), were separated using a separation buffer of 50 m*M* NH_4OAc, containing 10% MeOH and 1% AcOH at a pH of 4.1 (see Figure 12A). However, using a separation buffer of 50 m*M* NH_4OAc

FIGURE 10. Coating capillary with parent drug, and separation of mifentidine (MIF) plus seven synthetic standards/putative metabolites (MIF-AMINE, MIF-AZO, MIF-AMINE-OH, MIF-AMIDE, MIF-AZOXY, MIF-UREA, and MIF-NITRO) by CZE (monitored at 214 nm). Analysis performed on a Beckman P/ACE 2100. Separation buffer consisted of 50 m*M* NH_4OAc containing 30% MeOH and 1% AcOH (pH 4.2); capillary, 75-μm I.D. × 50 cm (effective length), 57 cm (total length); voltage, 20 kV at 25°C. Two-second pressure injection, sample in MeOH:H_2O (1:1). (A) Capillary was washed sequentially with 1 *M* NaOH (5 min), H_2O (5 min), and separation buffer (5 min) prior to sample injection onto the capillary. (B) Capillary was washed as described in (A), but then separation buffer containing 10 μ*M* mifentidine (MIF) was washed through the capillary for 5 min. The same separation buffer (containing 10 μ*M* MIF) was used in the electrophoresis run to separate components.

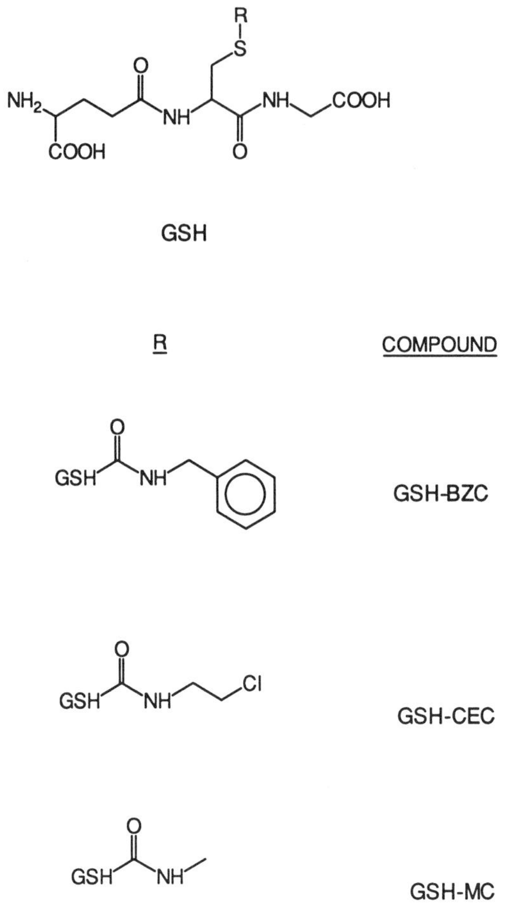

GSH

R

COMPOUND

GSH-BZC

GSH-CEC

GSH-MC

FIGURE 11. Structures of the carbamate thioester adducts of glutathione: GSH-BZC, GSH-CEC, and GSH-MC.

with 10% MeOH titrated to pH 8.9 with NH_4OH, a sizeable decrease in signal response was observed (see Figure 12B). On changing the separation buffer to 50 m*M* NH_4OAc, containing 10% MeOH and titrated to pH 10.5 with NaOH, no response was observed for analyte standards (Figure 12C). These results suggest that such putative phase II metabolites are base labile and are being degraded by separation buffer on the capillary. The lability of such glutathione adducts in basic conditions has been noted previously.[80]

2. Detection of Degradation Products by Capillary Zone Electrophoresis

Cefamandole is a widely prescribed second-generation cephalosporin antibiotic.[1] However, it is associated with a toxic reaction when alcohol is ingested after its administration. The reaction is similar to that seen with disulfiram (Antabuse™), a drug used in aversion therapy for alcoholics. It appears that the methyltetrazole-thiol (MTT), a degradation product of cefamandole, may play a role in inhibiting the enzyme aldehyde dehydrogenase, leading to a buildup of acetaldehyde in the body with a consequent systemic toxic reaction.[81] We are investigating the metabolism/chemistry of cefamandole in the presence of alcohols, such as methanol and ethanol, by CZE. A freshly prepared solution of cefamandole (1 mg/ml) in methanol, using 50 m*M* NH_4OAc containing 10% MeOH with 0.2 m*M* NH_4OH (pH 8.1) as a separation buffer, shows a single peak by CZE (see Figure 13A). However, if the drug is allowed to stand at room temperature in MeOH (or ethanol [EtOH] — results not shown), the appearance of two new components is readily detectable. These compounds appear ~20 to 30 min after addition of the alcohol, and are clearly present after 5 h, as shown in Figure 13B.

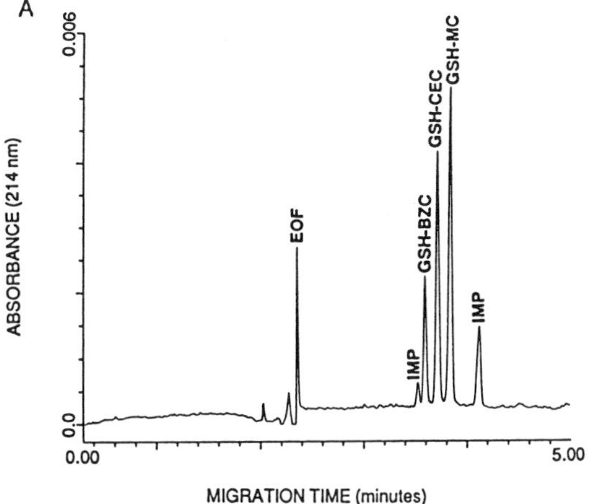

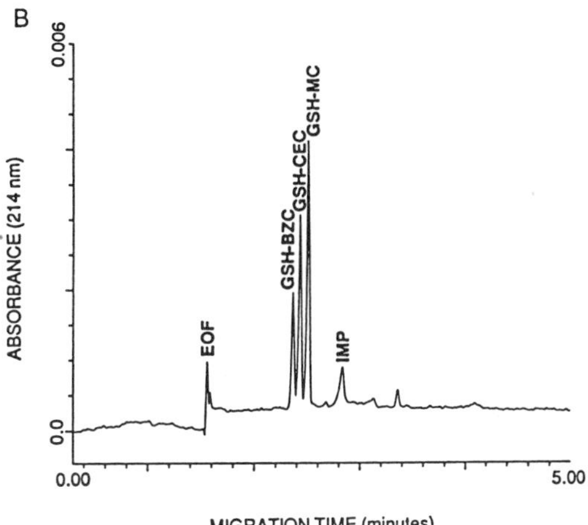

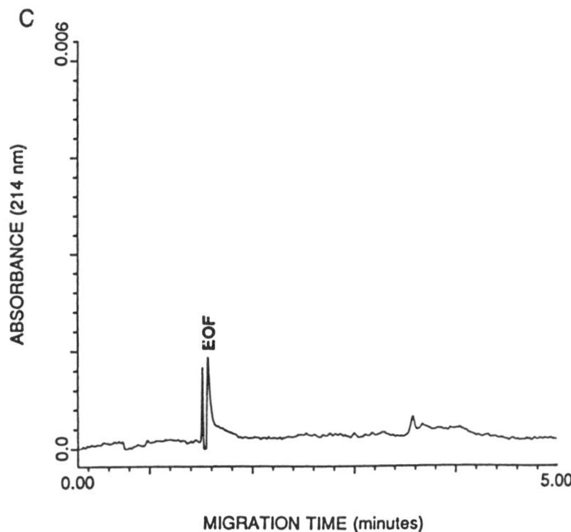

FIGURE 12. Effect of separation buffer pH: separation of three carbamate thioester adducts of glutathione by CZE (monitored at 214 nm). Analysis performed on a Beckman P/ACE 2100. Uncoated fused silica capillary, 50-μm I.D. × 20 cm (effective length), 27 cm (total length); voltage, 15 kV (25°C). One-second pressure injection, sample (250 pmol/μl) in separation buffer. (A) Separation buffer, 50 m*M* NH_4OAc containing 10% MeOH and 1% AcOH (pH 4.1); (B) separation buffer, 50 m*M* NH_4OAc containing 10% MeOH with NH_4OH to make pH 8.9; (C) separation buffer, 50 m*M* NH_4OAc containing 10% MeOH with NaOH to make pH 10.5. IMP, impurities.

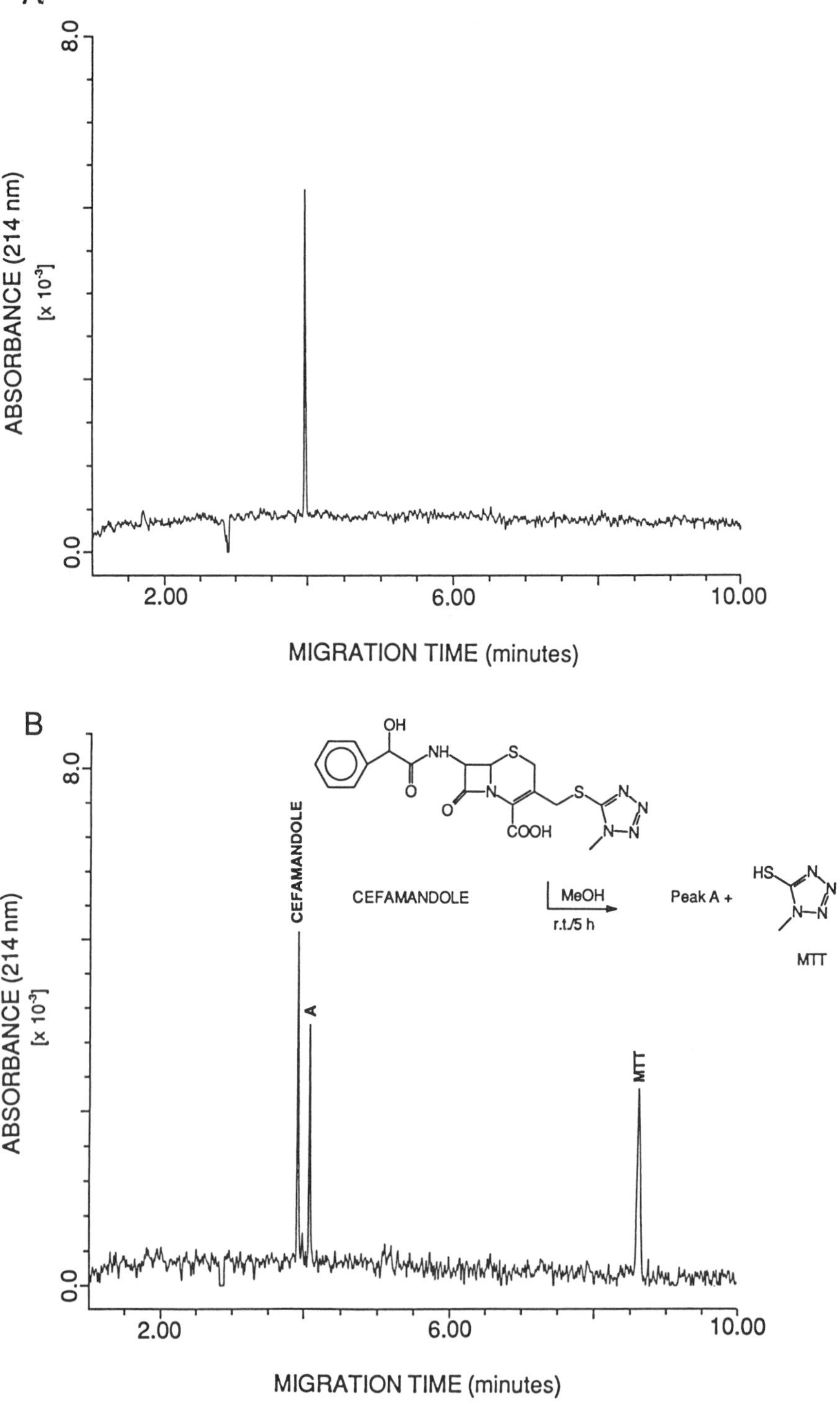

FIGURE 13. Analysis of the antibiotic cefamandole and its degradation products by CZE (monitored at 214 nm). Analysis performed on a Beckman P/ACE 2100. Separation buffer, 50 m*M* NH_4OAc containing 10% MeOH, with NH_4OH to adjust pH to 8.1, on an uncoated fused silica capillary (75-μm I.D. × 50 cm [effective length], 57 cm [total length]). Separation voltage, 30 kV at 25°C, with a 1-s pressure injection; sample in 100% MeOH. (A) Time zero, cefamandole dissolved in MeOH. (B) Cefamandole in MeOH after 5 h at room temperature. Peak A is still unidentified.

The late eluting response has been tentatively identified as MTT by spiking the 5-h sample with authentic material, and observing an increased response of the peak marked MTT in the electropherogram (result not shown).[82]

F. ANALYSIS OF BIOLOGICAL SAMPLES

1. Problems Associated with Complex Biological Matrices

Analysis by CE of drugs and their subsequent metabolites, derived from biological sources, involves either (1) quantitative studies, where it is important to determine the concentration of unmetabolized parent drug and/or a major metabolite present in the biological matrix (quantitative studies of pharmaceuticals and drug metabolites in biologically derived fluids by CE have not been widely reported,[18,24-26,34,35,40] and Silverman and Shaw discuss this in more detail in Chapter 10) or (2) isolation and structural characterization of metabolites formed. Many initial studies focus on the *in vitro* formation of metabolites, e.g., microsomal incubations, but ultimately it is important to understand metabolic pathways that occur *in vivo*.

The separation by CE of metabolites that are present in complex biological matrices, e.g., microsomal preparations, blood (plasma and serum), and urine, presents a considerable analytical challenge (see Chapter 20 for a more general discussion of sample matrix effects in CE). In part, this is due to the low concentration of analyte(s) present, as well as the complexity of the biological matrix, which generally contains a multitude of proteins, as well as a plethora of other components. Such proteins have been documented to cause two problems: (1) drug metabolites adhere to protein surfaces and, hence, cannot be separated by CE; and (2) protein introduced onto the capillary coats the wall, leading to analyte loss.

Various approaches have been adopted to overcome such problems, including traditional sample cleanup methods, such as liquid–liquid and/or solid-phase extraction (e.g., SepPak C_{18} or Bond Elut cartridges) prior to analysis by CE. We have used this approach to remove microsomal proteins from an *in vitro* incubation of haloperidol (discussed in the next section).[38] Roach and co-workers[35] have used a similar approach to measure methotrexate and 7-hydroxymethotrexate in human serum. Two groups have independently advocated the use of surfactants in direct injection analysis onto CE of antibiotics in human plasma.[13,18] They claim that the addition of surfactant (such as sodium dodecyl sulfate [SDS]) prevents both drug–protein interactions and adhesion of protein to the capillary wall.

It is clear that sample preparation prior to analysis by CE is an active area of research. Nonetheless, CE can now play a role in drug metabolism studies, and this is highlighted in examples given below.

2. Capillary Zone Electrophoresis in Metabolism

a. Species Difference

One area of particular interest in drug metabolism studies is the difference in metabolism of humans and various other animal species. Species-dependent differences in metabolism can occur in both phase I and II biotransformations and can be either qualitative and/or quantitative.[83] Such variations have been ascribed to differences in enzyme activity[3] and, more recently, to molecular aspects of gene evolution.[84] Clearly, the problems raised by the variation in metabolism by different species for new and clinically used drugs is important in understanding the toxicological and pharmacological activity of metabolites.

We investigated the *in vitro* metabolism of the neuroleptic drug haloperidol (see Figure 2 for structure) by both mouse and guinea pig hepatic microsomes,[38] using the previously described CZE conditions (see Figure 3). The electropherograms resulting from the analysis of the microsomal incubates after $ZnSO_4$ protein precipitation and solid-phase cleanup are shown in Figure 14A (mouse) and Figure 14B (guinea pig). Comparison of the relative migration times (metabolite:HAL) of metabolites to standards enabled us to tentatively identify five metabolites, namely, CPHP, HP^+, HTP, HTPNO, and FBPA (see Figure 2 for

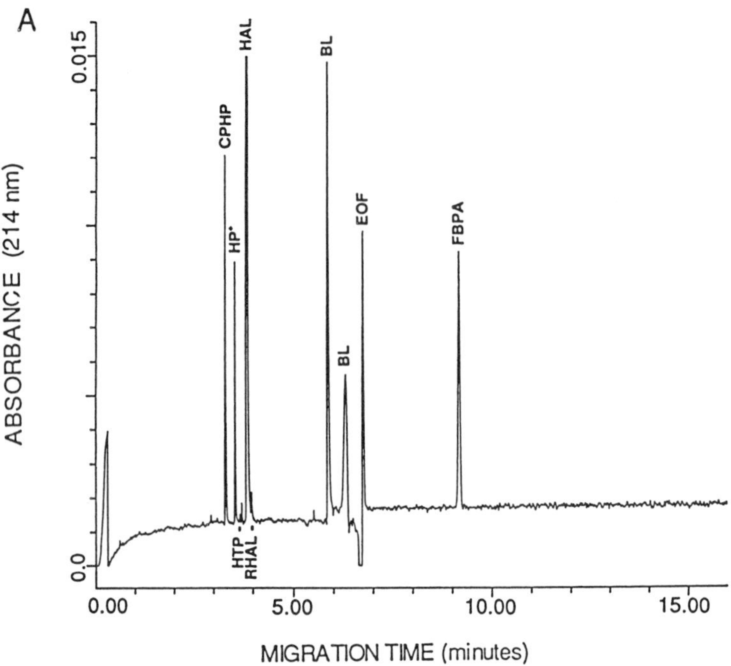

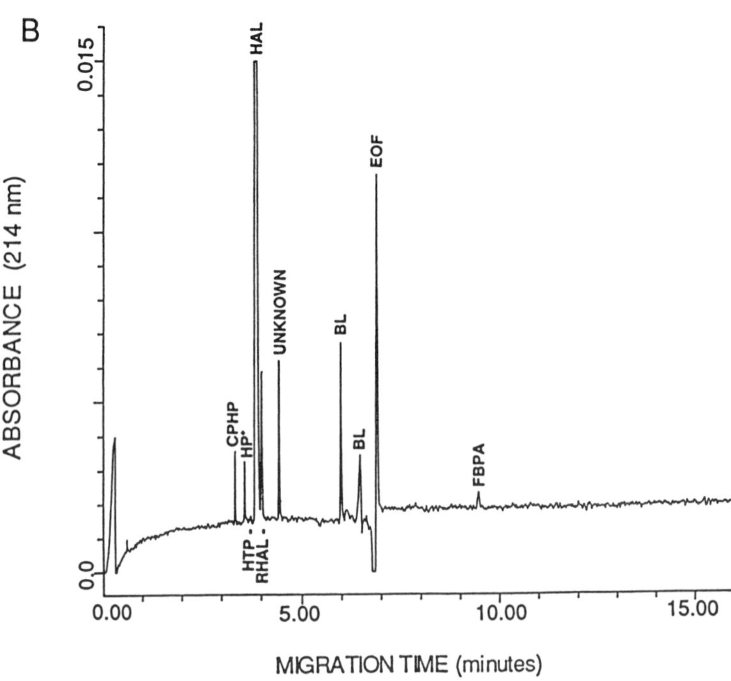

FIGURE 14. Analysis of hepatic microsomal incubations of the neuroleptic drug haloperidol (HAL) by CZE (monitored at 214 nm). Analysis performed on a Beckman P/ACE 2100. Separation buffer was 50 m*M* NH_4OAc containing 10% MeOH and 1% AcOH (pH 4.1) on an uncoated fused silica capillary (75-μm I.D. × 50 cm [effective length], 57 cm [total length]). Separation voltage, 30 kV at 25°C, with a 1-s pressure injection. (A) Hepatic mouse microsomes; (B) hepatic guinea pig microsomes. BL, Found in control microsomal incubation.

structures) from the mouse microsomal incubate (Figure 14A). The same five metabolites were also tentatively identified in the guinea pig microsomal incubate, as well as a further unknown metabolite (marked as UNKNOWN in Figure 14B). A clear difference in metabolism is exhibited by mouse and guinea pig in both a qualitative and quantitative manner. The induced mouse microsomes produce much more CPHP, HP^+, HTP, and FBPA than produced by the guinea pig microsomes. However, the guinea pig microsomal incubate contains much more HTPNO, as well as the UNKNOWN, as previously described by Fang and Gorrod.[65] Further structural studies are underway to characterize the unknown peak.

b. Chemical vs. Enzymatic Transformation

In the analysis of biotransformation pathways, it is important to distinguish between enzymatic and chemically medicated reactions. It is possible to rapidly distinguish this difference using CZE. We investigated the hepatic rat microsomal incubation of the H_2-antagonist mifentidine (see Figure 5 for structure) using CZE separation buffer conditions described in Figure 7. After a $ZnSO_4$ precipitation of protein and solid-phase extraction, three "metabolites," namely, MIF-AMINE, MIF-AMIDE, and MIF-UREA, were tentatively identified based on their relative migration times to synthetic standards (see Figure 15B).[71] However, analysis of a control incubate, using a heat-inactivated microsomal preparation, revealed the presence of both MIF-AMINE and MIF-AMIDE, as well as an unidentified component (marked C) in Figure 15A. This clearly demonstrates that MIF-AMINE and MIF-AMIDE are chemically produced, but not does not preclude this formation by enzymatic routes as well. At present, we are investigating the chemical mechanism of transformation of MIF to these two compounds.

c. **In Vivo** *Urine Analysis*

The ultimate goal in drug metabolism studies is to determine the metabolic fate of parent drug *in vivo*. Shown in Figure 16A and B are the electropherograms of a urine sample, after solid-phase extraction, from a healthy individual and a patient receiving a high, daily dose of haloperidol, respectively. Typically, 2 ml of urine was passed through a conditioned C_{18} Sep-pak cartridge, washed with 1 ml of H_2O, and eluted with 4 ml of MeOH. The organic solvent was removed by evaporation (under vacuum), and the sample was redissolved in 50 µl of MeOH:H_2O (1:1). It is clear that many more major components are present in the urine derived from the patient receiving haloperidol (Figure 16B) than in the profile from the healthy individual. However, this example serves to demonstrate the limitations of CE with UV as the method of detection. Although there are major differences in the two electropherograms, it is not possible to know if it is due to haloperidol-related phase I and/or phase II metabolites or some other unrelated physiological condition.

G. CAPILLARY ELECTROPHORESIS-MASS SPECTROMETRY AND CAPILLARY ELECTROPHORESIS-TANDEM MASS SPECTROMETRY

1. Introduction

A variety of detection devices other than conventional UV have been utilized in the analysis of small molecules (see Chapter 7 for more detailed discussion), including laser-induced fluorescence,[33-35] electrochemical,[73] and multiwavelength array UV[21,27,42] detectors. However, a limitation of these detectors is the lack of structural information obtained, particularly when compared to MS. Mass spectrometry, of all the detection systems presently available that analyze less than nanogram quantities, is most suited for acquiring structural data of unknowns. However, the use of on-line CE-MS and CE-MS/MS for the analysis of such molecules has not been fully exploited. Since most drug metabolism studies invariably involve

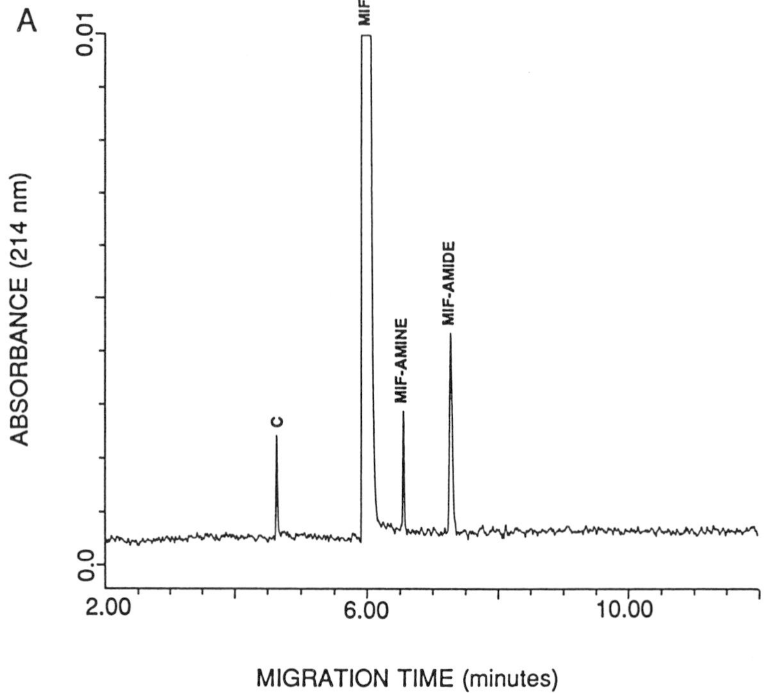

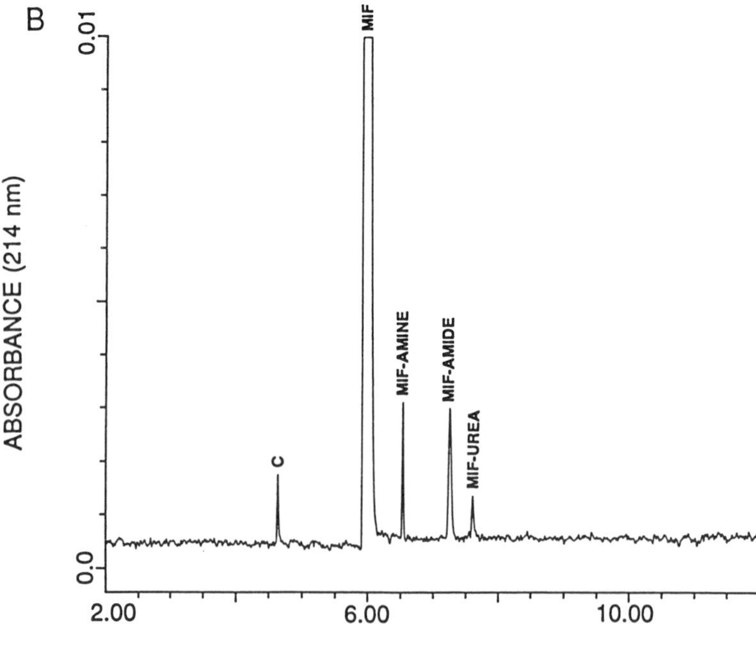

FIGURE 15. Analysis of the hepatic rat microsomal incubation of mifentidine (MIF). Injection and electrophoretic conditions identical to those described in Figure 4A. (A) Microsomes deactivated by boiling them in H_2O for 30 min; (B) active microsomes. (C) Unidentified component found in control microsome incubation.

structural determination of unknown metabolites (as demonstrated in the haloperidol urine study discussed above; see Figure 16B), it is clear that CE-MS and CE-MS/MS will play a major role in this arena in the future.

On the basis of the discussions of Smith et al. in Chapter 8 of this book and the work of Henion,[17, 50-53] it appears that electrospray ionization (ESI) and atmospheric pressure chemical ionization (AP/CI) will serve as the primary ionization techniques when coupled to CE for the analysis of small molecules and drug metabolites.

2. Metabolism Studies

a. In Vitro

The CE-ESI-MS ion chromatogram of seven haloperidol standards, including HAL, CPHP, HP^+, HTP, RHAL, HALNO, and HTPNO (see Figure 2 for structures), is shown in Figure 17.[64] The separation buffer was 50 m*M* NH_4OAc containing 10% MeOH and 1% AcOH, with a sheath liquid of isopropanol:H_2O:AcOH (60:40:1, v/v). All seven components were detected by the mass spectrometer, and the mass assignments are denoted in the caption to Figure 17.

It is possible, using the ESI source as a collision cell, to fragment precursor (molecular) ions obtained from the CE-ESI-MS interface (this is described in more detail by Smith et al. in Chapter 8) to produce CE-MS/MS spectra and, hence, structural information. A specific example of the fragmentation of the parent drug haloperidol in the ESI source is shown in Figure 18. Fragment ions (also known as product ions) detected at *m/z* 358, 165, and 123 can be readily ascribed to specific structural bond cleavages (see Figure 18).

A similar study was conducted on a hepatic guinea pig microsomal incubation of haloperidol, and five metabolites have been identified by CE-MS/MS studies, namely, CPHP, HP^+, HTP, RHAL, and FBPA (see Figure 2 for structures).[64] A sixth, unknown metabolite is still under investigation.

b. In Vivo

Johansson and co-workers have used CE-ESI-MS and CE-ESI-MS/MS to study the separation of a series of sulfonamides and benzodiazepines, as well as to investigate the *in vivo* metabolism of flurazepam.[17] Urine samples were collected 2 h after a 30-mg oral dose of flurazepam was administered, and the urine was subjected to β-glucuronidase treatment before analysis by CE-ESI-MS. These results are shown in Figure 19. They identified three metabolites, namely, didesethylflurazepam, monodesethylflurazepam, and *N*-1-hydroxyethylflurazepam. No ion corresponding to unmetabolized parent drug flurazepam was detected.

H. CAPILLARY ELECTROPHORESIS AND ANALYSIS OF DRUG–PROTEIN INTERACTIONS

The interaction of a drug with protein/enzyme is of fundamental importance in assessing the pharmacological properties of the parent drug or its active metabolite. At least two factors need to be considered.

1. If the concentration of unbound drug determines its pharmacological activity, then if the drug binds reversibly to proteins (particularly plasma proteins, such as albumin), an undetermined fraction of drug is removed from the active pool.
2. Many drug–enzyme interactions involve noncovalent binding of the drug.

At present, this area is in its infancy. Kraak et al.[85] have evaluated the use of CE for protein–drug-binding studies and discuss a variety of methodologies and their advantages and disadvantages. Chu and co-workers[86] are using what they term affinity CE to screen libraries of

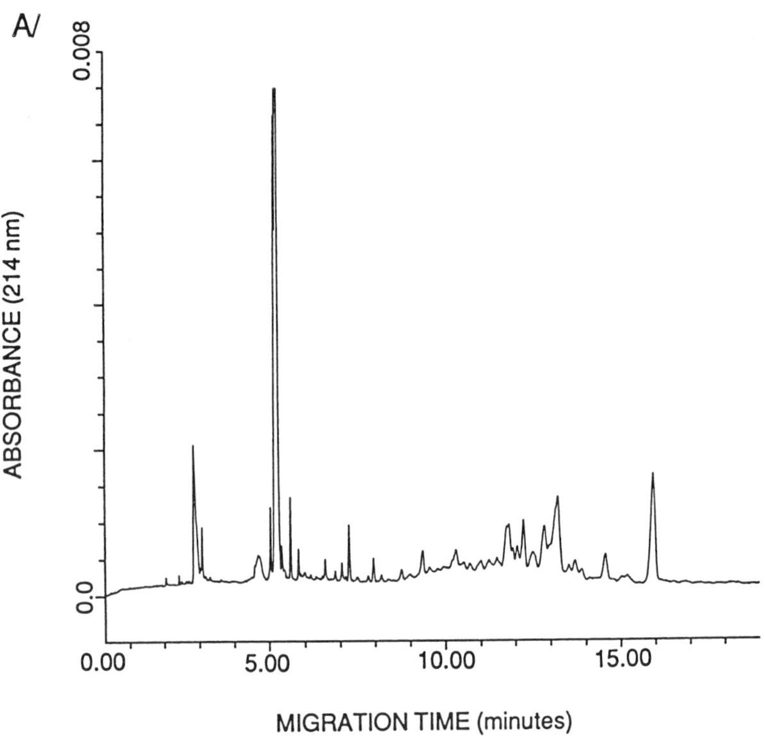
A/
0.008
0.0
ABSORBANCE (214 nm)
0.00
5.00
10.00
15.00
MIGRATION TIME (minutes)

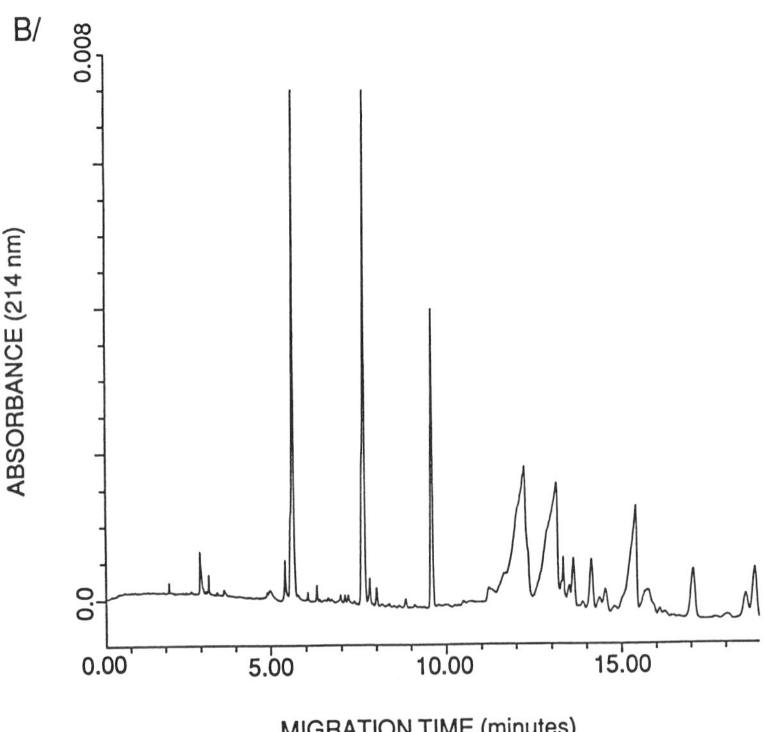
B/
0.008
0.0
ABSORBANCE (214 nm)
0.00
5.00
10.00
15.00
MIGRATION TIME (minutes)

FIGURE 16.

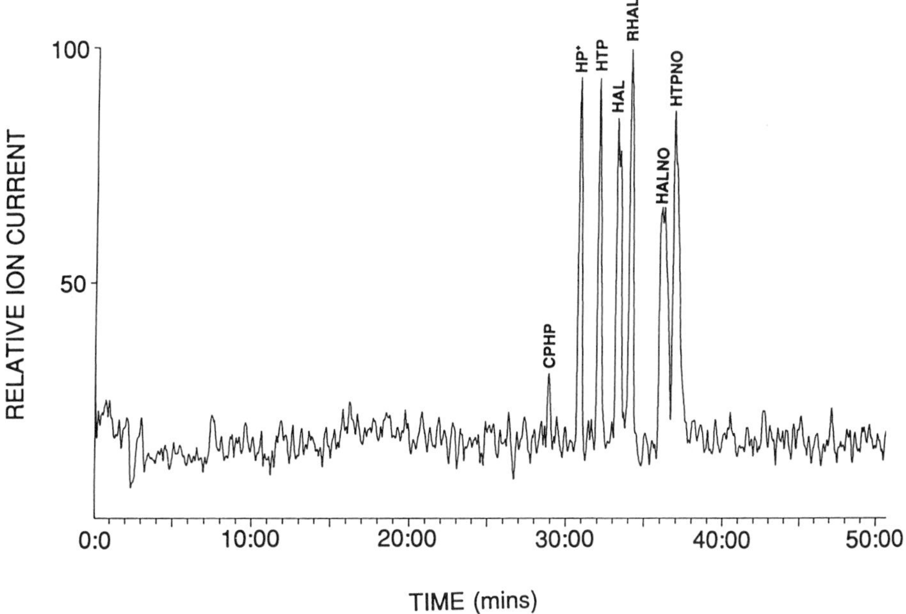

FIGURE 17. CE-MS analysis of six putative metabolites of the neuroleptic drug haloperidol (HAL). Compounds detected and identified were as follow: CPHP ($MH^+ = 212$), HP^+ ($MH^+ = 354$), HTP ($M^+ = 358$), HAL ($MH^+ = 376$), RHAL ($MH^+ = 378$), HALNO ($MH^+ = 392$), and HTPNO ($MH^+ = 374$). CZE separation on a Beckman P/ACE 2100 was effected using 50 m*M* NH_4OAc containing 10% MeOH and 1% AcOH (pH 4.1) as separation buffer at 15 kV, on an uncoated fused silica capillary (50 μm × 65 cm [total length]), using a 30-s pressure injection of sample. The sheath liquid for the ESI was isopropanol:H_2O:AcOH (60:40:1, v/v) and the ESI voltage was –3.4 kV. The scan range was *m/z* 420 to 125 at 3 s/decade at a resolution of ~1200 on a Finnigan MAT 900 mass spectrometer.

compounds to bind to soluble receptors, and Barker et al.[16] are investigating the use of CE to determine thermodynamic data for substrate binding interactions.

This is clearly an area of active development and will continue to grow.

III. FUTURE DIRECTIONS

A. CHIRAL DRUG METABOLISM

A rapidly expanding area of interest in the pharmaceutical industry is that of chiral drugs. At present, such agents are usually administered as racemic mixtures. However, recently, under pressure from new Food and Drug Administration (FDA) guidelines, companies are pursuing the development of single enantiomeric products.[87] On entering the body, such pharmaceuticals can, in addition to undergoing usual metabolic pathways, also undergo chiral inversion prior to metabolism. Detection of such inversion is important, since as one enantiomer may be the active agent, *in vivo* chiral inversion may lead to the formation of an enantiomeric toxic metabolite.

FIGURE 16. Analysis of urine samples: 2 ml of urine passed through a Sep-pak C_{18} cartridge and washed with 1 ml of H_2O. Components were eluted with 4 ml of MeOH, which was subsequently removed by evaporation under vacuum. The sample was redissolved in MeOH:H_2O (1:1) (50 μl). The samples were analyzed by CZE (monitored at 214 nm), using a separation buffer of 50 m*M* NH_4OAc containing 10% MeOH and 1% AcOH (pH 4.1) on an uncoated fused silica capillary (75-μm I.D. × 50 cm [effective length], 57 cm [total length]), at a voltage of 30 kV (25°C). Injection was for 1 s by pressure. (A) Urine from a healthy male on no medication; (B) urine from a male being treated daily with a high dose of haloperidol (HAL).

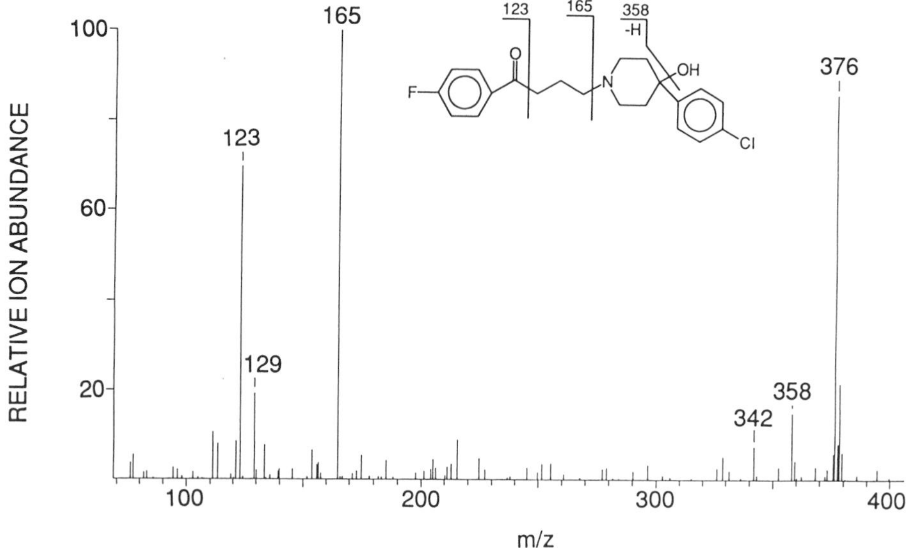

FIGURE 18. CE-MS/MS spectrum of the parent drug haloperidol (MH^+ = 376). Conditions were as described in Figure 17, except that the scan range was *m/z* 450 to 70 at 3 s/decade. Fragmentation was brought about by skimmer-induced dissociation in the ESI source.

As yet, while chiral separation of pharmaceuticals and small molecules by CE is being addressed (see Chapter 10), there have been few studies to separate chiral drug metabolites. In one study, Aumatell and Wells have demonstrated the separation of such metabolites by CE in urine for drug testing of athletes.[88] This research is, however, in its infancy and will require much more attention since we have found in our hands that chiral separations are not, as yet, routine, and replication of published methodology is difficult to achieve. Furthermore, in the arena of drug metabolism, where the identification of novel metabolites is of paramount importance, development of chiral CE-MS/MS approaches will be essential.

B. REACTIVE INTERMEDIATES

An important aspect in the determination of the metabolic pathway of a drug is the detection of reactive intermediates (transient metabolites). These compounds, like other metabolites, are usually of low abundance and are typically detected in time course studies of *in vitro* hepatic microsomal incubations. It should be noted that since such compounds are reactive, they are unlikely to be detected if cleanup procedures are employed prior to analysis. In this arena, CE offers enormous potential since it requires only small samples for analysis. Hence, changes in concentration of microsomal incubates on removal of samples can be avoided. However, currently, the preconcentration prior to CE is required to analyze such mixtures, and there are components of microsomal incubates that can adhere to the capillary and degrade its performance. Two approaches that may overcome these problems are presented below.

C. COMBINED LIQUID CHROMATOGRAPHY-CAPILLARY ELECTROPHORESIS

Combined LC-CE systems have already been demonstrated by Jorgenson et al.[89] and have been shown to possess the benefits of each individual component (e.g., high sample loading capacity for the LC and highly efficient separations for CE). Future developments of such equipment need to focus on detector systems that can yield structural information; hence, a

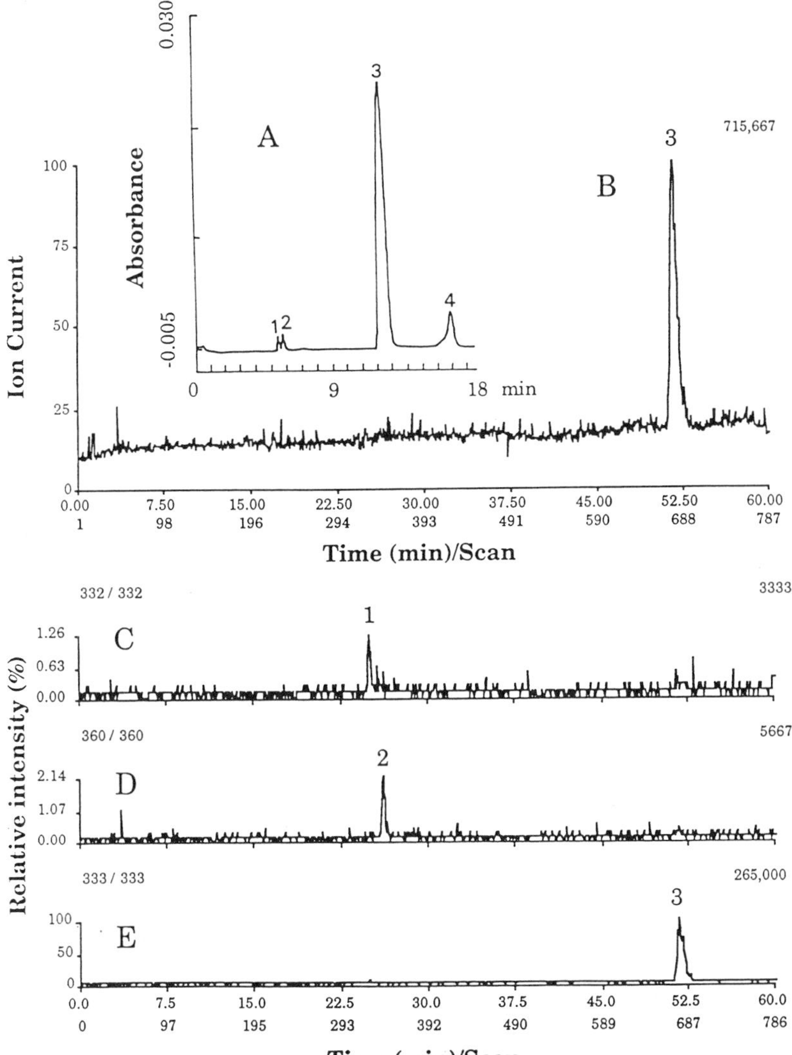

FIGURE 19. (A) CE-UV and (B) full-scan CE-MS total ion current and (C–E) extracted ion current profiles from the analysis of an extract of human urine collected 2 h after the administration of a 30-mg oral dose of flurazepam dihydrochloride. (A) CE-UV electropherogram at 254 nm, detection after 20 cm capillary length; (B) full-scan acquisition from m/z 250 to 400, 5 s/scan, detection after 100 cm capillary length. Peaks: 1, didesethylflurazepam, m/z 332 $[MH]^+$(C); 2, monodesethylflurazepam, m/z 360 $[MH]^+$(D); 3, *N*-1-hydroxyethylflurazepam, m/z 333 $[MH]^+$ (E); 4, "neutral fraction" in the urine extract. No peak corresponding to flurazepam (m/z 388) was observed. Injection volume, 5 nl; electrophoretic conditions consisted of an uncoated capillary (100 cm × 75 μm I.D.) and a separation buffer of 0.2 m*M* NH_4OAc adjusted to pH 1.3 with TFA, containing 15% (v/v) methanol; voltage was 26 kV. (Taken from Johansson, I. M., Pavelka, R., and Henion, J. D., *Anal. Chem.*, 559, 515, 1992. With permission.)

balance between the current stop flow LC-CE techniques and the now routinely obtained 4-ms separation of components by CE will be required to permit detection of components by MS.

D. COMBINED ELECTROCHROMATOGRAPHY

The technique of combined electrochromatography-CE is described in Chapter 5 of this book. It is currently not a popular technique, since capillaries are reported to be difficult to manufacture and peak broadening is reported to occur during the elution of analytes from the capillary. However, if these problems can be overcome, such techniques could become powerful in monitoring drug metabolite mixtures, since sample preconcentration and cleanup could be achieved on line.

E. DIRECT *IN VIVO* MONITORING OF DRUG METABOLISM

The development of microdialysis probes makes it increasingly likely that it will be possible to analyze metabolism products directly from humans in the near future. Such on-line coupling of patients to instrumentation, e.g., CE-MS, could potentially provide valuable pharmacokinetic and drug metabolism data. Hence, the clinician could, after administration of a drug, determine its efficacy with a particular patient and alter dosage immediately and, hence, reduce harmful side effects.

F. DRUG–PROTEIN INTERACTIONS

As discussed in an earlier section, drug–protein interactions are of importance in drug metabolism studies. However, little is known regarding how the drug and protein come to be in intimate contact, or how the drug is able to find the site of attachment on the protein. Capillary electrophoresis and CE-MS techniques could, with development, probe such interactions perhaps *in vivo* after administration of the therapeutic agent.

G. DETECTORS

It appears that CE-MS and CE-MS/MS will play an increasingly important role in the isolation and structural characterization of drug metabolites, particularly from *in vivo* sources. However, the use of LIF in clinical measurements where specific quantitation of a particular drug or metabolite is also an area of research that will receive considerable attention.

ACKNOWLEDGMENTS

We would like to thank Mrs. Val Langworthy for her considerable help in preparing all the figures and text in this chapter. We would also like to thank Dr. James Landers for his kind invitation to participate in this venture and his willingness to share his knowledge when we were first entering this arena.

REFERENCES

1. **Gilman, A. G., Rall, T. W., Nies, A. S., and Taylor, P., Eds.,** *The Pharmacological Basis of Therapeutics*, 8th ed., Pergamon Press (McGraw Hill), New York, 1990, p. 1811.
2. **Moffat, A. C., Ed.,** *Clarke's Isolation and Identification of Drugs*, 2nd ed., Pharmaceutical Press, London, 1986, p. 309.
3. **Gibson, G. G. and Skett, P.,** *Introduction to Drug Metabolism*, Chapman and Hall, London, 1986, p. 293.
4. **Pratt, W. B. and Taylor, P., Eds.,** *Principles of Drug Action. The Basis of Pharmacology*, 3rd ed., Churchill Livingstone, New York, 1990, p. 836.
5. **Benford, D. J., Bridges, J. W., and Gibson, G. G., Eds.,** *Drug Metabolism — From Molecules to Man*, Taylor Francis, London, 1987, p. 787.
6. **Dekant, W., Berthold, K., Vamvakas, S., Henschler, D., and Anders, M. W.,** Thioacylating intermediates as metabolites of *S*-(1,2-dichlorovinyl)-L-cysteine and *S*-(1,2,2-trichlorovinyl)-L-cysteine formed by cysteine conjugate β-lyase, *Chem. Res. Toxicol.*, 1, 175, 1988.
7. **Voelter, W. and Kronbach, T.,** High-performance liquid chromatographic comparison of *in vivo* and *in vitro* drug metabolism, *J. Chromatogr.*, 290, 1, 1984.
8. **Kuhr, W. G. and Monnig, C. A.,** Capillary electrophoresis, *Anal. Chem.*, 64, 389R, 1992.
9. **Landers, J. P., Oda, R. P., Spelsberg, T. C., Nolan, J. A., and Ulfelder, K. J.,** Capillary electrophoresis: a powerful microanalytical technique for biologically active molecules, *BioTechniques*, 14, 98, 1993.

10. **Wainwright, A.,** Capillary electrophoresis applied to the analysis of pharmaceutical compounds, *J. Microcolumn Sep.*, 2, 166, 1990.
11. **Nishi, H., Fukuyama, T., Matsuo, M., and Terabe, S.,** Effect of surfactant structures on the separation of cold medicine ingredients by micellar electrokinetic chromatography, *J. Pharm. Sci.*, 79, 519, 1990.
12. **Nishi, H., Tsumagari, H., Kakimoto, T., and Terabe, S.,** Separation of beta-lactam antibiotics by micellar electrokinetic chromatography, *J. Chromatogr.*, 477, 259, 1989.
13. **Nishi, H. and Terabe, S.,** Application of micellar electrokinetic chromatography to pharmaceutical analysis, *Electrophoresis*, 11, 691, 1990.
14. **Fujiwara, S. and Honda, S.,** Determination of ingredients of antipyretic analgesic preparations of micellar electrokinetic capillary chromatography, *Anal. Chem.*, 59, 2773, 1987.
15. **Nishi, H., Fukuyama, T., Matsuo, M., and Terabe, S.,** Separation and determination of lipophilic corticosteroids and benzothiazepin analogues by micellar electrokinetic chromatography using bile salts, *J. Chromatogr.*, 513, 279, 1990.
16. **Barker, G. E., Russo, P., and Hartwick, R. A.,** Chiral separation of leucovorin with bovine serum albumin using affinity capillary chromatography, *Anal. Chem.*, 64, 3024, 1992.
17. **Johansson, I. M., Pavelka, R., and Henion, J. D.,** Determination of small drugs by capillary electrophoresis — atmospheric pressure ionization mass spectrometry, *J. Chromatogr.*, 559, 515, 1991.
18. **Nakagawa, T., Oda, Y., Shibukawa, A., Fukuda, H., and Tanaka, H.,** Electrokinetic chromatography for drug analysis separation and determination of cefpiramide in human plasma, *Chem. Pharm. Bull.*, 37, 707, 1989.
19. **Atamna, I. Z., Janini, G. M., Muschik, G. M., and Issaq, H. J.,** Separation of xanthines and uric acids by capillary zone electrophoresis and micellar electrokinetic capillary electrophoresis, *J. Liq. Chromatogr.*, 14, 427, 1991.
20. **Weinberger, R. and Albin, M.,** Quantitative micellar electrokinetic capillary chromatography: linear dynamic range, *J. Liq. Chromatogr.*, 14, 953, 1991.
21. **Thormann, W., Meier, P., Marcolli, C., and Binder, F.,** Analysis of barbiturates in human serum and urine by high performance capillary electrophoresis-micellar electrokinetic capillary chromatography with on-column multi-wavelength detection, *J. Chromatogr.*, 545, 445, 1991.
22. **Salomon, K., Burgi, D. S., and Helmer, J. C.,** Separation of seven tricyclic antidepressants using capillary electrophoresis, *J. Chromatogr.*, 549, 375, 1991.
23. **Guzman, N. A., Ali, H., Moschera, J., Iqbal, K., and Malick, A. W.,** Assessment of capillary electrophoresis in pharmaceutical applications. Analysis and quantification of a recombinant cytokine in an injectable dosage form, *J. Chromatogr.*, 559, 307, 1991.
24. **Meier, P. and Thormann, W.,** Determination of thiopental in human serum and plasma by high-performance capillary electrophoresis-micellar electrokinetic chromatography, *J. Chromatogr.*, 559, 505, 1991.
25. **Soini, H., Tsuda, T., and Novotny, M. V.,** Electrochromatographic solid-phase extraction for determination of cimetidine in serum by micellar electrokinetic capillary chromatogram, *J. Chromatogr.*, 559, 547, 1991.
26. **Lloyd, D. K., Cypess, A. M., and Wainer, I. W.,** Determination of cytosine-β-D-arabinoside in plasma using capillary electrophoresis, *J. Chromatogr.*, 568, 117, 1991.
27. **Yeo, S. K., Lee, H. K., and Li, S. F. Y.,** Separation of antibiotics by high-performance capillary electrophoresis with photodiode-array detection, *J. Chromatogr.*, 585, 133, 1991.
28. **De Bruijn, E. A., Pattyn, G., David, F., and Sanda, P.,** Capillary electrophoresis of fluoropyrimidines, *J. High Resolut. Chromatogr.*, 14, 627, 1991.
29. **Swartz, M. E.,** Method development and selectivity control for small molecule pharmaceutical separations by capillary electrophoresis, *J. Liq. Chromatogr.*, 14, 923, 1991.
30. **Nishi, H. and Matsuo, M.,** Separation of corticosteroids and aromatic hydrocarbons by cyclodextrin-modified micellar electrokinetic chromatography, *J. Liq. Chromatogr.*, 14, 973, 1991.
31. **Guzman, N. A., Trebilcock, M. A., and Advis, J. P.,** The use of a concentration step to collect urinary components separated by capillary electrophoresis and further characterisation of collected analytes by mass spectrometry, *J. Liq. Chromatogr.*, 14, 997, 1991.
32. **Lee, K.-J., Heo, G. S., Kim, N. J., and Moon, D. C.,** Analysis of antiepileptic drugs in human plasma using micellar electrokinetic capillary chromatography, *J. Chromatogr.*, 608, 243, 1992.
33. **Swaile, D. F., Burton, D. E., Balchunas, A. T., and Sepaniak, M. J.,** Pharmaceutical analysis using micellar electrokinetic capillary chromatography, *J. Chromatogr. Sci.*, 26, 406, 1988.
34. **Burton, D. E., Sepaniak, M. J., and Maskarinec, M. P.,** Analysis of B_6 vitamers by micellar electrokinetic capillary chromatography with laser-excited fluorescence detection, *J. Chromatogr. Sci.*, 24, 347, 1986.
35. **Roach, M. C., Gozel, P., and Zare, R. N.,** Determination of methotrexate and its major metabolite 7-hydroxymethotrexate using capillary zone electrophoresis and laser-induced fluorescence detection, *J. Chromatogr.*, 426, 129, 1988.
36. **Fujiwara, S. and Honda, S.,** Determination of cinnamic acid and its analogues by electrophoresis in a fused silica capillary tube, *Anal. Chem.*, 58, 1811, 1986.

37. **Tanaka, Y. and Thormann, W.,** Capillary electrophoretic determination of *S*-carboxymethyl-L-cysteine and its major metabolites in human urine: feasibility investigation using on-column detection of non-derivatised solutes in capillaries with minimal electroosmosis, *Electrophoresis*, 11, 760, 1990.
38. **Tomlinson, A. J., Benson, L. M., Landers, J. P., Scanlan, G. F., Fang, J., Gorrod, J. W., and Naylor, S.,** An investigation of the metabolism of the neuroleptic drug haloperidol by capillary electrophoresis, *J. Chromatogr.*, 1993, in press.
39. **Weinberger, R., Sapp, E., and Moring, S.,** Capillary electrophoresis of urinary porphyrins with absorbance and fluorescence detection, *J. Chromatogr.*, 516, 271, 1990.
40. **Guzman, N. A., Berck, C. M., Hernandez, L., and Advis, J. P.,** Capillary electrophoresis as a diagnostic tool: determination of biological constituents present in urine of normal and pathological individuals, *J. Liq. Chromatogr.*, 13, 3833, 1990.
41. **Tomita, M., Okuyama, T., Nigo, Y., Uno, B., and Kawai, S.,** Determination of glyphosate and its metabolite (aminomethyl)phosphonic acid in serum using capillary electrophoresis, *J. Chromatogr. Biomed. Appl.*, 571, 324, 1991.
42. **Wernly, P. and Thormann, W.,** Analysis of illicit drugs in human urine by micellar electrokinetic capillary chromatography with on-column fast scanning polychrome absorption detection, *Anal. Chem.*, 63, 2878, 1991.
43. **Wernly, P. and Thormann, W.,** Confirmation testing of 11-nor-Δ^9-tetrahydrocannabinol-9-carboxylic acid in urine with micellar electrokinetic capillary chromatography, *J. Chromatogr.*, 608, 251, 1992.
44. **Weinberger, R. and Lurie, I. S.,** Micellar electrokinetic capillary chromatography of illicit drug substances, *Anal. Chem.*, 63, 823, 1991.
45. **Olivares, J. A., Nguyen, N. T., Yonker, C. R., and Smith, R. D.,** On-line mass spectrometric detection for capillary zone electrophoresis, *Anal. Chem.*, 59, 1232, 1987.
46. **Smith, R. D., Olivares, J. T., Nguyen, N. T., and Udseth, H. R.,** Capillary zone electrophoresis-mass spectrometry using an electrospray ionization interface, *Anal. Chem.*, 60, 436, 1988.
47. **Smith, R. D., Barinaga, C. J., and Udseth, H. R.,** Improved electrospray ionization interface for capillary zone electrophoresis-mass spectrometry, *Anal. Chem.*, 60, 1948, 1988.
48. **Smith, R. D., Loo, J. A., Barinaga, C. J., Edmonds, C. G., and Udseth, H. R.,** Capillary zone electrophoresis and isotachophoresis-mass spectrometry of polypeptides and proteins based upon an electrospray ionization interface, *J. Chromatogr.*, 480, 211, 1989.
49. **Loo, J. A., Jones, H. K., Udseth, H. R., and Smith, R. D.,** Capillary zone electrophoresis-mass spectrometry with electrospray ionization of peptides and proteins, *J. Microcolumn Sep.*, 1, 223, 1989.
50. **Lee, E. D., Mueck, W., Henion, J. D., and Covey, T. R.,** On-line capillary zone electrophoresis in spray tandem mass spectrometry for the determination of dynorphins, *J. Chromatogr.*, 458, 313, 1988.
51. **Lee, E. D., Mueck, W., Covey, T. R., and Henion, J. D.,** Liquid junction coupling for capillary zone electrophoresis/ion spray mass spectrometry, *Biomed. Environ. Mass Spectrom.*, 18, 844, 1989.
52. **Mueck, W. M. and Henion, J. D.,** Determination of leucine enkephalin and methionine enkephalin in equine cerebrospinal fluid by microbore high-performance liquid chromatography and capillary zone electrophoresis coupled to tandem mass spectrometry, *J. Chromatogr.*, 495, 41, 1989.
53. **Johannson, I. M., Huang, E. C., Henion, J. D., and Zweigenbaum, J.,** Capillary electrophoresis for the characterization of peptides. Instrumental considerations for mass spectrometric detection, *J. Chromatogr.*, 554, 311, 1991.
54. **Harvey, D. J.,** Drug metabolism, pharmacokinetics, and toxicity, *Mass Spectrom.*, 9, 303, 1987.
55. **Harvey, D. J.,** The use of mass spectrometry in studies of drug metabolism and pharmacokinetics, *Mass Spectrom.*, 10, 273, 1989.
56. **Abramson, F. P.,** Mass spectrometry in pharmacology, in *Methods of Biochemical Analysis*, Vol. 34, *Biomedical Applications of Mass Spectrometry*, Suelter, C. H. and Watson, J. T., Eds., John Wiley & Sons, New York, 1990, chap. 5.
57. **Busch, K. L., Glish, G. L., and McLuckey, S. A.,** *Mass Spectrometry/Mass Spectrometry: Techniques and Applications of Tandem Mass Spectrometry*, VCH, Weinheim, Germany, 1988.
58. **Fenselau, C.,** Tandem mass spectrometry: the competitive edge for pharmacology, *Annu. Rev. Pharmacol. Toxicol.*, 32, 555, 1992.
59. **Naylor, S., Kajbaf, M., Lamb, J. H., Jahanshahi, M., and Gorrod, J. W.,** An evaluation of tandem mass spectrometry in drug metabolism studies, *Biol. Mass Spectrom.*, 22, 388, 1993.
60. **Baillie, T. A.,** Advances in the application of mass spectrometry to the studies of drug metabolism, pharmacokinetics and toxicology, *Int. J. Mass Spectrom. Ion Process.*, 118/119, 289, 1992.
61. **Henion, J. and Covey, T.,** The determination of drugs in urine by LC-MS and LC-MS/MS, in *Mass Spectrometry in Biomedical Research*, Gaskell, S., Ed., John Wiley & Sons, Chichester, 1986, chap. 26.
62. **Tomer, K. B. and Parker, C. E.,** Biochemical applications of liquid chromatography-mass spectrometry, *J. Chromatogr. Biomed. Appl.*, 492, 189, 1989.

63. **Brown, M. A.,** *Liquid Chromatography/Mass Spectrometry*, Vol. 420, *ACS Symposium*. American Chemical Society, Washington, D.C., 1990.
64. **Tomlinson, A. J., Benson, L. M., Johnson, K. L., and Naylor, S.,** Investigation of the metabolic fate of the neuroleptic drug haloperidol by capillary electrophoresis coupled with tandem mass spectrometry, *J. Chromatogr. Biomed. Appl.*, 1993, in press.
65. **Fang, J. and Gorrod, J. W.,** An HPLC method for the detection and quantitation of haloperidol and seven of its metabolites, *J. Chromatogr. Biomed. Appl.*, 614, 267, 1993.
66. **Steuer, W., Grant, I., and Erni, F.,** Comparison of high-performance liquid chromatography, supercritical fluid chromatography and capillary zone electrophoresis in drug analysis, *J. Chromatogr.*, 507, 125, 1990.
67. **Smith, R. D., Loo, J. A., Edmonds, C. G., Barinaga, C. J., and Udseth, H. R.,** New developments in biochemical mass spectrometry: electrospray ionization, *Anal. Chem.*, 62, 882, 1990.
68. **Foret, F., Szoko, E., and Karger, B. L.,** On-column transient and coupled column isotachophoretic preconcentration of protein samples in capillary zone electrophoresis, *J. Chromatogr.*, 608, 3, 1992.
69. **Gebauer, P., Thormann, W., and Boček, P.,** Sample self-stacking in zone electrophoresis: theoretical description of the zone electrophoretic separation of minor compounds in the presence of bulk amounts of a sample component with high mobility and like charge, *J. Chromatogr.*, 608, 47, 1992.
70. **Kajbaf, M., Jahanshahi, M., Pattichis, K., Gorrod, J. W., and Naylor, S.,** Rapid and efficient purification of cimetropium bromide and mifentidine drug metabolite mixtures derived from microsomal incubates for analysis by mass spectrometry, *J. Chromatogr. Biomed. Appl.*, 575, 75, 1992.
71. **Benson, L. M., Tomlinson, A. J., and Naylor, S.,** Analysis of free solution capillary electrophoresis conditions to analyze the metabolism of the H_2-antagonist drug mifentidine, *J. Chromatogr.*, 1993, submitted.
72. **Lui, J., Cobb, K. A., and Novotny, M.,** Separation of precolumn ortho-phthalaldehyde-derivatized amino acids by capillary zone electrophoresis with normal and micellar solutions in the presence of organic modifiers, *J. Chromatogr.*, 468, 55, 1988.
73. **Wallingford, R. A. and Ewing, A. G.,** Separation of serotonin from catechols by capillary zone electrophoresis with electrochemical detection, *Anal. Chem.*, 61, 98, 1989.
74. **Wallingford, R. A., Curry, P. D., Jr., and Ewing, A. G.,** Retention of catechols in capillary electrophoresis with micellar and mixed micellar solutions, *J. Microcolumn Sep.*, 1, 23, 1989.
75. **Jackson, R. C., Sebolt, J. S., Shillis, J. L., and Leopold, W. R.,** The pyrazoloacridines: approaches to the development of a carcinoma-selective cytotoxic agent, *Cancer Invest.*, 8, 39, 1990.
76. **Benson, L. M., Tomlinson, A. J., Reid, J. M., Walker, D. L., Ames, M. M., and Naylor, S.,** Study of in vivo pyrazoloacridine metabolism by capillary electrophoresis using isotachophoresis preconcentration in nonaqueous separation buffer, *J. High Resolut. Chromatogr.*, 16, 324, 1993.
77. **Eisenhauer, E. A., Weinerman, B. H., Kerr, I., and Quirt, I.,** Toxicity of oral *N*-methylformamide in three phase II trials: a report from the National Cancer Institute of Canada Trials Group, *Cancer Treat. Rep.*, 70, 811, 1986.
78. **Langdon, S. P., Chubb, D., Gescher, A., Hickman, J. A., and Stevens, M. F. G.,** Studies on the toxicity of the antitumor agent *N*-methylformamide in mice, *Toxicology*, 34, 173, 1985.
79. **Threadgill, M. D., Axworthy, D. B., Baillie, T. A., Farmer, P. B., Farrow, K. C., Gescher, A., Kestell, P., Pearson, P. G., and Shaw, A. J.,** Metabolism of *N*-methylformamide in mice: primary kinetic deuterium isotope effect and identification of *S*-(*N*-methylcarbamoyl)glutathione as a metabolite, *J. Pharmacol. Exp. Therapeut.*, 242, 312, 1987.
80. **Baillie, T. A.,** Personal communication, 1993.
81. **Lipsky, J. J.,** Ability of 1-methyltetrazole-5-thiol with microsomal activation to inhibit aldehyde dehydrogenase, *Biochem. Pharmacol.*, 38, 773, 1989.
82. **Veverka, K., Tomlinson, A. J., Benson, L. M., Lipsky, J. J., and Naylor, S.,** Unpublished work, 1993.
83. **Williams, R. T.,** Inter-species variations in the metabolism of xenobiotics, *Biochem. Soc. Trans.*, 2, 359, 1974.
84. **Nerbert, D. W. and Gonzales, F. J.,** P_{450} genes: structure, evolution and regulation, *Annu. Rev. Biochem.*, 56, 945, 1987.
85. **Kraak, J. C., Busch, S., and Pope, H.,** Study of protein-drug binding using capillary zone electrophoresis, *J. Chromatogr.*, 608, 275, 1992.
86. **Chu, Y.-H., Avila, L. Z., Biebuyck, H. A., and Whitesides, G. M.,** Using capillary electrophoresis to identify the peptide in a peptide library that binds most tightly to vancomycin, *J. Org. Chem.*, 58, 648, 1993.
87. **Stinson, S. C.,** Chiral drugs, *Chem. Eng. News*, 70(39), 46, 1992.
88. **Aumatell, A. and Wells, R. J.,** Chiral separation of D,L racemethorphan and D,L raceorphan in urine for drugs in sport testing, paper presented at the 5th Int. Symp. on High Performance Capillary Electrophoresis, Orlando, FL, January 25–28, 1993.
89. **Jorgenson, J. W., Moore, A. W., Larmann, J. P., and Lemmo, A. V.,** Advances in two dimensional separation by LC/CE, paper presented at the 5th Int. Symp. on High Performance Capillary Electrophoresis, Orlando, FL, January 25–28, 1993.

PART V
PRACTICAL AND THEORETICAL CONSIDERATIONS IN CAPILLARY ELECTROPHORESIS

Chapter 18

MODIFICATION OF CAPILLARIES AND BUFFERS FOR ENHANCED SEPARATIONS IN CAPILLARY ZONE ELECTROPHORESIS AND CAPILLARY ISOELECTRIC FOCUSING OF BIOPOLYMERS

Jeff R. Mazzeo and Ira S. Krull

TABLE OF CONTENTS

0-8493-8690-X/94/$0.00+$.50

I. INTRODUCTION

Electrophoretic separations of large, multiply charged species, such as proteins, in fused silica capillaries suffers from electrostatic adsorption of positively charged regions of proteins to the negatively charged silanol groups on the capillary wall. This adsorption leads to poor efficiencies of separation, nonreproducible migration times and peak areas and, in the worst case, inability to detect completely adsorbed material. Furthermore, adsorption of material at the head of the capillary can cause changes in the local charge density, leading to differential flow profiles and band broadening for nonadsorbed solutes.[1] Thus, the need for surface deactivation of the silica wall to obtain optimal separation of biopolymers has led to the development of capillary coatings and buffer modifiers in capillary zone electrophoresis (CZE) and capillary isoelectric focusing (CIEF). In this chapter, we review advances in these two areas, concentrating on protein analysis. We also describe specific methodologies that can be successfully employed for capillary modification leading to improved separation of proteins. Relevant applications will be given and future prospects discussed. Some of this information has appeared previously.[2]

As is often the case in the scientific literature, semantics can cause a great deal of confusion. The distinction between "permanently" modified capillaries and "dynamic" coating agents is not a clear one. In this chapter, *coated capillary* refers to a silica capillary whose inner surface has been modified by covalent attachment of the coating material. In cases where the coating agent is physically adsorbed to the capillary wall, and applied either as a wash step or as a component in the running buffer, we use the term *buffer additive*. To describe all approaches, we use the term *wall modification*. These distinctions are our own classification; we do not advocate that this style be adopted.

A. COATED CAPILLARIES IN CAPILLARY ZONE ELECTROPHORESIS

One method employed to overcome the adsorption problem of proteins in CZE is to perform the separations at extremes of pH.[3-5] At low pH (<3), ionization of silanols is minimal, leading to decreased ionic attraction.[4] In contrast, at pH values above the p*I* of the most basic protein being analyzed, all proteins will be negatively charged and repulsive forces will predominate.[3,5] These methods have shown great improvements in the separation of proteins by CZE. However, their major limitation is the poor stability of proteins at extremes of pH, which can lead to degradation into multiple forms. This can present problems in interpreting separation patterns and should be avoided. Obviously, one would prefer to operate in the more neutral pH range of 5 to 9.

An alternative approach is to cover the silica capillary wall with a noncharged coating, which will minimize nonspecific adsorption. The ionized silanols will then be inaccessible to positively charged regions of the protein, leading to better peak shapes, efficiencies, resolution, and reproducibility. Initial efforts at coating capillaries involved the use of traditional silane-coupling chemistry, using trifunctional silane reagents, which attached to the silica silanols via siloxane bonds, Si-O-Si. The immobilized silane reagent could then be reacted further to give the final coating material. Although these coatings have been successful in certain cases, they have one major limitation: poor hydrolytic stability at basic pH values, owing to the siloxane bonds used for immobilization. Many of the bonded capillaries developed recently use different attachment chemistries, which should lead to more hydrolytically stable coatings.

In evaluating coated capillaries for CZE of proteins, several factors should be considered:

1. Separation efficiency, which should be evaluated in terms of plates per meter, rather than just plates. In theory, plates should approach 1 to 2 million/m.

2. Recovery, which should approach 100% in the ideal case.
3. Reproducibility of migration time from run to run, as well as longer time periods, such as day to day.
4. Maintainence of electroosmotic flow (EOF).

Plates per meter is calculated by first measuring the theoretical plates (using chromatographic equations) for a standard, usually a basic protein such as lysozyme or cytochrome *c*, since these proteins adsorb strongly to an unmodified capillary. This number is divided by the effective column length in meters, where effective column length is the distance from injection to detection. Obviously, plates per meter can vary significantly for a given coating with variables such as pH, ionic strength, field, etc. Thus, comparing different coatings on this basis is not really accurate, as each coating was run with a different field, pH, and buffer. In this light, it would be useful if a standard CZE buffer and field could be designated that would be used to evaluate and compare different coatings.

Reproducibility is determined by measuring the migration time for several standard proteins over a given time period, with a minimum of five injections required. Percent relative standard deviation (%RSD or %CV) is the standard deviation of these measurements divided by the average times one hundred.

Where possible, these criteria have been used to evaluate the effectiveness of the coatings reviewed below. Point 1 is particularly important, since it should be an indicator of resolving power, while points 2 and 3 are good indicators of the quantitative aspects and stability of the coating. The importance of point 4 may be debated. There are some cases where elimination of EOF is desirable, such as with sieving of sodium dodecyl sulfate (SDS)–proteins and nucleic acids, as well as certain methods of performing CIEF. Furthermore, certain samples of limited complexity (in terms of the p*I* spread of the proteins) may also make elimination of EOF necessary. However, if no EOF is maintained by the coating, the alternative approach is to use low- or high-pH buffers to separate complex mixtures of proteins. Using these conditions, as indicated earlier, protein stability is an issue, and peak capacity will be less due to minimization of charge differences.

McCormick has reported the use of a polyvinylpyrrolidone (PVP)-coated capillary in conjunction with low-pH buffers for the separation of a model protein mixture.[4] 3-Methacryloxypropyltrimethoxysilane was used as the silane-coupling agent to the capillary wall, followed by addition of 1-vinyl-2-pyrrolidone, ammonium persulfate, and tetramethylethylenediamine (TEMED), leading to a linear coating of PVP. The 1-vinyl-2-pyrrolidone adds to the methacrylate group and polymerizes, which is initiated by persulfate radicals. This capillary was then used at low pH (pH 2) to separate 15 proteins scanning the p*I* range 4.5–11 (Figure 1). The peaks in Figure 1 exhibited efficiencies on the order of 700,000 plates/m. Interestingly, two of the proteins showed the presence of multiple peaks, which the author attributed to denaturation to substituent subunits. This is a perfect example of the problems encountered when working at extremes of pH for protein separations.

Bruin et al. have described three coated capillaries for CZE of proteins.[6,7] A polyethylene glycol (PEG)-coated capillary was prepared using 3-glycidoxypropyltrimethoxysilane as coupling reagent, followed by reaction with PEG (M_r 600), which added to the epoxy ring to give the final coating.[6] This capillary proved to be useful only at pH < 5, with good long-term stability. The authors concluded that some adsorption was still going on, since theoretical plate counts for several proteins were significantly lower than predicted. An epoxy diol coating was prepared using the above-mentioned silane reagent, followed by the addition of HCl, opening the epoxy ring and generating the diol.[7] This column showed behavior similar to the PEG-coated capillary. The third coating described used 3-aminopropyltriethoxysilane as silane reagent, with the immobilized primary amine reacted with maltose and sodium

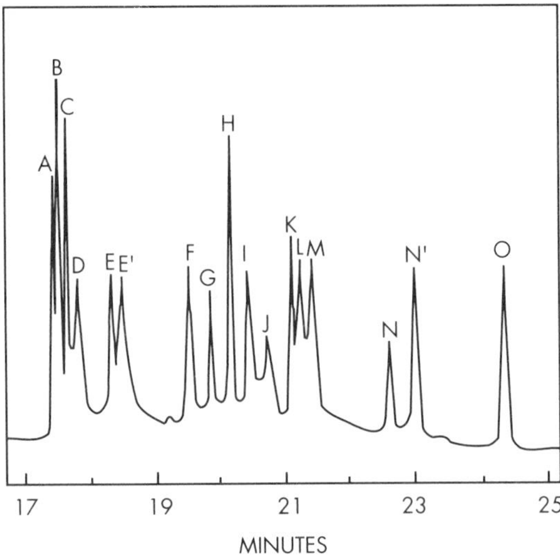

FIGURE 1. Capillary zone electrophoretic separation of protein mixture in PVP-modified capillary at pH 2.0. Voltage gradient from 5 to 25 kV in 150 s. Sample: (A) β-lactoglobulin B, (B) β-lactoglobulin A, (C) lysozyme, (D) albumin (human serum), (E) albumin (bovine serum), (F) cytochrome *c*, (G) trypsinogen, (H) myoglobin (whale), (I) transferrin, (J) conalbumin, (K) myoglobin (horse), (L) carbonic anhydrase B (bovine), (M) carbonic anhydrase A (bovine), (N) hemoglobin, (O) parvalbumin. (From McCormick, R. M., *Anal. Chem.*, 60, 2322, 1988. With permission.)

cyanoborohydride to give a maltose-coated capillary.[7] This coating proved stable up to pH 7, but, as with the previous two coatings from the same authors, the efficiency was much lower than predicted. Furthermore, the maltose coating proved susceptible to microbial attack.

An aryl pentafluoro-modified capillary was reported by Swedberg.[8] This coating was prepared using 3-aminopropyltrimethoxysilane as coupling agent, followed by reaction with pentafluorobenzoyl chloride, giving the final coating. This capillary proved quite effective at minimizing adsorption, as demonstrated by the separation of seven proteins in an uncoated capillary compared to the aryl pentafluoro-coated capillary (Figure 2). Efficiencies were on the order of 500,000 to 600,000 plates/m. The coating maintained a significant EOF, suggesting that ionized silanols were still present, which were shielded from interacting with proteins by the bulky aryl pentafluoro group. The reproducibility of migration time (within day) was 4% RSD in the worst case, less than 1% in the best case, and 7% on a day-to-day basis.

Towns and Regnier developed and rigorously characterized a polyethyleneimine-coated capillary for CZE of proteins.[9] Polyethyleneimine was first adsorbed to the silica wall, then cross-linked with ethyleneglycol diglycidyl ether and triethylamine. This coating maintains a constant positive charge over the pH range 2 to 12, thus pH can be changed to optimize selectivity without altering EOF. Because the wall is positively charged, the direction of EOF has been reversed, now moving from cathode to anode. This change in the direction of EOF requires that the detector be placed at the anodic end of the capillary, or that the polarity of the power supply be reversed. This coating is also different from those already described with respect to its thickness (~30 Å), as opposed to a monolayer type, described above. The authors concluded that this thick coating was necessary to mask all silanol groups from interacting with proteins. Positively charged proteins showed minimal adsorption, as demonstrated by the pH 7 separation of six basic proteins (p*I* > 7) (Figure 3). No data on efficiencies or reproducibility were reported.

The same authors also developed a C_{18}-coated capillary that was modified by the adsorption of nonionic surfactants.[10] The capillary was etched with 1 *M* NaOH, followed by reaction with octadecyltrichlorosilane, generating the C_{18} coating. A 0.5% by weight solution of the surfac-

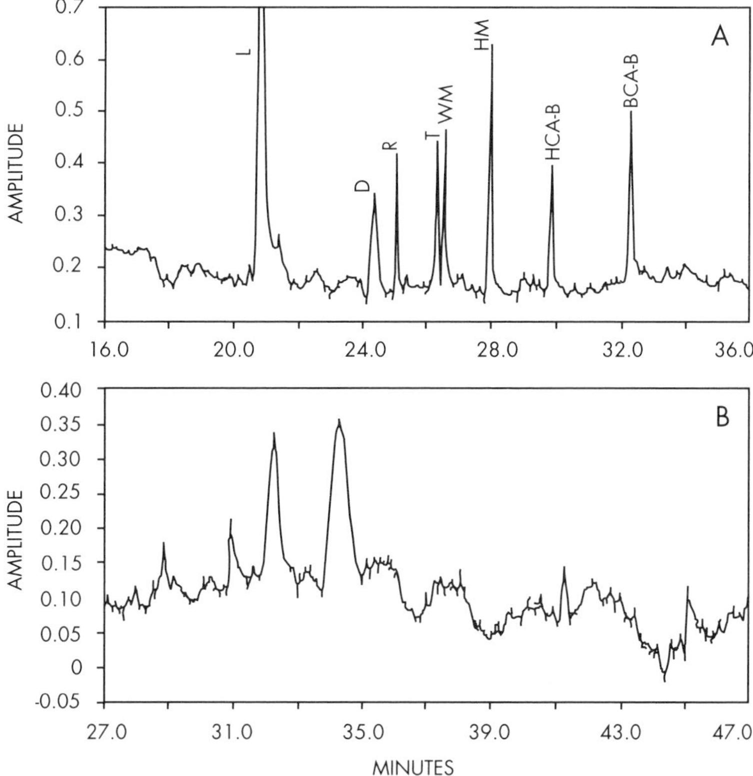

FIGURE 2. Elution profile of seven proteins and dimethyl sulfoxide (DMSO) on (A) an aryl pentafluoro-modified capillary and on (B) an untreated capillary. Buffer, pH 7; field, 250 V/cm. Sample: (L) lysozyme, (D) DMSO, (R) ribonuclease, (T) trypsinogen, (WM) whale myoglobin, (HM) horse myoglobin, (HCA-B) human carbonic anhydrase B, (BCA-B) bovine carbonic anhydrase B. (From Swedberg, S. A., *Anal. Biochem.*, 185, 51, 1990. With permission.)

tant, either a member of the Tween series (Tween-20, -40, or -80) or the Brij series (Brij-35 or -78), was pulled through the capillary for 2 h to obtain the surfactant-modified coatings. These coatings were found to give greater than 90% recovery for several standard proteins covering a wide range of p*I* values, and efficiencies on the order of 200,000 to 400,000 plates/m. Stability testing of the coatings over a 3-month period led to migration time reproducibility of 2.4% RSD. Separations of a model protein mix on Tween-20 and Brij-35 modified capillaries are shown in Figure 4. In a Brij-35-modified capillary, EOF was relatively constant with pH over a wide range.

Cobb, Dolnik, and Novotny reported a polyacrylamide-coated capillary similar to that described by Hjertén,[11] with the exception that the coating was attached to the silica wall through Si-C-Si bonding, as opposed to Si-O-Si.[12] This linkage results in greater hydrolytic stability compared to the siloxane. The silica wall was first chlorinated with thionyl chloride, followed by a Grignard reaction with vinyl magnesium bromide. The immobilized vinyl group was then reacted with acrylamide monomer and polymerizing agents ammonium persulfate and TEMED, resulting in a linear coating of polyacrylamide. The coating is of the monolayer type, which may leave some ionized silanols accessible to proteins. This chemistry was first reported for the modification of silica by Deuel et al.[13] and later by Pesek and Swedberg.[14]

This coating proved stable over the pH range 2 to 10.5, with elimination of EOF. The lack of EOF necessitated the use of low- or high-pH buffers for performing separations, so that all proteins traveled in the same direction. Nevertheless, impressive separations of model protein mixtures were obtained, as evidenced by the low-pH (2.7) separation of five model proteins in polyacrylamide-coated and uncoated capillaries (Figure 5). The polyacrylamide capillary

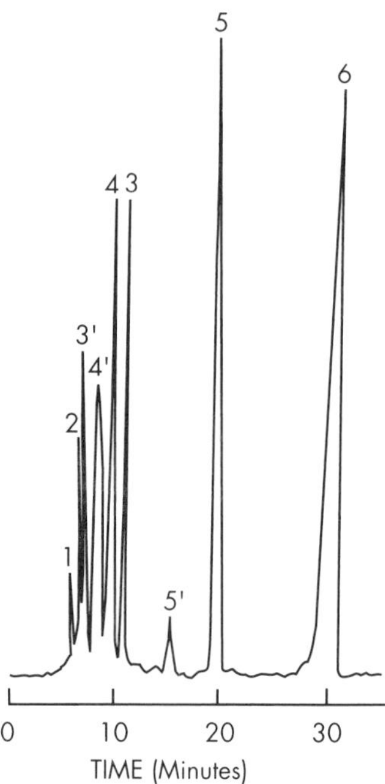

FIGURE 3. Capillary electrophoretic separation of model proteins on a polyethyleneimine-modified capillary. Buffer, pH 7.0; field, 250 V/cm. Sample: (1) mesityl oxide, (2) horse heart myoglobin, (3) ribonuclease, (4) chymotrypsinogen, (5) cytochrome *c*, (6) lysozyme. (From Towns, J. K. and Regnier, F. E., *J. Chromatogr.*, 516, 69, 1991. With permission.)

generated 200,000 to 300,000 plates/m, which, although impressive, was still lower than theoretically predicted values of 1 to 2 million plates/m. Thus, one must assume that some adsorption/interaction with the capillary wall was occurring. Migration time reproducibilities on the order of 0.3% RSD (within day) were obtained.

Nashabeh and El Rassi have described two different coatings for CZE of proteins,[15] referred to as "fuzzy" and "interlocked" coatings. For the fuzzy coating, a bottom layer of cross-linked glyceropropylpolysiloxane is covered by a hydrophilic polyether layer that may have one end in free solution or both ends cross-linked with the bottom layer. In the interlocked case, the polyether chain is bonded at both ends to the sublayer. In both cases, the polyether layer is polyethylene glycol of variable molecular weight. These coatings show more constant EOF with pH than an unmodified capillary. The capillaries exhibited average plate counts on the order of 150,000 plates/m, with reproducibility of migration time between columns of 2% RSD.

Table 1 summarizes the coatings described above.

Several other coated capillaries have been described for the analysis of proteins by CZE.[16,17] These include polymethylglutamate,[16] polyethylene glycol, and polyethyleneimine.[17] In general, they have shown improved performance over uncoated capillaries in terms of efficiencies, but not drastic improvement. Of the coatings described above, the nonionic surfactant-coated capillaries of Towns and Regnier give the best overall performance, in terms of efficiency, maintainence of EOF, reproducibility, and recovery. However, these coatings do not approach the 1 to 2 million plates/m predicted by theory, indicating that substantial interaction with the wall is still occurring.

It should be expected that many new coatings will be described in the near future, as much effort is being put forth in this area both by academic researchers as well as instrument companies. The ideal coating for CZE of proteins will consist of a hydrophilic, thick layer that shows minimal interaction with proteins and will also sterically hinder proteins from adsorbing

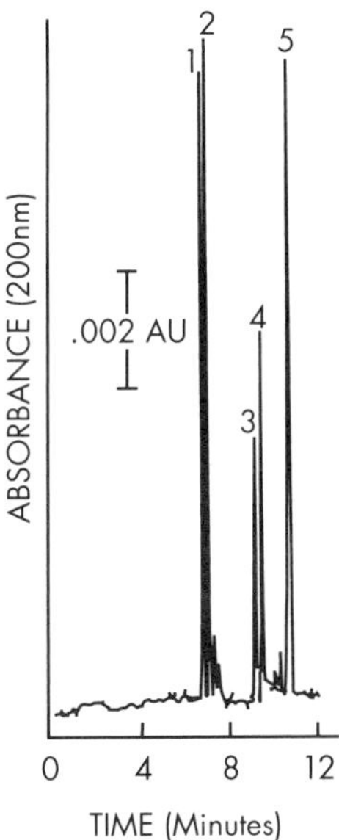

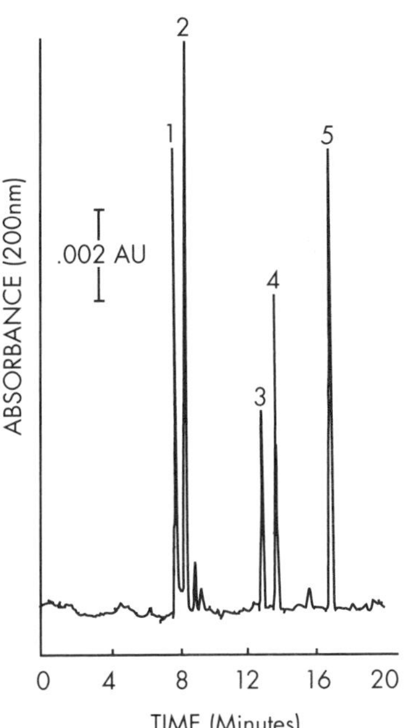

FIGURE 4. Separation of model protein mix in a Tween-20 (left) and a Brij-35 (right) surfactant-coated capillary. Buffer, 0.01 *M* phosphate, pH 7.0; field, 300 V/cm. Sample: (1) lysozyme, (2) cytochrome *c*, (3) ribonuclease A, (4) chymotrypsinogen A, and (5) myoglobin. (From Towns, J. K. and Regnier, F. E., *Anal. Chem.*, 63, 1126, 1991. With permission.)

to any remaining silanol groups on the capillary wall. It will also maintain some EOF such that basic and acidic proteins will be separable in the same run with one detector. It will have a wide range of pH stability, i.e., pH 2 to 12, good reproducibility from run to run and column to column, and a lifetime of at least several weeks. Until such a coating exists, researchers must find out which of the available coatings is best suited to their application.

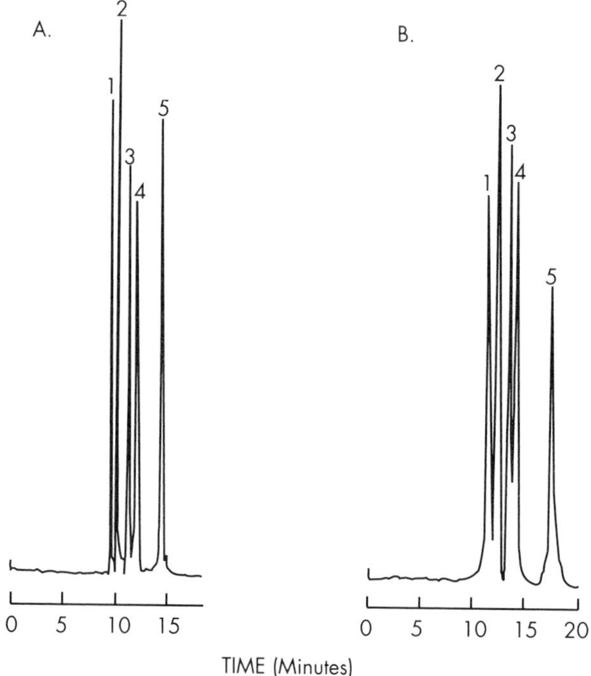

FIGURE 5. Capillary electrophoretic separation of model proteins in (A) a polyacrylamide-coated capillary and in (B) an uncoated capillary.[4] Buffer, pH 2.7. Field: (A) 367 V/cm and (B) 200 V/cm. Sample: (1) cytochrome *c*, (2) lysozyme, (3) trypsin, (4) trypsinogen, and (5) trypsin inhibitor. (From Cobb, K. A., Dolnik, V., and Novotny, M., *Anal. Chem.*, 62, 2478, 1990. With permission.)

TABLE 1
Various Coatings and Their Effects in Capillary Zone Electrophoresis of Proteins

Coating	pH stability	Affect on EOF	Plates/m	%RSD	Ref.
Polyvinylpyrrolidone			700,000		4
Polyethylene glycol	<5		300,000		6
Epoxy diol	<5		100,000	5	7
Maltose	<7		50,000	5	7
Aryl pentafluoro		Minimal	600,000	1	8
Polyethyleneimine	2–12	Reverses			9
Surfactant	4–11	Constant with pH	400,000	2.40	10
Polyacrylamide	2–10.5	Eliminates	300,000	0.30	11
Fuzzy/interlock	4–7.5	Constant with pH	150,000	2	15

B. BUFFER ADDITIVES FOR CAPILLARY ZONE ELECTROPHORESIS

Rather than use coated capillaries for CZE of proteins, it is possible to use buffer additives that will act as dynamic coatings, shielding the silica wall from interacting with proteins. Because the interaction between proteins and the capillary wall is primarily ionic, the ionic strength of the buffer can be increased, leading to decreased adsorption. This was demonstrated by Bushey and Jorgenson, who used high concentrations of zwitterionic salts for protein separations in CZE.[18] The advantage of using zwitterionic salts to decrease adsorption is that they can be added to the buffer in high concentration without generating excess current and Joule heat. Although the salts worked well for some proteins, very basic proteins, such as lysozyme, were not effectively separated. This approach has been commercialized by Millipore (Milford, MA), under the trade name Z1 Methyl Additive.

One way to overcome problems associated with basic proteins, such as lysozyme, is to change the immobile charge on the wall from negative to positive, which will cause repulsion of positive species. This was shown earlier with the coating of Towns and Regnier. Another approach was described by Wiktorowicz and Colburn,[19] who added, presumably, a cationic surfactant or polyethyleneimine to the capillary that coated the silica through adsorption, leading to a positively charged wall. This additive was rinsed through the capillary prior to analysis, and did not exist in the buffer or sample solutions. Separations of several basic proteins were demonstrated. The product is commercially available from Applied Biosystems (Foster City, CA) as MicroCoat.

In a similar approach, Emmer, Jansson, and Roeraade have used a cationic surfactant, FC-135, from 3M (Minneapolis, MN), in the CZE buffer to dynamically modify the capillary wall and make it positively charged,[20] thereby minimizing interaction of basic proteins with the wall. Efficiencies were on the order of 200,000 to 400,000 plates/m, and reproducibility of migration time was 0.5% RSD (within day), 1.8% day to day. Bullock and Yuan demonstrated that the addition of a basic compound, 1,3-diaminopropane, to the CZE running buffer greatly reduces the interaction of basic proteins with the wall at neutral pH.[21] Efficiencies of several hundred thousand plates per meter were obtained. Stover et al. previously described a similar method for improving the efficiency of histidine-containing compounds, adding putrescine to the buffer.[22]

A serious limitation of the techniques that rely on reversing the charge on the capillary wall is that adsorption of acidic analytes then becomes a problem. If the researcher is interested only in separating basic proteins, this is not a problem. However, separation of complex mixtures covering a wide p*I* range will lead to problems, not only for the acidic proteins, but also for basic ones, since the adsorption of any protein to the wall will create local changes in the capillary wall charge density and differential flow profiles.[1]

Gilges and co-workers have reported improved protein separations in CZE by the addition of 0.05% polyvinyl alcohol (M_r 15,000) to the running buffer.[23] Separations at pH 3 (Figure 6), led to average efficiencies for the five basic protein test solutes of about 700,000 plates/m. Reproducibility of migration time for 24 injections was 1.2% RSD. However, the authors reported difficulty in separating acidic proteins under the conditions used.

C. WALL MODIFICATIONS IN CAPILLARY ISOELECTRIC FOCUSING

The technique of capillary isoelectric focusing has been reviewed in Chapter 4 of this text. Wall modification in CIEF is necessary to both minimize adsorption, as well as eliminate/minimize EOF. If EOF is eliminated in CIEF, some means of mobilizing the focused zones past a stationary detection point must be used. Both salt mobilization and pressure mobilization have been described for cases where EOF was eliminated.[24-26] If some EOF is retained by the wall modification process, it can be used to mobilize focused zones.

1. Wall Modifications Eliminating Electroosmotic Flow in Capillary Isoelectric Focusing

There have been three coated capillaries described for CIEF that eliminate EOF. The first was the classic linear polyacrylamide coating of Hjertén.[11] The trifunctional silane reagent 3-methacryloxypropyltrimethoxysilane is first attached to the capillary wall, leading to an immobilized double bond, which is then reacted with acrylamide monomer in the presence of persulfate radicals and tetramethylethylenediamine (TEMED) to give a linear coating of polyacrylamide. This coating essentially eliminates EOF, and can be used successfully for both CIEF and CZE of proteins. Its main limitation is its poor hydrolytic stability at basic pH, due to the presence of siloxane (Si-O-Si), ester, and amide functionalities. Thus, with extended use, the coating is removed, leading to exposed silanol groups and EOF, as well as adsorption.

Kasper and Malera have reported the used of DB-17 (polyethylene glycol)-coated capillar-

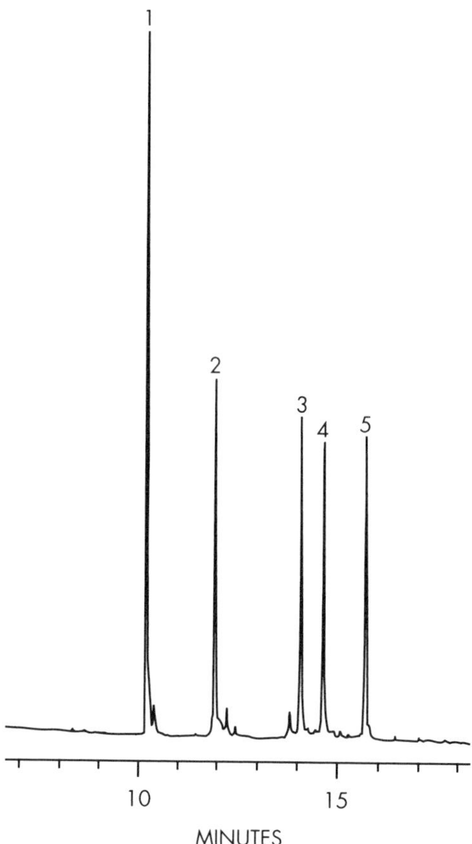

FIGURE 6. Capillary zone electrophoresis of basic proteins with polyvinyl alcohol added to the buffer. Buffer, 20 m*M* phosphate, 30 m*M* NaCl, pH 3.0, 0.05% (w/w) polyvinyl alcohol (M_r 15,000). Field, 357 V/cm. Sample: (1) cytochrome *c*, (2) lysozyme, (3) trypsin, (4) trypsinogen, and (5) chymotrypsinogen A. (From Gilges, M., Husmann, H., Kleemiß, M. H., Motsch, S. R., and Schomburg, G., *J. High Resolut. Chromatogr.*, 15, 452, 1992. With permission.)

ies from J&W Scientific, in conjunction with 0.05% methyl cellulose to abolish EOF for CIEF.[25] No long-term stability of this coating has been reported. Chen and Wiktorowicz have used DB-1 (dimethyl polysiloxane)-coated capillaries from J&W Scientific (Folsom, CA) with 0.4% methyl cellulose to eliminate EOF.[26] The stability of this coating was demonstrated by the linearity of p*I* vs. migration time plots over a 2-month period.

2. Wall Modifications Maintaining Electroosmotic Flow in Capillary Isoelectric Focusing

Mazzeo and Krull reported the use of methyl cellulose (viscosity of a 2% solution, 4000 cP; M_r 86,000) in uncoated capillaries for CIEF with EOF mobilization.[27] Subsequently, Thormann et al. described the use of hydroxypropylmethylcellulose in uncoated capillaries for performing CIEF.[28] In both cases, the cellulose derivatives adsorb onto the silica wall, increasing the viscosity and thereby decreasing EOF.

When performing CIEF with EOF mobilization, another important factor must be considered. Specifically, in an unmodified capillary, the extent of EOF strongly depends on the pH of the medium.[29] In CZE, this fact is tolerable, since a continuous buffer of constant pH is used to perform the separation. However, changing the pH to change the separation selectivity will

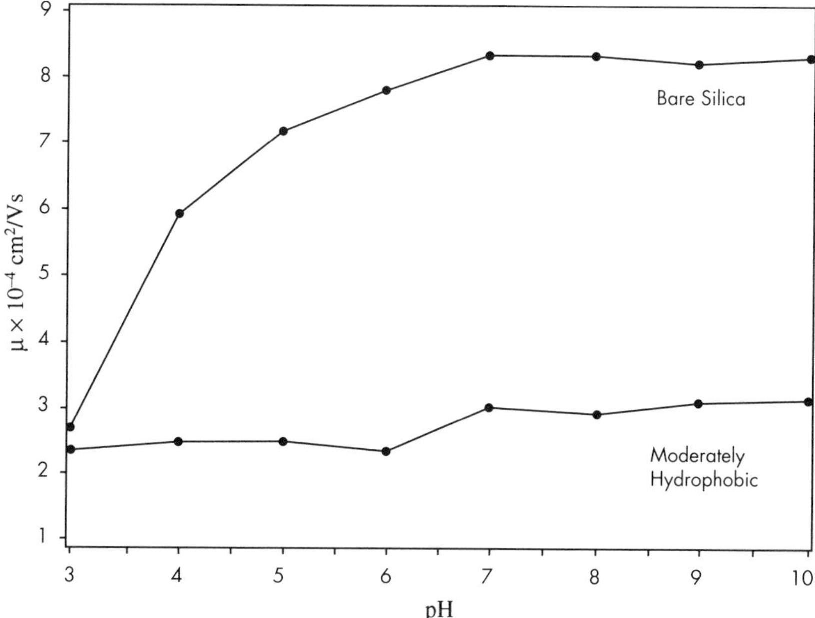

FIGURE 7. Electroosmotic mobility vs. pH for an uncoated and a C_8-coated capillary. (From Dougherty, A. M., Woolley, C. L., Williams, D. L., Swaile, D. F., Cole, R. O., and Sepaniak, M. J., *J. Liq. Chromatogr.*, 14, 907, 1991. With permission.)

also change the bulk flow, which Towns and Regnier have deemed a problem.[9] It would be preferable to be able to change the pH, and the selectivity, without changing the bulk flow.

In CIEF with EOF mobilization, the pH of the medium is not constant, but rather decreases from the cathode to the anode. Thus, in an uncoated capillary, this will lead to a nonconstant ζ potential at the wall. The low-pH region of the gradient will have a lower local ζ potential than the high-pH region, leading to a pressure difference between the two regions of the capillary and generation of laminar flow.[1] The problem becomes magnified as the run occurs, since the high-pH region of the gradient exits at the cathode side due to the EOF, while phosphoric acid is carried into the capillary at the anode. The net effect, as we have found, is to cause severe peak broadening for acidic proteins when CIEF with EOF mobilization is performed in uncoated capillaries.[30]

The obvious solution is to employ capillaries that have constant rates of EOF with changing pH. With this criterion in mind, we have found that commercially available C_8-coated capillaries from Supelco (Bellefonte, PA) (CElect H-150) are superior to uncoated capillaries for CIEF with EOF mobilization. Figure 7 shows the electroosmotic mobility dependence on pH in an uncoated capillary and the aforementioned C_8-coated capillary.[31] Below pH 7, the EOF drops substantially in the uncoated capillary, while staying reasonably constant in the C_8-coated capillary. Comparison of the two capillaries is shown in Figures 8 and 9, which show the CIEF separation of a standard protein mixture in an uncoated capillary (Figure 8) and the C_8-coated capillary (Figure 9). Note that for the basic standards, the performance of the two capillaries is similar, but for the acidic standards, the coated capillary shows superior performance. These results indicate that in order to successfully separate all proteins, basic and acidic, in one CIEF run, an uncoated capillary cannot be used.

However, there are certain additives that can be used to flatten the pH vs. EOF curve in an uncoated capillary. Organic solvents, such as methanol, have been shown to have this effect.[32] The methanol acts by hydrogen bonding to the capillary wall, suppressing the ionization of

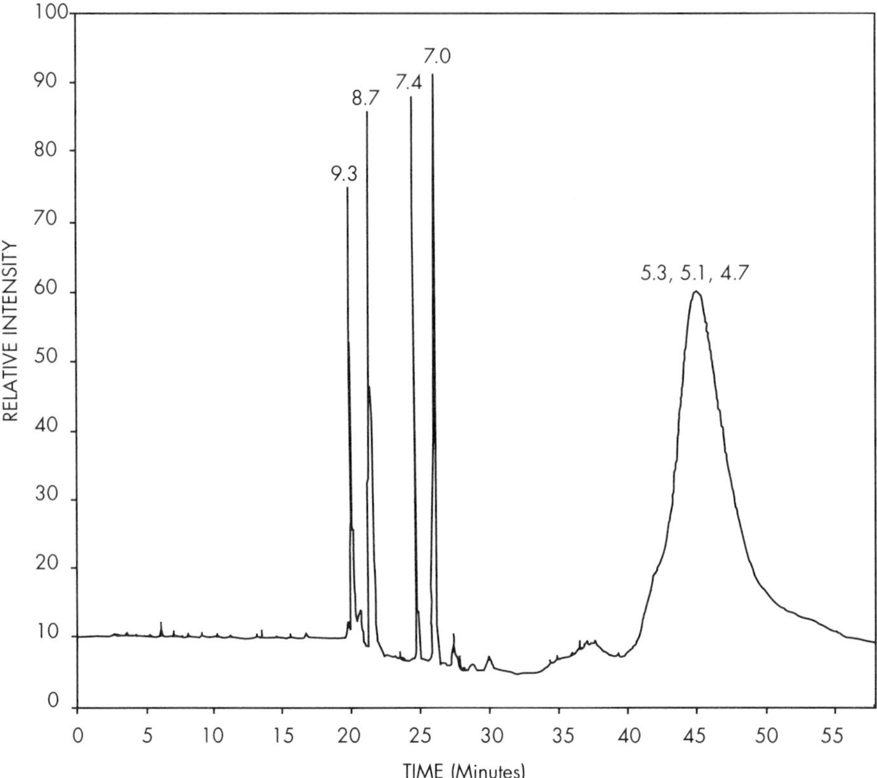

FIGURE 8. Capillary isoelectric focusing of acidic, neutral, and basic proteins in an uncoated capillary. Conditions: 50-μm I.D. capillary; distance from anode to detection, 40 cm; total distance, 60 cm; field, 400 V/cm; ultraviolet detection at 280 nm; 0.05 AUFS. Anolyte, 20 m*M* phosphoric acid; catholyte, 20 m*M* sodium hydroxide. Proteins dissolved at concentrations of 0.5 mg/ml in 5% Pharmalyte 3-10, 0.1% methyl cellulose, and 1.4% TEMED. Peak identification by isoelectric point. (From Mazzeo, J. R., Martineau, J., and Krull, I. S., *Methods*, 1993, in press. With permission.)

silanols and increasing the viscosity along the wall. We have not explored the use of methanol or other additives in CIEF with EOF mobilization, but it appears that they might solve the problem of separating acidic proteins in uncoated capillaries.

II. METHODOLOGY

A. CAPILLARY ZONE ELECTROPHORESIS APPROACHES

The exact method to use will depend greatly on the sample of interest, as well as the amount of time the researcher can invest. Many of the coatings that have been described above require much time and effort to prepare, and have been synthesized only by the original reporting group. There are some coatings that are commercially available, and others that can be prepared with some effort. From this standpoint, additives represent an attractive approach, since all of the additives that have been used for CZE of proteins are commercially available. The disadvantages of additives include possible interaction with the protein of interest, noise at the detector (especially in the case of mass spectrometry), and nongenerality.

Several additives are commercially available, marketed under the trade names Z1 Methyl Additive (Millipore) and MicroCoat (Applied Biosystems). The Z1 Methyl Additive is a zwitterionic species that may be added to the buffer to minimize wall interactions. MicroCoat is used to precondition the capillary and is not present in the buffer. It is applied as a rinsing

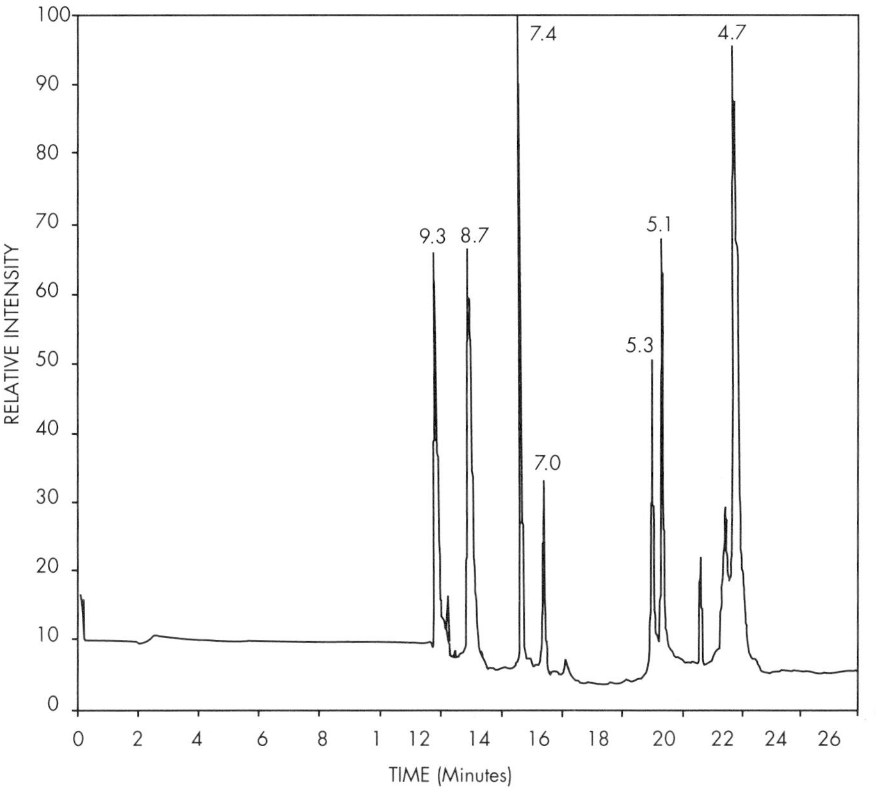

FIGURE 9. Capillary isoelectric focusing of acidic, neutral, and basic proteins in a C_8-coated capillary. All conditions as in Figure 7, except separation distance is 20 cm. Peak identification by isoelectric point. (From Mazzeo, J. R., Martineau, J., and Krull, I. S., *Methods*, 1993, in press. With permission.)

step prior to filling the capillary with buffer. The advantage here is that the additive cannot interact with the analyte, and will not cause excess noise at the detector. This additive reverses the charge on the wall from negative to positive. Thus, the direction of EOF will also be reversed. Furthermore, although the coating works well at pH 7 for basic proteins, acidic proteins would be expected to cause problems since they are oppositely charged from the wall. Both of these additives are supplied with detailed instructions for their optimum use.

When using additives in CZE, the following questions should be asked:

1. What is the p*I* range of the protein(s) of interest?
2. What is the pH stability of the protein?
3. Is the protein hydrophobic?

In cases where a wide range of p*I* values exists in the sample of interest, good peak shape of all species will best be achieved by working at low or high pH. In this case, pH stability must be considered. If at the extreme of pH some proteins still show broad peak shapes, additives should be tried that will minimize any interactions occurring. Try methyl cellulose or hydroxypropyl methyl cellulose (viscosity of a 2% solution, 4000 cP) at concentrations up to 0.2% in the buffer. At higher concentrations, size-based separation may take place. Alternatively, Z1 Methyl Additive at concentrations up to 1 *M* may be tried.

In cases where a single protein is being analyzed for degradation products, i.e., deamidation, phosphorylation, amino acid substitutions, etc., choice of conditions is more straightforward.

If the parent protein has an acidic p*I*, a buffer with near-neutral pH should suffice, with cellulose derivatives or Z1 methyl additive to minimize adsorption. For a protein with a basic p*I*, using MicroCoat to reverse the charge of the capillary wall and a buffer pH near neutral will generally give some separation. Alternatively, adding FC-135 to the buffer at concentrations of 0.05 to 0.1%, or 1,3-diaminopropane at 30 to 60 m*M*, may improve resolution and peak shape.

Permanently coated capillaries that are commercially available are few and far between. Supelco markets four different coated capillaries under the name CElect. They consist of a hydrophilic coating and moderate, medium, and highly hydrophobic coatings. They all maintain a significant EOF at typical buffer pH values, and also appear to minimize protein–wall interactions. The medium and highly hydrophobic coatings exhibit EOF that is nearly constant with pH.[31] The hydrophobic nature of some of the coatings may present problems with hydrophobic species, such as membrane proteins. Addition of a nonionic surfactant may alleviate this problem is some cases, but not all.

It may be possible to use the highly hydrophobic-coated capillary from Supelco to obtain a surfactant coating like that described by Towns and Regnier. A 0.5% (w/v) solution of the surfactant (i.e., Brij-35) in water should be pulled through the capillary for several hours and then removed by rinsing with water. Addition of a small amount of the surfactant to the running buffer (0.01%) may lead to improved separation efficiency.[10]

The linear polyacrylamide-coated capillary of Hjertén, which can be used for both CZE and CIEF of proteins, is commercially available from Bio-Rad (Richmond, CA). However, it is supplied in a cartridge that can be used only in the Bio-Rad CE instrument. Alternatively, the coating can be synthesized in the laboratory. The following procedure will outline that which has been recently reported.[33]

1. Preparation of Linear Polyacrylamide-Coated Capillaries

Fill the capillary with 0.1 *M* sodium hydroxide and remove after 3 h by rinsing with water. Fill the capillary with 0.1 *M* HCl and after 1 min remove by rinsing with water. Mix 30 μl of γ-methacryloxypropyltrimethoxysilane (available from Sigma [St. Louis, MO], LKB, and others) and 1 ml of 60% acetone (v/v) in water. Fill the capillary with this solution and allow to react overnight (12 h). Prepare a 3% (w/v) deaerated solution of acrylamide in water, which also contains ammonium persulfate (2 mg/ml) and TEMED (0.8 μl/ml). Remove the silane reagent from the capillary by rinsing with the acrylamide solution. Seal the capillary ends with rubber septa or modeling clay to avoid evaporation, and allow to react overnight. Rinse with water and the capillary is ready for use. It is recommended that detection windows be formed prior to the coating procedure, especially if a flame is used to remove the polyimide cladding.

To determine if the coating procedure was successful, measure the EOF in an untreated and coated capillary. This can be done by using a pH 7.0 phosphate buffer and injecting a solution of 0.1% (v/v) mesityl oxide, which will serve as a neutral marker. A peak should be seen for the untreated capillary within 15 min or so, depending on capillary length, applied field, and buffer concentration. For the linear polyacrylamide coated capillary, no peak should be seen even after 1 h, if the coating procedure was successful.

In CZE, buffer pH should be less than 9 for good long-term stability of the coating. Furthermore, base washing (i.e., 0.1 *N* sodium hydroxide) should be avoided. A common CZE buffer used with this coating is 20 to 100 m*M* phosphate, pH 2.5. Remember that this coating reduces EOF to unmeasurable levels, so that the pH of the buffer employed will depend on the sample to be analyzed. A complex protein sample with proteins covering a wide range of p*I* values will require the use of low pH, i.e., <4, in order to make all proteins positively charged, causing migration toward one electrode (cathode). For a single protein that may exhibit microheterogeneity, choose a pH close to the p*I* of the protein. The direction of migration will depend on the charge of the protein at the buffer pH.

B. CAPILLARY ISOELECTRIC FOCUSING APPROACHES

The procedure described above for the linear polyacrylamide coating can also be used to make capillaries for CIEF. An exact description of CIEF methodology will not be given here, as this is covered in Chapter 4. DB-1-coated capillaries (dimethylpolysiloxane) are commercially available from J&W Scientific. Since both of these coatings essentially eliminate EOF, some form of mobilization is required after focusing, either salt or hydrostatic.

For performing CIEF with EOF mobilization, uncoated capillaries can be used. However, to achieve successful separation of acidic proteins, a capillary is required that shows constant EOF with pH, as discussed above. The CElect H-150 medium hydrophobic coated capillaries from Supelco work well for this purpose. Methyl cellulose should still be added to the sample ampholyte mixture, since it appears to make the wall more hydrophilic, thereby minimizing adsorption.[34] For more specifics on performing CIEF, see Chapter 4.

III. APPLICATIONS

A. CAPILLARY ZONE ELECTROPHORESIS

Most of the current literature on CZE of proteins deals with development of conditions/coatings, using standard proteins. This is reflective of the problems with separating proteins successfully in CZE. However, there are a number of "real-world" applications of CZE to protein analysis that will be given here. We review only those applications that use a coating or buffer additive described above.

1. Capillary Zone Electrophoresis Applications with Coated Capillaries

The linear polyacrylamide coated capillary of Hjertén, commercialized by Bio-Rad, has been used in several CZE applications. Wu et al. performed CZE on human growth hormone at several pH values, with separation of the deamidated form from the parent realized at pH 6.5.[35] Gurley et al. separated histones using the linear polyacrylamide coating in a pH 2.5 phosphate buffer.[36] Ferranti et al. separated hemoglobin chains α, β, and γ in a pH 2.5 phosphate buffer.[37] Other applications include calmodulin,[38] membrane proteins,[39] and the proteins from the fluid lining of rat lungs.[40] Note that most of these applications have used pH 2.5 buffers, because the coating eliminates EOF. Again, we feel it is important to stress that at such a low pH, protein stability may be an issue. That is, are the resulting electropherograms really indicative of what was originally present in the sample?

2. Capillary Zone Electrophoresis Applications with Additives

The use of MicroCoat to obtain a charge-reversed capillary has been exploited for the analysis of recombinant chimeric glycoprotein[41] and polyethylene glycol-modified proteins.[42] On an uncoated capillary, the basic, chimeric glycoprotein showed severe adsorption, requiring the use of the charge-reversed capillary for successful separation. Polyethylene glycol-modified proteins have been studied with a MicroCoat-modified capillary.[42]

The Z1 Methyl Additive (trimethylammonium propanesulfonate) from Millipore has been used in the development of a method to determine the deamidation products of human insulin.[43] Under optimized conditions, migration time reproducibility less than 1% RSD was obtained. Neutral and acidic desamido insulin were separated from each other as well as from human insulin in less than 20 min. A direct comparison was made between the CZE assay and an ion-exchange high-performance liquid chromatography (HPLC) assay, in terms of the peak area percents of the neutral and acidic desamido insulin products. Good correlation between the two was realized. This type of quantitative protein assay is needed in order to increase the acceptance of CE methods for protein analysis.

Cellulose derivatives have been added to the running buffer in order to obtain improved separations of phosphorylated histone H1 variants,[44] and to monitor enzyme-labeled monoclonal antibody conjugates.[45] In these cases, it is believed that the cellulose derivatives hydrogen bond to the silanol groups on the capillary wall, generating a hydrophilic coating and shielding ionized silanols from interacting with proteins. The disadvantage here is the difficulty in working with solutions containing high concentrations of cellulose derivative, due to their high viscosity. This also must be considered when performing hydrodynamic injection.

B. CAPILLARY ISOELECTRIC FOCUSING APPLICATIONS

We refer the reader to Chapter 4 for applications of CIEF to protein analysis.

IV. FUTURE PROSPECTS

Capillary modification, whether with permanently coated capillaries or with dynamic coatings generated by buffer additives, is an area that will show substantial growth over the next 3 to 5 years. It does not appear, at this point, that there will be a single capillary coating that will be useful for all protein applications in CZE and CIEF. Rather, like chromatography, there will probably be many different types of coatings commercially available, with each useful for specific applications. Currently, the use of additives to generate dynamic coatings seems to be the most popular way to minimize protein–wall interactions, as opposed to permanently modified coated capillaries. That this is the situation is probably due to the unavailability of capillary coatings that show both reduced interaction as well as good long-term stability. Indeed, once coated capillaries become commercially available and are well characterized, they will gradually replace, to some extent, the use of additives. Until then, protein analysis by CZE and CIEF will require a certain degree of expertise, limiting its applicability. Expect many exciting developments in this area of capillary electrophoresis over the next few years.

REFERENCES

1. **Towns, J. K and Regnier, F. E.,** Impact of polycation adsorption on efficiency and electroosmotically driven transport in capillary electrophoresis, *Anal. Chem.*, 64, 2473, 1992.
2. **Mazzeo, J. R. and Krull, I. S.,** Coated capillaries and buffer additives for improved separation of proteins in capillary zone electrophoresis and capillary isoelectric focusing, *BioTechniques*, 10, 638, 1991.
3. **Lauer, H. H. and McManigill, D.,** Capillary zone electrophoresis of proteins in untreated fused silica tubing, *Anal. Chem.*, 58, 166, 1986.
4. **McCormick, R. M.,** Capillary zone electrophoretic separation of peptides and proteins using low pH buffers in modified silica capillaries, *Anal. Chem.*, 60, 2322, 1988.
5. **Zhu, M., Rodriguez, R., Hansen, D., and Wehr, T.,** Capillary electrophoresis of proteins under alkaline conditions, *J. Chromatogr.*, 516, 123, 1990.
6. **Bruin, G. J. M., Chang, J. P., Kuhlman, R. H., Zegers, K., Kraak, J. C., and Poppe, H.,** Capillary zone electrophoretic separation of proteins in polyethylene glycol-modified capillaries, *J. Chromatogr.*, 471, 429, 1989.
7. **Bruin, G. J. M., Huisden, R., Kraak, J. C., and Poppe, H.,** Performance of carbohydrate-modified fused-silica capillaries for the separation of proteins by zone electrophoresis, *J. Chromatogr.*, 480, 339, 1989.
8. **Swedberg, S. A.,** Characterization of protein behavior in high performance capillary electrophoresis using a novel capillary system, *Anal. Biochem.*, 185, 51, 1990.
9. **Towns, J. K. and Regnier, F. E.,** Polyethyleneimine-bonded phases in the separation of proteins by capillary electrophoresis, *J. Chromatogr.*, 516, 69, 1990.
10. **Towns, J. K. and Regnier, F. E.,** Capillary electrophoretic separations of proteins using nonionic surfactant coatings, *Anal. Chem.*, 63, 1126, 1991.
11. **Hjertén, S.,** High performance electrophoresis: elimination of electroendosmosis and solute adsorption, *J. Chromatogr.*, 347, 191, 1985.

12. **Cobb, K. A., Dolnik, V., and Novotny, M.,** Electrophoretic separations of proteins in capillaries with hydrolytically stable surface structures, *Anal. Chem.*, 62, 2478, 1990.
13. **Deuel, H., Wartmann, J., Hutschneker, K., Schobinger, U., and Gudel, C.,** *Helv. Chim. Acta*, 119, 1160, 1959.
14. **Pesek, J. J. and Swedberg, S. A.,** Allyl-bonded stationary phases as possible intermediates in the synthesis of novel high-performance liquid chromatographic phases, *J. Chromatogr.*, 361, 83, 1986.
15. **Nashabeh, W. and El Rassi, Z.,** Capillary zone electrophoresis of proteins with hydrophilic fused-silica capillaries, *J. Chromatogr.*, 559, 367, 1991.
16. **Bentrop, D., Kohr, J., and Englehardt, H.,** Poly(methylglutamate)-coated surfaces in HPLC and CE, *Chromatographia*, 32, 171, 1991.
17. **Huang, M., Vorkink, W. P., and Lee, M. L.,** Evaluation of surface-bonded polyethylene glycol and polyethyleneimine in capillary electrophoresis, *J. Microcolumn Sep.*, 4, 135, 1992.
18. **Bushey, M. M. and Jorgenson, J. W.,** Capillary electrophoresis of proteins in buffers containing high concentrations of zwitterionic salts, *J. Chromatogr.*, 480, 301, 1989.
19. **Wiktorowicz, J. E. and Colburn, J. C.,** Separation of cationic proteins via charge reversal in capillary electrophoresis, *Electrophoresis*, 11, 769, 1990.
20. **Emmer, A., Jansson, M., and Roeraade, J.,** A new approach to dynamic deactivation in capillary zone electrophoresis, *J. High Resolut. Chromatogr.*, 14, 738, 1991.
21. **Bullock, J. A. and Yuan, L. C.,** Free solution capillary electrophoresis of basic proteins in uncoated fused silica capillary tubing, *J. Microcolumn Sep.*, 3, 241, 1991.
22. **Stover, F. S., Haymore, B. L., and McBeath, R. J.,** Capillary zone electrophroresis of histidine-containing compounds, *J. Chromatogr.*, 470, 241, 1989.
23. **Gilges, M., Husmann, H., Kleemiß, M. H., Motsch, S. R., and Schomburg, G.,** CZE separations of basic proteins at low pH in fused silica capillaries with surfaces modified by silane derivatization and/or adsorption of polar polymers, *J. High Resolut. Chromatogr.*, 15, 452, 1992.
24. **Hjertén, S., Liao, J. L., and Yao, K.,** Theoretical and experimental study of high-performance electrophoretic mobilization of isoelectrically focused protein zones, *J. Chromatogr.*, 387, 127, 1987.
25. **Kasper, T. J. and Malera., M.,** Capillary electrophoresis: isoelectric focusing applications, paper presented at 27th Eastern Analytical Symposium, New York, 1988.
26. **Chen, S. M. and Wiktorowicz, J. E.,** Isoelectric focusing by free solution capillary electrophoresis, *Anal. Biochem.*, 206, 84, 1992.
27. **Mazzeo, J. R. and Krull, I. S.,** Capillary isoelectric focusing of proteins in uncoated fused silica capillaries using polymeric additives, *Anal. Chem.*, 63, 2272, 1991.
28. **Thormann, W., Caslavska, J., Molteni, S., and Chmelik, J.,** Capillary isoelectric focusing with electroosmotic zone displacement and on-column multichannel detection, *J. Chromatogr.*, 589, 321, 1991.
29. **Lambert, W. J. and Middleton, D. L.,** pH hysteresis effect with silica capillaries in capillary zone electrophoresis, *Anal. Chem.*, 62, 1585, 1990.
30. **Mazzeo, J. R., Martineau, J., and Krull, I. S.,** Performance of isoelectric focusing in uncoated and commercially available coated capillaries, *Methods*, 1993, in press.
31. **Dougherty, A. M., Woolley, C. L., Williams, D. L., Swaile, D. F., Cole, R. O., and Sepaniak, M. J.,** Stable phases for capillary electrophoresis, *J. Liq. Chromatogr.*, 14, 907, 1991.
32. **Schwer, C. and Kenndler, E.,** Electrophoresis in fused silica capillaries: the influence of organic solvents on the electroosmotic velocity and the ζ potential, *Anal. Chem.*, 63, 1801, 1991.
33. **Hjertén, S.,** Isoelectric focusing in capillaries, in *Capillary Electrophoresis: Theory and Practice*, Grossman, P. D. and Colburn, J. C., Eds., Academic Press, San Diego, 1992, p. 199.
34. **Mazzeo, J. R. and Krull, I. S.,** Unpublished results, 1993.
35. **Wu, S. L., Teshima, G., Cacia, J., and Hancock, W. S.,** Use of high-performance capillary electrophoresis to monitor charge heterogeneity in recombinant-DNA derived proteins, *J. Chromatogr.*, 516, 115, 1990.
36. **Gurley, L. R., London, J. E., and Valdez, J. G.,** High-performance capillary electrophoresis of histones, *J. Chromatogr.*, 559, 431, 1991.
37. **Ferranti, P., Malorni, A., Pucci, P., Fanali, S., Nardi, A., and Ossicini, L.,** Capillary zone electrophoresis and mass spectrometry for the characterization of genetic variants of human hemoglobin, *Anal. Biochem.*, 194, 1, 1991.
38. **Chan, K. J. and Chen, W. H.,** High performance capillary electrophoresis of calmodulin, *Electrophoresis*, 11, 15, 1990.
39. **Josic, D., Zeilinger, K., Reutter, W., Bottcher, A., and Schmitz, G.,** High-performance capillary electrophoresis of hydrophobic membrane proteins, *J. Chromatogr.*, 516, 89, 1990.
40. **Gurley, L. R., Buchanan, J. S., London, J. E., Stavert, D. M., and Lehnert, B. E.,** High-performance capillary electrophoresis of proteins from the fluid lining of the lungs of rats exposed to perfluoroisobutylene, *J. Chromatogr.*, 559, 411, 1991.

41. **Tsuji, K. and Little, R. J.,** Charge-reversed, polymer-coated column for the analysis of a recombinant chimeric glycoprotein, *J. Chromatogr.*, 594, 317, 1992.
42. **Cunico, R. L., Gruhn, V., Kresin, L., Nitecki, D. E., and Wiktorowicz, J. E.,** Characterization of polyethylene glycol modified proteins using charge-reversed capillary electrophoresis, *J. Chromatogr.*, 559, 467, 1991.
43. **Mandrup, G.,** Rugged method for the determination of deamidation products of insulin solutions by free zone capillary electrophoresis using an untreated fused-silica capillary, *J. Chromatogr.*, 604, 267, 1992.
44. **Lindner, H., Helliger, W., Dirschlmayer, A., Talasz, H., Wurm, M., Jaquemar, M., and Puschendorf, B.,** Separation of phosphorylated histone H1 variants by high-performance capillary electrophoresis, *J. Chromatogr.*, 608, 211, 1992.
45. **Harrington, S. J., Varro, R., and Li, T. M.,** High-performance capillary electrophoresis as a fast in-process control method for enzyme-labelled monoclonal antibody conjugates, *J. Chromatogr.*, 559, 385, 1991.

Chapter 19

THE IMPACT OF COLUMN TECHNOLOGY ON PROTEIN SEPARATIONS BY OPEN TUBULAR CAPILLARY ELECTROPHORESIS: PAST LESSONS, FUTURE PROMISES

Sally A. Swedberg

TABLE OF CONTENTS

0-8493-8690-X/94/$0.00+$.50

I. INTRODUCTION

A. OVERVIEW

Although two noteworthy reports appeared in the decade of the 1970s,[1,2] capillary electrophoresis (CE) enjoyed remarkable attention in the decade of the 1980s after the report by Jorgenson and Lukacs captured the attention of the analytical community with the implications of a high-efficiency, fully automated electrophoretic technique for macromolecular separations.[3] This chapter reviews the fundamental premises that indicated that CE would occupy a unique niche as a separation technique for macromolecules, and the potential impact of the open tubular capillary electrophoresis (OT CE) technique over the past decade for protein analysis. It is not meant as a comprehensive review of reports on column technology. Several previous reviews of that nature are available,[4-7] as well as Chapter 18 of this book. In this regard, this chapter is intended to impart an understanding of the initial presumptions and current realties of OT CE for protein analysis, in order to provide the reader with a vision of the future of this technique.

B. IN THE BEGINNING: OPEN TUBULAR CAPILLARY ELECTROPHORESIS AS AN EFFICIENCY-DRIVEN SEPARATION TECHNIQUE FOR PROTEINS

It is instructive to refer to an equation for the height equivalent to a theoretical plate (HETP), which describes the physical factors that limit efficiency in OT CE:[8]

$$H = \underbrace{\frac{l^2}{12L}}_{\text{injection plug term}} + \underbrace{\frac{2D_{\eta,T}}{V_{eof}}}_{\text{axial diffusion term}} + \underbrace{\frac{k'V_{eof}}{(1+k')^2}\left(\frac{r^2k'}{4D_{\eta,T}} + \frac{2}{k_d}\right)}_{\text{radial diffusion term}} \tag{1}$$

where H is the height equivalent to a theoretical plate (HETP; in micrometers), l is the injection plug length; L is the length of column to detection; $D_{\eta,T}$ is the diffusion coefficient of solute; k' is the capacity factor; V_{eof} is the electroosmotic flow (EOF) velocity; r is the internal capillary radius; and k_d is the first-order dissociation constant.

The first term for injection plug volume can be controlled experimentally so as not to be of significant impact.[9] With respect to the diffusion terms; the original presumption by Jorgenson and Lukacs was that the k' interaction term would be vanishingly small, so that the radial term would be zero.[10] This k' term is an indication of the degree of solute–surface interaction and, in an open tube with no column packing, it was presumed that this was a term that could be ignored. Originally, therefore, it was assumed that the efficiency-limiting term for OT CE would be the axial diffusion term.

If the diffusion coefficient for proteins is substituted into the expression, then through the relationship between HETP and the efficiency, N:

$$N = \frac{V_{eof}L}{2D_{\eta,T}} \tag{2}$$

under the assupmtion that V_{eof} is >> V_{epm}, where V_{epm} is the solute electrophoretic mobility, an efficiency of over 2 million theoretical plates could be achieved under these assumptions. This type of efficiency is currently achievable only for capillary gas chromatography (GC) of small solutes. Along with complete automation, and optical on-column detection providing the possibility of quantification, it appeared that OT CE had the potential to revolutionize the protein bench.

C. IN PURSUIT OF THE ASSUMPTIONS: THE ANTICHROMATOGRAPHY PRINCIPLE

In the OT CE experiment, although no column packing is used, it was realized at a very early stage that protein–capillary wall interactions severely degraded the efficiency and reproducibility of the technique. Jorgenson and Lukacs in 1983 remarked, "A better understanding of capillary surface modification will also be important, both for improved capillary surface deactivation and for better control over electroosmotic flow."[10] In 1986, after seeing limited success with dynamic deactivation, Lauer and McManigill concluded that stable wall coatings would be necessary to achieve, on the average, the 1 to 2 million theoretical plate efficiency for OT CE over a broad range of conditions.[11]

There are many attributes that are desirable in a bonded phase for protein analysis by OT CE. Although some specific requirements for a bonded phase have been enumerated in Chapter 18, and will be expanded on slightly in Section II, the surface must meet a special requirement in order to achieve the 1 to 2 million theoretical plate efficiency. Recalling that the assumption for dropping the radial term was a k' value of zero; this would be experimentally equivalent to creating a column packing for a liquid chromatography (LC) experiment where, given a heterogeneous mixture of proteins, all of the protein solutes would elute in the void volume. This type of material would then be ideal as a wall coating for the OT CE experiment. What is desirable for OT CE is, in effect, antichromatography.

In biophysics and bioengineering, there are many areas where knowledge of protein–surface interactions is crucial to advancing the frontier of science and/or technology. Within the area of biocompatible implant materials, many classes of materials that provide low and/or predictable protein–surface interactions have been identified. The next section reviews four bonded-phase surfaces reported in the literature for OT CE that are of particular relevance to the material covered in this chapter. These are surfaces that, based on knowledge from biophysics and bioengineering, are known surfaces that provide low protein–surface interactions.

II. BONDED-PHASE COLUMN TECHNOLOGY FOR OPEN TUBULAR CAPILLARY ELECTROPHORESIS OF PROTEINS

In Chapter 18 the authors outline criteria for evaluating coatings for OT CE. Additional criteria that are desirable for bonded phases in OT CE include the following.

1. Chemically stable wall coating; at least over the range of pH 4 to 7
2. Mechanically stable under high applied fields (200 to 600 V/cm) and appreciable liquid flow rates (0.5 to 2.0 mm/s)
3. From biophysical studies, a hydrated surface to provide reversible protein–surface interactions
4. A surface that can be readily replenished of any residual protein solute, using simple reagents and convenient wash routines

The first two criteria are requirements for long-term stability, while the last two criteria impact peak area and time reproducibility. Very simply, the surface must remain chemically constant to provide reproducibility, and deposited solute is a major source of irreproducibility in OT CE.[8,10,12]

The subject of EOF remains one of the unresolved issues concerning OT CE. The EOF arises as a result of a double-ion layer that forms at any solid–liquid interface (Figure 1).[13] In the case of silica, this is also influenced by partially ionized hydroxyl groups at the surface. However, any material in contact with water will form a ζ potential as a result of the double-

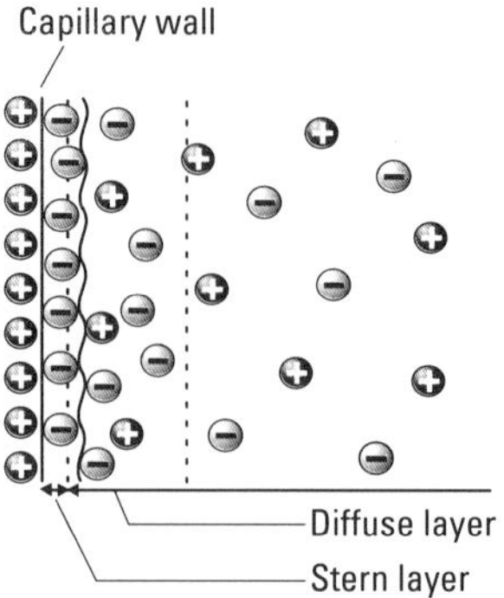

FIGURE 1. Surfaces in contact with an aqueous medium have a double-ion layer associated with the solid–liquid interface due to ionization, ion adsorption, and/or ion dissolution.

ion layer formation at the solid–liquid interface: to wit, Teflon has an appreciable EOF. Hjertén suggested that column technology with EOF would lead to irreproducible results.[12] In contrast, model studies, in conjunction with experimental data, suggest that the EOF component is not a source of irreproducibility, provided that the protein–surface interactions are small and reversible.[8,14] Discussion on this issue will be continued in Section III. For the purpose of this chapter, examples representing column technology with and without EOF will be given.

A. OPEN TUBULAR CAPILLARY ELECTROPHORESIS COLUMN TECHNOLOGY WITHOUT ELECTROOSMOTIC FLOW

1. Polyvinylpyrrolidinone

Polyvinylpyrrolidinone has been used successfully as a synthetic biopolymer and as a solid support for chromatography. It was first described as a bonded phase for OT CE by McCormick, and the surface produced encouraging results.[15] In Figure 1 of Chapter 18, an electropherogram of 15 proteins ranging from 14.5 to 66 kDa in size, and from p*I* 4.5 to 11, is shown. As discussed in Chapter 18, extremes in pH must be used in columns having no EOF. This is for the purpose of obtaining a single charge type in a mixture of proteins with diverse p*I* values for uniform migration past a single point of detection. In this example, the electrophoresis conditions used were a phosphate buffer (38.5 m*M* H_3PO_4–20 m*M* H_2PO_4) at pH 2. The intermediate synthetic layer is based on silylation. This is the most common type of primary surface activation for silica.[16] Its greatest disadvantage is that the pH stability for this type of chemistry falls in the pH range of 4 to 7. Fortunately, this is a pH range of considerable importance for the analysis of proteins. The average efficiency achieved in this example was about 200,000 theoretical plates.

2. Linear Polyacrylamide

Polyacrylamide, a synthetic hydrogel, has been the most popular medium for standard gel electrophoresis. Polyacrylamide has also been used as a biocompatible polymer in contact lenses, as well as a biocompatible implant material. In Figure 2, a pherogram of polypeptides in the range of 11 to 30 kDa, with p*I* values from 4 to 10.5, is shown. In this particular example, the primary activation of the silica surface is carried out so as to create an Si-C linkage. This linkage has been reported to provide greater pH stability than the Si-O-Si bond formed by conventional silylation.[17] Because of the absence of EOF acting as a mass transport system for a mixture of proteins of diverse charge type, a pH extreme is again required for homogeneity of charge. In this example, because of the putative expanded pH range of the chemical bond used in the intermediate phase, the conditions used for this separation included a 50 m*M*

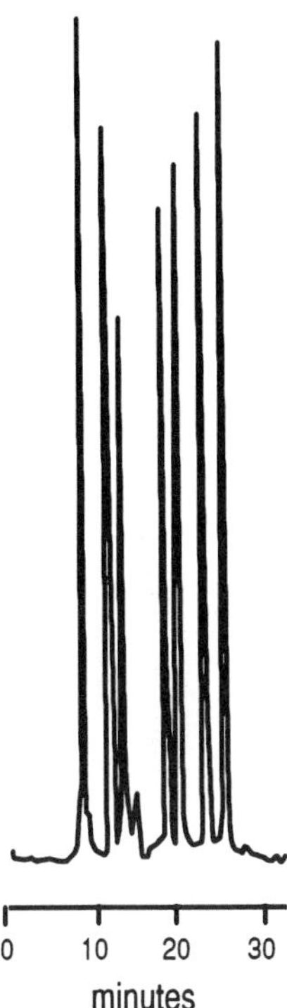

FIGURE 2. Without EOF: Separation of seven polypeptides, ranging from 11 to 30 kDa, and from p*I* 4 to 10.5, on an uncross-linked polyacrylamide-modified surface. Buffer, 50 m*M* GIAm–TEA, pH 9.5; L_{eff}, 45 cm; field, 333 V/cm. (From Cobb, K. A., Polnik, V., and Novotny, M., *Anal. Chem.*, 62, 2478, 1990. With permission.)

glutamate–triethylamine buffer at pH 9.5. The average efficiency in this separation was reported at about 200,000 theoretical plates.

B. OPEN TUBULAR CAPILLARY ELECTROPHORESIS COLUMN TECHNOLOGY WITH ELECTROOSMOTIC FLOW

1. Polyethylene Glycol

Polyethylene glycol (PEG) is considered an excellent material for biomedical implants due to its negligible protein adsorption characteristics. Shown in Figure 3 is the work by Bruin et al., using PEG-coated capillaries.[18] The proteins used in this example range in size between 14.5 and 30 kDa, and in p*I* from 7.5 to 10.8. Conditions used in this example were 30 m*M* phosphate buffer at pH 3.8. The average efficiency reported for this separation was about 100,000 theoretical plates.

2. Amphoteric Hydrogel

The example shown in Figure 4 is from a family of surface coatings that may be broadly termed as amphoteric hydrogels.[20,21] Synthetic hydrogels are a known class of important implant materials. One class of natural hydrogels consists of the proteins themselves. Other materials that have similar characteristics include certain types of ampholytes. One of the interesting characteristics of these types of surfaces is the predictable manner in which the EOF changes as a function of pH. By changing the pH, the direction and magnitude of the EOF may change (Figure 5). This gives a degree of control of resolution and elution order of solutes as a function of pH (Figure 6). The conditions under which the pherogram shown in Figure 6a was obtained included 10 m*M* citrate, pH 3.0. The EOF was anodic and –1.4 mm/s. In Figure 6b, the conditions under which this pherogram was obtained included 10 m*M* citrate,

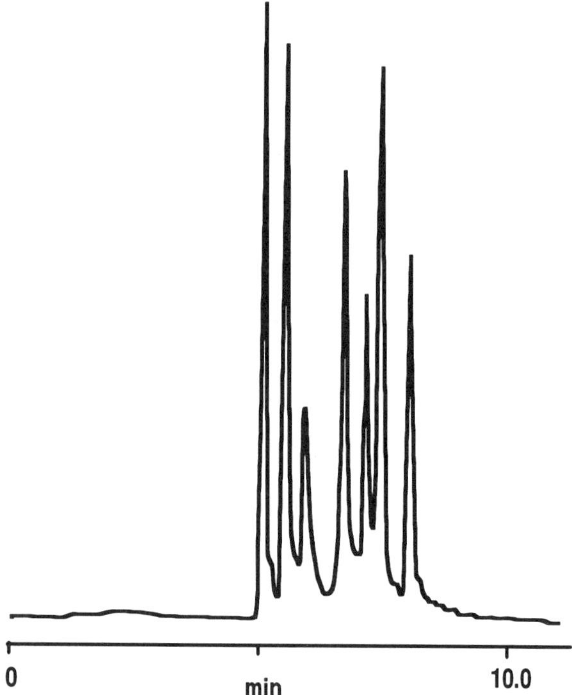

FIGURE 3. With EOF: Separation of seven polypeptides, ranging from 14.5 to 30 kDa, and from p*I* 7.5 to 10.5, on a polyethylene glycol surface. (From Bruin, G. J., et al., *J. Chromatogr.*, 471, 429, 1989. With permission.)

pH 6.5. The proteins used in this example range from 20.1 to 45 kDa in size, and from p*I* 4.6 to 9.5. Interestingly enough, although these conditions demonstrate good resolution of contaminant species, the efficiencies are on the order of 10K–50K plates.

III. LESSONS LEARNED FROM BONDED-PHASE COLUMN RESEARCH FOR OPEN TUBULAR CAPILLARY ELECTROPHORESIS

Based on the requirements that bonded phases should provide low adsorptive surfaces for proteins, the examples discussed above represent good prospects. All of the examples shown are separations of idealized, low molecular weight, single-stranded protein standards. Currently, few examples are available in the literature describing stable bonded-phase technologies for OT CE that have been used successfully to solve bioanalytical problems with samples truly indicative of those from a biosciences laboratory. This is a function of the lack of an information base describing what columns/electrophoresis conditions may produce stable, reproducible results, in conjunction with a lack of understanding from the user bench as to what types of bioanalytical problems OT CE can solve. This set of circumstances indicates the current nascent state of this evolving technology.

From 1985 to the present, a period during which the most intense research effort toward bonded-phase column technology for OT CE has been expended, there has been a considerable amount of knowledge gained concerning the potential for protein analysis by OT CE. The following section discusses three major issues that have been identified as a result of the research efforts on column technology for this technique: these are selectivity, EOF, and efficiency.

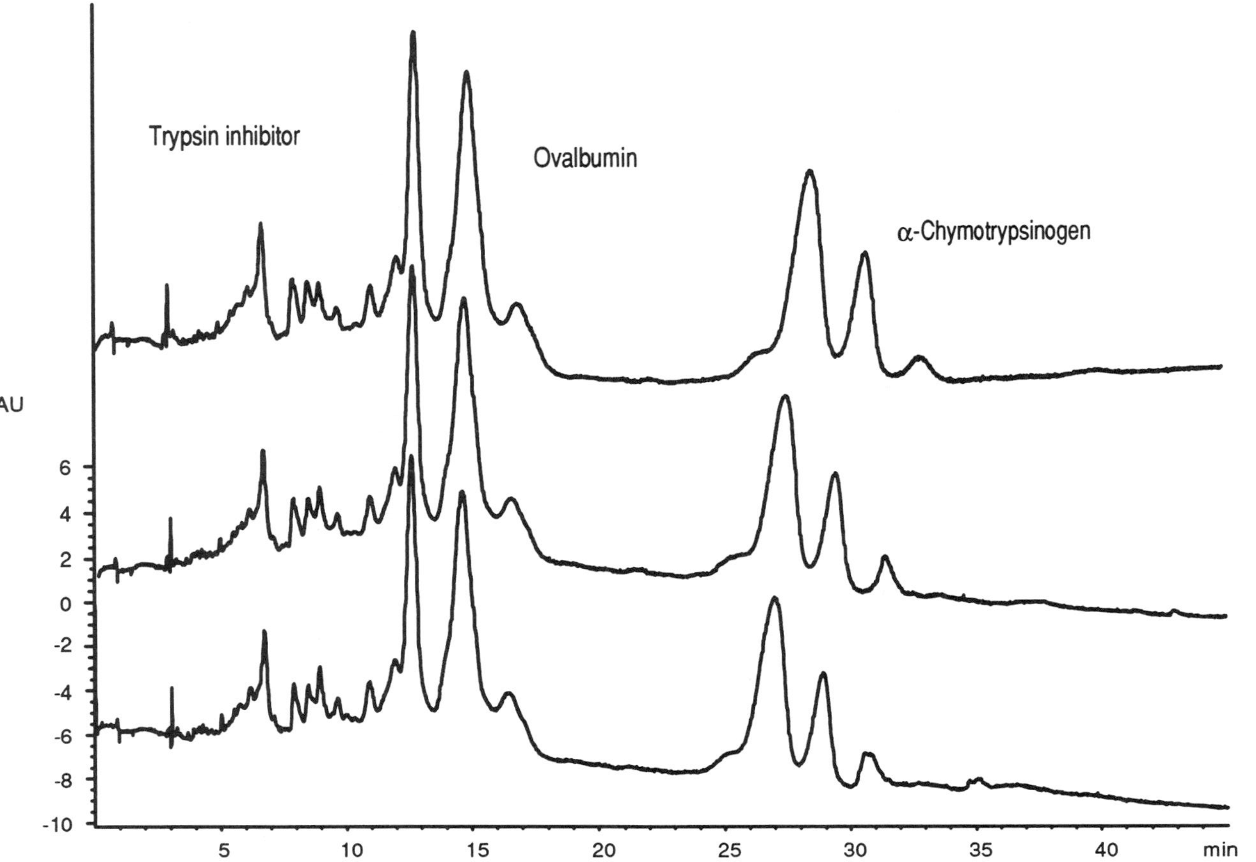

FIGURE 4. With EOF: Separation of three proteins by size (trypsin inhibitor [p*I* 4.6 and 20.1 kDa] vs. ovalbumin [p*I* 4.6 and 45 kDa]) and by charge (trypsin inhibitor and ovalbumin [p*I* 4.6] vs. α-chymotrypsinogen [p*I* 9.5]) on a bovine serum albumin-coated surface. Capillary, 25-μm I.D., 125-μm bubble; L_{eff}, 25 cm; field, –500 V/cm; EOF, –1.4 mm/s; buffer, 10 m*M* citrate, pH 3.0, 4 m*M* CHAPSO.

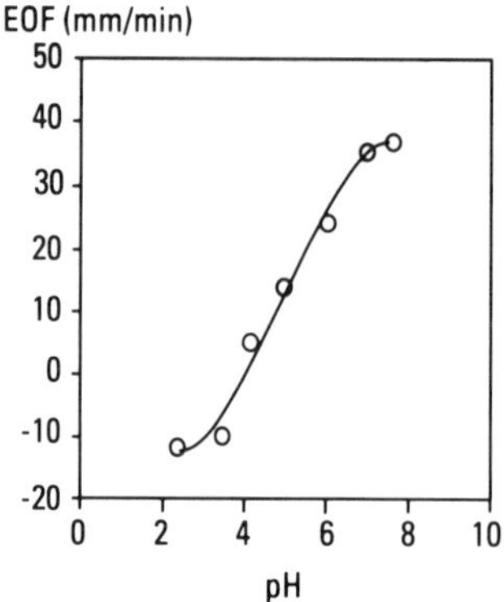

FIGURE 5. The electroosmotic flow changes in both magnitude and direction, as a function of pH, on an α-lactalbumin-modified surface.

A. SELECTIVITY IN OPEN TUBULAR CAPILLARY ELECTROPHORESIS

From the science of chromatography, the term *selectivity* has an exact definition. For chromatography, the selectivity, α, may be defined as

$$\alpha = \frac{k_2'}{k_1'} \tag{3}$$

where $k'_{2,1}$ is the capacity factor of solute species 2 and 1, respectively.

In this regard, the selectivity of the chromatographic system is the measure of the capability of such a system to resolve solute components in a heterogeneous sample mixture.

In OT CE, the selectivity may be defined as

$$\alpha = \frac{\mu_{epm2}}{\mu_{epm1}} \tag{4}$$

where $\mu_{epm2,1}$ are the electrophoretic mobilities of species 2 and 1, respectively.

This definition is strictly meant to imply electrophoretic selectivity. There are several factors that impact electrophoretic selectivity in OT CE, including pH, buffer strength and type, organic modifiers, etc. Whether or not surface modification could impact selectivity, by adding a chromatographic contribution to the term, has been the object of speculation.

Shown in Figure 7a and b are two examples from the literature that indicate that the surface plays a role in selectivity. The protein solutes used in these examples are single-stranded proteins for which the sequence and crystallographic data are known. Since all relevant information is known concerning hydrodynamic radius and charge, the elution order should be predictable. In both cases, numerous deviations occur based on what would be expected from the relevant protein data.

It may be argued that the conditions of sample preparation or analysis promote partial or total protein denaturation, aggregation, and/or possibly proteolytic degradation, and, hence, some ambiguity in the total interpretation of these results. In that regard, one study has been done under highly controlled conditions that strongly implicates that the surface is playing a role in selectivity. In Figure 8a and b, two pherograms are shown where the only difference in the two experiments is the surface of the capillaries.[21] Table 1 shows a selectivity difference for the two types of surfaces as determined in this study.

It is not surprising that surface chemistry may provide a degree of selectivity. The average efficiency in these examples indicates a k' component, according to the efficiency model given

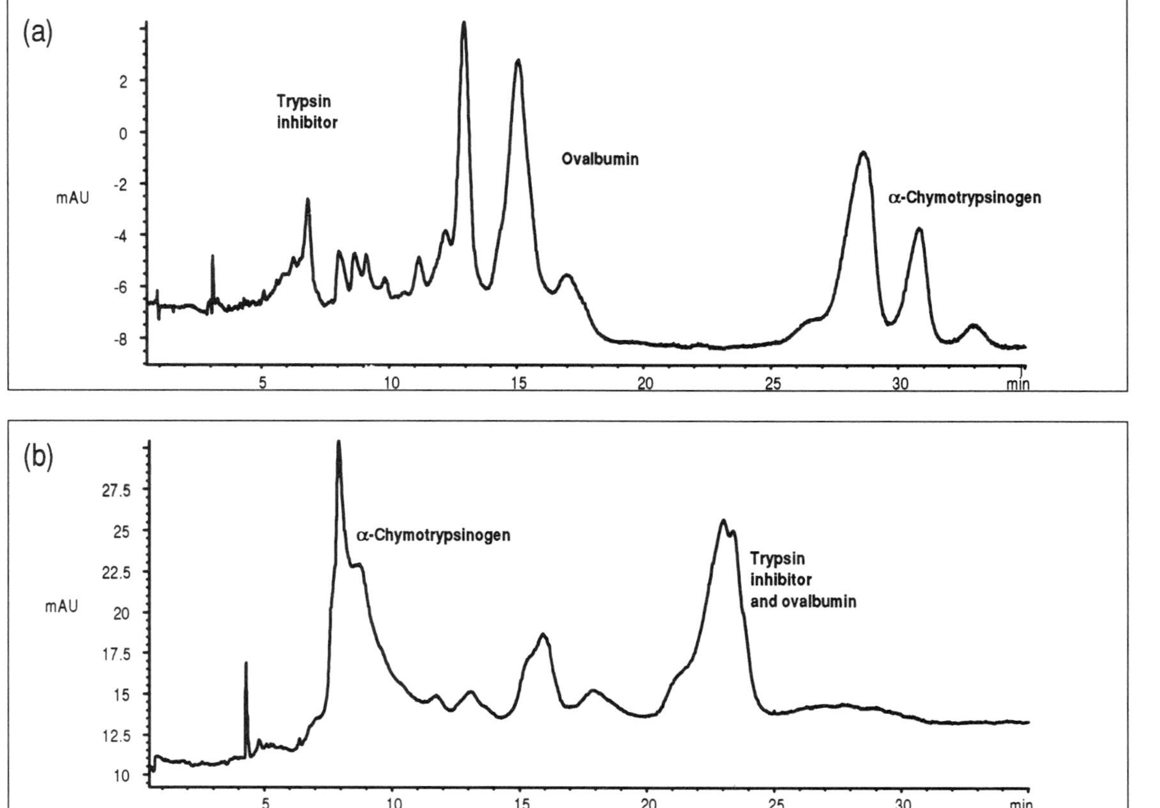

FIGURE 6. Control of resolution and elution order on BSA-deactivated columns. Electrophoretic selectivity can be changed by changing the buffer pH and voltage polarity for amphoteric hydrogel phases. For the proteins in (a) and (b), the resolution at pH 3.0 (a) is better. The elution order of species is reversed: low to high p*I* at low pH, and high to low p*I* at high pH. (a) Capillary, 25-µm I.D., 125-µm bubble; field, –500 V/cm; EOF, –1.4 mm/s; buffer, 10 m*M* citrate, pH 3.0, 4 m*M* CHAPSO. (b) Capillary, 25-µm I.D., 125-µm bubble; field, +500 V/cm; EOF, +1.0 mm/s; buffer, 10 m*M* citrate, pH 6.5, 4 m*M* CHAPSO.

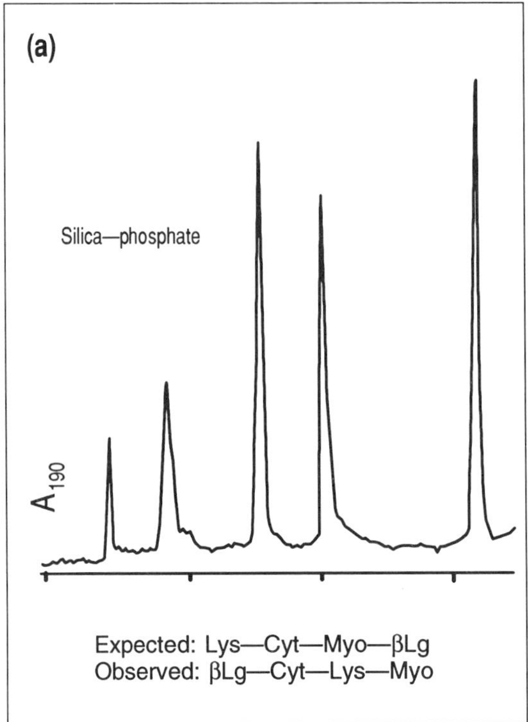

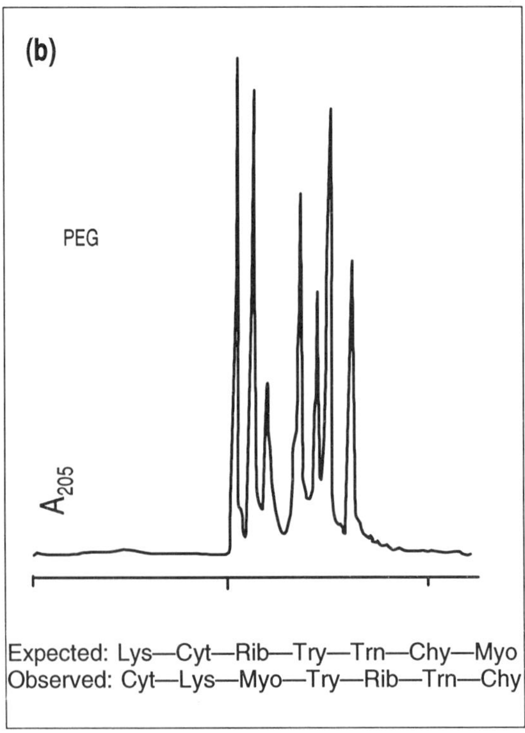

FIGURE 7. Selectivity in OT HPCE observed. Two examples from previous studies demonstrate the elution order of protein species, which are different than would be expected based on known size/charge relationships. ([a] From McCormick, R. N., *Anal. Chem.*, 60, 2322, 1988. With permission; [b] from Bruin G. J., et al., *J. Chromatogr.*, 471, 429, 1989. With permission.)

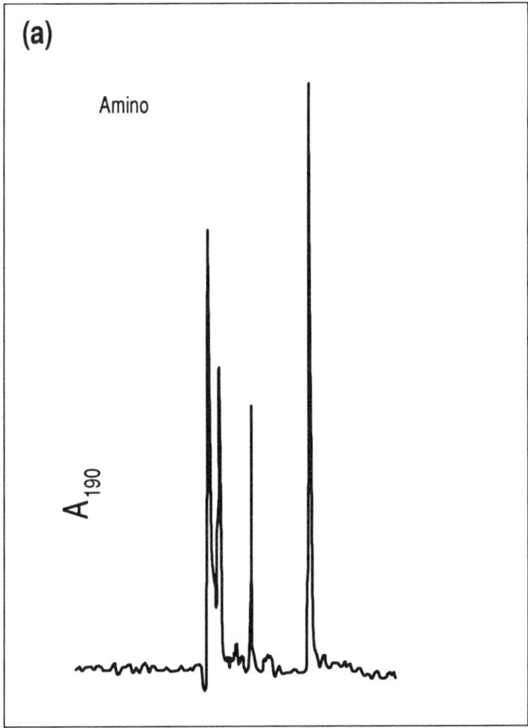

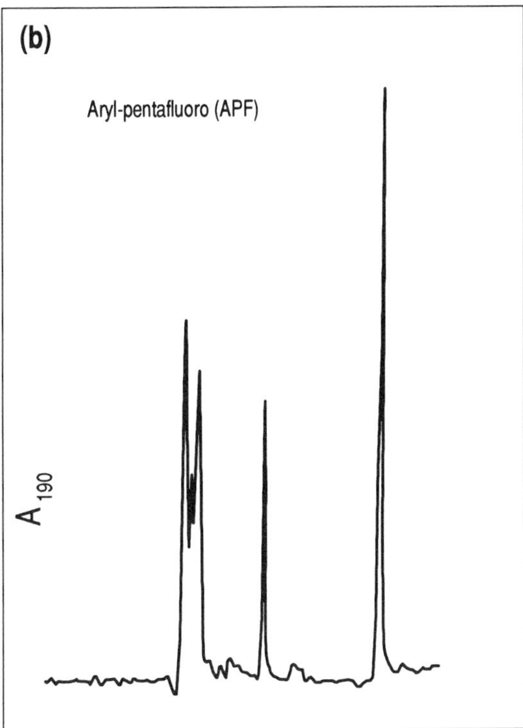

FIGURE 8. Selectivity in OT HPCE determined. Example of surface selectivity in a controlled experiment, where the only variable is the surface. In (a) the surface has been aminosilylated and in (b) half of the prepared silylated capillaries are processed to produce the arylpentofluoro (APF) surface. ([a and b] From Maa, Y.-F., et al., *J. High Resolut. Chromatogr.*, 14, 65, 1991. With permission.)

TABLE 1
Selectivity Difference for Two Surface Types in Open Tubular Capillary Electrophoresis

	Selectivity			
Surface	**C-R**	**C-M**	**C-D**	**D-M**
Amino phase	1.08	1.19	1.02	1.16
APF phase	1.11	1.27	1.02	1.25

Note: The values were calculated from the data in Figure 8a and b (Ref. 21), using Equation 4.

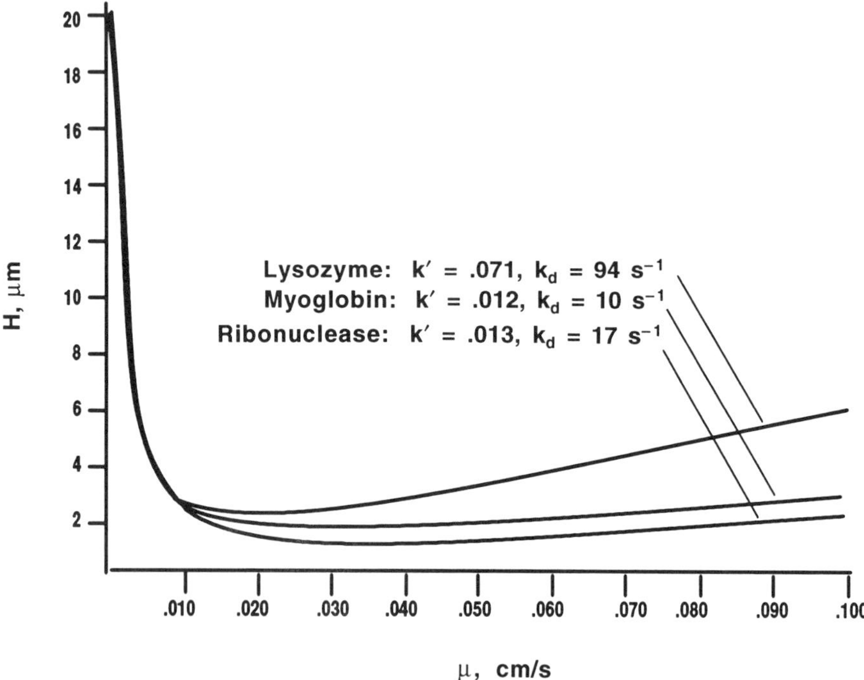

FIGURE 9. Curve fitting for three proteins analyzed by OT HPLC and OT HPCE. The three proteins were analyzed chromatigraphically under controlled conditions to determine independently the k' value under specified conditions on an APF capillary. The pherograms were analyzed for efficiency. The value of k' measured independently was inserted into Equation 1 and the k_d values were then determined.

in Equation 1. The data previously shown in Figure 4 also demonstrate that resolution in CE may not be dependent on efficiency alone.

B. THE ELECTROOSMOTIC FLOW IN OPEN TUBULAR CAPILLARY ELECTROPHORESIS

As mentioned previously, the concept that the EOF was a serious efficiency-degrading component and should, therefore, be eliminated, was first proposed by Hjertén. Model studies of a van Deemter relationship between HETP and EOF show that the EOF causes a serious drop in the efficiency of the OT CE method only if the k' value is appreciable (Figure 9). In this study, a stable bonded phase and electrophoresis conditions providing suitable reproducibility were used to demonstrate a good fit to Equation 1.[14] Additionally, since relevant protein separations in CE may not be simply efficiency driven, elimination of the EOF destroys a high-efficiency mass transport system for bringing species of dissimilar charge past a single point of detection.[3,8]

Furthermore, in Figure 10, two capillary systems using surfaces modified with linear polyacrylamide are shown.[17,19] In Figure 10a, the capillary was prepared so as to eliminate the EOF, and in Figure 10b, the EOF component is retained. The average efficiency in Figure 10a is slightly higher than the efficiency in Figure 10b, but the protein solutes in Figure 10b are baseline resolved in under 10 min vs. the 30-min analysis time for Figure 10a. This would seem to further illustrate that the EOF may be a very functional high efficiency mass transport system.

In addition, many important protein separations may not be efficiency driven. In Figure 11a and b, the separation of contaminants in two commercially available protein preparations is shown.[20] Under the conditions of this analysis, the flow component was appreciable (–1.4 mm/s). The CE separation of the contaminants in this analysis was comparable in resolving power to the state-of-the-art gradient elution high-performance liquid chromatography (HPLC) separation. These data clearly suggest that good resolution may be achieved under high flow conditions.

In column technology where the EOF is retained, one of the major issues remaining is control of the EOF as an additional mechanism for optimizing resolution. Figure 12 shows a graph of the resolution as a function of the EOF component. This graph clearly demonstrates why dynamic control over the EOF is desired in order to obtain optimal resolution of solute species.

One of the most elegant solutions to the problem of controlling the EOF magnitude has been championed by the early work of C. Lee of Iowa State University. Lee and co-workers have demonstrated an apparatus that may independently control the EOF component without impacting the solute mobility.[22,23] A schematic of the apparatus is shown in Figure 13. In this configuration, a high field is applied across the wall of the capillary. This field independently modulates the ζ potential at the solid–liquid interface. The EOF is related to the ζ potential through the following relationship:

$$\mu_{eof} = \frac{-\varepsilon\xi}{\eta} \tag{5}$$

where μ_{eof} is the electrophoretic mobility, ε is the permittivity, and η is the viscosity. The EOF is then independently controlled without influencing the solute mobility.

It is clear that the area of the external control of the EOF has important implications in providing control over the separation process in OT CE. This is discussed in detail in Chapter 22.

C. EFFICIENCY IN OPEN TUBULAR CAPILLARY ELECTROPHORESIS

Possibly the most important lesson learned from the past decade of effort on column technology for OT CE deals with the efficiencies that can be expected from OT CE.

Recall from Section I that one of the original boundary conditions assumed in order to reach the 1 to 2 million theoretical plate efficiency was that k' would be zero for all protein solutes. Figure 9 illustrates that k' does not need to be of great magnitude in order to have a significant impact on the efficiency in OT CE for macromolecular separation. In addition, several surfaces were discussed in Section II that represent, from the biophysical perspective, the most ideal surfaces for promoting low protein–surface interactions. Yet, such ideal surfaces fail to produce the 1 to 2 million theoretical plate efficiency. Table 2 shows how rapidly efficiency degrades as a function of increasing k'.

In hindsight, this is quite logical within the principles learned from separation science. In Section I the problem was expressed in terms of the experimental dilemma of the

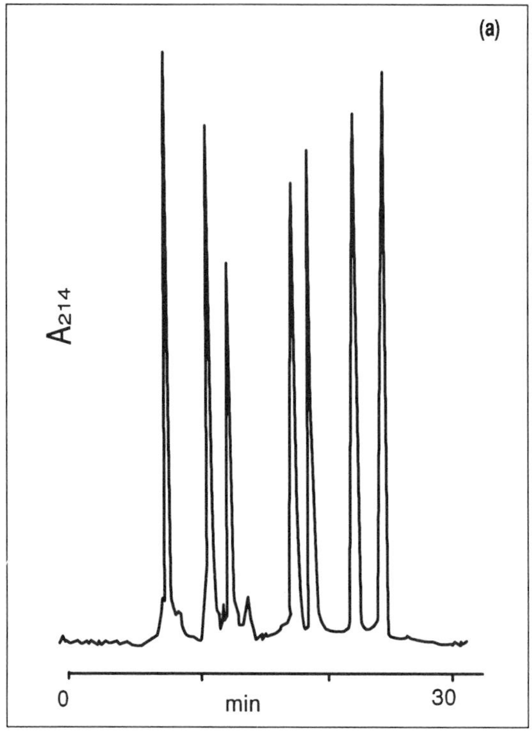

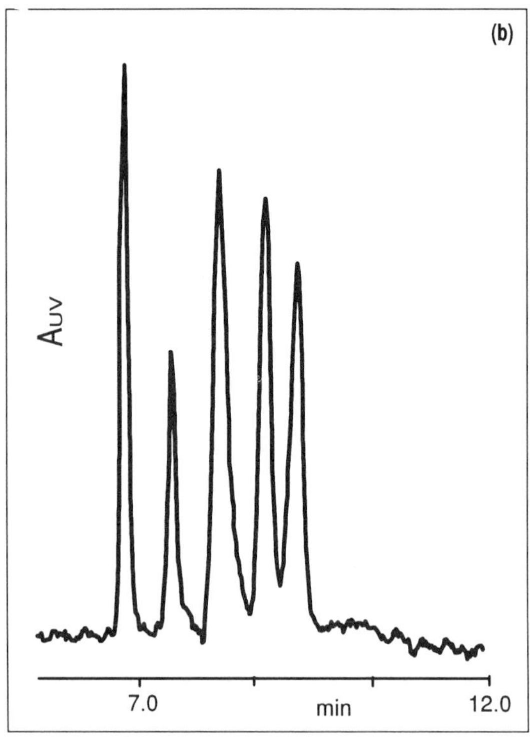

FIGURE 10.

antichromatography problem. From separation science there exists what is referred to as the central chromatography problem, which clearly states that, given a heterogeneous solute sample, there is not one set of chromatographic conditions that can separate all solutes in such a mixture. As well, the antichromatography problem, for the same reasons, is just as intractable. For CE, in corollary to one of the central premises of chromatography, it should be stated as the central antichromatography problem that, given a heterogeneous mixture of proteins, there is not one set of electrophoretic conditions that can separate all species (Table 3). It appears that, in addition to the electrophoretic selectivity that can be expected by changing pH, ionic strength, etc., there is a chromatographic selectivity that further complicates the separation process.

IV. A LOOK AT THE FUTURE OF PROTEIN SEPARATIONS FOR OPEN TUBULAR CAPILLARY ELECTROPHORESIS

To date, there are only a few examples of bioanalytical problems that are not only solved by OT CE, but that also offer an advantage over other techniques. However, from the biosciences bench, there appear to be two major areas emerging for which OT CE appears to be ideally suited for intact protein analysis: microheterogeneity/contaminant analysis in protein preparations in the late or final stages of purification, and the analysis of membrane proteins. The next two sections describe some of the progress in these two areas that suggests that OT CE will have a future in both areas.

A. MICROHETEREOGENEITY/CONTAMINANT ANALYSIS IN GLOBULAR PROTEIN PREPARATIONS

A growing number of reports from the biosciences and biopharmaceutical laboratories point to OT CE as a good technique for the analysis of globular protein microheterogeneity, and the protein contaminants in protein preparations. These analyses use a variety of approaches including unmodified fused silica, dynamically deactivated fused silica, and coated column technology.

Using unmodified fused silica, Holzman et al. reported the identification of isoforms of the prolyl isomerase, cyclophilin (Figure 14).[24] They compared the cloned protein to the bovine protein by gel electrophoresis and CE. One of the bovine fractions, showing a single band by isoelectric focusing (IEF), had two bands apparent by CE. Since OT CE separates by a completely different mechanism than either sodium dodecyl sulfate polyacrylamide gel electrophoresis (SDS PAGE) or IEF, this technique has the potential of providing information that cannot be obtained using traditional gel techniques.

Tran et al.[25] have reported separating isoforms of erythropoietin on a phospho-silicate modified surface (Figure 15). In light of the early work of McCormick,[15] it is now recognized that prolonged contact of the silica surface with phosphate salt results in the formation of a phospho-silicate surface that, in conjunction with high ionic strength buffers, may be useful for separating proteins of diverse types over a broad p*I* range.[26,27]

FIGURE 10. Electroosmotic flow: impact of efficiency in OT HPCE. Two uncross-linked polyacrylamide-deactivated surfaces are shown, one prepared to eliminate EOF (a) and one prepared to retain the EOF (b). (a) Polypeptides from 11 to 30 kDa and from p*I* 4 to 10.5 are separated. Buffer, 50 m*M* GIAm–TEA, pH 9.5; L_{eff}, 45 cm; field, 333 V/cm. (b) Polypeptides from 14.5 to 66 kDa and from p*I* 4.8 to 11 are separated. Buffer, 30 m*M* citrate, pH 3.0; L_{eff}, 26 cm; field, 250 V/cm. The analysis times of 30 min (a) vs. less than 10 min (b) are striking. ([a] From Novotny, M., et al., *Anal. Chem.*, 62, 2478, 1990. With permission. [b] From Engelhardt, H. and Kohr, H., *J. Microcolumn Sep.*, 3, 491, 1991. With permission.)

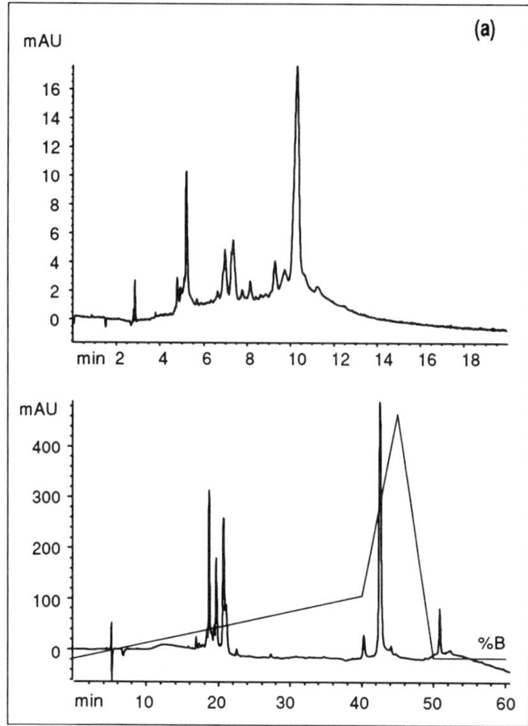

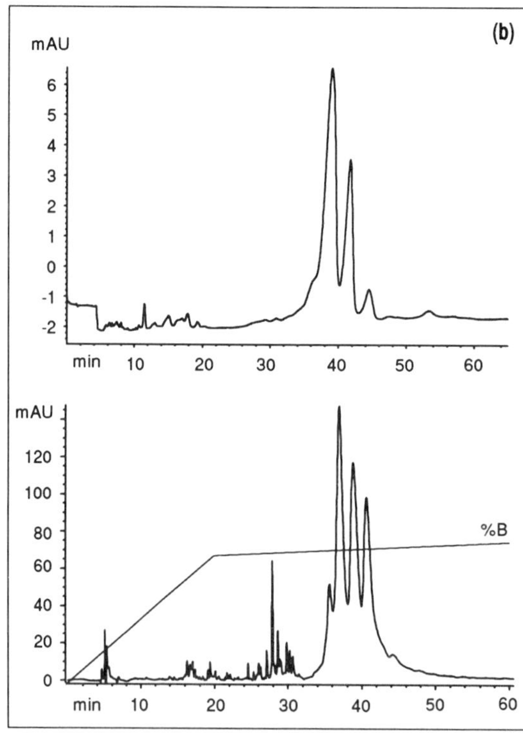

FIGURE 11. Comparison of components in lyophilized protein powders: CE (BSA surface deactivation) vs. HPLC separations of trypsin inhibitor (a) and α-chymotrypsinogen (b). (a) BSA capillary, 25-μm I.D. + 125-μm bubble; L_{eff}, 25 cm; field, −500 V/cm; EOF, −1.4 mm/s; buffer, 10 m*M* citrate, pH 3.0, 4 m*M* CHAPSO. Gradient elution: Vydac C_{18}, 250 × 2.1 mm. (b) BSA capillary, 75 μm; L_{eff}, 25 cm; field, −300 V/cm; EOF, −1.0 mm/s; buffer, 10 m*M* citrate, pH 3.0, 4 m*M* CHAPSO. Gradient elution: Vydac C_{18}, 250 × 2.1 mm.

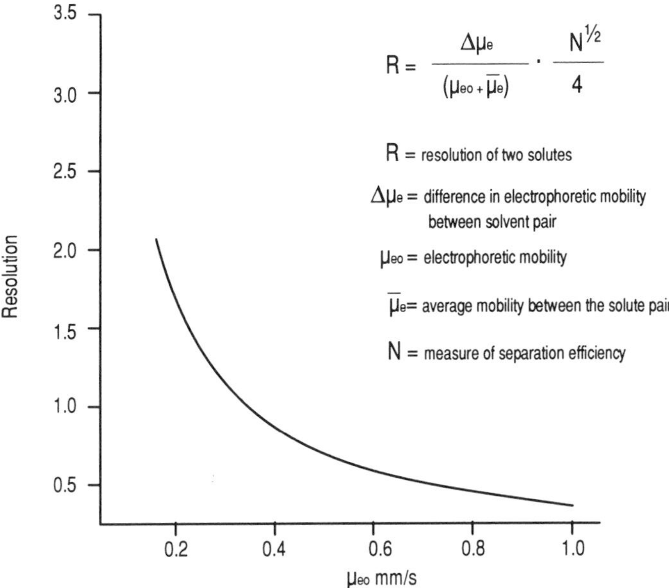

$$R = \frac{\Delta\mu_e}{(\mu_{eo} + \bar{\mu}_e)} \cdot \frac{N^{1/2}}{4}$$

FIGURE 12. The degradation of resolution as a function of EOF: the high-efficiency mass transport system, the EOF, may have an impact on solute resolution.

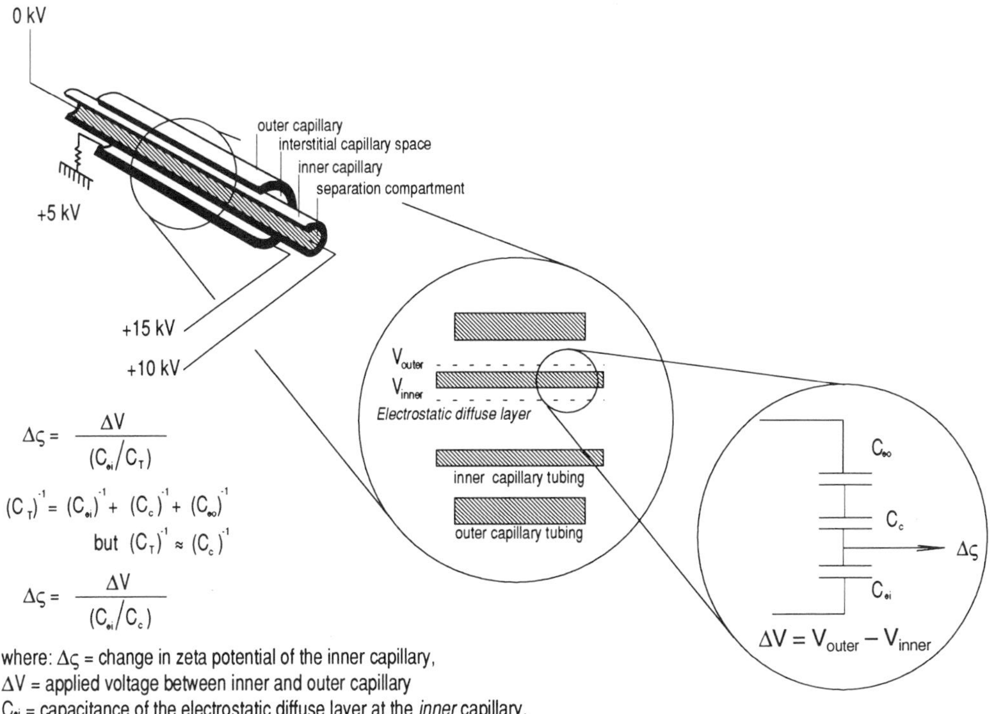

$$\Delta\varsigma = \frac{\Delta V}{(C_{ei}/C_T)}$$

$$(C_T)^{-1} = (C_{ei})^{-1} + (C_c)^{-1} + (C_{eo})^{-1}$$

$$\text{but } (C_T)^{-1} \approx (C_c)^{-1}$$

$$\Delta\varsigma = \frac{\Delta V}{(C_{ei}/C_c)}$$

FIGURE 13. Modulation of the EOF, using an externally applied field: the coaxial capillary configuration. The capillary configuration shown was used by Lee and co-workers to study the effects of an externally applied field on the EOF. (From Lee, C. S., McManigill, D., Wu, C.-T., and Patel, B., *Anal. Chem.*, 63, 1519, 1991. With permission.)

TABLE 2
The Impact of k' on Efficiency

$$H = \frac{l^2}{12L} + \frac{2D_{\eta,T}}{V_{eof}} + \frac{k'V_{eof}}{(1+k')^2}\left(\frac{r^2k'}{4D_{\eta,T}} + \frac{2}{k_d}\right)$$

(inj) (axial) (radial)

HETP (μm)

k'	Injection	Axial	Radial	Total	N
0.001	0.08	0.40	0.10	0.58	1.7×10^6
0.005	0.08	0.40	0.52	1.00	1.0×10^6
0.010	0.08	0.40	1.06	1.54	6.5×10^5
0.050	0.08	0.40	6.30	6.78	1.5×10^5
0.100	0.08	0.40	15.20	15.68	6.4×10^4

Note: The values were calculated for a 50-μm capillary, with 1 m to the point of detection, using a 1-mm injection plug, diffusion coefficient of 10^{-6} cm²/s, k_d of 10 s^{-1}, and V_{eof} of 0.5 mm/s.

TABLE 3
The General Elution Problem Revisited

For LC (general chromatography problem): Given a complex mixture of solutes with diverse k' values (1 to 10), there is not a single set of chromatographic conditions that can separate all components in such a mixture

For CE (general antichromatography problem): Given a complex mixture of proteins with diverse k' values (0.010 to 0.100), there is not a single set of electrophoretic conditions that can separate all components in such a mixture

Using covalently bonded column technology, separation of genetic variants of human hemoglobin,[28] and charge variants in recombinant DNA-derived proteins, have been reported (Figure 16).[29,30] Using surface deactivated columns to run an IEF separation, isoforms of transferrin have been reported.[31]

Finally, the CE separation of contaminants in commercial protein preparations, with the resolving power typically observed with gradient elution HPLC, has been shown. These separations were performed under low ionic strength conditions on a BSA surface deactivated column.[20] Compared to the separation run on the phospho-silicate–high ionic strength buffer approach, the BSA–low ionic buffer strength buffer system appears to resolve significantly more components in these preparations (Figure 17). Since it is known from biophysics that increasing ionic strength diminishes the electrophoretic mobilities, it is anticipated that increasing ionic strength may collapse the separation of species of similar charge and size. It is clear that evolving column technologies that allow good resolution at low ionic strengths will play an important role in microheterogeneity analysis by CE.

B. MEMBRANE PROTEIN ANALYSIS

The analysis of membrane proteins is difficult due to the relatively high surfactant concentrations required to stabilize these proteins *in vitro*. In addition, because of their hydrophobic nature, in combination with the high surfactant concentrations, many modes of HPLC are not compatible for high-resolution analytical separation of this class of proteins. Additionally, SDS-PAGE and IEF, which are heavily relied on for the key biomolecular signatures of size and charge for proteins, are frequently not compatible with the requirements of retaining

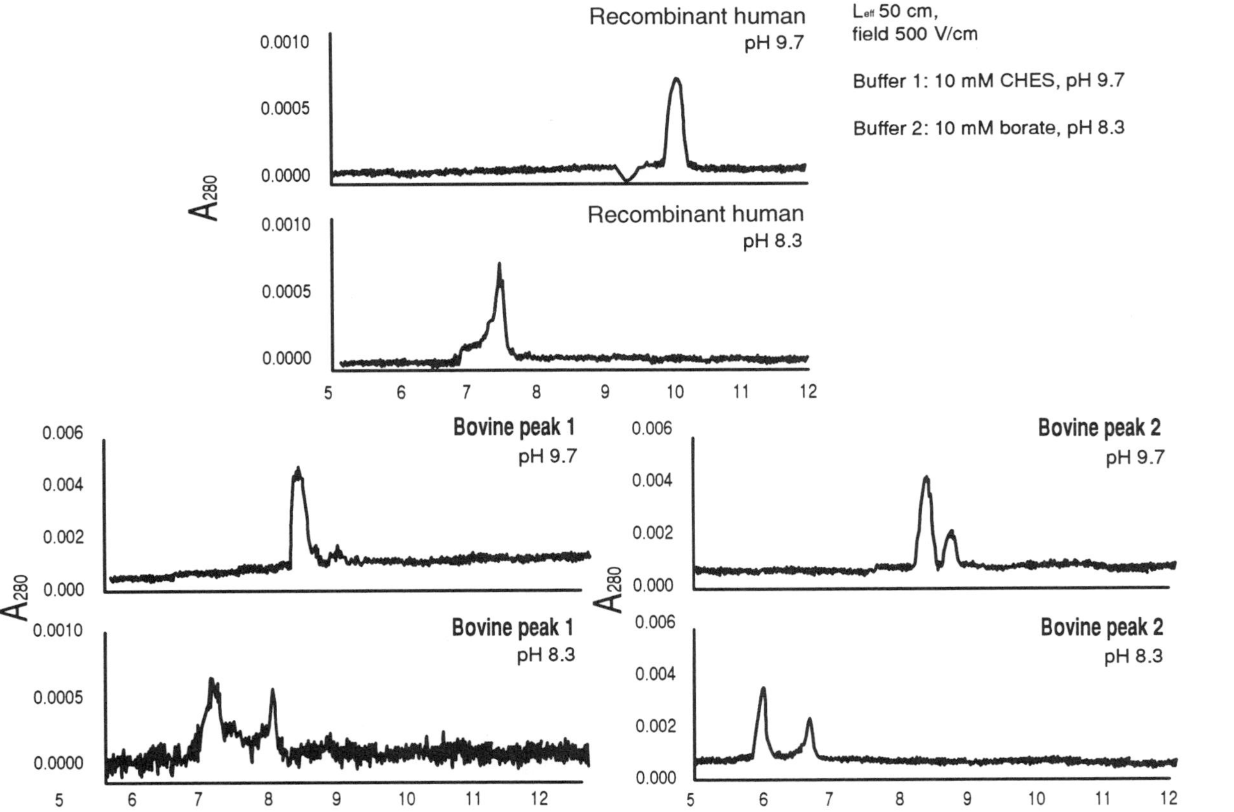

FIGURE 14. In the isoform analysis by CE of bovine cyclophilin, fraction 2 from the preparative separation column showed two species in the fraction while the IEF showed only a single band. Buffer 1, 10 m*M* CHES, pH 9.7; buffer 2, 10 m*M* borate, pH 8.3. L_{eff}, 50 cm; field, 500 V/cm. (From Holzman, T. F., et al., *J. Biol. Chem.*, 266, 2474, 1991. With permission.)

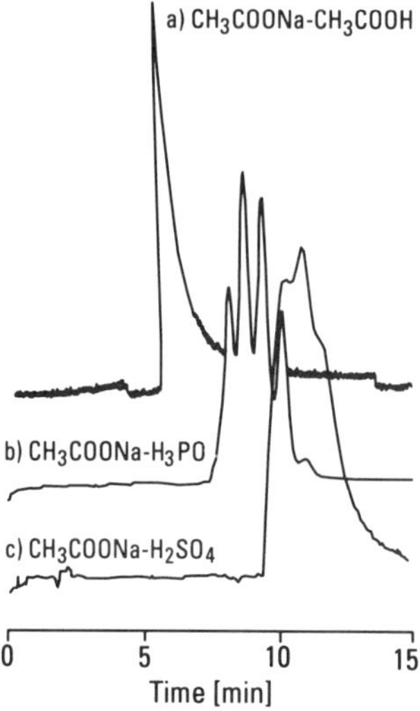

FIGURE 15. Capillary electrophoretic analysis of glycoforms: using a phospho-silicate-deactivated capillary, and a moderately high ionic strength (100 m*M* acetate–phosphate) glycoforms of the human recombinant erythropoietin are resolved. L_{eff}, 20 cm; field, 370 V/cm. (From Tran, A. D., et al., *J. Chromatogr.*, 542, 459, 1991. With permission.)

biologically active species of these proteins. Denatured proteins may carry a different net charge than the native-state protein. Denaturation may also result in the loss of bound prothetic groups. This may make interpretation of the gel results difficult or questionable.[32]

Open tubular capillary electrophoresis appears to be ideally suited for membrane protein aaysis. The typical surfactants used are either zwitterionic in nature, as in the CHAPS or CHAPSO series, or are nonionic, as in the alkyl glucoside or Triton series. In applications where SDS or CTAB are used, these charged surfactants tend to have an appreciable impact on the EOF. For the zwitterionic and nonionic surfactants, the net impact on the EOF is one of viscosity, not charge, and is usually not appreciable.

Figure 18 shows the separation of a nonbiologically active from a biologically active species in a preparation of bacteriorhodopsin. By using selective wavelength detection, the biologically active (retinal bound, 548 nm-absorbing) species can be distinguished from the nonbiologically active species. The SDS-PAGE results cannot distinguish biologically active from nonbiologically active species, as the biologically active form is denatured in SDS. Therefore, while not a complicated separation, it is a separation that currently can be done only by OT CE.

V. SUMMARY

Although once thought to be a separation technique that would have a level of efficiency currently paralleled only by the capillary gas chromatographic analysis of small solutes, it is now apparent that the 1 to 2 million theoretical plate efficiency for protein separations by CE is not, on the average, achievable. This is because very small solute–surface interactions have a large impact on the efficiency of the separation. In many literature examples, there exist elution order of protein solutes that are not predicted based on their charge and size. These data strongly suggest that surface selectivity may play a role in the separation process in the OT CE technique.

However, there are currently a growing number of examples of relevant protein separations done by CE that either cannot be done or are complementary to conventional separation techniques. The inherently distinct mode of separation offered by CE, coupled with the advantages of speed, automation, small sample size, and optical detection, are beginning to be shown, in the hands of biosciences users, as having a unique niche as a bioanalytical technique.

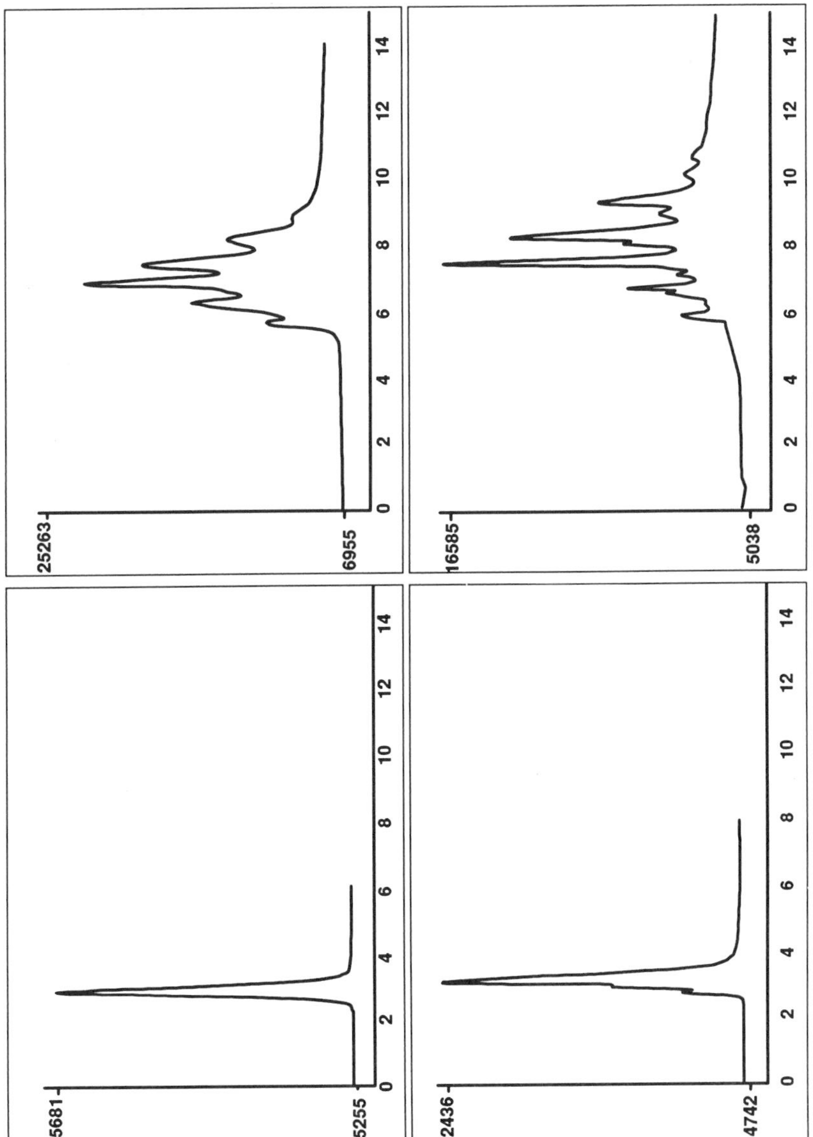

FIGURE 16. Using a surface-bonded deactivated phase, glycoforms of a soluble recombinant T4 receptor protein (rCD4) become resolved as a function of pH. (From Wu, S.-L., et al., *J. Chromatogr.*, 516, 115, 1990. With permission.)

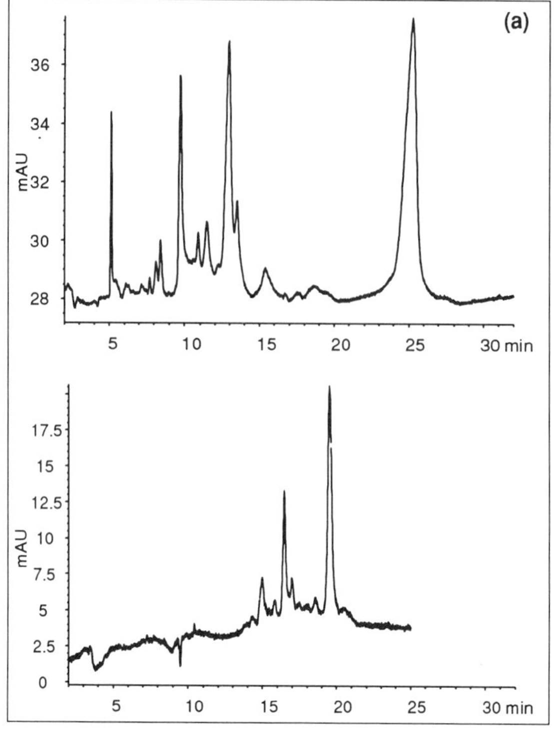

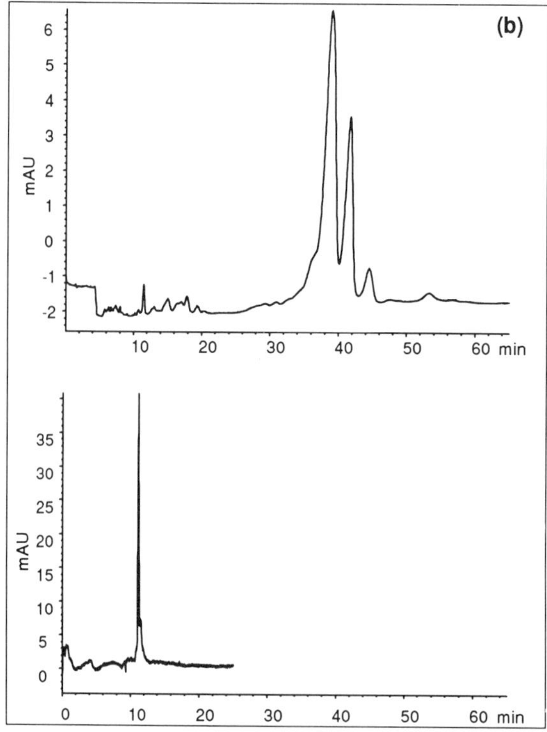

FIGURE 17. Two different capillary systems are used for the separation of lyophilized powders of α-chymotrypsinogen (a) and trypsin inhibitor (b). The BSA-deactivated column–low ionic strength buffer system (a) gives better resolution of species than the phospho-silicate-deactivated column-high ionic strength system (b).

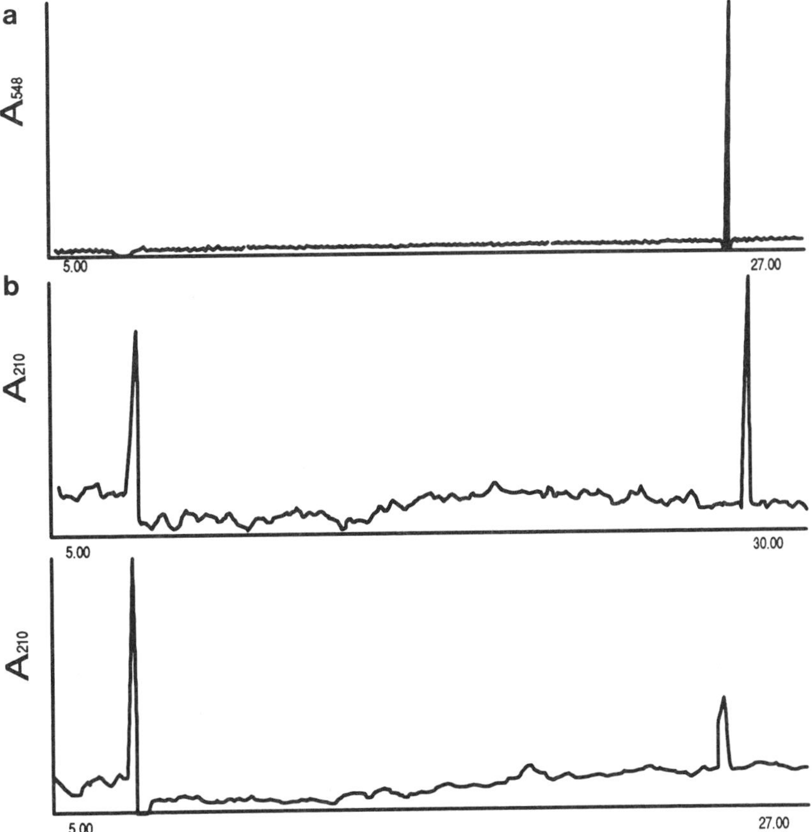

FIGURE 18. The biologically active species of bacteriorhodopsin have absorbance at 548 nm (retinal-bound forms) as well as in the far UV. Shown here in sequence is the 548-nm trace of the analysis of a preparation of bacteriorhodopsin (a) vs. the 210-nm trace (b). As the sample photobleaches the ratio of the retinal-bound form to the unretinylated form changes. Buffer, 10 m*M* OAC, pH 4, 18 m*M* CHAPSO; field, 200 V/cm; injection, 3 s at 200 V/cm; capillary, 92-mm effective length, 50-μm amino phase.

REFERENCES

1. **Virtanen, K.,** Zone electrophoresis in a narrow-bore tube employing potentiometric detection: a theoretical and experimental study, *Acta Polytech. Scand. Chem. Technol. Metall. Ser.*, 123, 1, 1974.
2. **Mikkers, F. E. P., Everaerts, F. M., and Verheggen, T. P. E. M.,** High performance zone electrophoresis, *J. Chromatogr.*, 169, 11, 1979.
3. **Jorgenson, J. W. and Lukacs, K. D.,** Zone elctrophoresis in open-tubular glass capillaries, *Anal. Chem.*, 53, 1298, 1981.
4. **Karger, B. L., Cohen, A. S., and Guttman, A.,** High performance capillary electrophoresis in the biological sciences, *J. Chromatogr.*, 492, 585, 1989.
5. **Goodall, D. M., Lloyd, D. K., and Williams, S. J.,** Current trends in capillary electrophoresis, *LC-GC*, 8(10), 788, 1990.
6. **Mazzeo, J. R. and Krull, I. S.,** Coated capillaries and additives for the separation of proteins by capillary zone electrophoresis and capillary isoelectric focusing, *BioTechniques*, 10(5), 638, 1991.
7. **Schomburg, G.,** Polymer coating of surfaces in column liquid chromatography and capillary electrophoresis, *Trends Anal. Chem.*, 10, 163, 1991.
8. **Swedberg, S. A.,** Characterization of protein behavior in high performance capillary electrophoresis using a novel capillary system, *Anal. Biochem.*, 185, 51, 1990.
9. **Huang, X., Coleman, W. F., and Zare, R. N.,** Analysis of factors causing peak broadening capillary zone electrophoresis, *J. Chromatogr.*, 480, 95, 1989.

10. **Jorgenson, J. W. and Lukacs, K. D.,** Capillary zone electrophoresis, *Science*, 222, 266, 1983.
11. **Lauer, H. H. and McManigill, D.,** Capillary zone electrophoresis of proteins in untreated fused silica tubing, *Anal. Chem.*, 58, 166, 1986.
12. **Hjertén, S.,** High performance electrophoresis: elimination of electroendosmosis and solute adsorption, *J. Chromatogr.*, 347, 191, 1985.
13. **Shaw, D. J.,** *Introduction to Colloid and Surface Chemistry*, 2nd ed., Butterworths, London, 1975, chap. 7.
14. **McManigill, D. and Swedberg, S. A.,** Factors affecting plate height in high performance zonal capillary electrophoresis, in *Techniques in Protein Chemistry*, Hugli, T., Ed., Academic Press, San Diego, 1989, p. 468.
15. **McCormick, R. M.,** Capillary zone electrophoretic separation of peptides and proteins using low pH buffers in modified silica capillaries, *Anal. Chem.*, 60, 2322, 1988.
16. **Unger, K. K.,** *Porous Silica*, Elsevier Scientific, Amsterdam, 1979, chap. 3.
17. **Cobb, K. A., Polnik, V., and Novotny, M.,** Electrophoretic separation of proteins in capillaries with hydrolytically stable surface structures, *Anal. Chem.*, 62, 2478, 1990.
18. **Bruin, G. J., Chang, J. P., Kuhlman, R. H., Zegers, K., Kraak, J. C., and Poppe, H.,** Capillary zone electrophoretic separations of proteins in polyethylene glycol-modified capillaries, *J. Chromatogr.*, 471, 429, 1989.
19. **Kohr, J. and Engelhardt, H.,** Capillary electrophoresis with surface coated capillaries, *J. Microcolumn Sep.*, 3, 491, 1991.
20. **Swedberg, S. A. and Herold, M.,** 1993, poster presentation, Protein Society Meeting.
21. **Maa, X.-H., Hyver, K. J., and Swedberg, S. A.,** Impact of wall modifications on protein elution in high performance capillary zone electrophoresis, *J. High Resolut. Chromatogr.*, 14, 65, 1991.
22. **Lee, C. S., Blanchard, W. C., and Wu, C.-T.,** Direct control of the electroosmosis in capillary zone electrophoresis by using an external field, *Anal. Chem.*, 62, 1550, 1990.
23. **Lee, C. S., McManigill, D., Wu, C.-T., and Patel, B.,** Factors affecting direct control of electroosmosis using an external electric field in capillary electrophoresis, *Anal. Chem.*, 63, 1519, 1991.
24. **Holzman, T. F., Egan, D. A., Edalji, R., Simmer, R. L., Helfrich, R., Taylor, A., and Burres, N. S.,** Preliminary characterization of a cloned neutral isoelectric form of the human peptidyl prolyl isomerase cyclophilin, *J. Biol. Chem.*, 266, 2474, 1991.
25. **Tran, A. D., Park, S., Lisi, P. J., Huynh, O. T., Ryall, R. R., and Lane, P. A.,** Separation of carbohydrate-mediated micro/heterogeneity of recombinant human erythropoietin by free solution capillary electrophoresis, *J. Chromatogr.*, 542, 459, 1991.
26. **Chen, F.-T. A.,** Rapid protein analysis by capillary electrophoresis, *J. Chromatogr.*, 559, 445, 1991.
27. **Chen, F.-T. A., Kelly, L., Palmeri, R., Biehler, R., and Schwartz, H.,** Use of high ionic strength buffers for the separation of proteins and peptides with capillary electrophoresis, *J. Liq. Chromatogr.*, 15, 1143, 1992.
28. **Ferranti, P., Malorni, A., Pucci, P., Fanali, S., Nardi, A., and Ossicini, L.,** Capillary zone electrophoresis and mass spectrometry for the characterization of genetic variants of human hemoglobin, *Anal. Biochem.*, 194, 1, 1991.
29. **Frenz, J., Wu, S.-L., and Hancock, W. S.,** Characterization of human growth hormone by capillary electrophoresis, *J. Chromatogr.*, 480, 379, 1989.
30. **Wu, S.-L., Teshima, G., Cacia, J., and Hancock, W. S.,** Use of high-performance capillary electrophoresis to monitor charge heterogeneity in recombinant-DNA derived proteins, *J. Chromatogr.*, 516, 115, 1990.
31. **Kilar, F. and Hjertén, S.,** Fast and high-resolution analysis of human serum transferrin by high-performance isoelectric focusing in capillaries, *Electrophoresis*, 10, 23, 1989.
32. **Miercke, L.,** personal communication.

Chapter 20

EFFECTS OF SAMPLE MATRIX ON SEPARATION BY CAPILLARY ELECTROPHORESIS

Zak K. Shihabi and L. Liliana Garcia

TABLE OF CONTENTS

0-8493-8690-X/94/$0.00+$.50

I. INTRODUCTION

The main appeal of capillary electrophoresis (CE) for routine analysis is the simplicity of sample treatment. In many instances, no sample pretreatment is necessary other than dilution. Although the sample constitutes a very small portion of the overall volume in the capillary once injected (<1%), the matrix (composition) of the sample has profound effects on the separation by CE. These effects are caused by the nature of the current conductance, where any slight change or difference in the ionic strength along the path of the current will have a significant effect on the electroosmotic flow (EOF), the migration of the ions, the bandwidth, and, consequently, on the overall separation. This phenomenon is unlike high-performance liquid chromatography (HPLC), where the sample matrix has little or no effect on the separation.[1]

Samples obtained from biological fluids and industrial sources often have a complex matrix. They may contain many compounds at concentrations several thousand times greater than the substances of interest. For example, theophylline, a drug commonly assayed in clinical laboratories, is present in serum at 10 mg/l in the presence of sodium at 10,000 mg/l and proteins at 100,000 mg/l. Such a complex matrix requires special maneuvers for a successful separation by CE. Complicating matters further, CE has a relatively low sensitivity, due to the short light pathway of the capillary, creating a need for clever manipulations of the sample to enhance the detection, similar to sample enrichment used in HPLC.

The effects of the sample matrix on CE separations have been evident since the birth of this technique with the early work of Mikkers et al.[2] These authors noticed that the theoretical plate numbers, which represent the separation efficiency, were much higher if the samples were dissolved in water compared to the electrophoresis buffer. In this chapter, the different factors related to the sample that can ultimately affect the separation, from both an efficiency and resolution perspective, are discussed in detail. These include ionic strength, the presence of proteins, pH, viscosity, and volume. In addition, how these effects can be manipulated to improve the (1) resolution, (2) signal, and (3) precision will be addressed.

II. MATRIX EFFECTS AND SEPARATION

Several factors related to the sample affect the separation in CE. Occasionally, in quest of speed or higher sensitivity, some of these factors (e.g., a large sample volume and/or high ionic strength) may be used in conjunction with inappropriate or suboptimal separation conditions such as a low ionic strength running buffer, a short capillary, or high voltage. The combination of such unfavorable conditions can yield a poor separation with unacceptable resolution, as demonstrated in Figure 1B.

A. SAMPLE IONIC STRENGTH

While high ionic strength buffers are good environments for proteins from a solubility perspective, they are deleterious as sample matrices for CE separations (see Chapter 2, Figure 6B). The most significant disadvantage of such conditions is the decreased "stacking" (described in Section II.F) that results. An excess of ions in the sample, especially with large sample volumes, is detrimental to the separation in CE (Figure 1B). Moreover, when using the electrokinetic injection mode with high ionic strength samples, as the sample is introduced, the excess ions in the sample are also introduced. This leads to less sample being introduced and, consequently, to a decrease in detector signal. When using the pressure injection mode, a high ionic strength sample matrix can lead to band broadening or multiple peaks (Figure 1B). Demorest and Dubrow[3] found that with a gel-filled capillary, the peak area decreased with an increase in the ionic strength of the sample. Later in this chapter (Section III) several approaches are described to overcome the deteriorating effects of salts in the sample.

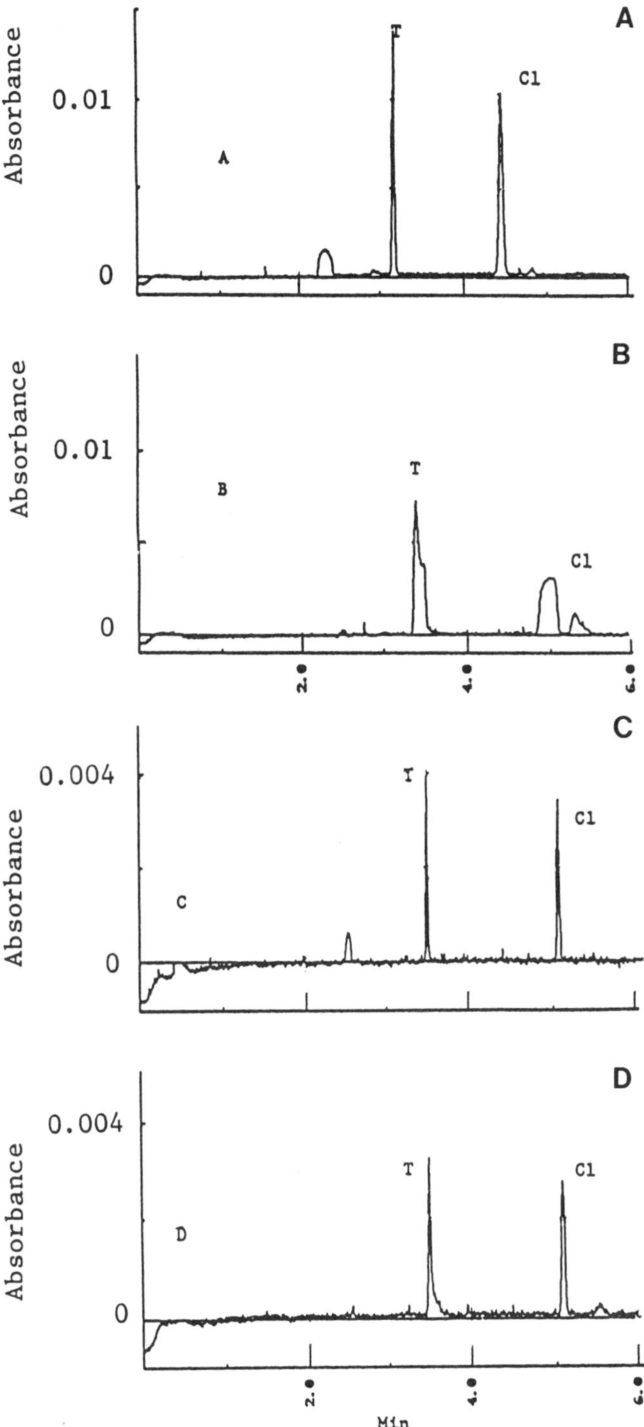

FIGURE 1. Effect of sample ionic strength and volume on the separation. Conditions: Beckman (Fullerton, CA) instrument equipped with a 50 μm × 40 cm capillary at 400 V/cm, 280 nm, 24°C, pressure injection, and a 300-mmol/l borate (pH 8.5) running buffer. (A) Theophylline (T; 20 mg/l) and chlorotheophylline (Cl; 20 mg/l) sample prepared in 30-mmol/l borate buffer with 15-s pressure injection. (B) Same conditions as in (A), but sample prepared in a 300-mmol/l borate buffer. (C) Same conditions as in (A), but sample injected for 3 s. (D) Same conditions as in (B), but sample injected for 3 s.

B. PROTEINS

At low concentrations and under appropriate conditions, proteins have little effect on CE separation. However, at high concentrations in the sample, proteins could mask the absorption of the compounds under study, affect their mobility by binding small molecules, and affect the viscosity of the sample. The most significant effect of proteins is to change the characteristics of the capillary wall. Cationic proteins tend to adsorb preferentially to the inlet side of untreated capillary wall, changing the zeta potential (ζ), which in turn affects the electroosmotic flow (v_{eo}):

$$v_{eo} = Ee\zeta/\eta$$

where E is the electric field strength, e is the fluid permittivity, and η is the viscosity.

Variations in the EOF along the capillary length cause an increase in band spreading and a decrease in peak symmetry.[4] When 2% of the capillary length is fouled with adsorbed proteins, the efficiency of the column decreases by 50%.[4] Since protein adsorption is a continuous process with each run, the reproducibility of the assay is affected greatly, as is discussed in Section V. This effect can be minimized by using either coated capillaries or micellar electrokinetic capillary chromatography (MECC). With the latter method, sodium dodecyl sulfate (SDS) enhances the solubility of the proteins.

C. pH

Changes in the sample pH will obviously affect the ionization of the sample components, and hence, their net charge. As a result, the migration rate and the solubility of many of the components change; the theoretical plate number, peak height, and peak area could also be affected. Sample concentration can be obtained through pH changes in the sample relative to the electrophoresis buffer[5] as is discussed in Section II.F. Also see Chapters 2, 5, and 12 of this book for more details on capillary preconcentration strategies.

D. VISCOSITY

As the viscosity of the electrophoresis buffer increases, the band diffusion decreases, leading to higher theoretical plate numbers and peak height. However, studies in this laboratory show that when the sample viscosity is increased, e.g., by addition of glycerol, the theoretical plate numbers decrease. The reasons for this observation are not clear; this is an area that certainly requires further study.

E. SAMPLE VOLUME

Unlike HPLC, the sample volume in CE contributes largely to the overall band diffusion. Because of the relatively low sensitivity of detection in CE, a high sample volume is desirable to increase the signal. Unfortunately, sample overloading can occur easily. This overloading occurs as a result of the inherent limitations of (1) the total sample volume injected (q_s) relative to the total volume of the capillary (q_c), which limits the maximum plate number (N_{max}):

$$N_{max} = 12(q_c/q_s)^2$$

and (2) the differences in the electrical conductivity between the sample and the running buffer (discussed in Section II.F).[6]

The theoretical plate numbers in CE depend greatly on the sample plug length (volume). Using β-hGH (biosynthetic human growth hormone) as a model compound, the theoretical plate numbers dropped from 800,000 to less than 10,000 by increasing the injection time from 0.2 to 15 s, with the highest theoretical plate number obtained at a sample volume close to zero.[7] An empirical formula to estimate the theoretical plate number relative to sample length is as follows:[8]

$$N = 12(L)^2/(W)^2$$

where L is the capillary distance from injection to detector and W is the injection plug length. The maximum allowable injection plug (W) can be expressed as[9]

$$W = (24DHt)^{1/2}$$

where D is the diffusion coefficient of the solute, H is the acceptable increase in plate height (defined as the length of the capillary divided by the theoretical plate number), and t is the migration time.

From the previous discussion, selecting an appropriate sample volume in CE is difficult. Low sample volumes offer very high theoretical plate numbers (Figure 1C and D) but at the same time they yield small detector signals. Consequently, the sample volume to be injected on the capillary is a tradeoff between the theoretical plate number and sensitivity. To obtain a minimum plate height, under nonstacking conditions,[9-11] the width of the band should be kept <1 mm,[8,10] which corresponds to an injection length of about 0.2% of the capillary length (to detector), i.e., 2 nl for a 50 μm × 50 cm capillary. Fortunately, increasing the sample volume fivefold to 1% of the capillary volume decreases the theoretical plate number only to approximately half. Hence, the rule of thumb is to keep the sample plug length less than 1% of the capillary length.

From a practical point of view, a simple plot of the injection time (or volume) vs. the peak height is very helpful in selecting the best injection time (or volume) for the separation. Such a plot is highly recommended when the method is initiated. Thereafter, peak height or peak area can be used for quantitation. In some instances, the peak area offers better reproducibility and linearity.[12]

F. STACKING AND FIELD-AMPLIFIED INJECTION

In polyacrylamide gel electrophoresis a large volume of sample can be preloaded on a small segment of a stacking gel with wide pores and/or with ions of different mobilities than that of the separating gel.[13] This causes a rapid migration of all of the molecules as a sharp band at the top of the separating gel, a phenomenon termed *stacking*. A similar effect can be induced in CE by preparing the sample in the same electrophoresis buffer but at a lower concentration. This causes the sample resistance and the field strength (in volts per centimeter) in the sample plug to increase, which in turn causes the ions to migrate rapidly and stack as a sharp band at the boundary between the sample plug and the electrophoresis buffer, with the positive ions lining up in front of the negative ones. Once in the electrophoresis buffer, the components of the sample migrate in different zones according to their charge/mass characteristics. As a result of this simple manipulation, the sample can be concentrated up to tenfold.

The amount of on-line sample stacking (or concentration) is proportional to the electrophoresis-to-sample buffer concentration ratio.[14] Theoretically, samples prepared in water should give the highest degree of stacking. However, since electroosmosis occurs much more rapidly in the diluted sample than the electrophoresis buffer, the mismatch in the flow rate causes a laminar flow inside the capillary that reduces the sharpness of the stacking process.[14,15] In addition, as a result of the high field effects, excessive heat can also be generated in the sample plug, causing band broadening[16] and thermal degradation of some components.[17]

Several investigators have used various mathematical models to describe stacking and band broadening during this process.[7,14,15] The stacking force, which is the difference between the field strength of the sample and that of the buffer, can be expressed as[7]

$$\text{Field strength} = (I/A)[(1/K_s) - (1/K_b)]$$

where I is the current, A is the capillary cross-sectional area, K_s is the specific conductivity of the sample, and K_b is the specific conductivity of the buffer.

The stacking effect can be utilized with both pressure (Figure 1A) and electrokinetic injections (field amplification injection). On the basis of this, higher theoretical plate numbers can be obtained if the sample buffer concentration is approximately tenfold lower than that of the electrophoresis buffer, the sample plug length is up tenfold the diffusion-limited peak width,[15] and a low potential is applied during the stacking process.[6,7] A less concentrated buffer or a longer sample plug will increase the laminar flow.

Chien and Burgi[18] have described a method of stacking by injecting a plug of water into the capillary before injecting the sample. They also have described another technique,[19] which further increases the sample volume for stacking by immediately reversing the polarity for a short period of time after the sample has been introduced into the capillary. In this case, the sample ions (i.e., the sample buffer), which can affect the separation, are removed first.

Stacking based on adjusting the pH of the sample has been described by Aebersold and Morrison.[5] They have concentrated peptides by dissolving the sample in a buffer 2 units above the net pI, so the peptides are negatively charged. As the potential is turned on, the peptides initially migrate toward the anode until they are stopped by the interface of the electrophoresis buffer, where they concentrate. After the short pH gradient of the sample dissipates in the electrophoresis buffer, the peptides become positively charged and migrate toward the cathode as a sharp zone. Using this method, a larger volume was introduced into the capillary, obtaining a fivefold concentration. There is a need, still, for other methods of stacking to concentrate the sample.

III. SAMPLE CLEAN-UP

A. DILUTION AND DIRECT SAMPLE INJECTION

As mentioned earlier, one of the main advantages of CE for routine analysis in industrial and clinical settings is the simplicity of sample introduction. In general, if the compound of interest has a strong absorptivity and is present in a high concentration relative to the interfering compounds, it may be detected easily without treatment. When the compound of interest is present in low concentrations in the presence of high concentrations of undesirable compounds, such as in samples obtained from serum, food, or industrial sources, a clean-up step(s) becomes necessary. Because of the matrix effects that have already been discussed, differences in migration time and peak height between the aqueous standards and the samples are commonly observed. Consequently, it is important in CE either to prepare the standards in the same matrix as that of the sample or to add the standards directly to the samples.

Several drugs, such as theophylline, caffeine,[20] and barbiturates,[21] have been successfully analyzed with direct serum injection by MECC. In this technique, the micelles solubilize the proteins. When standards were added directly to serum and the peak identity confirmed by spectral scanning, the results compared well with immunoassays. The simplicity and the low sample volume necessary for the assay render this method suitable for monitoring the levels of these drugs in pediatrics.

Several compounds, such as creatinine, uric acid, and phosphorus, have been analyzed by direct urine injection without pretreatment.[22,23] Several anions in urine, such as citrate, oxalate, and sulfate, have also been analyzed after appropriate dilution of the urine and optimization of CE conditions for each ion.[24] Several organic acids in food samples have been measured by indirect ultraviolet (UV) absorption after simple dilution.[25]

It is important to note that dilution does not clean up the sample, but it reduces the ionic strength of the matrix. Unfortunately, it also decreases the sample concentration, and thus the detector signal. To tolerate a higher ion concentration in the sample, the ionic strength of the buffer should be as high as possible so that the ratio of the electrophoresis/sample concentration is maintained at approximately ten for stacking.[26] For example, in order to analyze different drugs found in serum such as suramin,[12] pentobarbital,[27] iohexol,[28] a borate buffer of 200 to 500 mmol/l was used. The disadvantage of using such buffers, is that the high ionic strength leads to slower separations as a result of decreasing electroosmotic flow. The longer run times that result can be circumvented by using short capillaries (e.g., 25 cm) with narrow diameters (e.g., 25 or 50 μm). Such capillaries can also dissipate the generated Joule heat faster. As would be expected, the use of high buffer concentrations produces peaks with higher theoretical plate numbers, better resolution, and allows a larger amount of sample to be injected due to the enhanced "stacking effect." Vinther and Soeberg[29] found that high ionic strength buffers also decrease peak tailing due to a decrease in the radial pH gradients in the capillary. High-strength buffers used with narrow capillaries (25 μm) gave good protein separation with little or no protein adsorption and allowed the sample to focus.[30]

B. EXTRACTION

Traditional solvent extraction procedures are often used for the clean-up of small molecules, especially in gas chromatography (GC) and HPLC. Solid-phase methods, in particular the use of C_{18} cartridges, have become very popular in comparison to solvent extraction in the last decade because of the versatility of the methods and the wide choice of cartridge packings. Furthermore, the fractions from the HPLC can be used as clean-up for CE. In general, these methods are time consuming and require some skill, making them undesirable for routine assays. The different manufacturers of the solid-phase columns offer many helpful booklets, guidelines, and references on how to use these columns.

C. FILTRATION AND DIALYSIS

Large molecules can be separated from the small ones through special dialysis and filtration membranes, which are available with different molecular weight cut-off points. However, the use of small commercial dialysis cells or blocks (<500 μl) are more suitable for CE than the traditional bags. Under proper conditions, dialysis can be accomplished in 1 h.[31] Both sides of the dialysis chambers can be used for CE analysis based on whether the compound of interest has a high or low molecular weight. Small volumes of approximately 200 μl can also be filtered rapidly (1 to 7 min) through special filtration membranes in a microfuge at 15,000 *g*. Both of these techniques have been applied to the clean-up of urine proteins for analysis by CE. Figure 2 shows the electropherograms for a dialyzed and undialyzed urine sample. Addition of a concentrated solution of a high molecular weight compound, such as 20% polyethylene glycol 6000, to one side of the dialysis cell, can be used to concentrate proteins.

D. ORGANIC SOLVENT DEPROTEINIZATION

Acetonitrile has often been used in HPLC to remove serum proteins by addition of equal volumes of sample and acetonitrile. Experiments in this laboratory show that this ratio is not adequate to completely remove serum proteins for CE and that an acetonitrile:serum ratio of 3:2 (v/v) completely removes serum proteins.[28] In addition to removing proteins, the presence of acetonitrile in the sample for CE improves the solubility of some compounds and increases the resistance of the sample leading to stacking (Figure 3), thus allowing a larger volume of sample to be injected on the capillary.

1. A Method for Acetonitrile Deproteinization

Acetonitrile deproteinization is a very simple procedure. In practice 100 μl of serum is vortex mixed with 150 μl of acetonitrile (containing an internal standard) for 10 s, centrifuged in a microfuge at 15,000 *g* for 30 s and introduced into the capillary.

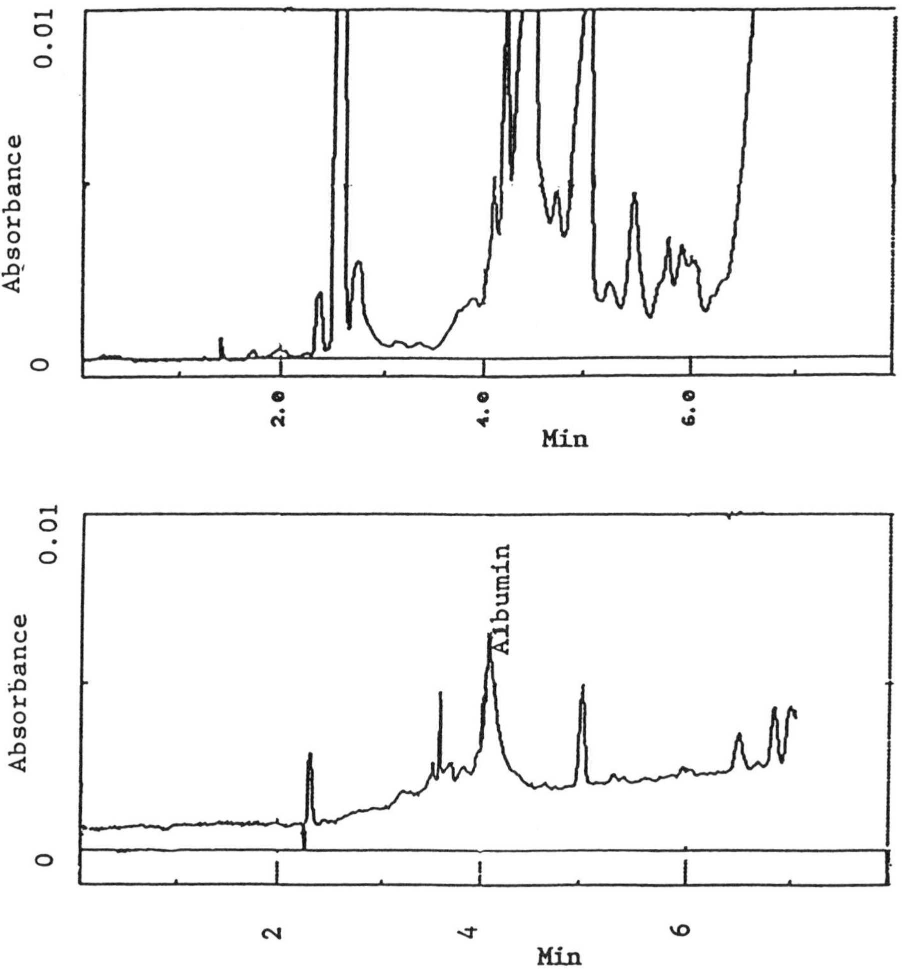

FIGURE 2. Effect of dialysis clean-up on urinary proteins. Conditions: Beckman instrument equipped with a 50 μm × 25 cm capillary at 250 V/cm, 214 nm, 24°C, 6-s pressure injection, and Beckman urine CE protein running buffer. *Top*: Undialyzed urine. *Bottom*: Dialyzed urine.

In this laboratory, several compounds, e.g., suramin,[12] pentobarbital,[27] and iohexol,[28] have been quantitated by CE after acetonitrile deproteinization. Ackerman et al.[32] determined 16 sulfonamides in meat after extraction with acetonitrile. Protein removal can also be accomplished by alcohols such as ethanol and methanol, and by acids such as perchloric and trichloroacetic acid. Precipitation with acids is not as desirable since it leads to the formation of an excess of salts. One of the advantages of alcohol precipitation is that proteins could also be directly dissolved in the appropriate buffers following precipitation and assayed by CE. For example, the globin chains of hemoglobin have been determined following acetone precipitation.[33]

E. DESALTING

Desalting is a difficult procedure to perform in routine assays. There are several methods for desalting, such as (1) dialysis as described above, (2) ion exchangers,[34] e.g., Chelex 100 and AG 50X2, and (3) reversing the polarity during CE as described in Section II.F.[18]

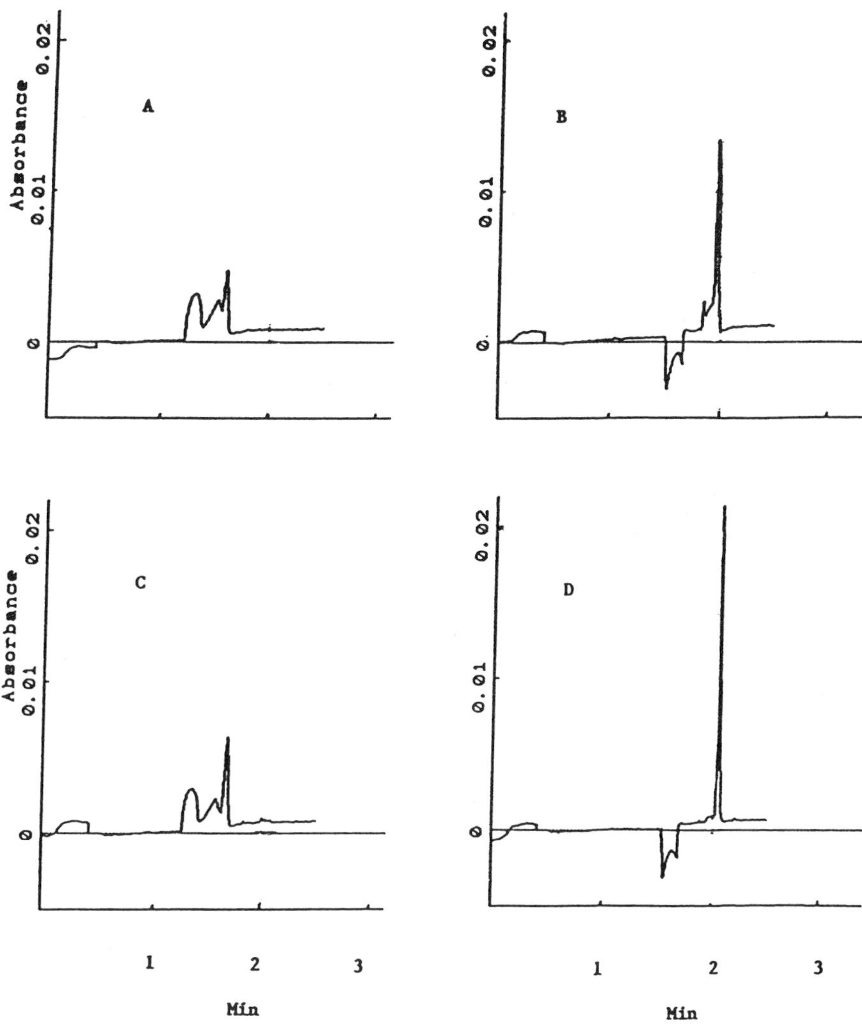

FIGURE 3. Effect of acetonitrile on sample stacking. Conditions: Beckman instrument equipped with a 50 μm × 25 cm capillary at 480 V/cm, 254 nm, 24°C, 10-s pressure injection, and 300-mmol/l borate (pH 8.5) running buffer. (A) Iohexol (16 mg/l) in water. (B) Iohexol in 60% acetonitrile. (C) Iohexol in 30-mmol/l borate buffer. (D) Iohexol in 60% acetonitrile and NaCl (60 mmol/l).

In general, the use of high ionic strength running buffers enables the direct analysis of samples containing a high salt concentration, eliminating in many instances the need for desalting.[25]

IV. ENHANCEMENT OF DETECTION

Because the light pathway in CE is relatively small, the detection limit in many cases is not adequate. Many methods for enhancing detection are described throughout this book (see Chapter 5), such as the use of sensitive detectors, using an antibody bound to a solid support, or chemical derivatization. A brief description of how the sample itself can be easily manipulated to enhance the detection is briefly presented.

A. STACKING AND FIELD-AMPLIFIED INJECTION

Stacking, as described earlier (Section II.F), is a simple and very common method for increasing the sensitivity of detection.

B. ACETONITRILE

The addition of acetonitrile to serum for deproteinization obviously represents a dilution of the sample. However, it has been observed[28] that under these conditions many compounds have a higher peak height and theoretical plate number than when dissolved in diluted buffers (Figure 3B). This can be explained by the high resistance of acetonitrile in the sample producing a stacking effect, which is slightly enhanced by the presence of serum ions (Figure 3D). This effect is more pronounced with short capillaries (<25 cm) and more dramatic with some compounds than with others. The exact mechanism involved in this phenomenon is not clear but deserves further study to investigate how the combination of salts, pH, voltage, and other factors enhances stacking.

C. PROTEIN PRECIPITATION

Some proteins can be precipitated with alcohols or acids and dissolved in small volumes of buffer to concentrate them with little or no denaturation, as described earlier. To quantitatively precipitate a protein at low concentration, another similar protein but at a higher concentration can be added, as a "carrier," to the sample. This will allow a few milligrams per liter to be precipitated and consequently concentrated. For example, urinary transferrin has been determined at levels of 1 to 5 mg/l by an immunoassay based on precipitation in the presence of added albumin.[35]

D. CAPILLARY ISOELECTRIC FOCUSING

In capillary isoelectric focusing the sample can be concentrated up to 100-fold, as is discussed in detail in Chapter 4 of this book.

E. TRANSIENT ISOTACHOPHORESIS

In many situations, the sample contains large amounts of salts that may interfere in the separation. By choosing the appropriate conditions and ions in the electrophoresis buffer and/or in the sample, a 50-fold increase in sensitivity can be obtained through a transient period of isotachophoresis during CE. In isotachophoresis, the sample is sandwiched between a fast leading ion and a slow terminating ion, while the sample components move as segments between these ions. Recently, Foret et al.[36] and others[37] have described several simplified approaches to perform a brief or transient (1 to 5 min) ITP before the separation conditions are switched to CE. This separation is termed *on-column transient ITP*.[36] For more details see Chapter 5.

V. SAMPLE MATRIX AND PRECISION

For implementing any analytical method into routine assays, precision is a very important consideration. Unfortunately, the effect of sample matrix on precision in CE has not been well documented. In general, when samples of biological origin are injected without extraction, the precision of the assay is poor. High protein content in samples, such as that found in serum, causes protein adsorption on the capillary walls, thus changing the EOF.[4] The continuous injection of such samples leads to a constant increase in migration time and band broadening. However, when the migration time is corrected based on an internal standard, the corrected standard deviation improves dramatically since the migration of the internal standard itself increases accordingly.

Whenever serum or unextracted samples are injected directly, a thorough wash of the capillary after each sample is necessary with either NaOH (0.1 *M*), phosphoric acid (0.1 *M*), or both[28,38] and an occasional or daily long wash of 15 to 30 min. These thorough washings are not necessary if the sample is extracted or has a clean matrix.[28] Chen et al.[30] found that high-strength buffers give good precision for migration (<1% relative standard deviation [RSD]) and peak area (<2.5%) because of the low protein adsorption to the uncoated capillary walls. In general, the precision in CE depends largely on having a clean capillary before each sample.[39]

VI. FUTURE PROSPECTS

This chapter has attempted to highlight the important role of the sample matrix in obtaining efficient, high-resolution separations with CE. Because of the relatively short light path in CE, sample concentration by different methods of stacking, such as by the use of lower ionic strength buffers, introduction of small volumes of water ahead of the sample, different sample pH or the addition of acetonitrile to the sample, is necessary. These methods can increase the sensitivity up to tenfold. However, devising other methods of stacking remains a challenge for the analytical chemist.

The use of high ionic strength running buffers is helpful in the separation of samples with excess concentration of salts. Other simple methods that can decrease the deleterious effect of salts in the samples would be welcomed. The addition of different compounds to the sample to increase the sensitivity and decrease the deleterious effects of salts, e.g., addition of special ions to combine CE and isotachophoresis in the same run, is worthy of further investigation.

REFERENCES

1. **Shihabi, Z. K.,** Review of drug analysis with direct serum injection on the HPLC column, *J. Liq. Chromatogr.*, 11, 1579, 1988.
2. **Mikkers, F. E. P., Everaerts, F. M., and Verheggen, Th. P. E. M.,** High-performance zone electrophoresis, *J. Chromatogr.*, 169, 11, 1979.
3. **Demorest, D. and Dubrow, R.,** Factors influencing the resolution and quantitation of oligonucleotides separated by capillary electrophoresis on a gel-filled capillary, *J. Chromatogr.*, 559, 43, 1991.
4. **Towns, J. K. and Regnier, F. E.,** Impact of polycation adsorption on efficiency and electroosmotically driven transport in capillary electrophoresis, *Anal. Chem.*, 64, 2473, 1992.
5. **Aebersold, R. and Morrison, H. D.,** Analysis of dilute peptide samples by capillary zone electrophoresis, *J. Chromatogr.*, 516, 79, 1990.
6. **Li, S. F. Y.,** Capillary electrophoresis. Principles, practice and applications, in *Journal of Chromatography Library*, Vol. 52, Elsevier Science, New York, 1992, p. 31–33.
7. **Vinther, A. and Soeberg, H.,** Mathematical model describing dispersion in free solution capillary electrophoresis under stacking conditions., *J. Chromatogr.*, 559, 3, 1991.
8. **Huang, X., Coleman, W. F., and Zare, R. N.,** Analysis of factors causing peak broadening in capillary zone electrophoresis, *J. Chromatogr.*, 480, 95, 1989.
9. **Grushka, E. and McCormick, R. M.,** Zone broadening due to sample injection in capillary zone electrophoresis, *J. Chromatogr.*, 471, 421, 1989.
10. **Terabe, S., Otsuka, K., and Ando, T.,** Band broadening in electrokinetic chromatography with micellar solutions and open-tubular capillaries, *Anal. Chem.*, 61, 251, 1989.
11. **Delinger, S. L. and Davis, J. M.,** Influence of analyte plug width on plate number in capillary electrophoresis, *Anal. Chem.*, 64, 1947, 1992.
12. **Garcia, L. L. and Shihabi, Z. K.,** Suramin determination by capillary electrophoresis, *J. Liq. Chromatogr.*, 16, 2049, 1993.
13. **Hames, B. D. and Rickwood, D.,** *Gel Electrophoresis of Proteins: A Practical Approach*, IRL Press, Washington D.C., 1981, pp. 7–10.

14. **Chien, R.-L. and Helmer, J. C.,** Electroosmotic properties and peak broadening in field-amplified capillary electrophoresis, *Anal. Chem.*, 63, 1354, 1991.
15. **Burgi, D. S. and Chien, R.-L.,** Optimization in sample stacking for high-performance capillary electrophoresis, *Anal. Chem.*, 63, 2042, 1991.
16. **Vinther, A. and Soeberg, H.,** Temperature elevations of the sample zone in free solution capillary electrophoresis under stacking conditions, *J. Chromatogr.*, 559, 27, 1991.
17. **Vinther, A., Soeberg, H., Nielsen, L., Pedersen, J., and Biedermann, K.,** Thermal degradation of a thermolabile *Serratia marcescens* nuclease using capillary electrophoresis with stacking conditions, *Anal. Chem.*, 64, 187, 1992.
18. **Chien, R.-L. and Burgi, D. S.,** Field amplified sample injection in high-performance capillary electrophoresis, *J. Chromatogr.*, 559, 141, 1991.
19. **Chien, R.-L. and Burgi, D. S.,** Sample stacking of an extremely large injection volume in high-performance capillary electrophoresis, *Anal. Chem.*, 64, 1046, 1992.
20. **Thormann, W., Minger, A., Molteni, S., Caslavska, J., and Gebauer, P.,** Determination of substituted purines in body fluids by micellar electrokinetic capillary chromatography with direct sample injection, *J. Chromatogr.*, 593, 275, 1992.
21. **Thormann, W., Meier, P., Marcolli, C., and Binder, F.,** Analysis of barbiturates in human serum and urine by high-performance capillary electrophoresis micellar electrokinetic capillary chromatography with on-column multi-wavelength detection, *J. Chromatogr.*, 545, 445, 1991.
22. **Guzman, N. A., Berck, C. M., Hernandez, L., and Advis, J. P.,** Capillary electrophoresis as a diagnostic tool: determination of biological constituents present in urine of normal and pathological individuals, *J. Liq. Chromatogr.*, 13, 3833, 1990.
23. **Guzman, N. A., Hernandez, L., and Advis, J. P.,** Effect of low temperature storage (on collected urine specimens) in the identification and quantitation of urinary metabolites by capillary electrophoresis, *J. Liq. Chromatogr.*, 12, 2563, 1989.
24. **Wildman, B. J., Jackson, P. E., Jones, W. R., and Alden, P. G.,** Analysis of anion constituents of urine by inorganic capillary electrophoresis, *J. Chromatogr.*, 546, 459, 1991.
25. **Kenney, F.,** Determination of organic acids in food samples by capillary electrophoresis, *J. Chromatogr.*, 546, 423, 1991.
26. **Garcia, L. L. and Shihabi, Z. K.,** Sample matrix effects in capillary electrophoresis. I. Basic considerations, *J. Chromatogr.*, 1993, in press.
27. **Shihabi, Z. K.,** Serum pentobarbital determination by capillary electrophoresis, *J. Liq. Chromatogr.*, 16, 2059, 1993.
28. **Shihabi, Z. K. and Constantinescu, M. S.,** Iohexol in serum determined by capillary electrophoresis, *Clin. Chem.*, 38, 2117, 1992.
29. **Vinther, A. and Soeberg, H.,** Radial pH distribution during capillary electrophoresis with electroosmotic flow. Analysis with high ionic strength buffers, *J. Chromatogr.*, 589, 315, 1992.
30. **Chen, F. A., Kelly, L., Palmieri, R., Biehler, R., and Shwartz, H.,** Use of high ionic strength buffers for the separation of proteins and peptides with capillary electrophoresis, *J. Liq. Chromatogr.*, 15, 1143, 1992.
31. **Shihabi, Z. K., Oles, K. S., McCormick, C. P., and Penry, J. K.,** Serum and tissue carnitine assay based on dialysis, *Clin. Chem.*, 38, 1414, 1992.
32. **Ackermans, M. T., Beckers, J. L., Everaerts, F. M., Hoogland, H., and Tomassen, M. J. H.,** Determination of sulfonamides in pork meat extracts by capillary zone electrophoresis, *J. Chromatogr.*, 596, 101, 1992.
33. **Ferranti, P., Malorni, A., Pucci, P., Fanali, S., Nardi, A., and Ossicini, L.,** Capillary zone electrophoresis and mass spectrometry for the characterization of genetic variants of human hemoglobin, *Anal. Biochem.*, 194, 1, 1991.
34. **Everaerts, F. M., Lemmens, A. A. G., Verheggen, Th. P. E. M., and Ogan, K.,** Recent developments in capillary isotachophoresis, in *Electrophoresis '86*, Dunn, M. J., Ed., VCH, Weinheim, Germany, 1986, pp. 117–131.
35. **McCormick, C. P., Shihabi, Z. K., and Konen, J. C.,** Microtransferrinuria and microalbuminuria: enhanced immunoassay, *Ann. Clin. Lab. Science*, 19, 444, 1989.
36. **Foret, F., Szoko, E., and Karger, B. L.,** On-column transient and coupled column isotachophoretic preconcentration of protein samples in capillary zone electrophoresis, *J. Chromatogr.*, 608, 3, 1992.
37. **Gebauer, P., Thormann, W., and Bocek, P.,** Sample self-stacking in zone electrophoresis. Theoretical description of the zone electrophoretic separation of minor compounds in the presence of bulk amounts of sample component with high mobility and like charge, *J. Chromatogr.*, 608, 47, 1992.
38. **Zhu, M., Rodriguez, R., Hansen, D., and Wehr, T.,** Capillary electrophoresis of proteins under alkaline conditions, *J. Chromatogr.*, 516, 123, 1990.
39. **Smith, S. C., Strasters, J. K., and Khaledi, M. G.,** Influence of operating parameters on reproducibility in capillary electrophoresis, *J. Chromatogr.*, 559, 57, 1991.

Chapter 21

TEMPERATURE CONTROL IN CAPILLARY ELECTROPHORESIS

Robert J. Nelson and Dean S. Burgi

TABLE OF CONTENTS

0-8493-8690-X/94/$0.00+$.50

I. INTRODUCTION

The measurement and/or control of temperature in many instrumental systems is often essential for a meaningful and reproducible result. Whereas liquid chromatography (LC) is frequently performed without the need to measure or optimize column temperature, the control of the buffer temperature in capillary electrophoresis (CE) is often of critical importance to success. The temperature of the buffer in the capillary affects most of the important parameters in the separation, and, therefore, even slight changes in buffer temperature can cause significant deviations in the results. This highlights the importance of understanding the factors that influence column temperature in CE.

Early work by Hjertén[1] described the basic theory of heat generation and dissipation in capillaries that were markedly larger than those typically used in CE today.[2-4] Other, more recent and varied work has extended the theory to narrower capillaries and different types of cooling systems.[5-16,41] Several researchers have shown examples of the importance to control column temperature since lack of control can lead to problems with injections,[17] sample stability,[5,17-18] quantitation,[12,19-21] and band broadening.[41] We suggest referring to these excellent publications for a more detailed discussion of the theory and derivation of the relationships shown below, since the focus here is on practical applications of the theory.

When considering the thermal aspects of a capillary electrophoresis experiment, the system can be simplified by focusing on three main characteristics: the ambient column temperature, heat generation rate, and heat dissipation characteristics of the cooling design. One can then optimize those parameters that are controllable, while having an understanding of the magnitude of the temperature difference between the buffer and the set temperature of the column thermostat.

II. HEAT GENERATION

A. THEORY

Several comprehensive works exist that detail the theory of heat generation and dissipation in capillary electrophoresis (CE).[1,5,7,8,10-13] Bello and Righetti[8] have developed a detailed and accurate computer model that estimates heat generation rates and buffer temperature given buffer conductivity and column length, inner diameter, and applied field. One can vary experimental parameters while the software accurately predicts the effect on column performance.

The heat that is generated is derived from the application of voltage to the system and the resulting electrical current flow. The amount of heat can be determined by casual observation of the voltage and current relationship displayed on most high-voltage power supplies and commercial instruments. The voltage can be thought of as the pressure applied to the ions to cause them to flow through the solution. The electrical current is simply the number of charges flowing past an observer located in the column, in coulombs per second, and can be thought of as the flow rate (assuming constant buffer conditions!). By utilizing the Ohm's law relationship, the applied power can be estimated as follows:

$$\text{Milliwatts of applied power} = (\text{microamps} \times \text{kilovolts}) \text{ or watts} = \text{volts} \times \text{amps} \quad (1)$$

For this example, if a system is running at 30 μA and 30 kV, the applied power would be 900 mW or 0.9 W. An Ohm's law plot, discussed briefly in Chapter 2, Section II.B, the relationship between voltage and current, can be easily and quickly constructed for a given buffer system. The data are generated by setting the applied voltage at 1- or 2-kV increments and observing the resulting current after stabilization (usually 1 min or so, depending on the type of cooling system and current level). A plot similar to that in Figure 1 can be constructed that is a characteristic of a CE system.

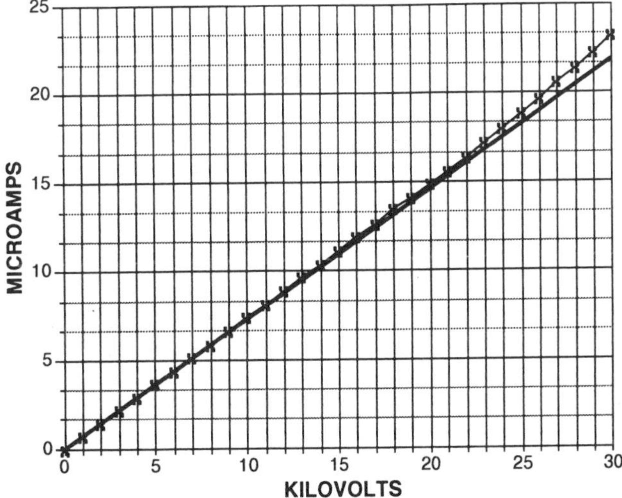

FIGURE 1. Ohm's law plot on the Spectra*PHORESIS* CE1000 (Thermo Separation Products, Fremont, CA) with forced air convection cooling, fresh 20 m*M* borate buffer (pH 9.0), 75 μm × 70 cm capillary column, at 25°C. The positive deviation of the column data × relative to an ideal resistor (solid line) is due to the temperature gradient between the column bore and the air temperature around the capillary. This gradient must develop to dissipate the ohmic heat. The lower the thermal conductivity of the cooling medium, or the slower the air velocity, the greater the positive deviation.

The straight line in Figure 1 represents an ideal resistor with no influence due to variance in temperature. This line is an extrapolation from the electrical current data at 1 kV and zero. The curved line shows the influence of the buffer inside the column warming due to the applied power. This warming occurs above the ambient temperature of the cooling system, the extent of which is determined by the type and design of cooling system, as is discussed in Section V.

Various problems can result from the application of more than a few watts to the system, as calculated in Equation 1. For example, McCormick[12] showed that the ohmic heating in the column by rapid application of the run voltage can expand the run buffer inside the column and expel an irreproducible volume of the sample during initial voltage application. The solution to the problem was to include a programmed voltage ramp at the beginning of the run so that the sample species had time to migrate into the capillary before thermal expansion of the buffer occurred. It should be noted that this expansion of the buffer can occur on any system at high power levels, even with the most efficient column cooling systems.

Another occasional problem relating to heat generation is the outgassing or boiling of the buffer when high power levels are utilized. The symptom of this problem is an instability or sudden drop in the current. This drop is sometimes difficult to observe if it happens during the initial ramp of the voltage. Care should be taken to note that the rate of current increase corresponds to that of the voltage ramp. If a small air or vapor bubble forms in the column, enough conductivity could remain as to appear as though a normal separation were underway. With experience, however, the discrepancy between the observed current and the expected current based on buffer composition and concentration should be evident. Degassing the buffer prior to use, even for only a few minutes in an ultrasonic bath, can lead to an extended range of applied powers without bubble or vapor formation. Interestingly, these vapor bubbles can be observed migrating through a column and can be mistaken as high-efficiency peaks (sometimes even appearing slightly tailed),[23] but they tend to lack expected reproducibility of better than 1 to 2% in migration time. The ability to record the current along with the detector signal is an invaluable diagnostic of a properly functioning system and can help to indicate these problems. Any drift in the current with time is directly related to changes in migration velocity of the different ions in solution through changing buffer concentration, composition, pH, or buffer temperature with time. Ion depletion and changes in pH of the run buffer can also be observed by monitoring the applied current or separation selectivity.[22]

B. FACTORS INFLUENCING HEAT GENERATION

The following sections contain a discussion of the factors that will have an influence on the applied power or heat generation in a CE experiment. Of note is that the greatest amount of heat is being generated in the center of the bore of the capillary. For the Joule heat to be dissipated, a temperature gradient from the center of the bore to the outer capillary surface must form that is proportional to the generated heat. There is a second important gradient from the outer surface of the capillary to the ambient environment of the cooling system that depends on the cooling system type and design. If the rate of heat generation can be minimized, then these gradients will likewise be reduced.

1. Buffer Composition

The buffer composition, as well as concentration, influences the buffer conductivity, and therefore the heat generated during a separation. One frequently asked question is which counterion to use. Potassium ions have a higher mobility than sodium and will tend to generate higher currents for the same molarity buffers; therefore they are not recommended for use when excessive heat generation is a problem. The ζ potential, and therefore conductivity, increases with size for these small cations in the order of $Li^+ < Na^+ < K^+ < Rb^+$, although using large organic counterions, such as histidine, can significantly reduce the electrical current.[6] It should be noted, however, that changing the counterion could also affect separation selectivity or adsorption of certain species to the surface of the capillary.

Zwitterionic buffers, or Good's buffers, for example, CHAPS, MOPS, MES, CAPS, and CAPSO, exhibit high buffer capacity while not contributing significantly to the conductivity of the system. Unfortunately, they tend to absorb ultraviolet (UV) light at 190 to 210 nm and may be an additional source of noise on the baseline when using low wavelengths for high-sensitivity applications. Nevertheless, for applications above 200 nm, one can use relatively high concentrations of these buffers while generating relatively low currents, thereby making it easy to work in the region of optimum efficiency around 1 to 2 W/m of generated heat.[14]

2. Applied Voltage

The applied voltage obviously has a direct relationship to the Joule heat generated by the applied power through the relationship in Equation 1. If the voltage is doubled, the applied power and heat generation will increase by a factor of 4. The applied voltage should be optimized for a given set of conditions to obtain the highest efficiency. The efficiency of the system can be sacrificed (due to excessive heat generation) for faster analysis times if there is sufficient resolution, and if the species of interest are not influenced by elevated buffer temperatures.

3. Column Length

The column length is indirectly related to the applied power. If column length is reduced by a factor of 2, the power will double at constant applied voltage. Conversely, if the column length is doubled at constant voltage the watts per meter decrease, as does the temperature gradient across the capillary bore,[5,10,13] yet the migration time will increase by a factor of 4. This may result in a very slight improvement in the resolution, although it is important to note that the resolution should not vary appreciably with column length at constant voltage, in accordance with the equations for resolution.[2,3] If the column length is doubled, the field should also be doubled (if possible) to obtain any appreciable improvement in resolution while only doubling migration time.

4. Column Internal Diameter

Since the cross-sectional area changes with the square of the radius, one will observe a doubling of the current (approximately) when changing from a 50-μm capillary to a 75-μm capillary, and a fourfold increase from 50 to 100 μm, or from 25 to 50 μm. Narrow bore capillaries have been used successfully with very high conductivity buffers in order to separate serum proteins[24] and other basic proteins,[25] for which adsorption to the wall of the capillary could be a problem. Perhaps the loss of sensitivity accompanying the use of narrow bore capillaries can be offset somewhat by the tremendous focusing effect achieved using 250 to 500 m*M* phosphate, sulfate, or borate buffers.

5. Buffer Viscosity

The buffer viscosity will affect the flow of ions in the buffer by increasing their resistance to migration in the column. The relationship of resulting current and viscosity is approximately linear, although this relationship depends greatly on the buffer additive and the range of viscosities studied. Large ions can flow quite rapidly in gel columns of extremely high viscosity. The addition of methanol or acetonitrile to the buffer will also reduce the heat generation by affecting the viscosity and/or the dielectric strength, thereby reducing the electrical current.

III. HEAT DISSIPATION

A. THEORY

Once heat is generated in the bore of the capillary it must be dissipated to the cooler environment around the column. Therefore, the temperature in the bore of the capillary will always be warmer than the cooling system temperature. The design of the cooling system will influence the difference between the ambient cooling system temperature and the temperature at the external surface of the column. Some magnitude of gradient, however, must always exist across the inner bore of the capillary and through the fused silica capillary wall to the cooling system in order to drive the heat from the point of generation to the cooling system.

An approximate temperature profile for a 50-μm I.D., 375-μm O.D. capillary is shown in Figure 2. In the capillary bore where heat is being generated, a roughly parabolic temperature profile exists. This is due to the heat being generated across the inner diameter of the capillary, but which is dissipated only at the interface between the bore and the inner wall of the capillary. Therefore, more heat is generated in the center of the capillary then near the inner wall, resulting in a higher temperature at the column center.

Interestingly, the magnitude of the temperature drop across the fused silica approximates that which is across only the polyimide coating in 50- and 75-μm capillaries. This suggests that the replacement of the polyimide coating with a more thermally conductive material should significantly improve the temperature drop across the capillary. However, it will probably be difficult to find a substitute that is thermally, yet nonelectrically, conductive and maintains the capillary flexibility and resistance to oxidation. The last temperature drop to the cooling system is the most easily influenced by the designer of CE instrumentation. With no cooling system this drop to ambient conditions could be on the order of several tens of degrees centigrade, which can cause a number of thermally related problems, for example boiling of the buffer, irreproducibility of migration time or peak area, or sample decomposition.

B. FACTORS INFLUENCING HEAT DISSIPATION

1. Cooling System Type

The cooling medium has a direct influence on the ability of the system to remove heat from the outer surface of the capillary. The thermal conductivity is the important property of the medium in contact with the capillary that determines its ability to remove the generated heat.

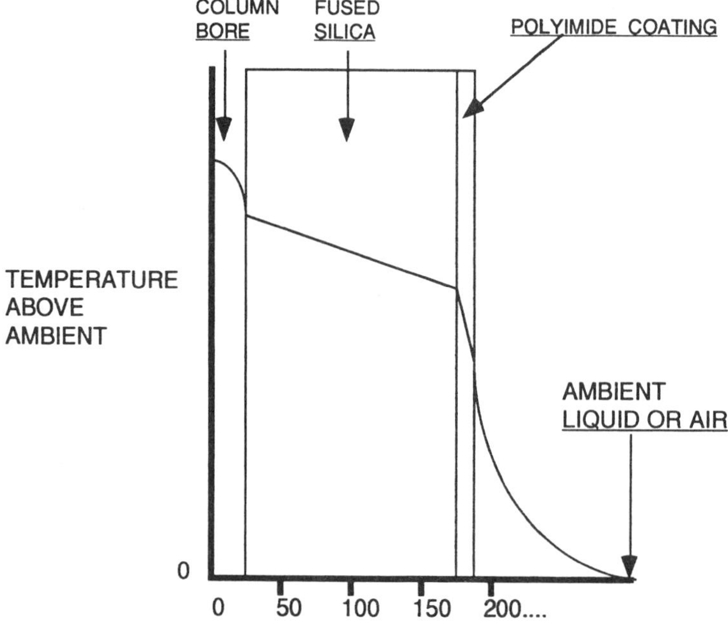

FIGURE 2. Column temperature profile for one radius of a 50-µm I.D., 375-µm O.D. capillary. Diagram predicts approximate temperature drops across each component in a fused silica capillary column typically used in capillary electrophoresis.

A metal has the highest thermal conductivity, although it is not practical to cool the capillary by placing it in direct contact with this material due to the close proximity of the high voltage, unless one is interested in attempting to control electroosmosis through the application of an exernal field to the plate as well. Ceramics such as beryllium oxides or alumina have good thermal conductivities as well as high electrical resistance, although ensuring good mechanical contact can prove difficult without the addition of a thermally conductive paste or liquid.[5] Circulating liquid coolants have superior thermal conductivities compared to air, although the design of a liquid-cooled system is usually somewhat more expensive than an air-cooled system. In addition, suitable nonelectrically conductive liquids tend to be expensive, viscous, harmful to the environment, or volatile depending on the substance of choice.

The velocity at which the liquid or air is passed over the outer capillary surface is also important for removing the heat from the capillary. High-velocity air cooling (10 m/s) can reduce the temperature difference between the buffer and the air bath to within a few degrees even at high power levels.[10] It is difficult to apply the same velocity of liquid flow over the capillary as can be attained with air cooling; however, this is compensated for by the higher thermal conductivity of a liquid and results in slightly better (lower) overall temperature rise in the buffer.

2. Cooling System Design

The design of the cooling system is also critical to its performance. One important consideration is the length of uncooled capillary between the buffer reservoir and the cooling system. In this region, significant heat can be generated that may offset the benefits of the the cooling medium described above. Ideally, the entire capillary, including the buffer vials, should be actively cooled to the same temperature. A design such as this would not contain any temperature gradients within the system and may improve reproducibility as well. Another consideration is the placement of the temperature sensors and the accuracy of the electronics employed for measurement and control of the system. Significant temperature fluctuations or poor instrument-to-instrument reproducibility could result from inadequate designs. However, existing commercial instruments that have thermostatted column designs seem to offer comparable levels of run-to-run reproducibility when the entire system is operating properly, regardless of cooling system type.

One additional advantage of an air-cooled design is that it is possible to run gradient temperature programs or, alternatively, to lower the temperature of the cooling system through feedback in response to the Joule heating from the applied power in order to maintain a constant resistance of the buffer in the capillary.[27] With this feature, one is assured that the average internal buffer temperature will be held nearly constant throughout the run.

3. Column Outer Diameter:Inner Diameter Ratio

Interestingly, a consequence of increasing the outside diameter of the column is an improvement in the thermal transfer of the heat to the cooling system.[5] If this is accomplished by the increase in the thickness of the fused silica, the result will be a decrease in the temperature drop from the buffer to the outside wall of the capillary. The primary reason for the improvement is that there is a concurrent increase in the surface area of the capillary in contact with the cooling system medium. Since fused silica has an acceptable thermal conductivity, increasing the surface area exposed to the cooling system tends to reduce the insulating effect of the polyimide coating by allowing more surface area for heat to be extracted by the cooling medium.

Similarly, one can increase the thickness of the polymer layer on the capillary and gain other improvements as well. It was noted earlier[5] that noise in the electrical current trace existed at high applied power levels, and Bello and Righetti[26] showed that this noise can translate into optical baseline noise. The primary benefit of increasing the polymer thickness was realized by reducing uneven, sporadic heat removal from the capillary at high applied power levels, even with high-velocity forced air cooling. This improvement was created by simply adding a Teflon tube as an insulator around the capillary to increase its surface area. Naturally, the average buffer temperature inside the capillary was higher, but the thermal noise picked up at the detector was significantly improved. To compensate for the increased buffer temperature, one could operate a cooling system at reduced temperatures. If operating at higher temperatures is not an issue, some benefit in optical baseline noise might be gained through this enhancement.

IV. ESTIMATING AVERAGE INTERNAL BUFFER TEMPERATURES

Often it is desirable to know the temperature of the buffer inside the capillary column. Without regard to buffer temperature, thermal breakdown of organic buffers can occur, yielding irreproducible migration times. In addition, thermally labile samples could give false peaks that might be misinterpreted not only as impurities, but also as the component of interest.[18] The change in pH as a function of temperature of the support buffer or sample buffer can either enhance or hinder the separation process.[28] Thus a complete understanding of the behavior of the separation buffer and sample under high electric fields is critical to a successful experimental design.

Terabe et al.[29] physically measured the temperature at the outside surface of the capillary column with a microthermocouple and estimated the internal temperature. An exact measurement of the capillary temperature was difficult due to only partial contact of the thermocouple with the curved capillary surface. Watzig measured the inside temperatures of the columns by placing a thermochromic solution of cobalt(II) chloride in the capillary and monitored the intensity of the spectral output from the solution.[30] Nelson et al.[5] have calculated the temperature, utilizing the deviation in the Ohm's law relationship described above. A computer simulation of the influence of temperature has been developed by Bello and Righetti.[8] Described below are some simple methods for calculating internal temperature in homogeneous as well as inhomogeneous environments.

A. ELECTROOSMOTIC FLOW METHOD

As seen in Equation 2, both the electroosmotic flow (μ_{eo}) and the mobility of an ion (μ_e) are dependent on the viscosity of the support buffer, therefore any change in the internal temperature will influence the migration time of the separation.

$$\mu_{eo} = \varepsilon\zeta/4\pi\eta$$
$$\mu_e = q/6\pi a\eta \tag{2}$$

where ε is the dielectric constant, ζ is the ζ potential, η is the viscosity, q is the charge of the molecule, and a is the effective radius of the molecule. For a first approximation, the viscosity of the solution is given in Equation 3:

$$1/\eta = Ae^{-B/T} \tag{3}$$

where B is determined from a plot of literature values for water ranging from 293 to 353 K and has the value of 1820.[6] Since μ_{eo} is inversely related to the viscosity, a ratio of a measured value of μ_{eo} at a low voltage to a measured value at a higher voltage should provide a method for calculating the temperature inside the column.

$$\mu_{eo1}/\mu_{eo2} = e^{-(B/T_1 - B/T_2)} \tag{4}$$

where subscript 1 is the lower voltage measurement and subscript 2 is the higher voltage measurement. It is assumed that ε and ζ are constant because the measurements are made at constant pH and buffer concentration, and the values do not change appreciably in the temperature range of most CE experiments.

If the lower voltage measurement is assumed to be at 298 K then Equation 4 becomes

$$T_2 = 1820/\ln(\mu_{eo1}) - \ln(\mu_{eo2}) + 6.11 \tag{5}$$

where T_2 is given in degrees Kelvin. Using Equation 5, one can calculate the temperature inside the capillary column by measuring the μ_{eo} at a low voltage, where one assumes there is no heating and then at the higher voltage, where one needs to know the temperature inside the column.[6] For example, if the μ_{eo} at 5 kV is 1.0×10^{-4} cm^2 V^{-1}s^{-1}, and μ_{eo} at 20 kV is 1.25×10^{-4} cm^2 V^{-1}s^{-1}, then

$$T_2 = 1820/[\ln(1.0 \times 10^{-4}) - \ln(1.25 \times 10^{-4}) + 6.11]$$
$$= 1820/(-9.21 + 8.97 + 6.11)$$
$$= 310 \text{ K or } 37°\text{C}$$

A linear plot can be obtained for the relationship between the calculated temperature and the applied power up to about 3 W/m. The slope of the line is related to the efficiency of heat removal and allows one to monitor the effectiveness of the cooling device used to remove heat during the separation. An approximation of internal temperature can be estimated as follows: natural convection is about 1.1°C per 0.1 W/m of power; forced air cooling is about 0.6°C per 0.1 W/m (e.g., about +6°C above the cooling system set temperature for 1 W/m of applied power) and liquid/solid state cooling is about 0.3°C per 0.1 W/m. Each cooling system can be calibrated to give the efficiency of heat transfer, using Equation 5.

B. CONDUCTIVITY, OHM'S LAW METHOD

An additional method for determining the internal temperature of the column during electrophoresis is to measure the relationship between temperature and conductivity of the buffer in the capillary column.[6,31,41] The conductivity of the solution, σ, is calculated from the measured current, i, at a given voltage, V:

$$\sigma = iA/VL \tag{6}$$

where A is the cross-sectional area of the column and L is the total length of the capillary. Since conductivity is inversely related to resistance, the Ohmic plot discussed above applies. The internal temperature of the a capillary column can be calculated using Equation 7.

$$T_2 = [(\sigma_2/\sigma_1) - 1/0.0205] + 25°\text{C} \tag{7}$$

Again, the buffer temperature is assumed for this example to be 25°C (or the set temperature of the cooling system) with a small applied voltage (1 to 2 kV), and the change in viscosity is linear with temperature. Generally, viscosity changes about 2%/1°C.

The average buffer temperature can be quickly calculated in front of the instrument by measuring the electrical current at 1 to 2 kV (e.g., 1 kV produces 4 μA). First calculate the expected current for an ideal resistor with no influence from temperature by solving for the current at a higher voltage (e.g., 4 μA/1 kV = x μA/25 kV ∴ x = 100 μA). Since the capillary is not an ideal resistor, the actual current will be higher than that just calculated due to the temperature increase by the buffer (see Figure 1). If the actual current at 25 kV is 120 μA, this indicates a 20% increase in current over the ideal resistor. Using the relationship of 2%/1°C, the average buffer temperature can be estimated to be about +10°C higher than the set temperature of the cooling system. The advantage of this method is that the estimation can be made using only two data points and two quick calculations.

As seen in Equation 2, a small change in the temperature during the electrophoresis run can greatly change the migration time of the sample both through the electroosmotic flow and the ion mobility. Stable temperature control during the separation is vital, and a close monitoring of the current will give valuable information on the behavior of the separation.

C. INHOMOGENEOUS HEATING

An emergence of on-column concentrating techniques for CE has occurred.[32-40] These techniques all employ the manipulation of the electric fields in the initial sample zone to stack the sample species against either a moving boundary (isotachophoresis) or a stationary boundary (field-amplified capillary electrophoresis). On occasion the electric fields in the sample regions can reach several hundred times those generated in the support buffer region. If the buffer system induces a large current flow during electrophoresis, the temperature inside the sample region can approach the boiling point of the sample buffer.[40] Under these conditions, thermally labile samples might decompose and generate a host of peaks that could be misinterpreted as impurities or even the component of interest.[18] Hence, care must be taken when sample concentration techniques are used to improve detection limits.

The conductivity in the sample plug under field amplified capillary electrophoresis conditions can be several times less than the conductivity of the support buffer. When this happens, the power induced in the sample region can be very large because of the electric field increase in this region.[40]

$$P_t = P_s + P_b \tag{8}$$

The power of the whole system is additive and consists of the power induced in the sample region and the power in the support buffer region. The power of each section is dependent on the conductivity of the section and the length of the section; thus Equation 8 becomes

$$P_t = i^2/A[x/\sigma_s + (1 - x)/\sigma_b] \tag{9}$$

where x is the length of the sample plug and σ_s and σ_b are the conductivities of the sample and run buffer, respectively. It is clear that if the sample completely fills the column, the total power induced is calculated from the current pulled and the voltage applied. The same condition exits for the column filled with the support buffer.

As discussed above and derived by Vinther and Soeberg,[40] the temperature of each section can be calculated because a linear relationship is assumed between the applied power and temperature. Thus Equation 9 will become

$$\begin{aligned} dT_b &= P_t x/\mathrm{a}[x + 1 - x(\sigma_b/\sigma_s)] \\ dT_s &= P_t x\sigma_b/\sigma_s/a[x + 1 - x(\sigma_b/\sigma_s)] \end{aligned} \tag{10}$$

Using Equation 10 with the sample in distilled water at $\sigma_s = 0.17$ mS/cm and the run buffer of 10 mM Tricine with 50 mM NaCl with a $\sigma_b = 4.96$ mS/cm, the temperature in the sample plug could possibly reach 99°C at only 15 kV.

V. CONTROL OF AMBIENT COLUMN TEMPERATURE

Controlling the ambient column temperature offers several advantages in CE. The migration time reproducibility and injection reproducibility can be improved (run to run, day to day, and laboratory to laboratory) if the column temperature is accurately and precisely controlled. In addition, the possibility of sample decomposition can be reduced by operating the system at lower temperatures. Conversely, increasing capillary temperature can shorten separations much the same as increasing the applied voltage. For samples where thermal decomposition is not an issue, and maximum voltage has been applied, each 10°C rise in ambient column temperature will shorten the analysis time by about 20%; thus, a 10-min analysis time becomes only 8 min.

A. MIGRATION TIME VARIABILITY

Since both the electrophoretic migration velocity as well as the electroosmotic flow velocity are affected by column temperature, the control of ambient conditions can have a significant influence on migration time variability. It is important to note that one should distinguish between run-to-run reproducibility (intraday reproducibility) and between day or between laboratory reproducibility (intraday or interlaboratory reproducibility). During the short time between injections for an intraday study, there may not be significant variation in temperature to greatly influence migration time reproducibility. Acceptable reproducibility can be obtained on manual systems that do not control ambient temperature, yet are able to produce migration time reproducibilities of less than 1% relative standard deviation (RSD) or coefficient of variation (CV). Between days or between laboratories, however, there can be a significant difference in ambient temperatures, and the use of a column thermostatting system can make a significant improvement in the coefficient of variation for a given study if one needs to compare results as in the pharmaceutical industry.

B. OPTIMIZING RESOLUTION

Of additional importance is the fact that relative migration times can be affected differently by changes in ambient temperature. Figure 3 indicates the change in migration order for different inorganic and organic compounds with increasing column temperature. Complete resolution of the different components occurs only at about 25 and 60°C. The elution order at 25°C is chloride, nitrite, sulfate, nitrate, and oxalate, while the elution order at 60°C is chloride, sulfate, nitrite, oxalate, and nitrate. The nitrite and nitrate move as a pair relative to the chloride, sulfate, and oxalate. The effect of capillary temperature on resolution is interesting, although not well explained at this time. The pH could be changing with temperature, or the sphere of hydration could be different for the species that are changing relative to the others. Nevertheless, more study is required to explain this effect further.

Guttman and Cooke[42] optimized the separation of DNA restriction fragments in the range of 20 to 50°C. Under constant current conditions, increasing the column temperature resulted in a maximum migration time for all DNA fragments and provided maximum resolution for the lower molecular weight fragments below about 200,000.

C. INFLUENCE ON INJECTION VOLUMES

Since the viscosity of the buffer in the capillary varies about 2%/1°C, the increasing injection volume with temperature should be taken into account during a temperature optimization study.[19] There are two methods to account for the increasing injection volumes with temperature. First, the resulting peak area can be normalized for decreasing viscosity by multiplying the area by the ratio of the viscosities. Second, one can simply inject 20% less with each 10°C temperature increase. Both methods should approximate the changes in buffer viscosity with increasing column temperature.

D. SAMPLE DECOMPOSITION

There have been several studies showing sample decomposition as a result of elevated column temperature.[5,17,18,40] The average buffer temperature can be estimated, as shown above, and can indicate whether there may be a problem with Ohmic heating in the system. In addition, the column temperature can be varied during optimization, and the influence on peak shape or area can be studied in order to observe any abnormalities.

E. COUNTERACTING BAND BROADENING

Several works have noted that Joule heating is responsible for increased bandwidth above an optimum applied voltage.[1-4,12-16] This is most likely due to one or more of the following factors in larger bore capillaries: increased diffusion due to overall elevated column temperatures, or radial gradients in temperature causing changes in pH, mobility, distortion of the flat zone profile, and convection. Gobie and Ivory[41] have shown an interesting method of counteracting the distortion of the zone profile in large bore capillaries by inducing an opposing laminar profile through the application of a small pressure to the outlet end of the capillary.

VI. CONCLUSIONS

The control of column temperature is critical to the overall performance of a capillary electrophoresis system. Since the temperature of the buffer affects most of the important

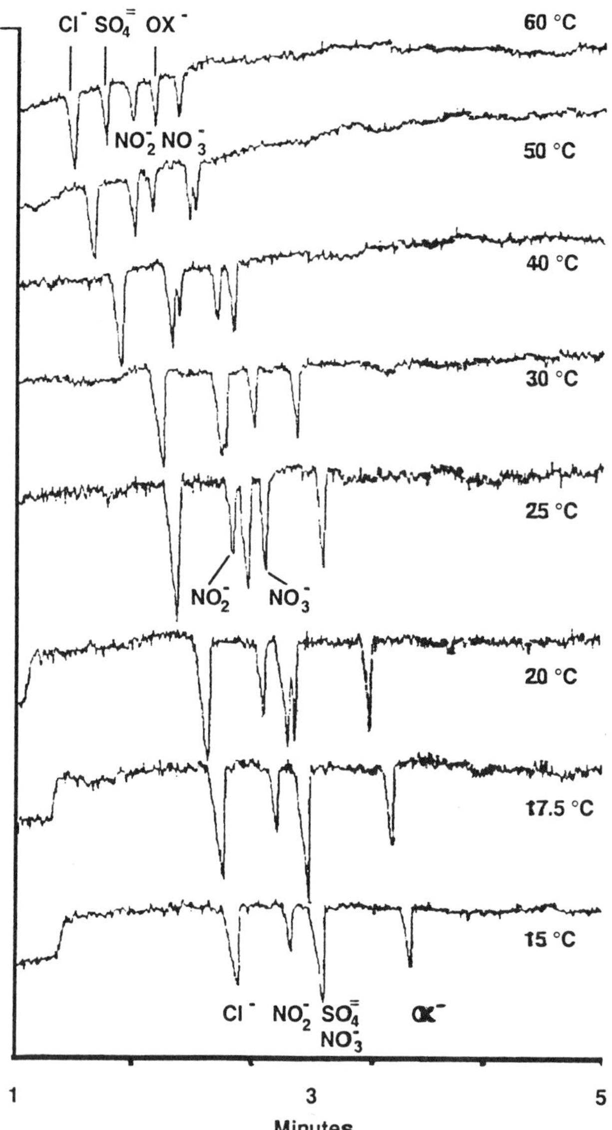

FIGURE 3. Separation of inorganic ions with an indirect absorbance buffer system. Buffer: 1,2,4,5-benzene tetracarboxylic acid (3 m*M*), diethylene triamine (2 m*M*), pH 9.6. Sample: 1 ppm each of sodium chloride, sodium sulfate, sodium nitrite, sodium nitrate, and oxalic acid. The separation was performed on a Spectra*PHORESIS* CE1000, –25 kV, in a 50 µm × 44 cm capillary column. Electrokinetic injection was performed at –5 kV for 2 s.

parameters in a CE separation it is vital not only to control the ambient temperature around the column, but also to rapidly remove the heat from the outer surface of the capillary. Temperature control also plays a significant role in the reproducibility of a method from day to day or laboratory to laboratory. The type of cooling system (air, liquid, or solid) is generally less important than its design (temperature accuracy, stability, and flow velocity or heat removal efficiency). Examples of easily estimating the average buffer temperature within the capillary have been shown. Finally, the column temperature should always be considered a variable in a separation that must be optimized with respect to speed, efficiency, or resolution.

REFERENCES

1. **Hjertén, S.,** Free zone electrophoresis, *Chromatogr. Rev.*, 9, 122, 1967.
2. **Jorgenson, J. W. and Lukacs, K. D.,** Capillary electrophoresis, *Science*, 222, 266, 1983.
3. **Gordon, M. J., Huang, X. H., Pentoney, S. L., and Zare, R. N.,** Capillary electrophoresis, *Science*, 242, 224, 1988.
4. **Karger, B. L., Cohen, A. S., and Guttman, A.,** High-performance capillary electrophoresis in the biological sciences, *J. Chromatogr. Biomed. Appl.*, 492, 585, 1989.
5. **Nelson, R. J., Paulus, A., Cohen, A. S., Guttman, A. B., and Karger, B. L.,** Use of Peltier thermoelectric devices to control column temperature in high-performance capillary electrophoresis, *J. Chromatogr.*, 480, 111, 1989.
6. **Burgi, D. S., Salomon, K., and Chien, R.-L.,** Methods for calculating the internal temperature of capillary columns during capillary electrophoresis, *J. Liq. Chromatogr.*, 14, 847, 1991.
7. **Bello, M. and Righetti, P. G.,** Unsteady heat transfer in capillary zone electrophoresis. I. A mathematical model, *J. Chromatogr.*, 606, 95, 1992.
8. **Bello, M. and Righetti, P. G.,** Unsteady heat transfer in capillary zone electrophoresis. II. Computer simulations, *J. Chromatogr.*, 606, 103, 1992.
9. **Terabe, S., Otsuka, K., and Ando, T.,** Band broadening in electrokinetic chromatography with micellar solutions and open-tubular capillaries, *Anal. Chem.*, 61, 251, 1989.
10. **Knox, J. H.,** Thermal effects and band spreading in capillary electroseparation, *Chromatographia*, 26, 329, 1988.
11. **Knox, J. H. and Grant, I. H.,** Miniaturisation in pressure and electroendosmotically driven liquid chromatography: some theoretical considerations, *Chromatographia*, 24, 135, 1987.
12. **McCormick, R. M.,** Capillary zone electrophoretic separation of peptides and proteins using low pH buffers in modified silica capillaries, *Anal. Chem.*, 60, 2322, 1988.
13. **Grushka, E., McCormick, R. M., and Kirkland, J. J.,** Effect of temperature gradients on the efficiency of capillary zone electrophoresis separations, *Anal. Chem.*, 61, 241, 1989.
14. **Sepaniak, M. J. and Cole, R. O.,** Column efficiency in micellar electrokinetic capillary chromatography, *Anal. Chem.*, 59, 472, 1987.
15. **Simpson, C. F. and Altria, K. D.,** Measurement of electroendosmotic flows in high-voltage capillary zone electrophoresis, *Anal. Proc.*, 23, 453, 1986.
16. **Altria, K. D. and Simpson, C. F.,** High voltage capillary zone electrophoresis: operating parameter effects on electroendosmotic flows and electrophoretic mobilities, *Chromatographia*, 24, 527, 1987.
17. **Rush, R. S., Cohen, A. S., and Karger, B. L.,** Influence of column temperature on the electrophoretic behavior of myoglobin and α-lactalbumin in high-performance capillary electrophoresis, *Anal Chem.*, 63, 1346, 1991.
18. **Vinther, A., Soeberg, H., Nielsen, L., Pedersen, J., and Biedermann, K.,** Thermal degradation of a thermolabile *Serratia marcescens* nuclease using capillary electrophoresis with stacking conditions, *Anal Chem.*, 64, 187, 1992.
19. **Rush, R. S. and Karger, B. L.,** Sample injection with P/ACE™ System 2000: importance of temperature control with respect to quantitation, TIBC-104, Beckman Instruments, Inc., Fullerton, California, 1990.
20. **Landers, J. P., Oda, R. P., Madden, B., Sismelich, T. P., and Spelsberg, T. C.,** Reproducibility of sample separation using liquid and forced air convection thermostated high performance capillary electrophoresis systems, *J. High Resolut. Chromatogr.*, 15, 517, 1992.
21. **Lookabaugh, M., Biswas, M., and Krull, I. S.,** Quantitation of insulin by capillary electrophoresis, *J. Chromatogr.*, 549, 357, 1991.
22. **Albin, M., Black, B., Wiktorowicz, J., Colburn, J., and Moring, S.,** Studies on reproducibility during automated electrophoresis analysis: evaporation and temperature control, Poster presentation, 3rd Int. Symp. on High-Performance Capillary Electrophoresis, San Diego, California, 1991.
23. **Nelson, R. J.,** Unpublished data, 1989.
24. **Chen, F. T. A., Liu, C. M., Hsieh, Y. Z., and Sternberg, J. C.,** Capillary zone electrophoresis in clinical chemistry, *J. Clin. Chem.*, 37, 14, 1991.
25. **Bushey, M. M. and Jorgenson, J. W.,** Capillary electrophoresis of proteins in buffers containing high concentrations of zwitterionic salts, *J. Chromatogr.*, 480, 301, 1989.
26. **Bello, M. and Righetti, P. G.,** Submitted for publication, February 1993.
27. **Weinburger, S. R. and Mills, J. L.,** Thermal control for capillary electrophoresis apparatus, U.S. Patent 5,066,382, 1991.
28. **Whang, C.-W. and Yeung, E. S.,** Temperature programming in capillary zone electrophoresis, *Anal. Chem.*, 64, 452, 1992.

29. **Terabe, S., Otsuka, K., and Ando, T.,** Micellar electrokinetic capillary chromatography, *Anal. Chem.*, 57, 834, 1985.
30. **Watzig, H.,** *Chromatographia*, 33, 445, 1992.
31. **Lukacs, K. D.,** Theory, instrumentation, and applications of capillary zone electrophoresis, Ph.D. thesis, University Microfilms International, Ann Arbor, Michigan, 55, 1983.
32. **Chien, R.-L. and Burgi, D. S.,** On-column sample concentrating using field amplification in CZE, *Anal. Chem.*, 64, A492, 1992.
33. **Kaniansky, D. and Marak, J.,** On line coupling of capillary isotachophoresis with capillary zone electrophoresis, *J. Chromatogr.*, 498, 191, 1990.
34. **Dolink, V., Cobb, K. A., and Novotny, M.,** Capillary zone electrophoresis of dilute samples with isotachophoretic preconcentration, *J. Microcolumn Sep.*, 2, 127, 1990.
35. **Foret, F., Sustacek, V., and Bocek, P.,** On-line isotachophoretic sample preconcentration for enhancement of zone detectability in capillary zone electrophoresis, *J. Microcolumn Sep.*, 2, 229, 1990.
36. **Foret, F., Szoko, E., and Karger, B. L.,** On-column transient and coupled column isotachophoretic preconcentration of protein samples in capillary zone electrophoresis, *J. Chromatogr.*, 608, 3, 1992.
37. **Debets, A. J. J., Mazereeuw, M., Voogt, W. H., Van Iperen, D. J., Lingeman, H., Hupe, K.-P., and Brinkman, U. A. Th.,** Switching valve with internal micro precolumn for on-line sample enrichment in capillary zone electrophoresis, *J. Chromatogr.*, 608, 151, 1992.
38. **Gebauer, P., Thormann, W., and Bocek, P.,** Sample self-stacking in zone electrophoresis, *J. Chromatogr.*, 608, 47, 1992.
39. **Jandik, P. and Jones, W. R.,** Optimization of detection sensitivity in capillary electrophoresis of inorganic anions, *J. Chromatogr.*, 546, 431, 1991.
40. **Vinther, A. and Soeberg, H. J.,** Temperature elevations of the sample zone in free solution capillary electrophoresis under stacking conditions, *J. Chromatogr.*, 559, 27, 1991.
41. **Gobie, W. A. and Ivory, C. F.,** Thermal model of capillary electrophoresis and a method for counteracting thermal band broadening, *J. Chromatogr.*, 516, 191, 1990.
42. **Guttman, A. and Cooke, N.,** Influence on temperature on high performance capillary gel electrophoresis, *J. Chromatogr.*, 559, 285, 1991.

Chapter 22

CONTROL OF ELECTROOSMOTIC FLOW IN CAPILLARY ELECTROPHORESIS

Takao Tsuda

TABLE OF CONTENTS

0-8493-8690-X/94/$0.00+$.50

I. INTRODUCTION

In capillary zone electrophoresis (CE), electroosmotic flow (EOF) is generated when three essential conditions are satisfied: (1) an electrovoltage is applied along a column axis, E_z, (2) charges are present on the inner surface of the column, and (3) a conductive solution is present within the capillary column. The phenomenon of electroosmotic flow has been studied by numerous workers.[1-38] During the infancy of CE development, the effects of pH, temperature, current, flow profile, electrolyte concentration, column material, and surface adsorption problems on electroosmotic flow were examined extensively.[1-7] The use of a radial applied voltage across the capillary column has advanced the understanding of electroosmotic flow as an electrical phenomenon and has peaked the interest of the CE community as a result of its potential role in controlling the efficiency of CE separations.[21-27]

II. INNER SURFACE OF THE CAPILLARY COLUMN

Since a glass surface has three to seven silanol groups per 10 Å^2,[36] it is easy to understand that the surface will have a charge due to the presence of SiO^- resulting from the dissociation of SiOH. While the concentration of SiO^- groups depends strongly on pH, the total charge on the surface cannot be explained solely by the dissociation of silanol groups.[7,30] The surface itself has charge as a result of the nature of the glass itself. This might be explained simply by the fact that everything on the earth is placed in an electrostatic field, which induces a surface charge on almost all materials. For example, glass, polytetrafluoroethylene, and poly(fluoroethylene)propylene have negative surface charges,[7] while charcoal has a positive charge. Hence, with capillaries made of these materials, an electroosmotic flow will be generated provided the three conditions described above are met.

Induction of surface charge also occurs when an electrovoltage is applied through the cross-sectional direction of the column, E_x. The interfaces between the inner wall and the buffer present inside the column, the silica capillary and the external polyimide coating, and the coating and the external medium act as a series of capacitors. The buffer, the column wall itself, the polyimide coating, and the external medium act as a series of resistors, and a charge can be induced by E_x at the inner solid surface.[21-29,34]

Even without E_x, a radial applied voltage will be generated merely by applying voltage along the length of the column (z axis), because at every point along the capillary, z_i, current will be emitted from the outside surface of the capillary to the surrounding medium, such as air or water, through the glass wall (its resistance is assumed to be 10^{15} ohms [Ω]).[27] At high applied potential gradients of 500 to 1000 V/cm of E_z, this phenomenon can be observed in that dust is attracted to the outer surface of the capillary. As dust particles have a charge opposite to that on the capillary wall, current will flow each time a dust particle contacts the surface. Thus, the applied potential along the column induces a potential gradient across the capillary column, which generates the charge on the inner surface of the capillary. This indicates that a potential gradient will be generated between the solid surface and the buffer, i.e., a zeta (ζ) potential.

When Hayes and Ewing[27] coated the outside of a capillary column with a thin film of conductive metal or ion-exchange resin, they found that rapid electroosmotic flow is observed when the metal sheath was connected to ground, because a strong radial electric field was generated between the inside and outside of the capillary. Therefore, it is clear that the glass wall, its polyimide coating (to prevent breakage of the capillary tubing), and both the inside and outside surfaces are working as a series of capacitors and/or resistors, and a small electric current will exist between the inner and outer surfaces of the capillary.

III. CHARGE AND THE POTENTIAL DISTRIBUTION AT THE SOLID–LIQUID INTERFACE

To understand the origin of electroosmotic flow, it is necessary to briefly discuss electrical potential and the physical environment of the interface. The discussion in this section follows the approach of R. J. Hunter.[34]

The electrostatic potential, ψ, near an isolated charged object *in vacuo* is defined by Coulomb's law. If the object is a sphere of radius r, the potential at a distance a from the surface is given by

$$\psi = Q/[4\pi\varepsilon_0(r + a)] \tag{1}$$

where Q is the total charge on the surface and ε_0 is the permittivity of free space.[34] The potential very close to the surface of the object (less than 100 Å) is impossible to determine, as the measurement is affected by the interaction with the charge on the object. When $r = 1$ cm, the potential *in vacuo* is constant up to 100 μm from the surface of the object, and equal to

$$\psi_o = Q/(4\pi\varepsilon_0 r) \tag{2}$$

where ψ_0 is the potential at the surface of the object. The potential, ψ, decreases with increased distance from the sphere.[34]

In capillary electrophoresis, the cylindrical solid inner surface is immersed in an electrolyte solution. Although the geometry is cylindrical, we may begin by studying a flat plate model. The simplest model is attributed to Helmholtz, in which both layers of charge at the interface between two phases (solid surface and liquid) are fixed in parallel planes to form a molecular condenser, called an electrical double layer (Figure 1).

Although the charges on a solid surface may be assumed to lie in a plane, the electrical force on the ions in solution compete with thermal diffusive forces, forming a diffuse double layer.[39,40]

Using the above model, the fundamental electrostatic equation for the system may be expressed as

$$\nabla^2\psi = -(4\pi\varepsilon_0)^{-1}[(4\pi\rho)/D] \tag{3}$$

where D is the dimensionless dielectric constant or relative permittivity ($D = \varepsilon/\varepsilon_0$, where ε is the permittivity of the dielectric) and ρ is the volume density of charge, given by

$$\rho = \sum_i n_i z_i e \tag{4}$$

where z is the valence of an ion of species i, and n_i is the number of ions of type i per unit volume, which is given by

$$n_i = n_i^0 \exp(-z_i e\psi/kT) \tag{5}$$

where $\psi = 0$ and $n_i = n_i^0$.

Substituting Equation 5 and Equation 4 into Equation 3, we obtain the complete Poisson–Boltzman equation:

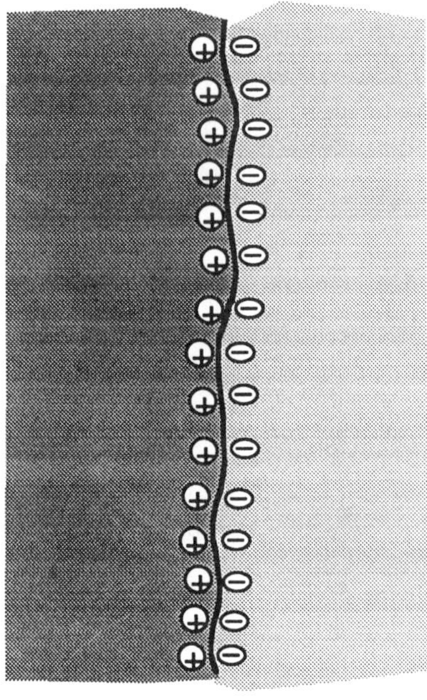

A

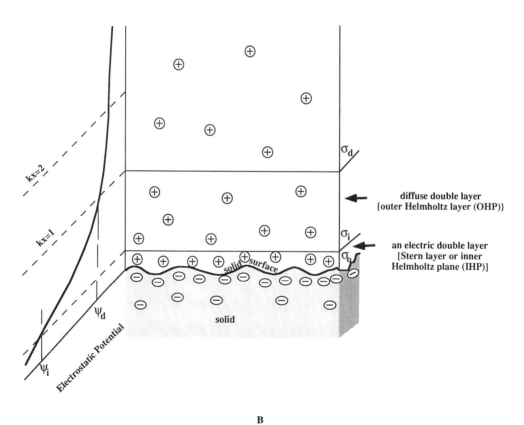

B

FIGURE 1. (A) Electrical double layer. (B) Schematic expression for the distribution of ions at and near a solid surface, and for the variation of electric potential.

$$\nabla^2\psi = d\psi^2/dx^2 = -(4\pi\varepsilon_0)^{-1}(4\pi/D)\sum_i n_i^0 z_i e \exp(-z_i e\psi/kt) \tag{6}$$

The equation may be simplified, as ψ varies only in one dimension, x. As ψ is small in the double layer (i.e., $z_i e\psi << kT$), we may use the exponential expansion, $e^x = 1 - x$. Utilizing the law of preservation of electroneutrality in the bulk electrolyte, Equation 6 becomes

$$d^2\psi/dx^2 = \kappa^2\,\psi \tag{7}$$

where

$$\kappa = \left[\frac{e^2 \Sigma n_i^0 z_i^2}{\varepsilon kt}\right]^{1/2} \tag{8}$$

Equation 7 can be solved to

$$d\psi/dx = -\kappa\psi \tag{9}$$

since $d\psi/dx = 0$ and $\psi = 0$ in the bulk solution, far from the solid surface where $x = \infty$.

A second integration of Equation 9 using the boundaries of the surface, where $x = 0$ and $\psi = \psi_{0,}$ gives

$$\psi = \psi_0 e^{-\kappa x} \tag{10}$$

This solution approximates the potential distribution near the wall. The general Equation 6 can be integrated, treating electrolytes as if they were symmetrical, to give

$$d\psi/dx = -[(2\kappa kt)/(ze)]\sinh[(ze\psi)/(2kt)] \tag{11a}$$

The integrated form of Equation 11a may be expressed in several forms, the most compact of which is[34]

$$\tanh(z\psi^*/4) = \tanh(z\psi^*/4)\exp(-kx) \tag{11b}$$

where $\psi^* = e\psi/kt$ is the dimensionless potential parameter, which we will refer to as the reduced potential. At 25°C, $\psi^* = 1$ when $\psi = 25.7$ mV. Figure 2 shows the electrical potential $z\psi^*$ in the double layer vs. κx according to Gouy[39] and Chapman.[40] The distance $1/\kappa$ is referred to as the thickness of the double layer (namely $\kappa x = 1$), although the potential is reduced to 2% of its original value at the extreme edge. The potential drop across the inner region of the double layer (inner Helmholtz plane [IHP], or Stern layer) is often around 0.1 V. Since this distance is less than 1 nm, the field strength is about 10^8 V m^{-1}, high enough to polarize molecules.

IV. CALCULATION OF ELECTROOSMOTIC FLOW VELOCITY

The theory for electroosmotic flow was first derived by Hunter[34] and Smoluchowski[42] for the movement of a liquid adjacent to a flat charged surface while under the influence of an electric field applied parallel to the interface. The velocity of the liquid in the direction parallel to the wall, v_z, rises from a value of zero in the inner Helmholtz plane to a finite maximum value, $v_{(osm)}$, at a distance $3/\kappa$ from the wall. The force on the unit volume of liquid, which is

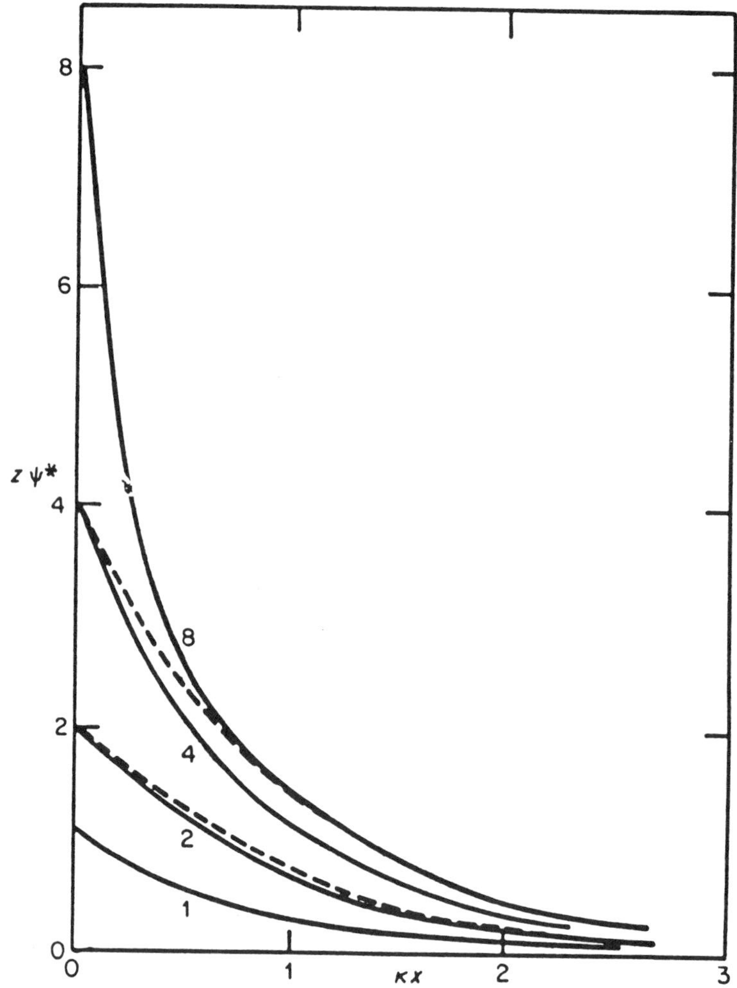

FIGURE 2. Electrical potential in the double layer (according to Gouy and Chapman). The full curves are from Equation 11b for various values of $z\psi_0$. The broken lines are from Equation 10 for $z\psi_0 = 2$ and 4. (From Overbeek, H. G. Th., in *Colloid Science,* Vol. 1, Kruy, H. R., Ed., Elsevier, Amsterdam, 1952. With permission.)

generated by the friction within the liquid, is equal to the force due to the attraction of ions under the applied voltage (Figure 3) and expressed by

$$E_z Q = E_z \rho A\, dx = \eta A(dv_z/dx)_x - \eta A(dv_z/dx)_{x+dx} \tag{12}$$

or

$$E_z \rho\, dx = -\eta(d^2 v_z/dx^2)dx \tag{13}$$

where E_z is the potential gradient applied parallel to the solid surface, Q is the number of charges in solution (ions), A is the unit area, and v_z the local electroosmotic velocity. The subscript z indicates the z axis, which is parallel to the surface.

Substituting for ρ, using the Poisson equation (Equation 3), we obtain

$$E_z(4\pi\varepsilon_0)(D/4\pi)(d^2 z/dx^2)dx = \eta(d^2 z/dx^2)dx \tag{14}$$

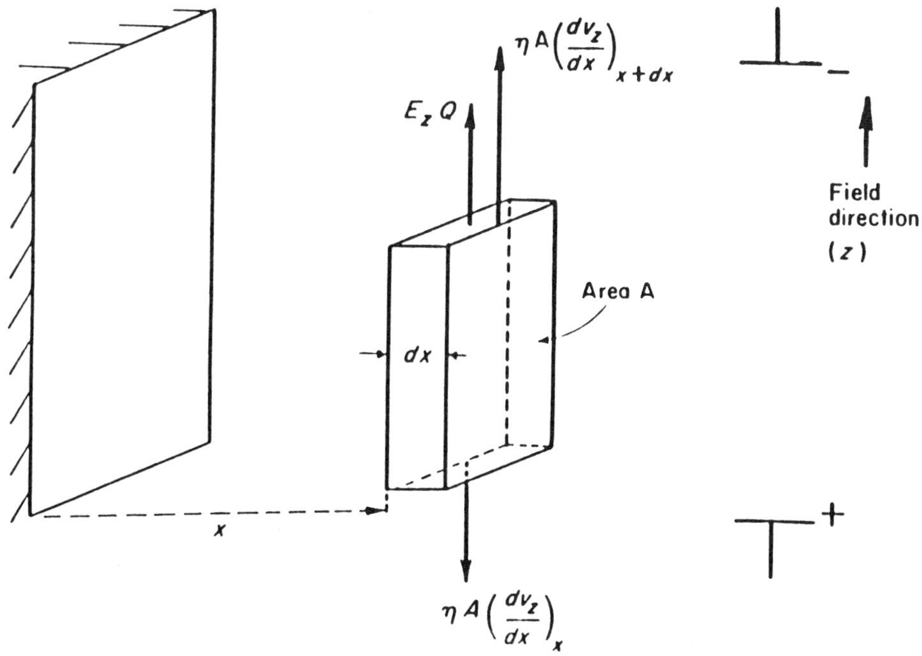

FIGURE 3. Forces on a volume of liquid of area *A* containing charge *Q*. (From Hunter, R. J., in *Zeta Potential in Colloid Science*, Academic Press, New York, 1981. With permission)

This equation may be evaluated between the limits of a distance far from the planar surface, where $\psi = 0$ and $v_z = v_{(osm)}$, to the inner Helmholtz plane, where $v_z = 0$ and $\psi = \zeta$. At the limits of the first integration, both $d\psi/dx$ and dv_z/dx are zero distant from the solid surface. Thus

$$v_{(osm)}/E_z = \mu_e = -4\pi\varepsilon_0[(D\zeta)/(4\pi\eta)] \tag{15}$$

$$= -(\varepsilon\zeta/\eta) \tag{16}$$

where μ_e is the electrophoretic mobility. Equation 16 is variously expressed[2,6,12,32,35,43] as

$$v_{(osm)} = [(DE_z)/(4\pi\eta)]\zeta \tag{17}$$

$$= [(\varepsilon\zeta E_z)/(4\pi\varepsilon_0\eta)] \tag{18}$$

Hunter (Ref. 34, p. 357) explains that there are two equations describing electroosmotic flow:

$$v_{(osm)} = (D\zeta)/(4\pi\eta) \tag{19a}$$

$$v_{(osm)} = (4\pi\varepsilon_0)[(D\zeta)/(4\pi\eta)] = (\varepsilon\zeta)/\eta \tag{19b}$$

The factor 4π appears because the earlier unit system was "unrationalized." After rationalization was achieved by defining the force between two charged particles in a dielectric, the factor 4π appears in formulas involving spherical symmetry and disappears from those involving flat plates. Whenever D appears in the old formulas, it is recommended that it be replaced by $4\pi\varepsilon_0 D$ or $4\pi\varepsilon$ (ε is equal to $\varepsilon_0 D$). D is dimensionless.

TABLE 1
Concentration of Electrolyte and Thickness of the Double Layer

c^a (mol l^{-1})	Thickness[b] (nm)
10^{-1}	1.04
10^{-2}	3.04
10^{-3}	9.62
10^{-4}	30.4

[a] Concentration of electrolyte.
[b] Values calculated using Equation 21.

V. CORRECTIONS DUE TO A CYLINDRICAL MODEL AND THE THICKNESS OF THE DOUBLE LAYER

For capillaries that have a cylindrical cross-section, one might assume the potential profile would be very different from that obtained with a flat plate model. The Poisson–Boltzmann equation (Equation 6) must satisfy a cylindrical model. Rice and Whitehead[35] found that the equations for the planar model must be modified for application to the cylindrical model by a correction factor, $F(\kappa r)$:

$$F(\kappa r) = 1 - \{[2I_i(\kappa r)]/[\kappa r I_0(\kappa r)]\} \quad (20)$$

The value of $F(\kappa r)$, plotted as a function of log κr (r = radius), is shown in Figure 4. The electroosmotic flow velocity in a cylindrical capillary is given by the expression

$$v_{(\mathrm{osm})} = [(\varepsilon\zeta E_z)/\eta]F(\kappa r) \quad (21)$$

The thickness of the double layer is dependent on the concentration of the electrolytes, through their effect on the parameter κ, which can be written (for water at 25°C)[34] as

$$\begin{aligned}\kappa &= [(2000F^2)/(\varepsilon_0 DRT)]^{1/2} I^{1/2} (\mathrm{nm})^{-1} \\ &= 3.288 I^{1/2}\end{aligned} \quad (22)$$

where F is the Faraday of charge (the magnitude of charge on 1 mol of electrons = 96,485 C) and R is the gas constant per mole. The ionic strength, I (moles per liter), is obtained from

$$I = 0.5\Sigma c_i z_i^2$$

where c_i is the concentration of ion i in moles per liter. The calculated values for the thickness of the double layer are given in Table 1. At $I = 10^{-2}$ mol l^{-1}, the value of $1/\kappa$ is 3.04 nm; at $I = 10^{-4}$ mol l^{-1}, $1/\kappa$ is 30.4 nm. The value of the geometric factor shown in Figure 4 deviates from 1 when κr is less than 10; if the concentration of electrolyte is 10^{-4} mol l^{-1}, r = 304 nm when κr = 10.

The typical capillary is 25 to 100 μm in inner diameter. It is not necessary to use the geometric factor to correct the equation for electroosmotic flow as the thickness of the double layer is less than 1 μm in general. However, with very narrow tubing, less than 5 μm in inner diameter, the flat geometry model no longer holds, and one should be careful to assess the effect of the double layer and its flow profile.

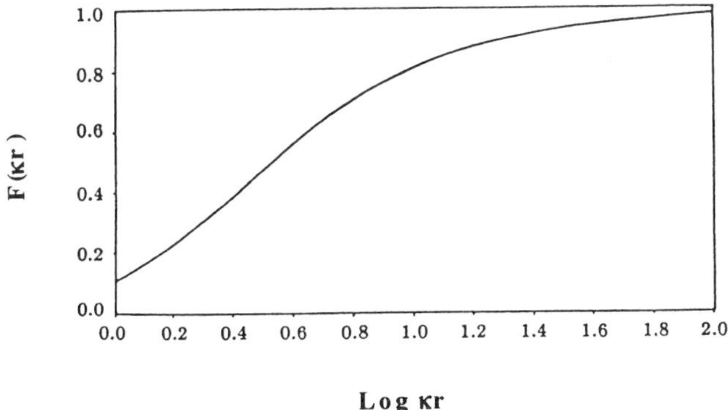

FIGURE 4. Corrective factor due to cylindrical geometry. (From Rice, C. L. and Whitehead, R., *J. Phys. Chem.*, 69(11), 4017, 1965. With permission.)

VI. APPLICATION OF AN EXTERNAL ELECTRIC FIELD TO CONTROL ELECTROOSMOTIC FLOW

Lee et al.[21] introduced a method to control electroosmotic flow with application of an external radial electric field to the fused silica capillary column. This method has been investigated further,[22-25] and confirmed by the studies of Ghowsi and Gale,[26] Hayes and Ewing,[27] and Hayes et al.[28] When a radial electric field is applied between the inside and outside surfaces of the capillary, both the wall of the glass capillary, C_c, and the interface between the inner surface of the glass tubing and the buffer, C_{ei}, act as capacitors. The capacitance is a function of charge density of the inner surface, σ_0, and is obtained from Gouy–Chapman theory:[34]

$$
\begin{aligned}
C &= d\sigma_0/d\psi = \varepsilon\kappa\cosh(z\psi_0^*/2) \\
&= 228.5zc^{0.5}\cosh(19.4z\psi_0)\ \ (\mu\text{F cm}^{-2})
\end{aligned}
\tag{23}
$$

where the unit dimension of c is moles per liter and that of ψ_0 is volts. Equation 23 may be modified for the capacitance of an electrostatic diffuse layer at the inner capillary surface–aqueous interface for a capillary column length l. The surface area is $\pi d_{id}l$ cm^2, ψ_0 is replaced with ζ, and units of electrolyte concentration are changed from moles per liter to moles per cubic centimeter:

$$C_{ei} = 7.23 \times 10^{-3}c^{1/2}2\pi r_{id}l\cosh(19.46\zeta)\ \ (\text{F cm}^{-2}) \tag{24}$$

where r_{id} is the inner radius of the capillary. The capacitance of the wall of the capillary tubing, C_c, is given by

$$C_c = \varepsilon_c 2\pi r_{id}l/[\ln(r_{od}/r_{id})] \tag{25}$$

where ε_c is the electrical permittivity of the silica tubing and r_{od} is the outer radius of the capillary tubing.[22]

The total capacitance of three capacitors in series, C_t, is given by

$$C_t^{-1} = C_{ei}^{-1} + C_c^{-1} + C_{out}^{-1} = C_c^{-1} \tag{26}$$

where C_{out} is the capacitance at the interface of the outside capillary surface and the surrounding medium.

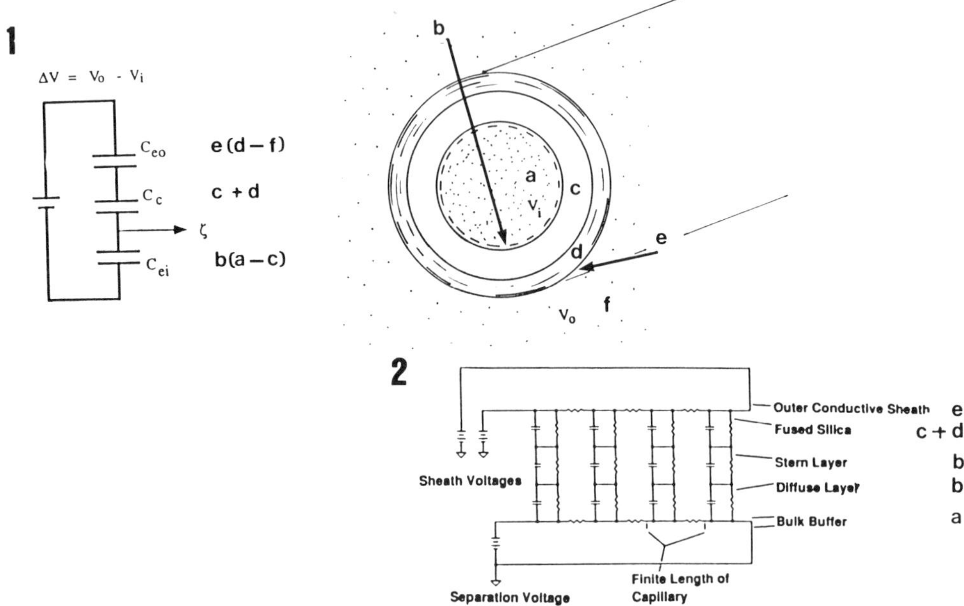

FIGURE 5. Models of capacity and/or resistance for CZE with axial applied voltage. (1) Capacitor model. (From Lee, C. S., McManigill, D., Wu, C.-T., and Patel, B., *Anal. Chem.*, 63, 1519, 1991. With permission.) (2) Capacitor and resistor model. (From Hayes, M. A. and Ewing, A. G., *Anal. Chem.*, 84, 512, 1992. With permission.)

As potential drop is inversely proportional to the capacitance, ζ is given as follows:

$$\zeta : E_x = C_t : C_{ei}$$
$$\zeta = E_x/(C_c/C_{ei}) \tag{27}$$

where E_x is the radial potential gradient applied.

The capacitor model proposed by Lee et al.[22,23] and the model including both capacitance and resistance described by Hayes and Ewing[27] are shown in Figure 5. As the radial applied voltage changes the surface charge density, σ_s, and zeta potential, ζ, it alters the electroosmotic mobility, μ_e. The value of the ζ potential could be estimated from Equation 27, if one approximated C_{ei} and C_c, or one could experimentally determine these values. Lee et al.,[21-23] Wu et al.,[24] Wu and Lee,[25] Hayes and Ewing,[27] and Ewing et al.[28] have presented the experimental results obtained by radial applied voltage. Resistance of the double layer is quite low compared to quartz (10^{15} Ω),[27] and the capacitance may be estimated at 5 to 25 μFarads (μF).

As depicted in Figure 6, an analytical column with 50-μm I.D., 150-μm O.D., and 25-cm length was placed in an acrylic box directly above six polonium-doped radioactive strips spaced 1 cm apart, which acted as an ionizing unit.[25] Each strip contained 250 μCuries (μCi) of polonium. The air within the box was ionized by the α particles emitted by the polonium, and had an electrical resistance of 10^9 Ω/cm. Reservoir 1 was filled with 10 m*M* phosphate buffer, pH 3; reservoir 2 contained 20 m*M* phosphate buffer at the same pH. The changes in the electroosmotic mobility resulting from various radial potential gradients were measured in the presence of a constant internal electric field strength of 220 V/cm. As shown in Figure 7, the electroosmotic mobility varied from 2.1×10^{-4} cm^2 V^{-1} s^{-1} at –8 kV to -1.2×10^{-4} cm^2 V^{-1} s^{-1} at 9 kV radial applied potential. The flow rate was assigned to be positive when the direction of flow was toward the cathodic end of the internally applied field, E_z. The direction

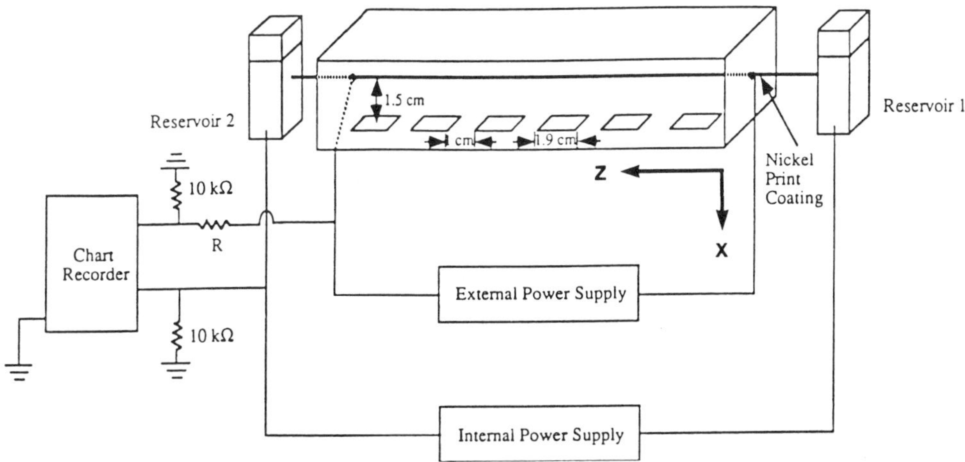

FIGURE 6. Air in a box is ionized by α particles. Nickel print coatings on the fused silica capillary work as electrodes for external power supply. (From Wu, C.-T. and Lee, C. S., *Anal. Chem.*, 64, 2310, 1992. With permission.)

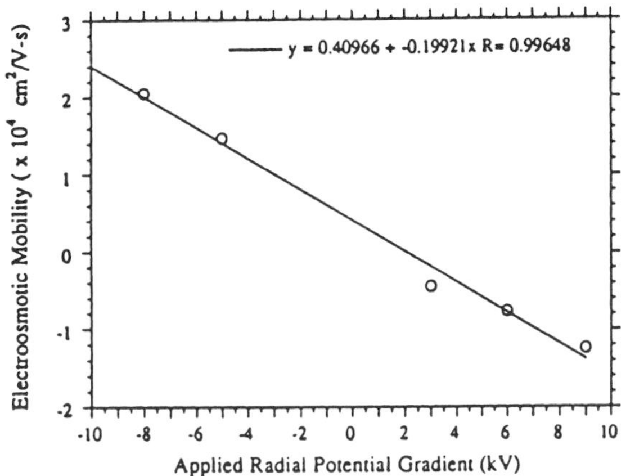

FIGURE 7. Plot of the electroosmotic mobility against the applied radial potential gradient. The experimental data, determined with 10 m*M*/20 m*M* phosphate buffer solution (pH 3), 50 μm I.D. capillary tubing, and the current-monitoring method are shown as the open circles. (From Wu, C.-T. and Lee, C. S., *Anal. Chem.*, 64, 2310, 1992. With permission.)

of electroosmotic flow is toward the anodic end when the ζ potential at the capillary–aqueous interface is changed from negative to positive by the application of a strong positive gradient across the capillary wall.[22]

The change in the magnitude and the polarity of the ζ potential has a multitude of possibilities, including the enhancement of the separation of peptides, and the reduction of solute adsorption on the capillary wall due to electrostatic repulsion.[24]

VII. ESTIMATION OF CHARGE DENSITY ON THE GLASS SURFACE

The estimation of charge density on the glass surface is the key to the solution of the equation relating ψ, ζ, capacitance, the thickness of the double layer, and electroosmotic mobility. It is clear that the charge density may be controlled by radial applied voltage. The charge density is also affected by its chemical environment, such as pH. The silanol groups on a glass surface of unit area dissociate,

$$[\mathrm{SiOH}]_s \xleftrightarrow{K} [\mathrm{SiO}^-] + [\mathrm{H}^+] \tag{28}$$

where K is the equilibrium constant, subscript s indicates the glass surface, and $[H^+]$ is the bulk buffer hydronium ion concentration. The ionized silanol groups generate a surface charge density, σ_{SiO^-}, which is related to solution pH by the following relationships:

$$\begin{aligned}\sigma_{SiO^-} &= [\mathrm{SiO^-}]_s\{[\mathrm{SiO^-}]_s/([\mathrm{SiOH}]_s + [\mathrm{SiO^-}]_s)\} \\ &= f'/(1 + [\mathrm{H^+}]/K)\end{aligned} \tag{29}$$

where $f' = [\mathrm{SiOH}]_s + [\mathrm{SiO^-}]_s$. The equilibrium constant, K, is assumed to be in the range of 10^{-5} to 10^{-7}, and the density of silanols is three to seven per 10 Å^2.[36] When there are five silanol groups per 10 Å^2, the number of silanols is equal to $5 \times 10^{14}/\mathrm{cm}^2$, and σ_{SiO^-} is equal to $2.5 \times 10^7/\mathrm{cm}^2$ at pH 7 (K assumed to be 10^{-7}).

The charge density at the inner surface of the glass capillary due to radial applied voltage, σ_{vs}, is given from the capacitance model,[11-15,17,18] and uses the following relations:

$$\sigma_{vs}\pi d_{id} l = C_{ei}\zeta = C_c E_x = C_t E_x$$

Therefore

$$\begin{aligned}\sigma_{vs} &= C_c E_x (\pi d_{id} l)^{-1} \\ &= E_x(\varepsilon_c/r_{id})[\ln(r_{od}/r_{id})]^{-1}\end{aligned} \tag{30}$$

where r is the radius of the capillary.

The total charge density, σ_t, is assumed to be given by the summation of the charges due to silanol groups and radial applied voltage, as follows:

$$\sigma_t = \sigma_{vs} + \sigma_{SiO^-} \tag{31}$$

This assumes that the electrostatic charges and the ionization of surface hydroxyl groups have no interaction between one another. Based on the work of Ewing et al.,[28] the relation of σ_t with $v_{(osm)}$ is derived as follows:

$$v_{(osm)} = E_z(\varepsilon_b/\eta)\exp(-\kappa x)(2kT/e)\sinh^{-1}\{(\sigma_{vs} + \sigma_{SiO^-})[500\pi/\varepsilon_b cRT]\}^{1/2} \tag{32}$$

where subscript b refers to buffer. Although σ_{vs} is originally defined as the charge density of the double layer,[19] it might be equated to the surface charge because of the electroneutrality of the system. The relation between calculated values obtained by Equation 32 and radial applied voltage is shown in Figure 8A.[27,28] The radial applied voltage is not effective above pH 4; at pH less than 4, the radial applied voltage has considerable effect on $v_{(osm)}$. The experimental values obtained with radial applied voltage are plotted in Figure 8B.[28] Values of $v_{(osm)}$ had to be increased by a factor of 3.9 to match the magnitude of the experimental data points. However, the trend predicted by Equation 32 is followed by the experimental values.

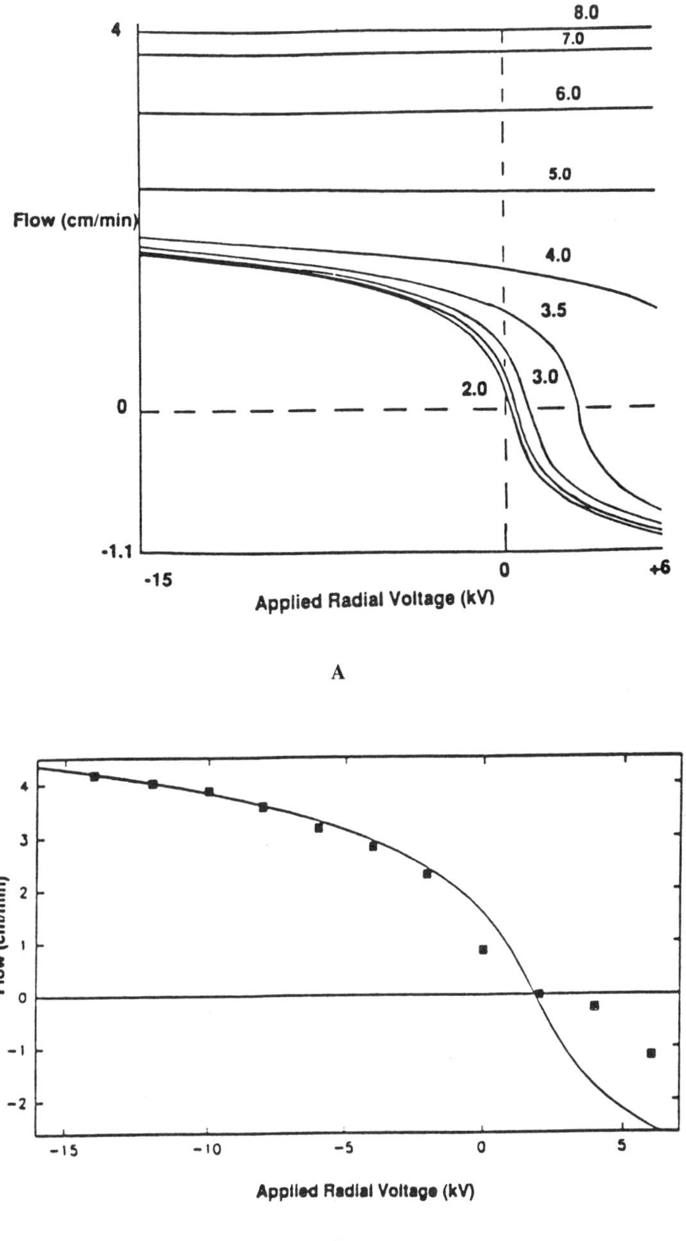

FIGURE 8. (A) A plot of Equation 32 for a series of buffer pH values. Separation potential, 140 V/cm; capillary, 50 cm × 10-μm I.D. × 150-μm O.D.; temperature, 298 K; concentration, 1 m*M*; dissociation constant, 1×10^{-7}; width of counterion 1.5×10^{-10} m; dielectric coefficient, 78; viscosity 0.00089 kg/(m · s); surface silanol group concentration, 5×10^{14} groups/cm^2. (From Hayes, M. A., Kheterpal, I., and Ewing, A. G., *Anal. Chem.*, 65, 27, 1993. With permission.) (B) Comparison of experimental data and theory from Equation 32 at pH 3.00. Each box represents the average of several neutral marker experiments, with $n = 1$ for data at −14 and 0 kV and $n = 3$ or more for all others. The solid line is theoretical data from Equation 32 employing the same variables as in the experiments (multiplied by 3.9 to match the magnitude; see text). Experimental conditions: separation potential, 140 V/cm; capillary, 50 × 10-μm I.D. × 144-μm O.D.; concentration, 1 m*M* phosphate buffer; detection, 214 nm. (From Hayes, M. A., Kheterpal, I., and Ewing, A. G., *Anal. Chem.*, 65, 27, 1993. With permission.)

VIII. ESTIMATION OF ZETA (ζ) POTENTIAL BY CAPILLARY ELECTROPHORESIS

ζ potential can be estimated experimentally by using Equation 16, 24, or 27. Although these equations contain several other physical parameters that should be estimated or calculated, we can select experimental conditions to eliminate some variables and, with several different experimental runs, it is possible to cancel the unknown parameters. From Equation 16 we can obtain following equation:

$$\zeta = -v_{(osm)}\eta(E_z\varepsilon)^{-1} \tag{33}$$

As η for pure water at 25°C is 8.95×10^{-3} P or 8.95×10^{-4} SI units (N m^{-2} s) and $\varepsilon = \varepsilon_0 D = 8.854 \times 10^{-12} \times 79$ F m^{-1}, Equation 33 becomes

$$\begin{aligned} \zeta \text{ (in mV)} &= -1.28 \times 10^{9}\, v_{(osm)}E_z^{-1} \text{ (m s}^{-1} \times \text{V}^{-1}\text{ m)} \\ &= -1.28 \times 10^{5}\, v_{(osm)}E_z^{-1} \text{ (cm s}^{-1} \times \text{V}^{-1}\text{ cm)} \end{aligned} \tag{34}$$

If $v_{(osm)}$ is 1×10^{-4} $cm^2 s^{-1}V^{-1}$, ζ is -12.8 mV.[34]

Lee et al.[22] calculated the ζ value by using Equation 27. They obtained a value of –18 mV for a capillary of 75-μm I.D., 375-μm O.D., containing 1.5 m*M* phosphate buffer (pH 5.0).

IX. CHEMICAL MODIFICATION OF THE INNER SURFACE BY ADSORPTION OF SURFACE-ACTIVE REAGENTS AND IONIC RESINS

Electroosmotic flow has the following relation:

$$v_{(osm)} = (f''n_{ins}eE_z)/(z\eta c^{0.5}) \tag{35a}$$

$$v_{(osm)} = (eA_3n_{ins}i)/(zc^{0.5}) \tag{35b}$$

where f'' is a constant, n_{ins} is the charge on the inner surface per unit area, and i is the current density.[10,12,19] From Equation 35, the electroosmotic flow velocity depends on the surface charge (n_{ins}) and its direction is dependent on the sign of the electric charge (e).

On the inner wall of the capillary, there are two kinds of surface charges: (1) surface charges induced naturally, σ_{nat}, due to the electrostatic field of nature, and (2) surface charges coming from the dissociation of –SiOH groups, σ_{SiO^-}. These surface charges may be covered partially or completely with surface-active reagents, such as cetyltrimethylammonium bromide (CTAB)[10,12] or tetrabutylammonium hydroxide (TBA).[20]

From the relationship between concentration of these surface-active reagents and electroosmotic flow velocity shown in Figure 9, it is clear that the original surface has a certain number of negative charges, σ_{nat}, which decrease as increasing amounts of CTAB or TBA are adsorbed on the surface of the glass wall (from the increased concentrations of CTAB or TBA in solution). The sign of the surface charge is changed from negative to positive at a concentration of 3.5×10^{-4} mol l^{-1} CTAB. The number of positive charges increase until complete coverage of the surface occurs. On the other hand, when we use a strong surface-active reagent of negative charge, such as sodium dodecyl sulfonate, the number of negative charges on the inner surface of capillary column increase by its adsorption.

Therefore, we can control the amount of the surface charge and its sign by the addition of surface-active reagents. This is often done in capillary electrophoresis to obtain short elution

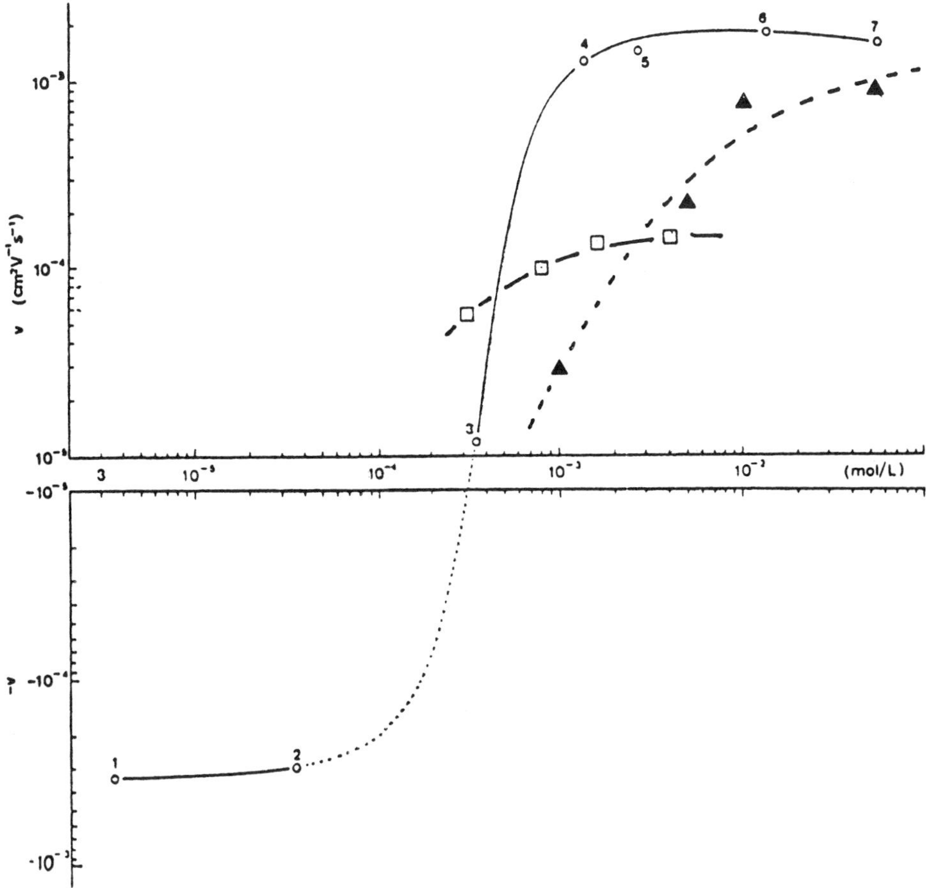

FIGURE 9. Variation of electroosmotic flow with the concentration of cetyltrimethylammonium bromide. (O)v; electroosmotic flow velocity. Horizontal line is the concentration of CTAB. The buffer for points 1–3 was 0.02 *M* 2-(*N*-morpholine) ethanesulfonic acid-L-histidine aqueous solution with 0.5% ethylene glycol (pH 6.2). The buffer for points 4–7 was 0.01 *M* phosphate buffer with 5% ethylene glycol (pH 8.3). FEP (0.2 or 0.3 m*M*) and a fused silica capillary (50-µm I.D.) were used for points 1–3 and 4–7, respectively. (From Tsuda, T., *J. Liq. Chromatogr.*, 12, 2501, 1989. With permission.) (▲) pH 3, 10 m*M* phosphate buffer containing 10% acetonitrile, 10-mm fused silica capillary. (Adapted from Pfeffen, W. D. and Yeung, E. S., *Anal. Chem.*, 62, 2178, 1990.) (❑) pH 7, 10 m*M* phosphate buffer, 10-mm fused silica capillary. (Adapted from Pfeffen, W. D. and Yeung, E. S., *J. Chromatogr.*, 557, 125, 1991.)

times for negative ions since both solute mobility and electroosmotic flow are in the same direction, and to decrease the adsorption of protein samples by the use of electrostatic repulsion between the surface and solute.[24,30,31]

Towns and Regnier[30] used the nonionic surfactants Tween and Brij to modify the capillary surface. A capillary was modified with octadecylsilane, and then a micellar solution of surfactant was pulled through the capillary for 2 h to ensure complete coating. The surface was covered both octadecylsilane (ODS) and nonionic surfactant adsorpted on the surface or partitioned with ODS. As the surface is covered with Brij/ODS, the electroosmotic velocity is decreased, especially at high pH, due to the coverage of the original charges and the loss of free silanol groups.[30] Polyethyleneimine (PEI), which acts as a polycation, has also been used to coat capillary tubing,[31] and imparts a positive charge to the capillary surface. Towns and Regnier[31] used a capillary that was partially coated with PEI (15 cm of a 100-cm length); the elution time of a neutral marker vs. capillary length is shown in Figure 10. The capillary partially coated with PEI (line A) shows a longer elution time compared to that of the uncoated capillary (line B). However, line A demonstrates the same behavior as line B when all of the

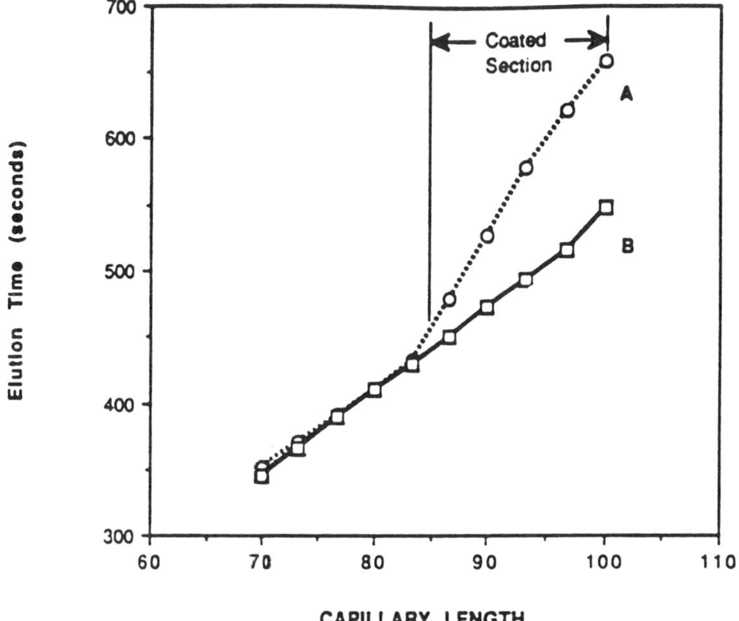

FIGURE 10. Elution time of a neutral marker (methyl oxide) vs. capillary length for (A) partially coated polyethyleneimine (M_r 20,000) (PEI-200) capillary and (B) uncoated capillary. Capillary A has the first 15 cm coated with PEI-200; the remaining length is uncoated. Sections (3.3 cm) of a 100-cm capillary were removed between runs with a corresponding drop in voltage to keep the field strength at 300 V/cm. Conditions: 50-μm I.D. capillary; 0.01 *M* phosphate buffer, pH 7.0; detection at 254 nm with window 35 cm from cathode. (From Towns, J. K. and Regnier, F., *Anal. Chem.*, 64, 2473, 1992. With permission.)

capillary coating is removed. The theoretical plate number obtained with the partially coated capillary is lower compared to that obtained without coating. These results indicate that the overall electroosmotic flow is toward the cathode, although the local electroosmotic flow in the portion of the capillary coated with PEI is toward the anode. Therefore, the flow profile is not equal throughout the entire capillary.

X. ELECTRIC CURRENT AS AN INDEPENDENT FACTOR

Electrovoltage, E_z, is dependent on RI, where R and I are electric resistance and current, respectively. R is dependent on the temperature of the buffer. Therefore, constant voltage operation means that current is always changing during the course of experiments.

In capillary electrophoresis, high voltage is applied along a capillary between 10 and 200 μm in diameter. For typical experiments, the current passing through the capillary tube produces heat, which inevitably causes a rise in temperature. A 45°C rise was observed in a 50-μm I.D. capillary, at 410 V/cm with a current of 50 μA,[16] and a 12° rise recorded in less than 10 s at 330 V/cm and 30 μA.[17] The production of heat is an inevitable result of applying high voltage, and causes the elevation of the buffer temperature. Therefore, it is very important to determine factors that are independent of the temperature of a buffer. One parameter is current density, which has been shown to be independent of temperature and geometry.[6,7,9,12] Thus, the equation, which shows a direct relationship between current density and electroosmotic flow, has been derived.[7,12]

As E is equal to i/k_c, where i denotes current density (A cm^{-2}) and k_c indicates specific conductance (Ω^{-1} cm^{-1}), Equation16 can be written as

$$v_{(osm)} = (i\varepsilon_0 D\zeta)(k_c\eta)^{-1} \tag{36}$$

D and ζ vary only slightly with temperature[1,39] and the variation is such that their product is essentially independent of temperature,[1,40] and ε_0 is constant. Accordingly, Equation 36 can be approximated as

$$v_{(osm)} = A_1(i/k_c\eta) \tag{37}$$

where A_1 is a constant. The product of k_c and η, the Walden product, is independent of temperature.[41] Thus Equation 37 becomes

$$v_{(osm)} = A_2 i \tag{38}$$

where A_2 is a constant and equal to

$$A_2 = (\varepsilon_0 D\zeta)(k_c\eta)^{-1}$$

From Equation 38, electroosmotic flow is seen to be dependent only on current density, not on temperature.

The relation between applied voltage and electric current is, at first, linear and then becomes nonlinear, as shown in Figure 11. This behavior is assumed to result from the increase in temperature of the buffer solution, the change in electrical resistance, and/or a change in the kinetics of electrophoresis.[9,41] There cannot be a linear relationship between electroosmotic flow and applied voltage.[7] However, when the electric current, I, or current density, i, is used as the independent variable instead of E, a linear relationship between electroosmotic flow and i is obtained, as shown in Figure 12.[7] This linear relationship has been determined experimentally by Tsuda et al.,[6,7,9,12] Terabe et al.,[16] and Huang et al.[17] The relationship between $v_{(osm)}$ and electric field strength is not linear.[6,7,9,12,16-18]

XI. EFFECT OF CONCENTRATION OF ELECTROLYTE

The concentration of electrolyte affects many factors: (1) in Equation 4 charge density per unit volume is directly related to concentration; (2) the thickness of the double layer of Equation 23 is proportional to the square root of ionic strength; (3) capacitance of the inner interface of Equation 24 is related to the square root of the ion concentration; (4) as ζ potential is directly proportional to the capacity at the inner interface, electroosmotic flow is inversely proportional to the square root of the concentration of electrolytes from Equations 22, 27, and 16. From Equation 35 the ion concentration is inversely proportional to electroosmotic flow.

There is another way to express the relation between electroosmotic flow and the concentration of electrolyte. The variables in Equation 36 that depend on the concentration of electrolyte are D, ζ, k_c, and η, and we assume that the product of D and ζ is constant.[1,40] For calculating approximate values of k_c and η with different concentrations of electrolyte, c (mol/l), many equations have been proposed. By Falkenhagen and Dole,[47]

$$\eta_{solu}/\eta_{water} = 1 + Ac^{0.5} \tag{39}$$

From Walden,[48]

$$k_c = (k_{c,inf})(1 + \mathrm{B}c^{0.5})^{-1} \tag{40}$$

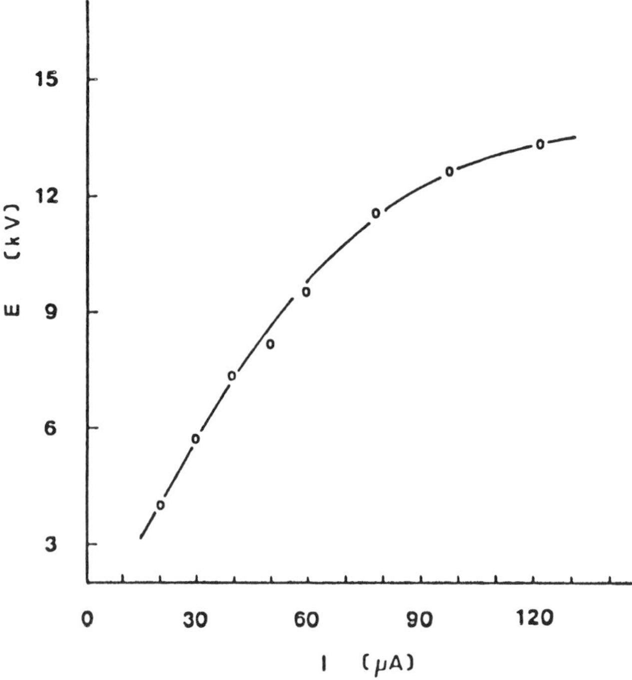

FIGURE 11. Relation between applied voltage, *E*, and electric current, *I*, under the conditions of zone electrophoresis. Capillary tubing: Pyrex glass capillary (422 mm × 85–μm I.D.) connected with fused silica capillary (142 × 195-μm I.D.). Solvent: 0.02 *M* phosphate buffer, pH 7, with 0.5% ethylene glycol. (From Tsuda, T., Nomura, K., and Nakagawa, G., *J. Chromatogr.*, 248, 241, 1982. With permission.)

where *A* and *B* are constants, and $k_{c,inf}$, is the value of k_c at infinite dilution of electrolyte, and η_{solu} and η_{water} are the viscosity of solution and viscosity of water, respectively. The product of k_c and η might be approximated by using Equations 39 and 40:

$$k_c\eta = 1 + (A - B)c^{1/2} + (A - B)Bc \tag{41}$$

Substituting Equation 41 into Equation 36 yields

$$v_{(osm)} = A_2 i[1 + (A - B)c^{1/2} + (A - B)Bc]^{-1} \tag{42}$$

where A_2 is a constant. The experimental results relating $v_{(osm)}$ and the concentration of electrolyte from studies by Tsuda et al.,[6,9] Fujiwara and Honda,[51] and Van Orman et al.[38] generally support Equation 42.

XII. EFFECT OF pH ON ELECTROOSMOTIC FLOW

The effect of pH is theoretically described by Hayes et al.[28] They proposed Equation 32, which includes the contribution from dissociated silanols on the glass inner surface, σ_{SiO^-}. However, their estimation does not correspond well with the obtained experimental values.[27,28] It is possible that their estimation of other factors and/or the additive law of charge densities are not valid. There are several experiments relating pH dependency to electrophoresis.[7,12,28,30,31,38] Values for the pK_a of silanols on the glass capillary surface range from 1.5 to 10; the most reasonable values are 5 to 7 and 10 for pure silicic acid.[49,50] Using Equation 32 we calculated pK_a values for silanols on a fused silica surface,[52] and

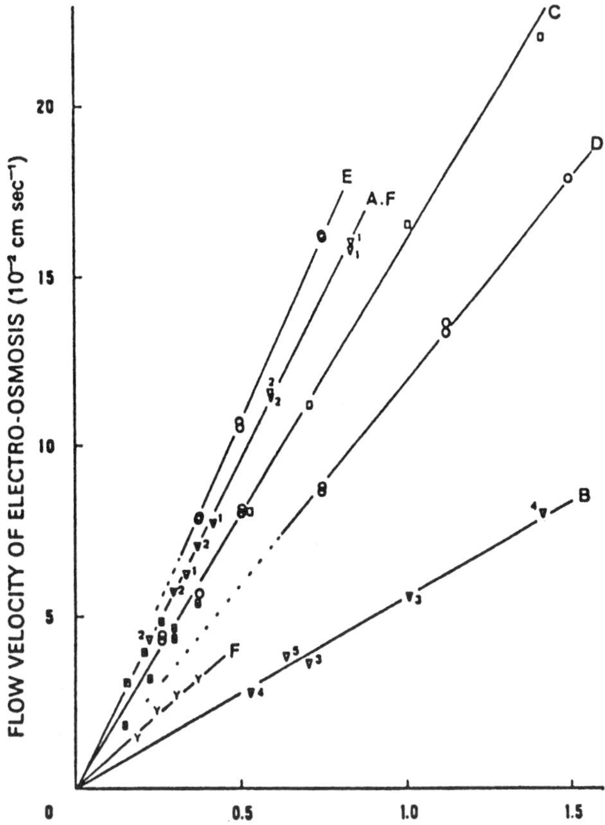

FIGURE 12. The relationship between electroosmotic flow velocity and electric current density. A and B, Pyrex glass capillary (∇) 88 (1), 208 (2), 60 (3), 85 (4), and 200 μm (5) inner diameter; C–E, fused silica capillaries (O) 195, 92, and 72 μm inner diameter, respectively; F, FEP with 323-μm I.D. (Y) and 500-μm I.D. (◪); C also shows (❏ and ▤) combined tubings of Pyrex glass and fused silica capillary, 85- and 195-μm I.D. (▤) and 208- and 195-μm I.D. (❏), respectively. The solution was 0.02 *M* phosphate buffer, pH 7, with 0.5% ethylene glycol, except in the case of B, where it was 0.05 *M* Na_2HPO_4. Solutes were benzene or pyridine. (From Tsuda, T., Nomura, K., and Nakagawa, G., *J. Chromatogr.*, 248, 241, 1982. With permission.)

have obtained 4.3 and 5 from the experimental data of Hayes and Ewing[27] and Towns and Regnier,[30] respectively.

As shown in Figure 13, at pH higher than 8 most of the silanol groups on the inner surface of an uncoated capillary are dissociated, and the velocity of electroosmotic flow is not dependent on pH. Under pH 4, $v_{(osm)}$ is not dependent on pH because most of the silanols are not dissociated. Therefore, $v_{(osm)}$ varies between pH 4 and 8. It is necessary to note that the pK_a of surface silanols is always dependent on the surface conditions and the buffer employed.

XIII. EFFECT OF THE COMBINATION OF TWO DIFFERENT CAPILLARY COLUMNS ON LOCAL ELECTROOSMOTIC FLOW VELOCITY

It is quite interesting how the apparent electroosmotic flow, $v_{(app)}$, behaves in a capillary column constructed of two different capillary materials. Tsuda et al.[7] examined the $v_{(osm)} - i$ relationship for Pyrex glass and fused silica tubing connected end to end with the 5-mm × 0.2-mm I.D. piece of poly(tetrafluoroethylene) tubing. The relationship obtained is the same as that given by the fused silica capillary of 195-μm I.D., shown by line C in Figure 12. It is

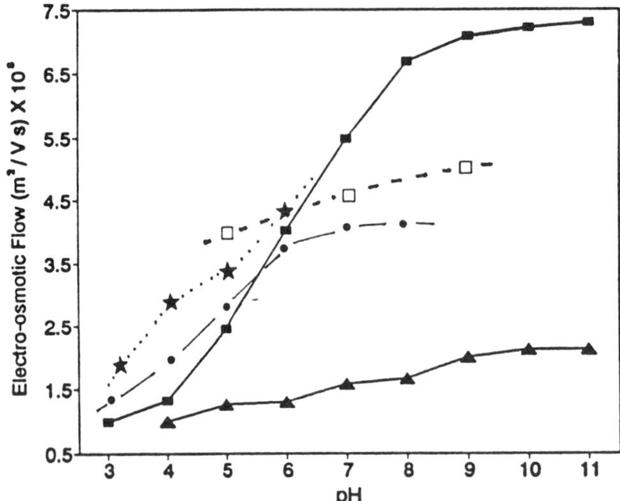

FIGURE 13. Dependence of electroosmotic flow ([m²/V · s] × 10⁸) on pH for (■) uncoated capillary; (▲) Brij-35/alkylsilane-coated capillary (from Towns, J. K and Regnier, F., *Anal. Chem.*, 63, 1126, 1991. with permission); (❑) uncoated (adapted from Tsuda, T., Nomura, K., and Nakagawa, G., *J. Chromatogr.*, 248, 241, 1982); (★) uncoated, theoretically estimated value, and (●) experimental value (adapted from Davies, J. T. and Riedeal, E. K., in *Interfacial Phenomena*, Academic Press, New York, 1963).

clear that the total electroosmotic flow in such a combination is controlled by the tubing that gives the lower electroosmotic flow.[7]

More recently, Towns and Regnier[31] used a capillary column in which 15% of the total length was coated with polycation (PEI) while the remaining 85% was uncoated. When the portion of the capillary coated by PEI was 15%, the $v_{(app)}$ was 83.6% of $v_{(osm)}$ for a 100% uncoated capillary column. From their experimental results obtained with different percentages of PEI, $v_{(app)}$ was found to be proportional to the ratio of length of PEI coating to the total column length, shown in Figure 10.[31] Although the electroosmotic flow in the PEI-coated portion is toward the anode and in the uncoated portion is toward the cathode due to the different local signs of surface charge, total electroosmotic flow is toward the cathode. This means that the main portion of the capillary dominates in determining electroosmotic flow even when there is partial flow against it. The theoretical plate number of a neutral maker is dramatically decreased by the combination of capillary surfaces (Figure 14). Therefore, the local flow profile may be very different of that in a uniform column. The flow profile in this case will be discussed in the next section.

On the basis of the above results,[7,31] we should be very careful about assumptions regarding the uniformity of the capillary inner surface. Unevenness will generate local flow effects,[1] affect the local electroosmotic flow profile, and cause the broadening of peaks. Consequently, a decrease of theoretical plate number will also be observed.

XIV. ELECTROOSMOTIC FLOW PROFILE

The electroosmotic flow profile has been discussed under the conditions of electroosmotically driven open-tubular liquid chromatography,[2,5,6,9,19,53,54] capillary zone electrophoresis (CZE)[4,7,12,14,15] and electroosmotically driven electrochromatography (ELC).[20] The electroosmotic flow profile is different from the parabolic laminar velocity profile of pressurized flow. The former flow profile is much flatter than the latter, thus very narrow peaks are obtained in electroosmotically driven chromatography and CZE. In early experiments, Pretorius et al.[2] and Tsuda et al.[6] obtained 10- and 30-fold less band broadening in ELC than expected with

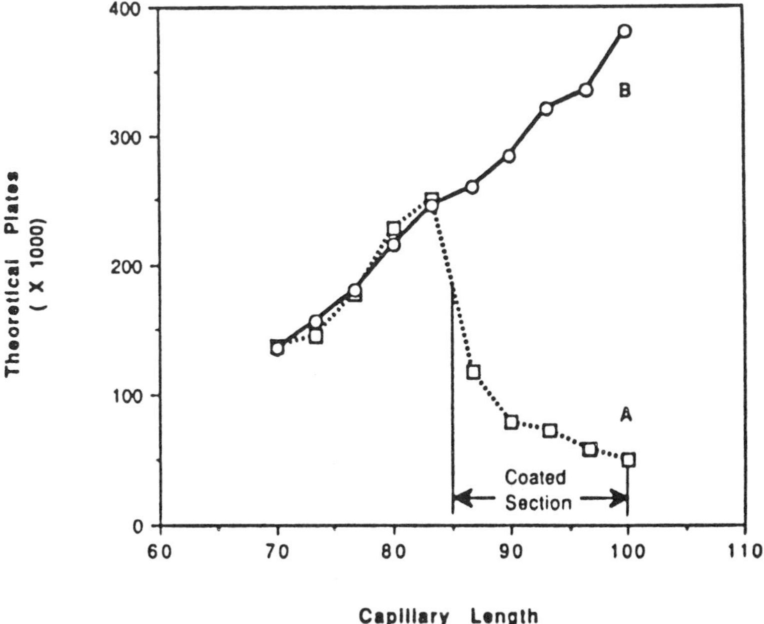

FIGURE 14. Plate number of mesityl oxide vs. capillary length for (A) partially coated PEI-200 capillary, and (B) uncoated capillary. Conditions are identical to those described in Figure 10. (From Towns, J. K. and Regnier, F., *Anal. Chem.*, 64, 2473, 1992. With permission.)

pressurized flow. Pretorius et al.[2] suggested that the profile is flat except in the region of the diffuse double layer near the column inner wall, where the flow profile has a quadratic velocity profile owing to the friction experienced by the viscous liquid flowing by the wall. Jorgenson and Lukacs[4] assumed that the zone broadening in CZE was generated only by axial molecular diffusion of solute and medium. Tsuda et al.[6] found that it was difficult to explain the experimental results solely on the basis of axial molecular diffusion, and proposed that the flow profile might be a combination of plug and Poiseuille flow. Rice and Whitehead,[35] Martin and Guiochon,[14] and Martin et al.[15] proposed the equations describing the electroosmotic flow profile as a combination of plug and Poiseuille flow. They studied the flow profile for pressurized flow under applied electrovoltage in a capillary tubing.

The average flow profile, u, can be computed as follows:

$$u/v_{(osm)} = F(\kappa r) \tag{43}$$

where $F(\kappa r)$ is given in Equation 20. Martin and Guiochon[14] use an adjustable parameter, ρ_d, which can be related to the thickness of the double layer relative to the column radius. Boundary values of $\rho_d = 1$ correspond to the usual Poiseuille flow (parabolic flow) and $\rho_d = 0$ to plug flow. One may estimate ρ_d:

$$\rho_d = (3/\kappa r)\{1 + 1/(4\kappa r) + 1/[4(\kappa r)^2] + 19/[64(\kappa r)^3]\} \tag{44}$$

Their result (dashed curve) is shown in Figure 15, with $\kappa r = 10$ and $\rho_d = 0.3083$.[14]

One typically uses a fused silica capillary of 50- to 100-μm I.D. and 10 m*M* buffer solution. Under these experimental conditions the diffusion double layer is assumed to be 30 to 300 nm; the value of κr is 830 to 83 for a 50-μm I.D. capillary column. Therefore, under normal conditions, the region of parabolic flow is limited and the flow profile is assumed to be a plug. Tsuda et al.[55] observed the zone profile by using a rectangular capillary, 50 × 1000 μm, with

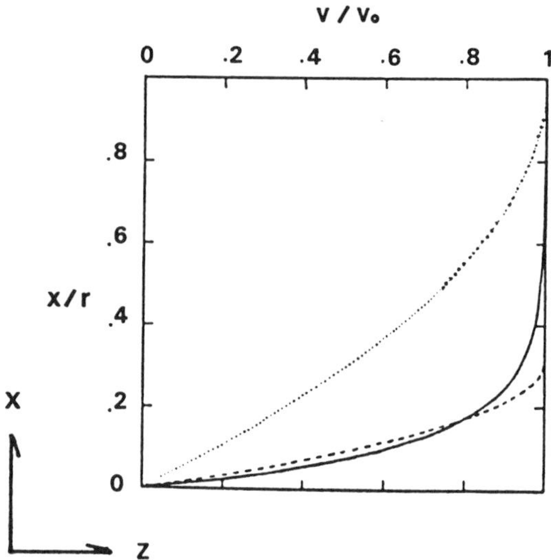

FIGURE 15. Velocity profile vs. relative distance from column axis: solid curve, electroosmotic flow; dashed curve, partially parabolic flow; $\kappa r = 10$ and $\rho = 0.3083$. The dotted curve represents the classic Poiseuille parabolic profile ($\rho = 1$). (From Martin, M. and Guiochon, G., *Anal. Chem.*, 56, 614, 1984. With permission.)

the aid of a microscope-video system. A rectangular capillary was used because the flat sides cause less distortion of the observed zone front than do the more common round walls of a cylindrical capillary.[55,56] The colored solute (Rhodamine 590, in methanol) was continuously introduced into a rectangular capillary and the zone front was tracked with a microscope for observation of the zone front profile.[55,57] The zone front was observed through the 1000-μm wide section of capillary.

The photographs and computer images of the zone front taken from the video tapes are shown in Figure 16 and 17, respectively. The length between each line on the *x* axis correspond to 59 μm. The period between pictures is 1 s. The advance of the zone front between two pictures can be measured from photographs by matching the positions of the stationary marks on the rectangular capillary. The zone front shown in Figure 16 is flat at the center with small bends at both ends of the 1000-μm axis, near the 50-μm plates. The progress of the zone front, recorded on video tape, was analyzed by computer software. The resulting image of the advancing zone front is shown in Figure 17. The 4 images in Figure 17 are successive 4-s spans, with the 1-mm axis divided into 17 sections. Both edges in the *z* direction of the image correspond to zone fronts. The advance of the zone front along the *z* axis is summarized numerically in Table 2. The average numbers at each *x* section were similar. The flow profile is pluglike in the center part of the capillary from sections 1 to 15. Unfortunately, the numerical values of the progress at sections 0 and 16 were not stable enough to estimate. To examine the profile in the vicinity of the 50-μm plate, we focused on the movement of the zone front with high magnification. One of the photographs obtained is shown in Figure 18. The zone front near the wall, namely near sections 0 and 16 of the *x* axis, has traveled further than the zone front in the central portion of the capillary. It was concluded that the advances of the central portion are similar in Figures 16 and 17 and Table 2, and the zone front at the edges of the 1000-μm axis in Figures 17 and 18 are ahead compared with the central portion. This phenomenon was observed in every experiment that we recorded.

We suggest the following. The edge portion of the front advances during the initial period, immediately after application of voltage on the rectangular capillary following the smooth operation of changing the reservoir from pure methanol to the methanol containing Rhodamine 590. From then on, the zone maintains equal speed at every point along the *x* axis. The center

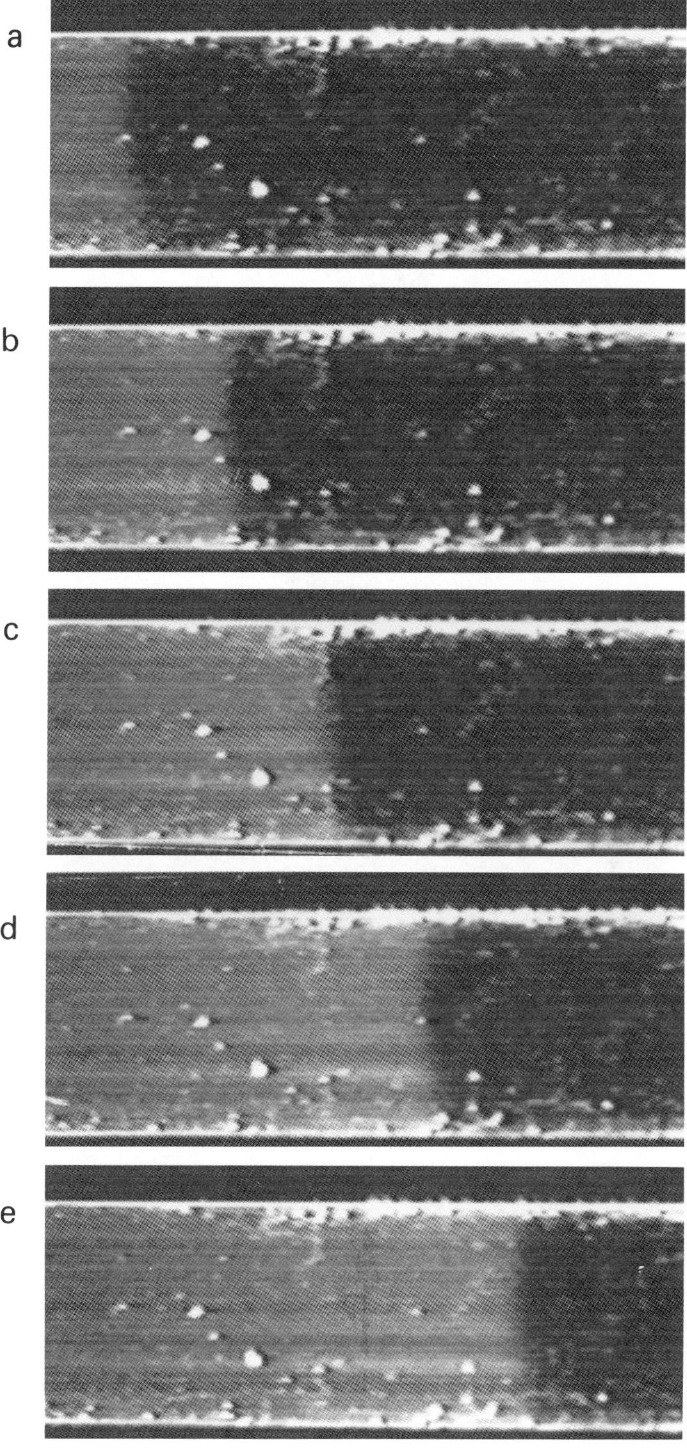

FIGURE 16. Photographs of zone front electroosmotic flow. Colored sample solution: 0.1 m*M* Rhodamine 590 in methanol. Applied voltage and current, 1.59 kV and 0.12 μA, respectively. The period between photographs is 1 s. The five photographs (a–e) are a series. (From Tsuda, T., Ikedo, M., Jones, G., Dadoo, R., and Zare, R. N., *J. Chromatogr.*, 632, 201, 1993. With permission.)

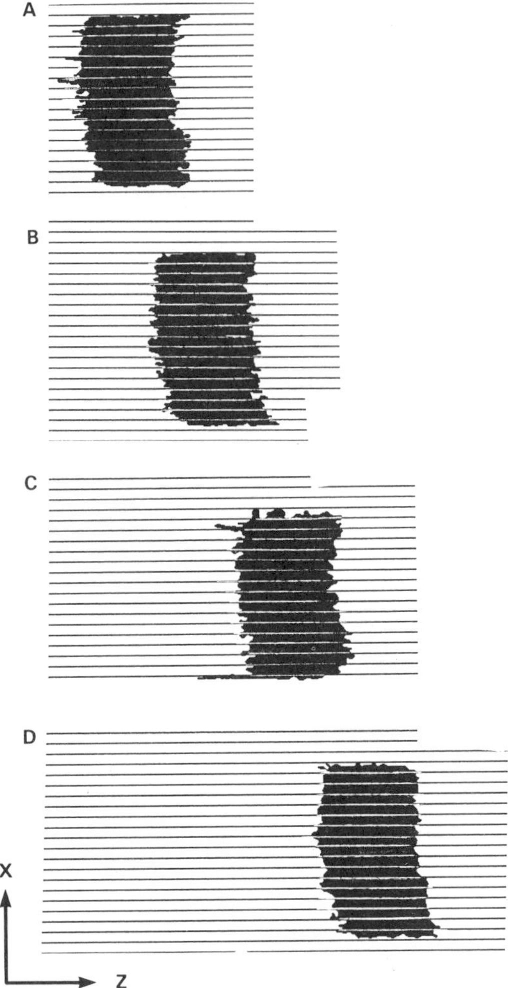

FIGURE 17. Computed images of zone front on the z axis over a 1-s span. The four images (A–D) are a series. Each image shows advances in the z axis over 1 s. The two ends of an image on the z axis correspond to zone fronts 1 s apart. (From Tsuda, T., et al., *J. Chromatogr.*, 632, 201, 1993. With permission.)

part lags behind edges. The flow profile observed is very different from the suggestions proposed previously. To our knowledge, this is the first time that the zone front pattern of electroosmotic flow has been directly observed.

This unique result could be explained in two ways. First, when the electroosmotic flow exits at the outlet of the capillary into the reservoir, some pumping power is required. Because there is not enough potential gradient to accept the flow coming from the column, pumping power due to electroosmotic flow is necessary. This causes the zone front in the central portion of the capillary to be forced back due to the hydrostatic back pressure of the reservoir.

A second explanation might be as follows. The distribution of positive and negative ions are such that the abundance of positive ions is extremely high in the region of the double layer up to $3/\kappa$, with the abundance of positive ions decreasing toward the center of the capillary tubing. In other words, the negative ions are concentrated at or near the center of capillary and their abundance in the region of the double layer is extremely small or none. The total electrocharge balance must be maintained over the whole xy axial range. The central part has a flow direction opposite to that near the wall regions. "Near the wall" does not mean in the

TABLE 2
Advances of the Zone Front in z Axis over a 1-s Period at x Section[a]

Number of y section	Experiment number 1	2	3	4	Average value[b]
0	—	8.3	—	7.5	—
1	7.8	7.6	8.0	7.0	7.6
2	7.2	8.3	7.6	7.2	7.78
3	8.0	6.9	8.4	7.8	7.78
4	8.1	7.8	8.0	7.1	7.75
5	8.0	8.3	6.7	8.5	7.88
6	7.6	8.0	7.4	7.8	7.7
7	7.0	8.2	7.8	7.4	7.63
8	7.7	8.0	6.5	8.5	7.68
9	8.1	8.1	7.4	7.6	7.8
10	7.5	8.4	8.0	7.5	7.85
11	8.0	7.7	7.6	7.7	7.75
12	7.4	7.6	8.5	7.0	7.63
13	7.8	7.5	8.2	7.5	7.75
14	6.8	8.0	8.3	6.9	7.5
15	6.0	8.6	7.6	7.7	7.48
16	—	8.0	—	—	—

[a] One unit of x axis = 57.8 μm.
[b] Mean of average value and its standard deviation are 7.69 and 0.12, respectively.

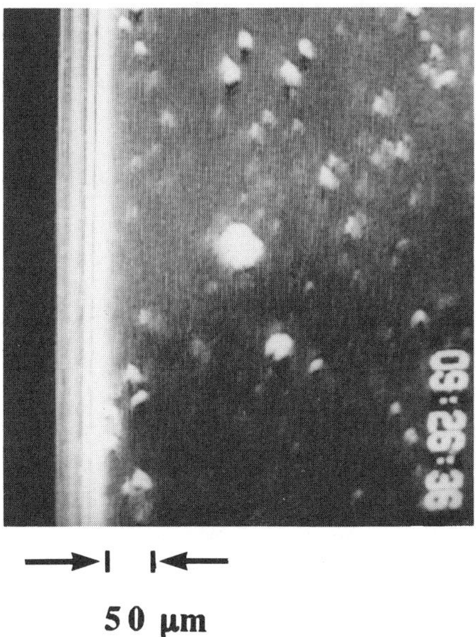

FIGURE 18. Zone front near the 50-μm plate. Experimental conditions as in Figure 16. (From Tsuda, T., Ikedo, M., Jones, G., Dadoo, R., and Zare, R. N., *J. Chromatogr.*, 632, 201, 1993. With permission.)

region of the diffuse double layer; it may be 10/κ or more. The local apparent flow velocity at the center region will be

$$v_{(app)center} = v_{(osm)edge} - v_{(ion)center}$$

Therefore the center region will trail behind the edge.

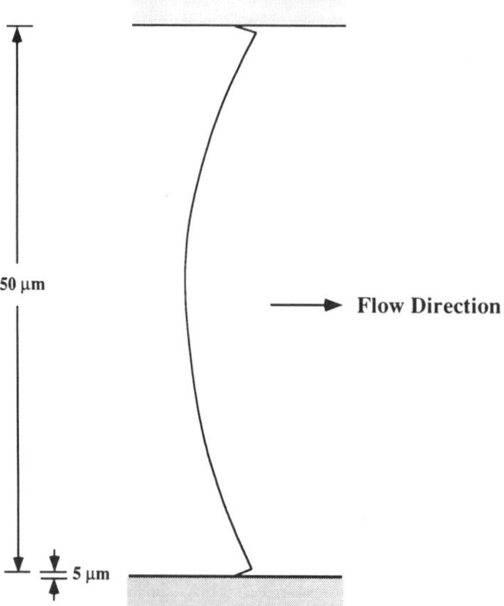

FIGURE 19. Flow profile for 50-μm fused silica capillary with 0.5 μm of the diffuse double layer. (From Tsuda, T., unpublished data, 1993.)

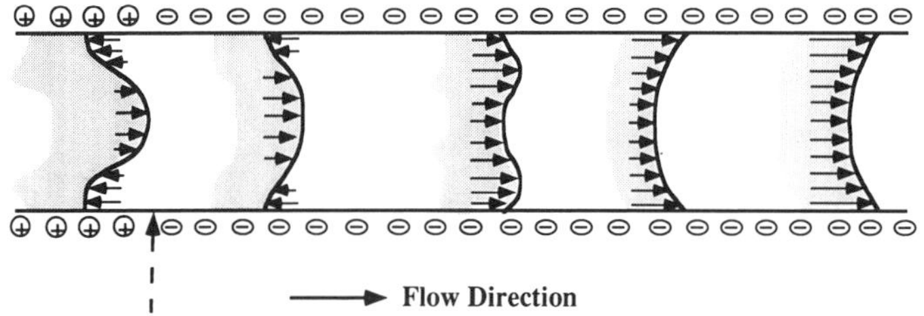

FIGURE 20. Proposed flow profile in combined capillary. (From Tsuda, T., unpublished data, 1993.)

The zone front in a capillary column with circular cross-section of 50- to 100-μm I.D. will appear different from the zone front shown in Figure 17, because the geometric dimensions are different from the rectangular capillary used, especially the length of the central portion. From the experimental results, we propose the zone front profile in a 100- to 50-mm I.D. capillary column with circular cross-section would appear as shown in Figure 19. The central portion of the zone front is retarded compared with the edges, as with a rectangular capillary.

The electroosmotic flow profile in the presence of ionic species is quite interesting. Tsuda et al.[55] observed that positive ionic solutes interact with the wall, which has negative charges. The edge of the zone front precedes ahead of the center part. Negatively charged solutes are repulsed from the wall and they are forced to concentrate in the center of the capillary, whereby the central portion precedes the edge of the frontal zone. These phenomena are explained by electrostatic repulsion or attraction. From these results we are able to imagine that the unequal distribution of ionic solutes or polarized neutral solutes (the physical basis of electroosmotic flow) will be generated by the solid surface charge.

Although the local flow profile is dependent on the local surface charge, in the case of combined capillary columns it depends also on total column conditions. In the column in which 15% in its length has positive surface charge, the flow profile may be very complex.[31] In Figure 20 we diagram the possible frontal profiles of electroosmotic flow at several local positions in a fused silica capillary with positively and negatively charged surfaces.

XV. CONCLUDING REMARKS

From the discussion presented above, one can readily see that we have only begun to delve into a true understanding of the basis for electroosmotic flow and the means to control it. As the depth of our knowledge increases, we should be able to devise practical methods of controlling electroosmotic flow, to enhance the separation of closely related molecules, or to enable the rapid determination of anionic solutes.

ACKNOWLEDGMENTS

The author would like to thank Robert P. Oda of the Mayo Clinic for his efforts in the final preparation of this chapter. His careful reading of the chapter and editorial comments were appreciated. The author would like to thank T. Yoshida for his helpful discussion.

REFERENCES

1. **Hjertén, S.,** Free zone electrophoresis, *Chromatogr. Rev.*, 9, 122, 1967.
2. **Pretorius, V., Hopkins, B. J., and Schieke, J. D.,** A new concept for high-speed liquid chromatography, *J. Chromatogr.*, 99, 23, 1974.
3. **Mikkers, F. E. P., Everaerts, F. M., Verhaggen, P. E. M.,** High-performance zone electrophoresis, *J. Chromatogr.*, 169, 11, 1979.
4. **Jorgenson, J. W. and Lukacs, K D.,** Zone electrophoresis in open-tubular glass capillaries, *Anal. Chem.*, 53, 1298, 1981.
5. **Jorgenson, J. W. and Lukacs, K D.,** High-resolution separations based on electrophoresis and electroosmosis, *J. Chromatogr.*, 218, 209, 1981.
6. **Tsuda, T., Nomura, K., and Nakagawa, G.,** Separation of organic and metal ions by high-voltage capillary electrophoresis, *J. Chromatogr.*, 264, 385, 1983.
7. **Tsuda, T., Nomura, K., and Nakagawa, G.,** Separation of organic and metal ions by high-voltage capillary electrophoresis, *J. Chromatogr.*, 264, 385, 1983.
8. **Tsuda, T., Nakajawa, G., Sato, M., and Yagi, K.,** Separation of nucleotides by high-voltage capillary electrophoresis, *J. Appl. Biochem.*, 5, 330, 1983.
9. **Tsuda, T.,** Electroosmotic flow for liquid chromatography and zone electrophoresis, *J. Chem. Soc. Jpn. Chem. Indust. Chem.*, 937, 1986.
10. **Tsuda, T.,** Modification of electroosmotic flow with cetyltrimethylammonium, *J. High Resolut. Chromatogr. Chromatogr. Commun.*, 10, 622, 1987.
11. **Tsuda, T.,** Elecrochromatography and the advantage and disadvantage of microchromatography, *Shimadju Sci. Instrum. News*, 28, 14, 1987.
12. **Tsuda, T.,** Electroosmotic flow and electric current in capillary electrophoresis, *J. Liq. Chromatogr.*, 12, 2501, 1989.
13. **Tsuda, T.,** *Chromatography*, Maruzen, Tokyo, 1989.
14. **Martin, M. and Guiochon, G.,** Axial dispersion in open-tubular capillary liquid chromatography with electroosmotic flow, *Anal. Chem.*, 56, 614, 1984.
15. **Martin, M., Guidochon, G., Walbroehl, Y., and Jorgenson, J. W.,** Peak broadening in open-tubular liquid chromatography with electroosmotic flow, *Anal. Chem.*, 57, 559, 1985.
16. **Terabe, C., Otsuka, K., and Ando, T.,** Electrokinetic chromatography with micellar solution and open-tubular capillaries, *Anal. Chem.*, 57, 834, 1985.
17. **Huang, X., Coleman, W. F., and Zare, R. N.,** Analysis of factors causing peak broadening in capillary zone electrophoresis, *J. Chromatogr.*, 48, 95, 1989.
18. **Huang, X., Gordon, M. J., and Zare, R. N.,** Current monitoring method for measuring the electroosmotic flow rate in capillary zone electrophoresis, *Anal. Chem.*, 60, 1837, 1988.
19. **Pfeffen, W. D. and Yeung, E. S.,** Open-tubular liquid chromatography with surfactant- enhanced electroosmotic flow, *Anal. Chem.*, 62, 2178, 1990.
20. **Pfeffen, W. D. and Yeung, E. S.,** Electroosmotically driven electrochromatography of anions having similar electrophoretic mobilities by ion pairing, *J. Chromatogr.*, 557, 125, 1991.

21. **Lee, C. S., Blanchard, W. C., and Wu, C.-T.,** Direct control of the electroosmosis in capillary zone electrophoresis by using an external electric field, *Anal. Chem.,* 62, 1550, 1990.
22. **Lee, C. S., McManigill, D., Wu, C.-T., and Patel, B.,** Factors affecting direct control of electroosmosis using an external electric field in capillary electrophoresis, *Anal. Chem.*, 63, 1519, 1991.
23. **Lee, C. S., Wu, C.-T., Lopez, T., and Patel, B.,** Analysis of separation efficiency in capillary electrophoresis with direct control of electroosmosis by using an external electric field, *J. Chromatogr.,* 559, 133, 1991.
24. **Wu, C.-T., Lopes, T., Patel, B., and Lee, C. S.,** Effect of direct control of electroosmosis on peptide and protein separations in capillary electrophoresis, *Anal. Chem.*, 64, 886, 1992.
25. **Wu, C.-T. and Lee, C. S.,** Ionized air for applying radial potential gradient in capillary electrophoresis, *Anal. Chem.*, 64, 2310, 1992.
26. **Ghowsi, K. and Gale, R. J.,** Field effect electroosmosis, *J. Chromatogr.,* 559, 95, 1991.
27. **Hayes, M. A. and Ewing, A. G.,** Electroosmotic flow control and monitoring with an applied radial voltage for capillary zone electrophoresis, *Anal. Chem.*, 84, 512, 1992.
28. **Hayes, M. A., Kheterpal, I., and Ewing, A. G.,** Effects of buffer pH on electroosmotic flow control by an applied radial voltage for capillary zone electrophoresis, *Anal. Chem.*, 65, 27, 1993.
29. **Davies, J. T. and Riedeal, E. K.,** in *Interfacial Phenomena*, Academic Press, New York, 1963.
30. **Towns, J. K and Regnier, F.,** Capillary electrophoretic separations of proteins using nonionic surfactant coatings, *Anal. Chem.*, 63, 1126, 1991.
31. **Towns, J. K. and Regnier, F.,** Impact of polycation adsorption on efficiency and electroosmotically driven transport on capillary electrophoresis, *Anal. Chem.*, 64, 2473, 1992.
32. **Delahay, P.,** *Double Layer and Electrode Kinetics,* Wiley Interscience, New York, 1965.
33. **Deyl, Z., Ed.,** Electrophoresis, *J. Chromatogr. Library*, 18, 1972.
34. **Hunter, R. J.,** *Zeta Potential in Colloid Science*, Academic Press, New York, 1981.
35. **Rice, C. L. and Whitehead, R.,** Electrokinetic flow in a narrow cylindrical capillary, *J. Phys. Chem.*, 69(11), 4017, 1965.
36. **Unger, K. K.,** Porous silica, *J. Chromatogr. Library,* 6, 1979.
37. **Fujiwara, S. and Honda, S.,** Effect of addition of organic solvent on the separation of positional isomers in high-voltage capillary zone electrophoresis, *Anal. Chem.*, 59, 487, 1987.
38. **Van Orman, B. B., Liversidge, G. G., McIntire, G. L, Olefirowicz, T. M., and Ewing, G.,** Effects of buffer composition on electroosmotic flow in capillary electrophoresis, *J. Microcolumn Sep.,* 2, 176, 1990.
39. **Gouy, G.,** *J. Phys. Chem.*, 9, 457, 1910.
40. **Chapman, D. L.,** A contribution to the theory of electrocapillarity, *Phil. Mag.,* 25, 475, 1913.
41. **Overbeek, H. G. Th.,** in *Colloid Science*, Vol. 1, Kruy, H. R., Ed., Elsevier, Amsterdam, 1952.
42. **Smoluchowski, M.,** in *Handbuch de Electrizitat und des Magnetismus* (Graetz), Vol. II, Barth, Leipzig, 1921, p. 366.
43. **Wallingford, R. A. and Ewing, A. G.,** Capillary electrophoresis, *Adv. Chromatogr.*, 29, 1, 1989.
44. **Hodgman, C. H.,** *Handbook of Chemistry and Physics*, CRC Press, Cleveland, OH, 1958.
45. **Watanabe, I., Ui, N., and Nakamura, M.,** Temperature dependence of the electrophoretic mobility of horse serum albumin, *Phys. Colloid Chem.*, 54, 1366, 1950.
46. **Horvath, A. L.,** *Handbook of Aqueous Electrolyte Solutions*, Ellis Horwood, Chichester, 1985, chap. 2, p. 11.
47. **Falkenhagen, H. and Dole, M.,** *Phys Z.,* 30, 611, 1929.
48. **Walden, P.,** *Z. Phys. Chem.*, 108, 341, 1924.
49. **Nawrocki, J.,** How to count strong adsorption sites by gas chromatography, *Chromatographia,* 24, 527, 1987.
50. **Jennings, W.,** *Analytical Gas Chromatography*, Academic Press, Orlando, FL, 1987.
51. **Fujiwara, S. and Honda, S.,** Determination of cinnamic acid and its analogues by electrophoresis in a fused silica capillary tube, *Anal. Chem.*, 58, 1811, 1986.
52. **Tsuda, T. and Kitagawa, S.,** unpublished data, 1993.
53. **Knox, T. H. and Grant, I. H.,** Miniaturization in pressure and electroendosmotically driven liquid chromatography: some theoretical considerations, *Chromatographia,* 24, 135, 1987.
54. **Yamamoto, H., Baumann, J., and Erni, F.,** Electrokinetic reversed-phase chromatography with packed capillaries, *J. Chromatogr.,* 593, 313, 1992.
55. **Tsuda, T., Ikedo, M., Jones, G., Dadoo, R., and Zare, R. N.,** Observation of flow profile in electroosmosis in a rectangular capillary, *J. Chromatogr.*, 632, 201, 1993.
56. **Tsuda, T., Sweedler, J. V., and Zare, R. N.,** Rectangular capillaries for capillary zone electrophoresis, *Anal. Chem.*, 62, 2149, 1990.
57. **Taylor, G.,** Dispersion of soluble matter in solvent flowing slowly through a tube, *Proc. R. Soc. London, Ser. A,* 219, 186, 1953.
58. **Huang, X., Luckey, J., Gordon, M. L., and Zare, R. N.,** Quantitative analysis of low molecular weight carboxylic acids by capillary zone electrophoresis/conductivity detection, *Anal. Chem.*, 61, 766, 1989.
59. **Tsuda, T.,** unpublished data, 1993.

CONCLUDING REMARKS

Chapter 23

FUTURE PROSPECTS FOR CAPILLARY ELECTROPHORESIS

James P. Landers and Thomas C. Spelsberg

TABLE OF CONTENTS

0-8493-8690-X/94/$0.00+$.50

I. INTRODUCTION

The acceptance of capillary electrophoresis (CE) as a viable analytical technique has begun. This is forcefully emphasized by the preceding 22 chapters, as well as the specialized volumes in the *Journal of Chromatography* covering the international symposiums on CE over the past 5 years. These not only emphasize the theoretical aspects but also the versatile nature of this technique as evidenced by its application to a multitude of analytical separations of importance in a wide spectrum of scientific disciplines. While the techniques currently available in CE offer a diversity of analyses, it is clear that its full potential is yet to be realized. Further developments in many areas of the CE field are underway and these include instrumentation, chemistry for enhanced detection and resolution, and applications.

II. INSTRUMENTATION

A. DETECTION

Improvements in detection will most probably involve modification of, or addition to, existing hardware. Improved ultraviolet (UV) detectors with greater capabilities such as spectral analysis (diode array detection) and enhanced detection limits are already becoming available. Non-UV and interface techniques such as fluorescence, chemiluminescence, conductivity, radioactivity, and CE-mass spectrometry (MS) systems have been established and some are commercially available. Of these, the use of laser-induced fluorescence (LIF) detection holds much promise. The sensitivity attainable with the LIF detection of fluorescent analytes is unprecedented, extending into the zeptomole (10^{-21}) to attomole (10^{-18}) range (10^{-12} to 10^{-9} *M*). As an extension of this detection approach, LIF microscopy has the potential for sensitivity at the 10- to 100-molecule level and, perhaps, down to single-molecule detection. This is yet to be fully explored.

B. MINIATURIZATION

In addition to the obvious advantages of CE for rapid, efficient, high-resolution analysis of microliter volume samples, other benefits result directly from its simplistic design. One of these is the potential for microminiaturization of the instrument. Using capillaries with a 10-μm I.D. and only a few centimeters in length, it has been shown that extremely high electric fields (5000 V/cm) can be used to facilitate "fast capillary zone electrophoresis (CZE)." [1] An example is given in Figure 1, where Jorgenson and co-workers' combined the use of a high electrical field and short capillary length to yield the separation of four fluorescein-labeled amino acids on the millisecond time scale. As an alternative to making capillaries with extremely small internal diameters, it has been shown that "micromachining" can be utilized to fabricate capillary channels in a "glass chip" that will suffice adequately for electrophoretic separations.[2] The very small internal diameters circumvent any thermal problems when high electrical fields are applied (2500 V/cm). The successful miniaturization of a CE system presents a number of possibilities, including the potential use as a microsensor system.[3]

C. CAPILLARY ELECTROPHORESIS IN TWO-DIMENSIONAL SYSTEMS

Like other separation techniques, CE has been shown to be amenable to interfacing with other analytical systems. As part of an on-line "two-dimensional configuration," CE has been shown to easily function as a first or second dimension. CE-MS, the first successful interfacing of CE with another analytical system, is an excellent example of CE as part of a two-dimensional system. In this case, separation by CE (which provides quantitative and qualitative information in the first dimension) is followed by mass spectrometric analysis, which provides a second quantitative dimension (mass or structural information) about the separated species. As dis-

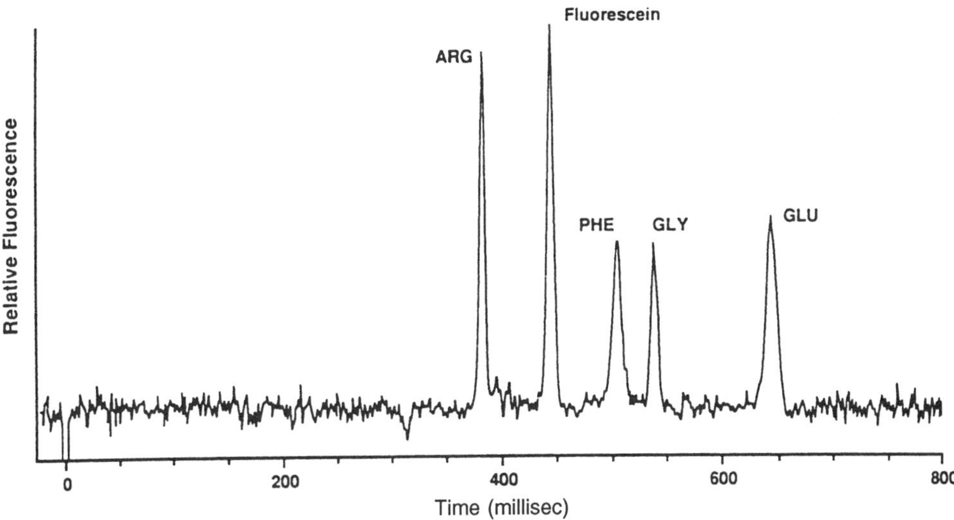

FIGURE 1. Fast CZE separation of fluorescein-tagged amino acids. Separation was performed in a 6-μm capillary having a $length_{tot}$ of 8 cm and a $length_{eff}$ of 0.5 cm at 20 kV in 10 m*M* phosphate buffer, pH 6.8. Injection was by pressure for a duration of 5 ms. Amino acid concentrations were 1.3×10^{-6} *M* for FTC-arginine, FTC-phenylalanine, and FTC-glycine and 3.8×10^{-6} *M* for FTC-glutamic acid. Excitation/emission was 488/520 nm. (Provided by Dr. J. Jorgenson, Chemistry Department, University of North Carolina.)

cussed in detail in Chapter 8 by Smith and co-workers, this makes for a tremendously powerful combination and is presently being explored by a number of researchers.

The utilization of CE as the second dimension in a two-dimensional array presents a new problem. Unlike CE-MS, where the speed of CE analysis is not a crucial factor (although very fast separations can be a problem), the analysis by CE as an on-line second dimension must be as fast or faster than the interval of sampling (sampling time) from the first dimension. Jorgenson's group, which has pioneered approaches for decreasing CE analysis time, has developed a system whereby CE analysis is completed on the low-second time scale. This makes sampling from a first-dimensional liquid chromatography (LC) separation possible.[4] Figure 2 shows the two-dimensional (reversed-phase liquid chromatography) RPLC-CE analysis of fluorescently labeled peptides from a tryptic digestion of horse heart cytochrome *c*. They have furthered this two-dimensional approach with the utilization of fast CZE. RPLC carried out over 40 to 50 min is sampled for fast CZE analysis completed on the millisecond time scale. This system will undoubtedly be extremely useful for analyses such as peptide mapping. It is clear that utilization of CE as part of a two-dimensional array will allow for the resolution of complex mixtures to approach that offered by standard two-dimensional systems such as isoelectric focusing/sodium dodecyl sulfate-polyacrylamide gel electrophoresis (IEF/SDS-PAGE).

Other multidimensional systems will surely be developed. It is reasonable to anticipate, for example, that a three-dimensional system consisting of RPLC-CE-MS is soon to follow: this would be tremendously powerful analytical tool.

D. EXTERNAL CONTROL OF ELECTROOSMOTIC FLOW

One aspect of CE that has gained much attention is the potential for controlling electroosmotic flow (EOF) externally. When working in bare fused silica capillaries, EOF varies proportionately with pH — extremely low EOF at low pH and substantial EOF at neutral and higher pH values. This can be disadvantageous when the EOF resulting from the choice of buffer system is not amenable to optimal resolution of sample components, i.e., at high pH EOF is too rapid, leading to poor resolution, or at low pH EOF is too slow, leading to

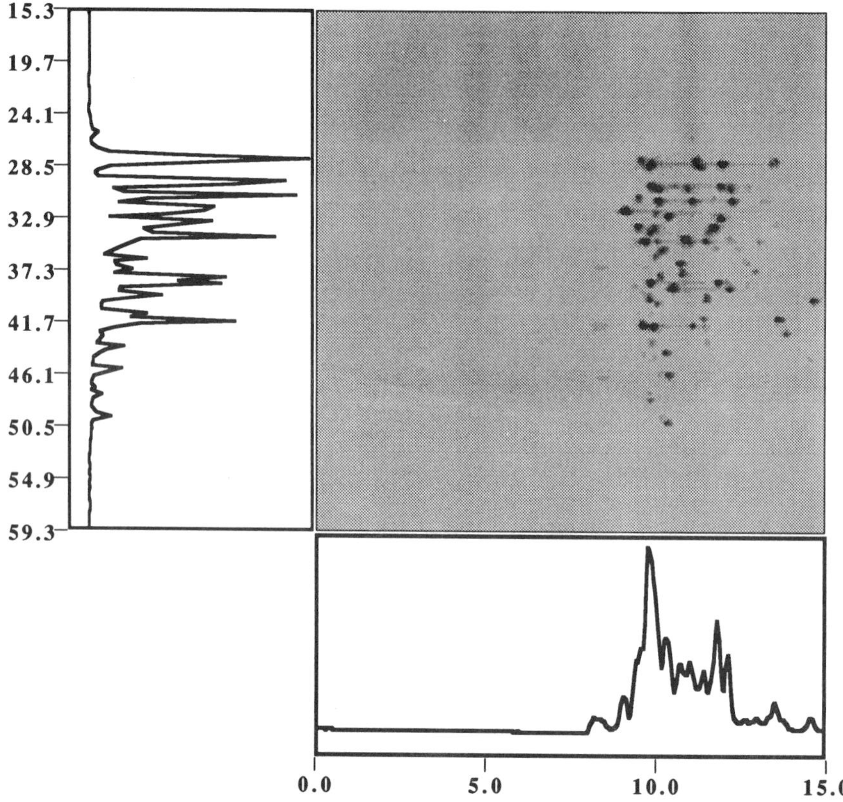

FIGURE 2. Two-dimensional RPLC-CE analysis of peptides from a tryptic digest of cytochrome *c*. RPLC was performed on a 2.1 × 150 mm Zorbax 300 SB-C8 column with 20 to 70% acetonitrile ramp over 50 min. CE analysis in a 1.2-cm $\text{length}_{\text{eff}}$ × 10-μm capillary at 20 kV in 10 m*M* phosphate buffer, 20 m*M* triethylamine, pH 11.0. Injection was by pressure for a duration of 5 ms. 0.3 × 10^{-6} *M*. Detection was at 214 nm. Numbers on both axes represent time in minutes. (From Larmann, J. P., et al., *Electrophoresis,* 14, 439, 1993. With permission.)

inconveniently long analysis times. Altering EOF presently requires the utilization of other means, such as varying ionic strength or using coated capillaries.

It is for this reason that having the ability to control EOF externally has captured the attention of the CE community. The work of Lee et al.[5] and others (discussed in Chapter 22 by Tsuda) shows that application of a radial potential gradient across the capillary alters the zeta (ζ) potential and, hence, the EOF in capillary electrophoretic separations. While this approach seems workable, it is limited to the use of acidic pH. It is clear that this problem will be circumvented with innovative designs. One such alternative approach that has shown promise is "flow counterbalanced CE".[6] With this system, an external pressure in the opposite direction of the EOF is used to "counterbalance" the flow from electroosmosis. This approach is extremely attractive since control of EOF is independent of pH.

III. CHEMISTRIES

There are few, if any, areas of the CE field that do not encompass a "chemistries" approach. However, some problems unique to CE have dictated the requirement for some new innovative approaches.

A. TRACE LEVEL DERIVATIZATION OF SAMPLE COMPONENTS

The tremendous sensitivity associated with LIF-CE presents the potential for identifying analytes present in complex sample matrices at very low concentrations. However, for

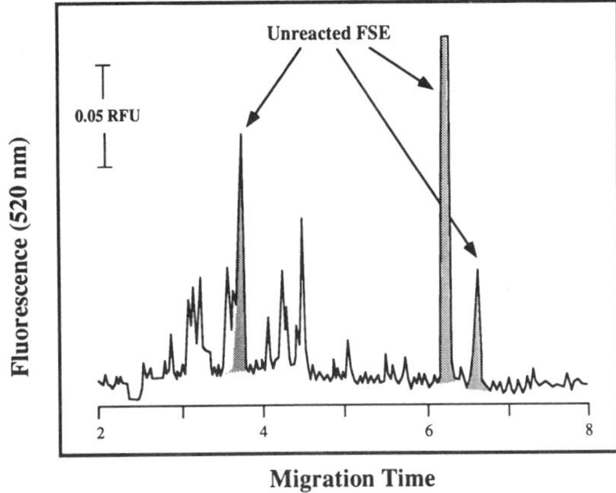

FIGURE 3. Zeptomole-level detection of fluorescently tagged peptides. CE separation of fluorescein succinimidyl ester-derivatized peptides from a tryptic digest of horse heart cytochrome *c*. Separation was performed in a 57 cm × 50 μm capillary at 25 kV in 100 m*M* buffer, pH 8.3. Sample was 10 ng/ml in water. Injection was by pressure for a duration of 1 s — total on-column mass was 400 × 10^{-21} mol.

detection of analytes that are not fluorescent, pre-column derivatization procedures (e.g., amine or sulfhydryl specific) have been described. At the present time, the limits for adequate specific labeling of trace quantities of a substance in a mixture is 1 to 2 orders of magnitude higher than the limits of detection achievable with LIF-CE (10^{-12}–10^{-9} *M*). Amine-specific labeling of peptides from a horse cytochrome *c* tryptic digest present in solution at concentrations less than 10 to 100 n*M* is extremely inefficient, and virtually nonexistent at less than 10 n*M*.[7] This is exacerbated by contamination of the desired labeled products with unreacted fluor or reaction by-products (Figure 3). Since it is unlikely that the unreacted fluor can be quantitatively removed by "clean-up" procedures with analyte concentrations at this low level, the design and synthesis of reactive species that fluoresce only once reacted (e.g., CBQCA for amines) will be of critical importance. It is clear that innovative alternatives will be necessary for solving these problems. One approach may be to concentrate the analytes of interest within a mixture into a micelle so that their effective concentration is greater, therefore enhancing the efficiency of the labeling reaction.[8] Alternatively, the low efficiency of labeling may be circumvented by not labeling at all, and detecting native fluorescence. Lee and Yeung[9] have demonstrated this to be a viable option with the laser-induced native fluorescence detection of specific proteins in single human erythrocytes.

B. CHEMICAL AND PHYSICAL GELS

Advances with gel buffer methods and new coating technologies for capillaries will be of great importance for the development of new applications. Capillary gel electrophoretic (CGE) analysis of DNA, both single and double stranded, has surpassed the performance of standard slab systems. On the other hand, denaturing protein analysis by capillary gel electrophoresis has progressed, but has not yet matched the resolution offered by standard SDS-polyacrylamide slab gel systems. It will be only a matter of time, however, before the CE technology will be optimized to complement, and eventually replace, traditional slab gel procedures by improving analysis speed and quantitation.

C. CAPILLARY COATINGS

Coatings, either permanent (bonded phase) or replaceable (dynamic), will permit the use of simpler buffer systems and should improve the reproducibility of the technique through minimizing analyte wall interactions. This area of CE is likely to advance at a rapid pace, since many manufacturers, both instrument and chemical suppliers, are dedicated to the research and development of new coatings for specific CE applications (involving both protein or DNA separations).

D. VOLATILE MICELLAR ELECTROKINETIC CAPILLARY CHROMATOGRAPHY (MECC) SYSTEMS

As described in Chapter 3, MECC has been shown to be applicable to a broad range of analytes in diverse sample matrices. The clear advantage of MECC is that the combination of electrophoresis and chromatography (partitioning of analytes into a micellar phase) allows for the resolution of neutral compounds that otherwise may not be possible with CE. While this mode of CE has been shown to resolve analytes of general biological interest, it is particularly useful for the analysis of a variety of organic compounds including aromatic hydrocarbons. For these types of samples, interfacing with MS could provide useful second-dimensional information. For example, Figure 4 of Chapter 3 shows the MECC separation of nine aromatic hydrocarbons, analytes that are suitable for identification by MS. However, it is not possible at the present time to interface MECC with MS, due to the nonvolatility of the micellar phase. It will therefore be important to develop micelle-forming components of sufficient volatility to permit interfacing with MS or a "separation trap" that would allow MS analysis of the volatile components without obstruction by the nonvolatile micellar constituents.

IV. BIOMEDICAL APPLICATIONS

A. FORENSIC DNA ANALYSIS

Forensic DNA analysis represents an exciting and challenging new area for CE. McCord and co-workers[10,11] have shown that physical gel systems can provide adequate resolution and reproducibility for the analysis of forensic DNA, using polymerase chain reaction (PCR)-amplified fragment length polymorphisms. They show that fluorescence detection is particularly useful for these types of samples with LIF-CE sensitivity in the 500-pg/ml range and resolution in the 4- to 6-bp range.[11] Clearly, the potential of CE for these types of analysis has only begun to be tapped.

B. CLINICAL

As clearly defined in Chapter 16, the rapid and reproducible analysis of extremely low volume samples makes CE ideal for clinical use. Yet, despite these characteristics, the acceptance of CE by clinical practitioners has been slow. One of the major barriers with transplanting CE from the research environment to the clinical laboratory is throughput. Even though some commercial CE units have the ability to analyze a large number of samples in an automated fashion, analysis is "sequential" and not "simultaneous." Therefore, for routine analyses, multicapillary "clinical CE" units will need to be devised so that simultaneous multiple sample analysis is possible. As described in Chapter 16, this has begun.

Capillary electrophoresis will also find a use in the clinical laboratory with specialized analysis (i.e., nonroutine tests). For example, determination of the presence of specific DNA fragments by "sequence-specific recognition"[12] presents the potential of CE for diagnosis of certain disease states. Capillary electrophoresis has also been shown to be useful for the monitoring of therapeutic drugs[13] and, as described in Chapter 17, for analysis of drug metabolites that are often indicative of metabolic dysfunction. In addition to identifying the presence of components foreign to the system, it also has potential for identifying a normal component of the system that is altered in structure. A good example is the analysis of abnormal hemoglobins associated with α-thalassemias.[14]

One area that shows tremendous promise is the utilization of antibodies to detect ultratrace levels of specific antigens in complex mixtures. Karger's group has shown preliminary evidence that this is possible with fluorescently tagged monoclonal antibodies.[15]

C. PROBING THE CONTENTS OF SINGLE CELLS

As should be clear from Chapter 15, the potential exists for analysis of the components of single cells. One of the obvious advantages of this technology would be the potential insights it would provide about cell-to-cell differences in heterogenous cell population from a single tissue. When studying the cellular components/mechanisms with a specific tissue or cells in culture, the biochemist is always cognizant of the fact that, although the sample is homogeneous from a tissue perspective, the population is heterogeneous at the cellular level. As a result, studies of cellular biochemical systems must always be interpreted as representing the average response from the mixed population. With CE, the possibility exists for dissecting the various characteristics associated with each cell in a heterogeneous population into individual responses.

V. OTHERS

A. FORENSICS

The success of CE for the analysis of small organic ions has made it perfectly suited for certain forensic analyses. Hargadon and McCord[16] have demonstrated the utility of CE for the analysis of low-explosive residues. They have shown that the short analysis times make CE ideal for the analysis of forensically relevant inorganic ions. The results were found to either complement or better those obtained with standard ion chromatography. Weinberger and Lurie[17] and Lurie[18] have described the forensic potential of CE for the analysis of illicit drugs. They show that MECC is ideal for the single-run analysis of both acidic and neutral components in illicit drug mixtures and is superior to separation obtained with high-performance liquid chromatography (HPLC).

B. PREPARATIVE CAPILLARY ELECTROPHORESIS

Fraction collection is an integral part of most biomolecular analyses. Capillary electrophoresis has been shown to be amenable to fraction collection,[19] although this is not done with the same ease as HPLC. Improvements in collection processes are needed for CE to be universally accepted as preparative method. Some commercial manufacturers have taken a dramatic step in making fraction collection and postseparation analysis easier, through modifications that allow for collection of separated proteins on a membrane. Further developments of this nature will be required.

C. MOLECULAR BIOLOGY

Within the field of molecular biology, new techniques are constantly evolving that require electrophoretic analysis. As discussed in Chapter 14, the routine use of CE for the sequencing of genes is not only feasible but appears to be superior to standard techniques. However, before this technique gains widespread acceptance, throughput will have to be improved, perhaps with a multichannel detection system. In other areas of molecular biology, the potential extension of CE to existing labor-intensive protocols has been demonstrated. For example, CE has been shown to have the potential to simplify a standard assay used for the analysis of gene promoter activity.[20] Other techniques that have potential for CE adaptability include (1) detection of single-stranded conformational polymorphisms to determine point mutations in genetic sequence, (2) temperature gradient electrophoresis to determine changes in DNA melting temperature due to sequence variations, and (3) various hybridization regimes designed to show DNA–DNA or DNA–protein interactions by a shift in electrophoretic mobility of the hybrid.

VI. CONCLUDING REMARKS

The developments in various areas of the CE field will ultimately lead to enhanced application of the technique to current separation and analytical problems. These developments include faster analysis times, amino acid analysis, standard ion analysis, intermolecular association studies, such as antibody–antigen complexation, receptor–ligand binding analysis, and enzyme–substrate kinetic analysis. One can envision that as the methods for analysis of complex mixtures become more standardized, automated instruments could be designed for a variety of diagnostic, biomedical, and forensic applications. In conjunction with development of specialized detection systems (electrochemical, radiometric, MS), new applications should be forthcoming, enhancing detection limits and pioneering new horizons.

REFERENCES

1. **Monning, C. A. and Jorgenson, J. W.,** On-column sample gating for high speed capillary zone electrophoresis, *Anal. Chem.*, 63, 802, 1991.
2. **Harrison, D. J., Seiler, K., Fan, Z., and Mann, A.,** Integration of capillary electrophoresis, sample injection, and flow injection analysis techniques on a chip using micromachining. Presented at the 4th Int. Symp. on High Performance Capillary Electrophoresis, Orlando, Florida, 1993.
3. **Hergenroder, R., Jacobson, S. C., Koutny, L. B., Warmack, R. J., and Ramsey, J. M.,** Microchip capillary electrophoresis. Presented at the 4th Int. Symp. on High Performance Capillary Electrophoresis, Orlando, Florida, 1993.
4. **Larmann, J. P., Jr., Lemmo, A. V., Moore, A. W., Jr., and Jorgenson, J. W.,** Two dimensional separations of peptides and proteins by comprehensive liquid chromatography-capillary electrophoresis, *Electrophoresis*, 14, 439, 1993.
5. **Lee, C. S., Blanchard, W. C., and Wu, C.-T.,** Direct control of the electroosmosis in capillary zone electrophoresis by using an external electric field, *Anal. Chem.*, 62, 1550, 1990.
6. **Culbertson, C. T. and Jorgenson, J. W.,** Flow counterbalanced capillary electrophoresis. Presented at the 4th Int. Symp. on High Performance Capillary Electrophoresis, Orlando, Florida, 1993.
7. **Landers, J. P., Oda, R. P., Ulfelder, K. J., and Spelsberg, T. C.,** A comparison of fluorescent derivatives for peptide and DNA analysis. Presented at the 4th Int. Symp. on High Performance Capillary Electrophoresis, Orlando, Florida, 1993.
8. **Sweedler, J. V., Timperman, A. T., Tracht, S. E., Fuller, R. R., and Cruz, L.,** Trace level derivatization schemes for laser-induced fluoresence detection in capillary electrophoresis. Presented at the 4th Int. Symp. on High Performance Capillary Electrophoresis, Orlando, Florida, 1993.
9. **Lee, T. T. and Yeung, E. S.,** Quantitative determination of native proteins in individual human erythrocytes CZE with laser-induced fluorescence detection, *Anal. Chem.*, 64, 3045, 1992.
10. **McCord, B. R., Jung, J. M., and Holleran, E. A.,** High resolution capillary electrophoresis of forensic DNA using a non-gel sieving buffer, *J. Liq. Chromatogr.*, 1993, in press.
11. **McCord, B. R., McClure, D. M., and Jung, J. M.,** Capillary electrophoresis of PCR-amplified DNA using fluorescence detection with an intercalating dye, *J. Chromatogr.*, 1993, submitted.
12. **Baba, Y., Tsuhako, M., Sawa, T., and Akashi, M.,** Sequence-specific recognition of DNA by capillary affinity gel electrophoresis using novel gel columns. Presented at the 4th Int. Symp. on High Performance Capillary Electrophoresis, Orlando, Florida, 1993.
13. **Wolfisberg, H., Schmutz, A., and Thorman, W.,** Assessment of automated capillary electrophoresis and micellar electrokinetic capillary chromatography for therapeutic drug monitoring. Presented at the 4th Int. Symp. on High Performance Capillary Electrophoresis, Orlando, Florida, 1993.
14. **Zhu, M., Wehr, T., Levi, V., Rodriquez, R., Shiffer, K., Montrose, S., and Cao, Z. A.,** Capillary electrophoresis of abnormal hemoglobins associated with α-thalassemias. Presented at the 4th Int. Symp. on High Performance Capillary Electrophoresis, Orlando, Florida, 1993.

15. **Karger, B. L., Foret, F., Schmalzing, D., Shimura, K., and Szoko, E.,** New approaches to trace analysis in capillary electrophoresis. Presented at the 4th Int. Symp. on High Performance Capillary Electrophoresis, Orlando, Florida, 1993.
16. **Hargadon, K. A. and McCord, B. R.,** Explosive residue analysis by capillary electrophoresis and ion chromatography, *J. Chromatogr.*, 602, 241, 1992.
17. **Weinberger, R. and Lurie, I. S.,** Micellar electrokinetic capillary chromatography of illicit drug substances, *Anal. Chem.*, 63, 823, 1991.
18. **Lurie, I. S.,** Micellar electrokinetic capillary chromatography of the enantiomers of amphetamine, methamphetamine and their hydroxyphenethylamine precursors, *J. Chromatogr.*, 605, 269, 1992.
19. **Rose, D. J. and Jorgenson, J. W.,** Fraction collector for capillary zone electrophoresis, *J. Chromatrogr.*, 438, 23, 1988.
20. **Landers, J. P., Schuchard, M. D., Subramaniam, M., Sismelich, T. P., and Spelsberg, T. C.,** High-performance capillary electrophoretic analysis of chloramphenicol acetyl transferase activity, *J. Chromatogr.*, 603, 247, 1992.

APPENDICES

Appendix 1

DESCRIPTION OF THE ELECTROPHORETIC PROCESS

TABLE OF CONTENTS

0-8493-8690-X/94/$0.00+$.50

I. MOBILITY OF AN ANALYTE IN AN ELECTRICAL FIELD

A charged particle in solution will become mobile when placed in an electric field. The velocity, v_i, acquired by the solute under the influence of the applied voltage H, is the product of μ_{app}, the apparent solute mobility, and the applied field E ($E = H/L$, where L is the length of the field).[1]

$$v_i = \mu_{app}E \tag{1}$$

II. SHAPE OF THE ANALYTE ZONE

Under the initial conditions of electrophoresis, the boundary between the buffer solution and the solute mixture forms a zone of infinitesimal thinness at right angles to the direction of applied current and migration. As migration proceeds, this initially sharp boundary will undergo a progressive deterioration in shape. The most important influence on this process is diffusion, as the initial conditions impart a severe concentration gradient across the zonal boundary. By applying Fick's second law of diffusion, Weber[2] has shown that the variation of solute concentration in the direction of migration is given by the equation

$$C_{x,t} = \left[\frac{k}{2(D_i\pi t)^{1/2}}\right] e^{\frac{-(x-v_i t)^2}{4D_i t}} \tag{2}$$

where C is the solute concentration at a distance x from the initial position after time t, v_i is the electrophoretic velocity of the solute, D_i is the diffusion coefficient of the solute, and k is a constant. By integration, initially with the boundary conditions at $t = 0$, at distant $x = \infty$, $C = 0$, and a second time with $t > 0$, $x = 0$ and $dC/dx = 0$, we obtain a mathematical description of the concentration profile. As the solute migrates a distance x from the origin after time t, the zone may be described by a Gaussian curve.

The characteristics of a Gaussian profile describe the peak maximum and the peak width. The maximum is dependent on the initial concentration of solute. The width depends on the length of time from initial conditions and the diffusion constant, D_i. The width may be given by the distance between the inflection points, in our case, $2/(2D_i t)^{1/2}$. From analogy with probability calculations, the width of a Gaussian curve is termed the standard deviation, σ:

$$\sigma = (2D_i t)^{1/2} \tag{3}$$

and the square of standard deviation is called the variance,

$$\sigma^2 = 2D_i t \tag{4}$$

Remembering that

$$t = L_d/v_i = L_d L_t/\mu_{app}V = L_d/\mu_{app}E \tag{5}$$

where L_d is the distance from the origin to the detector, and L_t is the total length of the capillary (i.e., the length of the applied electric field) we may substitute into Equation 4, to obtain

$$\sigma^2 = 2D_i\, L_d/(\mu_{app}E) \tag{6}$$

One should remember that the discussion to this point has dealt with diffusion only as a dispersive phenomenon affecting peak shape. Other factors that contribute to what is observed on the electropherogram as band broadening, the practical result of variance, are discussed later (Section IV, Sources of Variance).

III. RESOLUTION AND EFFICIENCY

The simplest way to characterize the separation of two components is to divide the difference in migration distance by the average peak width to obtain resolution (Res):

$$\text{Res} = 2(x_{i2} - x_{i1})/(w_1 + w_2) \tag{7}$$

where x_i is the migration distance of the analyte i, and the subscript 2 denotes the slower moving component, and w is the width of the peak at the baseline.[3] We can readily see that the position of a peak, x_i, is determined by the electrophoretic mobility. The peak width, w, is determined by diffusion and other dispersive phenomena. For two neighboring peaks, $w_1 = w_2$, and

$$\text{Res} = (x_{i2} - x_{i1})/w_2 \tag{8}$$

From the equation describing a Gaussian curve, the two peaks touch at baseline when $\Delta x_i = w_2 = 4\sigma$, and Res = 1, or

$$\text{Res} = \Delta x_i/4\sigma \tag{9}$$

Remembering that distance is equal to velocity times time ($x_i = v_i t$) and substituting for σ from Equation 3 into Equation 9, we obtain

$$\text{Res} = (\Delta\mu_{\text{app}}E)t/4(2D_{i,\text{avg}}t)^{1/2} \tag{10}$$

where $\Delta\mu_{\text{app}}$ is the difference in apparent electrophoretic mobility of the two solutes and $D_{i,\text{avg}}$ is the average diffusion of the two solutes.

To obtain a measure of efficiency for the process, we use probability theory.[4] For a random walk process of length L, made of n steps, the variance is given by

$$\sigma = l(n)^{1/2} \tag{11}$$

where l is the length of each step. If each step is independent of any other step, each contributes to the total variance of the process:[5]

$$\sigma_{\text{tot}}^2 = \Sigma\sigma_i^2 \tag{12}$$

Substituting from Equation 11 and rearranging,

$$1/n = L^2/\sigma_{\text{tot}}^2 = N \tag{13}$$

The number of steps in the random process, n, is inversely related to the number of theoretical plates, N, a measure of efficiency for the process.

We may substitute for L and σ in Equation 13 to express

$$N = (\mu_{avg}E)^2t^2/(2D_it) \quad (14)$$

where μ_{avg} is the average mobility of the two solutes. By comparing the expression for Res (Equation 10) with the definition of N (Equation 14), we obtain an expression relating resolution to the number of theoretical plates,

$$\text{Res} = (1/4)(\Delta\mu_{app}/\mu_{avg})N^{1/2} \quad (15)$$

If Res = 1, then

$$N = 16/(\Delta\mu_{app}/\mu_{avg})^2 \quad (16)$$

Another expression for N is derived from Equation 13, using the width at half-height of a Gaussian peak,

$$N = 5.54(L/w_{1/2})^2 \quad (17)$$

where 5.54 = 8 ln 2, and $w_{1/2}$ is the peak width at half-height.[6]

The utility of Equation 15 is that it permits one to assess independently the two factors that affect resolution, selectivity and efficiency. The selectivity is reflected in the mobility of the analyte(s), while the efficiency of the separation process is indicated by N.

At this point, it is important to point out that it is misleading to discuss theoretical plates in electrophoresis. The concept is a carryover from chromatographic theory, where a true partition equilibrium between two phases is the physical basis of separation. In electrophoresis, separation of the components of a mixture is determined by their relative mobilities in the applied electric field, which is a function of their charge, mass, and shape. The theoretical plate is merely a convenient concept to describe the analyte peak shape, and to assess the factors that affect separation.

IV. SOURCES OF VARIANCE

While N is a useful concept to compare the efficiency of separation among columns, or between laboratories, it is difficult to use to assess the factors that affect that efficiency. This is because it refers to the behavior of a single component during the separation process, and is unsuited to describing the separation of two components or the resolving power of a capillary. A more useful parameter is HETP, the height equivalent of a theoretical plate,[4]

$$\text{HETP} = L/N = \sigma^2_{tot}/L \quad (18)$$

The HETP might be thought of as the fraction of the capillary occupied by the analyte. It is more practical to measure HETP as an index of separation efficiency, rather than N, as the individual components that contribute to HETP may be individually evaluated and combined to determine an overall value. A consideration of all the factors influencing σ^2_{tot} should include not only diffusion, but also differences in mobility or diffusion generated by Joule heating, the reality that the sample is not introduced as a thin disk but as a plug of finite dimensions, and interaction of analytes with the capillary wall. Each of these factors is addressed separately below, in a simplistic manner. Theoretical derivation is given only to emphasize the importance of various parameters, and how they contribute to the overall result. References shall direct those desiring such information to the appropriate literature.

Two studies have addressed the variance due to a variety of sources. Huang et al.[7] investigated small molecules; Jones et al.[8] studied proteins as well as amino acids and a neutral marker. The two studies demonstrate that the band broadening observed in capillary electrophoresis is in excess over calculated values of diffusion and analyte interaction. Huang et al.[7] and Jones et al.[8] attribute the excess variance to sample introductory practices.

A. VARIANCE CAUSED BY TEMPERATURE

Current passing through a conducting solution generates heat. A simple way of looking at the problem is to compare the equivalent expressions for the applied potential, H, and heat generation, W,

$$H = i/\kappa\pi r^2 \tag{19}$$

$$W = i^2/\kappa(\pi r^2)^2 \tag{20}$$

where i is the current through the electrolyte solution, πr^2 the cross-sectional area of the capillary, and κ is the specific conductance of the buffer. By combining Equations 19 and 20, to take the ratio we obtain

$$H/W = (i/\kappa\pi r^2)[\kappa(\pi r^2)^2/i^2] = \pi r^2/i \tag{21}$$

From Equation 21, it can be readily seen that the reduction of heat production can be achieved by reducing the current density in the capillary. This may be accomplished by increasing the cross-sectional area of the capillary or, preferably, by reducing the current. The latter is preferred, as increasing the diameter of the capillary results in a reduction of the surface-to-volume ratio, and leads to less efficient heat dissipation. The choices to reduce the current lie in running the separation at a lower voltage, or reducing the ionic strength of the separation buffer.

To develop the quantitative expression for thermal heating, we need to describe the electrophoretic front, and its behavior under a thermal gradient.[9-13] According to the Poiseuille equation, which describes the parabolic flow due to pressure,

$$v_z = \Delta P r^2/8L\eta \tag{22}$$

Joule heating of the electrolyte solution creates a similar flow profile, the equivalent expression being

$$v_z = v(1 + E^2\Lambda C_b B r^2/4k_b T^2)[1 - (r_x/r)^2] \tag{23}$$

where $E^2\Lambda$ is the rate of heat generation per unit volume, C_b is the buffer concentration, k_b is the thermal conductivity of the electrolyte solution, Λ is the equivalent conductance of the electrolyte solution, T is the absolute temperature, r_x is the radial position (which varies from 0 at the center of the capillary to r at the capillary wall), and B is a buffer-related viscosity constant.[8] The variance caused by dispersion in a parabolic velocity profile is given by

$$\sigma_T^2 = 2D_i t + (r^6E^4\, C_b^2 B^2\Lambda^2 t)/[24D_i(8k_bT^2 - E^2\Lambda C_b B r^2)^2] \tag{24}$$

For most capillary applications, where the radius is small (25 to 50 μm), and E is in kilovolts, $E^2\Lambda C_b B r^2$ is $<<$ $8k_bT^2$ and the second term reduces to

$$\sigma_T^2 = r^6 E^4 C_b^2 B^2 \Lambda^2 t / 1536 D_i k_b^2 T^4 \tag{25}$$

The strong dependence of σ_T^2 on the radial dimension of the capillary (r^6) and the field strength (E^4) demonstrate the importance of performing high-voltage electrophoretic separations in narrow bore capillaries. It also highlights the necessity of obtaining efficient capillary cooling, to prevent thermal effects from affecting not only sample lability, but also solute mobility. For small molecules with relatively larger diffusion constants (on the order of 10^{-9} m^2/s) Grushka et al.[14] have calculated a maximum radius of 65 μm with a 0.1 *M* buffer solution at 30 kV, and 130 μm with 10 m*M* buffer for a less than 5% increase in dispersion. With larger molecules, having diffusion constants on the order of 10^{-10} m^2/s, these dimensions were halved. Jones et al.[8] have introduced a novel method to study Joule heating effects and nonideal plug flow contributions to analyte dispersion, with polarity reversal at constant voltage. By recording the variance (band broadening) of the peaks over time and plotting the slope of the obtained line vs. applied potential, these authors develop a measure of nonideal behavior. A comparison of charged analyte behavior with neutral solute variance (which should be unaffected by electrophoretic mobility effects of thermal heating, but affected equally by the temperature-dependent effects on diffusion and viscosity) allows an estimate of the Joule heating effect. With adequate thermostatting, one need not worry about thermal dispersion under normal operating conditions (see Ref. 15), but one needs to be aware of how temperature generated within the capillary affects the separation and resolution of the analytes.

B. VARIANCE CAUSED BY FINITE SAMPLE INTRODUCTION VOLUME

In Section I, we assume the solute was initially present as a thin disk that dispersed into a peak. In reality, the sample is introduced into the capillary as a cylindrical plug. Sternberg[5] derived the commonly used equation describing variance due to sample introduction,

$$\sigma_{\text{int}}^2 = l_{\text{int}}^2 / 12 \tag{26}$$

where l_{int} is the length of the sample introduction plug. This formula assumes that the sample is introduced as a rectangular plug, which is a close approximation for capillary electrophoresis. As the plug has a finite volume, it also has measurable length. Huang et al.[7] and Jones et al.[8] have investigated the relative contribution of introduction plug length (by hydrodynamic or electrokinetic methods) to the observed peak width, and concluded that l_{int}/L_d of less than 3% does not lead to excessive band spreading. Only at very small plug length (< 200 μm) does one observe variance due to diffusion of the sample plug; typical sample introduction plug lengths of 300 to 1500 μm obscure this effect. Huang et al. and Jones et al. conclude that sample introduction volume, *i.e.*, plug length, is the most significant factor in excessive band broadening observed in capillary electrophoresis.

C. VARIANCE CAUSED BY ANALYTE–WALL INTERACTIONS

Since the interactions of analyte and capillary wall, or components within the sample solution are numerous, complex, and sample specific, no systematic theoretical treatment for this problem has been developed. To obtain estimates of the relative contribution to overall variance, Jones et al.[8] used a voltage interruption/polarity reversal method. Huang et al.[7] attribute part of the excessive peak width observed in their experiments to analyte–wall interaction. One may minimize such interactions by the appropriate choice of buffer pH, ionic strength, or buffer additives (see section on methods development — Chapter 2, Section IV.C).

The variance of multiple dispersive phenomena on the analyte may be summed as

$$\sigma_{\text{tot}}^2 = \sigma_{\text{diff}}^2 + \sigma_T^2 + \sigma_{\text{int}}^2 + \sigma_{\text{wall}}^2 \tag{27}$$

To enable the reader to visualize the magnitude of the contribution made by each of these variances, we include some typical numbers.

For a small molecule, $D_i = 10^{-5}$ cm²/s
Time of analysis might be 10 min (600 s)
Introduction plug of 1.2 nl into a 50 μm × 47 cm capillary, 40 cm to the detector, results in a plug length of 0.6 cm

$$\sigma^2_{diff} = 12 \times 10^{-3}$$

$$\sigma^2_T = \sigma^2_{diff} = 12 \times 10^{-3}$$

$$\sigma^2_{int} = 3.0 \times 10^{-2}$$

$$\sigma^2_{tot}\text{ (exp. observed)} = 5 \times 10^{-2}$$

$$\sigma^2_{tot}\text{ (calculated)} = 1.2 \times 10^{-2} + 1.2 \times 10^{-2} + 3.0 \times 10^{-2} + \sigma^2_{wall} = 5.4 \times 10^{-2}$$

These figures indicate σ^2_{int} is by far the largest contributor to the band spreading.

REFERENCES

1. **Jorgenson, J. W. and Lukacs, K. D.,** Zone electrophoresis in open tubular glass capillaries, *Anal. Chem.,* 53, 1298, 1981.
2. **Weber, R.,** Concerning theories of electrophoretic separations in porous tubes, *Helv. Chim. Acta,* 36, 424, 1953.
3. **Snyder, L. R. and Kirkland, J. J.,** *Introduction to Modern Liquid Chromatography,* 2nd ed., Wiley Interscience, New York, 1979.
4. **Giddings, C. J.,** Generation of variance, "theoretical plates," resolution, and peak capacity in electrophoresis and sedimentation, *Sep. Sci.,* 4, 181, 1969.
5. **Sternberg, J. C.,** Extracolumn contributions to chromatographic band broadening, *Adv. Chromatogr.,* 2, 206, 1966.
6. **Szepesy, L.,** *Gas Chromatography,* CRC Press, Boca Raton, Florida, 1970.
7. **Huang, X., Coleman, W. F., and Zare, R. N.,** Analysis of factors causing peak broadening in capillary zone electrophoresis, *J. Chromatogr.,* 480, 95, 1989.
8. **Jones, H. K., Nguyen, N. T., and Smith, R. D.,** Variance contributions to band spreading in capillary zone electrophoresis, *J. Chromatogr.,* 504, 1, 1990.
9. **Coxon, M. and Binder, M. J.,** Radial temperature distribution in isotachophoresis columns of circular cross-section, *J. Chromatogr.,* 101, 1, 1974.
10. **Martin, M. and Guiochon, G.,** Axial dispersion in open-tubular capillary liquid chromatography with electroosmotic flow, *Anal. Chem.,* 56, 614, 1984.
11. **Brown, J. F. and Hinckley, J. O. N.,** Electrophoretic thermal theory. II. Steady-state radial temperature gradients in circular section columns, *J. Chromatogr.,* 109, 218, 1975.
12. **Hinckley, J. O. N.,** Electrophoretic thermal gradients. I. Temperature gradients and their effects, *J. Chromatogr.,* 109, 209, 1975.
13. **Hjertén, S.,** Zone broadening in electrophoresis with special reference to high-performance electrophoresis in capillaries: an interplay between theory and practice, *Electrophoresis,* 11, 665, 1990.
14. **Grushka, E., McCormick, R. M., and Kirkland, J. J.,** Effect of temperature gradients on the efficiency of capillary zone electrophoresis separations, *Anal. Chem.,* 61, 241, 1989.
15. **Landers, J. P., Oda, R. P., Madden, B., Sismelich, T. P., and Spelsberg, T. C.,** Reproducibility of sample separation using liquid or forced air convection themostated high performance capillary electrophoresis, *J. High Resolut. Chromatogr.,* 15, 517, 1992.

Appendix 2

CALCULATIONS OF PRACTICAL USE

TABLE OF CONTENTS

0-8493-8690-X/94/$0.00+$.50

In this section, we take a typical electropherogram (Figure 1) and, utilizing the data obtained from System Gold version 7.1 (Figure 1, inset), take the interested reader through a variety of calculations of practical use.

I. MOBILITY

The velocity of an analyte, v_i, is the distance traveled during the time of electric field application. Velocity is related to mobility, μ_i, by field strength, $E = H/L$. Thus,

$$v_i = \mu_{app}H/L = \mu_{app}E$$

The apparent mobility μ_{app} measured from an electropherogram is the sum of the mobility of the analyte and that due to electroosmotic flow:

$$\begin{aligned}\mu_{app,i} &= \mu_{ep} + \mu_{eo}\\ &= L_dL_t/t_{app,i}H\\ &= L_dL_t/H\ (1/t_{app,i} - 1/t_{ref}) + \mu_{ref}\end{aligned}$$

where L_d is length to the detector, L_t is total length of the capillary, and, if the reference peak is a neutral marker, $\mu_{ref} = 0$.

To calculate μ_{ref} for a charged reference peak, the above equation may be used, substituting $t_{app,i}$ for the charged reference and t_{ref} for the neutral marker, with $\mu_{ref} = 0$.

EXAMPLE

From Figure 1, $L_d = 40$ cm, $L_t = 47$ cm, and $t_1 = 1.723$ min:

$$\begin{aligned}v &= L_d/t_1 = 0.4\text{ m}/(1.723\text{ min})(60\text{ s/min})\\ &= 0.003869\text{ m/s}\\ &= \mu_{app}E = \mu_{app}H/L_t\end{aligned}$$

or

$$\begin{aligned}\text{apparent mobility} = \mu_{app,1} &= vL_t/H\\ &= (0.003869\text{ m/s})(0.47\text{ m})/25\text{ kV}\\ &= (0.001818/25) = 7.273 \times 10^{-4}\text{ cm}^2/(\text{V} \cdot \text{s})\end{aligned}$$

Similarly for the other peaks,

$$\mu_{app,i} = (0.4)(0.47)/(60)(25 \times 10^3)t_i = 12.533 \times 10^{-4}/t_i\text{ cm}^2/(\text{V} \cdot \text{s})$$

$$\mu_{app,2} = 12.533 \times 10^{-4}/2.642 = 4.744 \times 10^{-4}\text{ cm}^2/(\text{V} \cdot \text{s})$$

$$\mu_{app,3} = 12.533 \times 10^{-4}/2.818 = 4.448 \times 10^{-4}\text{ cm}^2/(\text{V} \cdot \text{s})$$

$$\mu_{app,4} = 12.533 \times 10^{-4}/2.924 = 4.286 \times 10^{-4}\text{ cm}^2/(\text{V} \cdot \text{s})$$

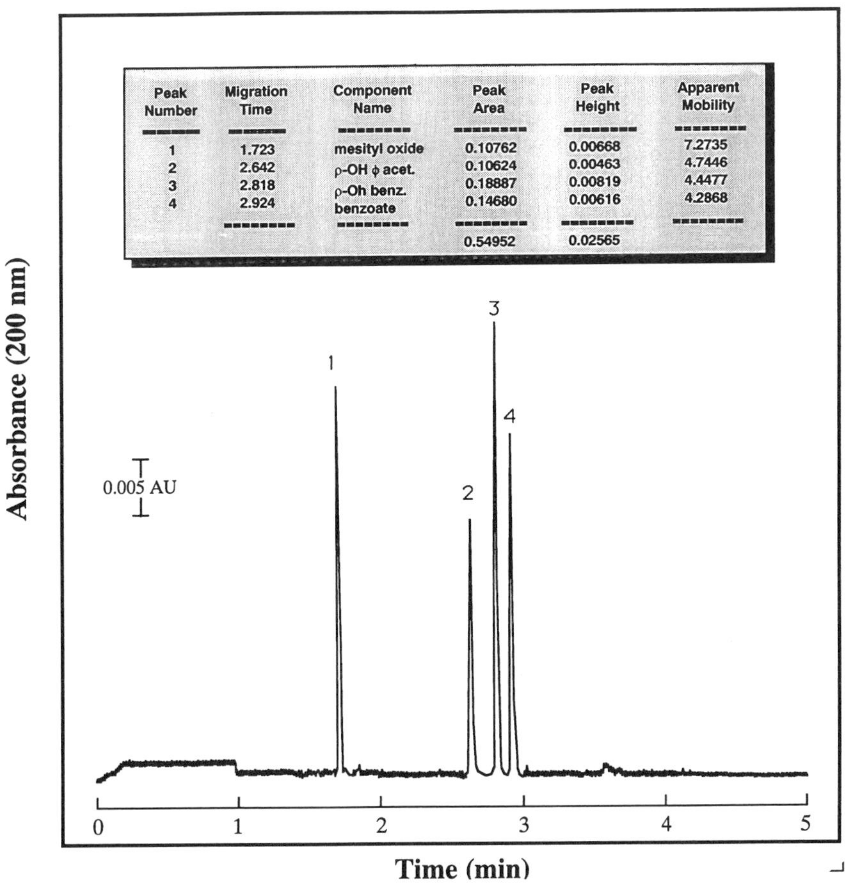

Peak Number	Migration Time	Component Name	Peak Area	Peak Height	Apparent Mobility
1	1.723	mesityl oxide	0.10762	0.00668	7.2735
2	2.642	ρ-OH φ acet.	0.10624	0.00463	4.7446
3	2.818	ρ-Oh benz.	0.18887	0.00819	4.4477
4	2.924	benzoate	0.14680	0.00616	4.2868
			0.54952	0.02565	

FIGURE 1. Simple separation of four small molecules for calculation of practical use. Separation of mesityl oxide (a neutral marker), *p*-hydroxyphenylacetic acid, *p*-hydroxybenzoic acid, and benzoic acid (in borate, pH 8.3; 3-s pressure [0.5 psi] injection). Analysis was carried out on a Beckman P/ACE System 2050. Separation conditions: capillary, 50 μm × 40 cm (effective length), 47 cm total length bare fused silica; *T*, 28°C; voltage, 25 kV; buffer, 100 m*M* borate, pH 8.3; detection, 200 nm.

Using

$$\mu_{app,i} = L_d L_t / H(1/t_{app,i} - 1/t_{ref}) + \mu_{ref}$$

we may make the same calculation using $\mu_{ref} = \mu_{eo} = 7.273 \times 10^{-4}$ cm²/(V · s):

$$\mu_{app,i} = [(0.4)(0.47)/25 \times 10^3](1/t_{app,i} - 1/1.723) + 7.273$$

$$\mu_{app,2} = 12.533(1/2.642 - 1/1.723) + 7.273 \times 10^{-4}$$

$$= 12.533(0.3785 - 0.5804) + 7.273 \times 10^{-4}$$

$$= -2.530 + 7.273 \times 10^{-4} \text{ cm}^2/(\text{V} \cdot \text{s}) = 4.743 \times 10^{-4} \text{ cm}^2/(\text{V} \cdot \text{s})$$

$$\mu_{app,3} = 12.533 \times 10^{-4}(0.3549 - 0.5804) + 7.273 \times 10^{-4}$$

$$= -2.826 \times 10^{-4} + 7.273 \times 10^{-4} = 4.448 \times 10^{-4} \text{ cm}^2/(\text{V} \cdot \text{s})$$

$$\mu_{\text{app},4} = 12.533 \times 10^{-4}(0.3420 - 0.5804) + 7.273 \times 10^{-4}$$

$$= -2.988 + 7.273 \times 10^{-4} = 4.285 \times 10^{-4} \text{ cm}^2/(\text{V} \cdot \text{s})$$

The electrophoretic mobility, μ_{ep}, may be calculated from

$$\mu_{\text{app},i} = \mu_{\text{ep}} + \mu_{\text{eo}}$$

as peak 1 in Figure 1 is a neutral marker, $\mu_{\text{app},1} = \mu_{\text{eo}} = 7.273 \times 10^{-4}$ cm^2/(V · s)

$$\mu_{\text{ep},2} = 4.744 - 7.273 = -2.529 \times 10^{-4} \text{ cm}^2/(\text{V} \cdot \text{s})$$

$$\mu_{\text{ep},3} = 4.448 - 7.273 = -2.825 \times 10^{-4} \text{ cm}^2/(\text{V} \cdot \text{s})$$

$$\mu_{\text{ep},4} = 4.286 - 7.273 = -2.987 \times 10^{-4} \text{ cm}^2/(\text{V} \cdot \text{s})$$

II. CORRECTED PEAK AREA

Integrators typically present peak area in dimensions of (response)(time). This calculation is a simple transformation to obtain area in (response)(width).

$$A_{\text{corr}} = A_i \text{ (mAU} \times \text{min)} v_i$$

$$= A_i \text{ (mAU} \times \text{min)} L_d \text{ (cm)}/t_m \text{ (min)}$$

$$= A_i \text{ (mAU} \times \text{cm)}$$

This can usually be achieved through the data collection and software system.

EXAMPLE

$$A_{\text{corr},1} = (0.10762)(40)/1.723$$

$$= 2.498 \text{ mAU cm}$$

Similarly,

$$A_{\text{corr},2} = (0.10624)(40)/2.642 = 1.608$$

$$A_{\text{corr},3} = (0.18887)(40)/2.818 = 2.681$$

$$A_{\text{corr},4} = (0.14680)(40)/2.924 = 2.008$$

Dividing the corrected area by peak height gives peak width, which approximates the peak width at half-height, which may be used to calculate N (see calculations, Section V).

$$w_{1/2,1} = 2.498 \text{ mAU}/0.00668 \text{ µAU} = 374 \text{ µm}$$

$$w_{1/2,2} = 1.608 \text{ mAU}/0.00463 \text{ µAU} = 347 \text{ µm}$$

$$w_{1/2,3} = 2.681 \text{ mAU}/0.00819 \text{ µAU} = 327 \text{ µm}$$

$$w_{1/2,4} = 2.008 \text{ mAU}/0.00616 \text{ µAU} = 326 \text{ µm}$$

III. QUANTITY OF SAMPLE INTRODUCED INTO THE CAPILLARY

We assume

$$Q = (\text{volume})_{\text{int}}(\text{concentration})$$

$$= \pi r^2 l[C_i]$$

where r is the internal radius of the capillary, and l the length of the sample plug, and $[C_i]$ is the concentration of the sample.

A. HYDRODYNAMIC INJECTION

$$Q = \pi r^2[\Delta P r^2 t_{\text{int}}/(8\eta L)][C_i]$$

where ΔP is the pressure difference, r is the capillary inner radius, t_{int} is the introduction time, η is the viscosity of the sample solution, and L is the length of the column.

EXAMPLE

Assuming typical values for the constants,

Column dimensions: 50-μm I.D. × 47 cm, 40 cm to the detector
$\eta = 0.9548$ cP $= 9.548 \times 10^{-4}$ N s m^{-2} at 22°C
$\Delta P = 0.5$ psi $= 3.435 \times 10^3$ N m^{-2}
$t_{\text{int}} = 3$ s

Then

$$\text{vol}_{\text{int}} = (3.1416)(25 \times 10^{-6})^2\{[(3.435 \times 10^3)(25 \times 10^{-6})^2 3]/[8(9.548 \times 10^{-4})0.47]\}$$

$$= (19.635 \times 10^{-10})(17.94 \times 10^{-4})$$

$$= 3.523 \times 10^{-12} \text{ m}^3 = 3.523 \text{ nl}$$

which correlates with $\simeq$ 1.2 nl/s injection.

B. ELECTROKINETIC INTRODUCTION

$$Q = [\pi r^2(\mu_{\text{app}})V_{\text{int}}t_{\text{int}}/L][C_i]$$

where μ_{app} is the mobility of the analyte, μ_{eo} is the electrosmotic mobility, and V_{int} is the introduction voltage.

EXAMPLE

Again assuming typical values,

Column dimensions: 50-μm I.D. × 47 cm, 40 cm to the detector
$t_{\text{int}} = 20$ s
$\mu_{\text{app}} = 4.744 \times 10^{-8}$ m^2 V^{-1} s^{-1}
$\mu_{\text{eo}} = 7.273 \times 10^{-8}$ m^2 V^{-1} s^{-1}
$V_{\text{int}} = 1.5$ kV

Then

$$\begin{aligned} vol_{int} &= (3.1416)(25 \times 10^{-6})^2[(4.744 \times 10^{-8})(1.5 \times 10^3)20/(0.47)]\ m^3 \\ &= (19.635 \times 10^{-10})(302.81 \times 10^{-5})\ m^3 \\ &= 5.945 \times 10^{-15}\ m^3 = 5.945\ nl \end{aligned}$$

However, if we calculate the volume of fluid injected, which is due solely to μ_{eo},

$$\begin{aligned} vol_{int} &= (3.1416)(25 \times 10^{-6})^2[(7.273 \times 10^{-8})(1.5 \times 10^3)20/(0.47)]\ m^3 \\ &= (19.635 \times 10^{-10})(464.2 \times 10^{-5})\ m^3 \\ &= 9115 \times 10^{-15}\ m^3 = 9.115\ nl \end{aligned}$$

With electrokinetic introduction, the amount of sample injected is proportional to the analyte mobility. It is typically used when the sample has greater mobility than electroosmotic flow. If one has a highly mobile analyte in relatively low ionic strength sample buffer ($1/x$ times separation buffer) one may load an x-fold greater portion onto the column by electrokinetic introduction without adversely affecting separation resolution. In the above example, if the sample were in 0.1X separation buffer, we could have loaded the equivalent of 59.5 nl of sample.

IV. RESOLUTION

Using Equation 3 from Chapter 2 for resolution, we obtain

$$Res = 2(x_{i2} - x_{i1})/(w_1 + w_2)$$

EXAMPLE

$$\begin{aligned} Res_{2-1} &= 2\ (2.818 - 2.642)/(0.02295 + 0.02306) \\ &= 0.352/0.04601 \\ &= 7.650 \end{aligned}$$

$$\begin{aligned} Res_{3-2} &= 2(2.924 - 2.818)/(0.02306 + 0.02383) \\ &= 0.212/0.04689 \\ &= 4.521 \end{aligned}$$

Using the criterion that peaks are resolved when Res = 1, we could state that all three peaks are resolved.

V. EFFICIENCY

From the above example, we may calculate the efficiency (from Equation 17) of the separation of the peaks in Figure 1.

$$N = 5.54(L_d/w_{1/2})^2$$

where L_d, the length to the detector, is 40 cm, and the peak width is (peak area/peak height), both of which must be in the same units, either in time (min) or distance (cm). We shall calculate the efficiency using both time and distance.

A. EXAMPLE 1

For time calculations,

$$N_i = 5.54(L_d/w_{1/2})^2 = 5.54[t_i/(\text{area}_i/\text{peak height}_i)]^2$$

From the data in the inset of Figure 1, peak area is given in mAU × min, and peak height is in AU. Thus,

$$N_1 = 5.54\{(1.732)/[0.10762/(0.00668)(1000)]\}^2$$

$$= 5.54(1.732/0.01611)^2 = 5.54(106.95)^2 = 63{,}368$$

$$N_2 = 5.54(2.642/0.02295)^2 = 5.54(115.14)^2 = 73{,}445$$

$$N_3 = 5.54(2.818/0.02306)^2 = 5.54(122.20)^2 = 82{,}728$$

$$N_4 = 5.54(2.942/0.02383)^2 = 5.54(122.70)^2 = 83{,}406$$

B. EXAMPLE 2

For distance calculations,

$$N_1 = 5.54(L_d/w_{1/2})^2 = 5.54[L_d/(\text{area}_{\text{corr}}/\text{peak height})]^2$$

From calculations, Section II, corrected peak area above,

$$w_{1/2,1} = 2.498 \text{ mAU}/0.00668\ \mu\text{AU} = 374\ \mu\text{m}$$

$$w_{1/2,2} = 1.608 \text{ mAU}/0.00463\ \mu\text{AU} = 347\ \mu\text{m}$$

$$w_{1/2,3} = 2.681 \text{ mAU}/0.00819\ \mu\text{AU} = 327\ \mu\text{m}$$

$$w_{1/2,4} = 2.008 \text{ mAU}/0.00616\ \mu\text{AU} = 326\ \mu\text{m}$$

Therefore,

$$N_1 = 5.54(40/0.374)^2$$

$$= 5.54(106.95)^2 = 63{,}370$$

Likewise,

$$N_2 = 5.54(40/0.347)^2 = 73{,}616$$

$$N_3 = 5.54(40/0.327)^2 = 82{,}896$$

$$N_4 = 5.54(40/0.326)^2 = 83{,}405$$

VI. JOULE HEATING

To calculate the Joule heating of a buffer, one should run an Ohm's law plot as outlined in Section IIIB, under Buffers. To calculate watts, we use the following:

Watts/m = (voltage)(amperage)/(column length, cm)1000

EXAMPLE

For 100 m*M* borate, pH 8.3: (25 kV)(13.17 μA)/(57 cm)1000 = 0.58 W/m
For 100 m*M* CAPS, pH 11.0: (25 kV)(134 μA)/(57 cm)1000 = 5.88 W/m
For 20 m*M* CAPS, pH 11.0: (25)(30.03)/57,000 = 1.32 W/m
For 100 m*M* PO_4, pH 2.5: (25)(229.5)/57,000 = 10.06 W/m
For 50 m*M* PO_4, pH 2.5: (25)(116.9)/57,000 = 5.13 W/m
For 25 m*M* PO_4, pH 2.5: (25)(70.03)/57,000 = 3.07 W/m

Appendix 3

TROUBLESHOOTING

Table 1 summarizes problems commonly encountered in capillary electrophoretic experiments, and possible remedies.

TABLE 1
Troubleshooting Commonly Encountered Capillary Electrophoresis Problems

Problem	Cause	Remedy
I. Peak Problems		
A. No peaks observed		
1. Baseline on scale	Inappropriate data collection scale	–Reset scale for appropriate absorbance parameter
	Inappropriate detector range	–Reset detector range
	Inappropriate detector wavelength	–Reset wavelength
	Separation time too short	–Lengthen analysis time –Lengthen capillary
	Current normal?	–Go to Section II.A
	Flow through capillary?	–Go to Section V
	Sample okay?	–Check sample level –Check for air bubble in bottom of sample vial –Increase sample introductory time –Check sample caps for leakage –Go to Section V
	Using voltage introduction	–Go to Section II.A
2. Baseline off scale	Offset baseline	–Rezero detector
	Inappropriate detector wavelength	–Reset wavelength
B. Peaks present		
1. Too many peaks	Random peaks may be caused by: microbubbles	–Warm buffer to room temperature –Reduce voltage
	Solid contaminants in sample	–Filter sample (0.45-μm pore size filter)
	Solid contaminants in buffer	–Filter buffer
	Residue from previous analysis	–Wash capillary
	Sample degradation	–Replace sample; check temperature of capillary chamber
2. Too few peaks	Proper wavelength?	–Reset detector wavelength
	Sufficient time?	–Lengthen analysis time –Lengthen capillary
	Proper current?	–Go to Section II.A
	Voltage introduction?	–Go to Section II.A
	Correct temperature?	–Reset temperature control
	Analyte–wall interactions	–Wash capillary –Check running conditions
	Fluid flow okay?	–Go to Section V
	Separation time too short	–Lengthen analysis time –Lengthen capillary –Reduce EOF

0-8493-8690-X/94/$0.00+$.50

TABLE 1 (Continued)
Troubleshooting Commonly Encountered Capillary Electrophoresis Problems

Problem	Cause	Remedy
	Similar charge/mass ratio (not resolving components)	–Alter buffer pH –Reduce EOF –Try alternative separation mode — MECC, CITP, CIEF
3. Distorted peak shape		
a. Peaks low	Proper wavelength?	–Reset wavelength
	Proper range on detector?	–Reset detector
	Proper integration?	–Reset integrator parameters
	Proper sampling?	–Longer sample introduction time
b. Peaks flat-topped	Sample too concentrated	–Reduce introduction time –Dilute sample
c. Peaks tail	Current too high	–Ionic strength of separation buffer too high. Check composition; dilute –Reduce voltage
	Sample buffer ionic strength too high	–Dilute sample buffer
	Analyte–wall interactions	Modify separation buffer: –Increase ionic strength –Add organic solvents –Add cationic compounds –Add zwitterions (phosphorylethanolamine) –Use coated capillary –Consider different CE mode (e.g., MECC)
II. Instrumental Problems		
A. Current		
1. Fluctuating current during separation	Vial levels low	–Replenish buffer
	Loose electrode connections	–Tighten connections
	Plugged capillary	–Rinse/replace
2. No current	Safety interlock off	–Reset interlock
	Plugged capillary	–Wash capillary
	Broken capillary	–Replace capillary
	Buffer depletion	–Replace buffers
	Empty capillary	–Fill reservoirs
3. Abrupt loss of current during separation	Short in system (buffer on reservoir cap)	–Dry cap
	Plugged capillary	–Rinse/replace
B. Baseline drift	Contaminated capillary	–Wash capillary
	Contaminated aperture on detector	–Clean aperture
	Bad capillary alignment	–Check alignment; realign or replace
	Detector instability	–Give adequate warm-up time –Replace lamp
	Unstable capillary temperature	–Check oven/bath temperature –Replace thermostat
	Unstable room temperature	–Stabilize room temperature –Deflect drafts off instrument –Move instrument
C. Data analysis		
Peaks observed — not analyzed	Inappropriate integration parameters	–Reset integrator attenuation

TABLE 1 (Continued)
Troubleshooting Commonly Encountered Capillary Electrophoresis Problems

Problem	Cause	Remedy
III. Sample Introduction		
A. Electrokinetic		
No peaks	Anodic sample	–Reverse polarity –Change separation buffer pH
	Sample ionic strength too high	–Dilute sample buffer
	Sample pH too high	–Adjust pH
	Sample ionic strength too low	–Raise ionic strength
B. Hydrodynamic		
No peaks	Poor seal	–Replace cap
	Pinched pressure/vacuum line	–Replace line
	Depleted pressure source	–Replace
	Anodic sample	–Reverse polarity –Change separation buffer pH
Irreproducible peak height/area	Poor seal	–Replace cap
	Pinched pressure/vacuum line	–Replace line
	Sample matrix volatility	–Decrease percentage volatile solvent
IV. Poor Quantitative Reproducibility		
A. Migration time		
Unstable temperature	Ionic strength too high?	–Dilute buffer
	Voltage too high?	–Decrease voltage
	Temperature too high?	–Decrease thermostatted temperature (see Section II.B)
	Sample matrix ionic strength too high	–Dilute sample
Others	Buffer depletion	–Replenish with fresh buffer
	Analyte–wall interactions	–See Section I.B.3.c
	Buffer siphoning	–Adjust level in inlet/outlet reservoirs[a]
	Contaminated capillary	–Rinse capillary extensively
	Inadequate rinse steps	–Increase rinse time with rinse solution/buffer
B. Peak height/area	Analyte–wall interactions	–See Section I.B.3.c
	Current instability	–See Section II.A.1
V. Capillary-Associated Problems[b]		
A. No peaks — proper flow through capillary	Anodic sample	–Reverse polarity
	Sample buffer too viscous	–Dilute sample
B. Reduced flow through capillary	Partially plugged capillary	–Wash capillary. Check sample, buffers for particulates; filter if necessary.
	Pinched pressure/vacuum line	–Replace line
	Depleted pressure source	–Replace
C. No flow	Broken capillary	–Replace capillary

[a] This is unlikely to be of significance with narrow diameter capillaries (i.e., <75 μm) since the hydrostatic head differential must be in range of 5 cm of water in order for this phenomenon to be an appreciable factor.

[b] Be confident that there is adequate capillary flow. This can be accomplished as described in Chapter 2, Section II.C.4.a. Alternatively, one may check the flow by pressure rinsing and observing droplets forming at the other end. If this must be done manually, a 6-cc syringe with a yellow Eppendorf pipette tip will produce enough pressure to create one drop every 20 to 25 s, with a 50 μm x 57 cm capillary.

Appendix 4

CAPILLARY ZONE ELECTROPHORESIS

Buffer	Additives	Analytes/Sample	Ref.
Ions			
5 m*M* Chromate or phthalate, pH 10.00	0.5 m*M* Nice-Pak OFM Anion-BT	Organic/inorganic ions	Romano et al., 1991
5 m*M* Chromate, pH 8.0 5 m*M* Chromate, pH 11.0 (LiOH)	0.4 m*M* OFM Anion-BT None	Organic/inorganic ions	Jones and Jandik, 1992
30 m*M* creatinine, pH 4.8 (acetic acid)	8 m*M* hydroxyisobutyric acid	Metal ions	Lu and Cassidy, 1993
100 m*M* Borate	50 m*M* Tetrabutyl ammonium hydroxide	Inorganic/organic ions	Avdalovic et al., 1993
Small Molecules/Ions			
50 m*M* Tetraethylammonium perchlorate	10 m*M* HCl in acetonitrile	Organic bases	Walbroehl, 1984
20–125 m*M* Sodium phosphate, pH 2.5, 7.0–9.2	None	Cinnamic acid and analogs	Fujiwara and Honda, 1986
50 m*M* Sodium phosphate, pH 7.0	50% Acetonitrile (v/v)	Substituted benzoic acid isomers	Fugiwara, 1987
50–100 m*M* Sodium acetate, pH 3.9–4.5	0.1% Hydroxypropyl cellulose (v/v)	Isotopic benzoic acids	Terabe et al., 1988
16 m*M* Sodium sulfate, 5 m*M* MES, pH 6.7	30% Methanol (v/v)	Methotrexate analysis	Roach et al., 1988
20–125 m*M* Sodium phosphate, pH 2.5, 7.0–9.2	None	Pharmaceuticals	Altria and Simpson, 1988
25 m*M* Tetrahexylammonium perchlorate	50% Acetonitrile (v/v)	Neutral organic molecules, e.g., polycyclic aromatic hydrocarbons	Walbroehl & Jorgenson, 1988
25 m*M* MES, pH 5.5–5.65	10–20% 2-Propanol (v/v)	Catecholamine analysis	Wallingford & Ewing, 1989
10 m*M* MES/His, pH 6.0	0.5 m*M* Tetradecyltrimethyl-ammonium bromide	Carboxylic acids	Huang et al., 1989
5 m*M* Sodium borate, pH 9.0	2% SDS (w/v), 0–1% Ethylene (v/v) diamine, and 5% ethylene glycol (v/v)	Polyamines	Tsuda et al., 1990
100 m*M* CAPS, pH 10.5	None	Nucleotides	Nguyen et al., 1990
20 m*M* Sodium citrate or 100 m*M* acetic acid, pH 2.0 or 2.9	None	Shellfish poisons	Thibault et al., 1991
150 m*M* Sodium dihydrogen phosphate, pH 2.98 or 5.98	Replacing water with D_2O, pD=2.98 or 5.98	Aniline derivatives	Okafo et al., 1991
20 m*M* Phosphoric acid/20% KOH, pH 7.00	None	Cimetidine in pharmaceutical preparations	Arrowood & Hoyt, 1991
10 m*M* sodium tetraborate-10-water and 50 m*M* boric acid or 40 m*M* sodium acetate, pH 8.5	Acetonitrile:water 50:50	cis/trans Isomers of butenedioic and retinoid acids	Chadwick & Hsieh, 1991

0-8493-8690-X/94/$0.00+$.50

50 m*M* CAPSO and 12.5 m*M* NaOH, pH 9.55	None	Tricyclic antidepressants	Salomon et al., 1991
20 m*M* Sodium phosphate or 3–5 m*M* imidazole or 50 m*M* acetic acid, pH 4.5–7.0	0.05% Ethylene glycol (v/v) or Pyrex glass tubing	Organic/inorganic cations and anions	Beck & Engelhardt, 1992
100 m*M* Borate, pH 8.3	None	CAT enzyme substrates and products	Landers et al., 1992
20 m*M* Imidazolium acetate, pH 7.0	None	Sulfonamides	Ackermans et al., 1992
20–50 m*M* Phosphate, pH 6.8	0.02% hydroxypropylcellulose (w/v) or 0.15% CPDAPS (w/v) and 20% methanol (v/v) or 0.5% polybrene (w/v)	Cefixime and its metabolites	Honda et al., 1992
6 m*M* Sorbate, pH 12.1	None	Carbohydrates	Vorndran et al., 1992
100 m*M* Borate, pH 8.4 100 m*M* Tricine, pH 8.4	None None	cis-diol containing compounds Norepinephrine, dopamine and metabolites	Landers et al., 1992
8 m*M* Sodium carbonate, 10 m*M* NaOH, pH 12.0	None	Simple carbohydrates	O'Shea et al., 1993
Peptides			
150 m*M* Phosphate, pH 3.0	None	Angiotensin II octapeptides	McCormick, 1988
10 m*M* Tricine, pH 8.0–8.1	5.8–45 m*M* Morpholine and 20 m*M* NaCl or KCl	LGH tryptic digest	Neilsen et al., 1989
100 m*M* Phosphate, pH 2.5	30 m*M* $ZnSO_4$	DL-His-DL-His	Mosher, 1990
25 m*M* Tris/25 m*M* Phosphate, pH 7.05	50 m*M* HTAB	Angiotensin analogs	Liu et al., 1990
0.5 mol/l Acetic acid, pH 2.6	None	Di- and triglycine, synthetic growth hormone-releasing peptide	Prusik et al., 1990
250 m*M* Borate, pH 7.0	1% Ethylene glycol (v/v) and 7% acetonitrile (v/v)	Proteinase-digested horse myoglobin	Tanaka et al., 1991
20 m*M* Citric acid, pH 2.5	None	Motilin and synthetic peptides	Florence et al., 1991
20 or 150 m*M* Sodium dihydrogen phosphate, pH 2.93, 7.93, or 7.95	Replacing water with D_2O,pD=2.93, 7.93, or 7.95	Simple peptides, tryptic digest of calcitonin, glucagon, and cytochrome *c*	Camilleri et al., 1991
40–80 m*M* Tris and tricine, pH 8.1–8.2	None	β-Casein tryptic and ACTH-endoproteinase Arg C digests	Krueger et al., 1991

25 m*M* Phosphate, pH 2.2 (KOH)	None	Adrenocorticotropic hormone (ACTH) peptide fragments	Van de Goor et al., 1991
20 m*M* Formate, pH 3.8	Alanine		
20 m*M* ε-Aminocaproate, pH 4.4 (Acetic acid)	None		
20 m*M* Histidine, pH 6.2	MES		
40 m*M* Imidazole, pH 7.5	MOPS		
100 m*M* Borate, pH 8.3 (KOH)	None		
50 m*M* Phosphate, pH 2.5	40% Acetonitrile (v/v)	Multiple antigen peptides	Tanaka et al., 1991
50 m*M* Sodium dihydrogen phosphate, pH 3.93	0.1% TFA (v/v) and 0.05% hydroxymethylpropyl cellulose (v/v)	α and β species of CGRP	Saria, 1992
20 m*M* Citrate buffer, pH 2.50	None	Peptide monomers & dimers	Landers et al., 1993
50 m*M* Formic acid, pH 2.5	10 m*M* sodium chloride	Basic peptides	Gaus et al., 1993
Peptides (Coated Capillaries)			
150 m*M* Phosphoric acid, pH 1.5 in polyvinylpyrrolidone-coated capillary	None	Dipeptides	McCormick, 1988
Proteins			
20 m*M* borate, pH 8.25 10 m*M* Tricine, pH 8.22	None 20 m*M* KCl	Model proteins	Lauer & McManigill, 1986
10 m*M* Tricine, pH 8.0	5.8 m*M* Morpholine and 20 m*M* NaCl	Biosynthetic human insulin derivatives and growth hormone	Nielsen et al., 1989
50 m*M* Glutamine-triethylamine, pH 9.5	Vinyl-bound polyacrylamide-coated capillaries	Model proteins	Cobb et al., 1990
50 m*M* Sodium phosphate, pH 7.0	Applied Biosystems coating reagent	Multiacetylated histone H4 proteins	Wiktorowicz & Colburn, 1990
50 m*M* Sodium borate decahydrate, pH 10.0	None	Human serum proteins	Gordon et al., 1991
50 m*M* Tricine, pH 8.0 50 m*M* MES, pH 6.0	30% Methanol (v/v) 30% Ethylene glycol (v/v)	Glycoproteins — recombinant human erythropoietin	Tran et al., 1991
50 m*M* Sodium tetraborate, pH 8.3	25–50 m*M* LiCl	Fluorescamine-labeled interferon and other proteins	Guzman et al., 1992
100 m*M* Borate, pH 9.0	1 m*M* Diaminobutane	Ovalbumin	Landers et al., 1992
20 m*M* Phosphate, pH 3.0	30 m*M* NaCl, 0.05% polyvinyl alcohol (w/w)	Basic proteins	Gilges et al., 1992

Proteins in Coated Capillaries			
10 m*M* Phosphate, pH 7.0 in capillaries coated with either Tween 20/alkylsilane or Brij 35/alkysilane	None 0.001% Brij 35	Basic proteins Acidic proteins	Towns & Regnier, 1991
10 m*M* Citrate, pH 3.0 in BSA-coated capillary	4 m*M* CHAPSO	Standard proteins	Swedberg, 1993
30 m*M* Citrate, pH 3.0 in a polyacrylamide-coated capillary (uncrosslinked)	None	Standard proteins	Kohr and Engelhardt, 1991
TES (NaOH), pH 7.0	0.2% SDS Linear acrylamide (CGE)	Proteins (3–205 kDa)	Werner et al., 1993
Oligonucleotides/DNA			
50 m*M* Phosphate, pH 7.0	5% Ethylene glycol (v/v)	Oligo dT ladder	Kasper et al., 1988
1 m*M* Borate, pH 9.1 (CGE)	30% Hydrolink (polymerized)	Restriction fragments	
100 m*M* Trizma 100 m*M* Boric acid, pH 8.7 (CGE)	Celsium hydroxide 0.1 m*M* EDTA 0.5% Hydroxyethylcellulose	Forensic DNA	McCord et al., 1993

MICELLAR ELECTROKINETIC CAPILLARY CHROMATOGRAPHY

Buffer	Additives	Applications	References
Small Molecules			
100 m*M* Phosphate, pH 7.0	25 m*M* CM-β-cyclodextrin	Cresol isomers	Terabe et al., 1983
25 m*M* Tetraborate/50 m*M* phosphate, pH 7.0	1 m*M* SDS	Phenols	Terabe et al., 1984
50 mM Phosphate/125 m*M* tetraborate, pH 7.0 100 m*M* Tris-HCl, pH 7.0	50 m*M* SDS 50 m*M* DTAB	PTH-amino acids	Otsuka et al., 1985
20–50 m*M* Phosphate, pH 7.0–8.0	10–50 m*M* SDS	Pharmaceuticals	Fujiwara and Honda, 1987
10 m*M* Disodium phosphate and 6 m*M* borate, pH 7.0–9.0	50 m*M* SDS or 50 m*M* dodecyltrimethylammonium chloride or 50 m*M* STS	Phenols and polycyclic aromatic hydrocarbons	Burton et al., 1987
10 m*M* Phosphate/6 m*M* borate, pH 7.0	10 m*M* SDS	Catechols and catecholamines	Wallingford and Ewing, 1989
20 m*M* Phosphate and 20 m*M* borate, pH 9.0	50 m*M* SDS and 20–60 m*M* tetralkylammonium salts	Vitamins and pharmaceuticals	Nishi et al., 1989
29 m*M* Phosphate-borate, pH 9.0	50 m*M* sodium cholate	Corticosteroids	Nishi and Terabe, 1990

20 m*M* Sodium dihydrogen phosphate and 20 m*M* borate, pH 9.2	100 m*M* SDS and 10% acetonitrile (v/v)	Nucleosides and nucleotide-3-monophosphates	Lecuq et al., 1991
50 m*M* Phosphate, pH 6.0	50 m*M* SDS and 5% 2-propanol (v/v)	Creatinine and uric acid and polycyclic aromatic hydrocarbons	Mikaye et al., 1991
50 m*M* Phosphate and 100 m*M* borate, pH 8.09	8 m*M* α-cyclodextrin, 1 m*M* β-cyclodextrin, and 1 m*M* γ-cyclodextrin	Plant growth regulators	Yeo et al., 1991
2.5–5.0 m*M* Borate, pH 7.8–8.9	10–50 m*M* SDS	Organic gunshot and explosives	Northrup et al., 1991
10 m*M* Borate-phosphate, pH 8.7	50 m*M* SDS, 6*M* urea, 20% methanol (v/v)	Benzothiazole sulfenamides	Nielsen and Mensink, 1991
8.5 m*M* phosphate and 8.5 m*M* borate, pH 8.5	85 m*M* SDS and 15% (v/v) acetonitrile	Acidic and neutral heroin impurities	Weinberger and Lurie, 1991
100 m*M* Borate and 50 m*M* phosphate, pH 7.6	30 m*M* SDS, 3% 2-propanol (v/v) or 3 m*M* γ-cyclodextrin	Vitamins	Ong et al., 1991
10 m*M* Borate, pH 9.5	75 m*M* SDS and 10% methanol (v/v)	Hydroquinone and related molecules in skin toning cream	Sakadinskaya et al., 1992
50 m*M* Phosphate and 100 m*M* borate, pH 6.0 or 7.0	3 m*M* β-cyclodextrin or 2 m*M* γ-cyclodextrin and 10 m*M* SDS	Sulfonamides and polycyclic aromatic hydrocarbons	Ng et al., 1992
10 m*M* Phosphate and 10 m*M* borate, pH 9.0	100 m*M* SDS and 20% methanol (v/v)	Enantiomers of amphetamine, methamphetamine, and their hydroxyphenethylamine precursors	Lurie, 1992
18 m*M* Sodium tetraborate and 30 mM phosphate, pH 7.0	50 m*M* CTAB	Glucosinolates and their desulfo-derivatives	Morin et al., 1992
60 m*M* Tris/phosphate, pH 2.5	20 m*M* Cyclodextrin (α,β,γ)	Basic drugs	Nielen, 1993
10 m*M* sodium phosphate 6 m*M* Borate, pH 9.2	10% acetonitrile 75 m*M* SDS	Morphine-3-glucuronide	Wernly et al., 1993
Peptides and Proteins			
10 m*M* Tris and 10 m*M* disodium, pH 7.05	50 m*M* Dodecyltrimethyl ammonium bromide	Angiotensin analogs	Novotny et al., 1990
50 m*M* Borate and 20–25 m*M* Tris and 10–25 m*M* disodium hydrogenphosphate, pH 9.50 or 7.05	50 m*M* SDS or 20 m*M* β-cyclodextrin and 1% THF (v/v) or 15% methanol (v/v) or additives such as 0.05 *M* HTAB or 2–50 m*M* DTAB	Derivatized peptides and angiotensin analogs	Lui and Novotny, 1990
10 m*M* Sodium phosphate 6 m*M* sodium borate	7 m*M* cyclodextrin (α, β, γ), acetonitrile (0–15%) 50 m*M* SDS, 50 m*M* Desoxycholate	Mycotoxins	Holland and Sepaniak, 1993
Oligonucleotides			
5–20 m*M* Tris and 5 m*M* phosphate, pH 7–8	2–7 *M* urea, 50 m*M* SDS, and 3 m*M* Mg or Cu or Zn	Bases, nucleotides, and oligonucleotides	Cohen, 1987
25 m*M* Phosphate and 50 m*M* borate, pH 7.0	100 m*M* SDS	Restriction fragments	Kasper et al., 1988
100 m*M* Tris-borate, pH 8.1	0.1% SDS (v/v), 2.5 m*M* EDTA, and 7 *M* urea	Restriction fragments	Cohen et al., 1988

REFERENCES

Ackermans, M. T., Beckers, J. L., Everaerts, F. M., Koogland, H., and Tomassen, T. J. H., Determination of sulphonamides in pork meat extracts by capillary zone electrophoresis, *J.* Chromatogr., 596, 101, 1992.

Altria, K. and Simpson, C. F., Analysis of some pharmaceuticals by high voltage capillary zone electrophoresis. *J. Pharm. Biochem. Med. Anal.*, 6, 801, 1988.

Arrowood, S. and Hoyt, A. M., Determination of cimetidine in pharmaceutical preparanons by capillary zone electrophoresis, *J. Chromamatogr.*, 586, 177, 1991.

Audalovic, N., Pohl, C. A., Rocklin, R. D., and Stillian, J. R., Determination of cations and anions by capillary electrophoresis combined with suppresed conductivity detection, *Anal. Chem.*, 65, 1470-1475, 1993.

Beck, W. and Engelhardt, H., Capillary electrophoresis of organic and inorganic cations with indirect UV detection, *Chromatographia,* 33, 313, 1992.

Burton, D. E., Sepaniak, M. J., and Maskarinec, M. P., Evaluation of the use of various surfactants in micellar electrokinetic capillary chromatography, *J. Chromatographic Sci.*, 25, 514, 1987.

Camilleri, P., Okafo. G. N., Southan, C., and Brown, R., Analytical and micropreparative capillary electrophoresis of peptides from calcitonin, *Anal. Biochem.* 198, 36, 1991.

Chadwick. R. R. and Hsieh, J. C., Separation of cis and trans double bond isomers using capillary zone electrophoresis, *Anal. Chem.*, 63, 2377, 1991.

Cobb. K. A., Dolnik, V., and Novotny, M., Electrophoretic separations of protein in capillaries with hydrolytically stable surface structures, *Anal. Chem.* 62, 2478, 1990.

Cohen, A. S., High performance capillary electrophoresis of bases, nucleosides and oligonucleotides: retention manipulation via micellar solutions and metal additives, *Anal. Chem.*, 59, 1021, 1987.

Cohen, A. S., Najarian, D., Smith, J. A., and Karger, B. L., Rapid separation of DNA restriction fragments usiny capillary electrophoresis, *J. Chromatogr.*, 458, 323, 1988.

Florance, J. R., Konteatis, Z. D., Macielag, M. J., Lessor, R. A., and Galdes, A., Capillary zone electrophoresis studies of motilin peptides, effect of charge, hydrophobicity, secondary structure and length, *J Chromatogr.*, 559, 391, 1991.

Fujiwara, S., Effect of addition of organic solvent and the separation of positional isomers in high voltage capillary zone electrophoresis, *Anal. Chem.* 59, 487, 1987.

Fujiwara, S., and Honda, S., Determination of ingredients of antipyretic analgesic prepartations by micellar electrokinetic capillary chromatography, *Anal. Chem.*, 59, 2773, 1987.

Fujiwara, S. and Honda, S., Determination of cinnamic acid and its analogues silica capillary tube. *Anal. Chem.*, 58, 1811. 1986.

Gaus, H.-J., Beck-Sickinger, A. G., and Bayer, E., Optimization of capillary electrophoresis of mixtures of basic peptides and comparison with HPLC, *Anal. Chem.*, 65, 1399-1405, 1993

Gilges, M., Hasmann, H., Kleemib, M. H., Motsch, S. R., and Schomburg, G., CZE separations of basic proteins at low pH in fused silica capillaries with surfaces modified by silane derivitization and/or adsorption of polar polymers, *J. High Resol. Chromatogr.*, 15, 452, 1992.

Gordon, M. J., Zare, R. N., Lee, K. J., and Arias, A. A., Protocol for resolving protein mixtures in capillary zone electrophoresis, *Anal. Chem.*, 63, 69, 1991.

Guzman, N. A., Moschera, J., Bailey, C. A., Iqbal, K., and Malick, A. W., Assay of protein drug substances present in solution mixtures by fluorescence denvitisation and capillary electrophoresis, *J. Chromatogr.*, 598, 123, 1992.

Holland, R. D. and Sepaniak, M. J., Qualitative analysis of mycotoxins using micellar electrokinetic capillary chromatography, *Anal. Chem.*, 65, 1140-1146, 1993.

Honda, S., Taga. A., Kakehi, K., Koda, S., and Okamoto, Y., Determination of cefixime and its metabolites by high performance capillary electrophoresis, *J. Chromatogr.* 590, 364, 1992.

Huang, X., Luckey, J. A., Gordon, M. J., and Zare, R. N., Quantitative analysis of low molecular weight carboxylic acids by capillary zone electrophoresis/conductivity detection, *Anal. Chem.*, 61, 766, 1989.

Jones, W. R. and Jandik, P., Various approaches to analysis of difficult sample matrices of anions using capillary ion electrophoresis, *J. Chromatogr.*, 608, 385-393, 1992.

Kasper, T. J., Melera, M., Gozel, P., and Brownlee, R. G., Separation and detection of DNA by capillary electrophoresis, *J. Chromatogr.*, 458, 303, 1988.

Kohr, H. and Englehardt, H., Capillary electrophoresis with surface coated capillaries, *J. Microcolumn Sep.*, 3, 491, 1991.

Krueger, R. J., Hobbs, T. R., Mihal. K. A., Therani, J., and Zeece. M. G., Analysis of endoproteinase ArgC action on adrenoconicotrophic hormone by capillary electrophoresis and reverse phase HPLC, *J. Chromatogr.*, 543, 451, 1991.

Landers, J. P., Schuchard, M. D., Subramaniam, S., Sismelich, T. and Spelsberg, T.C., High-performance capillary electrophoresis analysis of chloramphenicol acetyl transferase activity, *J. Chromatogr.*, 603, 247-257, 1992.

Landers, J. P., Oda, R. P., Madden, B. J., and Spelsberg, T.C., High-performance capillary electrophoresis of glycoproteins: The use of modifiers of electroosmotic flow for analysis of microheterogeneity, *Anal. Biochem.*, 205, 115-124, 1992.

Landers, J. P., Oda, R. P., and Schuchard, M. D., Separation of boron-complexed diol compounds using high-performance capillary electrophoresis, *Anal. Chem.*, 64, 2846-2851, 1992.

Landers, J. P., Oda, R. P., Liebenow, J. A., and Spelsberg, T. C., Utility of high performance capillary electrophoresis for monitoring peptide homo- and hetero-dimer formation, *J. Chromatogr.*, in press, 1993.

Lauer, H. H. and McMangill, D., Capillary zone electrophoresis of proteins in untreated fused silica tubing, *Anal. Chem.*, 58, 166, 1986.

Lecoq, F., Leuratti, C., Marafante, E., and DiBase, S., Analysis of nucleic acid derivatives by micellar electrokinetic capillary chromatography, *J. High Resolut. Chromatogr.*, 14, 667, 1991.

Liu, Y. and Chan. K. F. J., High performance capillary electrophoresis of gangliosides. *Electrophoresis*, 12, 402, 1991.

Liu, J. and Novotny, M., Capillary electrophoretic separation of peptides using micelleforming compounds and cyclodextrins as additives, *J. Chromatogr.*, 519, 189, 1990.

Lu, W. and Cassidy, R. M., Evaluation of ultramicroelectrodes for the detection of metal ions separated by capillary electrophoresis, *Anal. Chem.*, 65, 1649-1653, 1993.

Lurie, I. S., Micellar electrokinetic capillary chromatography of the enantiomers of amphetamine, methamphetamine and their hydroxyphenethylamine precursors, *J. Chromatogr.*, 605, 269, 1992.

McCord, B. R., Jung, J. M., and Holleran, E. S., High resolution CE of forensic DNA using a non-gel sieving buffer, *J. Liquid Chromatogr.*, 16 (9&10), 1963-1981, 1993.

McCormick, R. M., Capillary zone electrophoresis of peptides and proteins using low pH buffers in modified silica capillaries, *Anal. Chem.*, 60, 2322, 1988.

Mikaye, M., Shibukawa, A., and Nakasgawa, T., Simultaneous determination of creatinine and uric acid in human plasma and urine by micellar electrokinetic chromatography, *J. High Resolut. Chromatogr.*, 14, 181, 1991.

Morin, Ph., Villard, F., Quinsac, A., and Dreux, M., Micellar electrokinetic capillary chromatography of glucosinolates and desulfoglucosinolates with a cationic surfactant, *J. High Resolut. Chromatogr.*, 371, 1992.

Mosher, R. A., The use of metal ion-supplemented buffers to eliminate the resolution of peptides in capillary zone electrophoresis, *Electrophoresis*, 11, 765, 1990.

Neilsen, R. G., Sittampalam, G. S., and Richard, E. C., Capillary zone electrophoresis of insulin and growth hormone, *Anal. Biochem.*, 177, 20, 1989.

Ng. C. L., Lee, N. H., and Li, S. F. Y., Systematic optimisation of capillary electrophoretic separation of sulphonamides, *J. Chromatogr.*, 598, 133, 1992.

Nguyen, A. L., Luong, J. H. T., and Masson, G., Determination of nucleotides in fish tissue using capillary electrophoresis, *Anal. Chem.*, 62, 2490, 1990.

Nielen, M. W. F., Chiral separation of basic drugs using cyclodextrin-modified capillary zone electrophoresis, *Anal. Chem.*, 65, 885-893, 1993.

Nielsen, M. W. and Mensink, M. J. A., Separation of benzothiazole sultenamides using micellar electrokinetic capillary chromatography, *J. High Resolut. Chromatogr.*, 14, 417, 1991.

Nielsen, R. G., Riggin, R. M., and Richards, E. C., Capillary zone electrophoresis of peptide fragments from trypsin digestion of biosynthetic human growth hormone, *J. Chromatogr.*, 480, 343, 1989.

Nishi, H. and Terabe, S., Applications of micellar electrokinetic chromatography to pharmaceutical analysis, *Electrophoresis*, 11, 691, 1990.

Nishi, H., Tsumagari, N., and Terabe, S., Effects of tetraalkylammonium salts on micellar electrokinetic chromatography of ionic substances, *Anal. Chem.*, 61, 2434, 1989.

Northrop, D. M., Martine, D. E., and MacClehan, W. A., Separation and identification of organic gunshot and explosive constituents by micellar electrokinetic capillary chromatography, *Anal. Chem.*, 63, 10338, 1991.

Novotny, M. V., Cobb., K. A., and Liu, J., Recent advances in capillary electrophoresis of proteins, peptides and amino acids, *Electrophoresis*, 11, 735, 1990.

Okafo, G. N., Brown, R., and Camilleri, P., Some physico-chemical properties that make D.O-based buffer solutions useful media for capillary electrophoresis, *J. Chem. Soc. Chem. Commun.*, 864, 1991.

Ong, L. P., Ng, C. L., Lee, N. H., and Li, S. F. Y., Determination of antihistamines in pharmaceutical by capillary electrophoresis, *J. Chromatogr.*, 588, 335, 1991.

O'Shea, T. J., Lunte, S. M., and LaCourse, W.R., Detection of carbohydrates by CE with pulsed amperometric detection, *Anal. Chem.*, 65, 948-951, 1993.

Otsuka. K., Terabe, S., and Ando, T., Electrokinetic chromatography with micellar solutions: separation of phenylthiohydantoin amino acids, *J. Chromatogr.*, 332, 219, 1985.

Prusik, Z., Kasicka, V., Mudra, P., Stepanek, J., Smekal, O., and Hlavacek, J., Correlation of capillary zone electrophoresis with continuous flow zone electrophoresis: application to the analysis and purification of synthetic growth hormone releasing peptide, *Electrophoresis*, 932, 1990.

Roach, M. C., Gozel, P., and Zare, R. N., Determination of methotrexate and its major metabolite, 7-hydroxymethotreaxate using capillary zone electrophoresis and laser induced fluorescence detection, *J. Chromatogr.*, 426, 129, 1988.

Romano, J., Jandik. P, Jones, W. R., and Jackson. P. E., Optimisation of inorganic capillary electrophoresis for the analysis of anionic solutes in real sample, *J. Chromatogr.*, 545, 411, 1991.

Sakudinskaya, I. K., Desirodero, C., Nardi, A., and Fanali, S., Micellar electrokinetic chromatographic study of hydroquinone and some of its ethers. Determination of hydroquinone in skin toning cream, *J. Chromatogr.*, 596, 95, 1992.

Salomon, D. R. and Romano, J., Applications of capillary ion electrophoresis in the pulp and paper industry, *J. Chromaragr.*, 602, 219, 1992.

Saria, A., Identification of alpha and beta series of calcitonin gene-related peptide in the rat amygdala after separation with capillary zone electrophoresis. 573, 219, 1992.

Swedberg, S., The impact of column technology on protein separations in OT capillary electrophoresis: Past lessons, future promises, *CRC Handbook of Capillary Electrophoresis: A Practical Approach*, chapter 19, 1993.

Tanaka. H. W., Kanabe. T., Yameda, Y., and Semba, T., Capillary electrophoretic monitoring tor C-terminal fragment identification, *J. High Resolut. Chromatogr.*, 14, 491, 1991.

Terabe, S., Yashima, T., Tanaka. M., and Araki, M., Separation of oxygen isotopic benzoic acids by capillary zone electrophoresis based on isotope effects on the dissociation of the carboxyl group, *Anal. Chem.*, 60, 1673, 1988.

Terabe, S., Otsuka, K., Ichikawa. K., Tsuchiva, and Ando, T., Electrokinetic separations with micellar solutions and tubular capillaries, *Anal. Chem.*, 56, 113, 1984.

Terabe. S., Ozuki, H., Otsuka. K, and Ando, T., Electrokinetic chromatograhy with 2-0-carboxymethyl-_-cyclodextrin as a moving stationary phase, *J. Chromatogr.*, 332, 211, 1983.

Thibault, P., Pleasance, S., and Laycock, M. V., Analysis of paralytic shellfish poisons by capillary electrophoresis, *J. Chromatogr.*, 542, 483, 1991.

Towns. J. K. and Regnier, P. E., Capillary electrophoretic separations of proteins using non-ionic surfactant coatings, *Anal. Chem.*, 63, 1126, 1991.

Tran, A. D., Pak, S., Lisi, P. J., Huynh, O. T., Ryall, R. R., and Lane, P. A., Separation of carbohydrate-mediated microheterogeneity of recombinant human erythropoietin by free solution capillary electrophoresis, *J. Chromatogr.*, 542, 459, 1991.

Tsuda, T., Kobayashi, Y., Hori, A., Matsumoto, T., and Suzuki, O., Separation of polyamines in rat tissue by capillary electrophoresis, *J. Microcol. Sep.*, *2*, 21, 1990.

Van de Goor, T. A. A. M., Janssen. P. S. L., Van Nispen, J. W., Van Zeeland. M. J. M., and Everaerts, F. M., Capillary electrophoresis of peptides: analysis of adrenocorticotropic hormone-related fragments, 545, 379, 1991.

Vorndran, A. E., Oefner, P. J., Scherz, H., and Bonn, G. K., Indirect UV detection of carbohydrates in capillary zone electrophoresis, *Chromatographia*, 33, 1992.

Walbroech, Y. and Jorgenson, J. W., Capillary zone electrophoresis of neutral molecules by solvophobic association with tetraalkylammonium ions. *Anal. Chem.*, 58, 479, 1988.

Walbroehl, Y., On-column UV absorption detector for open tubular capillary zone electrophoresis, *J. Chromatogr.*, 315, 135, 1984.

Wallingford, and Ewing, A. G., Separation of serotonin from catechols by capillary zone electrophoresis using electrochemical detection, *Anal. Chem.*, 61, 98, 1989.

Weinberger, R. and Lurie. I. S., Micellar electrokinetic capillar chromatography of illicit drug substances, *Anal. Chem.*, 63, 827, 1991.

Werner, W. E., Demorgst, D. M., Stevens, J., and Wiktorewicz, J. E., Site dependent separation of proteins denatured in SDS by CE using a replaceable sieving matrix, *Anal. Biochem.*, 212, 253-258, 1993.

Wernly, P. and Thormann, W., Determination of morphine-3-glucuronide in human urine by capillary zone electrophoresis and micellar electrokinetic capillary chromatography, *J. Chromatogr.*, 616, 305-310, 1993.

Wiktorowicz, J. E. and Colburn, J. C., Separation of cationic proteins via charge reversal in capillary electrophoresis, **Electrophoresis,** 11, 769, 1990.

Yeo. S. K., Ong, C. P., and Li, S. F. Y., Optimisation of high performance capillary electrophoresis of plant growth regulators usiny the overlapping resolution mapping scheme, *Anal. Chem.,* 63, 2222, 1991.

INDEX

INDEX

D

E

I

K

L

M

N

O

P

Q

R

U

V

W

Y

Z